A Panoramic View
of Riemannian Geometry

Marcel Berger

A Panoramic View of Riemannian Geometry

With 424 Figures

Marcel Berger

Institut des Hautes Études Scientifiques
IHES
35 route de Chartres
91440 Bures-sur-Yvette, France
e-mail: mberger@ihes.fr

Library of Congress Control Number: 2003059588

Mathematics Subject Classification (2000): 53Axx, 53Bxx, 53Cxx, 37Cxx, 51Kxx

Corrected 2nd printing 2007
ISBN 978-3-540-65317-2 Springer Berlin Heidelberg New York

This work is subject to copyright. All rights reserved, whether the whole or part of the material is concerned, specifically the rights of translation, reprinting, reuse of illustrations, recitation, broadcasting, reproduction on microfilm or in any other way, and storage in data banks. Duplication of this publication or parts thereof is permitted only under the provisions of the German Copyright Law of September 9, 1965, in its current version, and permission for use must always be obtained from Springer. Violations are liable for prosecution under the German Copyright Law.

Springer is a part of Springer Science+Business Media

springer.com

© Springer-Verlag Berlin Heidelberg 2003

The use of general descriptive names, registered names, trademarks, etc. in this publication does not imply, even in the absence of a specific statement, that such names are exempt from the relevant protective laws and regulations and therefore free for general use.

Typesetting and production: LE-TeX Jelonek, Schmidt & Vöckler GbR, Leipzig
Cover design: Erich Kirchner, Heidelberg

Printed on acid-free paper   41/3180/YL - 5 4 3 2 1 0

Heinz Götze gewidmet

This book is a tribute to the memory of Dr. Heinz Götze who dedicated his life to scientific publishing, in particular to mathematics. Mathematics publishing requires special effort and talent.
Heinz Götze took up this challenge for almost 40 years with his characteristic energy and enthusiasm.

Preface

Riemannian geometry has become an important and vast subject. It deserves an encyclopedia, rather than a modest-length book. It is therefore impossible to present Riemannian geometry in a book in the standard fashion of mathematics, with complete definitions, proofs, and so on. This contrasts sharply with the situation in 1943, when Preissmann's dissertation 1943 [1041] presented all the global results of Riemannian geometry (but for the theory of symmetric spaces) including new ones, with proofs, in only forty pages.

Today, besides its basic use for the formulation of the theory of general relativity and the theory of Hamiltonian dynamical systems, the greatest achievement of Riemannian Geometry is the very recent proof of the Poincaré conjecture by Perelman, following a direction introduced and persued by Richard Hamilton, see more in sections 11.4.3, 14.4 and chapter 16 (p. 723).

Moreover, even at the root of the subject, the idea of a Riemannian manifold is subtle, appealing to unnatural concepts. Consequently, all recent books on Riemannian geometry, however good they may be, can only present two or three topics, having to spend quite a few pages on the foundations. Since our aim is to introduce the reader to most of the living topics of the field, we have had to follow the only possible path: to present the results without proofs.

We have two goals: first, to introduce the various concepts and tools of Riemannian geometry in the most natural way; or further, to demonstrate that one is practically forced to deal with abstract Riemannian manifolds in a host of intuitive geometrical questions. This explains why a long first chapter will deal with problems in the Euclidean plane. Second, once equipped with the concept of Riemannian manifold, we will present a panorama of current day Riemannian geometry. A panorama is never a full 360 degrees, so we will not try to be complete, but hope that our panorama will be large enough to show the reader a substantial part of today's Riemannian geometry.

In a panorama, you see the peaks, but you do not climb them. This is a way of saying that we will not prove the statements we quote. But, in a panorama, sometimes you can still see the path to a summit; analogously in many cases we will explain the main ideas or the main ingredients for the proof.

We hope that this form of presentation will leave many readers wanting to climb some peak. We will give all the needed references to the literature as the introduction and the panorama unfold. For alpinists, the equivalent of such a book will be the *refuge de haute-montagne* (the base camp) where you need to spend the night before the final climb. In the worst (we might say, the grandest) cases, like in the Himalayas, a climber has to establish as many as five base camps. The scientific analogue is that you need not only books, but also original articles.

Even without proofs or definitions, some of the peaks lie very far beyond. Distant topics will be mentioned only briefly in chapter 14. The judgement that a peak lies far away is personal; in the present case, we mean far from the author. His writing a book on Riemannian geometry does not indicate that he is an expert on every topic of it, especially the recent topics.

One may ask why we study only two objects: Euclidean domains with boundary, and Riemannian manifolds without boundary. There is a notion of Riemannian manifold with boundary, but in the Euclidean domain the interior geometry is given, flat and trivial, and the interesting phenomena come from the shape of the boundary. Riemannian manifolds have no boundary, and the geometric phenomena are those of the interior. Asking for both at the same time risks having too much to handle (however see §§14.5.1).

The present text is an introduction, so we have to refrain from saying too much. For example, we will mainly consider compact Riemannian manifolds. But noncompact ones are also a very important subject; they are more challenging and more difficult to study.

We will conform to the following principles:

- This book is not a handbook of Riemannian geometry, nor a systematic awarding of prizes. We give only the best recent results, not all of the intermediate ones. However, we mention when the desired type of results started to appear, this being of historical interest and at the same time helping the reader to realize the difficulty of the problem. We hope that those whose results are not mentioned will pardon us.
- We present open problems as soon as they can be stated. This encourages the reader to appreciate the difficulty and the current state of each problem.

Since this text is unusual, it is natural to expect unusual features of presentation. First, references are especially important in a book about mathematical culture. But there should not be too many. Generally, we will only give a few of the recent references. From these, the interested reader will be able to trace back to most of the standard sources. When we are considering very basic notions (like that of manifold or billiard) we will typically give many references. The reader might prefer to work with one more than another. Second, since we will not give formal definitions in the text, we thought the reader might find it useful to have most of them collected in the final chapter.

Some words about organization: first, the immensity of the field poses a problem of classification; in our division into chapters, necessarily arbitrary, we did not follow any logical or historical order. We have tried to follow a certain naturalness and simplicity. This explains why many recent discoveries, like those concerning the isoperimetric profile, the systolic inequalities, the spectrum, the geodesic flow and periodic geodesics come before a host of discoveries relating the topology of the underlying manifolds with various assumptions on curvature, although the latter results came to light much earlier than the former.

Second, our treatment of topics is certainly uneven, but this reflects the tastes and knowledge of the author. Disparities appear in the choice of results presented and in what we will offer as ideas behind the proofs. We apologize for that. For example, everything concerning bundles over Riemannian manifolds, especially spin bundles and spin geometry, will be very sketchy.

We hope that despite these weaknesses, the present book will bring pleasure and be of help to professional Riemannian geometers as well as those who want to enter into the realm of Riemannian geometry, which is an amazingly beautiful, active and natural field of research today. The reader who finds this book worthwhile will be interested in reading Dillen & Verstraelen 2000 [449].

Acknowledgements

I was able to write this book with enthusiasm thanks to the Università di Roma "La Sapienza", the Indian Institute of Technology at Powai–Bombay, the University of Pennsylvania and the Zürich Polytechnicum, where I was invited to give lectures, in Rome in 1992, in Bombay in 1993, in Philadelphia in 1994, and in Zürich in the winter semester of 1995–1996. These four departments permitted me to give lectures entitled "Topics in Riemannian geometry" where I covered a lot but with almost no proofs, only sketching ideas and ingredients. I want to thank them for having allowed me to give lectures which were not set in a classical frame. Many thanks also to the people who greatly helped me to write the surveys Berger 1998 [171], 2000 [172], 2003 [173]. They are too many to be thanked individually (their names are listed in Berger 1998 [171]), although I make an exception for Shanta Shrinivasan who wrote a first draft of my Bombay lectures.

I am deeply indebted to Dr. Benjamin McKay for taking on and carrying out the difficult task of language editing and typesetting the manuscript, inserting all the figures as he did so. Finally my special thanks go to Springer's mathematical editorial for agreeing to embark on this extraordinary project. Personal thanks go to my old friends Dr. Joachim Heinze and Dr. Catriona Byrne. Finally Mss. Susanne Denskus and Leonie Kunz had the hard time to completely put the manuscript in its final form.

Conventions of This Book

- We will generally assume (with the notable exception of all of chapter 1) that all manifolds are **compact** and **without boundary**.
- Einstein summation convention will be used.

- The lemmas, theorems, propositions, questions, corollaries, etc. are all numbered with a single sequence, to make it easier to find them; most mathematics books use separate sequences for the lemmas, theorems, etc.

Bures-Sur-Yvette, France,
June 2003 and May 2007   *Marcel Berger*

Contents

1 **Euclidean Geometry** 1
  1.1   Preliminaries 2
  1.2   Distance Geometry.. 2
        1.2.1   A Basic Formula... 2
        1.2.2   The Length of a Path 3
        1.2.3   The First Variation Formula
                and Application to Billiards..................... 4
  1.3   Plane Curves.. 9
        1.3.1   Length 9
        1.3.2   Curvature 12
  1.4   Global Theory of Closed Plane Curves................... 18
        1.4.1   "Obvious" Truths About Curves
                Which are Hard to Prove 18
        1.4.2   The Four Vertex Theorem 20
        1.4.3   Convexity with Respect to Arc Length 22
        1.4.4   Umlaufsatz with Corners 23
        1.4.5   Heat Shrinking of Plane Curves................... 24
        1.4.6   Arnol'd's Revolution in Plane Curve Theory 24
  1.5   The Isoperimetric Inequality for Curves................... 26
  1.6   The Geometry of Surfaces Before and After Gauß 29
        1.6.1   Inner Geometry: a First Attempt 30
        1.6.2   Looking for Shortest Curves: Geodesics 33
        1.6.3   The Second Fundamental Form
                and Principal Curvatures 45
        1.6.4   The Meaning of the Sign of K 52
        1.6.5   Global Surface Geometry 55
        1.6.6   Minimal Surfaces 58
        1.6.7   The Hartman-Nirenberg Theorem
                for Inner Flat Surfaces 62
        1.6.8   The Isoperimetric Inequality in $\mathbb{E}^3$ à la Gromov 63
                1.6.8.1   Notes 66
  1.7   Generic Surfaces 66
  1.8   Heat and Wave Analysis in $\mathbb{E}^2$ 70
        1.8.1   Planar Physics 70
                1.8.1.1   Bibliographical Note..................... 71

		1.8.2	Why the Eigenvalue Problem?..................... 71

- 1.8.2 Why the Eigenvalue Problem?..................... 71
- 1.8.3 Minimax... 75
- 1.8.4 Shape of a Drum 78
  - 1.8.4.1 A Few Direct Problems 79
  - 1.8.4.2 The Faber–Krahn Inequality............. 81
  - 1.8.4.3 Inverse Problems 83
- 1.8.5 Heat 87
  - 1.8.5.1 Eigenfunctions 90
- 1.8.6 Relations Between the Two Spectra 91
- 1.9 Heat and Waves in $\mathbb{E}^3, \mathbb{E}^d$ and on the Sphere 94
  - 1.9.1 Euclidean Spaces 94
  - 1.9.2 Spheres.. 95
  - 1.9.3 Billiards in Higher Dimensions 97
  - 1.9.4 The Wave Equation Versus the Heat Equation...... 98

2 Transition 101

3 Surfaces from Gauß to Today 105
- 3.1 Gauß.. 105
  - 3.1.1 Theorema Egregium 105
    - 3.1.1.1 The First Proof of Gauß's Theorema Egregium; the Concept of ds^2 106
    - 3.1.1.2 Second Proof of the Theorema Egregium ... 109
  - 3.1.2 The Gauß–Bonnet Formula and the Rodrigues–Gauß Map 111
  - 3.1.3 Parallel Transport 113
  - 3.1.4 Inner Geometry 116
- 3.2 Alexandrov's Theorems 120
  - 3.2.1 Angle Corrections of Legendre and Gauß in Geodesy....................................... 123
- 3.3 Cut Loci... 125
- 3.4 Global Surface Theory...................................... 131
  - 3.4.1 Bending Surfaces 131
    - 3.4.1.1 Bending Polyhedra 132
    - 3.4.1.2 Bending and Wrinkling with Little Smoothness 133
  - 3.4.2 Mean Curvature Rigidity of the Sphere 134
  - 3.4.3 Negatively Curved Surfaces 135
  - 3.4.4 The Willmore Conjecture 136
  - 3.4.5 The Global Gauß–Bonnet Theorem for Surfaces 136
  - 3.4.6 The Hopf Index Formula......................... 139

4 Riemann's Blueprints 143
4.1 Smooth Manifolds 143
4.1.1 Introduction 143
4.1.2 The Need for Abstract Manifolds 146
4.1.3 Examples 149
4.1.3.1 Submanifolds 151
4.1.3.2 Products 151
4.1.3.3 Lie Groups 152
4.1.3.4 Homogeneous Spaces 152
4.1.3.5 Grassmannians over Various Algebras 153
4.1.3.6 Gluing 156
4.1.4 The Classification of Manifolds 157
4.1.4.1 Surfaces 158
4.1.4.2 Higher Dimensions 159
4.1.4.3 Embedding Manifolds in Euclidean Space 161
4.2 Calculus on Manifolds 162
4.2.1 Tangent Spaces and the Tangent Bundle 162
4.2.2 Differential Forms and Exterior Calculus 166
4.3 Examples of Riemann's Definition 172
4.3.1 Riemann's Definition 172
4.3.2 Hyperbolic Geometry 176
4.3.3 Products, Coverings and Quotients 183
4.3.3.1 Products 183
4.3.3.2 Coverings 184
4.3.4 Homogeneous Spaces 186
4.3.5 Symmetric Spaces 189
4.3.5.1 Classification 192
4.3.5.2 Rank 193
4.3.6 Riemannian Submersions 194
4.3.7 Gluing and Surgery 196
4.3.7.1 Gluing of Hyperbolic Surfaces 196
4.3.7.2 Higher Dimensional Gluing 198
4.3.8 Classical Mechanics 199
4.4 The Riemann Curvature Tensor 200
4.4.1 Discovery and Definition 200
4.4.2 The Sectional Curvature 204
4.4.3 Standard Examples 207
4.4.3.1 Constant Sectional Curvature 207
4.4.3.2 Projective Spaces $\mathbb{KP}^n$ 209
4.4.3.3 Products 209
4.4.3.4 Homogeneous Spaces 210
4.4.3.5 Hypersurfaces in Euclidean Space 211

- 4.5 A Naive Question: Does the Curvature Determine the Metric? 213
  - 4.5.1 Surfaces 214
  - 4.5.2 Any Dimension 215
- 4.6 Abstract Riemannian Manifolds 216
  - 4.6.1 Isometrically Embedding Surfaces in $\mathbb{E}^3$ 217
  - 4.6.2 Local Isometric Embedding of Surfaces in $\mathbb{E}^3$ 217
  - 4.6.3 Isometric Embedding in Higher Dimensions 218

5 A One Page Panorama 219

6 Metric Geometry and Curvature 221
- 6.1 First Metric Properties 222
  - 6.1.1 Local Properties 222
  - 6.1.2 Hopf–Rinow and de Rham Theorems 226
    - 6.1.2.1 Products 229
  - 6.1.3 Convexity and Small Balls 229
  - 6.1.4 Totally Geodesic Submanifolds 231
  - 6.1.5 Center of Mass 233
  - 6.1.6 Examples of Geodesics 235
  - 6.1.7 Transition 238
- 6.2 First Technical Tools 239
- 6.3 Second Technical Tools 248
  - 6.3.1 Exponential Map 248
    - 6.3.1.1 Rank 250
  - 6.3.2 Space Forms 251
  - 6.3.3 Nonpositive Curvature 254
- 6.4 Triangle Comparison Theorems 257
  - 6.4.1 Bounded Sectional Curvature 257
  - 6.4.2 Ricci Lower Bound 262
  - 6.4.3 Philosophy Behind These Bounds 267
- 6.5 Injectivity, Convexity Radius and Cut Locus 268
  - 6.5.1 Definition of Cut Points and Injectivity Radius 268
  - 6.5.2 Klingenberg and Cheeger Theorems 272
  - 6.5.3 Convexity Radius 278
  - 6.5.4 Cut Locus 278
  - 6.5.5 Blaschke Manifolds 285
- 6.6 Geometric Hierarchy 286
  - 6.6.1 The Geometric Hierarchy 289
    - 6.6.1.1 Space Forms 289
    - 6.6.1.2 Rank 1 Symmetric Spaces 289
    - 6.6.1.3 Measure Isotropy 289
    - 6.6.1.4 Symmetric Spaces 290
    - 6.6.1.5 Homogeneous Spaces 290

	6.6.2	Constant Sectional Curvature 290	
		6.6.2.1 Negatively Curved Space Forms in Three and Higher Dimensions 292	
		6.6.2.2 Mostow Rigidity 293	
		6.6.2.3 Classification of Arithmetic and Nonarithmetic Negatively Curved Space Forms........................... 294	
		6.6.2.4 Volumes of Negatively Curved Space Forms 295	
	6.6.3	Rank 1 Symmetric Spaces 295	
	6.6.4	Higher Rank Symmetric Spaces.................. 296	
		6.6.4.1 Superrigidity 296	
	6.6.5	Homogeneous Spaces 296	

7 Volumes and Inequalities on Volumes of Cycles 299

7.1 Curvature Inequalities 299

7.1.1 Bounds on Volume Elements and First Applications . 299
- 7.1.1.1 The Canonical Measure 299
- 7.1.1.2 Volumes of Standard Spaces 303
- 7.1.1.3 The Isoperimetric Inequality for Spheres ... 304
- 7.1.1.4 Sectional Curvature Upper Bounds 305
- 7.1.1.5 Ricci Curvature Lower Bounds 308

7.1.2 Isoperimetric Profile 315
- 7.1.2.1 Definition and Examples 315
- 7.1.2.2 The Gromov–Bérard–Besson–Gallot Bound . 319
- 7.1.2.3 Nonpositive Curvature on Noncompact Manifolds................. 322

7.2 Curvature Free Inequalities on Volumes of Cycles 325

7.2.1 Curves in Surfaces 325
- 7.2.1.1 Loewner, Pu and Blatter–Bavard Theorems 325
- 7.2.1.2 Higher Genus Surfaces 329
- 7.2.1.3 The Sphere.............................. 336
- 7.2.1.4 Homological Systoles 338

7.2.2 Inequalities for Curves........................... 340
- 7.2.2.1 The Problem, and Standard Manifolds 340
- 7.2.2.2 Filling Volume and Filling Radius 342
- 7.2.2.3 Gromov's Theorem and Sketch of the Proof 344

7.2.3 Higher Dimensional Systoles: Systolic Freedom Almost Everywhere............... 348

7.2.4 Embolic Inequalities 353
- 7.2.4.1 Introduction 353
- 7.2.4.2 The Unit Tangent Bundle 357
- 7.2.4.3 The Core of the Proof 359
- 7.2.4.4 Croke's Three Results 363
- 7.2.4.5 Infinite Injectivity Radius 366
- 7.2.4.6 Using Embolic Inequalities 367

8 Transition: The Next Two Chapters ... 369
- 8.1 Spectral Geometry and Geodesic Dynamics ... 369
- 8.2 Why are Riemannian Manifolds So Important? ... 372
- 8.3 Positive Versus Negative Curvature ... 372

9 Spectrum of the Laplacian ... 373
- 9.1 History ... 374
- 9.2 Motivation ... 375
- 9.3 Setting Up ... 376
  - 9.3.1 Xdefinition ... 376
  - 9.3.2 The Hodge Star ... 378
  - 9.3.3 Facts ... 380
  - 9.3.4 Heat, Wave and Schrödinger Equations ... 381
- 9.4 Minimax ... 383
  - 9.4.1 The Principle ... 383
  - 9.4.2 An Application ... 385
- 9.5 Some Extreme Examples ... 387
  - 9.5.1 Square Tori, Alias Several Variable Fourier Series ... 387
  - 9.5.2 Other Flat Tori ... 388
  - 9.5.3 Spheres ... 390
  - 9.5.4 $\mathbb{KP}^n$ ... 390
  - 9.5.5 Other Space Forms ... 391
- 9.6 Current Questions ... 392
  - 9.6.1 Direct Questions About the Spectrum ... 392
  - 9.6.2 Direct Problems About the Eigenfunctions ... 393
  - 9.6.3 Inverse Problems on the Spectrum ... 393
- 9.7 First Tools: The Heat Kernel and Heat Equation ... 393
  - 9.7.1 The Main Result ... 393
  - 9.7.2 Great Hopes ... 396
  - 9.7.3 The Heat Kernel and Ricci Curvature ... 401
- 9.8 The Wave Equation: The Gaps ... 402
- 9.9 The Wave Equation: Spectrum & Geodesic Flow ... 405
- 9.10 The First Eigenvalue ... 408
  - 9.10.1 λ_1 and Ricci Curvature ... 408
  - 9.10.2 Cheeger's Constant ... 409
  - 9.10.3 λ_1 and Volume; Surfaces and Multiplicity ... 410
  - 9.10.4 Kähler Manifolds ... 411
- 9.11 Results on Eigenfunctions ... 412
  - 9.11.1 Distribution of the Eigenfunctions ... 412
  - 9.11.2 Volume of the Nodal Hypersurfaces ... 413
  - 9.11.3 Distribution of the Nodal Hypersurfaces ... 414
- 9.12 Inverse Problems ... 414
  - 9.12.1 The Nature of the Image ... 414
  - 9.12.2 Inverse Problems: Nonuniqueness ... 416
  - 9.12.3 Inverse Problems: Finiteness, Compactness ... 418

		9.12.4 Uniqueness and Rigidity Results 419
		9.12.4.1 Vignéras Surfaces 420
	9.13	Special Cases... 421
		9.13.1 Riemann Surfaces............................... 421
		9.13.2 Space Forms 424
		9.13.2.1 Scars 426
	9.14	The Spectrum of Exterior Differential Forms 426

10 Geodesic Dynamics 431
   10.1 Introduction 432
   10.2 Some Well Understood Examples 436
        10.2.1 Surfaces of Revolution............................ 436
              10.2.1.1 Zoll Surfaces 436
              10.2.1.2 Weinstein Surfaces 440
        10.2.2 Ellipsoids and Morse Theory 440
        10.2.3 Flat and Other Tori: Influence of the Fundamental
               Group.. 442
              10.2.3.1 Flat Tori................................ 442
              10.2.3.2 Manifolds Which are not Simply Connected 443
              10.2.3.3 Tori, not Flat........................... 445
        10.2.4 Space Forms 446
              10.2.4.1 Space Form Surfaces..................... 446
              10.2.4.2 Higher Dimensional Space Forms 448
   10.3 Geodesics Joining Two Points 449
        10.3.1 Birkhoff's Proof for the Sphere 449
        10.3.2 Morse Theory 453
        10.3.3 Discoveries of Morse and Serre 454
        10.3.4 Computing with Entropy 456
        10.3.5 Rational Homology and Gromov's Work 458
   10.4 Periodic Geodesics 461
        10.4.1 The Difficulties 461
        10.4.2 General Results 463
              10.4.2.1 Gromoll and Meyer...................... 463
              10.4.2.2 Results for the Generic ("Bumpy") Case 465
        10.4.3 Surfaces 466
              10.4.3.1 The Lusternik–Schnirelmann Theorem 466
              10.4.3.2 The Bangert–Franks–Hingston Results 468
   10.5 The Geodesic Flow....................................... 471
        10.5.1 Review of Ergodic Theory of Dynamical Systems ... 471
              10.5.1.1 Ergodicity and Mixing 471
              10.5.1.2 Notions of Entropy 473
   10.6 Negative Curvature 478
        10.6.1 Distribution of Geodesics 481
        10.6.2 Distribution of Periodic Geodesics 481
   10.7 Nonpositive Curvature................................... 482

10.8	Entropies on Various Space Forms		483
	10.8.1	Liouville Entropy	485
10.9	From Osserman to Lohkamp		485
10.10	Manifolds All of Whose Geodesics are Closed		488
	10.10.1	Definitions and Caution	488
	10.10.2	Bott and Samelson Theorems	490
	10.10.3	The Structure on a Given S^d and $\mathbb{KP}^n$	492
10.11	Inverse Problems: Conjugacy of Geodesic Flows		495

11 Best Metrics ... 499

11.1	Introduction and a Possible Approach		499
	11.1.1	An Approach	501
11.2	Purely Geometric Functionals		503
	11.2.1	Systolic Inequalities	503
	11.2.2	Counting Periodic Geodesics	504
	11.2.3	The Embolic Constant	504
	11.2.4	Diameter and Injectivity	505
11.3	Least curved		506
	11.3.1	Definitions	506
		11.3.1.1 $\inf \|R\|_{L^{d/2}}$	506
		11.3.1.2 Minimal Volume	507
		11.3.1.3 Minimal Diameter	507
	11.3.2	The Case of Surfaces	508
	11.3.3	Generalities, Compactness, Finiteness and Equivalence	509
	11.3.4	Manifolds with $\inf$ Vol (resp. $\inf \|R\|_{L^{d/2}}$, $\inf$ diam) $= 0$	511
		11.3.4.1 Circle Fibrations and Other Examples	511
		11.3.4.2 Allof–Wallach's Type of Examples	513
		11.3.4.3 Nilmanifolds and the Converse: Almost Flat Manifolds	514
		11.3.4.4 The Examples of Cheeger and Rong	514
	11.3.5	Some Manifolds with $\inf$ Vol > 0 and $\inf \|R\|_{L^{d/2}} > 0$	515
		11.3.5.1 Using Integral Formulas	515
		11.3.5.2 The Simplicial Volume of Gromov	516
	11.3.6	$\inf \|R\|_{L^{d/2}}$ in Four Dimensions	518
	11.3.7	Summing up Questions on $\inf$ Vol, $\inf \|R\|_{L^{d/2}}$	519
11.4	Einstein Manifolds		520
	11.4.1	Hilbert's Variational Principle and Great Hopes	520
	11.4.2	The Examples from the Geometric Hierarchy	524
		11.4.2.1 Symmetric Spaces	524
		11.4.2.2 Homogeneous Spaces and Others	524
	11.4.3	Examples from Analysis: Evolution by Ricci Flow	525

		11.4.4	Examples from Analysis: Kähler Manifolds 526

- 11.4.4 Examples from Analysis: Kähler Manifolds 526
- 11.4.5 The Sporadic Examples 528
- 11.4.6 Around Existence and Uniqueness 529
  - 11.4.6.1 Existence 529
  - 11.4.6.2 Uniqueness 530
  - 11.4.6.3 Moduli 531
  - 11.4.6.4 The Set of Constants, Ricci Flat Metrics ... 532
- 11.4.7 The Yamabe Problem 533
- 11.5 The Bewildering Fractal Landscape of $\mathcal{RS}(M)$ According to Nabutovsky 534

12 From Curvature to Topology 543
- 12.1 Some History, and Structure of the Chapter 543
  - 12.1.1 Hopf's Inspiration 543
  - 12.1.2 Hierarchy of Curvatures 546
    - 12.1.2.1 Control via Curvature 546
    - 12.1.2.2 Other Curvatures 547
    - 12.1.2.3 The Problem of Rough Classification 548
    - 12.1.2.4 References on the Topic, and the Significance of Noncompact Manifolds 549
- 12.2 Pinching Problems 549
  - 12.2.1 Introduction 549
  - 12.2.2 Positive Pinching 552
    - 12.2.2.1 The Sphere Theorem 552
    - 12.2.2.2 Sphere Theorems Invoking Bounds on Other Invariants 557
    - 12.2.2.3 Homeomorphic Pinching 558
    - 12.2.2.4 The Sphere Theorem with Lower Bound on Diameter, and no Upper Bound on Curvature 562
    - 12.2.2.5 Topology at the Diameter Pinching Limit .. 565
    - 12.2.2.6 Pointwise Pinching 567
    - 12.2.2.7 Cutting Down the Hypotheses 567
  - 12.2.3 Pinching Near Zero 568
  - 12.2.4 Negative Pinching 569
  - 12.2.5 Ricci Curvature Pinching 571
- 12.3 Curvature of Fixed Sign 576
  - 12.3.1 The Positive Side: Sectional Curvature 576
    - 12.3.1.1 The Known Examples 576
    - 12.3.1.2 Homology Type and the Fundamental Group 580
    - 12.3.1.3 The Noncompact Case 583
    - 12.3.1.4 Positivity of the Curvature Operator 588
    - 12.3.1.5 Possible Approaches, Looking to the Future 590

		12.3.2	Ricci Curvature: Positive, Negative and Just Below . 593

- 12.3.2 Ricci Curvature: Positive, Negative and Just Below .. 593
- 12.3.3 The Positive Side: Scalar Curvature 599
  - 12.3.3.1 The Hypersurfaces of Schoen & Yau 600
  - 12.3.3.2 Geometrical Descriptions 601
  - 12.3.3.3 Gromov's Quantization of K-theory and Topological Implications of Positive Scalar Curvature 602
  - 12.3.3.4 Trichotomy 603
  - 12.3.3.5 The Proof 603
  - 12.3.3.6 The Gromov–Lawson Torus Theorem 604
- 12.3.4 The Negative Side: Sectional Curvature............ 605
  - 12.3.4.1 Introduction 605
  - 12.3.4.2 Literature 605
  - 12.3.4.3 Quasi-isometries 606
  - 12.3.4.4 Volume and Fundamental Group 609
  - 12.3.4.5 Negative Versus Nonpositive Curvature 612
- 12.3.5 The Negative Side: Ricci Curvature 613

12.4 Finiteness and Collapsing 614
- 12.4.1 Finiteness 614
  - 12.4.1.1 Cheeger's Finiteness Theorems 614
  - 12.4.1.2 More Finiteness Theorems 618
  - 12.4.1.3 Ricci Curvature 622
- 12.4.2 Compactness and Convergence 624
  - 12.4.2.1 Motivation 624
  - 12.4.2.2 History 624
  - 12.4.2.3 Contemporary Definitions and Results 625
- 12.4.3 Collapsing and the Space of Riemannian Metrics.... 630
  - 12.4.3.1 Collapsing 630
  - 12.4.3.2 Closures on a Compact Manifold 634

13 Holonomy Groups and Kähler Manifolds 637
13.1 Definitions and Philosophy 637
13.2 Examples 639
13.3 General Structure Theorems 641
13.4 Classification 643
13.5 The Rare Cases.. 646
- 13.5.1 G_2 and Spin(7) 646
- 13.5.2 Quaternionic Kähler Manifolds 647
  - 13.5.2.1 The Bérard Bergery/Salamon Twistor Space of Quaternionic Kähler Manifolds.... 649
  - 13.5.2.2 The Konishi Twistor Space of a Quaternionic Kähler Manifold 651
  - 13.5.2.3 Other Twistor Spaces 651
- 13.5.3 Ricci Flat Kähler and Hyper-Kähler Manifolds 652
  - 13.5.3.1 Hyperkähler Manifolds 652

13.6	Kähler Manifolds		654
	13.6.1	Symplectic Structures on Kähler Manifolds	655
	13.6.2	Imitating Complex Algebraic Geometry on Kähler Manifolds	655

14 Some Other Important Topics ... 659

- 14.1 Noncompact Manifolds ... 660
  - 14.1.1 Noncompact Manifolds of Nonnegative Ricci Curvature ... 660
  - 14.1.2 Finite Volume ... 661
  - 14.1.3 Bounded Geometry ... 661
  - 14.1.4 Harmonic Functions ... 662
  - 14.1.5 Structure at Infinity ... 662
  - 14.1.6 Chopping ... 662
  - 14.1.7 Positive Mass ... 662
  - 14.1.8 Cohomology and Homology Theories ... 663
- 14.2 Bundles over Riemannian Manifolds ... 663
  - 14.2.1 Differential Forms and Related Bundles ... 663
    - 14.2.1.1 The Hodge Star ... 664
    - 14.2.1.2 A Variational Problem for Differential Forms and the Laplace Operator ... 664
    - 14.2.1.3 Calibration ... 666
    - 14.2.1.4 Harmonic Analysis of Other Tensors ... 667
  - 14.2.2 Spinors ... 668
    - 14.2.2.1 Algebra of Spinors ... 668
    - 14.2.2.2 Spinors on Riemannian Manifolds ... 669
    - 14.2.2.3 History of Spinors ... 669
    - 14.2.2.4 Applying Spinors ... 670
    - 14.2.2.5 Warning: Beware of Harmonic Spinors ... 670
    - 14.2.2.6 The Half Pontryagin Class ... 670
    - 14.2.2.7 Reconstructing the Metric from the Dirac Operator ... 671
    - 14.2.2.8 $Spin^c$ Structures ... 671
  - 14.2.3 Various Other Bundles ... 671
    - 14.2.3.1 Secondary Characteristic Classes ... 672
    - 14.2.3.2 Yang–Mills Theory ... 672
    - 14.2.3.3 Twistor Theory ... 672
    - 14.2.3.4 K-theory ... 673
    - 14.2.3.5 The Atiyah–Singer Index Theorem ... 674
    - 14.2.3.6 Supersymmetry and Supergeometry ... 674
- 14.3 Harmonic Maps Between Riemannian Manifolds ... 674
- 14.4 Low Dimensional Riemannian Geometry ... 676
- 14.5 Some Generalizations of Riemannian Geometry ... 676
  - 14.5.1 Boundaries ... 676
  - 14.5.2 Orbifolds ... 677

XXII Contents

- 14.5.3 Conical Singularities ... 678
- 14.5.4 Spectra of Singular Spaces ... 678
- 14.5.5 Alexandrov Spaces ... 678
- 14.5.6 CAT Spaces ... 680
  - 14.5.6.1 The CAT(k) Condition ... 680
- 14.5.7 Carnot–Carathéodory Spaces ... 681
  - 14.5.7.1 Example: the Heisenberg Group ... 681
- 14.5.8 Finsler Geometry ... 682
- 14.5.9 Riemannian Foliations ... 683
- 14.5.10 Pseudo-Riemannian Manifolds ... 683
- 14.5.11 Infinite Dimensional Riemannian Geometry ... 684
- 14.5.12 Noncommutative Geometry ... 685
- 14.6 Gromov's *mm* Spaces ... 685
- 14.7 Submanifolds ... 690
  - 14.7.1 Higher Dimensions ... 690
  - 14.7.2 Geometric Measure Theory and Pseudoholomorphic Curves ... 691

15 The Technical Chapter ... 693
- 15.1 Vector Fields and Tensors ... 693
- 15.2 Tensors Dual via the Metric: Index Aerobics ... 696
- 15.3 The Connection, Covariant Derivative and Curvature ... 697
- 15.4 Parallel Transport ... 701
  - 15.4.1 Curvature from Parallel Transport ... 703
- 15.5 Absolute (Ricci) Calculus and Commutation Formulas: Index Gymnastics ... 704
- 15.6 Hodge and the Laplacian, Bochner's Technique ... 706
  - 15.6.1 Bochner's Technique for Higher Degree Differential Forms ... 708
- 15.7 Gauß–Bonnet–Chern ... 709
  - 15.7.1 Chern's Proof of Gauß–Bonnet for Surfaces ... 710
  - 15.7.2 The Proof of Allendoerfer and Weil ... 711
  - 15.7.3 Chern's Proof in all Even Dimensions ... 713
  - 15.7.4 Chern Classes of Vector Bundles ... 714
  - 15.7.5 Pontryagin Classes ... 715
  - 15.7.6 The Euler Class ... 715
  - 15.7.7 The Absence of Other Characteristic Classes ... 716
  - 15.7.8 Applying Characteristic Classes ... 716
  - 15.7.9 Characteristic Numbers ... 716
- 15.8 Examples of Curvature Calculations ... 718
  - 15.8.1 Homogeneous Spaces ... 718
  - 15.8.2 Riemannian Submersions ... 719

16 Poincaré's Conjecture Solved by Riemannian Geometry ... 723

References 725

Acknowledgements 791

List of Notation 793

List of Authors 799

Subject Index 813

1 Old and New Euclidean Geometry and Analysis

Contents

1.1	**Preliminaries**		**2**
1.2	**Distance Geometry**		**2**
	1.2.1	A Basic Formula	2
	1.2.2	The Length of a Path	3
	1.2.3	The First Variation Formula and Application to Billiards	4
1.3	**Plane Curves**		**9**
	1.3.1	Length	9
	1.3.2	Curvature	12
1.4	**Global Theory of Closed Plane Curves**		**18**
	1.4.1	"Obvious" Truths About Curves Which are Hard to Prove	18
	1.4.2	The Four Vertex Theorem	20
	1.4.3	Convexity with Respect to Arc Length	22
	1.4.4	Umlaufsatz with Corners	23
	1.4.5	Heat Shrinking of Plane Curves	24
	1.4.6	Arnol'd's Revolution in Plane Curve Theory	24
1.5	**The Isoperimetric Inequality for Curves**		**26**
1.6	**The Geometry of Surfaces Before and After Gauß**		**29**
	1.6.1	Inner Geometry: a First Attempt	30
	1.6.2	Looking for Shortest Curves: Geodesics	33
	1.6.3	The Second Fundamental Form and Principal Curvatures	45
	1.6.4	The Meaning of the Sign of K	52
	1.6.5	Global Surface Geometry	55
	1.6.6	Minimal Surfaces	58
	1.6.7	The Hartman-Nirenberg Theorem for Inner Flat Surfaces	62
	1.6.8	The Isoperimetric Inequality in $\mathbb{E}^3$ à la Gromov	63
1.7	**Generic Surfaces**		**66**
1.8	**Heat and Wave Analysis in $\mathbb{E}^2$**		**70**
	1.8.1	Planar Physics	70
	1.8.2	Why the Eigenvalue Problem?	71
	1.8.3	Minimax	75

	1.8.4	Shape of a Drum	78
	1.8.5	Heat	87
	1.8.6	Relations Between the Two Spectra	91
1.9	**Heat and Waves in $\mathbb{E}^3, \mathbb{E}^d$ and on the Sphere**	**94**
	1.9.1	Euclidean Spaces	94
	1.9.2	Spheres	95
	1.9.3	Billiards in Higher Dimensions	97
	1.9.4	The Wave Equation Versus the Heat Equation	98

1.1 Preliminaries

Some (not all) of the following generalizes to any dimension; this will be left to the reader. We will only give references for special topics. General references could be Berger and Gostiaux 1988 [175], Coxeter 1989 [409], do Carmo 1976 [451], Klingenberg 1995 [816], Spivak 1979 [1155], Sternberg 1983 [1157], and Stoker 1989 [1160]. For those who like computer programming, Gray 1998 [584] will be of interest. We will assume elementary calculus and also that functions are differentiable as often as needed.

1.2 Distance Geometry

1.2.1 A Basic Formula

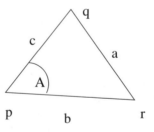

Fig. 1.1. A triangle in $\mathbb{E}^2$

In the Euclidean plane $\mathbb{E}^2$ we have distances between points and angles between smooth curves wherever the curves meet. The equation

$$a^2 = b^2 + c^2 - 2bc \cos A \tag{1.1}$$

should be kept in mind, where a, b, c are the pairwise distances between three points, as in figure 1.1 and A is the angle opposite the side of length a. The triangle inequality in $\mathbb{E}^2$:

$$a \leq b + c$$

is strict, in that equality forces the three points to be colinear, with p between q and r. Equation 1.1 is basic in telemetry, needed when your instant camera measures distances for you. To know the distance between any two of the points, you only need the angle (a local measurement) and two side lengths, not all three.[1] We will encounter other spaces like this later on. Think of mountains in your way along one side, as in figure 1.2. We use equation 1.1 in everyday life, in primitive geodesy, astronomy, automatic focusing, etc.

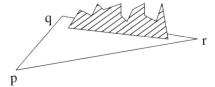

Fig. 1.2. One side is hard to measure

1.2.2 The Length of a Path

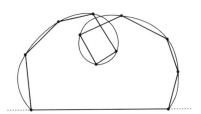

Fig. 1.3. Polygonal approximation

A *path* between two points is a continuous curve between those points. Its *length* is the supremum of the total lengths of polygons inscribed in the curve. The supremum need not be finite, e.g. a fractal curve like a snowflake. But if the curve, $c(t)$, is smooth enough[2] say with derivative $c'(t)$, then it has finite length $\int \|c'(t)\|$. The strict triangle inequality above ensures that the shortest path between two points is exactly the straight line interval between them.

For any metric space, one can define the length of a path as a supremum as above, and then consider the infimum of path lengths between two points.

---

[1] However, keep in mind the difficulty of making local measurements; angles between little pieces of straight lines are very sensitive to mismeasurement.

[2] More precisely, rectifiable; see Wheeden & Zygmund 1977 [1258].

4   1 Euclidean Geometry

If all points are connected by finite length paths, then this infimum is another metric on the same metric space. In general, it is a different metric, as in the case of the sphere in $\mathbb{E}^3$. The new metric is called the *inner metric*, and if the initial metric coincides with the inner metric, then the space is called a *length space*. Length spaces are becoming more important today. For example, see Burago, Gromov and Perelman 1992 [282] and Gromov 1999 [633].

1.2.3 The First Variation Formula and Application to Billiards

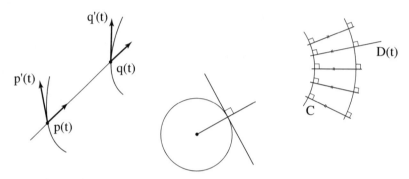

Fig. 1.4. Two points travelling through the plane

Imagine that we have two points $p(t)$ and $q(t)$ running along two curves in the Euclidean plane, as in figure 1.4 and we watch how the distance $d(p(t), q(t))$ varies with time t. The first variation formula is:

$$\frac{d}{dt} d(p(t), q(t)) = (q'(t) - p'(t)) \cdot \frac{q(t) - p(t)}{\|q(t) - p(t)\|} \qquad (1.2)$$

Consequently, if we have a one parameter family of straight lines, then any two curves which are perpendicular to the lines must be at constant distance; for example, circles about the origin. So the tangent line to a circle is orthogonal to the radius.

A more elaborate application of this equation, which will be important later on, is the *billiard table*. Suppose that we have a convex domain in the plane, say D, and consider a small particle (think of a small ball, of infinitesimal radius) flying inside D with no force applied to it. The particle will travel on a straight line with constant speed through the interior of D, and when it strikes the boundary ∂D it will bounce and travel on a new straight line. The new straight line path and the old one will be reflections of one another about the normal line of the curve ∂D at the point of impact. We will refer to this as *mirror bouncing*, since we see the same trajectories if we have light rays travelling inside a curved mirror in the shape of ∂D. For the moment, we will suppose that ∂D is differentiable and strictly convex. We will call our particle a *billiard ball*. The book of Tabachnikov 1995 [1175] is a complete reference on billiards; some other references are Sinai 1990 [1141], Katok et. al. 1986 [788], Arnoux 1988 [70], Veech 1989 & 1991 [1208, 1209], Katok 1987 [785], and Gutkin 1996 [672].

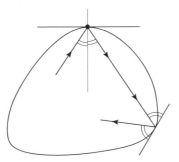

Fig. 1.5. A billiard ball

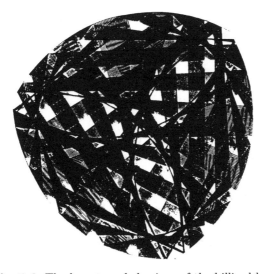

Fig. 1.6. The long term behaviour of the billiard ball

What happens to these trajectories over large periods of time? As we will see in chapter 10, this question is fundamental in rational mechanics and almost every field of physics. It may happen that a trajectory is periodic, returning eventually to its original position and velocity. The search for periodic trajectories is a corner stone in the study of long duration dynamics.

We devise the length counting function to try to count periodic trajectories, setting $CF(L)$ to be the number of periodic trajectories of length less than L. Let us now see how to exhibit some periodic trajectories.

In 1913, G. D. Birkhoff remarked that the first variation formula yields many periodic trajectories. Pick any number P of points of ∂D, in a fixed order (not necessarily consecutive along ∂D) as in figure 1.7, turning around the boundary Q times. Draw the polygon connecting those points in that order. Using the first variation formula, we find that mirror bouncing is exactly the condition that the polygon has critical length under perturbing its vertices along ∂D. But now for each choice of P and Q, we can take the polygon of greatest length, which exists by compactness of the space of polygons with given P and Q values. N.B. this construction does not yield all of the periodic trajectories. We will return to billiards in §§1.8.6.

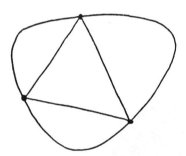 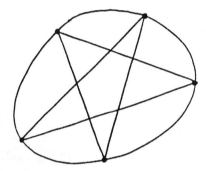

Fig. 1.7. Two of Birkhoff's configurations

Let us take our domain D to be a convex polygon, a case we expect to be simpler than a general convex curve. We shall be careful to eliminate trajectories which strike a vertex of the boundary. It is still an open question if there is even one periodic trajectory. This question is solved for polygons with angles rational multiples of π (see below), but is open for general polygons, even triangles. If the triangle is not obtuse, I leave it to you to find a (classical) periodic trajectory (hoping that you are not obtuse either). Everyone expects that infinitely many periodic trajectories exist, as for rational angled polygons and convex domains.

Rational angled polygons and convex domains are the only domains for which periodic billiard trajectories are well understood. To the rational angled polygons, one can associate a compact space, a generalized Riemannian manifold, locally Euclidean but for $2\pi k$ singularities, k an integer. See Arnoux 1988 [70] for more information. One knows that $CF(L)$ is asymptotically quadratic in L, as $L \to \infty$. See Vorobets, Gal'perin, and Stëpin 1992 [1224] and Gal'perin and Chernov 1991 [545].

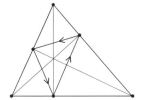 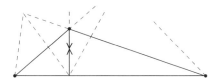

Fig. 1.8. Two triangular billiard tables

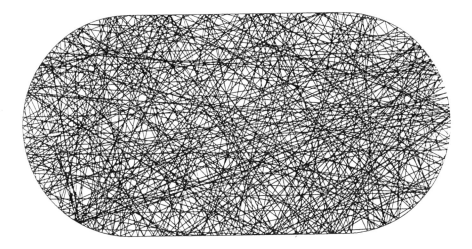

Fig. 1.9. The Bunimovitch Stadium

In concave domains, beams of light or pencils of trajectories are dispersed when reflected from the boundary, as in figure 1.10 on the following page and this implies *ergodicity* (see §§10.5.1.1 for the definition). See Katok et. al. 1986 [788] for the proof. Finding a simple proof is an open problem. However, a good deal is known about concave domains; for example let us just mention that $CF(L)$ is asymptotically exponential in L. Also see §§1.8.6 below for a heuristic relation between concave billiard tables, statistical mechanics, and manifolds with negative curvature.

For both rational angled polygons and convex domains, rigorous proofs of existence of periodic trajectories are quite difficult. There is a third case, a generalization of the *Bunimovitch stadium* which is not completely mysterious; see section 5.3 of Tabachnikov 1995 [1175] for a modern proof that there are periodic trajectories, and also see the references therein. In this case we have ergodicity and $CF(L)$ is exponential in L. These examples are convex (but with flat parts) and came as a shock to the ergodic community in 1979 when most people were convinced that some concavity was required for ergodicity, or if you prefer, that convex domains had too much focusing to achieve ergodicity.

Defocusing is apparent in figure 1.10. In §§1.4.3 we will see an interesting curvature condition needed to construct these examples.

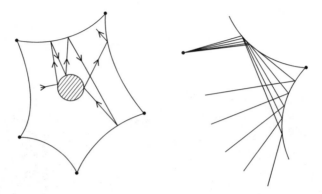

Fig. 1.10. Spreading rays in a nonconvex billiard table

A strictly convex billiard table can not be ergodic, because its flow has caustics (at least if its boundary is six times differentiable, or perhaps less—the optimal result is open), by a fundamental result of Lazutkin, discussed on page 92. But it is an open problem whether there are smooth ergodic billiard tables (the Bunimovitch and Wojtkowski examples are not smooth, having flat parts joined to curved parts with nonvanishing curvature).

One does not know how general these examples are. Let us mention two general results. The first, due to Katok, asserts that for polygonal billiards, $CF(L)$ grows more slowly than exponentially. See Katok's 1987 [785] and also the nice survey of Gutkin 1996 [672]. The proof is extremely interesting and consists of two parts. The first part studies the trajectories using successive reflexions on the sides, unwrapping the trajectory to make it a straight line on the plane. But this is not enough: the second part resorts to a strong theorem in ergodic theory connecting topological and metric entropy. We will define entropy in §§§10.5.1.2.

The second result is a density theorem in the set of all convex billiard tables, found in Arnoux 1988 [70]. For concave billiard tables, the length counting function $CF(L)$ is very well understood. One has the precise asymptotic expression

$$CF(L) \sim \frac{\exp hL}{hL}$$

as $L \to \infty$ with h the entropy, which is known to be positive. For information on higher dimensional billiards, see §§ 1.9.3 on page 97.

1.3 Geometry of Plane Curves and Two Dimensional Point Kinematics

1.3.1 Length

We will briefly recall some classical observations concerning plane curves, but also some less classical ones. Our style will probably appear different from standard expositions. This is intentional, in order to introduce the viewpoint of Riemannian geometry. So even if you think you know everything about plane curves and point kinematics, you might do well to glance at what follows.

We have a plane curve c with a point $c(t)$ running along it, a function of a parameter t, as smooth as needed (i.e. differentiable as many times as needed). The velocity of this curve c is $c'(t)$. As we have already noted, the length of c from $t = a$ to $t = b$ is

$$\int_a^b ||c'(t)|| \, dt .$$

This gives a distance on the curve, which is the time needed to traverse from $c(a)$ to $c(b)$ with unit speed. This inner geometry is the same as the geometry of an interval. By *inner* we mean that you care only about what happens on the curve itself, and not what happens outside. Consequently, all curves having the same length are *isometric*, namely they are the same as metric spaces. Equivalently, there is a one to one map which is an *isometry* between them (i.e. preserves distances). Moreover all of them are isometric to an interval of the same length on the Euclidean line $\mathbb{E}^1$. Surfaces will be completely different; see §§1.6.1.

Alternately, look at the triangle inequality. For the inner metric of a plane curve, at least locally,

$$d(p, r) = d(p, q) + d(q, r)$$

for any three points p, q, r with q between p and r. On a circle, between is a local notion, as you see in figure 1.11. Or think about the word "between". If we call a curve *closed* when it is periodic, then such a curve has a total length, and two closed curves are isometric exactly when they have the same length.

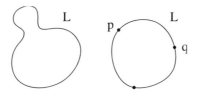

Fig. 1.11. These two curves are isometric

An important remark is in order: the attentive reader might have been puzzled, even outraged, that we never distinguished between a curve as a map $t \to c(t)$ from an interval of the real line, and its image as the set of points $c(t)$ when t ranges through the values of that interval. It is not easy to have a clear view of these two notions, and the relations between them; they are obscure even in many books. We refer the interested reader to chapter 8 of Berger and Gostiaux 1988 [175] for a very detailed exposition of this point. Now we will just say a few words about this distinction.

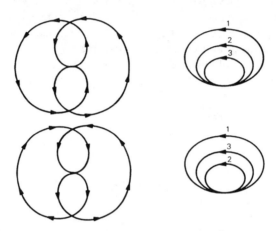

Fig. 1.12. Geometric images you can travel through on different routes

Maps $t \mapsto c(t)$ from an interval of $\mathbb{E}^1$ into $\mathbb{E}^2$ should be called *parameterized curves*. Most often only *regular curves* are allowed, i.e. the speed $c'(t)$ is never zero. Note that here, and also in the sequel, we use two different notions of how quickly something moves: the *speed* $\|c'(t)\|$ which is the norm of the *velocity* $c'(t)$ (see Feynman, Leighton and Sands 1963 [516], page 9-2). We skip intermediate developments (again, see Berger and Gostiaux 1988 [175]) and jump directly to the concept of a *geometric curve*: a line in the plane (not necessarily a straight line). By this we mean a subset of the plane which, near any point, can be reshaped into a straight line, i.e. taken by diffeomorphism of an ambient region of the plane into a straight line. In modern mathematical jargon, such an object is called a *one dimensional submanifold of the plane*. The link between the two notions is that the image set made by all of the $c(t)$ is always, but in general *only locally*, a geometric curve, and conversely a geometric curve admits regular parameterizations, in particular by arc length (i.e. with constant unit scalar speed). In the sequel, we will most often leave it to the reader to understand if parameterized or geometric curves are considered.

A word about orientations of curves: parameterized curves are automatically oriented, while geometric curves can be given two orientations. When you change parameterizations of a curve, two invariants appear: the first is

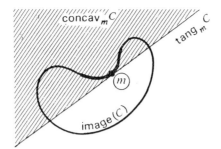

Fig. 1.13. Concavity and tangent line

the *tangent line* to which every $c'(t)$ belongs. The second is defined only for *biregular* curves, i.e. curves for which the acceleration $c''(t)$ is not proportional to the velocity $c'(t)$. The acceleration will then remain in the same open half plane under any reparameterization, and this half plane is called the *concavity* of the curve. It does not depend on orientation. If you parameterize the curve by arc length, the two orientations yield two opposite unit velocity vectors, but one acceleration vector, which is then attached canonically to the curve.

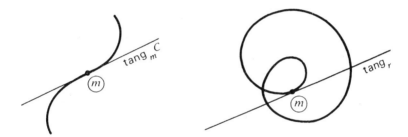

Fig. 1.14. Tangent lines

Everything said here is valid with obvious modification for curves in any Euclidean space $\mathbb{E}^d$.

A closed (periodic) plane curve is said to be *simple* if it is a one-to-one map up to the period (this for a parameterized curve). For a geometric curve, *simple* means a curve which has the global topology of a circle (in the jargon, it is a differentiable embedding in the plane of the circle seen as an abstract one dimensional manifold). We also will assume the speed never vanishes, indeed that it has unit speed.

Do not think that everything is known today about plane curves. A famous problem is to prove that in any plane closed curve, one can inscribe a square. For this problem, and other open ones, see the book Croft, Falconer & Guy 1994 [410]. The oldest unsolved problem in plane geometry is the *equichordal problem*, from 1916: prove that no plane curve can have two equichordal points, i.e. points so that all chords through them have constant length. The problem is studied extensively in note 6.3, chapter 6 of Gardner 1995 [548]. Do not miss the recent revolution in the theory of plane curves: see the end of §1.4.

1.3.2 Curvature

We now look at the outer geometry of plane curves. The inner geometry of curves does not differ from that of straight lines, but the geometry is radically different if we look at the way a curve sits in the plane. We are going to introduce a concept of *curvature* which measures how much a curve differs from a straight line. For a curve seen as a kinematic motion, curvature is directly linked with the *acceleration* vector $c''(t)$. For a geometric curve, we can introduce it by looking at the variation of the length of the curves drawn at a constant distance, called the *parallel curves* (recall the first variation equation 1.2 on page 4). If the absolute value of the infinitesimal change of length of these equidistant curves, close to a point $m = c(t)$, is the same as for a circle of radius r we say that the curve c has *radius of curvature* r at $m = c(t)$, and that its curvature is $K = 1/r$. The formula to compute curvature is

$$K(t) = \frac{1}{r(t)} = \frac{|\det(c'(t), c''(t))|}{\|c'(t)\|^3}$$

where t is any parameter, and boils down to

$$K(t) = \frac{1}{r(t)} = \|c''(t)\|$$

when t is a unit speed parameterization (i.e. arc length parameterization). Of course a circle of radius R has constant curvature equal to $1/R$. Another way to look at it is the following: the circle $C(t)$ which is defined as tangent to the curve at the point $c(t)$ and has radius $r(t)$ is the circle which has the most intimate contact with the curve (technically, the contact is third order, meaning that the curve and the circle have the same first three derivatives at that point). This circle is called the *osculating circle* to the curve at $m = c(t)$.

To rest a little, we mention an interesting fact rarely found in textbooks. When at a point m of the curve c, the curvature varies, that is to say $dK/dt(m) \neq 0$, then the osculating circle at m crosses the curve (guess where it is inside, and where outside). A baffling consequence is that the osculating circles of of a curve whose curvature is never critical never intersect one another. See the picture in figure 1.15. This picture gives two counterexamples. First, consider the unit tangent vectors to this family of circles. We get a continuous vector field, alias a differential equation in the plane. Now at every point of the curve there are at least *two* different integral curves of that vector field: namely the curve and the circle. The moral is that uniqueness of solutions of ordinary differential equations does not follow only from continuity. Second, consider the envelope of this family of circles. In many books it

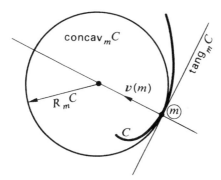

Fig. 1.15. Finding tangent line, concavity and curvature

is "proven" that the envelope is the limit of the so-called characteristic point, the point of intersection of close-by curves. This is wrong (unless we invoke "complex circles").

There is another geometric definition of the curvature. Parameterize the curve by arc length, so that $\|c'(t)\| = 1$ identically. Then $c'(t)$ is a unit vector, running through the unit circle of the plane. The curvature is the speed of its rotation. More precisely, one needs first to orient the plane, and choose an oriented direction in the plane. Then one can define an angle $\alpha(t)$ for $c'(t)$ which is a real number, defined only up to integer multiples of 2π. The curvature is $K(t) = d\alpha/dt$.

To get rid of the 2π ambiguity, there is an important fact which we are going to use below. For regular curves, when t runs through the interval of definition, one can follow $\alpha(t)$ by continuity and then define a map $t \mapsto \alpha(t)$ into $\mathbb{R}$. Note that if the curve is closed with period T then $\alpha(t+T)$ will in general differ from $\alpha(t)$ by an integral multiple of 2π.

We saw above that straight lines are the shortest paths between two points. Let us look at this again, now in a more sophisticated way. We start with a curve c with ends $p = c(a)$ and $q = c(b)$. Suppose it is as short as any curve can be with these extremities. Consider any one parameter family of curves c_μ with the same ends, with $c = c_0$, and compute their lengths. The fact that the length should be a minimum implies in particular that the first derivative of the length as a function of the parameter μ has to vanish. Computation yields

$$\frac{d \operatorname{length}(c_\mu)}{d\mu}\bigg|_{\mu=0} = -\int_a^b \langle f(s), c''(s) \rangle \, ds \qquad (1.3)$$

where the vector valued function $f(s)$ (which we can choose to be orthogonal to the curve everywhere) estimates the derivative of the normal displacement of the family of curves. This is also a first variation formula. Note that this formula is nothing but the integrated version of the definition we previously gave for curvature. A classical calculus trick (which is not obvious) proves that if we want all of these derivatives to vanish, namely the above integral

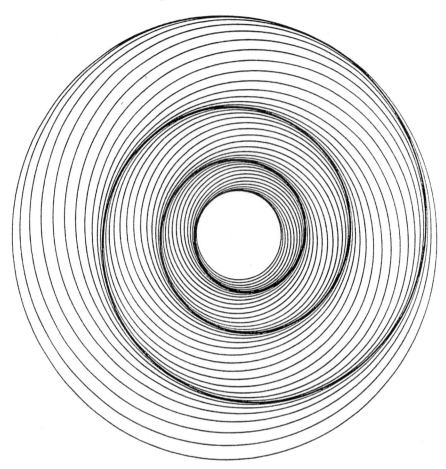

Fig. 1.16. Picture drawn for us by Etienne Ghys

to be 0 whatever the function f vanishing at $s = a$ and $s = b$, then $c''(t) = 0$ identically, and in particular our curve should have everywhere vanishing curvature. Curves with everywhere vanishing curvature are, of course, straight lines. Beware: what we have proven up to now by this method is that shortest paths can only be found among straight lines. Not every zero of the derivative of a function is a minimum. For a local minimum, a sufficient condition is to have positive second derivative. But we still have to search for an absolute minimum. The above proof is interesting because it is the simplest model of a field called *the calculus of variations*, and gives a general scheme to start searching for shortest paths. We will use it soon in §1.6 but principally in chapter 6. To conclude this story for the Euclidean plane, we know that if shortest paths exist, they should be straight lines. Then we check directly as above.

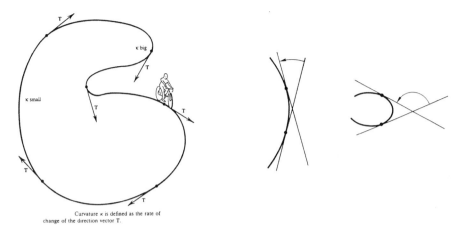

Fig. 1.17. Different definitions of curvature

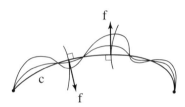

Fig. 1.18. Planar calculus of variations

There is a notion of *algebraic curvature* k as opposed to the scalar curvature K, linked by $K = |k|$. This is defined as soon as the curve is regular and oriented and the plane is also oriented, or (equivalently) if one side of the geometric curve is chosen, or (equivalently) if one of the two possible unit normal vectors is chosen. This side need not be the concavity side. Denote by $n(t)$ the unit normal vector you have chosen. Remember now that $c''(t)$ is an invariant. The algebraic curvature of c at t is the real number $k(t)$ so that

$$c''(t) = k(t)n(t)$$

A nice theorem about algebraic curvature:

Theorem 1 (More tightly curved curves) *If two curves c_1, c_2 are arc length parameterized, and have respective curvatures satisfying $k_1(t) \geq k_2(t)$ and start at the same initial point in the same direction, then the curve c_1 lies entirely inside the curve c_2.*

(because it is "more curved" as intuition tells you). For a proof see Chern 1989 [366] and exercise 7, section 5-7 of do Carmo 1976 [451]. Beware that you need simple curves if you want a global result. If not, you just get a local one. The

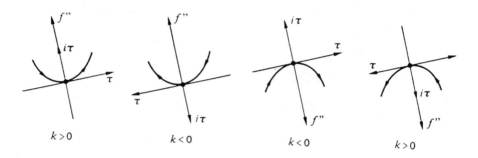

Fig. 1.19. Picturing algebraic curvatures

above theorem is one of the simplest ones in *control theory*. You will see many, many others later on. Note the following amusing consequences. Consider a simple closed plane curve, and the point where the curvature is maximum. Then the osculating circle at that point can keep rolling inside the curve all the way around. Similarly, the curve itself can roll inside its osculating circle of minimum curvature.

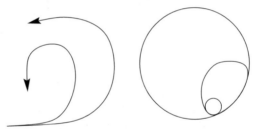

Fig. 1.20. (a) One is more curved than the other (b) The little circle can roll inside the closed curve, and the closed curve can roll inside the big circle

Let us make more precise our claim above that curvature is the only invariant of plane curves. More precisely the algebraic curvature is a characteristic invariant: if in the above inequality for curvatures we have equality ($k_1(t) = k_2(t)$ for every t) then the curves c_1 and c_2 are the same. Moral: if you know the value of the curvature as a function of the arc length then you know the curve completely. If the curves do not have the same initial point and direction, this just means that they are deduced one from the other by a Euclidean motion. The simplest example: a curve with non-zero constant curvature is part (or the whole) of a circle.

A spectacular application of this property of arcs of circle is the manufacture of a perfect straightedge. Take two pieces (of anything suitable for the job) and rub one against the other. The above theorem tells you that such a rubbing will eventually carve the two bounding curves into perfect circles. Now take three pieces, and rub them two by two. The boundaries will be circles, but the pieces will have to exhibit opposite convexity, i.e. equal algebraic curvature with opposite sign. For three, this is not possible unless the common curvature is zero. Naively, one might imagine that a laser can draw straight lines, but this is impractical. Today, lasers are used only to check the accuracy of the straightedge, after it is manufactured following the above recipe; once the laser has reported its evaluation of the straightedge, corrections are carried out by skilled hands using emery powder. Before the invention of the laser, accuracy was tested with interferences.

The same technique is still the only one available to make "perfect" planes: rub three surfaces against one another, and the result has to be a plane. To prove it, use the fact which we will quote on page 47 that pieces of a surface, all of whose points are umbilics, are pieces of spheres. To check accuracy, one doesn't use a laser, but drags a straightedge around on the surface, or a comparison plane, or applies an interferometer.[3] Rubbing only two pieces will yield pieces of a sphere, and is still the only method to sculpt spherical lenses.

In the same spirit, to fashion "perfect" balls, (which have thousands of practical applications, for example as ball bearings), one method was to put approximate balls inside a box, and stir them. Proof that the limit object is a sphere is not as elementary as proofs of the efficacy of the previous production methods, and was only recently established in Andrews 1999 [52]. Since 1907, balls for ball bearings have been produced by rolling rough balls between two plates with toroidal concentric groves and random indentations. See Berger 2003 [173] or technical books for more details.

For the story of curves in $\mathbb{E}^3$ see Berger and Gostiaux 1988 [175], do Carmo 1976 [451], or Stoker 1989 [1160]. We will just mention briefly what happens there. This time there are two characteristic invariants, curvature and torsion, but the curves need to be biregular, i.e. the acceleration should never vanish. Things also extend similarly to any $\mathbb{E}^d$, this time with $d-1$ invariants: see Spivak 1979 [1155]. But the first invariant is always the curvature and is defined simply as the norm $\|c''(t)\|$ for any arc-length parameterization. Only straight lines have everywhere zero curvature. For the kinematician this is the old fact that points with no force applied to them move along straight lines and at constant speed.

Take helices for example. They can be characterized as the curves with constant curvature and torsion. A practical application, in the spirit of the one just above: graduate a straightedge to make ruler; this is equivalent to

---

[3] An *interferometer* is a device which uses the interference of two waves (radio, acoustic, or light waves will do) to make very precise distance measurements.

making a protractor, dividing a circle in equal parts, etc. It is clear that you can do it if you can produce a screw (a helix). This is even simpler than a straightedge. Just start with an approximately good pair of a bolt and a nut and rub one against the other. At the end we will get a perfect bolt, and a perfect nut, because the common boundary will be made up of a union of curves with necessarily constant metric invariants, hence easily seen to be helices.

1.4 Global Theory of Closed Plane Curves

1.4.1 "Obvious" Truths About Curves Which are Hard to Prove

Systematic references for what follows are Stoker 1989 [1160], do Carmo 1976 [451], Berger and Gostiaux 1988 [175], among others. See also the new events in Arnol'd 1994 [65] and other references in §§1.4.6.

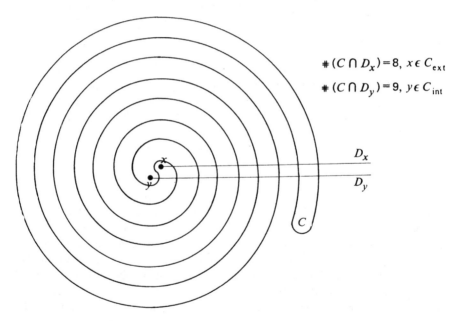

Fig. 1.21. A trick to find when a point belongs to the interior or to the exterior of a curve

Theorem 2 (Jordan curve theorem) *A simple closed plane curve has an interior and an exterior.*

This implies the more common statement that a simple plane closed curve separates the plane in two regions. More precisely the plane with the curve deleted from it has two connected components and moreover one is bounded whilst the other is not. The bounded one is of course the one called the interior. An immediate corollary is that a simple closed plane curve has a given side, the *interior* one, and by the above this implies that its algebraic curvature is defined, independently of an orientation of the curve and an orientation of the plane. More generally, closed plane curves have a well defined algebraic curvature. Note again that the interior side need not be concave. From the viewpoint of algebraic topology the interior is homeomorphic to a disk and so it is simply connected.

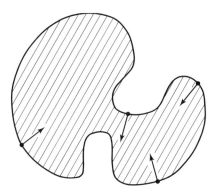

Fig. 1.22. Inner normals

This first result is valid even for curves which are not smooth but only continuous. The next result we will cite needs two derivatives and claims that any simple plane closed curve satisfies

$$\int k(s)\,ds = 2\pi \ . \tag{1.4}$$

The proof cannot avoid using the so-called Umlaufsatz (the turning tangent theorem) which asserts intuitively that the total rotation of the velocity (or of the tangent line to the curve) is equal to 2π. If you visualize this velocity vector as a point of the unit circle of $\mathbb{E}^2$ it means that when the parameter t runs through a complete period of definition of the curve the continuous determination $\alpha(t) \in \mathbb{R}$ which was considered above runs exactly from 0 to 2π. This does not mean that $\alpha(t)$ is monotonically increasing: it means that the unit circle is at the end algebraically covered once. Beware that the same point of the unit circle can be obtained many times but that, for example, if you come back a second time at a value previously passed you will have to come again a third time to "erase" it. To make all this mathematically precise one needs the notions of universal covering, simple connectivity, etc. The Umlaufsatz was essentially known to Riemann but a rigorous proof is pretty hard (try one if not convinced). Simplicity of the curve is necessary here. For closed (periodic) curves the *turning number*, defined as the number of rotations of the tangent line, hence

$$\text{turning number} = \frac{1}{2\pi} \int k(s)\,ds$$

can be any integer including 0. See the pictures below as well as Berger & Gostiaux 1988 [175] for an elementary proof and interesting theorems about that number. Also see §§1.4.4.

A basic remark is in order: the turning tangent theorem is a very particular case of the Gauß–Bonnet theorem given in §§3.1.2 for general surfaces, when moreover the curve can admit corners. See §§1.4.4.

The Umlaufsatz has many other corollaries. The first is a surprising explicit tube formula. Consider a simple plane closed curve c and take the tube around it of radius ε. Beware that the boundary of this tube might not be smooth if ε is too large. But for ε small enough,

$$\text{area of this tube} = 4\pi\varepsilon \cdot \text{length}(c)$$

Please check this on some examples. The length of the inner boundary curve is

$$\text{length}(c) - 2\pi\varepsilon$$

An amusing consequence: consider the belt highway of a big town and a car running through it completely. Is it more economical to drive on the inside lane rather than the outside? Not much, because the above shows that the difference of elapsed distance is always $2\pi\varepsilon$ where ε is the distance between the two lanes (say something like less than 100 meters at the end), this being true whatever the shape and the total length of the belt.

The second is a convexity result: suppose a simple closed plane curve satisfies $k(t) < 0$ for every t. Then its interior is a convex set. Try to prove it your own way before dismissing this result.

1.4.2 The Four Vertex Theorem

On a simple closed plane curve the curvature $k(t)$ is a continuous function (a periodic one). So it admits at least one maximum and one minimum. At such a point the derivative $k'(t)$ vanishes. Points where $k'(t) = 0$ are called *vertices* (think of ellipses if you are interested in the etymology). So any closed plane curve has at least two vertices. Now try to draw a curve with only two vertices: you will have a hard time (this is no reason not to try it seriously) because the four-vertex theorem asserts precisely that every closed plane curve has at least four vertices. In fact the proof of Osserman (see Osserman 1985 [983]) shows that generically it has at least six vertices. Ellipses have exactly four vertices. The converse of the four-vertex theorem was proven only recently in Gluck 1971 [569]: a function on a circle can be realized as the curvature

1.4 Global Theory of Closed Plane Curves   21

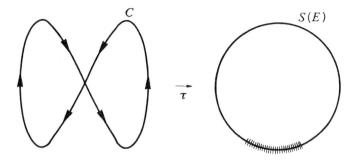

Turning number equal to 0

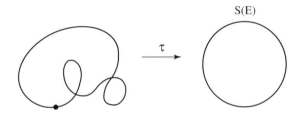

Turning number equal to 1

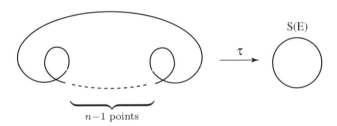

Turning number equal to n

Fig. 1.23. Turning numbers equal to 1, 0 and n

of a closed curve exactly if it admits at least two maxima separated by two minima.

Recently the four vertex theorem was applied to physics as follows. In a planar world with no gravitational forces, a plane convex body D stands in equilibrium between two liquids of different capillary constants and with a straight line common boundary. From the physics, at both of the two points of separation the angle θ between the tangent to the curve and the line of separation of the two liquids satisfies

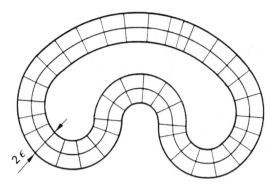

Fig. 1.24. The belt highway theorem

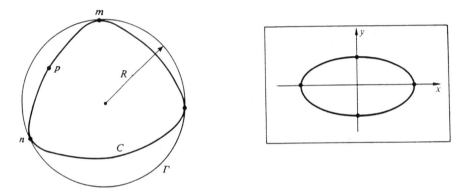

Fig. 1.25. A curve with six vertices

$$\cos\theta = \frac{\gamma_{13} - \gamma_{23}}{\gamma_{12}}$$

where γ_{ij} is the energy per unit area between phases i and j, the three phases here being 1 for the first liquid, 2 for the second liquid and 3 for the body. Using the four-vertex theorem and being quite tricky one can prove that there are at least four equilibrium positions, two of them moreover being stable ones (and two unstable): see Raphaël, di Meglio, Berger & Calabi 1992 [1048] or the latest on the story in Berger 1996 [168].

1.4.3 Convexity with Respect to Arc Length

Very recently an interesting local condition on a curve appeared in the literature, namely that the radius of curvature should be a concave function of the arc length:

$$\frac{d^2 r}{ds^2} \leq 0.$$

1.4 Global Theory of Closed Plane Curves

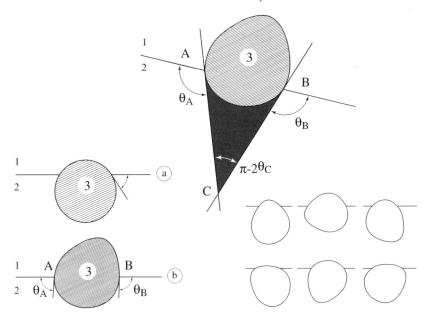

Fig. 1.26. (a) Angles of contact. (b) A spherical particle and convex particle at a fluid–fluid interface. (c) A convex particle with six equilibrium positions

As an exercise compute which parts of an ellipse obey it. This inequality allows one to build convex billiard tables (see §§1.2.3 above, and Wojtkowski 1986 [1274]) for which almost every nonperiodic trajectory is everywhere dense in space as well as in phase (they are called *ergodic*). The inequality controls geometric optics: it insures divergence after two reflections (see figure 1.27).

Fig. 1.27. Convexity with respect to arc length

1.4.4 Umlaufsatz with Corners

An important remark is in order, which might have already occured to the reader. Consider a triangle in the plane and think of it "*à la Umlaufsatz*", that is to say: drag the unit tangent vector along as you traverse each side. As a vector it keeps being constant but when you arrive at a vertex of angle A you have to turn by exactly $\pi - A$. Coming back to the origin you have finally turned from

$$(\pi - A) + (\pi - B) + (\pi - C) = 3\pi - (A + B + C) = 3\pi - \pi = 2\pi$$

As you might guess there is a formula covering both cases, namely for curves with reasonable singularities. We leave to you to write it, because we will give a much more general one in lemma 18 on page 112.

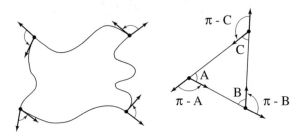

Fig. 1.28. Umlaufsatz with corners

1.4.5 Heat Shrinking of Plane Curves

Recently the global theory of plane curves became richer with results on the following evolution problem (sometimes also called the *heat shrinking curve equation*). This is one instance of the modern invasion of geometry by dynamical ideas, to be compared also with the invasion by iteration and combinatorics. Consider a simply closed plane curve and try to make it more like a circle by a systematic dynamical flow. The idea is to deform by moving the curve along its normal direction at a speed proportional to the algebraic curvature. This is natural; at points where the curve has large curvature, one should reshape it more. This leads to a parabolic partial differential equation, which one can call a heat equation, because it turns out to be very similar to the standard heat equation as in §1.8. But it is very hard to prove that evolution is possible, first for a short time and then forever. Moreover one can prove that the curve, suitably normalized, converges to a circle as expected. A good reference is the survey of Linnér 1990 [871] and Gage and Hamilton 1986 [537]. Related problems were studied by Andrews 1996 [51] and 1999 [52]. A practical application: when one drops some liquid on a hot plate of metal, the drop evolves according to this heat equation; see Wang 2002 [1242].

1.4.6 Arnol′d's Revolution in Plane Curve Theory

Is the global theory of curves just beginning? We alluded above to the turning number of a plane closed curve. The Whitney–Grauenstein theorem asserts

1.4 Global Theory of Closed Plane Curves

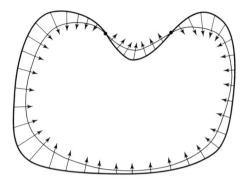

Fig. 1.29. A curve undergoing heat shrinking

that two immersed (not embedded in general) plane curves with the same turning number can be deformed one into the other through proper immersions. So one can say that the turning number is a characteristic invariant.[4] Until lately, that theorem remained isolated, along with the four vertex one, as well as a mysterious formula of Fabricius–Bjerre 1962 [500] relating the number of double tangents to the numbers of inflexion points and crossings (see section 9.8 of Berger & Gostiaux 1988 [175]). Very recently, Vladimir I. Arnol'd started a revolution when studying plane curves, hammering out a general frame to encompass these results. This is a very active field of research today. There is a price to pay of course. It is quite reasonably expensive. The curve has to be considered together with its tangent lines, so that the object to study is the set of all oriented tangent directions to the Euclidean plane, an object of dimension three (not two) and inside it the curve consisting of the tangents of a given plane curve. The three dimensional space has the topology of the inside of a torus; an immersed circle in the plane will lift to a knotted closed curve inside that torus and then one will appeal both to the theory of knots and the field of contact geometry, intimately linked to symplectic geometry, both very active fields of research. Deformations of curves will then be interpreted as wave fronts in a geometrical optics language, following flows given inside that torus by a more or less canonical vector field. Some references: the initial bomb was Arnol'd 1994 [65], then Aicardi 1995 [13], Ferrand 1997 [513], and Chmutov & Duzhin 1997 [374]. One can find an informal presentation of these theories, along with a discussion of the open problems in the theory of real algebraic plane curves, in chapter V of the work in progress Berger 2003 [173].

---

[4] For the historian, it was discovered by Pinkall (see Karcher & Pinkall 1997 [782]) that the Whitney–Grauenstein theorem appears in Boy 1903 [249] as a footnote. We will meet Boy's article again on page 136.

1.5 The Isoperimetric Inequality for Curves

The isoperimetric inequality for closed plane curves tells us that among all such curves bounding regions of fixed area, the circle, and only the circle, is the shortest. Or, among closed curves with fixed length, the circle encloses the greatest area. Explicitly, if A is the enclosed area, and L the perimeter,

$$L^2 \geq 4\pi A$$

with equality only for circles. This was known to the Greeks, but the history of the proof is fantastic. Let us just say that the first solid proof was written by Schwarz around 1875; references for this history are Osserman 1978 [982] and section 1.3 of the article by Talenti in Gruber and Wills 1993 [661]. One might also consult chapter 12 of Berger 1987 [164]. For the plane, and arbitrary dimensions as well, there is the excellent reference Burago & Zalgaller 1988 [283]. To our knowledge, it is the only source of not just one but all of the classical proofs of the isoperimetric inequality, handling the cases of equality with minimal regularity assumptions on the boundary.

Do not be surprised that we are going to spend quite a lot of time on isoperimetric inequalities. Generalizations of them to Riemannian manifolds are very important, and surprisingly recent. We believe that it is useful to explain the inequality in the plane because the general case is quite complicated.

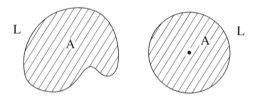

Fig. 1.30. Two planar domains

One of the most natural proofs should be one using Stokes' theorem. The desired inequality is a relation between the area, which is an integral over the interior of the region, and the perimeter: an integral over the boundary. Stokes' theorem is a basic equality between integrals over domains and other integrals over their boundaries. Astonishingly, it was not until 1957 that Knothe found a proof of the isoperimetric inequality using Stokes' theorem.[5] This proof remained unnoticed for a long time, and was brought to light by Misha Gromov, who communicated it to the author in the 1980's. One can find it in Berger 1987 [164] and also in chapter VI of Sakai 1996 [1085].

---

[5] Peter Petersen points out that Schmidt's proof in Schmidt 1939 [1105] uses Stokes' theorem, even if obliquely.

1.5 The Isoperimetric Inequality for Curves

Since closed curves are periodic objects, an analyst or a physicist will naturally use harmonic analysis, i.e. Fourier series. Hurwitz established just such a proof in 1901 (see section 7.4 of Groemer's contribution to Gruber & Wills 1993 [661]), and it is quite simple. Contrary to the Gromov–Knothe proof, it has the major drawback that, at least up to today, nobody has been able to extend it to higher dimensions using the natural extension of Fourier series, *spherical harmonics*. See §1.9 and Gruber & Wills 1993 [661] for more on spherical harmonics.

The other most natural proof should be as follows: pick up a curve which realizes the minimum of the ratio L^2/A and just prove that it is a circle. But in exchange it is the most expensive proof, because one needs very hard results from analysis to ensure first the existence of such a curve and second that such a minimizing curve is differentiable. But this is known now and more, in any dimension and also in general Riemannian manifolds (with the proviso that the minimizing object is differentiable only almost everywhere, but that is enough as we will see in §§§7.1.2.2). As for plane curves, let us consider our minimizing curve c. Follow the same scheme as above when we proved that straight lines minimize length among curves with fixed ends. Again we employ the (sophisticated but fruitful) notion of curvature. Here we have to express that the length is critical, among a family of curves deforming the given one, as long as all of the curves enclose the same area. The derivative is easily computed (using the first variation formula 1.3 on page 13) to show that the algebraic curvature satisfies:

$$\int_a^b k(t)\,f(t)\,dt = 0$$

for every function $f(t)$ such that

$$\int_a^b f(t) = 0\,.$$

The classical Lagrange multiplier technique implies immediately that $k(t)$ has to be constant (a necessary, but perhaps not sufficient, condition). But we know that only circles have constant curvature (prove it directly). This comes from the above statement to the effect that curvature is a characterizing invariant. We will see in section §1.6 that this technique, with one trick more, extends to the isoperimetric inequality for surfaces in $\mathbb{E}^3$, and in §§§7.1.2.2 we will watch it extend to Riemannian manifolds with Ricci curvature bounded from below.

The reader will ask now what were the older, classical proofs. Proving that only circles can achieve equality turns out to be difficult in every demonstration of the inequality, which is otherwise simple (this for general dimension, see §§1.6.8). But for the plane case, there is a very tricky and completely elementary geometric "quadrilateral" argument due to Steiner which can be seen for example in the pictures below (see also Berger 1987 [164]).

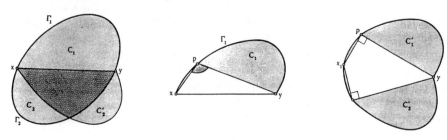

Fig. 1.31. Steiner's quadrilateral argument

Two proofs are classical and not too difficult: the one based on Steiner symmetrization and the one based on the Brunn–Minkowski theorem. We refer the reader to Berger 1987 [164] for both of them; see also the entry of Talenti in Gruber & Wills 1993 [661] and the very comprehensive reference Burago & Zalgaller 1988 [283]. Steiner symmetrization is too important, powerful and geometric an idea to be concealed. Briefly it goes like this. To every plane domain D and every straight line d we associate the symmetrization $d(D)$ of D with respect to d as depicted in figure 1.32. Slice D along every perpendicular line to d and then slide every slice in order that its middle sits on d. This yields by construction a domain which is symmetric with respect to d. One shows quite easily that the operation $D \mapsto d(D)$ keeps areas constant and can only lessen the perimeter. With some general topology one then concludes the inequality using the fact that disks are the only domains which are symmetric with respect to every direction.

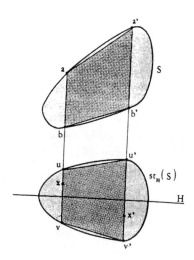

Fig. 1.32. Steiner symmetrization

Performing line symmetrizations with respect to suitably chosen directions is interesting to visualize: the convergence to a circle is very fast. We leave the interested and capable reader to see this on a computer.

Quite recently a very expensive proof has arisen, based on the evolution equation mentioned in §§1.4.5. Under this deformation the initial curve will look more and more like a circle; one discovers that, under the evolution, the isoperimetric ratio does not increase. See Linnér 1990 [871] for this. This proof is not really interesting when applied to a planar curve—its true purpose lies in the fact that it also works for Riemannian surfaces to yield the existence of simple periodic geodesics, a hard and expensive topic; see a lot about this in §10.4.

1.6 The Geometry of Surfaces Before and After Gauß

Unlike for curves, we will only define geometric surfaces in the space $\mathbb{E}^3$ (see why for yourself). Geometric curves are everywhere locally differentiable nice one-dimensional subsets in $\mathbb{E}^2$, while surfaces are the nice subsets of $\mathbb{E}^3$, nice in the sense that there is a local map from $\mathbb{E}^3$ to itself sending the local piece of the surface onto an open subset of a plane. That map must be differentiable, one to one and with a differentiable inverse. The jargon is: surfaces are two dimensional submanifolds of $\mathbb{E}^3$. Equivalently near any point, M can be written as $f^{-1}(0)$ where f is a smooth locally defined function $f : U$ open $\subset \mathbb{E}^3 \to \mathbb{R}$ whose derivative never vanishes. The fact that we do not introduce parameterized surfaces does not prevent our using local parameterizations of geometric surfaces, also called local charts. Moreover, unless explicitly stated, we require our surfaces to be connected. Alexandrov gave a very brief and accessible account of the theory of surfaces in chapter 7 of Aleksandrov, Kolmogorov & Lavrent'ev 1956 [15], translated into English in Aleksandrov, Kolmogorov & Lavrent'ev 1999 [16]; also see Morgan 1998 [936]. Monge gave a terse explanation of surface theory, translated to English in Struik 1969 [1165]. For an exposition of the submanifolds of $\mathbb{E}^d$ see Berger & Gostiaux 1988 [175].

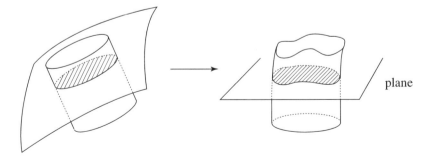

Fig. 1.33. Straightening a piece of a surface

1.6.1 Inner Geometry: a First Attempt

Following the scheme we used for plane curves we first study the *inner geometry* of our surface. The way that M sits in $\mathbb{E}^3$ will be studied next. Imagine people living on the surface M and ignoring things which happen outside of it. Those people walk, and they are interested in the distance needed to travel from one point p to another point q of M. Mathematically, we consider a curve drawn on M—it is forbidden to leave M and to go, even for one second, into the outside world. Because such a curve c lies in $\mathbb{E}^3$ it has a length, and of course the surface's inhabitants want to find the shortest path from p to q. We define the distance from p to q in M as the infimum of the lengths of all curves going from p to q in M, and denote it by $d_M(p,q)$ (or $d(p,q)$ if there is no need to be more precise):

$$d(p,q) = d_M(p,q) = \inf\{\text{length}(c) \,|\, c \text{ is a curve in } M \text{ from } p \text{ to } q\}. \quad (1.5)$$

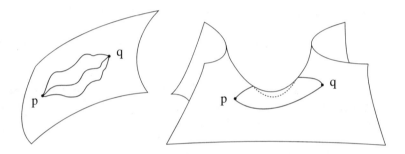

Fig. 1.34. Curves drawn on a surface

It is easy to see that $d = d_M$ is a metric on the set M (since M is connected). This metric is sometimes called the *inner* or *intrinsic* metric (see again §§1.2.2). Beware that d_M is not what is called the *induced metric* on M from that of $\mathbb{E}^3$. The only surfaces M for which the induced metric and the inner metric coincide are portions of a plane. This inner metric is determined by the Euclidean structures on the collection of the tangent planes $T_m M$ when m runs through M, as we will see.

As a first example, consider the unit sphere $S^2 \subset \mathbb{E}^3$. How do we compute $d(p,q)$? We should find the shortest paths. Assume we know, or have made the right guess: the shortest paths on spheres are arcs of great circles (i.e. the intersections of the sphere by planes through its center). From this the distance is easily computed:

$$d(p,q) = \arccos\langle p, q\rangle \in [0, \pi].$$

To prove that the shortest paths are indeed great circular arcs, we take such an arc, less than half of a great circle, and arrange by rotation that it be

1.6 The Geometry of Surfaces Before and After Gauß

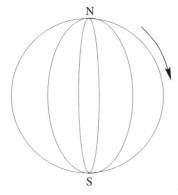

Fig. 1.35. Latitude grows at unit rate along longitude lines

a longitude line, and that it touches neither the north nor the south pole; see figure 1.35. The latitude (expressed in radians) increases at unit rate as we traverse the longitude line at unit speed, i.e. change in latitude is arc length. One now needs to see only that the rate of increase of latitude along any unit speed curve is slower than unit rate (i.e. slower than arc length) except where the curve is tangent to a longitude line (due south). We conclude that total arc length must be greater than difference in latitude, except along longitude lines. If we have a longitude line, and any other path with the same end points, then the other path has the same total change in latitude between its end points, and so must have greater length. This is an important idea because it is the basic scheme of the proof that geodesics locally minimize length on arbitrary surfaces; see equation 3.8 on page 116.

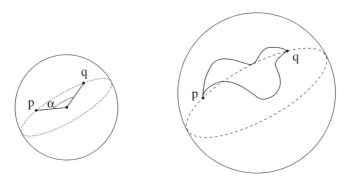

Fig. 1.36. Curves on spheres

The *principal formula of spherical trigonometry* provides the sphere an analogue of the Euclidean formula 1.1 on page 2:

$$\cos a = \cos b \cos c + \sin b \sin c \cos A \qquad (1.6)$$

32   1 Euclidean Geometry

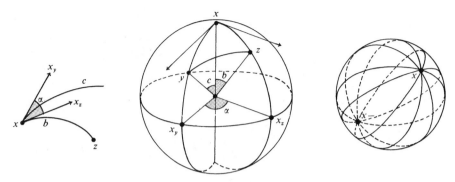

Fig. 1.37. Spherical trigonometry

for every spherical triangle $\{p,q,r\}$ with side lengths $a = d(q,r), b = d(r,p), c = d(p,q)$ and where A is the angle at the vertex p. It is a good exercise to extend this formula to a sphere of radius R, let R go to infinity and prove that the formula converges to the Euclidean one. Another exercise is to check that the sphere has the strict triangle inequality, which easily comes out of equation 1.6. Conclusion: shortest paths on the sphere are the arcs of great circles provided we keep them of length smaller than or equal to π. In particular, shortest paths between two given points are unique with the sole exception of *antipodal* points. The word "principal" is used because, as with equation 1.1 on page 2, all the other formulas for spherical triangles can be deduced from this one by pure trigonometrical computations.

Another consequence of the formula is that the sphere is never, even locally, isometric to a part of the Euclidean plane $\mathbb{E}^2$. From the inner geometry distance measurements, we can compute angles. But then the angle sums of a spherical triangle will be too large to be those of a planar triangle. Most surfaces are not even locally isometric to $\mathbb{E}^2$ (see §§3.1.1 for proof). The case of curves was radically different, remember §§1.3.1. This explains why cartography is a whole world in itself; if interested see chapter 18 of Berger 1987 [164].

Before leaving the sphere, recall an old formula whose importance for the sequel is not to be underestimated. The formula says that for a spherical triangle T with angles A, B, C its area is given by

$$\mathrm{Area}(T) = A + B + C - \pi. \tag{1.7}$$

The real history of this formula seems to have come to light only very recently. It was discovered by Thomas Harriot (1560-1621) in 1603 and published (perhaps rediscovered) in 1629 by Albert Girard (1595-1632). See references on page 55 of Ratcliffe 1994 [1049], a fascinating and extremely informative book.

Note 1.6.1.1 (Space forms) Returning to formulas 1.6 on the previous page and 1.1 on page 2 we see that spheres share with Euclidean spaces the

1.6 The Geometry of Surfaces Before and After Gauß

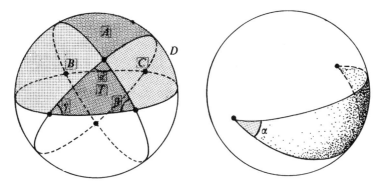

Fig. 1.38. Spherical areas and angle sums

basic philosophy of §§1.2.1. The tempting question to find all such spaces was in the minds of many mathematicians starting in the second half of the 19th century and thereafter. See §§4.3.2, note 4.3.2.3 on page 182 and §§6.3.2 for the continuation of this story. ♦

1.6.2 Looking for Shortest Curves: Geodesics

Let us talk to somebody who does not know how to find the shortest paths on the sphere. We want to apply the first variation technique of §§1.3.2 and find the formula which is the analogue of equation 1.3 on page 13. Because we can displace curves only under the proviso that the curves are traced entirely on our surface M, only the tangential part of a variation can be chosen freely. At a point $c(t)$ of the considered curve pick up a unit vector $n(t)$ tangent to the surface and at the same time normal to the curve. Note that (as for plane curves) there are two possibilities up to ± 1. We parameterize c by its arc length s.

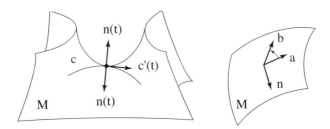

Fig. 1.39. M is oriented and $\{a, b, n\}$ is positively oriented

The first variation of the lengths of a one parameter family of curves c_α around $c = c_0$ is

$$\left.\frac{d\,\text{length}\,(c_\alpha)}{d\alpha}\right|_{\alpha=0} = -\int_a^b f(s)\,\langle c''(s), n(s)\rangle\,ds \qquad (1.8)$$

where the variation of the curves is

$$\left.\frac{\partial c_\alpha}{\partial \alpha}\right|_{\alpha=0} = f(s)n(s)$$

We give a name to the important quantity $\langle c''(s), n(s)\rangle$, calling it the *geodesic curvature* of c. It will be denoted by $k_g = k_g c$. Geodesic curvature changes sign when a different choice is made for the unit normal vector. The *scalar geodesic curvature* is just its absolute value $|k_g|$. From equation 1.8 we see that

Lemma 3 *The geodesic curvature of a curve on a surface depends only on the inner metric of the surface, not on how the surface is embedded in $\mathbb{E}^3$.*

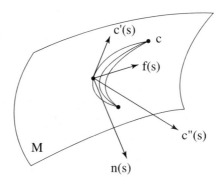

Fig. 1.40. Variation of a curve on a surface

As before we deduce from this formula that the curve c can be shortest only if its geodesic curvature vanishes identically: $k_g(t) = 0$ for every t. This amounts exactly to ask the curve c to have acceleration normal to the surface. This condition was known to Jacob Bernoulli I, who taught it to his student Euler in the beginning of the 18th century. It appeared for the first time in written form in Euler's work. Such curves are called *geodesics* of the surface. This justifies the name of geodesic curvature because k_g is a measure of how the curve under consideration differs from a geodesic. Unless otherwise stated, geodesics will always be parameterized by arc length.

In an intuitive sense, to follow a geodesic is just walking straight ahead in front of you. A more picturesque (approximate) way to get geodesics is to roll along the surface a small buggy made of two wheels of equal radii joined by a short axle, as in figure 1.41 on the facing page. We leave to the reader the proof of this fact. The kinematic interpretation of geodesics is

1.6 The Geometry of Surfaces Before and After Gauß

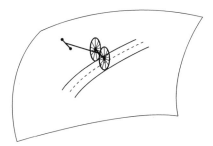

Fig. 1.41. A cart traverses geodesics

that they are the trajectories followed by a point moving on the surface with no force applied to it. In some sense you should feel no inner acceleration. Shortest paths are necessarily geodesics but geodesics are not in general the shortest paths between points, as we will see later. We will call a path a *segment* if it is the shortest path between its end points (some books call these *minimal geodesics*). Geodesics are segments only between points which are close enough to one another (think of the sphere). This comment is valid in any Riemannian manifold; see §§6.1.1. The distance between end points of a segment is exactly the length of the segment. To give the reader a flavor for the general Riemannian manifold, we study the theory of geodesics and of the shortest paths in a somewhat detailed way. This because there is not much difference in difficulty between surfaces and general Riemannian manifolds.

There are extremely few surfaces on which geodesics (and a fortiori shortest routes) can be more or less explicitly determined. But one can get a reasonably general statement first as follows. We want to find out if geodesics exist. For a general surface the simplest possible chart is the one given when one considers the surface (locally) as the graph of a function, and this is always possible (perhaps after relabeling the Cartesian coordinates). So we write M locally as the image of the map

$$(x, y) \mapsto (x, y, F(x, y)) \tag{1.9}$$

where (x, y) runs through some open set in $\mathbb{E}^2$.

Our curve will be c given by

$$t \mapsto (x(t), y(t), F(x(t), y(t)))$$

and we just write that its acceleration $c''(t)$ is normal to M at every point. Computation yields the equations

$$x''(t) + P \frac{(x')^2 R + 2x'y'S + (y')^2 T}{1 + P^2 + Q^2} = 0 \tag{1.10}$$

$$y''(t) + Q \frac{(x')^2 R + 2x'y'S + (y')^2 T}{1 + P^2 + Q^2} = 0 \tag{1.11}$$

where for simplicity we set

$$P = \frac{\partial F}{\partial x} \quad Q = \frac{\partial F}{\partial y}$$

$$R = \frac{\partial^2 F}{\partial x^2} \quad S = \frac{\partial^2 F}{\partial x \partial y} \quad T = \frac{\partial^2 F}{\partial y^2}$$

and general theorems about differential equations tell us immediately that geodesics exist with any given initial conditions, at least for small values of t, and are unique under these conditions. In particular there is one and only one geodesic parameterized by arc length, starting from a given point $m \in M$ and with given unit length velocity vector in the tangent plane to M at m.

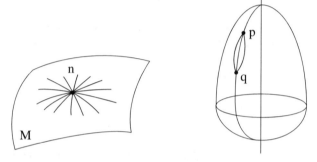

Fig. 1.42. (a) Geodesics through a point. (b) Meridians are geodesics by symmetry

From the above, general topology implies quite easily (using Gauß's lemma on page 116) that geodesics are shortest paths locally and that for any two sufficiently close points on M there is one and only one shortest path. But to find it explicitly is almost always impossible. Moreover globally uniqueness is false as already seen on the sphere. Let us now mention a few cases where geodesics are exceptionally easy to find.

The first case is when one has a plane of symmetry for the surface, because then the local uniqueness implies that the intersection with that symmetry plane is always a geodesic. Example: the great circles of the sphere are exactly its geodesics. For a surface of revolution every meridian will be a geodesic.

Surfaces of revolution are those obtained by rotating a curve around an axis. Geodesics on surfaces of revolution are a step easier to find because they satisfy a conservation law. The conservation law was discovered by Clairaut around 1730. We use two facts: the first is that the normal straight line to a surface of revolution always meets the axis of revolution. The second is that under projection to a plane orthogonal to the axis of revolution, the acceleration of the projection is the projection of the acceleration. This implies, in kinematic language, that the projected curve is a motion under a central force because the straight line containing the acceleration vector always goes through the origin of the plane. Then it is classical that such a motion obeys

1.6 The Geometry of Surfaces Before and After Gauß

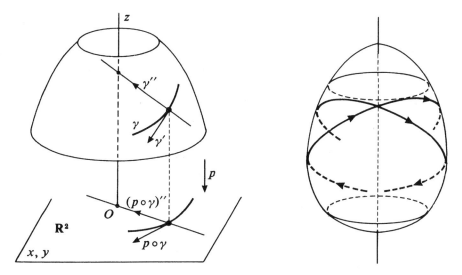

Fig. 1.43. Geodesics on surfaces of revolution

the law of area and from there the geodesic can be uncovered just by finding a primitive of a given function. For details and examples, see Berger & Gostiaux 1988 [175], Klingenberg 1995 [816] or do Carmo 1976 [451] and also §§10.2.1.

For example, the global behavior of geodesics is very simple when the meridian is convex: geodesics move between two parallels of the surface, parallels which moreover are at the same distance from the axis. In particular the behavior of a geodesic when the time goes to infinity is completely known. One can also find the periodic ones. For a torus of revolution we suggest that you study the different kinds of geodesics, or see Bliss 1902 [208].

The second type of surface whose geodesics are easily determined is a quadric. Quadrics are the surfaces of $\mathbb{E}^3$ given as the points where a quadratic form vanishes. The simplest ones are the compact ones, i.e. the ellipsoids:

$$\frac{x^2}{a^2} + \frac{y^2}{b^2} + \frac{z^2}{c^2} = 1$$

When $a = b = c$ we have a sphere; when only two of the three are equal we are on an ellipsoid of revolution and the determination of geodesics is easy to carry through completely (but the geodesics are given by elliptic functions).[6] But for a, b, c all distinct, it is much harder. Jacobi succeeded in 1839 in integrating the equations for geodesics of ellipsoids by a quite difficult trick; explicit expression requires hyperelliptic functions. Details in textbook form are in Klingenberg 1995 [816] along with a lot of information about ellipsoids, but note that some statements are imprecise.

---

[6] Do not be afraid of elliptic functions. They are just the "classical functions" which come next after polynomials, rational fractions, exponentials and logarithms, trigonometric functions and their combinations.

Jacobi's motivation to find ellipsoidal geodesics has some historical interest. He was asked to do this job by Weierstraß for the following reason. At that time, the earth was considered to be an ellipsoid of revolution, whose eccentricity was quite well measured. However, geographers found discrepancies and the conjecture was that the real shape is that of an ellipsoid with three distinct axis lengths. So there was enormous interest in finding the geodesics. The picture is now back to an ellipsoid of revolution, but with a little flattening around the poles.

The geometric version of Jacobi's result is too nice to be concealed. It is explained in Arnol'd 1996 [66] pp. 469–479. First we introduce the one parameter family of quadrics called *homofocal* to the above ellipsoid (or *confocal*). They are defined by the equation

$$\frac{x^2}{a^2+\lambda} + \frac{y^2}{b^2+\lambda} + \frac{z^2}{c^2+\lambda} = 1$$

and are pictured in figure 1.44 on the facing page. A line will usually be tangent to exactly two quadrics of this family, at two different points of the line. Given any quadric of this family, we take the lines which are tangent to it at one point and to the original ellipsoid somewhere else. These lines form a geodesic on the ellipsoid, and all geodesics arise this way. As a consequence, a geodesic on the ellipsoid will either be periodic or everywhere dense in a region given by the corresponding quadric. Thus the global behavior of geodesics is completely understood. In particular, for example, there are infinitely many periodic geodesics, coming in one parameter families (bands). Note, for the geometer, that the intersection curves of the ellipsoid with the confocal quadrics are exactly the curvature lines of the ellipsoid: those lines are the integral curves of the two line fields given by the principal directions and play an important role in surface theory (see §§1.6.3). The family of confocal quadrics gives the ellipsoid nice coordinates which are basic for various studies, see Morse & Feshbach 1953 [945] or Courant & Hilbert 1989 [407] for example.

But the story of the shortest paths on ellipsoids is not finished, even when they are of revolution. Here is why. We continue to suppose that our axes have three distinct lengths. On our ellipsoid there are four points called umbilic points; they come in two pairs p, p' and q, q'. These four points are the intersection of the ellipsoid with the quadric of the confocal family which has degenerated into a hyperbola located in the $x = 0$ plane. Every geodesic from one umbilic goes after a time T to the other point of the pair. The time T does not depend on the initial direction because of the first variation formula for surfaces (see §§3.1.4). Caution: if you take a second lap for time T you will come back to your original point, but in general with a different direction, unlike on the sphere. This is of course completely proven. The complete picture of such a geodesic, infinite in both directions, is quite subtle: it will be used for an interesting application in §10.9.

1.6 The Geometry of Surfaces Before and After Gauß   39

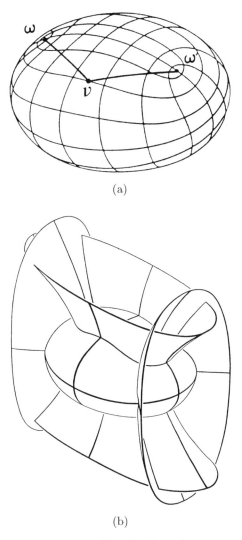

Fig. 1.44. Homofocal quadrics

If p is not one of the four umbilics then associated to it is its *cut locus* (see §3.3 and §6.5 for the general definition of the cut locus and details). The cut locus of a point p is the closure of the set of points which can be joined to p by more than one segment. For nonumbilic points p of the ellipsoid, it was claimed by Braunmühl 1878 that the cut locus of a point p, as a subset of the ellipsoid, is homeomorphic to a compact interval (see Braunmühl 1882 [257]). The two extremities are joined to p by a unique shortest path, while points in the interior of this interval are joined to p by exactly two shortest paths. There is still no complete proof of this assertion, despite von Mangoldt

1881 [893]. Very recently, Sinclair & Tanaka 2002 [1142][1143] wrote software to study the cut loci of surfaces; moreover they proved the conjecture for the particular case of an ellipsoid of revolution for points sitting on the equator.

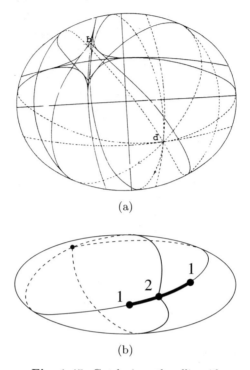

Fig. 1.45. Cut loci on the ellipsoid

Some clever readers will have been wishing to ask the author for some time why he did not address two natural questions. The first: can we extend geodesics indefinitely? The second: is there always at least one shortest path between any pair of points of a surface? Both answers are no in general; the simplest counterexample killing both questions is a plane with one point deleted.

The answers to these questions on general surfaces are neither easy to find nor to state. What is happening in the above example of a punctured plane is that a line running through this deleted point will arrive at it as to a boundary point. So we hope that things will be alright for surfaces with no boundary. Indeed they will, but for boundary in the sense of the inner metric of the surface. For example, if you draw in the plane a curve of infinite length, and then produce in three dimensional Euclidean space the "cylinder" on that curve (see figure 1.49 on page 42), i.e. the lines through the curve which are perpendicular to the plane, then on that surface every geodesic is indefinitely extendable and any two points can be joined by a unique geodesic

1.6 The Geometry of Surfaces Before and After Gauß   41

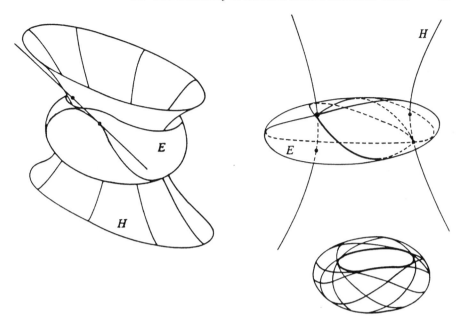

Fig. 1.46. Geodesics on the ellipsoid

which is moreover the shortest path between those points. However, in $\mathbb{E}^3$ this cylinder can have a boundary. The jargon answer is: the surface should be a complete metric space when endowed with the inner metric forced on it by its embedding in $\mathbb{E}^3$. The simplest case is when the surface is compact: then it is also complete in the above sense. We leave the subject here because we will study it intensively in the general case of Riemannian manifolds in §6.1.

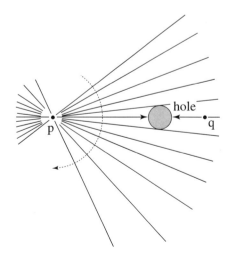

Fig. 1.47. There is no shortest path from p to q

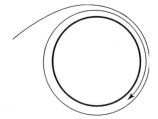

Fig. 1.48. Curve asymptotic to a circle

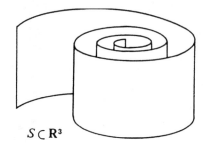

$S \subset \mathbf{R}^3$

Fig. 1.49. A complete surface, not closed

Another important question: do our geodesics provide the shortest passage? Assume the surface M is complete and pick up a geodesic c (parameterized by arc length) starting from a point $p = c(0) \in M$. We know that we can extend it indefinitely and that at first (for small t) it will be the shortest route between $c(0)$ and $c(t)$. But there is no reason that this will remain true for larger t; think for example of any compact surface M. People like to control things, analysts like to have a priori inequalities. Here we would like to specify a positive real number I (attached to a given surface M of course) such that any geodesic of length smaller than I is a segment. As we will see later (in §6.5) for the general compact Riemannian manifold, the best possible I always exists (is positive) and has a name because of its basic importance: it is called the *injectivity radius* of M, written $\mathrm{Inj}\,(M)$. For the unit sphere its value is π. Even for ellipsoids it is not easy to compute. Note that it can fail to exist (more precisely, it can be zero) for complete noncompact surfaces as shown in figure 1.50 on the facing page. Even for surfaces of revolution, cut loci and injectivity radius are impossible to find explicitly. A few exceptions are tori of revolution and Zoll's surfaces; see Besse 1978 [182] for Zoll's ones. For tori of revolution, the result is not found in textbooks. From what will be said in chapter 6 it can be left to the reader as a hard exercise, say already only for the case where the point is sitting in the outer meridian.

In §§1.2.3 we were interested in periodic trajectories of billiard balls, for example as a kinematic problem. Running along geodesics on a given surface is also a type of kinematics. Moreover it is a paragon for kinematics because many mechanical situations, as soon as they involve only two parameters,

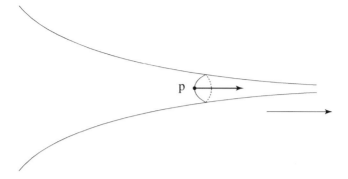

Fig. 1.50. $\operatorname{Inj}(M) = 0$

have motions given exactly as the geodesics of a surface. Note, to be more careful, that the surface under consideration should most often be an abstract surface (not one embedded in $\mathbb{E}^3$) and that will one of our important motivations to extend the notion of surface and of geometry to abstract Riemannian manifolds. In any event we now play billiards on a surface M in $\mathbb{E}^3$ with no boundary and so we look for periodic geodesics and ask ourselves the standard questions: are they many of them, or is there at least one? How many of them have length smaller than a given length L (this is called the *length counting function* and denoted by $CF(L)$).

We look first at the sphere and discover that all geodesics are periodic with the same period equal to 2π. Given a general surface, assuming moreover that it is strictly convex, even Poincaré could not prove there is at least one periodic geodesic on it. Birkhoff established it in 1917. Proof that there is an infinity for every surface had to wait until 1993; prior to that date it was not known even for strictly convex surfaces. We will come back to this more seriously in chapter 10.

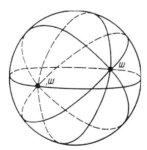

Fig. 1.51. Geodesics on the sphere

Be careful about infinity: we want an infinite number of *geometrically distinct* geodesics. As a kinematic motion, running twice along a periodic geodesic

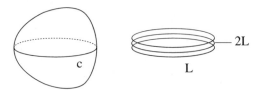

Fig. 1.52. Counting periodic geodesics

is different from running only once, but it is not geometrically distinct. For example when working on counting functions one should be careful to distinguish between the counting function for geometric periodic geodesics and that for parameterized ones. The question is difficult because the standard ways to prove existence of periodic geodesics consider them as motions. We will devote an entire chapter to this (and more) in the case of a general compact manifold, namely chapter 10.

We come back to the sphere. A typical mathematically minded question: are spheres the only surfaces all of whose geodesics are periodic? In kinematics this is the case for one of the simplest systems, the harmonic oscillator: one considers in Euclidean space a point which is attracted to the origin by a force positively proportional to the distance. Then every trajectory is an ellipse and thus periodic. So our question (at least as regards surfaces) asks for the harmonic oscillators of Riemannian geometry.

The question was taken up by various people around the end of the last century. They worked only with surfaces of revolution because their geodesics are easily computed as we saw above. The question was settled by Zoll 1903 [1310] when he proved that there are many surfaces of revolution all of whose geodesics are periodic and which are not spheres.[7] These surfaces are not all convex and most remarkably some of them are even real analytic.[8]

This does not end the question. In fact mathematicians like to classify, if possible completely, objects with a given property. So we want to find all surfaces which are harmonic oscillators. Only partial results are known today. Moreover to have a clear view of the problem it is better to work with abstract Riemannian surfaces. We then refer the reader to §10.10 where we will see the current state of the art.

Let us come back again to the inner geometry of the sphere. Not only are all geodesics periodic, all of those starting from a given point pass after time π through the same point, its antipode. Are spheres the only surfaces having that property? This time the answer conforms to intuition: only spheres can do

---

[7] Some people say *closed* instead of *periodic*, which can be ambiguous since it could also be used to mean only that a geodesic comes back to the same point, but not with the same direction; such a geodesic will not usually be periodic, and will be called a *geodesic loop*.

[8] *Real analytic* functions are those equal to their convergent Taylor series, in an open set about each point. The definition of real analytic surfaces is analogous.

1.6 The Geometry of Surfaces Before and After Gauß   45

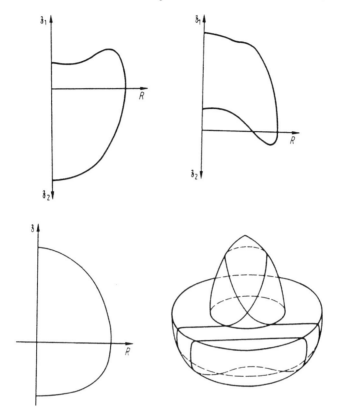

Fig. 1.53. Zoll's surfaces with concave parts

this. The question was raised by Blaschke in the 1930's but solved only in 1961 by Leon Green, see Green 1963 [586]. . For all the above questions the standard reference is Besse 1978 [182], but it is now outdated, so see §10.10. We will just prepare the reader for the first variation formula for general Riemannian manifolds to be uncovered in §§6.1.1: this formula shows immediately that in an antipodal manifold the running time from one point to its antipode has to be a constant (for example, π for unit spheres).

We stop here because we need more powerful tools to study the inner metric, namely those invented by Gauß. They will provide the content of chapter 3.

1.6.3 The Second Fundamental Form and Principal Curvatures

We now are interested in how a surface M sits in $\mathbb{E}^3$. We will generalize the notion of curvature for plane curves seen in §§1.3.2, and erect an object called the *second fundamental form*, which will tell us how much the surface differs from its tangent plane.

46   1 Euclidean Geometry

We will look at the curvature of various curves sitting in M, parameterized by arc length. The first result on these curvatures is Meusnier's theorem which goes back to the end of the 18th century. To a curve c, to be studied at the point $m = c(t)$, we attach the curve d which is the cross section of M by the plane containing both the tangent $c'(t)$ to c at m and the normal straight line to M at m. Orient M near m (i.e. pick a unit normal vector to M at m), and orient d to agree with the orientation of c at m where c and d are tangent. In general, the acceleration vector $c''(t)$ is not normal to M at m, but makes an angle $\alpha \in [0, \pi/2]$ with it. Meusnier's theorem relates the two geodesic curvatures $k_c(t)$ and $k_d(t)$:

$$k_c(t) = k_d(t) \arccos \alpha.$$

Note those are essentially the lengths of the accelerations, not the radii of curvature, explaining why things are not too intuitive. But for osculating circles this relation is the one we expect: the osculating circle of c is the projection of the osculating circle of d on the plane determined by $c'(t)$ and $c''(t)$. It helps to draw curves on a sphere (see figure 1.54).

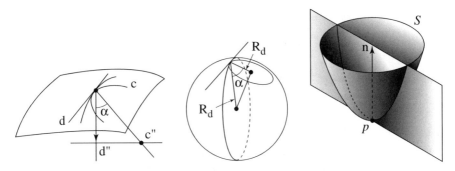

Fig. 1.54. Meusnier's theorem

Conclusion: to know the curvature of any curve, we need only know the curvatures of the normal cross sections of M.

The next theorem organizes these curvatures algebraically. The curvature of a normal section d at $m \in M$ depends obviously only on the unit tangent vector $u = d'(t)$. When u runs through the unit circle in the tangent plane $T_m M$ to M at m, the curvature k_u turns out to be a quadratic form in u, denoted by II_m and called the *second fundamental form* of M at m. With respect to the Euclidean norm on $T_m M$ this form has principal directions (which do not depend on the unit normal vector choice) and principal values for these principal directions. We denote them by k_1 and k_2 and call them the *principal curvatures* of M at m. Keep in mind our choice of an orientation for M: if we reorient M, then the sign of II_m flips, and so do the signs of k_1 and k_2, although the principal directions are unchanged. We now have on $T_m M$ two quadratic forms: the first one, the Euclidean norm induced by the embedding of $T_m M$ in $\mathbb{E}^3$ (which then depends only, so to speak, on the inner geometry) and the second, denoted by II which depends on the outer geometry (i.e. on the way that M sits in $\mathbb{E}^3$). The second fundamental form measures the defect of M to be a plane.

It turns out that for most standard examples, the elements of the second fundamental form are quite easy to compute (contrary to the inner geometry, where we had trouble computing distances and geodesics). For a unit sphere, at every point $k_1 = k_2 = 1$ (if the unit normal vector is the inward pointing one) and so the principal directions are not defined. For any surface the points where $k_1 = k_2$ are called *umbilics*. This is not a contradiction with the same word used for ellipsoids in §§1.6.2 because in fact these points of ellipsoids are also umbilics in the present sense. It is not hard to see that a surface all of whose points are umbilics is a part (or the totality) of a sphere or plane. But there is a conjecture of Carathéodory (made in the early 1920's) still open:

Conjecture 4 (Carathéodory) *Every compact surface in $\mathbb{E}^3$ which is homeomorphic to a sphere has at least two umbilics.*

This was proven between 1940 and 1959 by different authors (Hamburger, Bol, Klotz 1959 [818]) for surfaces which are strictly convex and real analytic. The proof is very hard and it seems that rigor was varying. For the state of affairs today see Scherbel 1993 [1102] and Gutierrez & Sotomayor 1998 [671]. It is natural to apply Hopf's index formula (see equations 15.9 on page 709 and 15.12 on page 713)

$$\chi(M) = \sum_x \text{index}_x(\xi) \quad (1.12)$$

where ξ is a vector field on M with a finite number of singular points. If M is topologically the sphere S^2 then one has $\chi(M) = 2$. Now at nonumbilic points of M one has the two principal directions. By continuity and orientation one hopes to use them to define a vector field, with singularities only at umbilics. In fact one can first extend Hopf's result to fields of directions; then indices are no longer necessarily integers but in general only half integers (rational numbers with denominator 2): for example the four umbilics of an ellipsoid have index equal to $1/2$. This does not prove that there are at least two umbilics, because one umbilic might have an index equal precisely to 2. The essence of the result above consists in proving, when the surface is moreover real analytic, that the index of an umbilic cannot be larger than one. The proof is extremely involved.

For very recent contributions to the subject, see Rozoy 1990 [1073] and Smith & Xavier 1999 [1150].

More generally for surfaces of revolution, using a symmetry argument, one sees immediately that the principal directions at a point $m \in M$ are the tangents to the meridian and the parallel through m. Then one of the principal

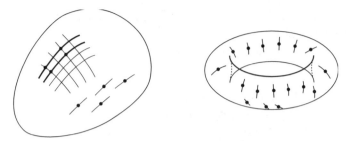

Fig. 1.55. (a) Impossible everywhere on a topological sphere. (b) Possible on a topological torus

curvatures is the curvature of the meridian, while the other is the inverse of the distance between m and the point of the axis of revolution where the normal to M at m meets this axis.

For a cone, a cylinder and more generally for the envelope of a one parameter family of planes things are as follows.[9] Such surfaces contain through every point m a *generatrix* which is a portion of a straight line and also a geodesic. Along a generatrix the tangent plane does not turn, i.e. the unit normal vector is constant. The generatrix is always a principal direction and the associated principal curvature vanishes.

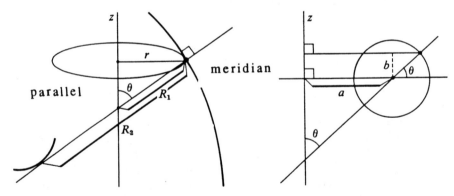

Fig. 1.56. The principal curvatures of a surface of revolution and the particular case of a torus of revolution

We leave the reader to compute the elements of the second fundamental form for ellipsoids. More generally principal curvatures are easy to compute either when M is given as the level set of a function or by some parameterization: see one case on page 52.

To every quadratic form on Euclidean space are attached two basic invariants: its trace and its determinant. It is thus most natural to attach to a

---

[9] These surfaces are called *developable* for a reason to be seen below in §§1.6.7.

1.6 The Geometry of Surfaces Before and After Gauß

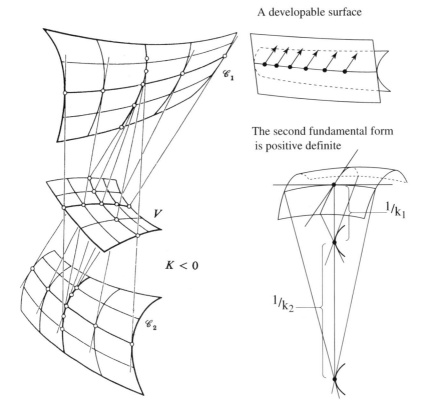

Fig. 1.57. The second fundamental form has opposite signs

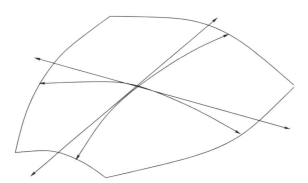

Fig. 1.58. The second fundamental form is positive definite

surface M the two following invariants:

$$K = k_1 k_2 \quad \text{the } total \text{ or the } Gau\ss!curvature$$
$$H = \tfrac{k_1+k_2}{2} \quad \text{the } mean\ curvature$$

50   1 Euclidean Geometry

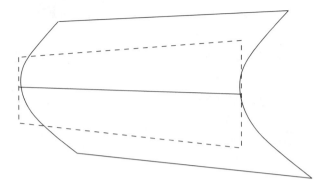

Fig. 1.59. A developable surface

Note that K does not depend on the choice of unit normal, while H changes sign if we change the unit normal. The reason for the name of Gauß will appear fully in the sequel. As a first example one sees from the above that K vanishes identically on developable surfaces. Obviously K and H are constant on spheres and planes. But there are many surfaces, different from spheres, with either H or K constant (not both). We will come back to this later on in §3.4. For the moment, we just mention the simplest examples of these. Surfaces of revolution with constant H or K can be explicitly determined, for the reason that (from what was said above about their principal curvatures) one can reduce the condition to an ordinary differential equation for their meridian curve. The shapes of some such surfaces are given in figure 1.60.

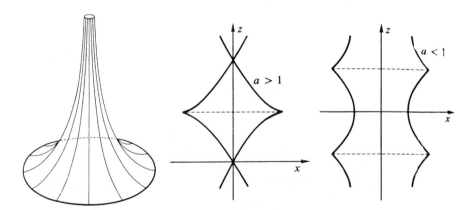

Fig. 1.60. (a) Beltrami's trumpet. (b), (c) Meridians of surfaces of revolution of constant negative Gauß curvature

Note that there are singularities (except for the standard sphere, and the standard cylinder). This is compulsory because there are theorems which say

1.6 The Geometry of Surfaces Before and After Gauß   51

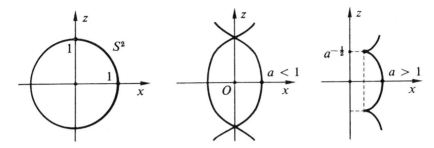

Fig. 1.61. Constant positive Gauß curvature surfaces of revolution

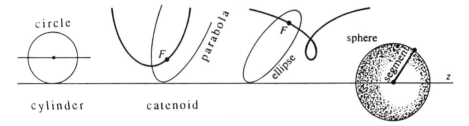

Fig. 1.62. Surfaces of revolution with constant mean curvature

that closed surfaces of either constant mean or constant Gauß curvature must be spheres; see §§3.4.1 and §§3.4.2. Note that these theorems are hard to prove. They were obtained only at the turn of the century. But the philosophy is that local pieces of surfaces of revolution are then easy to find. Now take a small cap of sphere or take some piece out of a tennis ball. Intuitively you might think it is easy to deform it in many ways keeping either K or H constant, because we have lot of freedom and are asking only for one condition. Rotational symmetry is exceptional. Indeed this flexibility with fixed mean or Gauß curvature turns out to be true for "general" surfaces, e.g. when K is not zero, but one needs extremely hard analysis to prove it. Moreover it is false for some surfaces which have exceptionally flat points. Indeed, some surfaces with such a strange flat point on them can not be bent without stretching (i.e. deformed through isometries other than rigid motions), even after cutting away all but an arbitrarily small neighborhood of that point.

For K a positive constant, it is natural to ask for closed surfaces, because positive K makes the surface shaped a little like an ellipsoid at each point. But how about $K \equiv -1$? It should be shaped more like a saddle. We know from §§1.6.4 that if it is complete for the inner metric, then such a surface cannot be compact. But is it possible to find an inner-complete surface in $\mathbb{E}^3$ with $K \equiv -1$? The answer is no, and this was proved by Hilbert in 1901. The proof is hard. Hilbert proved even more: the answer is still no even if you permit the surface to cross itself as in figure 1.63 on the following page. Such surfaces are called *immersions* as opposed to *embeddings*, see Berger &

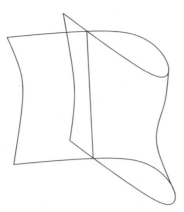

Fig. 1.63. An immersion

Gostiaux 1988 [175], do Carmo 1976 [451], Gallot, Hulin, & Lafontaine 1990 [542] for the jargon. Hilbert's theorem is of fundamental historical importance. It explains why hyperbolic geometry has to be defined abstractly, and can never be obtained as the inner geometry of a surface in $\mathbb{E}^3$. The interested reader can find its proof in section 5.11 of do Carmo 1976 [451] or in chapter 5 of volume III of Spivak 1979 [1155]. A recent survey of the state of affairs for the various generalizations and conjectures connected with Hilbert's result is presented in chapter II of Burago & Zalgaller 1992 [284].

Note the following kinematic interpretation of the meridian of surfaces of revolution with H constant (see figure 1.62 on the previous page), called *Delaunay surfaces*. They are obtained by rolling a conic in a plane along one of its tangents and the desired curve is that described during that rolling by one of its foci. This is not only a graceful depiction; there are deeper things behind it in connection with harmonic maps (see §14.3 and V.3 of Eells & Rato 1993 [480]).

We mention that computing K and H is easy even in the more general form
$$(u,v) \mapsto (x(u,v), y(u,v), z(u,v)).$$
When M is locally the graph $(x,y) \mapsto (x, y, F(x,y))$ of a function F, then K is
$$K = \frac{RT - S^2}{(1 + P^2 + Q^2)^2} \tag{1.13}$$
(with notation as in equation 1.10 on page 35).

1.6.4 The Meaning of the Sign of K

A first interpretation of K concerns the outer geometry. To say that $K(m) > 0$ is to say that the principal curvatures are not zero and of the same sign.

1.6 The Geometry of Surfaces Before and After Gauß

You can deduce easily that the surface M is locally strictly convex at m, and therefore that locally all points of M stay on one side of the tangent plane $T_m M$ with only the point m being in $T_m M$. If $K(m) < 0$ the principal curvatures are non-zero and of different signs. Locally the surface is close to a hyperbolic paraboloid. This, for example follows from the Taylor expansion for M around m:

$$(x,y) \mapsto \left(x, y, \frac{1}{2} k_1(m) x^2 + \frac{1}{2} k_2(m) y^2 + o\left(x^2 + y^2\right) \right) \qquad (1.14)$$

in a chart where $m = (0,0,0)$, where $T_m M$ is the plane $z = 0$ and the x and y axes are the principal directions. A similar expansion clarifies the positive case.

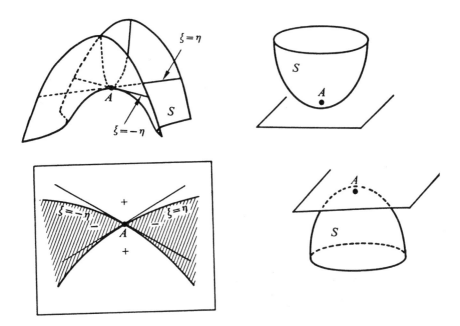

Fig. 1.64. (a) $K < 0$ (b) $K > 0$

It is important to remark that if $K(m) = 0$ nothing can be said about the shape of M locally at m without further assumptions. In some sense, every shape is possible. Consider some of the possibilities presented in figure 1.65 on the next page.

The second meaning will be treated in the most conceptually appropriate manner later on, but we can now give just the first zest of it. Consider a surface having the property that $K \geq 0$ everywhere. In particular, it is locally convex. Thinking, for example, of the spherical case, we might guess that if we have a triangle with sides a, b, c and an angle A, the basic formula 1.1 on

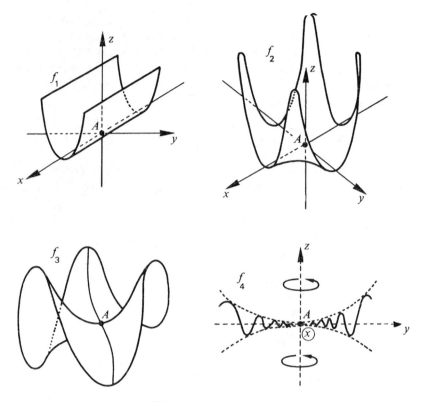

Fig. 1.65. Flat points

page 2 is bent into an inequality:

$$a^2 \leq b^2 + c^2 - 2bc\cos A \qquad (1.15)$$

We will see in §3.2 that such an inequality is always valid under the sole assumption that $K \geq 0$ everywhere on M (provided, of course, that M is inner-complete).

How about $K \leq 0$? This time the surface is in some sense the opposite of convex[10] in the sense that every point (at least when $K < 0$) is a saddle point. The reader can guess that at saddle points geodesics diverge, and possibly confirm her impression on some explicit examples. We will see in §3.2 that this intuition is correct, namely that for triangles on surfaces enjoying $K \leq 0$ everywhere one always has

$$a^2 \geq b^2 + c^2 - 2bc\cos A. \qquad (1.16)$$

But beware that in the negative case, unlike the positive one, one has to restrict to small triangles, more precisely to triangles which can be filled up

---

[10] *Also* the opposite of concave, since concave versus convex is just a change of orientation, which doesn't affect K.

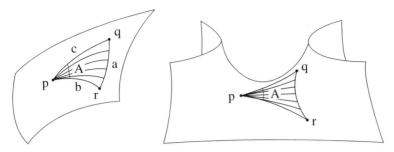

Fig. 1.66. Triangles on surfaces with positive and negative Gauß curvature

by a nice piece of the surface. Figure 1.67 shows a counterexample, the angle at the vertex being equal to π and the distance between q and r being as small as we like, for large b and c. This counterexample is linked with the notion of injectivity radius introduced on page 42. For the proof of these two theorems see §3.2.

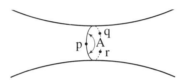

Fig. 1.67. Angle $A = \pi$

1.6.5 Global Surface Geometry

Let us mention some global results on surfaces, to give the reader an inkling of the subject. But only a few, much less comparatively than we did for plane curves in §1.4. There are two reasons: first, surfaces are more complicated objects than curves. Secondly, surfaces $M \subset \mathbb{E}^3$ are not our aim, which is Riemannian geometry. Surfaces are just used here as an introduction.

One open problem is the two umbilics conjecture mentioned above. Another open problem is *Alexandrov's conjecture*. Consider all compact surfaces with $K \geq 0$ and look for a kind of isoperimetric inequality, better an *isodiametric inequality*, specifying the maximal area of a surface with given inner diameter. The diameter of a metric space is the supremum of the distances between any two of its points. The inner diameter of a surface is its diameter for its inner metric. Check that the sphere is certainly not the best (just bend it). The conjecture is that the best bound is for the (flat) double disk for which, if the diameter is equal to D, the area is equal to $\pi D^2/2$. Note: this a bound which is not realizable by a smooth surface, but probably only by an object which is a limit of smooth surfaces with better and better ratios for Area$/D^2$, as in figure 1.68 on the next page. See partial results in Croke 1988 [417].

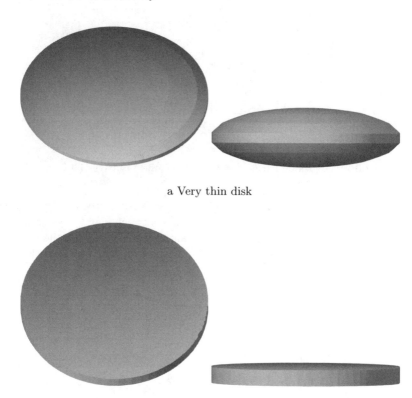

a Very thin disk

b Double flat disk

Fig. 1.68. The limit of a flattening disk

A simple theorem:

Theorem 5 *On any compact surface there is at least one point m with $K(m) > 0$.*

You will enjoy finding the proof for yourself. Another result (which to our knowledge appears in no modern textbook and was told us by Gromov) extends the rolling theorem 1 on page 15.

Theorem 6 (Rolling surface theorem) *Take a compact surface M and let $\sup k$ be the largest of the absolute values of its principal curvatures. Then a sphere of radius $\sup k$ can roll everywhere inside M. And the analogous result for the infimum value of the principal curvatures. Consequence: every closed surface can be squeezed between two spheres, one of radius $1/\sqrt{\inf k}$ and the other of radius $1/\sqrt{\sup k}$.*

1.6 The Geometry of Surfaces Before and After Gauß

We now explain Hadamard's theorem, which is the analogue for surfaces of the plane curve global convexity theorem we described in §1.4.

Theorem 7 (Hadamard, 1898) *Let M be a compact surface with positive Gauß curvature everywhere. Then M is the boundary of a convex body and consequently as a topological surface is a sphere.*

Proof. Remember the proof in the case of the plane. What is needed is a result ensuring here that there is only one point of M where the unit inner normal vector has any particular value (the existence of one being trivial). In the plane case this intermediate result was a consequence of the Umlaufsatz. Here things are simpler. The idea is to introduce the normal map, most often called the Gauß map. It is a map going from the surface M into the unit sphere $S^2 \subset \mathbb{E}^3$ obtained, as you have already guessed, by attaching to the point m the unit inner normal vector to M at m. The positivity of K (its convexity if you prefer) ensures that this map is a covering map. Now algebraic topology says that a sphere cannot have a covering map unless the map has a continuous inverse, because the sphere is simply connected (see §§4.1.3 for example).

Note 1.6.5.1 Various people discovered more or less recently that this map appeared explicitly before Gauß in the work of Rodrigues in 1815, so we propose to call it in this book the *Rodrigues–Gauß map*. Rodrigues used it to prove, among other things, that the integral of the curvature of an ellipsoid is equal to 4π (i.e. without any explicit computation). See references in §3.4.5.
♦

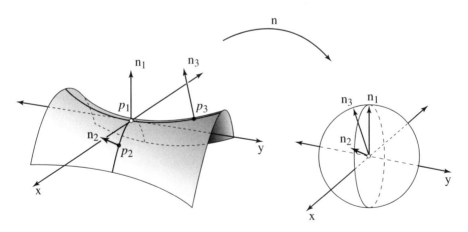

Fig. 1.69. The Rodrigues-Gauß map

Note 1.6.5.2 For closed (but not necessarily simple) curves in the plane, it is possible to have examples where the curve is everywhere locally strictly convex but is not the boundary of a convex domain of the plane. The turning number can be any integer as we saw on page 20. In $\mathbb{E}^3$ things are simpler: consider an immersion in $\mathbb{E}^3$ of an abstract surface M (for the jargon, if needed, any differential geometry book will do[11]). It is clear how to define the Gauß curvature K of an immersed surface. Assume that $K > 0$ everywhere. Then this immersion has to be one-to-one, i.e. an embedding, and of course the image is the boundary of a convex body; in particular, topologically M is a sphere. This follows with no change from the argument above because the Gauß map is still defined and is still a covering of S^2 by S^2. See more details in §§3.2.1. ◆

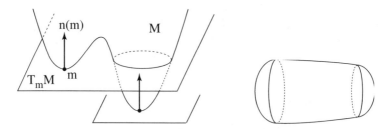

Fig. 1.70. (a) If M were not entirely on one side of $T_m M$. (b) Flat parts are permitted

Note 1.6.5.3 It seems intuitive that for a compact surface in $\mathbb{E}^3$ the condition $K \geq 0$ should be enough to insure that it is a topological sphere which is the boundary of a convex body: why not admit "flat" parts, or flat points, as in figure 1.70? You expect a sphere, but proof had to wait for Chern and Lashof in 1958. For references see Berger & Gostiaux 1988 [175] section 11. 3. 2, do Carmo 1976 [451] 5. 6-B and Burago & Zalgaller 1992 [284], chapter I, for generalizations. ◆

1.6.6 Minimal Surfaces

In this section and the next one we are going to address the natural question

Question 8 *What are the surfaces for which one of the two fundamental invariants K or H vanishes identically?*

We begin with the mean curvature $H = (k_1 + k_2)/2$. The mean curvature happens to have a very simple geometrical meaning: exactly the extension to

---
[11] Except, perhaps, this one.

1.6 The Geometry of Surfaces Before and After Gauß

surfaces of the curvature of plane curves when this curvature is seen as yielding the first variation for the length of curves (see equation 1.3 on page 13).

More precisely consider a surface M, a piece Ω of it and a one parameter family of neighboring surfaces, parameterized by a parameter t, and given by a normal variation function f. Then

Lemma 9 *The first derivative at M of the area of the piece of surface that Ω stretches into is*

$$\frac{d}{dt}\operatorname{Area}(\Omega_t) = 2\int_\Omega H(m)f(m)\,dm \qquad (1.17)$$

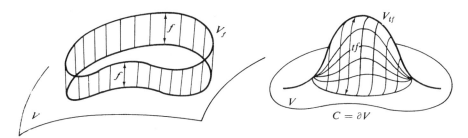

Fig. 1.71. Variation of a piece of a surface

We will use this first variation formula in §§1.6.8 and in §§3.4.2. It implies that the surfaces with $H \equiv 0$ everywhere are in some sense candidates for minimizing area. For this reason, surfaces with $H \equiv 0$ are called *minimal surfaces*. Consider the so-called *Plateau problem*: given a curve C in $\mathbb{E}^3$ find a surface S whose boundary is C and which has smallest area. This you can solve experimentally by dipping your curve, as realized by a piece of wire, in soapy water (these are not soap bubbles which correspond to surfaces for which H is not zero but only constant, see §§1.6.8). The study of minimal surfaces is a mathematical topic in itself which has been extremely active since the 1960's. Before then, there were only a few results, due to the lack of analysis tools. We refer to the two complementary bibles (and of course the references there): Dierkes, Hildebrandt, Küster, & Wohlrab 1992 [444, 445] and Nitsche 1989 [968], and to the excellent survey Meeks 1981 [911], and the recent Osserman 1997 [984]. Some minimal surfaces are drawn in figure 1.72 on the next page. Minimal surfaces can sometimes be found by an evolution flow; see Chopp 1993 [376] and Wang 2002 [1242].

Examples of minimal surfaces with no boundary condition are easy to build up; $H \equiv 0$ is a weak condition in that case. To get examples systematically there is a trick linking minimal surfaces with complex functions of a one variable: see section 10.2.3.6 of Berger & Gostiaux 1988 [175].

Minimal surfaces of revolution are straightforward to produce and analyze, because we saw in §§1.6.3 how to find principal curvatures of surfaces of

60   1 Euclidean Geometry

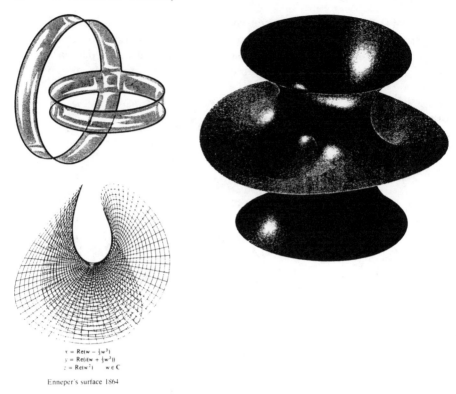

$x = \mathrm{Re}(w - \tfrac{1}{3}w^3)$
$y = \mathrm{Re}(i(w + \tfrac{1}{3}w^3))$
$z = \mathrm{Re}(w^2)$ $w \in \mathbb{C}$

Enneper's surface 1864

Fig. 1.72. Some minimal surfaces

revolution. From this, an exercise will prove that surfaces which are at the same time minimal and of revolution are necessarily *catenoids*, namely those whose meridian is the graph of the cosh (hyperbolic cosine) function. It is also easy to prove that the only ruled minimal surfaces are *helicoids* and planes. A surface is *ruled* when it is the set made up by a one parameter family of straight lines called the *generatrices* of the surface; beware that for a general ruled surface the tangent plane $T_m M$ turns when m moves along a generatrix. A helicoid is the image of a straight line moving under a spiral staircase motion. Computing the principal curvatures is especially easy for ruled surfaces. For catenoids and helicoids, if needed, see section 4.2 of do Carmo 1976 [451]. An important technique in thinking about mean curvature is the *mean curvature flow*; see §§1.4.5, Ecker 2001 [477], Chopp 1993 [376] and Chen & Li 2001 [361].

We can slice a catenoid to wrap it into a helicoid, as in figure 1.73 on the facing page. This is a famous example of a one parameter family of surfaces (a deformation) having two properties. The first is that every surface in the family is minimal. The second is that all the inner metrics of these surfaces are isometric under a suitable point to point correspondence (this property to be

1.6 The Geometry of Surfaces Before and After Gauß   61

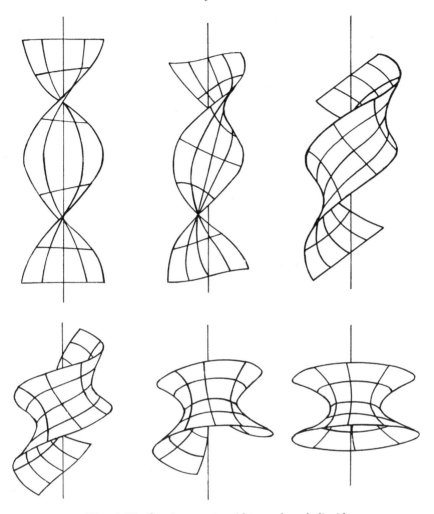

Fig. 1.73. Cutting a catenoid to make a helicoid

an isometric deformation is sometimes called bending [without stretching]). The fundamental theorem of Gauß which is going to be the keystone of chapter 3 implies then that the invariant K is preserved by isometries. But $H = 0$ throughout the bending, hence the two principal curvatures are preserved since their two values are $\pm\sqrt{-K}$.

Note 1.6.6.1 Recalling that the curvature of a plane curve, given as a function of the arc length, characterizes the curve up to Euclidean motion, could have suggested to you that the knowledge of the two principal curvatures will do the same for surfaces. This is false. However, if one knows the second fundamental form, that is, not only the principal curvatures but also the principal directions, then the surface is determined up to rigid motion. In the family above these principal directions are not preserved by the isometric correspondence, they turn. For demonstration of the above see for example the theory of these families in 3.5 of Dierkes, Hildebrandt, Küster, & Wohlrab 1992 [444, 445]. ♦

1.6.7 The Hartman-Nirenberg Theorem for Inner Flat Surfaces

We want to study surfaces M with $K \equiv 0$. We saw in §§1.6.3 that this is always the case for developable surfaces, i.e. for the envelope of a one parameter family of planes. Let us not forget the special case of a piece of plane for which the second fundamental form vanishes identically. One can build very wild examples of surfaces by attaching to pieces of planes pieces of cones and pieces of cylinders. This can be done as smoothly as we like using functions like $\exp(-1/x^2)\sin(1/x)$, etc. Or just smoothly bend the unbound upper corner of this page; now notice that you still have freedom to bend the unbound lower corner with your other hand. Such a surface will have both planar points (where $k_1 = k_2 = 0$) and non-planar points where one of the principal curvatures is non zero.

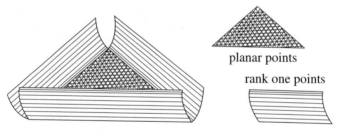

Fig. 1.74. Surfaces with $K = 0$

Locally at a non-planar point it is easy to prove (see section 5-8 of do Carmo 1976 [451] for details) that a surface with vanishing Gauß curvature is a piece of a developable one and second, that its inner metric is locally isometric to that of the Euclidean plane (this will also follow from §§3.1.4). To get the generatrices, just follow the principal direction associated with zero principal curvature. A similar assertion being of course true at planar points, finally surfaces with $K \equiv 0$ have everywhere the same inner geometry locally as the Euclidean plane. Now how about a global statement? Things are quite easy in the real analytic case, but for milder smoothness hypotheses, the question was settled only in 1959 by Hartman and Nirenberg. They prove moreover that if M is complete then it has to be a complete cylinder (i.e. the lines perpendicular to a plane, passing through a plane curve). The starting idea, which is hard to work out completely and correctly, is that a family of generatrices will develop singularities if extended far enough unless they are parallel. See do Carmo 1976 [451], section 5-8 for a very detailed and lucid exposition. Do not think mathematicians are now completely happy, they never are; see for example the very recent Fuchs & Tabachnikov 1999 [526].

1.6.8 The Isoperimetric Inequality in $\mathbb{E}^3$ à la Gromov

We now follow up our work on the planar case in §1.5. We consider domains $D \subset \mathbb{E}^3$ with compact boundary $M = \partial D$, and we ask

Question 10 *What is the best possible dimensionless ratio*
$$\frac{\mathrm{Area}(M)^3}{\mathrm{Vol}(D)^2}$$
for the area of $M = \partial D$ with respect to the volume of D, and which domains achieve this lower bound?

The expected answer is that round balls (with sphere as boundary) achieve the bound, and nothing else does.

Theorem 11 (Isoperimetric inequality in $\mathbb{E}^3$) *Among all domains $D \subset \mathbb{E}^3$ with twice continuously differentiable boundary, those which minimize the ratio*
$$\frac{\mathrm{Area}(\partial D)^3}{\mathrm{Vol}(D)^2}$$
are precisely the Euclidean balls.

Note 1.6.8.1 The plane is the only surface where uniqueness is easy to prove. There is a good reason for this: starting in dimension three, uniqueness in the isoperimetric problem cannot be true in general. For consider an object (which will turn out below to be a sphere) achieving the best possible ratio, and just add "hair" to it, that is to say, curves, etc. Then both the volume of the domain and the area of the boundary will not change. So uniqueness can be expected only under an additional hypothesis; the standard ones are either convexity or smoothness. From these remarks, and from what follows, one can see why the isoperimetric inequality—an amazing beautiful, useful and simply stated result—is usually absent from mathematics curricula. ◆

For a spectacular application of the isoperimetric inequality to convex bodies and the behavior of functions on spheres of high dimension (namely *Dvoretzky's theorem* and its proof), see the note 7.1.1.2 on page 305 and the references therein.

In every dimension, the proof by Stokes' theorem using the Knothe-Gromov trick, the one using Steiner's symmetrization and the one using the Brunn-Minkowski inequality are valid to get the inequality. It is always more

64   1   Euclidean Geometry

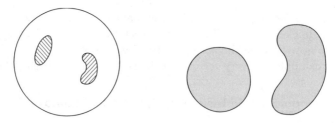

Fig. 1.75. (a) Isoperimetric inequality on the sphere. (b) Isoperimetric inequality in $\mathbb{E}^3$

difficult to prove that only spheres achieve equality, i.e. uniqueness. We now know how to prove the inequality and uniqueness using a technique found by Paul Levy in the 1930's. It remained buried up to 1980 when Gromov used it at full strength as follows. This proof is the more powerful one (see §1.5 and §§§7.1.2.2) and at the same time the most expensive. This is because it starts by asserting that there exist optimal domains D and that they have a boundary which, as a subset $M = \partial D$ of $\mathbb{E}^3$, is a smooth surface everywhere. This is known from geometric measure theory, although the required results emerged only in the 1960's (see §§14.7.2). Note that in higher dimensions everything works the same, with the proviso that geometric measure theory tells us that the solution in general will have singularities, but at most of codimension 6, which implies that they are of zero measure, and then the scheme below applies perfectly. A technical point: using the first variation equation 1.2 on page 4, we see that the point on M closest to a point of D will be a point for which the supporting tangent cone is contained in a half-space: this implies that such a point is regular (see §§§7.1.2.2 and the references there for more details).

The idea is fill up the inside of D starting from points of M and following the inner normal straight line. To be brief, we just speak of the normal $N_m M$ to M at m. Every point $d \in D$ is on at least one normal, namely take for $m \in M$ a point which is as close to d as possible. Then $d \in N_m M$.

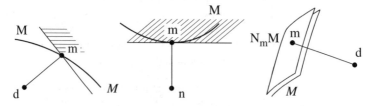

Fig. 1.76. Not a minimum; a local minimum

The infinitesimal volume element at distance t from m along a normal $N_m M$ is easily calculated to be

1.6 The Geometry of Surfaces Before and After Gauß

$$(1 - k_1(m)t)(1 - k_2(m)t)$$

with the effect that

$$\text{Vol}(D) = \int_M \int_0^{\text{Cut}(m)} (1 - k_1(m)t)(1 - k_2(m)t) \, dt \, dm \tag{1.18}$$

where Cut(m) is the *cut value*, namely the length of the interval along the normal to M coming out of m at which we should stop in order to fill up D completely (see figure 1.77).

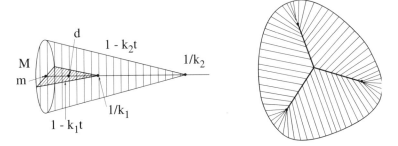

Fig. 1.77. Filling up a domain in the plane with normals from the boundary

The minimum property of M, in conjunction with the first variation formula 1.17 on page 59, shows that the mean curvature H is a constant, say h. From the elementary geometric-arithmetic mean inequality

$$xy \leq (x+y)^2/4$$

it follows that the integrand is bounded from above by $(1-ht)^2$. A elementary geometric trick shows that the value Cut(m) at which we stop integrating along $N_m M$ never exceeds $1/k_1(m)$ if $k_1(m) \leq k_2(m)$. This trick is essential and involves the notion of focal point (see Berger & Gostiaux 1988 [175], 2.7.11 or Gray 1990 [583], 8.1). It will be used at least two times below, in §§6.5.2 and in §§§7.1.2.2. In conclusion one obtains

$$\text{Vol}(D) \leq \frac{h}{3} \text{Area}(M = \partial D) \tag{1.19}$$

The proof is now completed using the fact that Area$(\partial D)^3/\text{Vol}(D)^2$ is also a minimum. One computes the first variation of that quotient for the normal variation $f = $ constant. This yields

$$\text{Vol}(D) = \frac{h}{3} \text{Area}(\partial D).$$

Equality in equation 1.19 is now forced and tracing our steps backwards we find that $k_1 = k_2$ everywhere on M, i.e. all points are umbilics; we have seen already that this forces M to be a sphere.

1.6.8.1 Notes

A first remark is in order for the attentive reader. Putting together the mean curvature rigidity of the sphere mentioned in §§3.4.2 and the nonexistence of singularities in dimensions below 6 the above proof seems useless. But it is not that simple. First, the fact that surfaces of constant mean curvature are spheres is an extremely difficult one to prove; see §§3.4.2. Moreover, the above technique extends word for word in every dimension, and in general Riemannian manifolds with Ricci curvature bounded from below (see §§§7.1.2.2). So it is extremely powerful.

The reader might be interested to know that Schmidt 1939 [1105] proved that there is an isoperimetric inequality for domains of S^2, the extremal figures being the spherical caps. He used symmetrization, but the above technique works as well. Schmidt proved his inequality in any dimension and not only for spheres but also for hyperbolic spaces: see §§4.3.2. Note also that our method "à la Gromov" works on spheres (and in any dimension) with easy changes in the integrand (the infinitesimal volume element) of 1.18 on the preceding page. You can guess which change to make if desired from theorem 111 on page 314. Steiner symmetrization also works because these spaces admit a lot of hyperplane symmetries (see theorem 40 on page 208). Also see §§§7.1.1.3.

One can be excused for the incorrect impression that the geometry of spheres is completely mastered today. Let us mention a famous open question. Girard's formula 1.7 on page 32 implies that a spherical triangle all of whose angles are rational multiples of π has an area which is rational multiple of the total area of the sphere. Starting in dimension three, i.e. studying spheres $S^d \subset \mathbb{E}^{d+1}$, the question is to compute the volume of a simplex (the direct generalization of triangle in higher dimensions) as a function of the so-called dihedral angles, namely the angles along edges between the two adjacent faces. Schläfli found a formula in 1858, but one is unable to deduce from it the same rationality conclusion. The common view is that this time (in dimensions at least three) rational dihedral angles will not imply rational volume. But nobody can prove it today. For more on this in a general setting involving also volumes of simplices in hyperbolic spaces, see the survey Vinberg 1993 [1222] and the references there, as well as the references quoted in section 10.6 of Ziegler 1995 [1306].

Finally one can consider the volume of tubes around surfaces in $\mathbb{E}^3$ (comparing with §1.4 above). For this question, which has a nice answer and is of historical importance, see Gray 1990 [583], entirely devoted to tubes. With much less material, see chapter 6 of Berger & Gostiaux 1988 [175]. We will meet tubes again in §15.7.

1.7 Generic Surfaces

We want to say that a property of a surface (or more general mathematical object) is *stable* if the property continues to hold after small perturbation,

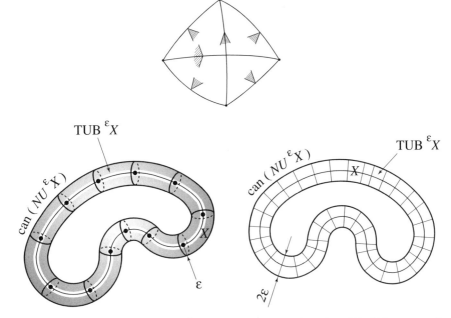

Fig. 1.78. (a) A tetrahedron in S^3 with its six dihedral angles. (b) Tubes around curves

and say it is *generic* if it always holds after (large or small) random perturbation. For example, positivity of Gauß curvature is stable, but not generic. Vanishing of Gauß curvature (inner flatness) is neither stable nor generic. The property of a parameterized smooth curve being immersed is generic. We will say that a generic parameterized smooth curve is immersed. Any property of a surface that can be observed by actual measurements must be stable, since the measurements will not be perfectly precise. If the property is generic, then we don't even have to make a measurement; we can expect it beforehand with near certainty.

Let us be more precise about these perturbations. Cut a sequence of points out of a line. Since the line is not countable, there will still be points left over, not belonging to the sequence. Similarly, if we cut a sequence of smooth curves out of a surface, there will be points left over. Call a subset of a topological space *skinny* if it is closed and nowhere dense. For example, a curve in a surface. A subset of a topological space is called *meager* if it is the union of a sequence of skinny sets, and *residual* if it is the complement of a meager set, i.e. the residue left over from cutting out a sequence of skinny sets. In our example of cutting curves out of a surface, each curve is skinny, the union of the curves is meager, and the complement residual. A topological space is said to be *fat* if it is not a meager subset of itself. A complete metric space is fat (the first skinny set in a sequence misses some closed ball, and the next skinny set misses a smaller closed ball inside the previous ball, etc. converging down to a point or a closed ball of points). In particular, a finite set is fat. Similarly, a complete Frechet space is fat. A residual subset of a fat space is fat. A property of elements of a fat topological space is *generic* if it holds on a residual subset.

Whitney 1955 [1260] proved that the generic C^3 map from a surface to the plane has only cusp and fold singularities. The fold is the map

$$(x, y) \mapsto (x^2, y)$$

while the cusp is the map

$$(x, y) \mapsto (x^3 + xy, y)$$

and to say that these are the singularities of a map means that near any singular point, there are local coordinates (x, y) on the surface, and (u, v) on the plane, so that the map is represented in this way. These maps are drawn in figures 1.79 on the facing page and 1.80 on the next page. In each map, the singular points form a smooth curve. The image of this curve on the plane forms another curve, smooth except possibly with corners, the curve of singular values. One may strengthen this result (only slightly) to examine the preimages of singular values: generically, the preimage will contain at most two singular points, and if there are two, then the two corresponding tangent lines of the curve of singular points map to distinct tangent lines on the plane. Thus every compact smooth surface is obtained by stitching together finitely many pieces of cloth (i.e. regions of the plane) each of which has smooth edges, and finitely many corners. In effect, the curve of singular values is a sewing pattern for the surface (as in fashion magazines); compare figure 1.81 on page 70. But some of the corners of the cloth must be smoothed out (look at the cusp), while others (corresponding to double singular values) must remain corners when stitched together. Keep this in mind when we discuss triangulations of surfaces and the Gauß–Bonnet theorem.

Now consider the Euclidean geometry of a surface in $\mathbb{E}^3$. A point of positive Gauß curvature is called an *elliptic point*. Near an elliptic point, any tangent line to the surface will have only first order contact (since there are no zeros of the second fundamental form). At a point of negative Gauß curvature (called a *hyperbolic point*) such a line will have first order contact, except along the two asymptotic directions (null directions of the second fundamental form), where the tangent line achieves at least second order contact. These asymptotic directions form the tangents to the *asymptotic lines*. At points of vanishing Gauß curvature (*parabolic points*) we have collision of asymptotic directions, forming a single tangent line, acheiving at least second order contact. The elliptic and hyperbolic points form open sets, and generically (in the natural topology on smooth surfaces) they are bounded by a smooth curve of parabolic points. Inside the region of hyperbolic points is a curve consisting of those points where the asymptotic lines have an inflection. For a little more detail,

1.7 Generic Surfaces   69

Fig. 1.79. A fold

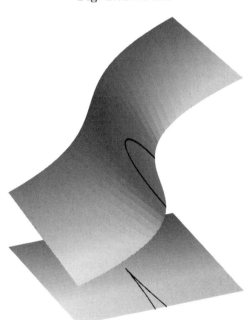

Fig. 1.80. A cusp

see Arnol'd 1992 [63]. For a lot more, especially on the singularities of wave fronts emanating from the surface, and on the umbilics of generic surfaces, see

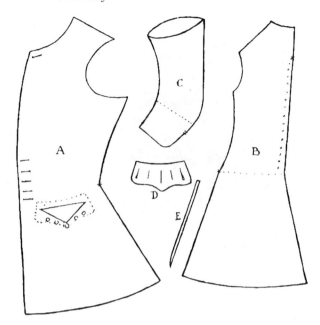

Fig. 1.81. A sewing pattern

Porteous 1994 [1039]. A brief discussion of the same topics will be found in Berger 2003 [173]. Generic surface geometry has been applied to the study of retinal imaging and the melting of ice and deforming of landscapes by wind and rain; see Donati & Stolfi 1997 [460].

1.8 Heat and Wave Analysis in $\mathbb{E}^2$

1.8.1 Classical equations of physics for a plane domain

We come back to the physics of a compact domain D in the Euclidean plane $\mathbb{E}^2$, thinking some more about billiard balls. In §§1.2.3 we studied the classical mechanics of a particle moving inside D. Now we want to study physics related to continuum and quantum mechanics. Namely, we consider the equations called respectively the *heat equation*, the *wave equation* and the *Schrödinger equation*.

These equations can be simply written using the notation

$$\Delta f = -\frac{\partial^2 f}{\partial x^2} - \frac{\partial^2 f}{\partial y^2}$$

for a function $f : D \to \mathbb{R}$. The operator $f \mapsto \Delta f$ is called the *Laplacian*. The heat equation describes what happens when you start with a given distribution of heat on D at a given time, and ask how the heat has spread at a later time.

The heat equation for the function f describing the temperature $f(x, y, t)$ at time t at the point (x, y) of D is

$$\Delta f = -\frac{1}{\nu}\frac{\partial f}{\partial t}$$

where ν is the *conductivity* of the material.

For the wave equation, you consider the cylinder built over D, and pour a thin layer of water in it. Then you try to describe the motion of the surface of the water.[12] The wave equation for the height $f(x, y, t)$ of the water after time t at the point $(x, y) \in D$ is

$$\Delta f = -\frac{1}{c^2}\frac{\partial^2 f}{\partial t^2}$$

where c is the *speed of sound* in the fluid. The wave equation coincides with the *vibrating membrane equation* which describes the normal motions of the domain D when it is considered as a vibrating membrane, a "drum". It should be understood that the solutions of these equations in fact represent only a first order approximation of the motions under consideration. The wave equation also describes, in first approximation, the behavior of sounds in a flat object, see §§1.8.4.

The Schrödinger equation of a free particle uses complex valued functions. It is written

$$\frac{\hbar^2}{2m}\Delta f = i\hbar\frac{\partial f}{\partial t}$$

where $i = \sqrt{-1}$, $\hbar$ is *Planck's constant*, and m is the *mass* of our free particle. Henceforth, we will take the constants appearing in any of these equations to be equal to 1. This can be achieved by rescaling the time variable (which may sound bizarre to an engineer, but is natural to a mathematician).

1.8.1.1 Bibliographical Note

The subject of classical continuum physics and quantum mechanics, even in the plane, is immense, and so is the bibliography. We gave many references during our journey, but the reader will find a commented bibliography at the end of Bérard 1986 [135], unfortunately dating from 1985.

1.8.2 Why the Eigenvalue Problem?

We want to solve the equations above, under various conditions on the boundary ∂D of D and on the initial condition at time $t = 0$. To solve such an equation depending both on the time t and the point $m = (x, y)$, our first clue is to

---

[12] The wave equation is just a first approximation. For better approximations, one needs to work much harder. For example, see Greenspan 1978 [598]. To our knowledge, no one has ever considered how to extend Greenspan's work to curved surfaces, or to Riemannian geometry.

Fig. 1.82. The first two eigenfunctions of a square

use the fact that (roughly by the Stone–Weierstraß approximation theorem) we need only consider product functions $f(m,t) = g(m)h(t)$. Subsequently, one constructs series of them (as in Fourier series theory). For example, look at the heat equation. We find that g and h must satisfy:

$$\frac{\Delta g}{g} = -\frac{h'}{h}$$

where h' is the usual derivative in t.

Since the first fraction depends only on the point m and the second only on the time t their common value has to be a constant, call it λ. A function $g : D \to \mathbb{R}$ such that $\Delta g = \lambda g$ is called an *eigenfunction* of D and λ is called an *eigenvalue* of D. Of course, one should specify a boundary condition. Once these eigen-objects are calculated, we are almost done. Just note that the time dependence will give

$$h(t) = \begin{cases} e^{-\lambda t} & \text{for the heat equation} \\ e^{i\lambda t} & \text{for the Schrödinger equation} \\ e^{it\sqrt{\lambda}} & \text{for the wave equation.} \end{cases}$$

Physically, for a vibrating membrane, the product motions $g(m)h(t)$ are the *stationary* ones; they are the ones one can observe through *stroboscopy*.[13]

For all of these equations, the set of the eigenvalues λ of the Laplacian on D is an essential piece of data. For simplicity's sake, we will work only with the boundary condition $f(m,t) = 0$ for every $m \in D$ and every time t; this is often called the *Dirichlet problem*. We will call the set of the corresponding

---

[13] Recall that a stroboscope is an instrument which periodically flashes light, used for studying periodic motion.

eigenvalues the *spectrum* of D (for the Dirichlet problem) and denote it by Spec (D). It is the analogue for quantum mechanics of the set of lengths of periodic trajectories in D which we met in §§1.2.3 for classical mechanics. This set of lengths is sometimes called the *length spectrum* of D.

The eigenvalues represent—in some sense— the energy levels of the domain D, so that the spectrum is of primary importance. We will see that the spectrum is much better understood that the length spectrum (see §§1.2.3) because there are systematic tools to address to it. Physicists are extremely interested in these energy levels: how do they look as a subset of $\mathbb{R}$? The first and simplest number to estimate is the analogue here of the counting length function introduced in §§1.2.3 for classical mechanics. Written $N(\lambda)$, it is defined to be the number of eigenvalues of Δ (for the Dirichlet problem on D) which are not larger than λ:

$$N(\lambda) = \# \{\lambda_i \leq \lambda\}$$

This function, and its vague family resemblance to its cousin $CF(L)$, is far from understood (see §§1.8.6).

Affinities between these two spectra are still emerging from the shadows. They provoke curiosity in the mathematician, drawing comparison between a geometric notion and a notion coming from analysis. Some physicists explore this topic these days because it is one of the simplest models where one can test "explicitly" various conjectures on relations between classical and quantum mechanics.

In the sequel, we will amply revisit this question on compact Riemannian manifolds: see chapters 9 and 10. Treat our discussion here as an introduction and a motivation for later. Physicists are also interested in abstract Riemannian manifolds, even though naively they may seem unphysically abstract, because some of them are much simpler to handle than all but a few bounded regions of Euclidean space. The simplicity comes from having no boundary, enabling us to ignore boundary conditions, which are often the source of subtleties. The relations between periodic motions and vibrations (or waves) of a domain D transcend the similarities in the spectra. In kinematics, we have trajectories (and not only their length), while in vibrations we have the eigenfunctions, of which we can retain the nodal curves (the points of the domain where an eigenfunction vanishes). When you vibrate a membrane and cover it with dust, the dust collects in the nodal curves, because the membrane stays still at those points—at the first order. For known relations between spectra, nodal curves and periodic motions, we refer the reader to the recent book of Sarnak 1995 [1095] and the references therein.

Let us look for the eigenfunctions. As in Fourier series theory, (which corresponds essentially to D being an interval in the line $\mathbb{E}^1 = \mathbb{R}$), any reasonable function f vanishing on ∂D can be expressed as a converging series whose entries are eigenfunctions of the Dirichlet problem for D. Notation: write the totality of the spectrum as

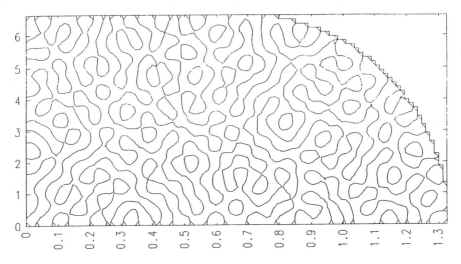

Fig. 1.83. A nodal set of Bunimovitch's stadium

$$\mathrm{Spec}\,(D) = \{\lambda_i\}_{i=1,2,\ldots}$$

and the set of the corresponding eigenfunctions as $\{\phi_i\}_{i=1,2,\ldots}$ normalized so that

$$\int_D \phi_i^2 = 1$$

for every $i = 1, 2, \ldots$ and

$$\int_D \phi_i \phi_j = 0$$

for $i \neq j$. Beware that eigenvalues can have multiplicity, the multiplicity of λ being the dimension of the space of eigenfunctions with eigenvalue λ. This dimension is always finite. When writing the spectrum as above, it is generally understood that multiple eigenvalues are repeated a number of times equal to their multiplicity. Let us clarify a minor point: not only do the eigenfunctions ϕ_i have to be normalized, but they must also be orthogonal to one another. This is automatic when their eigenvalues are different (exercise for the reader), but for multiple eigenvalues one needs to choose an orthonormal basis of a finite dimensional Euclidean vector space.

The Fourier-like decomposition says that every function f vanishing along D is equal to the following series:

$$f = \sum_{i=1}^{\infty} c_i \phi_i$$

where

$$c_i = \int_D f(m) \phi_i(m)\, dm$$

for each $i = 1, 2, \ldots$. When thinking of a vibrating membrane, the λ_i will be the frequencies of vibration, also called the *harmonics*; the lowest one is called the *fundamental tone*.

1.8.3 The First Way: the Minimax Technique

One way to hunt down the eigenfunctions is as follows. We get the first one by the so called *Dirichlet principle*: among functions vanishing along ∂D one looks for one minimizing the ratio

$$\frac{\int_D \|\operatorname{grad} f(x,y)\|^2 \, dx \, dy}{\int_D f^2(x,y) \, dx \, dy}$$

where $\operatorname{grad} f$ is the gradient of f:

$$\operatorname{grad} f = \left(\frac{\partial f}{\partial x}, \frac{\partial f}{\partial y} \right).$$

This ratio is called the *Dirichlet quotient*.

Suppose that we have a function f which minimizes this ratio. Computing the derivative of this ratio with respect to ε for a variation $f + \varepsilon g$ of the function f and using Stokes' theorem together with the Lagrange multiplier technique, reveals

$$\Delta f = \lambda_1 f$$

for some constant λ_1, and reveals that this λ_1 is the infimum of the Dirichlet quotient over all nonvanishing functions:

$$\lambda_1 = \inf \left\{ \frac{\int_D \|\operatorname{grad} f\|^2 \, dx \, dy}{\int_D f^2 \, dx \, dy} \,\Big|\, f \neq 0 \right\}. \tag{1.20}$$

This yields the first eigenfunction: $\phi_1 = f$, together with the first eigenvalue (which, by the way, is always of multiplicity one: you can prove it yourself). To dig up the next eigenvalue and its eigenfunctions, one applies the same trick, but restricting the set of functions f in consideration, allowing only functions which are orthogonal to the first eigenfunction:

$$\int_D f(x,y) \phi_1(x,y) \, dx \, dy = 0.$$

Proceed in this way, ad infinitum. See more on this in §9.4.

For rigorous proofs of the facts above, and also for facts which we did not quote explicitly, namely that the spectrum is a discrete, countable subset of the real line (of course made up entirely of positive numbers) and the fact that eigenvalues are of finite multiplicity, we refer the reader to Bérard 1986 [135], Chavel 1984 [325], Courant & Hilbert 1953 [406, 407], or any book on classical continuum physics or quantum mechanics. Note that those facts were rigorously proven only in the 1920's. If you look at a proof using Sobolev inequalities, keep in mind that the best current derivation of those inequalities uses the isoperimetric inequality.

There are very few examples where the spectrum or the eigenfunctions can be determined explicitly. Two old standards are rectangles and disks. In both cases, separation of variables disentangles the eigenfunctions. Using again the Stone–Weierstraß theorem, and because the boundary condition agrees with the separation, on a rectangle one need only look for product functions $f(x,y) = g(x)h(y)$, and there will be no other eigenfunctions. If the rectangle has side lengths a and b respectively, then the eigenfunctions are

$$\sin\frac{\pi m x}{a} \sin\frac{\pi n y}{b} \qquad (1.21)$$

where m and n are any integers, yielding the set of eigenvalues

$$\lambda = \pi^2 \left(\frac{m^2}{a^2} + \frac{n^2}{b^2}\right).$$

i.e. to obtain an eigenvalue at most λ we need m and n to be integer points inside a certain ellipse. We will see below that a simple expression yields an easy first order approximation of $N(\lambda)$ when $\lambda \to \infty$, but the second order term in λ is related to deep number theory and is still not completely understood today. It is believed that the number of integers m, n with

$$m^2 + n^2 \leq \lambda^2$$

is asymptotic to $\pi^2\lambda^2 + O\left(\lambda^{\varepsilon+1/2}\right)$ as $\lambda \to \infty$, for any $\varepsilon > 0$, but there is still no proof. This is called the Gauß circle problem; see §§1.8.5. However, it is known that $\pi^2\lambda^2 + O\left(\lambda^{1/2}\right)$ is too small.

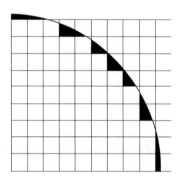

Fig. 1.84. The circle problem: find the integer points inside a circle

The eigentheory of the disk is treated by separating the variables using polar coordinates (ρ, θ) and looking for product functions $f(x,y) = g(\rho)h(\theta)$.

One calculates $h'' = -nh$ where n is any integer, and $g(\rho)$ satisfies the n-th Bessel equation. Everything is explicit, modulo knowing everything about Bessel functions; see any textbook in the realm of mathematical methods of physics, e.g. Morse & Feshbach 1953 [945], Courant & Hilbert 1953 [406, 407].

Be wary: vibrations of the disk are not easily grasped. Figure 1.85 is the intensity distribution of an eigenfunction of quite high frequency, taken from Sarnak 1995 [1095]. The very high concentration near the boundary is surprising. It explains the famous "whispering gallery effect". The picture is less surprising if you know that higher Bessel functions vanish to higher order at the origin.

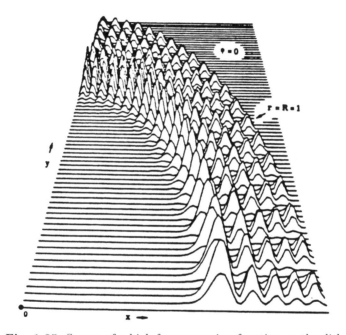

Fig. 1.85. Square of a high frequency eigenfunction on the disk

The eigenvalues (harmonics) of a vibrating string are just the integral multiples of the lowest one. This is never the case for a plane domain. Check it for yourself on the two above cases. This is why drums are never very musical, but the player tries when he beats a drum to generate a suitable mixture of various eigenfunctions, ideally only one. Figure 1.86 on the following page (picture and legend) is taken from Balian & Bloch 1970/1971/1972 [97, 98, 99, 100, 101], the pioneering papers which started the search (mentioned above) for relations between the two spectra. If you see some resemblance between the figure and certain musical instruments, it is not by chance.

To our knowledge the only other regions of the plane whose eigenfunctions, or even just eigenvalues, are explicitly computable are ellipses and isoceles,

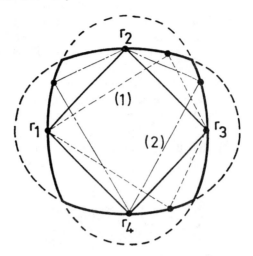

Fig. 1.86. From papers of Balian & Bloch

rectangular, equilateral and half equilateral triangles. The eigenvalues make themselves manifest as follows: for ellipses, separate variables using homofocal coordinates (see the figures in §§1.8.6 and the books of Courant & Hilbert 1953 [406, 407] and Morse & Feshbach 1953 [945]). For the triangles, use reflections on the boundary to extend the functions (vanishing at the boundary) to a square (for isoceles triangles) or to the whole plane with periodicity (for equilateral or half equilateral triangles). Periodic functions in the plane can be decomposed in Fourier series, but we will come back at length to this from the point of view of abstract manifolds. This will be achieved by replacing lattice-periodic objects in the plane by objects on an abstract (flat) torus: see §§4.3.3. Fourier analysis of periodic functions on the line or of functions on a circle are the same subject.

As a shocking example of our ignorance, one knows nothing about regular hexagons, not even the first eigenvalue. Try to understand why the reflection method does not work with regular hexagons. For more see Bérard 1980 [134], Pólya & Szegö 1951 [1037]. We did not really justify our use of the word "minimax;" it will be amply justified in §9.4.

1.8.4 Direct and Inverse Problems: Can One Hear the Shape of a Drum?

Mathematicians are never trumped when they cannot determine mathematical objects explicitly. Under their own steam, with aid from physicists, they look for various replacements like approximations, estimates of the counting function $N(\lambda)$, etc. There are two ways to think of this. Direct problems are of the type:

Question 12 *I know more or less the shape of the domain D; what can I deduce from that about its spectrum?*

Inverse problems work the opposite direction:

Question 13 *I know (more or less) the spectrum of a domain; what can I say about its shape?*

These two problems were addressed (more or less explicitly) in the 19th century and people have never stopped working on them. This book is an introduction to Riemannian geometry, and we will devote the whole of chapter 9 to these two problems in the Riemannian realm. Therefore, we will treat spectral geometry of Euclidean domains as mere motivation, and we will content ourselves with giving a few results, ideas and problems.

1.8.4.1 A Few Direct Problems

The main direct problem is the asymptotic expansion of $N(\lambda)$ and will be addressed in the next section in some detail because the technique extends automatically to Riemannian manifolds. A more naive problem is to consider the first eigenvalue λ_1. For a vibrating membrane, λ_1 is called the *fundamental tone*, for obvious musical reasons. In the heat equation, λ_1 gives the dominant information (together with the first eigenfunction), i.e. the mode that decays with slowest exponential rate. Everything else quickly disappears. Controls on λ_1 are of practical importance to avoid "resonances".

A natural question concerning the first eigenfunction ϕ_1 was open for a long time. Using Dirichlet's principle, it is easy to see that the first eigenfunction of any domain never vanishes, except at the boundary. (After flipping sign if needed, take it to be positive.) But, as a sort of control, one expects that the shape of the boundary has some influence on the first eigenfunction, more precisely on its level curves $\{f = t\}$ when t runs from 0 to the maximum of f. Assume, for example that the domain is convex: is every level curve convex also? Brascamp & Lieb 1976 [256] proved that this is so. A similar problem was solved by them in the same paper. We want to know a little about how λ_1 depends on D. They proved that λ_1 is a concave function when one performs Minkowski addition of two domains.[14]

Other direct problems: to give systematic lower or upper bounds on the λ_i depending on the domain. See Bérard 1986 [135], Chavel 1984 [325], Chavel 1993 [326] and §§1.8.5 for examples. We note here that the minimax technique (explained in equation 1.20 on page 75, see also chapter 9) shows that upper bounds are in general quite easy to obtain, because one just exhibits some suitable function, calculating its Dirichlet quotient. For a lower bound, one has

---

[14] A definition of Minkowski addition is presented in Berger 1994 [167] (11.1.3) or any book on convexity.

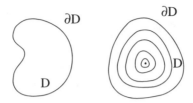

the level lines are all convex curves

Fig. 1.87. What are the level lines?

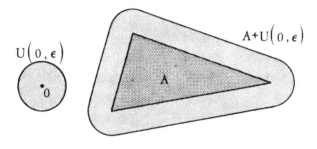

a A Minkowski sum in the plane

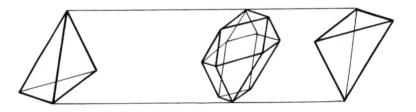

b A Minkowski sum in space

Fig. 1.88. Minkowski sums

to show an inequality valid for every function (in particular the eigenfunction under consideration). We will give the basic strategy later in this section.

Another natural idea when looking at the spectrum is to look at the ratio of the first two eigenvalues λ_1 and λ_2. It was conjectured in 1955 by Payne, Pólya and Weinberger that the ratio λ_2/λ_1 is strictly smallest on disks. This was proven only in 1992: see Ashbaugh & Benguria 1994 [72] for this question, and the recent Ashbaugh & Benguria 1995 [73].

1.8.4.2 The Faber–Krahn Inequality

The last direct problem we shall ponder, but not the least, is the Faber–Krahn inequality: among all domains with equal area the one with lowest λ_1 is the disk:
$$\lambda_1(D) \geq \lambda_1(D_*) \tag{1.22}$$
where D_* denotes a disk with the same area as D. Musically it is quite intuitive. For the heat equation, a body with longer boundary should lose heat to its environment more quickly.

The assertion was stated without proof by Rayleigh in 1877; for history and earlier results see Rayleigh 1894 [1053] or Pólya & Szegö 1951 [1037]. Faber and Krahn proved it independently in 1920. The proof below extends with no difficulty to any dimension.

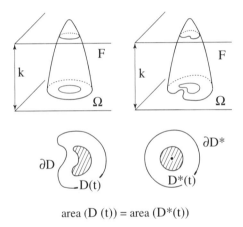

Fig. 1.89. Schwarz symmetrization

The underlying idea of the proof is too beautiful to be concealed. Moreover it is the archetype for the more general result which will see for Riemannian manifolds in chapter 9. Therefore, we are going to give a detailed proof. Here, we will symmetrize functions (as opposed to the geometric symmetrization of Steiner in §1.5) called *Schwarz symmetrization*. It is a kind of transplanting of functions defined on D to functions defined on a disk D_* such that Area(D_*) = Area(D). See figure 1.89. Pick up the graph of some function $f : D \to \mathbb{R}$ and slice it horizontally. Watch the level curves $f^{-1}(t)$ when t goes from zero to the maximum of f. We now build up a function $f_* : D_* \to \mathbb{R}$ by forcing its level curves to be circles centered at the center of D_*, so that each level curve surrounds a disk whose area is the same as the area surrounded by the corresponding level curve of f. Perhaps we can make this clearer: let $D(t) = f^{-1}(t, \sup f]$ and define f_* by the condition that the disk $D_*(t) =$

$(f_*)^{-1}(t, \sup f]$ is centered at the origin and that $\operatorname{Area} D_*(t) = \operatorname{Area} D(t)$. Note that $\sup f = \sup f_*$. By equation 1.20 on page 75 we only need to prove

$$\frac{\int_D \|\operatorname{grad} f\|^2 \, dx \, dy}{\int_D f^2 \, dx \, dy} \geq \frac{\int_{D_*} \|\operatorname{grad} f_*\|^2 \, dx \, dy}{\int_{D_*} f_*^2 \, dx \, dy}.$$

The basic idea is to replace the double integral in $dx \, dy$ by an integral first in dt and then along the level curves $f^{-1}(t)$. Figure 1.90 shows that

$$dx \, dy = \frac{ds \, dt}{\|\operatorname{grad} f\|} \tag{1.23}$$

where ds denotes the arc length differential along the level curves. Two basic tools of the theory of integration are used here: Fubini's theorem and the change of variables, giving

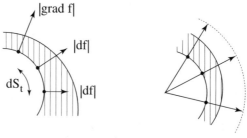

the variation of area goes as $\dfrac{1}{\operatorname{grad}(f)}$

Fig. 1.90. The variation of area across level curves

$$\operatorname{Area} D(t) = \int_t^{\sup f} dt \int_{f^{-1}(t)} \frac{ds}{\|\operatorname{grad} f\|} \tag{1.24}$$

The same equation applies to f_* and D_*, and because of the condition on the areas inside the level curves

$$\int_{f^{-1}(t)} \frac{ds}{\|\operatorname{grad} f\|} = \int_{f_*^{-1}(t)} \frac{ds}{\|\operatorname{grad} f_*\|}.$$

Applying equation 1.24 to D_* and f_* one gets

$$\int_D f^2 \, dx \, dy = \int_{D_*} f_*^2 \, dx \, dy.$$

To prove

$$\int_D \|\operatorname{grad} f\|^2 \, dx \, dy \geq \int_{D_*} \|\operatorname{grad} f_*\|^2 \, dx \, dy$$

1.8 Heat and Wave Analysis in $\mathbb{E}^2$

we compute the derivative with respect to t of the integral

$$G(t) := \int_D \|\operatorname{grad} f\|^2 \, dx \, dy$$

From equation 1.23 on the facing page one calculates

$$G'(t) = -\int_{f^{-1}(t)} \|\operatorname{grad} f\| \, ds$$

and the Schwarz inequality reads

$$\left(\int_{f^{-1}(t)} \|\operatorname{grad} f\| \, ds\right)\left(\int_{f^{-1}(t)} \frac{ds}{\|\operatorname{grad} f\|}\right) \geq \left(\int_{f^{-1}(t)} ds\right)^2$$

$$= \left(\operatorname{length} f^{-1}(t)\right)^2 \qquad (1.25)$$

The classical isoperimetric inequality applied to both $D(t)$ and $D_*(t)$ and their boundaries $f^{-1}(t)$ and $f_*^{-1}(t)$ give:

$$\operatorname{length} f^{-1}(t) \geq \operatorname{length} f_*^{-1}(t)$$

Write the analogue of 1.25 but for $D_*(t)$. We remark that $\|\operatorname{grad} f_*\|$ is constant on the circles by the circular symmetry, so now we have equality in the Schwarz inequality. This shows that $G'(t) \geq G'_*(t)$ for every t so that

$$\int_D \|\operatorname{grad} f\|^2 \, dx \, dy = G(0) \geq G_*(0) = \int_{D_*} \|\operatorname{grad} f_*\|^2 \, dx \, dy$$

since $G(\sup f) = G_*(\sup f_*) = 0$, giving the finishing touch.

We hid a technical difficulty: the level curves need not be smooth. This is not too hard to overcome (see the remarks in Bérard 1986 [135], IV.6).

The Faber–Krahn inequality can also be obtained by Steiner line symmetrization (see §1.5); this symmetrization respects the area and diminishes λ_1.

We could ask for more than the asymptotic expansion, namely how does the subset $\{\lambda_i\}_{i=1,2,\ldots} \subset \mathbb{R}^+$ look? Physicists are interested in that question. They would like a sort of random repartition around Weyl's asymptotic formula. We will say few words on this in §9.11, see also Sarnak 1995 [1095]. The spectrum is a mystery even for triangles in the hyperbolic plane. Hyperbolic triangles are presently the physicists' favourite specimens, because hyperbolic geometry (negative curvature) is more realistic than Euclidean geometry for systems of particles.

1.8.4.3 Inverse Problems

The importance of inverse problems cannot be underestimated. For plane domains they provide the simplest model of determining the structural soundness of an object by listening to its vibrations. Consider a vibrating object, anything from a space craft to a church bell. We need to test the strength of this object. Already in the middle ages, bell makers knew how to detect invisible cracks by sounding a bell (on the ground, before lifting it up to the belfry). How can one test the resistance to vibrations of large modern structures by non-destructive assays? This is a perplexing problem; to our knowledge there is no solid mathematical contribution on the subject. A small crack will not only change the boundary shape of our domain—one side of the crack will strike the other during vibrations, invalidating our use of the simple linear wave equation. On the other hand, presumably heat will not leak out of a thin crack very quickly, so perhaps the heat equation will still provide a reasonable approximation for a short time, unless cracks permeate the object, or take up significant volume. The effect on quantum mechanics is unclear. For references on nondestructive assays, see Bourguignon 1986 [238]. In the absence of mathematical results, engineers use both scale models and assays.

The spectrum determines (more or less) the length spectrum (for more precise statements, see the references). But we saw that periodic billiard motions are very difficult to study, and in particular recovery of something of the shape of a domain from its length spectrum is beyond reach today (see Guillemin & Melrose 1979 [669], from which we extract figure 1.91 on the facing page to help the reader feel the difficulties). There is basically only one exception: disks are characterized by their spectrum as will be proven twice in the next section. See also Guillemin & Melrose 1979 [668] for the spectral geometry of ellipses.

The Faber–Krahn inequality 1.22 on page 81 can also be seen as an inverse problem; another one is the following. Let us ask for more than only Faber–Krahn: their result tells us that in order for a domain to have a low fundamental tone, it should have large area. But this does not expose the shape of the domain. For example, consider rectangles where we have the explicit value $\lambda_1 = \pi^2 \left(1/a^2 + 1/b^2\right)$ for the fundamental tone. Then one sees that a large area ab with a small a will never yield a low fundamental tone. What one can feel is that both a and b should be large or, otherwise stated, the domain should contain in its interior a large disk (see figure 1.92 on the facing page). This was conjectured long ago, but proven only in 1977 by Hayman. We refer to Osserman 1977 [981] and Croke 1981 [412] for an improvement, and note that the optimal result is still an open question, namely the optimal constant and the corresponding shapes. In exchange, the minimum principle shows immediately that the fundamental tone of a domain is never larger than that of the maximal disk included in the domain.

Another long standing conjecture has been solved only recently. Consider the second eigenfunction ϕ_2 of a bounded plane domain D. What are the possibilities for its nodal lines? Using the minimax principle, Courant proved a long time ago that the nodal lines of the k-th eigenfunction ϕ_k cannot divide D into more than k different domains. Thus we have only two topological possibilities for ϕ_2, as in figure 1.93 on the next page.

1.8 Heat and Wave Analysis in $\mathbb{E}^2$ 85

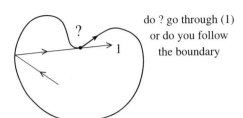

a very bad periodic trajectory

do ? go through (1) or do you follow the boundary

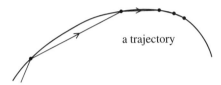

a trajectoryn accumulating by an infinite number of reflections to a point (this exists in strictly convex domains)

a trajectory

Fig. 1.91. From Guillemin & Melrose, 1979 [669]

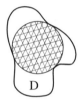

a large area need not to produce a low tune nor a large diameter, one needs to include a large disk

Fig. 1.92. Neither a large area nor a large diameter suffices to produce a low tone; one needs to include a large disk

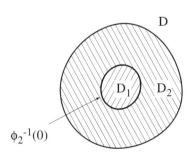
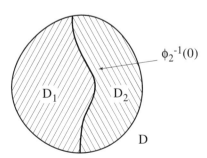

excluded now

Fig. 1.93. Topology of the nodal lines of ϕ_2

In 1974, Payne conjectured that the only possibility is the one with a transverse nodal line (check it for yourself on the examples you know). This has just been proven in Melas 1992 [912], but only for convex domains. It is interesting that the proof makes an explicit use of Hayman's result mentioned above.

We come back to the main question: is the knowledge of the totality of Spec (D) enough to fix the shape of D? Formally, are two domains D and D' for which Spec (D) = Spec (D') congruent, that is to say, is there an isometry of the Euclidean plane identifying D with D'? This is exactly the musical question of the title of this section. It is not absurd to expect that only a countable sequence of numbers might yield the complete knowledge of a domain; think for example of a periodic function determined by its Fourier coefficients. Formulated in 1882 by Schuster, the negative answer was not uncovered until 1992, in Gordon, Webb, & Wolpert 1992 [579]. The two plane domains shown in figure 1.94 have the same spectrum. Simpler and more general examples have been found in Buser, Conway, Doyle & Semmler 1994 [294].

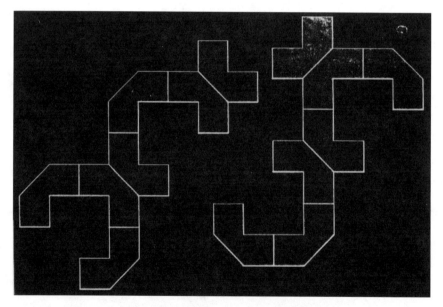

Fig. 1.94. Isospectral planar domains

You can never clip the wings of a mathematician; deciding the existence of isospectral domains does not end our inquiry. There is no uniqueness, but what is then the defect of uniqueness? Are there (up to congruence, of course) only a finite number of domains with identical spectra? Let us just mention that today there are compactness results (see Osgood, Phillips, & Sarnak 1988 [978]), but no example is known of a one parameter family of isospectral domains (not all congruent). For the theory of isospectral Riemannian manifolds, see §9.12.

The counterexamples pictured in figure 1.94 on the preceding page are not convex. This question is addressed in Gordon & Webb 1994 [578] where convex counterexamples are built in four dimensions. But there are still no planar convex isospectral domains known.

1.8.5 Second Way: the Heat Equation

Today, one of the simplest tools to study the spectrum and the eigenfunctions is the heat equation. It is not the most powerful (the wave equation is more powerful), but the technique is much simpler. Both techniques, incidentally, extend to Riemannian manifolds with little alteration.

Henceforth, all functions vanish at the boundary ∂D.

We start with the *fundamental solution* of the heat equation,[15] denoted by $K(m, n, t)$.
$$K : D \times D \times \mathbb{R}^+ \to \mathbb{R}.$$

It gives the solution $f(m, t)$ of the heat equation with given initial condition $f(m, 0) = g(m)$. The desired formula is the integral
$$f(m, t) = \int_D K(m, n, t) g(n)\, dn$$

and one can prove (see Bérard 1986 [135], Chavel 1984 [325] and also §9.7) that
$$K(m, n, t) = \sum_{i=1}^{\infty} \phi_i(m) \phi_i(n) e^{-\lambda_i t}$$

The reader can check that this ensures that f will satisfy the heat equation, as long as the above series converges. The series is known to converge, but that is not so easy to prove.

Another way to write the solution $f(m, t)$ with initial condition
$$g(m) = \sum_{i=1}^{\infty} a_i \phi_i(m)$$

is
$$f(m, t) = \sum_{i=1}^{\infty} a_i \phi_i(m) e^{-\lambda_i t}.$$

Our scheme to study the spectrum $\{\lambda_i\}$ is as follows. First, integrate the trace $K(m, m, t)$ over D, which yields
$$\int_D K(m, m, t)\, dm = \sum_{i=1}^{\infty} e^{-\lambda_i t}.$$

---

[15] Also called the *heat kernel*.

It remains to get information on the diagonal values of K. To get some intuition, the fundamental solution can also be interpreted as follows: consider a singular initial data of heat at time zero consisting of a Dirac distribution located at the point m. Then $K(m,n,t)$ is the temperature after time t at the point n (see figure 1.95). In the case of the whole plane (no boundary D this time) this value is well known:

$$K_0(m,n,t) = \frac{1}{4\pi t} \exp\left(-\frac{\|m-n\|^2}{4t}\right). \tag{1.26}$$

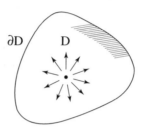

Fig. 1.95. It is infinitely cold outside; the particles diffusing heat away from the point m are not aware (at the first order at least) of the disaster that awaits them when they reach the boundary.

Note first that this implies that heat is diffused instantaneously, say with infinite speed, from any point (not very intuitive). Now, as Mark Kac says in Kac 1966 [775], in case there is a boundary ∂D, please do not worry. Your (correct) intuition is that the particles are not aware of the disaster that awaits them when they reach the boundary. It can be rigorously proven that their awareness is really not subtle, at least to the first order. The mathematical claim is that

$$K(m,m,t) \sim \frac{1}{4\pi t} \quad \text{as} \quad t \to 0$$

whatever the point m is. We won't give a proof of this asymptotic. Integrating this over D:

$$\sum_{i=1}^{\infty} e^{-\lambda_i t} \sim \frac{\text{Area}(D)}{4\pi t} \quad \text{as} \quad t \to 0$$

A very easy exercise will tell you that if you know the function

$$t \mapsto \sum_{i=1}^{\infty} e^{-\lambda_i t}$$

then you know Spec(D). But do not deduce from this that you can easily estimate the counting function $N(\lambda)$ when $\lambda \to \infty$. Nevertheless, this is possible, but it is quite a hard theorem called the Hardy–Littlewood–Karamata Tauberian theorem:

$$N(\lambda) \sim \frac{\text{Area}(D)}{4\pi}\lambda$$

(see Ivrii 1998 [761]).

This asymptotic formula was proven as early as 1911 by Hermann Weyl; but note that his proof was completely different, not employing the heat equation. His more intuitive technique can be found in Bérard 1986 [135]; it uses minimax properties. Weyl's innovation is to approximate the domain D by an aggregate of squares (of smaller and smaller size, as in figure 1.96) and to paste eigenfunctions on these squares. To these functions we apply the minimax procedure, and can bound the eigenvalues of D on both sides by a limit argument. This time one does not need any Tauberian theorem. The first outcome of the minimax procedure is the strict monotonicity principle:

Proposition 14 *Consider two domains* $D \subset D'$.

$$\lambda_1(D) \geq \lambda_1(D')$$

Proof. Take the first eigenfunction ϕ_1 of D and extend it to be zero on $D'-D$. You are done by applying Dirichlet's principle to D'.

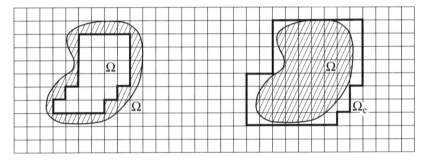

Fig. 1.96. Approximation by small squares

An immediate corollary is an asymptotic estimate for the eigenvalue λ_i:

$$\lambda_i \sim \frac{4\pi}{\text{Area}(D)}i \quad \text{as} \quad i \to \infty$$

A second immediate corollary of Weyl's result is that you can hear if a domain is a disk because the spectrum will tell you the area of D, and then you just apply the Faber–Krahn inequality 1.22 on page 81.

How about a second order approximation? In 1954, Pleijel got the next order approximation. In his paper Kac 1966 [775], Kac works quite hard to get the third term, guessing that the right formula should be:

90   1 Euclidean Geometry

$$\sum_{i=1}^{\infty} e^{-\lambda_i t} \sim \frac{\text{Area}(D)}{2\pi t} - \frac{\text{length}(\partial D)}{\sqrt{2\pi t}} + \frac{1}{6}(1-r)$$

where r is the number of holes inside D. The second term is Pleijel's. Note that can one hear the area and the perimeter of D, hence the isoperimetric inequality yields again the fact that disks are characterized by their spectrum.

Kac could only prove the third term for polygons. It was proven the next year, 1967, in the very general context of Riemannian manifolds with boundary by McKean and Singer in their fundamental paper of 1967 [910]. You will read much more about it in chapter 9.

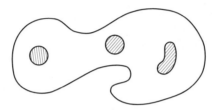

Fig. 1.97. One can hear the number of holes

The Tauberian theorem above shows that the first terms of $N(\lambda)$ and $\sum_{i=1}^{\infty} \exp(-\lambda_i t)$, can each be acquired from the other. But this does not apply to the following terms, so the preceding results say nothing concerning the second term of $N(\lambda)$. The second term of $N(\lambda)$ is still not completely understood today, even for squares; this is the famous *Gauß circle problem* in number theory (see Erdös, Gruber, & Hammer 1989 [491] and the references there on page 104, and also Bérard 1980 [133] for an introduction to the subject, and Huxley 1996 [752]). However, some results can be obtained, but one should use the wave equation instead. The fundamental solution is trickier than for the heat equation, but once it is understood, one purchases deeper results. In the case of Riemannian manifolds, optimal results can be obtained for the second order term. See chapter XXIX of Hörmander 1983 [738] and the very informative Trèves 1980 [1198, 1199], chapter 18. For more recent results and conjectures, see chapter 9.

1.8.5.1 Eigenfunctions

Another important question, having looked at eigenvalues, is to look at eigenfunctions, in particular where they vanish. The sets $\{\phi_i = 0\}$ are called the *nodal lines* of the domain. Nodal lines have been the subject of intensive studies on huge computers. They have many physical interpretations. Cover a vibrating membrane (a drum) with dust. Nodal lines are the places where the dust concentrates, precisely because (at least at the first order) in a given vibrating mode associated to an eigenfunction, nodal lines are the points which do not move. Put your finger at a point m on the membrane; then the only possible modes occuring in the vibration will be those whose nodal line contains the point m. Nodal lines very often have a surprisingly "regular" behavior. This is only partially explained today; see references in chapter 9. Nodal lines have practical consequence. An example: when autoworkers attach the motor or the body to the underframe, bolts (or, more often these days, point solderings) are carefully placed on nodal lines of the underframe.

Fig. 1.98. Nodal lines of ergodic domains have surprisingly regular behaviour

1.8.6 Relations Between the Two Spectra

A plane domain D has both a length spectrum (for classical billiards) and a quantum mechanical spectrum (for quantum billiards). Formulas relating the two spectra, when they exist, are called *Poisson type formulas*. We will see the archetype of them, the classical Poisson formula 9.14 on page 389 for flat tori. The task of unearthing their mutual relations is that of connecting classical and quantum mechanics. It is not surprising that physicists published the first papers on the subject: Balian & Bloch 1970/1971/1972/1974 [97, 98, 99, 100, 101]. They gave heuristic support to the conjecture that the spectrum determines the length spectrum. The other way around is more challenging. This is not surprising, since there are many more ways to address the spectrum than to address the length spectrum and the periodic trajectories. To our knowledge, this implication today is known only in the "generic" case; see Guillemin & Melrose 1979 [669]. We will not pursue this further, first because the case with boundary is much harder, and second, because we will study these relations on general Riemannian manifolds (but without boundary) in §9.9.

We mention an interesting fact, not well known, and basically the only result (with its extension in Colin de Verdière 1977 [390] to compact Riemannian manifolds) which goes from the length spectrum to the spectrum. It is due to V. F. Lazutkin and concerns smooth convex domains D of the plane. We recall first that the billiard trajectories inside an ellipse behave almost like those of a circle. A given trajectory remains constantly tangent to a homofocal ellipse (or hyperbola), as depicted in figure 1.99 on the facing page. By the way, it is unknown if only ellipses have this property. Back to an arbitrary smooth strictly convex domain: Lazutkin proved the existence of an infinite number of nested *caustics*. These are curves which are the envelopes of trajectories. Then he showed that to every caustic, there is an infinite sequence of numbers which approximate an infinite number of elements of the spectrum. The idea is beautiful: every tangent to the caustic defines an exterior half-plane. One then can perform, in a sense which has to be made precise, the sum of the waves sent outside from this half-plane. This sum is called a *quasimode* and will turn out to yield an approximation for a lot of eigenvalues. See chapters 9 and 10 for Riemannian manifolds, and Lazutkin 1993 [852] for a recent reference to Lazutkin's many results.

Very recently, "visual" relations may have been discovered between periodic motions and nodal lines. The pictures in figure 1.100 on the next page are quite fascinating; they are taken from Gutzwiller 1990 [673] (a very informative book). On one hand, Gutzwiller conjectured that there are mathematical relations underlying the apparent visual relations between the two pictures, a phenomenon he called *scarring*. On the other hand, in the so-called *arithmetic case*, Sarnak proved in Sarnak 1995 [1095] that there is *no* scarring; also see §§§9.13.2.1.

There are some abstract Riemannian manifolds whose periodic geodesics are easier to track, and from them one can then get information on the spectrum; see the book Buser 1997 [293] for an exemplary case, and chapter 10.

1.8 Heat and Wave Analysis in $\mathbb{E}^2$

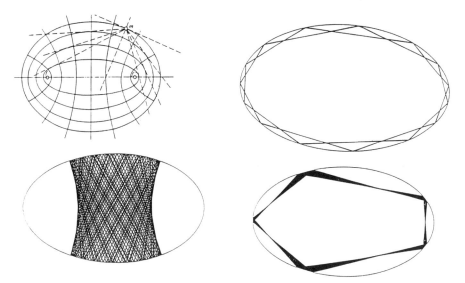

Fig. 1.99. Billiards in an ellipse

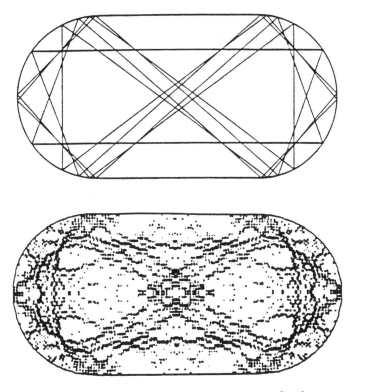

Fig. 1.100. Pictures from Gutzwiller 1990 [673]

1.9 Heat and Waves in $\mathbb{E}^3, \mathbb{E}^d$ and on the Sphere

1.9.1 Euclidean Spaces

In $\mathbb{E}^3$ or $\mathbb{E}^d$, there is little change from the plane, at the outset of our investigation. We replace Δ by:

$$\Delta = -\sum_{i=1}^{d} \frac{\partial^2}{\partial x_i^2} \tag{1.27}$$

where $\{x_i\}_{i=1,\ldots,d}$ are Cartesian coordinates on $\mathbb{E}^d$. Stokes' theorem, used in the derivation of Dirichlet's principle, remains valid here:

$$\int_D \langle \operatorname{grad} f, \operatorname{grad} g \rangle = \int_D f \Delta g.$$

Note here something that we skipped above, namely the intrinsic character of the Laplacian Δ with respect to the Euclidean geometry of $\mathbb{E}^d$. In the language of modern analysis, one can say that the symbol of Δ is $-\|\ \|^2$. And if you add that the subsymbol vanishes, this completely identifies the Laplacian to an analyst.

Phenomena including the nature of the spectrum, expansion of a function as a generalized Fourier series in a basis of eigenfunctions, the expression via eigenfunctions of the fundamental solution of the heat equation, and Weyl's asymptotic formula, are identical but for the obvious changes. For example, the asymptotic expansion for $N(\lambda)$ becomes

$$N(\lambda) \sim \omega_d \frac{\lambda^{d/2}}{(2\pi)^d} \operatorname{Vol}(D)$$

where ω_d is the volume of the unit ball in $\mathbb{E}^d$. From this one gets immediately

$$\lambda_i \sim \left(\frac{(2\pi)^d}{\omega_d \operatorname{Vol}(D)} i \right)^{d/2}$$

We will look for explicit spectra of some special domains. Parallelepipeds are left to the reader. Let us now look at balls in $\mathbb{E}^3$. We aim to progress as on on page 76, this time using spherical coordinates (ρ, θ, ϕ) where ρ is the distance to the origin, θ the latitude and ϕ the longitude, as in figure 1.101 on the next page. Recall that the change of coordinates is given by:

$$x = \rho \sin\theta \cos\phi$$
$$y = \rho \sin\theta \sin\phi$$
$$z = \rho \cos\theta.$$

We will now try to understand Δ on the spheres $\rho = \rho_0$.

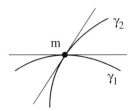

how to compute the Laplacian of f on S^2

Fig. 1.101. (a) Spherical coordinates. (b) How to compute the Laplacian on the sphere

1.9.2 Spheres

It is very instructive to try to write Δ in spherical coordinates. Do not despair if you are stuck, it is a classical trap. One needs a trick to get out of it. You will find it, for example, in Berger, Gauduchon, & Mazet 1971 [174] (chapter III, G.V), Chavel 1984 [325], Courant & Hilbert 1953 [406], or in any text treating mathematical methods of physics. Even if you succeed, this equation looks a mess. You can get out of that mess, but you will not see the *spherical harmonics* very clearly. Moreover, you will suffer, because the representation of the sphere via latitude and longitude is very poor at the poles; see the pictures in your favourite atlas, or chapter 18 of Berger 1987 [164].

Why are things so complicated conceptually compared to a disk in the plane, where we separated variables, and looked for vibrations $h'' = -nh$ (n an integer)? The disk vibrations were gotten first through standard vibrations on the circle (and then for the radial directions, we had the Bessel equation). But the philosophy is that function theory on a circle is the same as for periodic functions on the real line, solved by Fourier analysis. What can replace the circle's Fourier analysis when we are on the sphere?

Approach the intrinsic geometry of the sphere, and consider spherical coordinates rather in the form (ρ, m) where $\rho > 0$ and $m \in S^2$, and work now with the inner geometry of S^2, ignoring $\mathbb{E}^3$. We define the Laplace–Beltrami operator Δ (more precisely, Δ_{S^2} of S^2 if necessary to avoid confusion) for a function $f : S^2 \to \mathbb{R}$ by:

$$\Delta_{S^2} f(m) = -\left.\frac{d^2(f \circ \gamma_1)}{dt^2}\right|_{t=0} - \left.\frac{d^2(f \circ \gamma_2)}{dt^2}\right|_{t=0} \tag{1.28}$$

where γ_1, γ_2 are any two geodesics issuing from m at time $t = 0$, parameterized by arc length, and orthonormal at m. This is easily checked to be well defined (if you know some analysis, think of the symbol). Analogously, the reader can define Δ_{S^d}. A difficult computation yields, for a function $f : \mathbb{E}^d \to \mathbb{R}$

$$\Delta_{\mathbb{E}^d} f|_{S^{d-1}} = \Delta_{S^{d-1}} f - \frac{\partial^2 f}{\partial \rho^2} - (d-1)\frac{\partial f}{\partial \rho}$$

The eigenfunctions $\Delta f = \lambda f$ which are separated as $f(x,y,z) = g(\rho)h(m)$ must satisfy $\Delta_{S^2} h = \lambda h$, These functions on the sphere have been well known for a long time, and are called *spherical harmonics*. They are the subject of an immense literature, since they are next to the Fourier series for their importance in geometry and physics. They are the first examples of Fourier analysis on a Riemannian manifold considered abstractly, although this was not the way people looked at them originally. But we emphasized this aspect immediately, since we want to introduce you to Riemannian geometry and analysis in the general setting. For a spectacular geometric application of spherical harmonics to the extremely hard problem of balancing a large number of points evenly around the sphere, see Colin de Verdière 1989 [394] or the very good book Sarnak 1990 [1094]. For the use of spherical harmonics to estimate the kissing number (the maximum number of solid nonintersecting unit balls which can touch one given unit ball), see chapter 13 of Conway & Sloane 1999 [403]. We will meet the kissing number again on on page 583.

The eigenvalues of S^d are the integers $k(d+k-1)$, with the multiplicity

$$\frac{(d+2k-1)(d+k-2)!}{k!(d+k-3)!}$$

where k runs through the integers. References can be Berger, Gauduchon & Mazet 1971 [174], Chavel 1984 [325]. The miracle is that the harmonic polynomials in $\mathbb{E}^d$, when restricted to S^{d-1}, are eigenfunctions of S^{d-1} and, by Stone–Weierstraß, they are numerous enough to prove that there are no other ones.

In particular, the first eigenfunctions are especially appealing; they are the restrictions of the linear forms of $\mathbb{E}^d$, and their eigenvalue is $\lambda_1 = d$. Interpreted in spherical geometry, they are the cosine of the distance function to some point. We will use this heavily in §§12.2.5.

Let us finish things quickly by saying that, spherical eigenvalues and eigenfunctions being known, this in turn solves Dirichlet's problem for the vibrations of balls in $\mathbb{E}^3$ through Bessel type equations. Our analysis extends to any dimension d for balls in $\mathbb{E}^d$ via spherical harmonics for the spheres S^{d-1}. Again, for extremely few other domains D of $\mathbb{E}^d$ can the eigenvalues or the eigenfunctions be computed explicitly. Ellipsoids can be managed by the use of confocal quadrics; see Courant & Hilbert 1953 [406, 407], Morse & Feshbach 1953 [945]. And quite a recent result of Pierre Bérard works for a few very special polyhedra[16] known as Weyl type domains, corresponding to Lie groups; see Bérard 1980 [134]. See also Bérard & Besson 1980 [138].

We will not say much now about spherical harmonics since you are going to meet them often later on and because they are not our principal concern. For those who like recent results and open problems (yes, there are some, even for such old and much belabored subjects) we mention first that Yves Meyer says among other things the following. We consider the wave equation

---

[16] Polyhedra are rather called *polytopes* starting with dimension $d = 4$.

1.9 Heat and Waves in $\mathbb{E}^3, \mathbb{E}^d$ and on the Sphere

on spheres

$$\Delta_{S^2} f = -\frac{\partial^2 f}{\partial t^2}$$

as, for example, describing the evolution with time of a thin layer of water on the sphere. This would describe the motion of the sea, if there were no continents on the planet. It might not be a bad approximation, since oceans are the bulk of the Earth's surface. His first result backs intuition. Disturb the water with a big shock (a Dirac distribution, if you prefer) somewhere, say at the north pole. Then you will always find another big shock at the south pole after a time T which is no larger than 2π. But Meyer's second series of results oppose intuition: the big shock can move from the north to the south pole while remaining extremely small everywhere for all time between 0 and T, as in figure 1.102 on the next page. Worse: some apparently moderate shocks can, with larger and larger time, be very small almost everywhere and almost all the time, but at some times and some places can be as big as desired (everything there and then will break down). Meyer's results explained the following strange observation: quite recently a tidal wave was observed at Martinique, and came back there later, this being completely unnoticed everywhere else in between times.

1.9.3 Billiards in Higher Dimensions

The Faber–Krahn inequality extends to any dimension. Again, the proof involves the isoperimetric inequality (in any dimension). Contrarily, billiards in dimensions larger than two are still to be fully explored. There is one exception: billiards in concave regions, for which we refer to Katok, Strelcyn, Ledrappier & Przytycki 1986 [788]. Such regions have the best possible ergodic behavior.

But for nonconcave regions, even for polyhedra in three dimensional space, almost nothing is known concerning periodic motions or ergodicity, except for rectangular parallelepipeds, balls (of course) and the Bérard examples mentioned above. There is an unpublished result of Katok asserting that the length counting function is subexponential. But, even in the simplest possible cases, things are either not finished, or completely open. Let us mention only two cases. The first is that of the cube. Dynamics in a cube are comparable to dynamics in a square: a trajectory is periodic or everywhere dense. But how quickly a non-periodic trajectory darkens your computer screen is complicated. It depends on the simultaneous approximation of a pair of real numbers by rational numbers. This subject is partly open; in particular one does not know which triples of real numbers give the cubic analogue of the golden ratio (the trajectory in the square for which your computer screen gets dark most rapidly). But the coding of the faces successively hit by the ball is beginning to be unmasked: see Arnoux, Mauduit, Shiokawa, & Tamura 1994 [71]. For the regular tetrahedron, it seems that nobody today has the foggiest idea whether trajectories are ergodic or more like polygons with angles

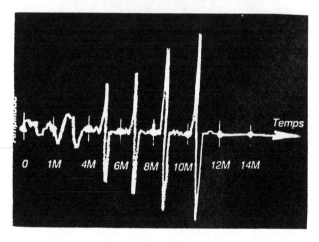

Fig. 1.102. Tidal waves returning to Martinique

rational multiples of π. The fact that there is no guess is surprising: today no one can program a computer to accurately simulate tetrahedral billiards. However, there are a few results yielding periodic trajectories in tetrahedra, based on the geometry of isometries of $\mathbb{E}^3$, but they are difficult to locate; Conway has some unpublished results, and there are published results of R. Hayward, M. Gardner, and D. Wells.

1.9.4 The Wave Equation Versus the Heat Equation

In any dimension, the wave equation is much harder to study that the heat equation. The drastic difference of behavior of waves according to the parity of the dimension of the space where waves propagate was observed long ago. We will come back to these two equations in chapter 9. Briefly put, the heat equation corresponds to the Laplace transform, while the wave equation corresponds to the more powerful Fourier transform.

2 Transition:
The Need for a More General Framework

Euclidean geometry and the geometry of surfaces in $\mathbb{E}^3$ that we looked at in the preceeding chapter turn out to be quite unsatisfactory for many reasons. We will review some of them here; they are not all logically related.

The first two problems with Euclidean geometry came to light in the mind of Gauß. Recall Euclid's fifth postulate (see Euclid *circa* 300 B.C.E.[499]):

Postulate 15 *If a straight line falling on two straight lines makes two interior angles on the same side with sum less than two right angles, the two straight lines, if produced indefinitely, meet on that side on which the angles are less than two right angles.*

By the end of the 18th century, Euclidean geometry's structure was unclear. Although many people (perhaps Euclid among them) had believed that Euclid's fifth postulate was a consequence of his four others, possibly with some additional postulate which would be simpler and more elegant, none had convincingly succeeded in finding a candidate additional postulate, or proving the redundancy of the fifth.[1] Discomfort with the fifth postulate arises from its complexity, and also because one can not verify it (or even test its approximate accuracy) by direct physical measurement, since the point where two lines meet could be very far off of your sheet of graph paper.

By 1817, Gauß had convinced himself that there must be other geometries satisfying Euclid's first four postulates (and, of course, satisfying the common notions), but not the fifth. He did not publish these ideas. One reason for his reticence is quite plausible: Gauß could not clearly define his new geometry, and he often described himself in letters to his friends as a "perfectionist". That was why he also delayed for years publishing his contributions to the theory of surfaces, as we will mention shortly.

Various partial constructions of non-Euclidean geometry appeared starting in 1829, in the writings of Lobachevskii, followed by a work of Bolyai. They

---

[1] Recall that Euclid introduced ten axioms for his geometry, five of which were called *postulates* while the other five were called *common notions*. The significance of this distinction is a matter of conjecture, but the postulates are clearly geometric, while the common notions are concerned with the nature of equations and inequalities generally, essentially defining $=$ and $<$. The common notions are assumed in any geometry.

satisfied the first four postulates, but not the fifth; however, they were not built on solid mathematics. The proper construction of non-Euclidean geometry demands abstraction, since (as Hilbert demonstrated much later) there are no candidates for such a geometry inside Euclidean space; this is our first motivation for Riemannian geometry; see §§1.6.3 and §§3.4.3.

Even if you understand surfaces in $\mathbb{E}^3$ pretty well, there are many reasons to work in dimensions 3, 4, etc. For example, to treat time as an additional coordinate, or to investigate the space of lines, the space of circles, the space of spheres, etc. There is no limit on the number of parameters, i.e. dimensions, needed to describe sets of geometric objects. Even if there is an elementary method to put such a geometric set into some Euclidean space $\mathbb{E}^n$, such an embedding may not be geometrically equivariant, i.e. symmetries of the geometric set might not extend to symmetries of Euclidean space.

Consider classical mechanics, starting with the notion of a rigid body. It has six parameters: three for the location of the center of gravity, and three to say how it has been rotated around that center. You can avoid working in a six dimensional space, because the center of gravity lives in a three dimensional Euclidean space, but what is the set of rotations, as a three dimensional object? How can one explore geometry on it? The reader may have struggled over mechanics texts and suffered through *Euler angles*, and to such a reader, the question does not appear superfluous. We would like a general framework, in which the motions of the rigid body will be geometrically meaningful curves.

More generally, we need to define the set of positions of a mechanical object, like a multiple pendulum, and see what the trajectories of this object are, as curves in that set. For a double pendulum, we will see below that the space of positions is a surface, in fact a torus, but that the motions of the pendulum are not the geodesics on the torus, for any torus inside $\mathbb{E}^3$. They will be geodesics only for an abstract Riemannian geometry.

In statistical mechanics, physicists consider systems of N particles, with N large. The space of positions is $3N$ dimensional, and to carry complete information about their motions, one needs to know their positions and velocities, so a $6N$ dimensional *phase space*. This is not simply $\mathbb{E}^{6N}$, because of collisions. Thus our third reason for abstraction is that we have already been manipulating such abstract spaces, and trying to pretend we were not (i.e. trying to parameterize them).

As we will see in theorem 27 on page 135, abstract Riemannian geometry of surfaces can only be represented locally, and only subject to some nondegeneracy hypothesis, by surfaces in $\mathbb{E}^3$. Moreover, this is special to two dimensional objects; for example, the geometry of a three dimensional smooth object in $\mathbb{E}^4$ is very special among three dimensional Riemannian geometries.

The program of abstract Riemannian geometry received two decisive thrusts, the first by Gauß in 1827, and the second by Riemann in 1854. In essence, Gauß realized the concept of inner metric of a surface in $\mathbb{E}^3$, and then realized that the *first fundamental form* of a surface (the Euclidean inner product restricted to its tangent spaces) and the inner geometry each determine the other. The length of a curve γ is measured by an integral:

$$\int \sqrt{I(\dot{\gamma})}\, dt$$

and this I is our first fundamental form, so this connects the inner geometry to a quadratic form I on tangent vectors. With this form, one can calculate angles, distances, etc., and strenuous algebraic manipulations uncover the Gauß curvature, as we will see.

Riemann forged two simultaneous innovations: first, he defined (not too rigorously) a differentiable manifold to be a set of any dimension n, where one can perform differential calculus, change coordinates, etc. In particular, one has differentiable curves, tangent vectors (velocities) of those curves, and a tangent space at each point (i.e. all possible velocities of any curves through that point). Then he asked that a geometry on a manifold be simply an arbitrary positive definite quadratic form on each of those tangent spaces, thought of as the analogue of Gauß's first fundamental form. One could use the same expression to define length of curves, look for shortest curves, etc.

We are concerned with the immense program to develop the properties of such a geometry, looking for invariants generalizing the invariant of Gauß, the Gauß curvature, and looking for extensions of the results of Gauß. Riemann managed only to find the right invariant generalizing Gauß curvature, now called the *Riemann curvature tensor*. Many mathematicians after Gauß and Riemann set down the foundations of Riemannian geometry. Technical tools such as parallel transport and absolute differential calculus were designed at the end of the 19th century, culminating in the work of the Italian school of Ricci and Levi-Civita, but the underlying concept of manifold was definitively understood only by Whitney in 1936.

In the sequel, we will first present the crux of Gauß's contributions, and also review more recent results. Gauß's accomplishment is effectively two dimensional Riemannian geometry, so that many parts of Gauß's work are covered by an exposition of the latter. But we think that developing the subject quite slowly, using concepts available to Gauß, will help the reader when we go on to general abstract Riemannian geometry, because two dimensional objects can be visualized and have technical peculiarities making them more tractable.

After surfaces, we will turn to general Riemannian manifolds. Our presentation will be in the following spirit: we will try to remain geometric for as long as possible, and also state results, including recent ones, not only without proof, but (even worse) hiding the techniques of the proofs. For example, in the first chapters we will assume that there is a parallel transport operator allowing us to drag tangent vectors along a curve; we will not even define these ideas. Nor will we enter into the hornet's nest of giving a definition of a Riemannian manifold. We are justified in taking this path: *parallel transport* and *manifold* had to wait until around 1900 and 1936 respectively to receive definitions. You should regard them as expensive.

Later on, as in chapter 1, we will undertake analysis on Riemannian manifolds. This is possible because the basic differential operator in Euclidean analysis, the Laplace operator, can be defined invariantly on a general Riemannian manifold, following a natural analogy. It is called the *Laplace–Beltrami operator*, but we will call it the *Laplacian* for short. A dramatic consequence (not mentioned in the written work of Riemann nor by generations of his successors, though likely in Riemann's imagination, since he often worked in mathematics with physics in mind) is that heat, wave and Schrödinger equations are born, out of pure geometry. This explains the popularity of Riemannian geometry today in many domains of physics. We will also see that Riemannian manifolds are the natural worlds in which to study Hamiltonian mechanics.

Along with introducing the classical perspective, we will treat modern views of Riemannian geometry. Counter to intuition, to study geometry on a Riemannian manifold, one has to work not only on the manifold itself, but to relate its geometry to various larger spaces (fiber bundles).

3 Surfaces from Gauß to Today

Contents

3.1 **Gauß** **105**
  3.1.1 Theorema Egregium 105
  3.1.2 The Gauß–Bonnet Formula
         and the Rodrigues–Gauß Map 111
  3.1.3 Parallel Transport 113
  3.1.4 Inner Geometry 116
3.2 **Alexandrov's Theorems** **120**
  3.2.1 Angle Corrections of Legendre and Gauß
         in Geodesy 123
3.3 **Cut Loci** **125**
3.4 **Global Surface Theory** **131**
  3.4.1 Bending Surfaces 131
  3.4.2 Mean Curvature Rigidity of the Sphere 134
  3.4.3 Negatively Curved Surfaces 135
  3.4.4 The Willmore Conjecture 136
  3.4.5 The Global Gauß–Bonnet Theorem for
         Surfaces 136
  3.4.6 The Hopf Index Formula 139

3.1 Gauß

3.1.1 Theorema Egregium

This theorem is baffling. Its Latin wording ("excellent theorem") was forged by Gauß because he was so excited about it. Gauß studied surfaces for years before discovering this result. It is the kind of theorem which could have waited dozens of years more before being discovered by another mathematician since, unlike so much of intellectual history, it was absolutely not in the air.

Theorem 16 (Theorema Egregium) *The total curvature $K = k_1 k_2$ of a surface depends only on its inner metric.*

It is not completely clear how Gauß found this: see the remarkable historical account of Gauß's results on surfaces in Dombrowski 1979 [455]. We will say a few words about this later on. To our knowledge there is no simple geometric proof of the theorema egregium today. The only geometric argument we know which is close to this is the one in page 195 of Hilbert & Cohn-Vossen 1952 [713]. The picture is given in figure 3.1.

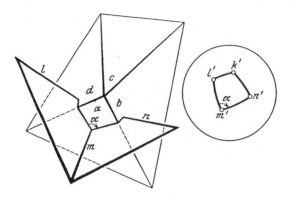

Fig. 3.1. Ideas from the proof of the theorema egregium

There are many different proofs of the theorema egregium. For them we refer to the standard texts, as for example: do Carmo 1976 [451], Klingenberg 1995 [816], O'Neill 1966 [974], Stoker 1989 [1160], Thorpe 1994 [1188], Sternberg 1983 [1157], Boothby 1986 [220]. We give two proofs, one of them rarely presented. The first proof below is the most standard and quite cheap modulo equation 1.13 on page 52. In case this formula is not assumed, the second proof we are going to give is no more complicated and as an advantage is deeper conceptually. Moreover, this second proof yields extra results immediately. It is stems out of Riemann's fundamental paper, and represents the gist of Riemann's generalization of Gauß curvature to any dimension. Since we will not give Riemann's proof in general dimensions, instead we offer it here in our discussion of surfaces.

3.1.1.1 The First Proof of Gauß's Theorema Egregium; the Concept of ds^2

Let us assume equation 1.14 on page 53 and prove that $k_1(m)k_2(m)$ depends only on the inner metric of the surface M near the point m. We first need to express the inner metric of a surface in coordinate language. In other words, we introduce the device refered to in the jargon as ds^2. Suppose that we have a coordinate system locally describing M:

$$(u,v) \mapsto (x = x(u,v), y = y(u,v), z = z(u,v))$$

To designate the inner metric is equivalent to designating the lengths of curves. If a curve is given by
$$t \mapsto (u(t), v(t))$$
then the composition of the two maps will be
$$c : t \mapsto (x(t), y(t), z(t))$$
which is a curve in $\mathbb{E}^3$ whose length can be computed as the integral of its speed:
$$\|\dot{c}(t)\| = \sqrt{\dot{x}(t)^2 + \dot{y}(t)^2 + \dot{z}(t)^2}$$
If we write subscripts for partial derivatives, e.g. $x_u = \partial x / \partial u$, then the chain rule yields:
$$\begin{pmatrix} \dot{x} \\ \dot{y} \\ \dot{z} \end{pmatrix} = \begin{pmatrix} x_u & x_v \\ y_u & y_v \\ z_u & z_v \end{pmatrix} \begin{pmatrix} \dot{u} \\ \dot{v} \end{pmatrix}$$
We collect the terms in $\|\dot{c}(t)\|$ in the following way:
$$\|\dot{c}\|^2 = E\dot{u}^2 + 2F\dot{u}\dot{v} + G\dot{v}^2 \tag{3.1}$$
where E, F, and G are functions of (u, v), depending only on the chart chosen to parameterize the surface. Today most people prefer to write this in differential notation:
$$ds = \|\dot{c}(t)\|\, dt$$
$$du = \dot{u}(t)\, dt$$
$$dv = \dot{v}(t)\, dt$$
$$ds^2 = E\, du^2 + F\, du\, dv + G\, dv^2 \tag{3.2}$$

Do not underestimate the next few lines; they are at the root both of Gauß's way of thinking and of Riemann's dramatic reversal. We analyze the mathematical meaning of the above. In order to know the inner metric (in a given chart) we need only know three functions E, F, G, but more conceptually to know a quadratic form (depending on two variables). Clearly the quadratic form determines the inner geometry, since we can use it to calculate integrals giving lengths of curves. The trick is to show that we can recover the quadratic form from the inner geometry.

The reader should take it as an exercise to prove that any smooth map between surfaces which preserves the lengths of curves must preserve the Euclidean inner product on the tangent spaces of the surfaces. This exercise has several parts: first, define what a smooth map should be between surfaces. For example, just ask for it to extend smoothly to some neighborhood in the ambient space. Now, define the derivative of such a map. Recalling the usual notion of derivative of a smooth map between Euclidean spaces, as a linear approximation, one wants to prove that the derivative of a map between surfaces is well defined as a linear map on tangent planes to the surface. Also, the reader should produce an example to become convinced that the derivative is not well defined in any other directions. Acclimatize yourself with what happens to derivatives when viewed through charts. Finally, if the map preserves length of curves, one shows (by looking at "small curves", i.e. taking a limit) that the derivative map preserves lengths of tangent vectors to parameterized curves, and so is an isometry. Moreover, now show that any smooth isometry of the inner metric must have the same property: it preserves lengths and so its derivative preserves the Euclidean quadratic form. From the above comments, the reverse is true as well: the Euclidean quadratic form, ds^2, is preserved by a smooth map precisely when the map is an inner metric isometry. Thus ds^2 is just a repackaging of the inner geometry.

One can (should!) quibble here. The isometries we spoke about are assumed smooth. We will simply take this smoothness requirement to be part of the inner geometry. This is not obviously consistent with our definition of inner geometry as a pure metric space notion. It is not obvious how to ensure that all isometries of the inner metric are smooth. The reader is encouraged to worry about this. The author will not. Proof of smoothness of isometries requires the Gauß lemma below, and consideration of the trigonometry of small triangles. For guidance, see Palais 1957 [991].

The idea that E, F, G compose a quadratic form is not surprising, since a quadratic form (symmetric matrix) in dimension two consists of three independent matrix elements. In the present case these three elements are in the basis made up by the two tangent vectors associated to x and y: namely their square norm and their scalar product. This quadratic form is nothing but the Euclidean structure of the tangent plane $T_m M$ to the surface M under consideration, this Euclidean structure being written in coordinates via the map $(u, v) \mapsto \mathbb{E}^3$.

We now want to find a Taylor expansion for ds^2 at some point $m = (0,0)$ in a suitable chart (x, y). If ds^2 has the form

$$ds^2 = dx^2 + dy^2 + O(x^2 + y^2)$$

(i.e. if ds^2 is Euclidean at the first order) then

$$K(m) = -12\left(E_{yy} - 2F_{xy} + G_{xx}\right) \qquad (3.3)$$

(here indices represent partial derivatives).

This is the proof, since everything involves only ds^2. The assertion is checked when using a chart as in equation 1.13 on page 52 and taking $m = (0, 0)$. Then P and Q vanish at m and equation 3.3 follows directly from equation 1.13.

Note 3.1.1.1 Gauß wanted so much to be sure of his theorem that he checked equation 3.3 for the most general chart. The explicit formula, which took him

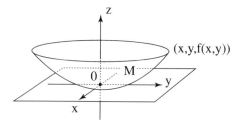

Fig. 3.2. Adapting coordinates to see Gauß curvature

five pages of computation, is given in Dombrowski 1979 [455]. We will see below how he might have guessed the result. ♦

Note 3.1.1.2 Formula 3.3 raises a paradox which is never mentioned in textbooks. The left hand side involves only the second derivatives of the equation defining the surface while the right hand side involves third derivatives. This topic is connected with the end of §4.5. Many people worked on this peculiarity, which is of some significance in the study of Riemannian geometry with minimal smoothness. Consult Weyl 1916 [1255], van Kampen 1938 [1205], Chern, Hartman, & Wintner 1954 [369], Klotz 1959 [818], and Burago & Zalgaller 1992 [284] on this point. Van Kampen treats the C^2 case. ♦

3.1.1.2 Second Proof of the Theorema Egregium

The chart given by taking our surface to be the graph of a function avoided some mess, since it was Euclidean at first order. But the inner metric enables us to define a canonical chart (locally) near any point. Such charts will turn out to be of fundamental importance for general Riemannian manifolds, so we will use the opportunity to introduce them on surfaces. We use a fact met in §§ 1.6.2 on page 33: a geodesic emanates from every point in every direction with any choice of constant speed. This gives a map from the tangent vectors sticking out of some point to the surface itself: just follow for a unit of time a constant speed geodesic with given initial velocity. Fixing any orthonormal basis of the tangent plane, this gives us a canonical local map from $\mathbb{E}^2$ into $\mathbb{E}^3$, a canonical chart called *normal coordinates*, which we write as u, v. Geometrically, geodesics through a point of the surface are represented by straight lines through the origin in the chart, and with an arc-length parameterization.

We want to compute ds^2 in that chart, up to the second order and for this we factor it through the graph chart from equation 1.9 on page 35. We have to find the second order terms of $x(u, v), y(u, v), z(u, v)$. For this we use the equations 1.10 and 1.11 on page 35 and moreover we rotate the x, y plane so that $S(m) = 0$. As in equation 1.14 on page 53

$$F(x,y) = \frac{1}{2}k_1 x^2 + \frac{1}{2}k_2 y^2 + o\left(x^2 + y^2\right)$$

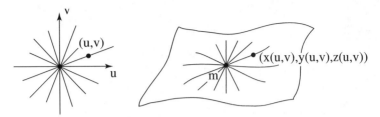

Fig. 3.3. Normal coordinates

with $k_1 = k_1(m)$ and $k_2 = k_2(m)$.

After some pain we find

$$x = u - \frac{1}{24}k_1 u \left(k_1 u^2 + k_2 v^2\right) + o\left(u^2 + v^2\right)$$
$$y = v - \frac{1}{24}k_2 v \left(k_1 u^2 + k_2 v^2\right) + o\left(u^2 + v^2\right)$$

and substituting yields

$$ds^2 = du^2 + dv^2 - \frac{1}{12}K(m)\left(u\,dv - v\,du\right)^2 + o\left(u^2 + v^2\right) \qquad (3.4)$$

We now explore consequences of equation 3.4. One sees clearly that K measures the defect of the inner metric of the surface M to be locally Euclidean (by this we mean isometric locally to $\mathbb{E}^2$; we will also use the word *flat* instead of locally Euclidean). We can now easily prove a claim we made in §§ 1.6.7 on page 62: our surface is locally Euclidean if and only if K vanishes identically. In general the knowledge of K is not enough to determine the inner metric of the surface: see §§4.5.1. But if there is a map between two surfaces which is an isometry for their inner metrics, then their Gauß curvatures must correspond under this map.

Suppose now that you live in M and want to discover if your geometry is Euclidean, or more, compute its Gauß curvature K. From the point m at which you want to compute $K(m)$, you look at the points which are at a small distance ε. They describe a curve which we can call the *circle* $C(m, \varepsilon)$ with center m and of radius ε. Calculate its length. Equation 3.4 gives you the answer:

Theorem 17 (Bertrand & Puiseux, 1848)

$$\operatorname{length} C(m, \varepsilon) = 2\pi\varepsilon - \frac{\pi}{3}K(m)\varepsilon^3 + o\left(\varepsilon^3\right)$$

so that you can get $K(m)$ as a limit. Actually, to be precise, we still have to see that the circles $C(m, \varepsilon)$ are represented in normal coordinates precisely by

$$C(m, \varepsilon) = \left\{u^2 + v^2 = \varepsilon^2\right\}\;;$$

a remarkable coincidence, matching exactly the Euclidean circle of the same radius in these coordinates. See §§ 3.1.4 on page 116 for this. Please check this formula directly on the sphere and also check that this confirms what we saw in §§ 1.6.4 on page 52.

$K(m) > 0$: length$(C(m;\varepsilon)) < 2\pi\varepsilon$ $K(m) < 0$: length$(C(m;\varepsilon)) > 2\pi\varepsilon$

Fig. 3.4. Small circles in differently curved surfaces

3.1.2 The Gauß–Bonnet Formula and the Rodrigues–Gauß Map

This is one of the deepest and hardest formulae to prove for surfaces. It will imply the theorema egregium by a limit argument, and can also be viewed as an integrated version of the theorema egregium. We refer the reader to Dombrowski 1979 [455] for what can be said about how Gauß guessed all of these things. We state the theorem as improved by Bonnet, Gauß having proved it only for geodesic polygons. There is no simple proof; it is always expensive and hard to do it correctly. The difficulty is the same as for the Umlaufsatz in § 1.4 on page 18 with the additional complication of carrying it out on a surface. One cannot get the formula only using Stokes' theorem (directly down on the surface) however tempting it may be (since we have an equality between an integral in a domain and one on its boundary). For various types of proof, see the set of references given in §§ 3.1.1 on page 105.

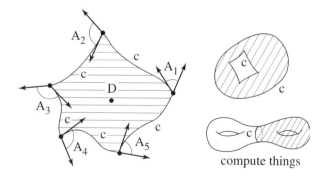

Fig. 3.5. The Gauß–Bonnet theorem

We consider on a surface M a domain D, which is the continuous image of a Euclidean disk (if you are familiar with a little algebraic topology, this amounts to asking D to be simply connected). This condition is sufficient but not necessary. The necessary and sufficient one is the existence on D of a continuous nowhere vanishing vector field. We ask moreover that D have as boundary a curve c which is smooth except at a finite number of points m_i. At those points there are well-defined exterior angles which we denote by α_i (see the picture in figure 3.5 on the preceding page). From remarks made on page 34 we infer that c has well defined algebraic geodesic curvature k_g, away from the points m_i.

Lemma 18 (Local Gauß–Bonnet)

$$\int_D K\, dA + \int_c k_g\, ds + \sum_i \alpha_i = 2\pi$$

We did not say what the integral is. On any surface there is a canonical measure dA. You can find yourself its expression in any chart. In the ds^2 jargon of equation 3.2 on page 107 the quantity to integrate is

$$dA = \sqrt{EG - F^2}\, du\, dv.$$

Check also that the formula can be false if the domain D is not simply connected. We saw on on page 34 that k_g depends only on the inner metric. So take the limit of smaller and smaller domains. Then K depends only on the inner metric. It seems that is the way Gauß guessed the theorema egregium statement (but it was not the solid proof he wanted).

Let us check first that equation 18 is a generalization of two formulas we met before. The first for a simple closed curve in the plane: in the plane $K \equiv 0$ and, if the curve is smooth everywhere, we get the Umlaufsatz of § 1.4 on page 18. If the curve has corners, we get an extension of the Umlaufsatz. Assume now we are on the sphere S^2 and consider a geodesic triangle. Then $K \equiv 1$ and we get Girard's formula 1.7 on page 32. On a surface where $K \equiv -1$ (this will be hyperbolic geometry, see §§ 4.3.2 on page 176 and equation 4.22 on page 180) as in figure 1.60 on page 50, geodesic triangles T verify

$$\text{area}(T) = \pi - A - B - C.$$

The proof of the Gauß–Bonnet theorem is never simple; in fact it is subtle and tricky. By its very essence, it cannot be derived directly using Stokes' theorem 34 on page 171, even if that is what you were expecting, since the formula is a relation between the inside of a domain and its boundary. Various proofs have been devised; see the many books on Riemannian geometry we have already mentioned. In 1944, Chern was the first to find a conceptual proof; the trick is to lift things to the unit tangent bundle (the set of unit length tangent vectors). One can then cleverly apply Stokes' theorem. This is explained with some detail in § 15.7 on page 709. We suggest also that the reader find counterexamples when D is not simply connected.

The Rodrigues–Gauß map was already introduced in note 1.6.5.1 on page 57. We fix a continuous choice of a unit normal N along M, forming a map $N : M \to S^2$. The Jacobian of N (easy to calculate) is equal to $k_1 k_2$. Hence

Proposition 19 *If we denote by $d\sigma$ the canonical measure of S^2, then $K\,dA$ is the pullback measure:*
$$K\,dA = N^* d\sigma\,.$$

The choice of the normal is irrelevant since K is invariantly defined. To develop your intuition, check that this agrees with the sign of K. It seems that Gauß looked quite carefully at the map N to get his results. He noted moreover that this map is very natural in astronomy. One can now interpret the Gauß–Bonnet formula as

$$\int_D N^* d\sigma = 2\pi - \int_C k_g\,ds - \sum_i \alpha_i \qquad (3.5)$$

where the left hand integral can be interpreted as the algebraic measure of the image of D under the Rodrigues–Gauß map.[1] The sign of K corresponds to whether N respects or reverses orientation. The proof of equation 3.5 consists in integrating the infinitesimal equation $dN = K\,ds$, the difficulty being to see what is happening at the boundary, because we need more than just Stokes' theorem.

3.1.3 Another Look at the Gauß–Bonnet Formula: Parallel Transport

We come back to the inner metric and ignore again the outside world. In the Euclidean plane, if you have a closed curve and pick some vector at one of its points you can transport it along the curve by just asking the vector to be constant. And when you come back home, your vector is unchanged. You probably know that this is not possible on surfaces; think of the sphere. The various tangent spaces to a surface can not be all at once identified.

We want to define a notion of parallel transport on a surface depending only on its inner metric. This will turn out to be possible in the sense that we will be able to identify the various tangent planes to a given surface, but only along a given curve. To do this, we reverse the idea of defining curvature of a plane curve by the rate at which its tangent line turns, as in §§ 1.3.2 on page 12. Curvature measures the defect of a curve from a straight line (or, on

---

[1] By algebraic measure, we mean (as we discussed in studying turning numbers) that we have to count with opposite signs when we are on regions that get mapped with reversed orientation.

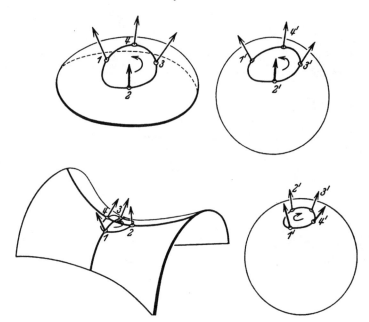

Fig. 3.6. The Rodrigues–Gauß map is not too easy to visualize, especially when $K < 0$. The Rodrigues–Gauß map preserves orientation at an elliptic point, and reverses it at a hyperbolic point

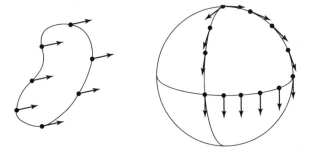

Fig. 3.7. Euclidean case: one comes back to the same vector. On a sphere, you come back to a different vector in general

a surface, defect from a geodesic). So we say that a vector field $X(s)$ along a given curve c is parallel if, first, it is of constant norm, and second, if one has for its angle $\alpha(s)$ with $c'(s)$:

$$\alpha'(s) = -k_g c(s) \qquad (3.6)$$

We say that $X(s)$ is the *parallel transport* of $X(0)$ along c.[2]

---

[2] The function k_g is the geodesic curvature introduced on page on page 34.

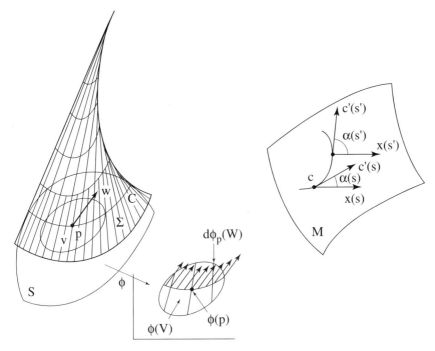

Fig. 3.8. Parallel transport

This is determined by the inner geometry, because the geodesic curvature is inner-invariant. Note that the geodesics are exactly the tangent-self-parallel curves. When you make a journey along a closed curve and follow a parallel vector field, at the end you will have a discrepancy from what you started with. The discrepancy (i.e. the angle of rotation) is measured in domains D parameterizable by a disk, with boundary $c = \partial D$, exactly by

$$\int_D K\,dA.$$

If you think of such a curve as a loop, this discrepancy is called the *holonomy* of the loop. In more general contexts of mathematics and physics it is an extremely important notion. See chapter 13 for holonomy.

There is a kinematic way to look at parallel transport. We start with surface M, a curve c in M, and some initial point m on c. Consider a fixed plane tangent to M at m, and make the surface M roll on this plane along the curve c, the surface M being considered as a rigid body. The point of contact in this rolling will describe a curve c^* in the fixed plane. The curvature of c^* is equal to the curvature of c and parallel vector fields $X(s)$ in M along c correspond to vector fields $X^*(s)$ along c^* which are just constant as vectors in the fixed plane.

3.1.4 Inner Geometry

The first result is due to Gauß. It is the exact analogue for surfaces of the first variation formula 1.2 on page 4. We have a one-parameter family of geodesics γ_t along which we pick two moving points $c(t) = \gamma_t(a(t)), d(t) = \gamma_t(b(t))$, as in figure 3.9. Call $\delta(t)$ the arc length from $a(t)$ to $b(t)$ along γ_t. Be aware that $\delta(t)$ is not in general the distance for the inner metric between the points $c(t)$ and $d(t)$. But it is if $\delta(t)$ is small enough (remember equation 1.8 on page 34).

$$\frac{d\delta}{dt} = d'(t) \left.\frac{\partial \gamma}{\partial s}\right|_{s=b(t)} - c'(t) \left.\frac{\partial \gamma}{\partial s}\right|_{s=a(t)} \tag{3.7}$$

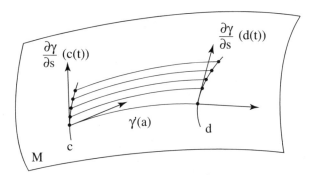

Fig. 3.9. Variation through geodesics

A first consequence:

Lemma 20 (Gauß) *Geodesics forming a smooth family and connecting two fixed points must all have the same length.*

This fact was employed for ellipsoids on page 38. A second consequence is that the circles $C(m;\varepsilon)$ of theorem 17 on page 110 have tangent orthogonal to their radius. This is often called the *Gauß lemma*. We can use it immediately as follows. Pick up a normal chart at a point m of a surface M and look at it in the associated polar coordinates. Then, for ρ small enough,

$$ds^2 = d\rho^2 + f^2(\rho, \theta)\, d\theta^2 \, . \tag{3.8}$$

We will soon relate f to the curvature K. For the moment, we claim that equation 3.8 implies that the inner distance from m to the point (ρ, θ) is always equal to ρ. Therefore the Euclidean disk $\{m \,|\, \rho(m) < \rho_0\}$ is always, for some $\rho_0 > 0$, also the metric disk for the intrinsic metric (not the metric induced by the embedding). The proof is elegant: you draw a curve from m to a point (ρ, θ). If it stays entirely in the domain of the chart, then it is obvious from equation 3.8 that its length is larger than or equal to ρ and that equality occurs only when the curve is the radial line $\theta = $ constant. This is because ρ increases at unit rate along radial lines, and more slowly in other directions. If the curve goes outside of the domain, it is "obviously" longer, but to prove it rigorously we have to appeal to the general topology result called the theorem of the custom passage: the curve must meet the boundary circle $C(m, \rho_0)$ so that its length is larger than $\rho_0 > \rho$.

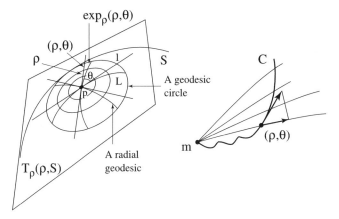

Fig. 3.10. Radial geodesics

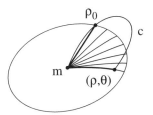

Fig. 3.11. The custom theorem: the length of c is larger than ρ_0

The second result, also due to Gauß, serves for surfaces the task that will later on be carried out by the notion of *Jacobi vector field*, and is a basic tool in Riemannian geometry: see § 6.3 on page 248. It is in some sense an integrated version of equation 3.4 on page 110. Namely, consider again a one parameter family of geodesics $\gamma(s, \alpha)$ where s is the arc length on each geodesic, and α is the parameter. We want to study the displacement vector $\frac{\partial \gamma}{\partial \alpha}$. Assume it is orthogonal to the geodesics so that it yields a function f along a chosen geodesic γ_0:

$$\frac{\partial \gamma}{\partial \alpha} = fn$$

where n is a unit normal vector field to γ_0. The *Gauß equation* is
$$f''(s) + K\left(\gamma\left(s,\alpha\right)\right) f(s) = 0 \tag{3.9}$$
in short: $f'' + Kf = 0$, where the derivatives are with respect to arclength s.

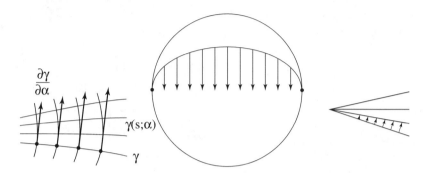

Fig. 3.12. A Jacobi field on the sphere: check that it is always of the form $f = a \sin s$

The Gauß equation is not too hard to prove, and it could have been expected that it was a second order linear differential equation. This because geodesics are determined as soon as a starting point and starting velocity are given. Or, if you prefer, the equations for geodesics (equations 1.10 and 1.11 from page 35) are second order. Differentiation with respect to a parameter linearizes these equations. What is important is the fact that in some sense K tells the whole story of geodesic spreading, to first order. This does not contradict the fact that K is not enough to know the metric, because equation 3.9 implicitly involves the knowledge of some geodesics. We say again here that K does not determine the inner metric up to isometry: on a generic surface, we can slide along the level sets of K with infinitely many degrees of freedom, while the metric geometry of a generic surface is absolutely rigid. We postpone examples to §4.5 because they are much better understood in the abstract frame of Riemannian manifolds.

If you know $K(x, y)$ in normal coordinates (as on page 109) then you know the metric completely.

However, there is one case where you need not know the geodesics: when K is constant. Locally, metrics of constant curvature are all known. Surfaces with the same constant curvature are locally isometric. If you use polar coordinates (ρ, θ), the three cases are

$$ds^2 = \begin{cases} d\rho^2 - \frac{1}{K}\sinh^2\left(\rho\sqrt{-K}\right) d\theta^2 & K < 0 \\ d\rho^2 + \rho^2 \, d\theta^2 & K = 0 \\ d\rho^2 + \frac{1}{K}\sin^2\left(\rho\sqrt{K}\right) d\theta^2 & K > 0 \end{cases} \tag{3.10}$$

The first is the hyperbolic plane, the second the Euclidean plane, and the third is the sphere of radius $K^{-1/2}$. We saw in the pictures in §§ 1.6.3 on page 45 examples of surfaces with $K \equiv -1$. Study carefully the remarks made in §§ 1.6.3 on page 45 and §§ 4.3.2 on page 176.

Gauß's equation proves quite easily the formulas 1.15 on page 54 and 1.16 on page 54. But we will prove much more in section § 3.3 on page 125. For the moment we leave to the reader to check, with equation 3.9 that, as in the pictures in figure 3.13, geodesics which start from the same point will converge if $K > 0$ and diverge if $K < 0$.

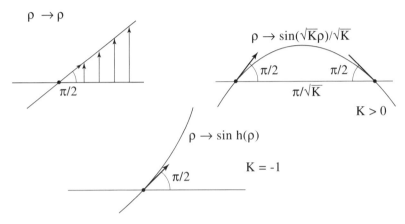

Fig. 3.13. Geodesics on surfaces of constant curvature

Comparing with Euclidean geometry, the third result is new to us (you can guess why after looking at it). It is called the *second variation formula*. In effect, we will try to pluck a guitar string, on a surface (see figure 3.14 on the next page). A second variation formula has substance only when the first variation is zero, i.e. around geodesics. It has primary importance if you want to look for shortest geodesics: at least locally you will not want a negative second derivative. Let $c(t, \alpha)$ be a one parameter family of curves around a geodesic $c(\cdot, 0) = \gamma$ parameterized by arc length, and consider the infinitesimal displacement $\frac{\partial c}{\partial \alpha}(t, \alpha = 0)$ along γ. If you assume it to be normal to the geodesic it is then given by a function f, multiplied by a unit normal vector field. Assuming that the ends of the curves occur at $t = a$ and $t = b$, and also that those ends traverse geodesics as we vary α (a particular case is when the extremities are fixed) then

$$\frac{d^2}{d\alpha^2} \text{length } c(\cdot, \alpha)\bigg|_{\alpha=0} = \int_a^b \left(\left(\frac{df}{dt}\right)^2 - Kf^2\right) dt \qquad (3.11)$$

Curves equidistant from a geodesic are concave when $K > 0$ and convex when $K < 0$, as in figure 3.15 on the following page. We will use this idea in-

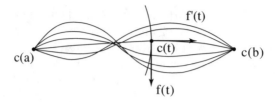

Fig. 3.14. Second variation: plucking a geodesic guitar string

tensively in § 3.3 on page 125 when studying whether geodesics are segments.[3] A nice application is Bonnet's theorem 21 on page 125.

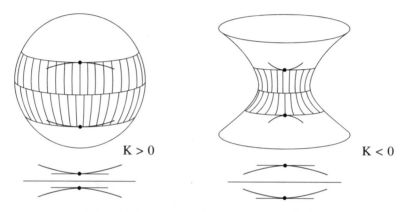

Fig. 3.15. Concavity/convexity of geodesics

If the curves drawn by the two extremities are no longer geodesics, the formula should be corrected by an "integrated" term which involves only the geodesic curvature of these two curves, see for example recent references such as Cheeger & Ebin 1975 [341], Gallot, Hulin, & Lafontaine 1990 [542], do Carmo 1992 [452], Chavel 1993 [326].

3.2 Alexandrov's Theorems and Gauß's Angle Correction

In §§ 1.6.4 on page 52 we claimed global metric inequalities for surfaces with $K > 0$ and other ones with $K < 0$. With equation 3.9 on page 118 we can obtain triangle inequalities with $\delta = \inf K$ and $\Delta = \sup K$. This was proven by Alexandrov[4] around 1940 and was extended Riemannian manifolds of any

---

[3] Recall that *segment* is the word for shortest geodesic, that is to say a geodesic with length equal to the distance between its end points.

[4] Alexandrov proved it even for surfaces which are not very smooth; Cartan had already proven it around 1930 for smooth surfaces with $\Delta = 0$

3.2 Alexandrov's Theorems and Gauß's Angle Correction 121

dimension by Rauch and Toponogov; see § 6.4 on page 257. We sketch here the proof of those inequalities on smooth surfaces, because it is the root of the later developments.

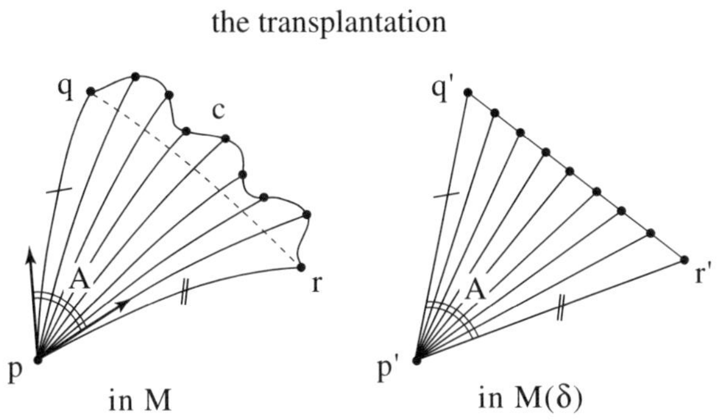

Fig. 3.16. Transplantation of triangles

We start with the assumption that $K \geq \delta$ everywhere on M. Assume first that the triangle $\{p, q, r\}$ under consideration is in the domain of a normal polar coordinate system like equation 3.8. We claim that if $K \geq \delta$ everywhere, and if $f \neq 0$ between 0 and ρ, then $f(\rho) \leq f_\delta(\rho)$ where f_δ is given by equation 3.10 on page 118 after we plug in $K = \delta$. Once that inequality is obtained, one uses the following transplantation. Consider in the model space $M(\delta)$ of constant curvature δ a triangle $\{p', q', r'\}$ with equal sides $pq = p'q', pr = p'r'$ and angle $A = A'$ at p and p' respectively. One draws the third side $q'r'$ and transplants it back to M. This gives a curve c in M from q to r with

$$\text{length}(c) \leq d_{M(\delta)}(q', r').$$

By the very definition of the inner metric of M one finds

$$d_M(q, r) \leq d_{M(\delta)}(q', r') \tag{3.12}$$

which is the control desired since $d_{M(\delta)}$ can be explicitly computed. When $\delta = 1$ this is formula 1.6 on page 31; for $\delta = 0$ it is formula 1.1 on page 2; and for $\delta = -1$ see equation 4.21 on page 180.

A more geometrical formulation of equation 3.12, which is in fact equivalent, is just to say that if you built a triangle $\{p', q', r'\}$ with its three sides equal to those of $\{p, q, r\}$, then the angles in M are greater than or equal to the respective ones in $M(\delta)$.

The inequality $f \leq f_\delta$ comes from the classical Sturm–Liouville technique, which applies to equations like 3.9. Mechanically it is intuitive: f and f_δ have identical initial values $f(0) = 0, f'(0) = 1, f_\delta(0) = 0, f'_\delta(0) = 1$. They represent the motion of a point along a line which is attracted to the origin

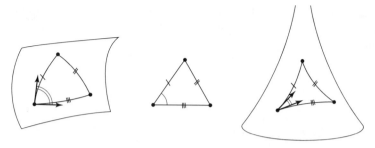

Fig. 3.17. Angles of triangles with equal side lengths

with a force K times the distance from the origin. See figure 3.18. In the motion under Kf, the force is always stronger than for the motion under δf. Since the initial conditions are the same, it is clear that the first motion will always be closer to the origin than the second one, at least before it returns to the origin.

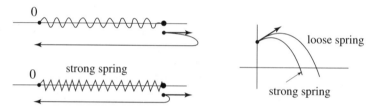

Fig. 3.18. Sturm–Liouville theory: springs in motion

The mathematical proof is very nice. We want to study the ratio f/f_δ. The derivative of this ratio has the same sign as $f'f_\delta - f'_\delta f$. We write the two equations $f'' + Kf = 0$ and $f''_\delta + \delta f_\delta = 0$ and calculate

$$\frac{d}{dt}(f'f_\delta - f'_\delta f) = (\delta - K)ff_\delta$$

(since the two terms $f'f'_\delta$ cancel each other). This is never positive, by assumption. This implies that the function $f'f_\delta - f'_\delta f$ is never positive, since it starts with value 0 at initial time. So the ratio f/f_δ never increases. But it starts at initial time with value one. So the ratio f/f_δ is always smaller than equal to 1, i.e. $f(t) \leq f_\delta(t)$.

The inequality 3.12 is still valid in the large for any triangle in any complete surface. To see it you cleverly decompose the large triangle into smaller ones to which the above applies.

For surfaces with $K < \Delta$, we get mutatis mutandis the inequality:

$$d_M(q, r) \geq d_{M(\Delta)}(q', r') \tag{3.13}$$

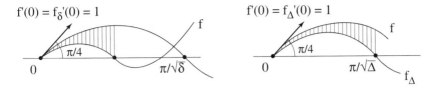

Fig. 3.19. Sturm–Liouville theory: Comparing functions

but only for triangles within a normal chart. The natural conjecture for a global result is false as already seen in figure 1.67 on page 55. The transplantation is now carried out from the surface under consideration to the model surface $M(\Delta)$.

Note 3.2.0.1 Inequality 3.12, namely Alexandrov's theorem for a lower bound on the curvature, has recently been the subject of great attention. Geometers always like to work on more and more general spaces, both for theory and applications. But a new concept should not be to general if one wants to penetrate its mysteries. It turns out, as discovered by various mathematicians, that metric spaces for which there is a fixed δ such that 3.12 holds have a geometry which is extremely reasonable. See §§ 14.5.5 on page 678 for these Alexandrov spaces. ♦

3.2.1 Angle Corrections of Legendre and Gauß in Geodesy

What follows now is a very interesting piece of history of mathematics. We took it from the fascinating reference Dombrowski 1979 [455], where the reader can find more details. As we said in chapter 2, Gauß wanted to undertake extremely precise geodesy. Here is a typical example of a geodesy problem. Assume $\{p, q, r\}$ is a triangle (not too large) on a surface S, and the triangle has side lengths a, b, c and angles A, B, C. In the Euclidean plane $\mathbb{E}^2$ draw a triangle $\{p', q,' , r'\}$ with the same side lengths a, b, c and denote its angles by A', B', C'. The geodesist wants to estimate the differences $A - A', B - B', C - C'$. In 1787, Legendre succeeded to compute these differences when the surface is a sphere of radius R, i.e. $K = 1/R^2$, and got the angle correction formula:

$$A = A' + \frac{1}{3}\sigma K + o\left(a^4 + b^4 + c^4\right) \tag{3.14}$$

and idem for B and C where σ denotes the area of the triangle $\{p, q, r\}$.

This agrees with our intuition: when $K > 0$, if you try to wrap a triangle coming from the Euclidean plane on such a surface you feel that operation will make the angles larger. We also suggest that you prove formula 3.14.

Our planet is a sphere only in first approximation. In fact it is closer to an ellipsoid of revolution, having an eccentricity for which one had already quite a good assessment in Gauß's time. On such an ellipsoid, Gauß wanted a better approximation than Legendre's spherical one. Using all of the machinery he had developed for the inner geometry of surfaces, and after quite hard labor, he obtained the following correction valid for any surface:

$$A = A' + \frac{\sigma}{12}\left(2K(p) + K(q) + K(r)\right) + o\left(a^4 + b^4 + c^4\right) \quad (3.15)$$

and idem for B, C, where $K(p), K(q), K(r)$ are the total curvatures at these points.

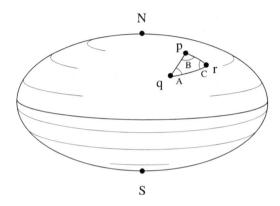

Fig. 3.20. Gauß's angle correction

Gauß immediately compared his correction with Legendre's where the surface is no longer a sphere with constant K but an ellipsoid of revolution, which is the next best approximation to our planet. Note that now the correction generally depends on the vertex (which was not the case for Legendre's correction). This means that you have to know the latitude of the three points. Then, as we mentioned in §§ 1.6.3 on page 45, the curvature is easily computed. You will notice for example that the correction is smaller than Legendre's when closer to the pole. The numerical results are as follows for a triangle explicitly surveyed by Gauß near Göttingen, and for the values admitted at that time for the eccentricity of our planet:

$$\begin{array}{l}\text{Legendre's common correction } 4.95116''\\ \text{Gauß's corrections} \qquad 4.95104'', 4.95113'', 4.95131''\end{array} \quad (3.16)$$

Gauß's comment on these in a letter to a friend is worth quoting:

> *In practice this is of course not at all important, because it is negligible for the largest triangle on earth that can be measured; however the dignity of science requires that we understand clearly the nature of this inequality.*
>
> C. W. F. Gauß

Even the accuracy of contemporary measurements does not alter this circumstance, since the imprecision of the instruments now available at metrology labs is of the order of one second; see Berger 2003 [173].

Note 3.2.1.1 In 1820, Gauß was hired as a map maker, (at the behest of King George III of Britain, who was the Elector of Hannover), so his interest in geodesy was also driven by an important problem: to survey Hannover (today a province of Germany) to within an accuracy sufficient to produce useful maps. To make a map is to capture the shape of a surface up to a certain accuracy, in other words up to a certain maximum magnitude of Gauß and mean curvature. ♦

3.3 Back to Metric Questions: Cut Loci and Injectivity Radius

We address the basic question: when is a geodesic a segment, i.e. when is its length equal to the distance between its end points? We saw in §§3.1.4 that this always the case when the end points are close enough. But we would like to control the distance in the large with the curvature. Our first result will be a negative one:

Theorem 21 (Bonnet, 1855) *If $K \geq \delta > 0$ then no geodesic of length $L > \pi/\sqrt{\delta}$ can be a segment.*

Proof. Apply the second variation formula 3.11 on page 119 to the function

$$f(s) = \sin\left(\frac{\pi s}{L}\right)$$

as in figure 3.21 on the next page and check that the second derivative in equation 3.11 is negative. Therefore, nearby that geodesic there are shorter curves with the same end points.

Turn this negative result in a positive one. Assume the surface M is complete and that $K \geq \delta > 0$ everywhere. The above result proves that the diameter of M is smaller than or equal to $\pi/\sqrt{\delta}$, and in particular M is compact.

Corollary 22 *For a surface like the one in figure 3.22 on the following page, going to infinity, e.g. an elliptic paraboloid, the curvature gets arbitrarily close to zero.*

Completeness is essential; take as a counterexample:

$$\mathbb{R} \times (-\pi/2, \pi/2)$$

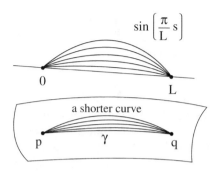

Fig. 3.21. Bonnet's theorem: finding a shorter curve

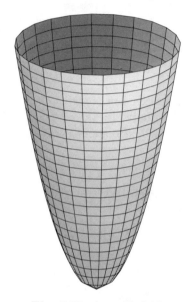

Fig. 3.22. A paraboloid

endowed with the metric

$$ds^2 = dx^2 + \cos^2 x \, dy^2$$

We now look at surfaces with $K < \Delta$. The reader will prove that for every geodesic of length smaller than or equal to $\pi/\sqrt{\Delta}$ the second variation (from equation 3.11 on page 119) is never negative. This is called *Wirtinger's inequality*, and nicely proven for example with Fourier series: any f with $f(0) = f\left(\pi/\sqrt{\Delta}\right) = 0$ (see figure 3.23 on the facing page) satisfies

$$\int_0^{\pi/\sqrt{\Delta}} \left(\left(\frac{df}{dt}\right)^2 - \Delta f^2 \right) ds \geq 0$$

and equality only for f proportional to $\sin\left(s\sqrt{\Delta}\right)$.

Fig. 3.23. Wirtinger's inequality

We recall here the definition of injectivity radius, already met on page 42:

Definition 23 *The injectivity radius of a metric space M is the largest number Inj such that any pair of points $\{p,q\}$ with $d(p,q) <$ Inj are joined by exactly one segment.*

The Wirtinger inequality, together with the picture, will suggest:

Theorem 24 (Klingenberg, 1959) *The injectivity radius of every compact surface is not smaller than the lesser of the two numbers: (1) $\pi/\sqrt{\sup K}$ and (2) half the length of the smallest periodic geodesic.*

This is the ultimate control for the geometry of a compact surface. The proof is not simpler for surfaces than for the general dimension, so we refer the reader to our later discussion in arbitrary dimensions of theorem 89 on page 272. Note that we can just omit $\pi/\sqrt{\sup K}$ in the case where we know that $\sup K \leq 0$. It is very easy to show the optimality of the result. Take a prolate ellipsoid of revolution to see the need for a curvature assumption (more on that shortly) and figure 3.24 for the periodic geodesic assumption.

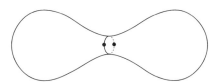

Fig. 3.24. A thin neck

We come back to the cut locus, which was introduced in §§ 1.6.2 on page 33 without any detailed definition. We consider it briefly here, because it will be studied in detail in §§ 6.5.4 on page 278. The *cut locus* was first considered by Poincaré in 1905; he named it the "ligne de partage". We consider a geodesic γ starting from a given point $m = \gamma(0)$ and run along it. We know that at the beginning we have $d(m,\gamma(s)) = s$, but also that this will not in general remain valid for every t. The *cut point* of γ is the last point of M on γ for which this remains true (it might happen that it is so to speak "rejected to

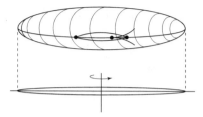

Fig. 3.25. Along the equator, Gauß curvature is very large, if the ellipsoid is very thin

infinity"). The cut locus of m, denoted by Cut-Locus(m), is the union of the cut points of the geodesics emanating from m.

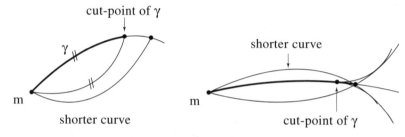

Fig. 3.26. Cut points

Besides the fact that it is impossible in general to compute geodesics or distances, the major source of difficulty in studying the cut locus is that there are two kinds of points on it. Most cut points are connected to m by two distinct segments. But some can be connected by only one segment. It can be proven (see later on) that the cut locus is always the closure of the points of the first type. For a generic surface, a cut locus will look like a graph. The ends of that graph are precisely the points with only one segment joining them to m. Beware that the picture of this graph can be horrible, for example an infinite graph with ends accumulating at one point. Systematic bad examples are given in Gluck & Singer 1979 [570] with quite simple surfaces, even surfaces of revolution. Cut loci of real analytic surfaces are better behaved, but in the Riemannian realm objects are essentially only smooth. See more on the cut locus in §§ 6.5.4 on page 278 for history and details, even in the case of surfaces.

Let us give a few examples and remark that basically no other ones are workable. The first is the sphere: the cut locus of every point is the antipodal point. The next natural surfaces are the quadrics, especially ellipsoids, but we saw already on page 39 that their cut loci are still not known today. There is only a conjecture, drawn in figure 3.30 on page 130. Cut loci are intractable even for ellipsoids of revolution. The pictures are different for prolate and oblate ones. In both cases we have drawn the conjectured envelope of the geodesics from some point of the equator: this envelope plays a basic role in Riemannian geometry. It is called the *conjugate locus*, and its points the *conjugate points* of the point we started from; the conjugate locus will appear again with a more formal definition 85 on page 268. For the poles, as for the sphere, the cut locus is reduced to the other pole. The results of Myers 1935 [953] are as follows: for a real analytic, simply connected surface, the cut locus of every point p is always a tree, and at every point q of the cut locus, the number of arcs in this tree is always equal to the number of segments between p and q. Moreover, the ends of the tree are necessarily conjugate points. The conjecture on ellipsoidal cut loci will follow from the following "statement" of Jacobi (known as *Jacobi's last "theorem"*): the envelope of the geodesics starting from a nonumbilic point has exactly four cusps. Since the geodesics of an ellipsoid are algebraic objects (given explicitly by hyperelliptic functions), one naturally feels that such a statement should be easy to prove, but in fact it is still unproven; see Arnold 1994 [65].

For generic surfaces, one should expect at least triple points on the cut locus.

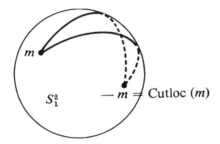

Fig. 3.27. Cut locus of the sphere

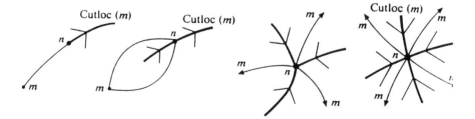

Fig. 3.28. Cut points of various orders

Fig. 3.29. Cut locus of the torus: the Dirichlet-Voronoi diagram

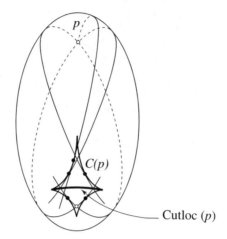

Fig. 3.30. The conjectured cut locus of an ellipsoid

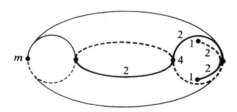

Fig. 3.31. The cut locus of a point located outside and on the equator of a torus of revolution. The numbers denote the number of connecting segments

Note 3.3.0.2 The sphere $S^2 \subset \mathbb{E}^3$ has a very special inner metric. The fact that the cut locus of any point is always reduced to a single point implies that its diameter is equal to its injectivity radius. The cut value is not only constant (which can be seen by applying the first variation formula) but as soon as you attain the cut value you also attained the diameter. In particular, the diameter is attained for every pair of antipodal points and is attained on every geodesic. We saw on page 51 that this geometry is characteristic of the sphere. But for abstract Riemannian manifolds, we will see on page 238 that there are manifolds other than spheres with the property that their diameter equals their injectivity radius, and in §§ 6.5.5 on page 285 that it is still unknown exactly which Riemannian manifolds enjoy this property. ♦

We will meet the cut locus for general Riemannian manifolds in §§ 6.5.4 on page 278.

3.4 Global Results for Surfaces in $\mathbb{E}^3$

3.4.1 Bending Surfaces

We pursue a little bit a topic we touched on page 51. First, the rigidity of the sphere: a compact surface with $K \equiv 1$ is a sphere. This was proven by Liebmann in 1899 and is not too hard. In 1927, Cohn-Vossen proved that two compact surfaces with positive Gauß curvature which are inner isometric are in fact congruent, i.e. there is a global isometry of $\mathbb{E}^3$ identifying the two surfaces. Thus, strictly convex surfaces are rigid. Every known proof is quite tricky. See for example Berger & Gostiaux 1988 [175] 11.14, Burago & Zalgaller 1992 [284] page 21, and Klingenberg 1995 [816] 6.2.8. Compact surfaces without everywhere positive curvature can be isometric without being congruent, as shown in figure 3.32.

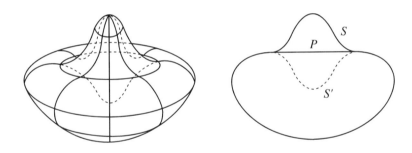

Fig. 3.32. Two isometric surfaces which are not congruent

Associated to that result and to the picture in figure 3.32, two questions are still open today. Note that the two surfaces shown in the picture are not real analytic. So the first question is to extend Cohn-Vossen's result to any pair of surfaces, provided both are real analytic. Apparently, there is no one parameter family of inner isometric surfaces connecting these two surfaces. The following question is also open: assume we have a one parameter family of surfaces M_α, all inner isometric to one another; such a family is called a *bending*. Are they necessarily congruent in $\mathbb{E}^3$? Audoly 1999 [89] presents a new approach to rigidity. All nondeformation results known today are for convex surfaces and noncongruent examples are smooth surfaces (infinitely differentiable). Locally, it looks obvious that any piece of surface can be deformed (and with a lot of parameters: take a tennis ball with some piece removed). But in fact constructing such deformations is unbelievably hard. We refer to the survey in Burago & Zalgaller 1992 [284]. We mention again that there are pointed surfaces which cannot be deformed no matter how small the neighborhood of the point.

3.4.1.1 Bending Polyhedra

The theory of bending polyhedra has developed differently from that of smooth surfaces. First, there is an analogue of the rigidity of convex surfaces, namely the celebrated theorem of Cauchy (proven when he was 24):

Theorem 25 (Cauchy, 1813) *Convex polyhedra are rigid.*

Two convex polyhedra which have a point correspondence which is an isometry when restricted to every face are necessarily congruent. One has to prove that the corresponding dihedral angles are equal, and the proof is extremely subtle; see for example 12.8.6 in Berger 1987 [164], or Stoker 1989 [1160], to which you might want to add Karcher 1968 [778].

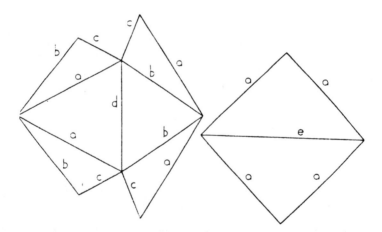

Fig. 3.33. Mountain and valley folds

There are examples of continuous deformations (bendings) of nonconvex polyhedra; see the survey on *Rigidity* by Robert Connelly, section 1.7 of Gruber & Wills 1993 [661]. They are not well understood, although Connelly, Sabitov & Walz 1997 [399] did prove that the volume bound by a polyhedron remains constant under bending. Our opinion is that the known examples of bendable polyhedra do not flex very much.

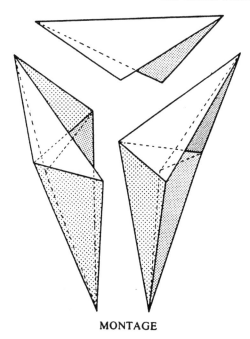

MONTAGE

Fig. 3.34. Steffen's flexible polyhedron. This picture is taken from Berger 1977 [156], and is missing from the English translation Berger 1987 [164]; after Klaus Steffen, drawing by Benoit Berger

3.4.1.2 Bending and Wrinkling with Little Smoothness

Differentiability assumptions are very important in Cohn-Vossen's rigidity theorem. We stated the result of Cohn-Vossen with the word "curvature", implicitly implying that our surfaces are at least twice differentiable. And in fact the result is deadly false without this assumption. A famous result of Nash and Kuiper (see theorem 47 on page 218) says that any continuously differentiable embedding of a surface in $\mathbb{E}^3$ which reduces distances can be approximated (in the topological sense, i.e. uniformly) by a continuously differentiable isometric embedding. The same way you can wrinkle (locally) a flat piece of paper (zero curvature, locally Euclidean metric) into a very finely corrugated one, preserving the locally Euclidean structure (and the zero curvature), one can locally wrinkle any piece of surface preserving the metric structure. The difficulty is to get continuously differentiable objects in the limit (and not fractals). One has to preserve a good ratio between "amplitude" and "frequency".

Recently, Bleecker 1997 [207] extended the Nash–Kuiper result to one parameter continuous deformations of surfaces. As a nice corollary, he got examples of isometric deformations of compact embedded surfaces in $\mathbb{E}^3$ which are volume increasing. This is to be compared with the Connelly–Sabitov–Walz result for flexible polyhedra which we met above. Bleecker heuristically describes the process as "finely corrugated wrinkling". It is to be compared with the Lohkamp result on negative Ricci curvature in §§ 12.3.5 on page 613.

But note also that the C^2 assumption can be replaced by convexity. The corrugation wrinkling process of Nash–Kuiper is typically not convex. We will just see mean curvature rigidity assuming convexity, a famous result of A.D. Alexandrov. See the basic reference on the subject: Pogorelov 1973 [1033]. For a viewpoint on nonsmooth bending, see Schlenker 1998 [1104].

3.4.2 Mean Curvature Rigidity of the Sphere

The sphere is also rigid for mean curvature: a compact surface with positive Gauß curvature and constant mean curvature is necessarily a sphere. There is a generalization of both mean and Gauß rigidity for the sphere. A surface M in $\mathbb{E}^3$ is called a *Weingarten surface* if there is some relation between its two principal curvatures, that is to say, there is a smooth function F of two variables such that

$$F(k_1(m), k_2(m)) = 0$$

for every $m \in M$. Constant Gauß or mean curvature surfaces are Weingarten. The classification of Weingarten surfaces is almost completely open today. The only general result concerning them is:

Proposition 26 *If M is a Weingarten surface with positive Gauß curvature, it is never possible for k_1 to have a maximum at the same point where k_2 has a minimum, unless M is a sphere.*

For proof, see the short survey in 11.18 of Berger & Gostiaux 1988 [175].

But we know much more about the rigidity of the sphere for mean curvature: in 1950, H. Hopf proved the rigidity of compact constant mean curvature surfaces without the $K > 0$ proviso, but still assuming that the surface is a topological sphere. In 1955, Alexandrov proved it without any topological restriction. The proof is extremely hard: it is a subtle blending of analysis and geometry. One can look at do Carmo 1976 [451] page 324 for the case of the sphere and to the beautiful book of Hopf 1989 [732] for the general case. From a practical point of view, Hopf's theorem proves that soap bubbles are necessarily round spheres. But be cautious on a subtle point which is missing in many texts: the physics tells us that soap bubbles are local minima for area among nearby surfaces with the same volume; in particular they have constant mean curvature, by equation 1.17 on page 59. The physics does not tell us that soap bubbles are absolute minima, or else theorem 11 on page 63 would suffice to ensure that soap bubbles must be spheres.

A dramatic historical development came in Wente 1986 [1254], where Wente constructed immersed (but not embedded, by the above) compact surfaces of constant mean curvature. It opened up an active field of research. See, for example, Pinkall & Sterling 1989 [1027]. A recent point of view is

Kamberov 1998 [776] where spinors come into the picture, see also Hitchin 1997 [722].

To see surfaces of constant mean curvature of different topologies (noncompact but with more "ends" than those of revolution seen above), it was necessary to wait until Kapouleas 1995 [777]. Today there is still no framework in which to complete the classification. The constructions of Kapouleas are based on fine analysis which makes it possible to make connect sums. One will find recent results and a bibliography in Mazzeo, Pacard & Pollack 2000 [905]. See many computer drawings in Große-Brauckmann 1997 [641] and Große-Brauckmann & Polthier 1997 [642].

In higher dimensions, Alexandrov's theorem is still valid for embedded hypersurfaces: only round spheres have constant mean curvature. Immersions are radically different: there are many nonspherical immersed topological spheres S^{d-1} of constant mean curvature in $\mathbb{R}^d$. See chapter VII of Eells & Ratto 1993 [480] for references; the topic is directly related to harmonic maps, about which a little bit will be said in § 14.3 on page 674.

3.4.3 Negatively Curved Surfaces

Theorem 27 (Hilbert, 1901) *No complete immersed surface in $\mathbb{E}^3$ has constant negative Gauß curvature.*

The proof is hard. If you look at figure 1.62 on page 51 you will see that the surfaces there with $K \equiv -1$ are smooth, but if you try to extend them to be complete, they have singularities like a cusp of revolution. Near that cusp, this subset of $\mathbb{E}^3$ is no longer a manifold. Hilbert's proof consists in showing that traveling far enough along a geodesic, one is forced to develop such a singularity. References are, for example: do Carmo 1976 [451] 5-11, Stoker 1989 [1160] VIII.16. Hilbert's theorem is historically important, since it proves that the hyperbolic plane cannot be constructed as a surface in $\mathbb{E}^3$. See in particular §§ 4.1.2 on page 146. This is very far from exhausting the subject of complete surfaces of negative curvature. See the recent survey in chapter II of Burago & Zalgaller 1992 [284]. This result of Hilbert was generalized in the 1950's by Efimov in a deep paper; see Milnor 1972 [926] and Burago & Zalgaller 1992 [284]: there is no immersion in $\mathbb{E}^3$ of a complete surface of Gauß curvature $K < K_0 < 0$ bounded from above by a negative constant.

Is there any complete, negative Gauß curvature surface in a bounded region of $\mathbb{E}^3$? Some people call the preceding question "Hadamard's conjecture" because Hadamard took nonexistence for granted. An analogous conjecture, of Calabi and Yau, was that there can be no complete minimal surface staying in a bounded region of $\mathbb{E}^3$. In Nadirashvili 1996 [964], both conjectures were killed. He constructed a surface living in a bounded region, which is both minimal and of negative Gauß curvature. The tool is the Weierstraß formula for minimal surfaces.

3.4.4 The Willmore Conjecture

We mentioned the *Carathéodory conjecture*, a famous and still open problem about umbilic points, in §§1.6.3. An equal famous open problem is *Willmore's conjecture*: for any torus M immersed in $\mathbb{E}^3$, Willmore 1971 [1271] believes that

$$\int_M H^2 \geq 2\pi^2$$

(where H is the mean curvature) and he has some ideas on which tori should achieve equality. Among tori of revolution, equality is achieved on precisely those which are conformally square. The remarkable fact (not hard to prove) is that this integral, of squared mean curvature, is invariant under not only rigid motions of $\mathbb{E}^3$, but under conformal transformations, such as

$$x \mapsto \frac{x}{\|x\|^2}.$$

Leon Simon proved that whatever the infimum of total squared mean curvature on immersed tori may be, it is achieved on some (topological) torus. But the value of the minimum is not known. This is quite atypical of variational problems; once a guess is available for the minimum, and one knows the existence of a minimum, then the problem usually solves itself. The conjecture has been proven among large classes of torus immersions. Recent references on Willmore's conjecture: Ammann 1999 [32], Li & Yau 1982 [863]. A strongly related conjecture is *Lawson's conjecture* that a minimal surface in S^3 which is topologically a torus must be a Clifford torus.

3.4.5 The Global Gauß–Bonnet Theorem for Surfaces

We now prove the global version of the Gauß–Bonnet local formula 18 on page 112. In all books, except Chavel 1984 [325], it is called the Gauß–Bonnet "theorem". One wonders why. This is quite surprising since the notion of characteristic did not exist during Bonnet's lifetime. See below for the Euler-Poincaré characteristic of a surface. It seems that there is no detailed historical study of the Gauß–Bonnet theorem. See Chern 1990 [367], but it has to be completed some day. We just provide a soupçon of information. For embedded surfaces in $\mathbb{E}^3$, the formula is proven in Dyck 1888 [467], with crucial help from Kronecker 1869 [835]. But Poisson already remarked in 1812 that the integral is constant under variation, and in 1815, Rodrigues used the Rodrigues–Gauß map (not published at that time) to prove the formula in some cases. Karcher & Pinkall 1997 [782] discovered very recently that the Gauß–Bonnet formula is completely proven, by triangulations (as it will be here) in Boy 1903 [249]. In the same article, Boy elegantly constructed an immersion of the projective plane in Euclidean space $\mathbb{E}^3$. The global Gauß–Bonnet formula appeared partially in Blaschke 1921 [203] page 109 and completely in Blaschke 1930 [205] page 167. But from Blaschke's way of writing, it is clear that it was a folk theorem in the twenties. It also appears in van Kampen 1938 [1205].

Our approach to the theorem can be followed without modification for abstract Riemannian surfaces (see chapter 4 on page 143) but we will ask our surface to lie in $\mathbb{E}^3$, just because of the title of this chapter. We consider some compact surface M in $\mathbb{E}^3$, and subdivide it into small triangles whose sides are geodesics. The smallness condition is that every such triangle is topologically the boundary of a disk (if you prefer, it can filled up without a hole). For the feasibility of this, see the notes 3.4.5.2 and 3.4.5.3 on page 139.

The idea is now to apply the Gauß–Bonnet formula 18 on page 112 to every such triangle and sum it over the total decomposition. Let $\{T_i\}$ be the collection of triangles, and denote by α_{ij} its angle at the vertex m_j. (N.B. we have switched from talking about exterior angles before to interior angles here.) For any T_i, formula 18 reads:

$$\int_{T_i} K \, dA = \sum_j \alpha_{ij} - \pi$$

where the sum is over the vertices of the triangle T_i.

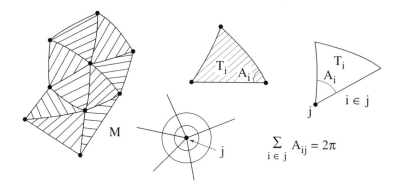

Fig. 3.35. Gauß–Bonnet applied to triangles

We sum over the triangles:

$$\int_M K \, dA = \sum_{i,j} \alpha_{ij} - \pi F$$

where F is the number of triangles (faces of the triangulation). Exchanging the order of summation and noting that

$$\sum_{i,j} \alpha_{ij} = 2\pi$$

at every vertex, we finally get

$$\int_M K\,dA = 2\pi V - \pi F$$

where V is the number of vertices. Let E be the number of sides (edges of the triangulation). Because every edge belongs to exactly two faces and every face has three edges, we finally get

$$\int_M K\,dA = 2\pi(V - E + F).$$

We assume that you know Euler's formula to the effect that $V - E + F$ depends only on the topology of the surface triangulated as above. It is called the *Euler characteristic* and is denoted by $\chi(M) = V - E + F$. It is also equal to the alternating sum of the Betti numbers

$$\chi(M) = b_0(M, \mathbb{Z}) - b_1(M, \mathbb{Z}) + b_2(M, \mathbb{Z}) \tag{3.17}$$

(see §§§4.1.4.2) Pictorially, for an orientable surface, $\chi(M)$ is related to the number γ of "holes," also called the *genus* of M (a sphere has no holes, a torus has one hole) by $\chi = 2(1 - \gamma)$. There is a classical theorem in algebraic topology, the classification of surfaces, which says that every compact surface in $\mathbb{E}^3$ is homeomorphic to one with some number γ of holes as drawn in figure 3.36. See §§§ 7.2.1.2 on page 329 and §§§ 4.1.4.1 on page 158. The existence of a triangulation is a connected question; see the notes 3.4.5.2 and 3.4.5.3 on the next page.

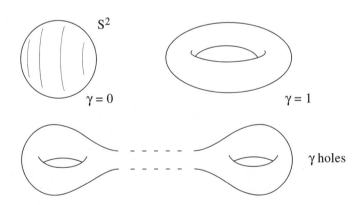

Fig. 3.36. A surface with γ holes

Finally, we get the

Theorem 28 (Global Gauß–Bonnet theorem)

$$\int_M K\,dA = 2\pi\chi(M)$$
$$= 4\pi(1 - \gamma) \qquad \text{if } M \text{ is orientable.}$$

3.4 Global Surface Theory   139

This formula is fundamental: the totality of the Gauß curvature does not depend on the way the surface sits in $\mathbb{E}^3$. A first consequence is a very weak form of Hadamard's theorem 7 on page 57. If $K > 0$ then the surface has to be topologically a sphere, by theorem 28. If our surface is a torus, then theorem 28 says that the set of points where K is positive is exactly compensated by the set of points where it is negative. You know that the total area of the sphere is equal to 4π, and this verifies the Gauß–Bonnet theorem on the sphere. You can also check it for surfaces of revolution, since K is easy to compute; see §§ 1.6.3 on page 45. Later on in equation 15.9 on page 709, we will see a third quantity equal to both sides of the Gauß–Bonnet formula. The formula will certainly be valid for abstract Riemannian surfaces, since we only worked with the inner metric. In particular, the Gauß–Bonnet theorem holds for unorientable compact surfaces (which can never be embedded in $\mathbb{E}^3$) and for negatively curved compact surfaces (which can never be isometrically immersed in $\mathbb{E}^3$).

Another application is described in the note 6.5.2.4 on page 275.

Note 3.4.5.1 There are global inequalities regarding curves on surfaces, e.g. between the area and lengths of periodic geodesics. But they are better treated in abstraction; see § 7.2 on page 325. ♦

Note 3.4.5.2 In the proof of global Gauß–Bonnet theorem, the condition that the sides of the triangulation are geodesics is not really needed. Even if an edge is not a geodesic, in each of the two triangles it belongs to, the curvature of an edge will have opposite signs. The integrals of the geodesic curvature along the edges will cancel each other in the sum. ♦

Note 3.4.5.3 It seems a naive question whether one can triangulate a surface. But a correct proof is hard. The first was by Radó in 1925; see Stillwell 1993 [1158], Massey 1991 [902] for references, almost the only accessible one being Ahlfors & Sario 1960 [12]. For recent references on surface triangulations, see Colin de Verdière 1991 [395] and Colin de Verdière & Marin 1990 [398]. However, there is a cheap way to construct a triangulation of any compact surface. It was proposed to us by Hermann Karcher and seems to have never appeared in the literature. On a compact surface M, choose any Riemannian metric. Then use the results of §§ 6.5.3 on page 278 to get a covering of M by a finite set of convex balls. Then play with the drawings in figure 3.38 on the following page and design the desired geodesic triangulation. ♦

3.4.6 The Hopf Index Formula

For surfaces M the characteristic $\chi(M)$ can also be computed by Hopf's formula for the sum of the indices of a field ξ of tangent vectors on M vanishing at finitely many points:

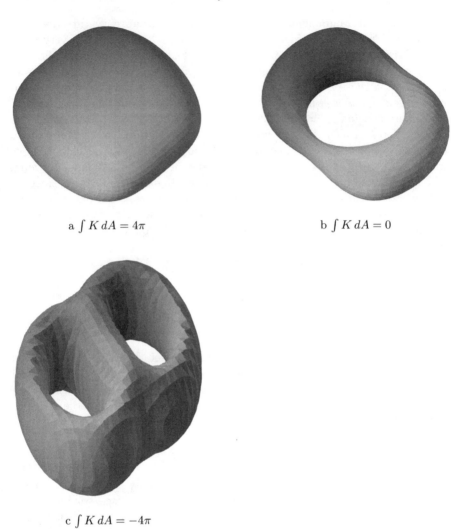

a $\int K\,dA = 4\pi$ b $\int K\,dA = 0$

c $\int K\,dA = -4\pi$

Fig. 3.37. Applying the Gauß–Bonnet theorem

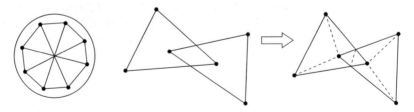

Fig. 3.38. Building a geodesic triangulation, following Hermann Karcher

$$\chi(M) = \sum_{x} \operatorname{index}_x(\xi) . \tag{3.18}$$

The index of such a vector field at one of the points x where it vanishes is an integer defined as follows. Since there are only finitely many zeros of ξ, we can draw a circle around x which contains no other zeros of ξ. Then, as we travel counterclockwise around this circle, the direction of ξ rotates some number of times, as in figure 3.39. This number of rotations is our index.

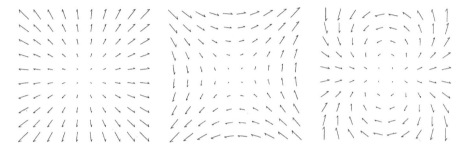

Fig. 3.39. Indices of vector fields

This is a particular case of a formula valid in any dimension: equation 15.9 on page 709. It was used already for the result on umbilics on page 47, and will be the key for a higher dimensional generalization of the Gauß–Bonnet theorem.

There is another way to look at theorem 28. Let us use the Rodrigues–Gauß map $N : M \to S^2$, here for the whole compact surface M. Then Gauß–Bonnet tells us that

$$\frac{1}{4\pi} \int_M K \, dA = \frac{1}{2} \chi(M)$$

while proposition 19 on page 113 identifies this ratio with

$$\frac{\int_M N^* d\sigma}{\int_{S^2} d\sigma} .$$

This quotient expresses the number of times (in total) that M covers S^2 through the map N. This is called the *degree* of the map N. For the general notion of degree, one can look at Petersen 1997 [1018], or Appendix A.6 of Berger & Gostiaux 1988 [175]. But a classical theorem of Hopf, valid in fact for any hypersurface M^d in $\mathbb{E}^{d+1}$ (where a generalized Gauß map exists analogously) says that:

$$\chi(M) = 2 \deg(N) \tag{3.19}$$

and this is again the Gauß–Bonnet theorem. For this see also equation 15.10 on page 712.

Here we bring to an end our study of the global properties of surfaces. Not that we have run the well dry, but primarily we are interested in the inner and not in the outer geometry. Secondly, some questions, even for surfaces in $\mathbb{E}^3$, can be considered in the realm of abstract Riemannian surfaces and thereby in fact better understood. A typical example is the Laplacian; the definition we gave in equation 1.28 on page 95 for the sphere works word for word for any surface M in $\mathbb{E}^3$. So we can consider heat, wave and Schrödinger equations on such a surface. But this is better done in the abstract context because it is much more general. More importantly, some inequalities for the spectrum and some purely geometric inequalities are valid for surfaces in $\mathbb{E}^3$, but equality is attained only for abstract surfaces, not embeddable in $\mathbb{E}^3$.

4 Riemann's Blueprints for Architecture in Myriad Dimensions

Contents

- 4.1 **Smooth Manifolds** 143
  - 4.1.1 Introduction 143
  - 4.1.2 The Need for Abstract Manifolds 146
  - 4.1.3 Examples 149
  - 4.1.4 The Classification of Manifolds 157
- 4.2 **Calculus on Manifolds** 162
  - 4.2.1 Tangent Spaces and the Tangent Bundle 162
  - 4.2.2 Differential Forms and Exterior Calculus 166
- 4.3 **Examples of Riemann's Definition** 172
  - 4.3.1 Riemann's Definition 172
  - 4.3.2 Hyperbolic Geometry 176
  - 4.3.3 Products, Coverings and Quotients 183
  - 4.3.4 Homogeneous Spaces 186
  - 4.3.5 Symmetric Spaces 189
  - 4.3.6 Riemannian Submersions 194
  - 4.3.7 Gluing and Surgery 196
  - 4.3.8 Classical Mechanics..................... 199
- 4.4 **The Riemann Curvature Tensor** 200
  - 4.4.1 Discovery and Definition 200
  - 4.4.2 The Sectional Curvature 204
  - 4.4.3 Standard Examples 207
- 4.5 **A Naive Question: Does the Curvature Determine the Metric?** 213
  - 4.5.1 Surfaces 214
  - 4.5.2 Any Dimension 215
- 4.6 **Abstract Riemannian Manifolds** 216
  - 4.6.1 Isometrically Embedding Surfaces in $\mathbb{E}^3$ 217
  - 4.6.2 Local Isometric Embedding of Surfaces in $\mathbb{E}^3$ 217
  - 4.6.3 Isometric Embedding in Higher Dimensions .. 218

4.1 Smooth Manifolds

4.1.1 Introduction

As we said in chapter 2, Riemann's construction of the Riemannian manifold consisted first in building the foundation of the smooth manifold. He then established on that foundation the concept of a Riemannian metric. In the first two sections we will present smooth manifolds, and thereafter define Riemannian metrics. The notion of smooth manifold is at the same time extremely natural and quite hard to define correctly. This notion started with Riemann in 1854 and was widely used. Hermann Weyl was the first to lay down solid foundations for this notion in 1923. The definition became completely clear in the famous article Whitney 1936 [1259].

A correct definition of smooth manifold is not too difficult today if one uses modern general topology and differential calculus. For precise definitions, the reader can consult almost any differential geometry text. Here are some: Berger & Gostiaux 1988 [175], Bishop & Crittenden 1964 [197], Boothby 1986 [220], do Carmo 1992 [452], Chavel 1993 [326], Gallot, Hulin, & Lafontaine 1990 [542], Hu 1969 [742], Sakai 1996 [1085], Spivak 1979 [1155], Thorpe 1994 [1188], Warner 1983 [1243]. But it still takes time. In 1927, Élie Cartan published a textbook on Riemannian manifolds. This book was followed in 1946 by a second edition, much enlarged: Cartan 1946 [321], and in English translation, with comments by R. Hermann: Cartan 1983 [319]. Cartan's was the only book on Riemannian geometry up to the 1960's. Then many books started to appear. Even in the second printing, Cartan preferred not to define manifolds precisely. Page 56 of the second edition reads:

> La notion générale de variété est assez difficile à définir avec précision.
>
> [The general notion of manifold is quite difficult to define with precision.]
>
> Élie Cartan

Only examples and considerations follow. We will take Cartan's approach, referring the reader to standard references for a modern exposition of the notion of smooth manifold. Do not think that Cartan was without a clear concept of manifold. In the same way, many people, typically Siegel, were writing with coordinates, matrices, etc. although familiar with abstract vector spaces, linear maps, etc.

A d dimensional *smooth manifold* is a topological space (most often connected) which is everywhere locally smoothly equivalent to $\mathbb{E}^d$. These local equivalences are called *charts* or *coordinate systems*, the essential condition being that when they overlap, two charts are related by a smooth diffeomorphism, i.e. a bijection which is a smooth map, as is its inverse. A map is *smooth* if it admits derivatives of any order. The pictures in figure 4.1 on the next page are *not* manifolds.

We will see below that smooth manifolds admit at every point a tangent space, which is a d dimensional real vector space. Differential and integral calculus (of any order) operate on such a manifold. In a sense, one can define smooth manifolds as the objects where calculus is possible (at least to the first order, if one wants an intrinsic calculus). We will often omit the word "smooth". It is important to know that in some special cases one has to work in

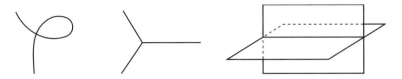

Fig. 4.1. *Not* manifolds

larger class of manifolds, of tensors, etc. For example the class C^1 is the class in which only continuous first derivatives are demanded; similarly, if we ask that the derivatives also are continuously differentiable, we are in the class C^2. A typical example: a Riemannian metric need only be C^2 to have well defined curvature. In §§12.4.2 we will meet the class $C^{1,\alpha}$ of metrics for which one needs the first derivative to satisfy a *Lipschitz condition* of order $\alpha \in (0,1)$. It is meaningless to ask that Riemannian metric tensors be smoother than the manifolds they live on. For example, for a metric to be C^2, it needs an underlying manifold of class C^3. It will always be understood that we work in the C^∞ class, for manifolds, for tensors, etc. unless explicitly stated. Manifolds whose coordinate changes belong only to the class C^0 of continuous maps will be called *topological manifolds*. We will eventually, but briefly, meet the notions of real analytic manifold, and analytic Riemannian metric; this means that these things are defined by means of real analytic functions. But, essentially, the Riemannian realm is that of indefinitely differentiable objects. A *diffeomorphism* between two manifolds is a smooth bijective map whose inverse is also smooth.

Note 4.1.1.1 We mention briefly the two technical points which make the correct notion of manifold very difficult. It is not too expensive to define a smooth manifold as a set covered by charts, which are smoothly related to one another where their domains overlap. But this won't work. The first sign of trouble is that such a manifold can be too large, i.e. not countable at infinity, e.g. the so-called *long line* or the Prüfer surface, which can be found in Berger & Gostiaux 1988 [175] or in Spivak 1979 [1155], volume I. The second problem is that it might fail to be separated i.e. might not be Hausdorff, as in figure 4.2.

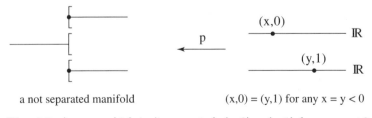

Fig. 4.2. A space which isn't separated: $(x, 0) \sim (x, 1)$ for any $x < 0$

So one has to add the two extra conditions of Hausdorffness and countability at infinity to prevent these two problems. Then a key theorem of Whitney, theorem 30 on page 161, says that if we add these two extra conditions to the definition of manifold, a manifold is the same thing as a submanifold of some $\mathbb{E}^d$; we will see later on that ultimately this point of view is not very illuminating. ◆

4.1.2 The Need for Abstract Manifolds

In chapter 2 we mentioned that one motivation for inventing abstract manifolds came from studying sets of geometric objects. Let us examine the simplest such set: the set of all straight lines through the origin of $\mathbb{R}^3$. We will denote this set by $\mathbb{RP}^2$ and call it the *real projective plane*. You might think of the sphere $S^2 \subset \mathbb{R}^3$ centered at the origin and associate to a line the points where it meets the sphere. The trouble is that there are two such points, so that you need to keep only "half" of the sphere. Restrict yourself to the northern hemisphere. Then there are still two intersection points of horizontal lines with the hemisphere, on the equator. And if you cut half of that equator off, you have a total mess. This piece of a sphere is not a nice surface anymore, at the equatorial points where the missing half of the equator meets the half still in place. Moreover, the construction is not equivariant; we have privileged some hemisphere. The original set of lines is acted on by the group of linear maps in an elementary way, while the chopped up sphere is not. We will see how to get out of this dilemma.

Examining the set of lines through the origin is not a mathematician's luxury. Making a colour involves mixing the three basic colours in correct proportions, and this is represented by a line through the origin in $\mathbb{R}^3$. Colour mixing is of vital importance for car makers, printers, graphic artists, etc. An objection could be that, the coefficients being positive, one may look at only the positive octant of S^2. But it turns out that one really needs to work in $\mathbb{RP}^2$, even if only in a part of it.

Another example of a set of geometric objects described in chapter 2 was the set of positions of a rigid body in $\mathbb{E}^3$. As we said before, you can first restrict consideration to a body rotating around a fixed point. We also mentioned *Euler angles*. Not only are the formulas complicated, but as when using latitude and longitude to describe the sphere, there are positions for which those angles are not well defined. We will refer to the set of rotations in $\mathbb{R}^3$ around a fixed point as $SO(3)$; of course it is a group, the *special orthogonal group*. To define Euler angles, an axis is chosen. But $SO(3)$ should look the same near any of its points. Hamilton made this homogeneity of $SO(3)$ manifest by applying the quaternions he had just discovered. Recall that the *quaternions* are $\mathbb{H} = \mathbb{R} \oplus \mathbb{R}^3$ with multiplication

$$(x_0, x) \cdot (y_0, y) = (x_0 y_0 - \langle x, y \rangle, x_0 y + y_0 x + x \times y) \ .$$

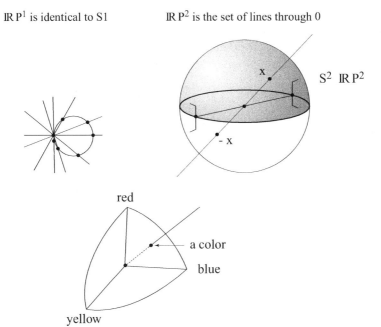

Fig. 4.3. (a) $\mathbb{RP}^1$ is identical to S^1. (b) $\mathbb{RP}^2$ is the set of lines through 0. (c) Colour mixing

where $\times$ is the cross product and $\langle,\rangle$ is the scalar product on $\mathbb{R}^3$. If $X = (x_0, x)$ then let $\bar{X} = (x_0, -x)$. Identify $\mathbb{R}^3 = 0 \oplus \mathbb{R}^3$ with the imaginary quaternions $x_0 = 0$. Unit length quaternions $Y = (y_0, y)$ act on imaginary quaternions $X = (0, x)$ by

$$X \mapsto YX\bar{Y}.$$

This gets the unit sphere $S^3 \subset \mathbb{H}$ to rotate $\mathbb{R}^3$. Just as for $\mathbb{RP}^2$, although the set of unit length quaternions form a three dimensional sphere S^3, there are two unit length quaternions $\pm Y$ giving the same rotation. The geometry will be clarified by the notion of covering space which we will soon consider.

Let us look at robots. The set describing the limb postures and locations of a robot is typically described by an abstract manifold. For the robot's moves to be graceful, not abrupt, the space of its states has to be a smooth manifold. In a similar story, previously we mentioned that in statistical mechanics, one has to work with the set made up by the positions of a large collection of particles. Because of collisions, this set is (not much) worse than a manifold: it has corners.

Finally, in classical mechanics, the space of configurations of a mechanical system is a manifold. In all but the simplest problems, it will have a boundary, etc. But the first thing to do for mastering the general situation is to start with problems in which the configurations form a manifold. We saw above the rigid body. A very simple example is the double pendulum. The configuration space is a two dimensional torus T^2, a surface we will see more of in the next section. You should really think of it abstractly, and not as embedded in $\mathbb{R}^3$: see §§§4.3.3.2. To sum up, let us quote

> There is no notion of acceleration on manifolds without a Riemannian structure.
>
> Vladimir I. Arnol'd

More on this in §15.5.

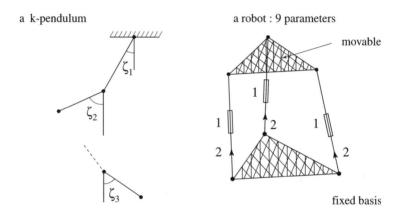

Fig. 4.4. A k-pendulum; A robot whose configurations depend on nine parameters. The number 1 on the three lines indicates a one-parameter gliding, while the number 2 at the base points indicates that the direction of the gliding bars depends on two parameters.

As we have said before, Hilbert's theorem 27 on page 135 gives a mathematical motivation for abstract geometry. We have complete surfaces in $\mathbb{E}^3$ with $K \equiv 1$ and with $K \equiv 0$. By Hilbert's theorem there is no complete surface with $K \equiv -1$. For many reasons it is extremely desirable to have a complete metric space everywhere locally isometric to a surface with $K \equiv -1$. We could ask for more: compact "surfaces" with $K \equiv 0$ or with $K \equiv -1$. They certainly do not exist in $\mathbb{E}^3$, because compact surfaces in $\mathbb{E}^3$ must have positively curved parts. For more about these important abstract constant curvature surfaces, called *space forms*, see §§6.3.2 and the note 4.1.4.1 on page 158. Note that the theory of space forms draws heavily on that of Lie groups; see §§§4.1.3.3.

A recent motivation for thinking of manifolds as a collection of charts arose in General Relativity. Even if our solar system can be covered by a single chart (since it pretty much lives in a three dimensional affine space), relativity specialists cover it by various charts (with laws for how to relate views of the same physical phenomena from different charts), typically with a global chart centered at a suitable place, a chart centered on the sun, and a chart centered on each planet; see Damour, Soffel & Xu 1991 [427].

4.1.3 Examples

Our first tool to construct examples of manifolds can be viewed two ways: discrete quotienting of a previously built manifold down to a smaller quotient manifold, or the unraveling of a smaller manifold into a larger covering manifold. The whole story is a straightforward generalization of the identification of the circle with the real numbers modulo 2π (or modulo anything else). The downward trip sees the circle as the quotient of the real numbers under the identification $x = x + 2\pi$. Going upward, we look at some map

$$\mathbb{R} \to S^1$$

given, for example, by

$$t \mapsto (\cos t, \sin t) \ .$$

Such a map should be a *covering map*, which demands what is pictured in figure 4.5. One of the essential properties of a covering is that one can lift a continuous map $T \to S^1$ of a simply connected space[1] T into a map $T \to \mathbb{R}$. This property has been used on pages 13 and 19 when studying the global properties of the curvature of plane curves.

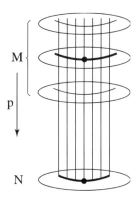

Fig. 4.5. Lifting a curve upstairs

We will be a little more specific. First one defines when a map $p : M \to N$ between two manifolds is a covering map. The condition is that the base N is covered by open sets U so that for each of these U the set $p^{-1}(U)$ is diffeomorphic to the product of U with a discrete set. Then one defines the *deck transformations* of a covering map $p : M \to N$ to be the maps $M \to$

---

[1] See page 150 for the definition of simple connectivity.

M commuting with p. The deck transformations form a discrete subgroup of the group of all diffeomorphisms of M (acting moreover with no fixed point). We call M a *covering space* of N. In our example, $\mathbb{R} \to S^1$, the deck transformations are the translations

$$x \to x + 2\pi k$$

for any integer k. Conversely a discrete group G of diffeomorphisms of a manifold M gives rise to a quotient set M/G and, if the action is well behaved, the quotient M/G will be a manifold and the quotient map $M \to M/G$ will be a covering map. If G is a finite group, then the number of points in any stalk $p^{-1}(m)$ is constant, called the *number of sheets* of the covering. See for example Berger & Gostiaux 1988 [175] 2.4. We just note here that compact manifolds are easier to handle. Be careful: a succession of covering maps can be subtle. The so-called *normal coverings* fit together well, but otherwise all bets are off. This point is most often omitted in books on algebraic topology. The only places we found it were in chapter 4, §19 page 163 of Dubrovin, Novikov & Fomenko 1985 [464], and in an exercise in Greenberg 1967 [587].

Any reasonable topological space, for example any manifold, has a *universal covering* which is uniquely determined by the condition that it is *simply connected*. Simple connectivity means that any loop is contractible to a point. Any connected covering of a simply connected manifold is a diffeomorphism (has only one sheet). We used this fact in the proof of Hadamard's theorem 7 on page 57. Finally, the group of deck transformations of the universal covering $\tilde{M} \to M$ of any manifold M is naturally identified with the fundamental group $\pi_1(M)$ of M which is the group made up by the set of loops traveling away from and returning to some chosen point, under continuous deformation keeping the point fixed. The specific choice of fixed point is inessential.

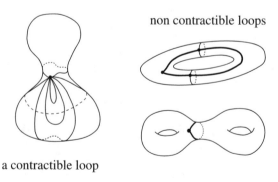

Fig. 4.6. (a) A contractible loop. (b) Noncontractible loops

A typical example: quotient the sphere S^d by the antipodal map. We saw in §§4.1.2 (in dimension $d = 2$) that the quotient is the set of straight lines through the origin in $\mathbb{R}^{d+1}$. It will be called in the sequel *real projective space*, denoted by $\mathbb{RP}^d$. For example, $\mathbb{RP}^3$ and $SO(3)$ are the same manifold, as we have seen via quaternions.

Another basic example is the higher dimensional generalization of $\mathbb{R} \to S^1$. This is the frame for Fourier series. We want to consider Fourier series in d variables. Consider functions on $\mathbb{R}^d$ which are periodic in every variable. So we have to quotient $\mathbb{R}^d$ by the group generated by the d translations moving one unit along each coordinate axis. The quotient is called the d dimensional torus and is denoted by T^d. As in the one dimensional case, its tremendous advantage over $\mathbb{R}^d$ is its compactness. The d dimensional torus can be viewed also as the product of d circles. It can also be seen as the unit cube with the opposite faces identified in an obvious manner. The two dimensional case is pictured in figure 4.7 and explains the name "torus".

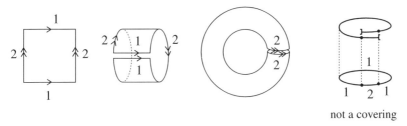

Fig. 4.7. (a) A torus (b) Not a covering

Beware that a map $f : M \to N$ between two manifolds of the same dimension, which is everywhere of maximal rank (hence a local diffeomorphism), need not be a covering map as seen in figure 4.7. It is a covering map when both M and N are compact (see e.g. 4.1.5 in Berger & Gostiaux 1988 [175]) or when both are complete Riemannian manifolds: see corollary 55 on page 228.

4.1.3.1 Submanifolds

A submanifold of a manifold is just another manifold living inside it, but the notion is subtle (just as in Euclidean space) since we can insist that the submanifold be *embedded*, or can allow it to cross itself, etc. (called an *immersed* submanifold). The relevant technical points are addressed in any differential geometry text. A submanifold of codimension one (one dimension less than the ambient manifold) is called a *hypersurface*.

4.1.3.2 Products

The simplest tool to make new manifolds from old is the product. If M and N are manifolds, the product set $M \times N$ is a manifold whose dimension is the sum of the dimensions of the original manifolds. The product of a line and a circle is an infinite cylinder. The product $S^2 \times S^2$ is more complicated, and geometrically extremely interesting; see the Grassmannian Grass $(2, 4)$ just below. And we will see (on page 579) that it remains a mystery for Riemannian geometers.

You can take the product of a manifold with a set, e.g. finite, which you can think of as a zero dimensional manifold. This just means taking a certain number of copies of the initial manifold. The result is never connected. Remember that we defined a covering $p : M \to N$ by the condition that locally on N, the inverse image $p^{-1}(N)$ is a product by a discrete set. The lifting property then appears to be quite natural.

4.1.3.3 Lie Groups

Lie groups are, by definition, the groups which are manifolds and for which the group operations are differentiable. The group operations are then automatically C^∞. They are pervasive not only in geometry but also in almost all fields of mathematics and of mathematical physics. The reason is simple: Lie groups are at the same time geometric objects and algebraic ones, so they promise a rich interplay.

The Platonic Lie group is $GL(n, \mathbb{R})$. Many of the subgroups of $GL(n, \mathbb{R})$ are Lie groups too. Indeed all closed or path connected subgroups of a Lie group are Lie groups. For example

$$O(n), SO(n), O(p, q), SO(p, q), U(p, q), SU(p, q)$$

and the torus $T^n = \mathbb{R}^n/\mathbb{Z}^n$. Covering spaces of Lie groups are Lie groups, e.g. Spin $(p, q) \to SO(p, q)$. We have seen the circle S^1 and the 3-sphere Spin $(3) = S^3 \subset \mathbb{H}$ clothed as Lie groups. The beautiful group G_2 is the symmetry group of the algebra of Cayley numbers; see the definition of Cayley numbers on page 154. Symmetry groups of differential equations and geometric structures on manifolds are usually (not always!) Lie groups. The abelian subgroup $\mathbb{Q} \subset \mathbb{R}$ is *not* a Lie group.

Some Lie groups are not subgroups of $GL(n, \mathbb{R})$, for any n, but they are rarely encountered. For example, the universal cover $\widetilde{GL}(n, \mathbb{R}) \to GL(n, \mathbb{R})$ is not a subgroup of $GL(N, \mathbb{R})$ for any N.

4.1.3.4 Homogeneous Spaces

Many geometric objects appear, not as a Lie group, but as a quotient of a Lie group by a subgroup. Such objects are called *homogeneous spaces*. Consider for example the set of all k dimensional vector subspaces of $\mathbb{R}^d$. We denote this set by Grass (k, d) and call it a *Grassmann manifold* or *Grassmannian*. It is also a quotient of Lie groups. Introduce the orthogonal group $O(d)$ of all linear isometries of $\mathbb{R}^d$. The Grassmannian Grass (k, d) is the quotient of the orthogonal group $O(d)$ by the subgroup $O(k) \times O(d-k)$ of the isometries which preserve the k-plane $\mathbb{R}^k \subset \mathbb{R}^d$ (and consequently preserve the orthogonal complement of this plane); i.e.

$$\text{Grass}(k,d) = O(d)/O(k) \times O(d-k).$$

Recall that a quotient of groups is not itself a group unless the subgroup which is the denominator of the quotient is a normal subgroup.

The name "homogeneous space" expresses their fundamental property: they look the same everywhere, since the numerator group acts transitively on them. The orthogonal group $O(d)$ consists of matrices whose rows are mutually orthonormal. How do we describe them? Pick the first row out of the unit sphere, then the second out of the part of the sphere which is perpendicular to the first, and so on. You see that there are $d-1$ degrees of freedom for the first choice, $d-2$ for the second, etc. Therefore

$$\dim O(d) = (d-1) + (d-2) + \cdots + 2 + 1 = \frac{d(d-1)}{2}.$$

The Grassmannian $\text{Grass}(k,d)$ is a manifold of dimension $k(d-k)$ because

$$\dim \text{Grass}(k,d) = \dim O(d) - \dim O(k) - \dim O(d-k)$$
$$= \frac{d(d-1)}{2} - \frac{k(k-1)}{2} - \frac{(d-k)(d-k-1)}{2}$$
$$= k(d-k).$$

In case $d=4$ and $k=2$ (this was our starting example) the result is a manifold of dimension $2 \times 2 = 4$. It can be shown that it is the same (as a manifold) as the quotient of $S^2 \times S^2$ by the double antipody. The product of spheres $S^2 \times S^2$ is diffeomorphic to the set of oriented 2 dimensional vector subspaces of $\mathbb{R}^2$: see Singer & Thorpe 1969 [1145].

So we need to define quotients of manifolds by equivalence relations whose equivalence classes are not discrete but instead have a positive dimension. The quotient will still be a manifold if the equivalence relation is not too wild. In the case G/H of a Lie group G divided by some subgroup H the manifoldness of the quotient is guaranteed as soon as the subgroup H is closed in G. Compact subgroups H will produce Riemannian geometry on the quotient; see §§4.3.4 and §§§4.4.3.4.

4.1.3.5 Grassmannians over Various Algebras

We finish our examples with a case you should progressively learn to visualize, so we had better introduce it now. We want to extend the notion of Grassmannian to complex geometry, that is to say, to study the set denoted by $\text{Grass}_\mathbb{C}(k,n)$, made up of all k dimensional complex vector subspaces of $\mathbb{C}^n$. It is called the *complex Grassmannian*. You will not be surprised that its complex dimension is $k(n-k)$ and that its ordinary (real) manifold dimension is then $2k(n-k)$. The case $k=1$ is the analogue of real projective space; we naturally denote it by $\mathbb{CP}^{n-1}$ and name it *complex projective space*: its real dimension is $2(n-1)$. The same construction can be performed over the quaternions $\mathbb{H}$ and we get the *quaternionic Grassmannian* $\mathrm{Grass}_{\mathbb{H}}(k,n)$ and the *quaternionic projective space* $\mathbb{HP}^{n-1}$. Guess the dimensions. These projective spaces will recur frequently throughout this book.

The Cayley numbers[2] $\mathbb{C}a$ give birth to only one space, called the *Cayley plane* and denoted by $\mathbb{C}a\mathbb{P}^2$; its (real) dimension is 16. A Cayley space of three or more dimensions cannot exist: three dimensions provide room to prove the Desargues theorem which in turn implies associativity of the base field: see 4.8.3 in Berger 1994 [167] for a set of references. Any Grassmannian, and in particular any projective space, can be written as a quotient of Lie groups for which we use the notation:

$$\mathbb{KP}^n = U\mathbb{K}(n+1)/U\mathbb{K}(n), \quad \mathbb{K} = \mathbb{R}, \mathbb{C}, \mathbb{H}, \mathbb{C}a \tag{4.1}$$

The letter U stands for "unitary" which means here the group of linear transformations of the spaces $\mathbb{C}^n$, $\mathbb{H}^n$, $\mathbb{C}a^3$ which respects the Euclidean structure of each of these spaces. But to write $\mathbb{C}a\mathbb{P}^2$ one needs the exceptional Lie group

---

[2] Recall that the *Cayley numbers* form an eight dimensional nonassociative algebra. We define them following Robert Bryant: let V be a seven dimensional real vector space, and $\phi \in \Lambda^3(V^*)$ a 3-form. Write $v \lrcorner \phi$ for $\phi(v, \cdot, \cdot)$. Define the symmetric pairing

$$(x,y) = (x \lrcorner \phi) \wedge (y \lrcorner \phi) \wedge \phi$$

for $x, y \in V$, so that

$$(,) : \mathrm{Sym}^2(V) \to \Lambda^7(V^*) \ .$$

Call ϕ positive if $(x,x) \neq 0$ for all $x \in V$. Any two positive 3-forms can be identified by a linear transformation. This pairing has a determinant (the reader should define this); it is not a number but:

$$\det(,) \in \mathrm{Sym}^9\left(\Lambda^7(V^*)\right) \ .$$

There is a unique volume form for which this determinant is 1, call it Ω, and define an inner product on V by $\langle,\rangle = \frac{(,)}{\Omega}$ and define $P : \Lambda^2(V) \to V$ by

$$\langle P(x,y), z \rangle = \phi(x,y,z) \ .$$

The Cayley numbers are $\mathbb{C}a = \mathbb{R} \oplus V$ with

$$(x, X) \cdot (y, Y) = (xy - \langle X, Y \rangle, xY + yX + P(X,Y)) \ .$$

For example, suppose that $V = \mathbb{R}^7$ with coordinates x^i. Write dx^{ijk} for $dx^i \wedge dx^j \wedge dx^k$. The 3-form

$$\phi = dx^{123} + dx^{145} + dx^{167} + dx^{246} - dx^{257} - dx^{347} - dx^{356}$$

is positive.

F_4, namely
$$\mathbb{C}\mathrm{a}\mathbb{P}^2 = F_4/\operatorname{Spin}(9).$$

The Cayley projective plane $\mathbb{C}\mathrm{a}\mathbb{P}^2$ is beautiful. In the Riemannian zoo, we like to call it the *panda*. The projective lines are eight dimensional spheres and verify the axioms of the projective plane. Cartan suspected its projective structure in Cartan 1939 [317], page 354. Freudenthal in 1951 made the first proper projective construction of $\mathbb{C}\mathrm{a}\mathbb{P}^2$, which was never published. For the intricate history of the panda, see chapter 3 of Besse 1978 [182].

We made the space $\mathbb{R}\mathbb{P}^d$ by quotienting the sphere S^d with the antipody. Instead of viewing $\mathbb{C}\mathbb{P}^n$ as a Grassmannian, we can use the same tack. We identify $\mathbb{C}^{n+1}$ with $\mathbb{R}^{2n+2}$ and look where a given complex line pierces the unit sphere $S^{2n+1} \subset \mathbb{R}^{2n+2}$. The intersection this time is no longer a pair of points but a circle, because the ambiguity is now exactly the set of complex numbers of norm one. This presents an important map called the *Hopf fibration*, pictured in figure 4.8.

Fig. 4.8. Two views of the $S^3 \to S^2$ Hopf fibration. The sphere S^3 has one point deleted, which is now the point at infinity of $\mathbb{R}^3$. The various two dimensional tori are associated to circles in S^2 with two exceptions for the degenerate tori (circles) which are associated to the two poles of S^2.

The inverse image of every point in $\mathbb{C}\mathbb{P}^n$ is a great circle sitting in S^{2n+1}. You can also see this as the quotient of this sphere by the action of the circle, treating the circle as the group of complex numbers of norm 1 (under multiplication). The case $n = 1$ is famous; $\mathbb{C}\mathbb{P}^1$ is the sphere S^2 (the Riemann sphere): see 4.3.7 and 18.9 in Berger 1994 [167], Berger 1987 [164] for more about this map $S^3 \to S^2$. Finally, associated to the quaternionic projective spaces, one constructs a fibration of S^{4n+3} by a family of 3 dimensional subspheres. The Cayley plane is much more subtle: for $n = 1$ we unearth a fibration $S^{15} \to S^8$ of S^{15} by 7 dimensional subspheres. But it can be proven that there is no well tempered map $S^{23} \to \mathbb{C}\mathrm{a}\mathbb{P}^2$. This reflects failure of associativity of Cayley numbers, but the proof is still one of the deepest theorems in algebraic topology: see Husemoller 1994 [751], chapters 14 and 15.

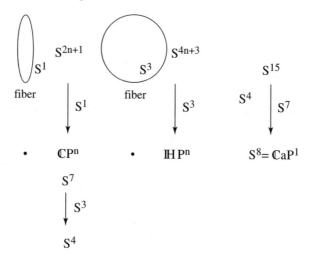

Fig. 4.9. Various Hopf fibrations

4.1.3.6 Gluing

There is a completely different manifold manufacturing method: gluing. We glued when building the torus by gluing opposite faces of a cube. The pictures in figure 4.10 on the facing page show quite a variety of gluing. We will meet them again in §§4.3.7.

Beware that gluing in higher dimension is subtle. For example if you glue two hemispheres of dimension 7 by using a clever trick to attach one equator to the other, then you can get smooth manifolds which are topologically but not smoothly identical to the usual $S^7 \subset \mathbb{E}^8$, as in figure 4.10 on the next page. Such exotic spheres were obtained for the first time by Milnor in 1956. More terrible animals appeared soon after: in 1961, Kervaire found a topological manifold, in the sense of having continuous charts, which cannot be given any smooth structure.

You should get used to looking at a manifold as an abstract object and also up to diffeomorphisms. For example, S^2 is considered as not necessarily the round sphere, and moreover up to diffeomorphism.[3]

Note 4.1.3.1 Gluing is a very particular case of a extremely important technique to build new manifolds from old ones. This technique, called *surgery*, will be presented in §§4.3.7. Surgery will enter naturally in studying manifolds with positive scalar curvature in §§12.3.3 and in our analysis of negative pinching in §§12.2.4; surgery for space forms will be considered in §§6.6.2 and will also enter in our study of systolic inequalities in §§7.2.3. ◆

---

[3] The group of diffeomorphisms of S^2, or of any other manifold, is a very large group.

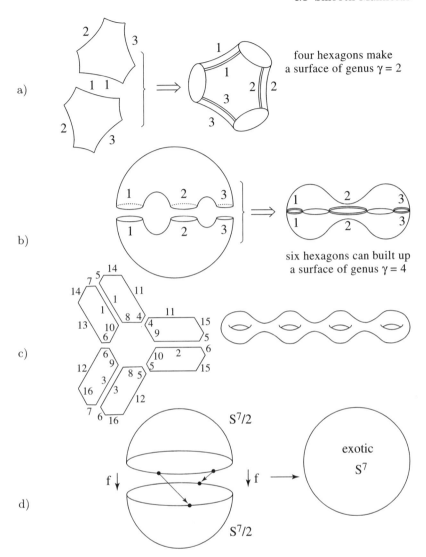

Fig. 4.10. (a) Two hexagons make a pair of pants (b) Two pairs of pants make a surface of genus $\gamma = 2$. (c) Six hexagons can build a surface of genus $\gamma = 4$ (d) Gluing an exotic sphere.

4.1.4 The Classification of Manifolds

This book, above all in chapter 12, will make free use of the foundations of algebraic and differential topology. It is very hard to give references which will cover everything needed, or even a large part. A general picture, in a historical context, is given in Dieudonné 1989 [448]; there is also the recent survey Novikov 1996 [970]. We will say a few words on the subject, in a simple minded manner, but in the spirit needed for the present book.

4.1.4.1 Surfaces

The classification of compact surfaces is a milestone. It says that the only compact surfaces with boundary, up to topological equivalence, are the ones depicted in figure 4.11, and their unorientable analogues. For example, no boundary curves and no holes gives the sphere, one hole gives the torus, etc. Topological equivalence of surfaces implies differentiable equivalence, a phenomenon which is exceptional to two dimensions. This classification is old, going back to Jordan and Möbius in the 1860's. But a complete proof had to wait for Kerékjártó in 1923. It is still quite lengthy to prove in complete detail, and for that reason is absent from most textbooks on algebraic topology. One can use Morse theory (a new idea) to simplify the job. References can be Gauld 1982 [552], Hirsch 1994 [715], Massey 1991 [902], Moise 1977 [932], Seifert & Threlfall 1980 [1119], Stillwell 1993 [1158], Wallace 1969 [1230].

Another major and very natural theorem is the existence of a triangulation of any surface. This looks obvious, but is not. A classical reference is Seifert & Threlfall 1980 [1119]. For more and more recent references, see the notes 3.4.5.2 on page 139 and 3.4.5.3 on page 139.

Fig. 4.11. An orientable surface with some holes and some boundary curves

Note 4.1.4.1 (Riemann surfaces) Let us temporarily borrow in advance some notions of Riemannian geometry to be introduced shortly. The Riemannian geometer needs to know that every compact surface admits a metric of constant curvature: positively curved on the sphere or projective plane, vanishing curvature (a flat metric) on the torus, and negatively curved on anything else. A surface with such a metric is called a *Riemann surface* as opposed to a mere Riemannian surface, i.e. surface with arbitrary Riemannian metric. We will keep company with these Riemann surfaces in §§6.6.1. Every negatively curved Riemann surface can be sewn out of right angled hexagons of the hyperbolic plane, as depicted in figure 4.10 on the previous page. Riemann surfaces appear naturally in the theory of one complex variable, and as algebraic curves. The first central problem concerning Riemann surfaces was to associate a Riemann surface to each complex curve. It was solved around 1900, and is called the fundamental theorem of conformal representation (theorem

70 on page 254). The complex geometry on the curve enables angle measurements, but not distance measurements. One needs to find a Riemann surface structure with matching angle measurements. It is easy to find a Riemannian metric that does this, so then we have only to deform it to have constant curvature. Angle preserving deformations of Riemannian metrics (called *conformal*) are given precisely by multiplying the original metric by a positive function. The constant curvature requirement imposes an elliptic differential equation on that function. See §§11.4.7. ♦

4.1.4.2 Higher Dimensions

A fundamental open problem of differentiable topology is to classify compact differentiable manifolds. No book covers everything known about this problem. There are prominent difficulties in the low dimensions 3 and 4; there is some hope that the Riemannian geometry will help to provide the classification in precisely these dimensions. We will now present four levels of ever finer classification, levels which coincide only for surfaces.

The first level is *homology* (and cohomology). One can define (simplicial) homology groups over any field by simplicial chains. This works for suitable topological spaces; here compact manifolds (differentiability is not necessary). The notations for homology and cohomology are $H_p(M, \mathbb{F})$ and $H^p(M, \mathbb{F})$ for the p dimensional homology and cohomology, where $\mathbb{F}$ is the coefficient field. The associated Betti numbers are

$$b_p(M, \mathbb{F}) = \dim_{\mathbb{F}} H_p(M, \mathbb{F}).$$

The de Rham theorem 32 on page 168 describes the cohomology of differentiable manifolds over the field $\mathbb{F} = \mathbb{R}$. In particular, with differential forms one can obtain the *Euler–Poincaré characteristic* which by definition is the alternating sum

$$\chi(M) = \sum_{p=0}^{d} (-1)^p b_p(M, \mathbb{R}).$$

As revenge, Morse theory (to be seen below in §§10.3.2) and Gromov's theorem in §§§12.3.1.2 for positive curvature both work over any field. One can also define homology and cohomology over the ring of integers $\mathbb{Z}$, instead of over fields. We just remark that to know the Betti numbers, even over all fields, is far from enough to know the manifold, even if the manifold is simply connected. The homology groups carry many other operations, like the cap-product, Massey triple products, Steenrod operations, etc. There are many other numbers attached to manifolds; the classical ones are the characteristic numbers (see a little bit about them in §15.7), the Euler-Poincaré characteristic being the simplest. They interest the Riemannian geometer because some of them can be computed with the curvature, via a far reaching generalization of the Gauß–Bonnet theorem 28, see §15.7.

The fundamental group $\pi_1(M)$ is only reflected in homology by its Abelianization:

$$H_1(M, \mathbb{Z}) \cong \pi_1(M)/[\pi_1(M), \pi_1(M)]$$

For example, there are many *homology spheres*, i.e. compact manifolds having the same homology groups as the standard sphere, but different topologically. This story is not finished today, most prominently at the π_1 level; a long standing conjecture of Novikov is proven every year for more and more groups, but not definitively for every possible group. It might well be false, and random groups may be employed to provide a counterexample; see the very end of Gromov 1999 [633].

The second level is the *homotopy type*: you consider two manifolds M and N homotopy equivalent if there exist two maps $f : M \to N$ and $g : N \to M$ so that both $f \circ g$ and $g \circ f$ are continuously deformable into respectively the identity map on M and on N (we say they are *homotopic to the identity*). In particular, the *homotopy groups* $\pi_k(M)$ of a manifold play the role at the homotopy level that the homology group played above. The "ratio" homotopy/homology is infinite: in any dimension larger than 2, there are infinitely many manifolds having the same homology but not homotopy equivalent. It is important to mention that with only differential forms, their exterior product, and the d operator between them, one can bring to light much more information than the real Betti numbers given by the de Rham theorem. This basic discovery was made in Sullivan 1977 [1166], to the effect that the weaker notion of *rational homotopy*, the set of $\pi_k(M) \otimes \mathbb{Q}$, can be recovered completely from the analysis of differential forms. We will meet this notion from the perspective of Riemannian geometry in the note 12.3.1.2 on page 591 and in §13.6. A book treatment of the subject is Bott & Tu 1982 [229].

The third level is *homeomorphy*: two manifolds M and N are *homeomorphic* if there is a bijective map $f : M \to N$ such that f and f^{-1} are both continuous maps. The diffeomorphy class was already defined above. The ratio homeomorphy/homotopy is also infinite in every dimension except two. For manifolds which are not simply connected, there are already homotopic, non-homeomorphic examples among three dimensional lens spaces, i.e. the finite quotients of the three dimensional sphere. This is a little more subtle. The theory of characteristic classes is an ideal tool for examining these lens spaces; Milnor & Stasheff 1974 [925] give an extremely lucid exposition; we will give short shrift to characteristic classes in §15.7. Examples of manifolds differing in homeomorphy but not homotopy arise already among S^3 bundles over S^4.

The fourth level is diffeomorphy. The problem is to classify the possible differentiable structures on a given topological manifold. The first discovery was that of unusual differentiable structures on spheres, called *exotic spheres*. We note here that, from the point of view of Riemannian geometry, namely what sort of curvature they can bear, these exotic spheres are not at all understood today; see §§12.2.2. However, the ratio diffeomorphy/homeomorphy is very satisfying when the dimension is larger than 4: it was a wonderful achievement of the 1970s that this ratio is always finite in dimensions above 5; basic texts are Hirsch & Mazur 1974 [716], Kirby & Siebenmann 1977 [805]. Moreover, these finite numbers can be computed explicitly using topological information on the manifold. Dimension 4 is an exception; counterexamples can even be taken among complex algebraic surfaces. See chapter 10 of Donaldson & Kronheimer 1990 [457]. For compact simply connected and oriented 4-manifolds, Poincaré duality shows that the entire homology is contained in the structure of $H_2(M, \mathbb{Z})$ (which is isomorphic to $H^2(M, \mathbb{Z})$) if one knows moreover the type (the signature) of the quadratic form

$$H^2(M, \mathbb{Z}) \times H^2(M, \mathbb{Z}) \to \mathbb{Z}$$

given by

$$\omega_0, \omega_1 \mapsto \int_M \omega_0 \wedge \omega_1 .$$

Definitive progress was made recently: see Donaldson & Kronheimer 1990 [457]. The problem is to find which such quadratic forms occur and who the corresponding manifolds are.

For dimensions 3 and 4, see §14.4; there is hope that Riemannian geometry will help pure topologists in this query in dimensions 3 and 4. We already gave some references for low dimensions: a very informative sketch is found in Hansen 1993 [681], Wall 1999 [1229], and Donaldson & Thomas 1990 [458, 459], Donaldson & Kronheimer 1990 [457], Morgan 1996 [939], Seiberg & Witten 1994 [1118], and for the link with Riemannian geometry: Anderson 1997 [47].

4.1.4.3 Embedding Manifolds in Euclidean Space

A natural question:

Question 29 *Are there more abstract manifolds than just the submanifolds of $\mathbb{E}^n$?*

Theorem 30 (Whitney 1936 [1259]) *Any manifold of dimension d can be embedded in $\mathbb{E}^{2d+1}$.*

An elementary proof for compact manifolds, and references, appear in 1.133 of Gallot, Hulin & Lafontaine 1990 [542].

Theorem 31 (Nash & Tognoli) *Any smooth compact manifold is diffeomorphic to a smooth real algebraic submanifold of some $\mathbb{E}^n$, i.e. to the set of solutions of a system of polynomial equations.*

For a modern proof see chapter 14 of Bochnak, Coste, & Coste-Roy 1998 [209].

4.2 Calculus on Manifolds

In short: on smooth manifolds, just as in $\mathbb{E}^d$, one can perform an intrinsic differential calculus of first order. If you are worried about calculus of higher order, which exists certainly in a vague sense since our manifolds are C^∞, one of the privileges of Riemannian geometry is that it permits an "absolute (or intrinsic) calculus" of any order. We will come back to absolute calculus later on in §15.5.

The simple reason behind the intrinsic first order calculus is that changes of coordinates are smooth by definition. The chain rule for the derivative of a differentiable function of many variables insures the invariance of objects of the first order:

$$(h \circ g)'(m) = h'(g(m)) g'(m) \qquad (4.2)$$

But at the second order everything breaks down:

$$(h \circ g)''(m) = h''(g(m))(g'(m), g'(m)) + h'(g(m))g''(m) \qquad (4.3)$$

because of the extra quadratic term.

For example, a function $f : M \to \mathbb{R}$ will have a differential df as in $\mathbb{R}^d$, but not a second differential $d^2 f$, which we would like to be a symmetric bilinear form. A curve will have a tangent vector (its velocity), but no intrinsic acceleration: it has an acceleration in every chart, but there is no invariance property. The marvel of the absolute calculus found by the Italian geometers at the turn of the century is that a Riemannian metric will give rise automatically and canonically to such a $d^2 f$ (see for example the definition of the Laplacian in §§9.3.1). A curve will have an acceleration (so that kinematics will make sense) and more: a calculus of any order; see §15.5.

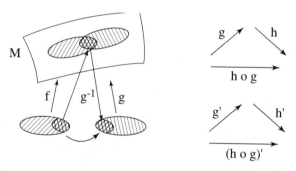

Fig. 4.12. The matching map $g^{-1} \circ h$ should be smooth

4.2.1 Tangent Spaces and the Tangent Bundle

Let us be more explicit about calculus on manifolds. Since our manifolds are abstract objects, *a priori* they do not have natural tangent spaces, as submanifolds of $\mathbb{E}^n$ do. Nonetheless, at any point m of a manifold M, one can attach in a canonical but abstract fashion its *tangent space* at m, denoted by $T_m M$, and its elements are called *tangent vectors*. The tangent vector is so fundamental that it is worth giving various equivalent definitions of it. First, just take equivalence classes of vectors $(a_1, \ldots, a_d)$ under the chain rule 4.2 on the preceding page. Second: a smooth parameterized curve in a manifold M is a smooth map from some interval of $\mathbb{R}$ into M (the notion of smooth maps being defined through charts), as pictured in figure 4.13. Take the equivalence classes of smooth parameterized curves in M passing through m at time $t = 0$ under the equivalence relation of having the same velocity at m in some (hence any) chart. By the chain rule, we find that the velocity is a tangent vector.

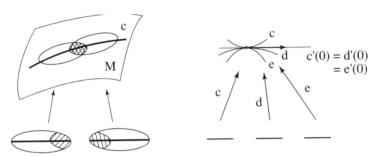

Fig. 4.13. (a) Coordinate charts. (b) Three curves with the same tangent vector

There is more than one way to employ the chain rule. Assume a curve c has velocity $\xi = c'(0)$ at $m = c(0)$. Then, for any function $f : M \to \mathbb{R}$, the chain rule says

$$df(\xi) = (f \circ c)'(0) . \tag{4.4}$$

Analysts and physicists like functions (observables) so we give now a definition of tangent vectors using only functions. Call a linear operation ξ a *derivation at* $m \in M$ if ξ takes smooth functions on M to real numbers, and satisfies Leibnitz's rule

$$\xi(fg) = g(m)\xi(f) + f(m)\xi(g) .$$

for any smooth functions $f, g : M \to \mathbb{R}$. A tangent vector ξ to a manifold M at a point m is just the same as a derivation at m. This abstract definition is quite new; it seems to have appeared in the literature only in the 1960's.

On $\mathbb{R}^d$, all of the tangent spaces $T_m \mathbb{R}^d$ are essentially the same, namely $\mathbb{R}^d$. If we form a quotient torus $T^d = \mathbb{R}^d / \Lambda$, then we still have an identification of all of the tangent spaces of the torus with $\mathbb{R}^d$. For a general manifold, there is no way to compare the tangent spaces at different points. See figure 4.14 on the next page. This was already the case in a somewhat restricted sense for a surface $M \subset \mathbb{E}^3$: compare with §§3.1.3, where tangent vectors could be compared but only along a given curve in the surface. In an abstract manifold this is impossible without some additional structure. A Riemannian metric will (partly) do it.

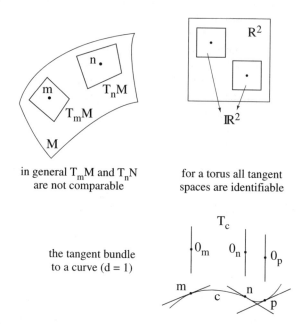

Fig. 4.14. In general, $T_m M$ and $T_n M$ are not comparable. For a torus, all tangent spaces are identifiable. The tangent bundle to a curve.

We cannot compare different tangent spaces to a manifold, but we can put them together into a bigger manifold, called the *tangent bundle* to M and denoted by TM. As a set, the tangent bundle is just the union of the tangent spaces $T_m M$, and the manifold structure is not too hard to get; see any of the texts quoted above. There is a canonical projection map $p: TM \to M$ which attaches to every tangent vector the point it sticks out of (its foot).

The notion of a smooth map $f: M \to N$ between two manifolds is just defined by pairs of charts, the coherence of the notion being insured by the definition of a manifold, and again the chain rule. The basic point is that such a map also has a *derivative* (sometimes called the *differential*). It has many possible notations: f', df, Tf. At every point, it can be defined in the same three ways that tangent spaces were defined: by curves, by functions and by charts. It gives rise to an elegant commutative diagram, drawn in figure 4.15 on the facing page.

We say that a smooth map $f: M \to N$ is an *immersion* if $df_m: T_m M \to T_{f(m)} N$ is an injective for all $m \in M$, and a *submersion* if df_m is surjective for all $m \in M$. The *rank* of $f: M \to N$ is defined to be the rank of df_m, assuming that this is constant. The preimage of a submanifold $S \subset N$ viewed through a submersion $f: M \to N$ is a submanifold $f^{-1}S \subset M$. The preimage

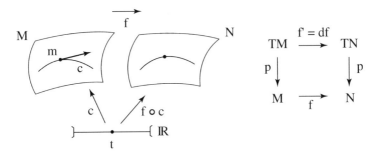

Fig. 4.15. Taking the first derivative

of a point under a constant rank map is also a submanifold, but the preimage of a submanifold might not be a submanifold. (Find an example.)

An important special case: when the target manifold is $\mathbb{R}$, in which case df is at every point $m \in M$ a linear map

$$df_m : T_m M \to \mathbb{R}$$

i.e. an element of the dual space $T_m^* M$ of the vector space $T_m M$. Again, the union of these dual spaces, for m running through M, makes up a manifold called the *cotangent bundle* $T^* M$ of M. For technical details, see your favourite text book of differential geometry. Note that the tangent and the cotangent bundles are different objects. In Riemannian geometry, they can be identified because there will be an additional Euclidean structure yielding a canonical duality between each tangent space and its dual space.

We do not introduce here a fundamental notion in differential geometry, that of vector field, this in order to remain geometric as long as possible; see §15.1 for this notion and its developments.

Just two examples: for a product manifold $M \times N$ the tangent space $T_{(m,n)}(M \times N)$ is canonically identified with the product $T_m M \times T_n N$. For a Lie group G, the tangent space $T_e G$ at the identity element $e \in G$ is called the *Lie algebra* of G and very often denoted by the corresponding Gothic letter $\mathfrak{g}$. Multiplication on the group G is reflected on $\mathfrak{g}$ in the quadratic terms of a Taylor expansion for the multiplication map $G \times G \to G$. These quadratic terms yield an operation on $\mathfrak{g}$, the *Lie bracket*

$$x, y \in \mathfrak{g} \mapsto [x, y] \in \mathfrak{g} .$$

This bracket is the value $[X, Y](e)$ at $e \in G$ of the two vector fields X and Y on G obtained by transporting x and y all over G by left translations. The Lie bracket is an antisymmetric map

$$\mathfrak{g} \times \mathfrak{g} \to \mathfrak{g}$$

satisfying the Jacobi identity:

$$[x,[y,z]] + [y,[z,x]] + [z,[x,y]] = 0$$

for every triple of elements $x, y, z \in \mathfrak{g}$. More precisely, one can first define one parameter subgroups in G, as smooth group homomorphisms

$$g : \mathbb{R} \to G .$$

Such a homomorphism, thought of as a curve, is determined by its velocity at the identity element: $x = g'(0)$. Define the *exponential map*

$$\exp : \mathfrak{g} \to G$$

by asking that

$$\exp x = g(1)$$

where $g(t)$ is the one parameter subgroup with initial velocity equal to x. The inverse of the exponential map is a natural system of local coordinates near $e \in G$. Moreover, left translating these coordinates around, we produce nice charts around every point, so that the transition functions are real analytic. Thus Lie groups are real analytic objects, and their multiplication and homomorphisms are real analytic.

The Lie bracket $[x, y]$ is the infinitesimal defect of commuting in G, calculated by taking the commutator $g(t)h(t)g^{-1}(t)h^{-1}(t)$ of two one parameter subgroups, and computing its second derivative at $t = 0$ where $g(t)$ and $h(t)$ are the one parameter subgroups whose velocities at $t = 0$ are x and y. Finally one can completely reconstruct the group structure of G (assuming that G is simply connected), out of the Lie algebra structure of $\mathfrak{g}$. If needed, see §15.1 for the definition of bracket of vector fields. Warner 1983 [1243] gives a very lucid exposition of the basics of Lie groups, but of course one can also look at the other references we will give at various places, e.g. when studying symmetric spaces in §§4.3.5.

4.2.2 Differential Forms and Exterior Calculus

The exterior differential calculus on $\mathbb{E}^d$ extends without difficulty to manifolds, because it only involves first derivatives (the exterior derivative). We briefly summarize here elementary exterior calculus on (smooth) manifolds. We assume the reader is familiar with exterior algebra on vector spaces and the exterior calculus for differential forms on $\mathbb{E}^d$. We just recall that exterior algebra on a vector space of dimension d exists only in degree from 0 to d and that in dimension 0 and d the exterior product is itself a one dimensional vector space.

So on any manifold M, we first introduce the p-exterior tangent bundle $\Lambda^p(T^*M)$ which is the collection

$$\Lambda^p(T^*M) = \bigcup_m \Lambda^p(T_m^*M)$$

with the obvious projection map. This can be made into a manifold as usual. A p-differential form α on M is a section (smooth) of that bundle, namely a collection $\{\alpha(m)\}_{m\in M}$ of $\alpha(m) \in \Lambda^p(T_m^*M)$, required to be smooth, which means for example that, transferred by any chart to $\mathbb{E}^d$ it is a smooth ordinary differential form. The set of these p-differential forms on M is denoted by $\Omega^p(M)$. Three operations defined on differential forms on $\mathbb{E}^d$ transfer to those defined on M:

1. *exterior product* (a.k.a *wedge product*):

$$\alpha \wedge \beta = (-1)^{pq} \beta \wedge \alpha$$

   where α is a p-form and β a q-form,
2. *inverse image* (a.k.a. *pullback*): $f^*\alpha$ of a p-form α on a manifold N by a smooth map $f : M \to N$, giving a p-form $f^*\alpha$ on M, and
3. exterior derivative: $d\alpha$, which is a $(p+1)$-form, when α is a p-form. Just as in $\mathbb{E}^d$, $d \circ d = 0$.

The exterior derivative gives rise to the sequence:

$$0 \longrightarrow \mathbb{R} \longrightarrow \Omega^0(M) \xrightarrow{d} \Omega^1(M) \xrightarrow{d} \Omega^2(M) \xrightarrow{d} \cdots$$

$$\cdots \xrightarrow{d} \Omega^p(M) \xrightarrow{d} \Omega^{p+1}(M) \xrightarrow{d} \Omega^{p+2}(M) \xrightarrow{d} \cdots$$

$$\cdots \xrightarrow{d} \Omega^{d-1}(M) \xrightarrow{d-1} \Omega^d(M) \xrightarrow{d} 0.$$

with $d \circ d = 0$.

The elements of $\Omega^0(M)$ are just the functions $f : M \to \mathbb{R}$ and on them the exterior derivative d is nothing but the differential df of f. The functions with $df = 0$ are precisely those which are constant on each component of M. And of course $d(d(f)) = 0$. But how about the converse? A differential form α is said to be *closed* if $d\alpha = 0$. Consider a closed 1-form α. Is there some function f such that $df = \alpha$? Such a function f need not exist. It must exist on spheres S^d when $d \geq 2$ but generally such a function will not exist on the circle. Any constant coefficient 1-form α on $\mathbb{E}^d$ will descend to the torus T^d, but if for example $\alpha = dx_1$ on $\mathbb{E}^d$, the function x_1 does not descend. In such a situation, mathematicians examine the failure for f to exist as follows. Introduce:

$$Z^p(M) = \{\alpha \in \Omega^p(M) \mid d\alpha = 0\} \tag{4.5}$$
$$B^p(M) = \{d\beta \mid \beta \in \Omega^{p-1}(M)\}$$
$$= d\Omega^{p-1}(M).$$
$$\tag{4.6}$$

The forms in $Z^*(M)$ are called *closed*, while those in $B^*(M)$ are called *exact*. The defect of exactness of a closed form occurs in the quotient space:

$$H_{dR}^p(M) = \frac{Z^p(M)}{B^p(M)} \tag{4.7}$$

called the *p-th de Rham cohomology group* of M.

If the manifold is connected, then clearly

$$H_{dR}^0(M) = \mathbb{R}\ .$$

Please find $H_{dR}^0(M)$ for manifolds which are not connected. There is no general description of the de Rham cohomology groups, except for the celebrated:

Theorem 32 (de Rham) *For any compact manifold M^d and any $0 \leq p \leq d$, the de Rham groups $H_{dR}^p(M)$ are isomorphic to the real cohomology groups $H^p(M^d, \mathbb{R})$.*

Such a theorem is plausible if one thinks of Stokes' theorem 34 on page 171. The cohomology groups $H^*(M, \mathbb{R})$ are, by definition, the dual vector spaces of the homology groups $H_*(M, \mathbb{R})$. We can get differential forms ω to eat compact submanifolds $S \subset M$ by

$$S \mapsto \int_S \omega$$

(defined to be 0 if the degree of ω is not the dimension of S). We have only to check that this integration operation is defined on the quotient $H_{dR}^*(M) = Z^*(M)/B^*(M)$ and that if S_0 and S_1 are homologous, then

$$\int_{S_0} \omega = \int_{S_1} \omega\ .$$

Both of these are direct consequences of Stokes' theorem. This is not a proof, since homology classes are not in general occupied by submanifolds; it is just a hint. The de Rham theorem is proven in many books; one very lucid presentation is in Warner 1983 [1243], and see also the recent Jost 2002 [768].

It is natural to ask if by chance there is a way to realize by a canonical choice of differential forms the elements of the $H_{dR}^p(M)$, which appear so far only as equivalence classes. We are looking for a canonical section

$$H_{dR}^p(M) \to \Omega^p(M)\ .$$

Such a section is not canonically determined on a manifold without any additional structure. Again one of the miracles of Riemannian analysis, the Hodge–de Rham theorem, will provide us with such a canonical section attached to each Riemannian metric on M. The corresponding forms are called *harmonic*; see theorem 405 on page 665 and §15.6 for more on this.

The top dimension $p = d$ is a little more complicated than $p = 0$, but not too much. It is linked with orientation as follows. A manifold is *orientable* when one can define a continuous choice of orientations for its various tangent spaces, or equivalently cover it with an atlas of coordinate charts such that all the coordinate transforms have positive determinant. Assume that the manifold M is orientable. Since orientation is connected to determinants and because d-forms transform like determinants, we can always build a d-form ω on M, vanishing nowhere. Such a form is called a *volume form*. The converse is also true. An unorientable manifold admits a unique two-sheet covering which is orientable. Canonical examples are the Möbius band and the real projective spaces $\mathbb{RP}^d$ when d is even. From this observation, one deduces without too much pain that for a connected manifold M the top cohomology is $H^d_{dR}(M) = \mathbb{R}$ when M is orientable and $H^d_{dR}(M) = 0$ otherwise. See for example Berger & Gostiaux 1988 [175], 7.2.1.

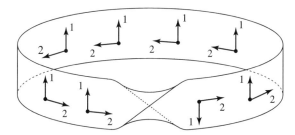

Fig. 4.16. The Möbius band is not orientable; its tangent bundle is not a product

The top dimensional cohomology is extremely important because as soon as M is oriented the *integral* $\int_M \alpha$ of any d-form α is defined. One can call it the *volume* of M for α. In particular, a volume form will define a measure on M, since any two d-forms are proportional by a function, due to the fact that the exterior power $\Lambda^d(\mathbb{R}^d)$ in top dimension is one dimensional. The above integral transforms accordingly when taking inverse images under diffeomorphisms $f : M \to N$.

$$\int_M f^*\beta = \int_N \beta, \quad \beta \in \Omega^d(N). \tag{4.8}$$

There is an interesting theorem about volume forms, simple to state but proven only recently. It says in essence that the structure given by a volume form is so weak that is has only one invariant: the above integral. More precisely

Theorem 33 (Moser 1965) *Let M be a compact manifold and α, β two volume forms. Then there exists a diffeomorphism*

$$f : M \to M$$

such that $f^*\alpha = \beta$ if and only if

$$\int_M \alpha = \int_M \beta .$$

For a proof, one can look at Berger & Gostiaux 1988 [175] 7.2.3. An oriented Riemannian manifold will have a canonical volume form.

The crowning result in the exterior differential calculus is Stokes' theorem. It relates the exterior differential, a concept from analysis, with a concept of geometry: the boundary of a domain. Most differential geometry books cover Stokes' theorem; Berger & Gostiaux 1988 [175] try to give a large number of its applications. If you know some algebraic topology, then cohomology groups are related to homology groups and these are defined using special cases of domains and their boundaries. So Stokes' theorem is the first evidence for the de Rham theorem, but the complete proof had to wait for de Rham in 1929. The proof waited quite a long time to be written neatly. A good reference for the proof is Warner 1983 [1243]. Remember that Stokes' theorem 34 on the next page yields an interesting proof of the isoperimetric inequality; see Berger 1987 [164] 12.11 or Berger & Gostiaux 1988 [175] 6.6.9 for details.

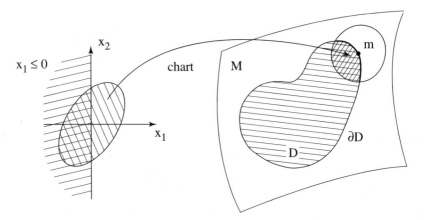

Fig. 4.17. A chart for a domain with smooth boundary

Stokes' theorem operates on a compact domain D in a manifold M. The domain should have a smooth boundary, which means that every point $m \in D$ is of two possible types. Either m is an interior point, or there is a smooth function x_1 defined on a neighborhood U of m having a nonzero differential dx_1 such that

$$D \cap U = \{ m \in U \mid x_1(m) \leq 0 \} .$$

This makes ∂D a smooth submanifold of M, of codimension 1, i.e. of dimension one less than that of M.

Theorem 34 (Kelvin–Stokes)

$$\int_D d\alpha = \int_{\partial D} i^*\alpha$$

for every $\alpha \in \Omega^{d-1}(M)$, where $i : \partial D \to M$ denotes the canonical injection. (More prosaically, one says that $i^*\alpha$ is the restriction of α to ∂D.)

The attentive reader should have been worrying: both integrals above need some orientation to be defined. So we should add that the manifold M is oriented (or at least has a chosen local orientation covering at least D). Then the basic remark is that ∂D inherits a canonical orientation from that of M, given geometrically by the inner side of D, and analytically by asking that dx_1 (locally) be used to orient the normal directions to ∂D, which will fit together with only one orientation of ∂D to produce the given orientation of M.

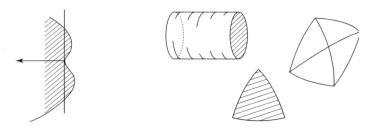

Fig. 4.18. Inner normal; Domains with reasonable singularities

Note 4.2.2.1 Thinking for example of the Gauß–Bonnet theorem 18 on page 112, domains with smooth boundary are too restrictive for many geometrical situations. Many important domains have corners. There are extensions of Stokes' theorem to domains with worse and worse boundary singularities. Some references: Dieudonné 1969 [446] chapter XXIV.14 or the very last section of Lang 1993 [846]. Think also of the isoperimetric inequality in §1.5. The ultimate goal is to define a notion for which the exterior derivative and the boundary of a domain are defined by asking Stokes' theorem to be true for them; this is a starting point in *geometric measure theory*. See 4.2 of Morgan 2000 [937] and the references there. ♦

Note 4.2.2.2 In the many classical applications of Stokes' theorem to analysis or physics in two or three dimensional Euclidean space (e.g. the Green's, Ostrogradski's, etc. formulas) a systematic use is made of the duality between forms and vectors, exterior products, etc. For example, the differential of a function can be identified with its gradient, and more generally a 1-form or 2-form with a vector field. The exterior product becomes the cross product, and the operation d then corresponds to div, grad or curl depending on what you apply it to. ♦

4.3 Examples of Riemann's Definition of a Geometric Object

4.3.1 Riemann's Definition

Gauß was interested in the inner geometry of a surface embedded in $\mathbb{E}^3$ and made profound use of the determination of the inner geometry by the collection of Euclidean structures on its tangent planes. In coordinate language, we wrote this Euclidean structure as ds^2. Riemann made a dramatic reversal: he defined a geometry as a ds^2 on an abstract manifold, i.e. as a collection of Euclidean structures on its abstract tangent spaces. Analytically, this is given by various ds^2 in various charts, but they have of course to match where the domains of two charts overlap.

You can think of such a Riemannian object as a field of ellipsoids: the ellipsoids built up by the vectors of norm equal to one. It is interesting to think about this with some care. Understand that it is, as it stands, too difficult a notion to work with directly. One can also consider a Riemannian object as a manifold which is everywhere infinitesimally Euclidean. Note also that an ellipsoid—in an affine space—has no privileged directions (of course, you should realize that is the case because we consider ellipsoids in affine, not in Euclidean spaces). Ellipsoids are all the same. For example, in the affine plane (do not put any Euclidean structure), a Riemannian structure is a field of ellipses, as in figure 4.19 on the next page. You will see immediately how hard it is to make any sense out of it.

This dramatic reversal was very innovative. Riemann lectured his new theory of geometry in 1854 in front of Gauß. After the lecture, Gauß was walking home with a friend and told him: "This lecture surpassed all my expectations, I am greatly astonished." Moreover these comments were made with the greatest appreciation and with an excitement rare for him. To be fair, it should be mentioned that in 1868 Helmholtz arrived at the same definition of a geometric object, though without being aware of Riemann's 1854 contribution, for the good reason that Riemann's work was only published in 1868 (after Riemann died in 1866).

A typical example is to consider the sphere S^2 (abstract or embedded, that is not the question) and put on every $T_m S^2$ some Euclidean structure, as in figure 4.19 on the facing page. Here, as in all the above, this collection should depend smoothly on the point m varying through S^2. The new geometry has in general nothing to do with the standard one on the sphere. A natural question is to ask if such a geometry can always be identified with one coming from a (probably wild) embedding of S^2 in $\mathbb{E}^3$ (not necessarily a "round" sphere). We will address this question below in §4.6.

The first new example of Riemann, namely the first solid construction of hyperbolic geometry, is so important that we will devote the next section to it. We now make precise a few immediate details. First, about notation. The Riemannian ds^2 will in general be denoted by g. It is a collection of positive

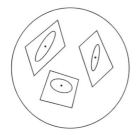 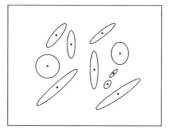

Fig. 4.19. An abstract ds^2 on S^2; A field of ellipses in the plane

definite quadratic forms $g(m)$ on $T_m M$ for m running through M. The total object will be denoted by (M, g) where M is the underlying manifold (the clay for our sculpture). If we write M^d this means that d is the dimension of M. In a coordinate chart $(x_1, \ldots, x_d)$, the standard notation in general dimension (replacing the E, F, G in dimension 2, as in 3.1 on page 107) is:

$$ds^2 = g = \sum_{i=1}^{d} g_{ij} dx^i dx^j \qquad (4.9)$$

where the g_{ij} are functions

$$g_{ij} = g_{ij}(x_1, \ldots, x_d)$$

In this expression for g, the g_{ij} are the scalar products of the vector fields $\frac{\partial}{\partial x_j}$ associated to the coordinate chart:

$$g_{ij} = g\left(\frac{\partial}{\partial x_i}, \frac{\partial}{\partial x_j}\right) \qquad (4.10)$$

Write for yourself how the g_{ij} behave under changes of charts.

The notation will imitate Euclidean geometry:

$$\begin{aligned} \|v\| &= \sqrt{g(v,v)} \quad \text{for the } norm \\ v \cdot w &= \langle v, w \rangle \\ &= g(v,w) \quad \text{for the } scalar\ product \end{aligned} \qquad (4.11)$$

when there is no need to mention g explicitly. In particular there is a well-defined angle sitting in $[0, \pi]$ for every pair $\{v, w\}$ of nonzero tangent vectors to M sticking out of the same point. The definition of an orthonormal tangent frame at a single point is now clear.

Riemann's idea is that a ds^2 gives a geometry as follows: we just ditto the inner geometry of surfaces (or even of Euclidean space). The length of a curve $c : [a, b] \to M$ is by definition:

$$\text{length}(c) = \int_a^b \|c'(t)\|\, dt \qquad (4.12)$$

and the distance between two points p, q of M is (cf. 1.5 on page 30):

$$d(p,q) = \inf_c \{\text{length}(c) \,|\, c \text{ is a curve from } p \text{ to } q\} \qquad (4.13)$$

as in figure 4.20.

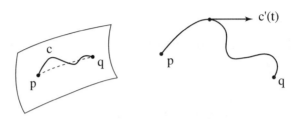

Fig. 4.20. Measuring length

This seems to yield trivially a metric on M as for surfaces in §§1.6.1. The two first axioms for a metric space are obvious: $d(p,q)$ is clearly symmetric and

$$d(p,q) + d(q,r) \geq d(p,r)$$

clearly, by gluing paths together, and employing the infimum in the definition of distance. The big and dangerous trap is the axiom

$$d(p,q) = 0 \quad \text{implies} \quad p = q \,.$$

Here one uses the Hausdorff (separated) nature of manifolds. The proof of this third axiom will be given in §§6.1.1. Oppositely, in the non-Hausdorff one dimensional example on page 145, the two end points are at zero distance from each other in any Riemannian metric.

Theorem 35 (Palais 1957 [991]) *The metric of a Riemannian manifold determines its structure as a manifold (the smooth charts) and its ds^2.*

We will henceforth call $g = ds^2$ on a manifold M a *Riemannian metric* on M, or sometimes just a *metric* (employing Palais' result to ensure that there is little ambiguity), and the couple (M, g) will be called a *Riemannian manifold*. Maps between Riemannian manifolds preserving their metrics are called *isometries*. Also see §14.5 for generalized Riemannian manifolds.

In only extremely special cases will we be able to compute the metric explicitly, i.e. find distances between points. One reason is that it is given by an infimum definition. It is not an algebraic object you can compute from the ds^2. On the other hand, the curvature invariants we are going to meet soon are eminently computable.

Isometries are of a different kind from general diffeomorphisms, in the sense that, for example, an isometry is completely determined as soon as two points and two orthogonal frames are asked to correspond under it, as in figure 4.21. A nice consequence is that the group of all isometries of a given Riemannian metric, which we will write

$$\mathrm{Isom}(M),$$

is always a Lie group, see 2.34 in Gallot, Hulin & Lafontaine 1990 [542]. Also chapter II of Kobayashi 1995 [826] and section III.6 of Sakai 1996 [1085].

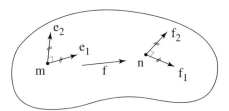

Fig. 4.21. An isometry f of a surface is uniquely determined as soon as equations $n = f(m)$, $f_1 = f'(m) \cdot e_1$ and $f_2 = f'(m) \cdot e_2$ are imposed.

We are interested only in the geometric "structure" of a Riemannian manifold, which is to say in a Riemannian metric up to isometries. What we will call a *Riemannian structure* on a given manifold M is an element of the quotient of of the set of all possible Riemannian metrics on M by the group of all diffeomorphisms of M. Let $\mathrm{Diff}(M)$ be the group of diffeomorphisms. For the total sets we will use the notations $\mathcal{RM}(M)$ and $\mathcal{RS}(M)$:

$$\mathcal{RM}(M) = \{\text{Riemannian metrics on } M\} \tag{4.14a}$$

$$\mathcal{RS}(M) = \{\text{Riemannian structures on } M\}$$
$$= \mathcal{RM}(M) / \mathrm{Diff}(M) \tag{4.14b}$$

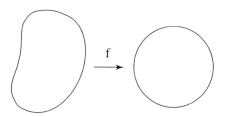

Fig. 4.22. Equivalence under diffeomorphisms: (M, f^*g) is the same as (M, g)

4.3.2 The Most Famous Example: Hyperbolic Geometry

One of the driving forces in mathematics has been to get rid of Euclid's fifth axiom. In the early 19th century, Gauß realized that there were other geometries than Euclidean ones verifying the first four axioms but not the fifth one. In the 1820's, Lobachevskii and Bolyai came to the same conclusion and started building a "new geometry". But what they developed was properly defined only locally (see page 50).

In his 1854 lecture, Riemann enlarged the reservoir of mathematical models and concepts, and used the new objects to solve a pending problem. He pointed out that the sphere of radius ρ has, in appropriate coordinates: [4]

$$ds^2 = \frac{dx_1^2 + \cdots + dx_n^2}{\left(1 + \frac{K}{4}\left(x_1^2 + \cdots + x_n^2\right)\right)^2} \tag{4.15}$$

where

$$K = \frac{1}{\rho^2}.$$

He then said explicitly that the geometry everybody was looking for, the *hyperbolic geometry*, is just the one defined by the same ds^2, but with $K < 0$ a negative constant, defined on the open ball

$$x_1^2 + \cdots + x_n^2 < \frac{4}{|K|}.$$

It could hardly be simpler. The interpretation of the negative constant K will be clarified hereafter.

This simple statement is dangerously misleading. Note for $K > 0$, i.e. on the sphere, the x coordinates are not defined at exactly one point of the sphere. So the sphere is not fully described by these coordinates. What Riemann is asserting is that for $K < 0$, this metric in these coordinates is complete (with the coordinates defined up for $|x|^2 < \frac{4}{|K|}$) so we aren't missing anything—we have parameterized the entire hyperbolic space.

This expression for the metric doesn't make clear the properties of the geometry, above all that the resulting Riemannian geometry is the most symmetrical one. Moreover, this representation renders obscure the fact that for positive K the above ds^2 is that of a sphere (with one point deleted). A word about the phrase "most symmetrical". Euclid's first four axioms were used to derive the congruence properties of Euclidean spaces: we have enough isometries of such a geometry to transform any pair of triangles one into the other as soon as they have three equal sides or two equal sides with an equal angle

---

[4] The coordinates are, for $y \in \mathbb{E}^{n+1}$ with $|y| = \rho$:

$$x = \left(\frac{y_1}{\rho - y_{n+1}}, \ldots, \frac{y_1}{\rho - y_{n+1}}\right).$$

4.3 Examples of Riemann's Definition

between them. It amounts to the same to ask for a formula generalizing 1.1 on page 2 and 1.6 on page 31. The essence of such a formula is to give the distance $d(q,r)$ as a function of $d(p,r), d(p,q)$ and the angle at p between the two segments (shortest connections) from p to q and p to r: look also at §§6.3.2. Those segments should exist by the axioms: there is one and only one line through two given points. This prevented people (before Felix Klein at the end of the 19th century) from realizing that the quotient of the sphere by the antipodal map places an extremely simple geometry on $\mathbb{RP}^2$ verifying Euclid's first four axioms but not the fifth one. Another reason for this realization coming so slowly is that it is natural to consider that in a good geometry looking roughly Euclidean, straight lines should be infinite.

When $K < 0$, the geometric space given by formula 4.15 is called the *hyperbolic space* of dimension d and of curvature K. We will denote it by $\text{Hyp}^d(K)$. These spaces are pervasive in almost every field of mathematics: algebraic geometry, number theory, differential geometry, complex variables, dynamical systems and in physics (ergodic theory, string theory, semiclassical asymptotics, etc.). For this reason we will now present three other models of hyperbolic space;[5] for simplicity's sake we only describe these models for the hyperbolic plane, with curvature $K = -1$. Let us write $\text{Hyp}^d = \text{Hyp}^d(-1)$. References for hyperbolic geometry: Vinberg 1993 [1221] section I, Benedetti & Petronio 1992 [129], Berger 1987 [164] chapter 19 if you like, Ratcliffe 1994 [1049] and for the hyperbolic plane only: Buser 1992 [292].

The simplest representation of the hyperbolic plane, from the Riemannian geometry viewpoint, is as one sheet of a two-sheeted hyperboloid. We consider in $\mathbb{R}^3$ the two-sheet hyperboloid

$$z^2 = 1 + x^2 + y^2$$

as in figure 4.23 on the next page, and let M be the sheet on which $z > 0$. Every tangent plane $T_m M$ to M has a positive definite quadratic form, given by restricting $x^2 + y^2 - z^2$ to it. So this construction induces on M an abstract Riemannian metric. The construction is invariant under the group of linear transformations which preserve the quadratic form $x^2 + y^2 - z^2$. It is easy to see that this group acts transitively on M and also transitively on pairs of tangent vectors to M making a given angle.

There is a preferred chart for M:

$$(u, v) \mapsto (\cos u \sinh v, \sin u \sinh v, \cosh v) \qquad (4.16)$$

in which:

$$\begin{aligned} ds^2 &= dx^2 + dy^2 - dz^2 \\ &= dv^2 + \sinh^2 v \, du^2 \end{aligned} \qquad (4.17)$$

But this is exactly equation 3.10 on page 118 so that this mysterious formula is now explained.

---

[5] and even a fifth model in §§4.3.4

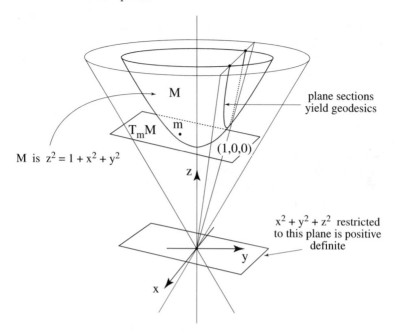

Fig. 4.23. A hyperboloid of two sheets as a model of the hyperbolic plane

The Klein model is a direct metric representation where the manifold M is the open unit disk $x^2 + y^2 < 0$. The metric $d(p, q)$ is defined in terms of the two points r and s where the (Euclidean) straight line through p and q meets the boundary circle of the disk:

$$d(p, q) = \frac{1}{2} |\log[p, q, r, s]| \tag{4.18}$$

where $[p, q, r, s]$ denotes the cross ratio.

The third model is Poincaré's and also defined in the open unit disk. There is only one (Euclidean) circle containing two given points p and q of the disk, and striking the boundary of the unit disk at right angles, as drawn in figure 4.24 on the next page. This circle strikes the boundary of the unit disk at two points, say r and s. Set

$$d(p, q) = \left| \log \left(\frac{pr}{ps} \cdot \frac{qs}{qr} \right) \right| \tag{4.19}$$

The fourth and last model is Poincaré's half-plane $y > 0$ in $\mathbb{R}^2$ with metric defined in the style of Riemann:

$$ds^2 = \frac{dx^2 + dy^2}{y} \tag{4.20}$$

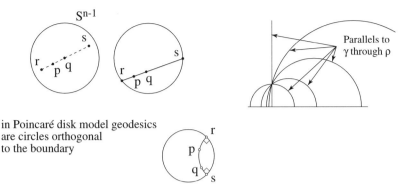

Fig. 4.24. In Klein's model, geodesics are straight lines; Poincaré half plane; In the Poincaré disk model, geodesics are circles orthogonal to the boundary

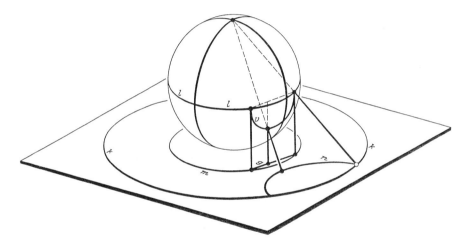

Fig. 4.25. How to pass from the Klein model to the Poincaré model

Needless to say, these four geometries are all isometric. Most books give fewer models (some give only one!). According the type of problem, there is always a more adapted model for the geometry or the computations you have to do.

There is one and only one shortest path between any two points p and q. In the hyperboloid model, it is given by the intersection of the sheet with various Euclidean planes through the origin. In Klein's model, geodesics are just exactly straight lines. In Poincaré's disk, the geodesics are the Euclidean circles orthogonal to the boundary. In the upper half plane model, the geodesics are the circles orthogonal to the x-axis. So in the first two models, "straight lines" coincide with ordinary straight lines. This is not the case in the last two, but in exchange, the last two models have hyperbolic angles the same as Euclidean angles. A correspondence preserving angles, extremely important in various domains (not only in cartography), is called *conformal*.

The universal formula for triangles $\{p, q, r\}$ with side lengths a, b, c and angles α, β, γ is

$$\cosh a = \cosh b \cosh c - \cos \alpha \sinh b \sinh c \tag{4.21}$$

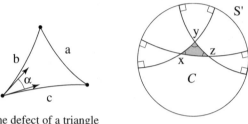

the defect of a triangle
in the Poincaré conformal model

Fig. 4.26. The defect of a triangle in the Poincaré conformal model

The defect of angles in triangles is given by:

$$\mathrm{area}\{p, q, r\} = \pi - \alpha - \beta - \gamma \tag{4.22}$$

The above formula is of course to be compared with the Harriot-Girard formula 1.7 on page 32.

Note 4.3.2.1 The geometry of hyperbolic spaces takes time to master. We just give a few hints. In contrast to Euclidean spaces, hyperbolic spaces of a given dimension are not all isometric to each other; they depend on a parameter K, the *curvature*. This should be compared with the fact that Euclidean spaces admit homothetic self-mappings (scalings). Hyperbolic spaces do not. Put the other way around, Euclidean spaces can be characterized by Euclid's first four axioms and the presence of these self-similarities. ♦

Note 4.3.2.2 Hyperbolic spaces have a lot of symmetries. They have totally geodesic subspaces of every dimension (analogous to lines, planes, etc. in Euclidean geometry), which are themselves hyperbolic spaces. Those of one less dimension we will call *hyperplanes*. For each hyperplane, there is a symmetry of the space fixing exactly that plane, a *hyperplane reflection*. Therefore an isoperimetric inequality is valid in hyperbolic spaces just as in Euclidean spaces and on spheres; see 10.2 of Burago & Zalgaller 1988 [283]. One may use a proof *à la Steiner*; see §1.5. But a proof *à la Gromov* will also work: see §§1.6.8 and §§§7.1.1.3. ♦

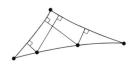 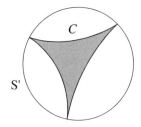

triangles are thin (compute the maximum in finction of K)

Fig. 4.27. An ideal triangle; Triangles are *thin* (compute the maximum as a function of K)

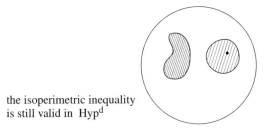

the isoperimetric inequality is still valid in Hyp^d

Fig. 4.28. The isoperimetric inequality is still valid in $\text{Hyp}^d(K)$

Formulas like 4.21 on the preceding page or 3.10 on page 118 show that for a triangle with a fixed vertex p and moving vertices q and r going to infinity, $d(q,r)$ becomes larger and larger. But (in contrast to Euclidean geometry) the straight line from q to r does not go to infinity—it stays at finite distance. Even the three vertices can go to infinity; these are called *ideal triangles*. Every triangle is universally thin, i.e. there is a fixed real number h such that, for every triangle, any point on any side is at a distance less than h from one of the other two sides. This is more and more important in contemporary geometry: see the very end of §§12.3.2 and Abresch & Gromoll 1990 [5] for Riemannian geometry and Gromov 1987 [622] for a completely different context.

Hyperbolic trigonometry can be quite subtle, but is a prerequisite to study Riemann surfaces. A very complete set of formulas and properties is presented in Buser 1992 [292]. Particularly important are the formulas for right angled hexagons. To test his force, the reader can try to prove the meeting property of the figure 4.29 on the next page, valid for any right angle hexagon.

The volume of any simplex in Hyp^d is no larger than a universal constant, denoted by $VRS(d)$; this is obvious in view of equation 4.22 and $VRS(2) = \pi$. This value is attained for *any* ideal triangle. In higher dimensions, that there is such a bound is not too hard to prove. What was discovered only in the late 1970s is that, when $d > 2$, this value is attained only for ideal and regular simplices; see references and proof in Benedetti & Petronio 1992 [129] or Ratcliffe 1994 [1049]. A simplex is said to be *regular* (think of Euclidean spaces, say $\mathbb{E}^3$) if any permutation of its vertices can be obtained by an isometry of the

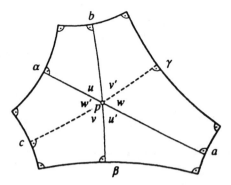

Fig. 4.29. A right angled hexagon in the hyperbolic plane and its three colliding codiagonals

whole space. In the present case, the simplex may extend to infinity, since the isometries extend to the ideal boundary $S(\infty) = S^{d-1}$ and moreover are determined by this extension. The isometry group of Hyp^d is called the *Möbius group*, and its action on S^{d-1} consists entirely of conformal transformations. The basic difference between the dimension $d = 2$ and the dimensions larger is that any diffeomorphism of the circle S^1 is conformal, and the Möbius group is a finite dimensional subgroup of the diffeomorphism group of the circle. For $d > 2$, the Möbius group (i.e. the action of the isometry group of hyperbolic space on the S^{d-1} sphere) is exactly the group of conformal transformations of the sphere S^{d-1}. This is the key to Gromov's proof of the Mostow rigidity theorem; see page 293, and see Ratcliffe 1994 [1049] for a detailed exposition. The ideal regular simplices appear also in theorem 273 on page 517. There is a subtle intertwining of asymptotics: in Hyp^d when you go to infinity some things go to the infinite exponentially (like the divergence of point based geodesics), but some others remain bounded (like the volume of simplices). Volumes of tetrahedra in Hyp^3 are still under study. The problem is to write down explicit formulas; see Cho & Kim 1999 [375].

Note 4.3.2.3 (Space forms again) We come back now to the subject of note 1.6.1.1 on page 32. Formula 4.21 is like formulas 1.1 on page 2 and 1.6 on page 31: it tells us that the distance between two points at given distances from a point m and seen at a given angle is some universal function of those two distances and the angle. We will see in §§6.3.2 a partial classification of all spaces with this property. ♦

Note 4.3.2.4 The end of this section will be devoted to various examples of Riemannian manifolds, some of which will be described in more detail later on. One important feature of the recent contributions in Riemannian geometry is the construction of various examples. These examples can be very sophisticated and profound. They serve very often as counterexamples to see which results can be expected and why some others can not. Otherwise stated, they serve to show that some theorems are the best possible. We will not present such sophisticated examples here, because there are too many of them, and because they need techniques not explained thus far in our story, and finally because we think they will be better appreciated in the various contexts in which they are used. Therefore they will appear progressively as the chapters go on. ♦

Any manifold carries some Riemannian metric. The two provisos of being Hausdorff and countable at infinity are necessary, as metric spaces always verify them. It is classical that they are sufficient, using partitions of unity; see most references on Riemannian geometry. Note that such *au hasard* Riemannian metrics are in some sense as common as functions. Finding special metrics on a given manifold is a natural task and will be the topic of chapter 11.

4.3.3 Products, Coverings and Quotients

4.3.3.1 Products

The product $(M \times N, g \times h)$ of two Riemannian manifolds (M,g) and (N,h) is exactly like the product of two Euclidean spaces. We define the metric $g \times h$ by the Pythagorean theorem: for a tangent vector to $M \times N$, say (v,w), we should have
$$\|(v,w)\|^2 = \|v\|_M^2 + \|w\|_N^2 \ .$$
The v and w components are often referred to as *horizontal* and *vertical*. One of the simplest examples is $S^2 \times S^2$. As an abstract manifold, it carries with it an open problem; see §12.1 and §§§12.3.1.1.

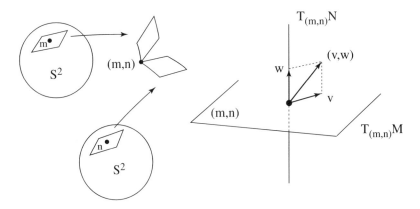

Fig. 4.30. A product of spheres; Riemannian metric of a product

When it is a product a Riemannian manifold is said to be *reducible*. This notion also exists locally. Theorem 56 on page 229 will pass from local to global reducibility (another de Rham theorem). We will see in theorem 394 on page 642 a canonical product decomposition into irreducible pieces.

4.3.3.2 Coverings

We return to the ideas we pursued in §§4.1.3 on coverings and quotients and mix it up with Riemannian metrics. When you have a Riemannian manifold (N, h) and a covering $p : M \to N$ then M inherits a Riemannian metric $p^*h = g$ by asking p to be a local isometry (or using the pull-back notion p^* for differential forms—the forms need not be exterior). This situation can be called a *Riemannian covering*. In particular, any Riemannian manifold has a well-defined universal Riemannian covering (necessarily simply connected; see page 150). The deck transformations are isometries. A Riemannian covering will inherit the curvature properties of the base. We will apply this idea many times, in particular in chapter 12. Conversely, given a covering map $p : M \to N$ where (M, g) is a Riemannian manifold and where all the deck transformations are isometries, there is a quotient Riemannian metric $h = g/p$ down on N such that p is everywhere a local isometry. This comes from the following general nonsense: in a quotient situation you can always take the quotient down of a structure upstairs provided this latter is invariant by the action giving the quotient.

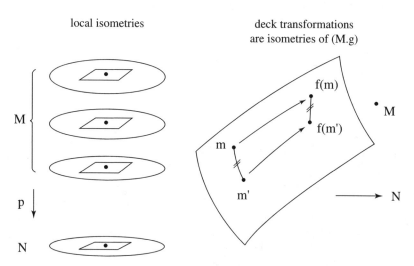

Fig. 4.31. Local isometries; deck transformations are isometries of (M, g)

Let us have examples of Riemannian coverings. The first is $p : S^d \to \mathbb{RP}^d$ where p is the antipodal map. The antipodal map preserves the canonical

Riemannian metric of S^d, so that we have now a canonical Riemannian metric on $\mathbb{RP}^d$. Check that distance in this metric is measured as follows: let D and E be two straight lines in $\mathbb{R}^{d+1}$ through the origin and v and w two unit vectors on them. Then the distance in our metric is

$$d(D, E) = \arccos |\langle v, w \rangle|$$

or, if you prefer, the angle between D and E which belongs to $[0, \pi/2]$.

Consider the example in §§4.1.3 of tori T^d, but extend it slightly by making the quotient of $\mathbb{R}^d$ not by the translations

$$x_i \to x_i + 1 \quad i = 1, \ldots, d$$

but a set of translations

$$v \to v + e_i \quad i = 1, \ldots, d$$

where $\{e_i\}_{i=1,\ldots,d}$ is any basis of $\mathbb{R}^d$. Equivalently, one can speak of a *lattice* Λ of $\mathbb{R}^d$. Any translation is an isometry of the canonical Euclidean structure of $\mathbb{R}^d$. On the manifold T^d we get a Riemannian metric g. It is always locally Euclidean; that is why the resulting Riemannian manifold is called a *flat torus*. N.B. in general two different bases (or lattices) can give nonisometric Riemannian structures on T^d. This is clear if you draw the *fundamental domain* made up of the points closer to 0 than to any e_i, as in figure 4.32 on the following page.

Recall that the fundamental domain is the closure of the set of points of the space which are closer to the origin than to any other point of the lattice. It is drawn by drawing all of the mediating hyperplanes of the pairs $(0, \lambda)$ where λ runs through the nonzero points of Λ.

The most symmetrical (T^2, g) we come up with via this construction is not the standard one (the square) but the one corresponding to the regular hexagon, which will be met again later on in §§§7.2.1.1. The regular hexagon is obtained for

$$e_1 = (1, 0) \quad \text{and} \quad e_2 = \left(\frac{1}{2}, \frac{\sqrt{3}}{2}\right).$$

We said above that hyperbolic geometry is encountered in many fields of mathematics and physics. In fact the most commonly studied objects are not regions of Hyp^2 but its compact quotients. There are many ways to find covering maps $p : \text{Hyp}^2 \to N$ (as a manifold, Hyp^2 is the same as $\mathbb{R}^2$) where N is a smooth manifold. The arithmetic approach consists in finding a discrete group of isometries of Hyp^2 which operates nicely enough. It is difficult to do but very important for applications. We refer for this to §§6.6.2 and the references there, in particular to the fine recent survey constituting section II of Vinberg 1993 [1221]. One can also look at the short survey Berger 1996 [168]. Soon below we will get such compact quotients by geometry using gluing. See also Ratcliffe 1994 [1049].

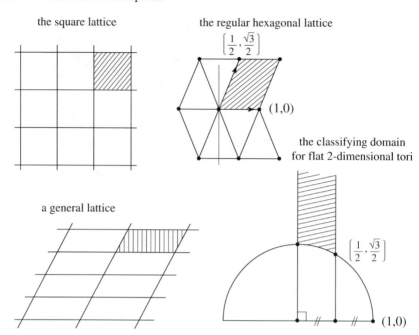

Fig. 4.32. The square lattice; the regular hexagon lattice; a general lattice; the classifying domain for flat two dimensional tori

4.3.4 Homogeneous Spaces

We saw in §§§4.1.3.3 that Lie groups are manifolds (by definition). Assume G is any Lie group and consider the *left translation map*

$$\lambda_h : G \to G$$
$$g \mapsto hg .$$

as in figure 4.33 on the next page. It is a diffeomorphism, so its differential $d\lambda_h$ is a linear isomorphism between the two vector spaces $T_g G$ and $T_{hg} G$. Use this remark as follows: pick any Euclidean structure (any positive definite quadratic form) on $\mathfrak{g} = T_e G$ (the tangent space to G at the identity element e) and transport it to $T_h G$ by demanding that $d\lambda_h$ be an isometry. Doing this for every $h \in G$, one gets a Riemannian metric on G which is invariant under left multiplication, called *left invariant*. In general it will not be right invariant, unless the adjoint representation leaves the initial quadratic form on $T_e G$ invariant. When G is compact such a bi-invariant Riemannian metric always exists and these are important examples of Riemannian manifolds: see §§4.4.3.

After real vector spaces, the simplest example of a Lie group is the Lie group structure on $G = S^3$, established by writing this sphere is the set of unit quaternions. Then the adjoint representation admits an invariant quadratic

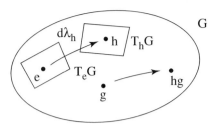

Fig. 4.33. Left translation

form at T_eG and you will not be surprised that by translations (left or right, one gets the same thing) one gets the standard "round" Riemannian metric on S^3. Now let q be any positive definite quadratic form on T_eG and diagonalize it with respect to the standard one; call $a \geq b \geq c$ the values of its principal axis. For various ratios between these three numbers one has on S^3 different left invariant metrics. In the case $a > b > c$ its geometry is still not understood; see note 6.5.4.1 on page 282.

One obtains quotient metrics from the preceding ones on $\mathbb{RP}^3 = SO(3)$, which is the quotient of S^3 by the antipodal map. The metrics thus derived are the ones whose geodesics describe the motions of a rigid body around its center of gravity, the coefficients a, b, c being directly related to the ellipsoid of inertia of the body. These motions are called *Poinsot motions*; see the beautiful rolling interpretation in the note 6.5.4.1 on page 282 and on page 200 in §§4.3.8.

We saw in §§4.1.3.4 that homogeneous spaces G/H are essential in geometry, being typical examples of the proper setting for sets of geometric objects of a given type. Assume one has a Riemannian metric g on G and that one wants to push it down to the quotient G/H. By the general nonsense quoted above, one needs g to be invariant under the action of H. More precisely, H is called the *isotropy group* of G/H and its action is to be considered on T_eG by the differentials of the right translations associated to the $h \in H$. A Lie group acting faithfully on a real vector space can leave invariant a positive definite quadratic form if and only if it is compact. So we are sure that homogeneous spaces G/H have invariant (homogeneous) Riemannian metrics as soon as H is compact. As examples, all the Grassmann manifolds of §§4.1.3.5, and in particular the $\mathbb{KP}^n$. Contrarily, if H is not compact there is no G invariant metric. The simplest example is the set of all plane straight lines. Try by hand to define on it a metric (i.e. a distance between two straight lines) invariant under Euclidean isometries, as in figure 4.34 on the following page. Convince yourself it is hopeless. The deep reason for this phenomenon is that the subgroup of Euclidean isometries leaving a given straight line invariant is not compact.

The examples above are homogeneous spaces G/H with the extra property that the action of H on T_eG is irreducible. This implies that a quadratic

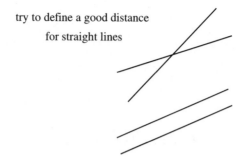

Fig. 4.34. Try to define a good distance for straight lines

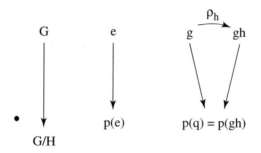

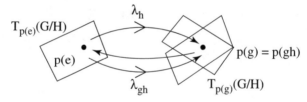

Fig. 4.35. For every $h \in H$ the derivative $d\rho_h$ should preserve the ds^2 on $T_{\rho(e)}(G/H)$

form on $T_e G$ invariant under adjoint action by H is unique up to a scalar (just use diagonalization of quadratics forms). Consequently the G invariant Riemannian metrics on these spaces are unique (up to a scalar) and we will use the word *canonical* for them. It is also in that way that one can get a fifth model for Hyp^d (see §§4.3.2) by defining it as a quotient space. Denote by $SO(d,1)$ the group of linear transformations of $\mathbb{R}^{d+1}$ which leave invariant the quadratic form

$$x_1^2 + \cdots + x_d^2 - x_{d+1}^2$$

and have positive determinant. Then

$$\mathrm{Hyp}^d = SO(d,1)/SO(d)$$

where $SO(d)$ is included in $SO(d,1)$ in the obvious way.

Even though there are plenty of them, homogeneous spaces do not constitute a deluge. I would not dare to say "classified", but there is a classification of maximal subgroups of the so-called *simple* Lie groups, at least at the Lie algebra level; see Dynkin 1952 [468, 469]. Also see Onishchik 1994 [977] and Gorbatsevich, Onishchik & Vinberg 1999 [574]. Let $\mathfrak{g}$ (resp. $\mathfrak{h}$) be the Lie algebra of G (resp. H); then the isotropy property of H reads: there is a direct sum decomposition $\mathfrak{g} = \mathfrak{h} \oplus \mathfrak{m}$ with the bracket condition $[\mathfrak{h}, \mathfrak{m}] \subset \mathfrak{m}$ and note that $\mathfrak{m}$ can be seen as the tangent space to G/H at the origin $m_0 =$ the neutral element coset eH. One is left (at least at the Lie algebra level) to look at the isotropy representation and its decomposition into irreducible pieces. To pass from Lie algebras to Lie groups is only a matter of finding discrete quotients. But practically the job of studying these homogeneous metrics is immensely more difficult than naively expected. We refer to §§15.8.1 for the formulas for their curvature and to §§§11.4.2.2 to see that even the Einstein homogeneous metrics are still far from being classified. Chapter 7 of Besse 1987 [183] is a good reference.

4.3.5 Symmetric Spaces

Between hyperbolic spaces and spheres on one hand and general homogeneous spaces on the other, there is an intermediate extremely important category, that of *symmetric spaces*, introduced by Élie Cartan in the late 20's. Their complete theory, including the classification, all due to Élie Cartan, is quite long in complete detail. The bible is Helgason 1978 [701]; see also Loos 1969 [878, 879]. Symmetric spaces are more or less completely treated in various Riemannian geometry books: section IV.6 of Sakai 1996 [1085], section 2.2 of Klingenberg 1995 [816], section 7.G of Besse 1987 [183], Eschenburg 1997 [495]. Besse 1987 [183], which is oriented toward Riemannian geometry, has complete and explicit tables giving various invariants. We now give a brief sketch of the theory, following the extremely efficient presentation of Cheeger & Ebin 1975 [341].

We define *locally symmetric spaces* as the Riemannian manifolds for which the local geodesic symmetry around any point is a local isometry. This sym-

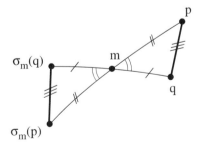

Fig. 4.36. The geodesic symmetry of a symmetric space

metry around m, denoted by I_m, is defined as the map changing $\gamma(t)$ into $\gamma(-t)$ for every geodesic γ through $m = \gamma(0)$, depicted in figure 4.36 on the previous page. If the manifold is complete, then one can define (thanks to the Hopf–Rinow theorem 52 on page 227) a global symmetry $I_m : M \to M$ and the manifold M is then called *symmetric* if all of the I_m are isometries. One says that (M, g) is a *symmetric space*. One can also say that I_m is the conjugate, via the exponential map, of the trivial symmetry $x \mapsto -x$ of $T_m M$.

As is true of most interesting mathematical objects, symmetric spaces can be defined by equivalent definitions not at all apparently related; here we have three equivalent properties. Let us settle everything in:

Theorem 36 (Élie Cartan) *Let M be a Riemannian manifold.*

1. *M is locally symmetric if and only $DR = 0$ where DR is the covariant derivative of the curvature tensor (see §15.5), i.e. the curvature is invariant under parallel transport. Every locally symmetric space is locally a part of a symmetric space.*
2. *If M is symmetric and simply connected then M is homogeneous: $M = G/H$ where G is a Lie group and H the isotropy group of some point $m_0 \in M$.*
3. *For a homogeneous symmetric space G/H, denote by I the symmetry around the origin eH. Then the map*

$$\sigma : G \to G$$
$$g \mapsto I \circ g \circ I$$

is an involutive group automorphism: $\sigma^2 = \mathrm{Id}_G$. The fixed point set H' of σ in G is closed, contained in H and H and H' have the same identity component.
4. *Conversely, let σ be an involutive automorphism of a Lie group G with fixed point subgroup H. Then σ induces a diffeomorphism I_0 of G/H. If there exists a left invariant metric on G/H which is invariant under I_0, then G/H is symmetric.*

The first part of property 1 is seen as follows: at every point $m \in M$ the tensor $DR(m)$ is of order 5 and the isometry I_m transforms it into

$$(-1)^5 DR(m) = -DR(m) .$$

Since I_m is an isometry, $DR(m) = -DR(m) = 0$. The converse comes from the Cartan philosophy on page 249. The metric around a point $m \in M$ is locally given by the solution of a Jacobi field with initial condition as in proposition 67 on page 249. But $DR = 0$ says exactly that the Jacobi field equation is an ordinary differential equation with constant coefficients, the constant value being the curvature tensor $R(m)$. The symmetry I_m around m just reverses the initial conditions, so that the solutions at time t have opposite values, hence the same norm. And this norm yields the metric: again see proposition 67 on page 249.

Keep in mind a two dimensional sphere with two points (or two disks) deleted. This is locally symmetric, as is its universal cover, a kind of snail shell of constant curvature (or onion peel), as in figure 4.37. But the snail shell is not isometric to a subset of the sphere. Indeed the snail shell has infinite diameter. Yet the sphere is the only complete surface of constant positive curvature. A similar phenomenon happens with a cone with the sharp point deleted. Therefore, locally symmetric spaces (even simply connected ones) are not always contained in symmetric spaces, or even in complete Riemannian manifolds.

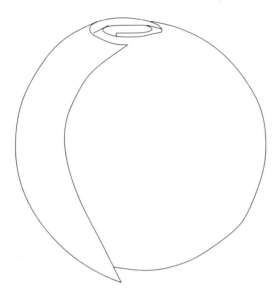

Fig. 4.37. Onion peel

Assertion 2 will come from the obvious local homogeneity; because this classification is essentially using only the Lie algebra properties of symmetric spaces, properties which are obtained from a local condition.

Composing various symmetries demonstrates local, and even global, homogeneity: one can reach every point from a given one. The claim 3 is trivial, the relation between H and H' coming from the fact that m is a isolated point of I_m. For 4 just define the symmetry around the coset $m = gH$ to be given by
$$I_m = \lambda_g \sigma \lambda_g^{-1}$$
where λ denotes left translation in G.

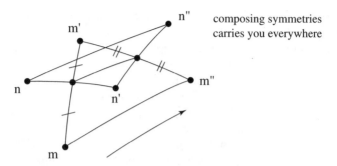

Fig. 4.38. Composing symmetries carries you everywhere

4.3.5.1 Classification

Who are those spaces? Are they many of them; are they fewer than arbitrary homogeneous spaces? The amazing fact discovered by Élie Cartan is that symmetric spaces are very rare: we describe now this classification. In theorem 36 above one does not see the Lie algebras appearing. We now use the language of decompositions we introduced on page 189. The presence of an involutive automorphism σ of G (with fixed point set H) says that the pair of Lie algebras $\mathfrak{g}, \mathfrak{h}$ satisfies the very strong conditions

$$[\mathfrak{h}, \mathfrak{m}] \subset \mathfrak{m} \quad \text{and} \quad [\mathfrak{m}, \mathfrak{m}] \subset \mathfrak{h}$$

for the direct sum

$$\mathfrak{g} = \mathfrak{h} \oplus \mathfrak{m}$$

of the total Lie algebra $\mathfrak{g}$ associated to the eigenvalues $+1$ and -1 of our involution. Note that $\mathfrak{m}$ corresponds bijectively to the tangent space of G/H at the base point eH. This strong Lie algebra condition enabled Cartan to get a complete classification of such pairs $\mathfrak{g}, \mathfrak{h}$, and thereafter a classification of the Riemannian symmetric spaces themselves. One assumes now simple connectivity and the global symmetry property. First one shows that symmetric spaces are uniquely decomposed as Riemannian products (see §§§4.3.3.1)

$$M = M_1 \times \cdots \times M_k$$

of irreducible symmetric spaces $M_i = G_i/H_i$, where the groups G_i are simple, plus the special case seen below of the groups themselves, plus an eventual Euclidean factor. The second basic fact is that, in the irreducible case, symmetric spaces occur in pairs $(G/H, G'/H)$, where the space G/H is compact and of nonnegative curvature, and the other space G'/H of the pair is of nonpositive curvature; technically G' is a noncompact real form of the complex Lie group associated to G and H is one of its maximal compact subgroups (these maximal compact subgroups being all conjugate, we can just write down any of them). So that finally one has to look for the involutive automorphisms of the various simple Lie groups. But Cartan realized that the above duality reduces the problem to finding the real forms of simple Lie groups, a job he already done in Cartan 1914 [312]. One has just to add the compact simple Lie groups G and their negative correspondents $G_\mathbb{C}/G$ where $G_\mathbb{C}$ is the complex form of G.

The final list of irreducible symmetric spaces is surprisingly small, showing finally that the geodesic symmetric condition is an extremely strong one. One finds first the various real simple Lie groups G which are symmetric spaces for themselves, written as $(G \times G)/G$ where the involution of $G \times G$ is the exchange of factors and the denominator G is the diagonal. In the nonnegative curvature category are the spheres, the Grassmann manifolds over the various fields $\mathbb{R}, \mathbb{C}, \mathbb{H}$ and add $\mathbb{C}a\mathbb{P}^2$. There are in addition only four other infinite series of spaces as well as finite exceptions in low dimensions. Those infinite series are all of interest, not only for themselves, but for various geometric situations e.g. symplectic geometry, Kähler geometry, since they are essentially the set of complex structures, the set of symplectic structures, the set of quadratic forms and the set of Hermitian forms on a vector space. We will meet them again in the future: first in §6.6 and then in §§12.3.1.1. There are also nonpositive curvature symmetric spaces associated to the above list by the pairing described before, among them the various hyperbolic spaces $\text{Hyp}_\mathbb{K}^n$.

For tables giving the classification, with moreover geometric interpretation, one can consult books on symmetric spaces, especially the tables 7.H and 10.K of Besse 1987 [183].

4.3.5.2 Rank

A basic notion for symmetric spaces is that of *rank*. One proves that, in a symmetric space G/H, there are flat totally geodesic submanifolds (see §§6.1.4), and that those of maximal dimension are all conjugate under the action of G, and that any tangent vector is contained in at least one such submanifold. Their common dimension is called the *rank* of the symmetric space. At the Lie algebra level, they are not mysterious; the corresponding linear subspaces $\mathfrak{t} \subset \mathfrak{m}$ are those on which the bracket vanishes identically: $[\mathfrak{t}, \mathfrak{t}] = 0$. In symmetric spaces of nonnegative curvature, these subspaces are tori. In simply connected, nonpositively curved symmetric spaces, they are Euclidean spaces. This completely solves the geodesic behavior of symmetric spaces: on compact symmetric spaces we have every geodesic sitting as a geodesic inside a flat torus. Thus every geodesic (continued indefinitely) is periodic or is everywhere dense in a totally geodesic flat torus. The compact symmetric spaces of rank equal to 1 are exactly the spheres and the $\mathbb{K}\mathbb{P}^n$. This is how Élie Cartan discovered that all of their geodesics are periodic and of same length. This notion of rank will reappear later on in various places.

In some sense, every question you can ask about symmetric spaces is answered. For example, computing the curvature, analyzing the geodesic behaviour in the small and in the large. This is exceptional in Riemannian geometry. A heuristic way to present symmetric spaces is to say that, in the hierarchy of geometries developed in §6.6, starting with Euclidean geometry, Riemannian symmetric spaces (and their associated generalized space forms) come just after hyperbolic and elliptic geometries, ranked by the quality of our understanding.

4.3.6 Riemannian Submersions

A *Riemannian submersion* is an often encountered generalization of the too strict notion of product seen in §§4.3.3. It consists in a map

$$p : (M, g) \to (N, h)$$

between two Riemannian manifolds which is infinitesimally a product along every fibre, as in figure 4.39. Let us make this more mathematical. Every point $n \in N$ admits a neighborhood U such that $p^{-1}(U)$ is a product manifold. Now at every point $m \in p^{-1}(n)$ we have the *vertical tangent space*, which is the tangent space $V_m M$ at m to the fiber $p^{-1}(n)$. But the Euclidean structure $g(m)$ provides us with a horizontal tangent space, namely the orthogonal complement $H_m M$ of $V_m M$. The differential dp of p restricted to $H_m M$ is a vector space isomorphism between $H_m M$ and $T_n N$ by construction. But both $H_m M$ and $T_n N$ are equipped with Euclidean inner products. We say that $p : (M, g) \to (N, h)$ is a *Riemannian submersion* when for every $n \in N$, and every $m \in p^{-1}(n)$, this vector space isomorphism is a Euclidean isomorphism.

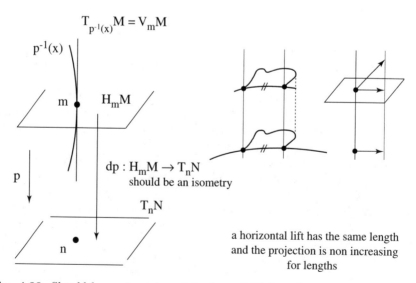

Fig. 4.39. Should be an isometry; A horizontal lift has the same length and the projection does not increase lengths

Remark that when the fiber is discrete (say of dimension 0) then we have nothing but a Riemannian covering as in §§§4.3.3.2. The case of a Riemannian product is also a very peculiar case, highly atypical. A surface of revolution provides a Riemannian submersion: the submersion goes from the surface to any meridian (check it). One can define objects of revolution in any dimension d; they have the spherical symmetry given by an action of the orthogonal group $O(d-1)$, namely

$$M = I \times S^{d-1}$$

where I is any interval of the real line. The base is still one dimensional. The metric can be written

$$g = dt^2 + \phi^2(t)\, ds_{d-1}^2$$

where ds_{d-1}^2 designates the standard metric of the sphere S^{d-1}.

Note 4.3.6.1 (Warped products) We will not use it explicitly, but the notion of *warped product* is important in various Riemannian situations, see also §§15.8.2. Keep in mind that the definition varies according to authors. In Petersen 1997 [1018] the term is reserved for the objects of revolution above, and the notion extended moreover to double warped products

$$M = I \times S^p \times S^q\;.$$

In Sakai 1996 [1085] a warped product is defined on any Riemannian product manifold $(M \times N, g \times h)$ by a numerical function $f : M \to \mathbb{R}$ forcing the modification of the metric $g \times h$ into

$$\|(v,w)\|^2 = \|v\|_M^2 + f\|w\|_N^2\;.$$

♦

Other examples of Riemannian submersions are the fibrations of §§§4.1.3.5:

$$S^{2n+1} \to \mathbb{CP}^n$$

and

$$S^{4n+3} \to \mathbb{HP}^n\;.$$

Basic examples are the tangent bundle TM and the unit tangent bundle UM of any Riemannian manifold. This metric on the unit tangent bundle will be very significant in chapter 10. First, we need to define the canonical Riemannian metric on TM and UM. One approach to the canonical Levi-Civita connection (see §15.3), generates a canonical horizontal subspace in the tangent space to TM at each point. Take the obvious inner product on the vertical space (the space tangent to the fiber). On the horizontal space, take the lift of the metric on M by the differential of the projection $TM \to M$. Define the metric on TM to be the Euclidean product, at every point, of these inner products. This is now a Riemannian submersion by the very definition. There is no substantial difference in constructing a Riemannian geometry for UM.

Among the most interesting Riemannian submersions are those all of whose fibers are totally geodesic (see §§6.1.4 for this notion). Then the fibers are mutually isometric, and the horizontal trajectories yield those isometries (see figure 4.40). In this special situation, the formulas 15.17 on page 721 for computing the curvature simply greatly. For a new viewpoint on Riemannian submersions, see Karcher 1999 [781].

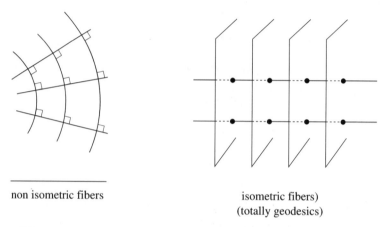

non isometric fibers   isometric fibers)
                                  (totally geodesics)

Fig. 4.40. Nonisometric fibers; Isometric fibers (totally geodesic)

4.3.7 Gluing and Surgery

Let us now consider gluing (see §§§4.1.3.6). There are two points here. First one should endow the pieces to be glued with a Riemannian metric: this requires a definition of Riemannian manifold with boundary, which we will not supply, allowing the interested reader to supply a rigorous definition. The second point is that the object constructed in the gluing operation should be a smooth manifold.

4.3.7.1 Gluing of Hyperbolic Surfaces

We consider only gluing of surfaces, for simplicity, and look at the pictures in figure 4.41 on the facing page. Assume that the pieces shown are domains of the hyperbolic plane Hyp^2. Trouble in matching arises first along the sides and then at the vertices. Along the sides: as a curve in a domain, each side has a geodesic curvature. Sides glued together should have the same geodesic curvature, but with opposite signs. The simple thing to do is to require that the geodesic curvature of each side be everywhere zero i.e. all sides are geodesics. And that turns out out to be enough in Hyp^2, because the isometry group of Hyp^2 contains a reflection with respect to any given straight line. At the vertices one needs only that the sum of the angles of the matching pieces add up to 2π.

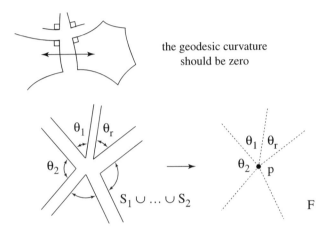

Fig. 4.41. The geodesic curvature should be zero

For example, it is an nice exercise in hyperbolic plane geometry to prove that there are (many) hexagons all of whose sides are straight lines and all of whose vertex angles are equal to $\pi/2$. With a pair of pants one can glue together (abstract) compact surfaces of any genus greater than one, to the effect that on any such orientable surface there exists at least one metric everywhere locally isometric to Hyp^2 (see figure 4.10 on page 157). Equivalently: the universal covering of any one of them, endowed with the metric pulled up by the covering, is isometric to Hyp^2. Note that this is weaker than the conformal representation theorem 70 on page 254. We will see how many such structures exist on a given surface in §§6.6.2. These results are extremely important, for example for physicists in the recently developed string theory whose starting point is to replace particles with curves. Instead of a path of a particle, this curve as time evolves will traverse a surface.

Any complex structure on a surface of genus greater than one is Hermitian for a unique Riemannian metric locally isometric to the hyperbolic plane. Geometrically this means that we are given an operator J on tangent vectors, determining a rotation, so that $J^2 = -1$ and we want the metric to measure the angle of the rotation to be $\pi/2$, and moreover to be locally isometric to Hyp^2.

Other examples are made from tilings of Hyp^2, typically with triangles with angles equal to $\pi/p, \pi/q, \pi/r$ where p, q, r are integers, or with regular

198   4 Riemann's Blueprints

n-sided polygons with vertex angle equal to $2\pi/n$. In both these cases the tiling is obtained by reflexions along the sides.

We will come back to these space forms in §§6.3.2 and §6.6, but also in §§10.2.4 and in the first part of chapter 12. Even more important than the existence of a constant curvature metric on any surface is the fundamental theorem of conformal representation theorem 70 on page 254 to the effect that one can find, for any metric g on a compact surface, a new one of the form fg (f a numerical function) which is of constant curvature.

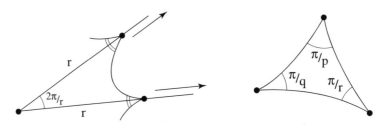

Fig. 4.42. In Hyp^2 there are n-gons which are tiling, this for every n; When $r \to \infty$, the angle goes to zero, so that for some r it attains the value $2\pi/n$

4.3.7.2 Higher Dimensional Gluing

In higher dimension, gluing is now also possible along hypersurfaces. But the condition for such hypersurfaces is extremely strong: since the flip (the symmetry) around the hypersurface should be a local isometry, the hypersurface under consideration should be totally geodesic to be sure to have no singularity in the metric. As to be seen in §§6.1.4, totally geodesic submanifolds of dimension larger than one (those of dimension one are just geodesics) do not exist in generic Riemannian manifolds. Notable exceptions are space forms (manifolds of constant curvature, see §6.6); applications of such gluings will be seen there and in §§12.2.4.

If one allows local modifications of the metrics around the hypersurfaces, then gluing is possible in great generality. Let us consider first the *trivial surgery*, namely the *connected sum* $M\#N$ of two Riemannian manifolds (M,g) and (N,h) of the same dimension. Pick up points p and q in M and N respectively and consider small balls B and B' around them. The connected sum consists in identifying the boundary spheres ∂B and $\partial B'$ by some diffeomorphism. Pictorially, one thinks more of a cylinder. Then it is clear that, using bump functions, one can put on $M\#N$ a Riemannian metric which coincides with g on $M\backslash B$ and on $N\backslash B'$. This kind of construction is obviously possible for any surgery of a manifold along any submanifold.

The challenge, later on, will be to do the above smoothing in order to preserve some given condition on the Riemannian structures, typically conditions on their curvature. Examples will occur in §§§12.3.1.1 and §§§12.3.3.2.

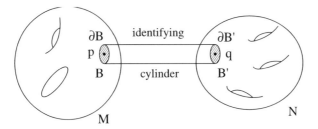

Fig. 4.43. Connected sum

4.3.8 Classical Mechanics

The last source of Riemannian geometry we will consider is classical mechanics. We follow the route initiated in §§4.1.2. Start with the configuration space. The first object is the kinetic energy of the physical system. It is a quadratic positive definite form on velocity vectors, so this exactly a Riemannian metric. For the double pendulum it is:

$$ds^2 = r_1^2 \, d\phi_1^2 + 2r_1 r_2 \cos(\phi_1 - \phi_2) \, d\phi_1 \, d\phi_2 + r_2^2 \, d\phi_2^2 \tag{4.23}$$

where ϕ_1 and ϕ_2 are the two angles defining the position and r_1 and r_2 are

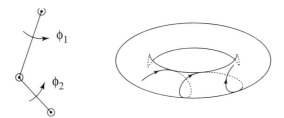

Fig. 4.44. The torus (not with the metric induced by the embedding in $\mathbb{E}^3$)

the lengths of the links. How now if there is gravity or more generally some potential energy field V on the pendulum? What we say now is in fact valid for any configuration space. Fix an initial value E for the energy, and then the motions of the system will be the geodesics of the metric

$$d\sigma^2 = (E - V) ds^2 \tag{4.24}$$

where ds^2 is the kinetic energy. You need to start with quite a large E if you want $E - V$ to remain positive during the motion. For the double pendulum, you have to start with large speed if you want both pendulums to turn around completely. If not, you will meet some "boundary" and return to the domain you just left. Riemannian geometry represents the classical mechanics of excited systems, while for near equilibrium systems, the kinetic energy is overwhelmed by potential energy, so that a different approach is required. This explains why economists rarely learn Riemannian geometry: economics is a total enigma away from equilibrium.

The above is completely general. For example, it applies to the motions of a rigid body, described above by the manifold $SO(3)$. If there is a potential, one has only to modify the Riemannian metric by 4.24. Practically, the result is a left invariant metric of the following nature: $SO(3)$ has at the origin a canonical Euclidean structure given by the adjoint representation. By left transport this will yield the standard "round" Riemannian metric on the special orthogonal group $SO(3)$ (i.e. the group of Euclidean 3-space rotations around a fixed point). This is the case for motions of a body whose ellipsoid of inertia is a sphere. In the general case, the metric to use is a general quadratic form, say with eigenvalues $\{a < b < c\}$ with respect to the canonical one, and then left transport it. This is called *le mouvement à la Poinsot* and we cannot resist giving Poinsot's geometric description of these moves, from Poinsot 1842 [1035]. The move of the rigid body is the same as if, fixed at its center, the ellipsoid of inertia is rolling without friction on a plane, the distance of the plane to the center depending on the initial conditions; see figure 6.62 on page 283. For a proof, see Appell & Lacour 1922 [57] tome II, page 219.

Conversely one can say that on a given Riemannian manifold a mass particle which is submitted to no force moves along geodesics. See chapter 10 for the global long term behavior of geodesics. For applications of Riemannian geometry to the two cases above, see the note 6.5.4.1 on page 282 as well as the book Arnol'd 1996 [66].

4.4 The Riemann Curvature Tensor

4.4.1 Discovery and Definition

Let (M, g) be a Riemannian manifold. The first thing to do is to look for segments i.e. for shortest connecting curves between two given points. We already studied segments for surfaces $M^2 \subset \mathbb{E}^3$, and we succeeded in a disguised way, essentially using the embedding to define the geodesic curvature and then proving by the calculus of variations that the shortest had to be geodesics, i.e. curves with vanishing geodesic curvature. To be a geodesic turned out to be a curve satisfying a differential equation of the second order (cf. 1.10 on page 35) in some chart. There is a similar story for abstract Riemannian manifolds, but it needs some quite hidden machinery, so we postpone it to the technical chapter 15 and only apply the main result: the shortest paths have to be chosen among geodesics, these being the curves which are solutions in any chart of a second order differential equation which can be written in a vector notation:

$$c'' = F(c, c') \qquad (4.25)$$

where F is quadratic in c'

4.4 The Riemann Curvature Tensor

We recall that the major difficulty here is that this has to be written in a chart (at least now) because in a manifold curves have well defined speed c' but no well defined acceleration c''. If this is puzzling you, let me add that what we said is that the equation 4.25 transforms when changing charts in a funny way. But that equation is invariantly attached to the ds^2 (which also transforms in a funny way) so that geodesics are well defined curves, not depending on the choice of a chart but really attached to (M, g).

This is a real difficulty. Evidence of this—a posteriori—is that it needed half a century to become clear. In 1869, Christoffel made a major contribution to clarify this matter, before the final touch by the Italian geometers at the beginning of the 20th century.

From equation 4.25 we retain that, for every point $m \in M$ and any tangent vector $v \in T_m M$, there is one and only one geodesic c with $c(0) = m$ and $c'(0) = v$. Beware that this geodesic may be defined only for a small interval of time t containing 0. Geodesics enjoy the property that their speed $\|c'(t)\|$ is constant. Most often in the sequel we will parameterize them by unit speed without mentioning it. Due to the compactness of the unit sphere in $T_m M$, there is a positive number t such that all the geodesics emanating from a given point $m \in M$ are defined at least in $[-t, t]$.

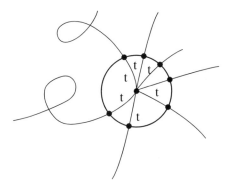

Fig. 4.45. Geodesics defined for time at least t

We now follow the route of the proof of the formula 3.3 on page 108. The above property of geodesics implies that near any point m of a Riemannian manifold, there are *normal coordinates* centered at m i.e. charts with domain in $\mathbb{R}^d$ centered at $(0, \ldots, 0)$ and such that the image under the chart of every straight line through $(0, \ldots, 0)$ is a geodesic of (M, g) through m. The second major contribution of Riemann in his 1854 address was to show that in normal coordinates centered at some point $m \in M$, ds^2 has an expansion which necessarily takes the form:

$$ds^2 = dx_1^2 + \cdots + dx_d^2 \qquad (4.26)$$
$$+ \frac{1}{12} \sum_{i,j,k,h} R_{ijkh}(m) \left(x_i\, dx_j - x_j\, dx_i\right)\left(x_k\, dx_h - x_h\, dx_k\right)$$
$$+ o\left(x_1^2 + \cdots + x_d^2\right)$$

Of course the hard work consists in showing that the second order terms are not made up by general quadratic terms but that they can be written in a expression involving only the various

$$\left(x_i\, dx_j - x_j\, dx_i\right)\left(x_k\, dx_h - x_h\, dx_k\right)$$

products.[6] The $1/12$ factor is written not only to match when $d=2$ with formula 3.3 but is in fact forced when we identify the curvature tensor with other definitions of the curvature tensor below in proposition 413 on page 700. Important note: we match formula 3.3 but only up to sign. In the literature some authors use also R with the opposite sign: this is conceptually not important, but be careful not to trip over signs.

Formula 4.26 can be interpreted by saying that a Riemannian metric is Euclidean at the zeroth order, which is obvious, but more: intrinsically it is also Euclidean to the first order, which is more surprising. One can also consider that a Riemannian metric does not have tensor invariants of the first order but only invariants starting at the second order, the second order one being precisely the curvature tensor. (The Levi-Civita connection, described in chapter 15, is a first order invariant but not a tensor.)

There are many other conclusions we can derive from equation 4.26. The first is in dimension 2: the Gauß curvature is defined for abstract surfaces with Riemannian metrics, even without embedding in $\mathbb{E}^3$. We will now refer to the Gauß curvature as simply the *curvature* of (M, g): it is a function $K : M \to \mathbb{R}$. For example in any two dimensional Riemannian manifold the Bertrand–Puiseux formula 17 is still valid and gives a very geometric way to feel about K.

The second is that there is an extension of the curvature (function) for Riemannian manifolds of any dimension but that notion is quite complicated. In fact it is so complicated that some of its aspects are still not understood today. The object which comes into the picture is a 4-tensor R, called the *Riemann curvature tensor*. This means that it is a smooth collection of quadrilinear forms $R(m)$ on the tangent spaces T_mM when m runs through M. This quadrilinear form can be defined intrinsically through coordinate changes but also by extending by linearity the object whose values on the basis $\{e_i = \partial/\partial x_i\}_{i=1,\ldots,d}$ are the $R(e_i, e_j, e_k, e_h) = R_{ijkh}$. Then one will write $R(x, y, z, t)$ applying the curvature tensor to tangent vectors $x, y, z, t \in T_mM$. By formula 4.26, the Riemann curvature tensor is invariant under isometries.

---

[6] See Spivak 1979 [1155], volume 2, for a guess as to how Riemann achieved this—he did not give any detailed computation in his text.

Third: the curvature appears as the "acceleration" of the metric, for example at the center of normal coordinates:

$$\partial_i \partial_j g_{kh} = \frac{1}{6} \left(R_{ikhj} + R_{ihjk} \right) . \tag{4.27}$$

One might hope to control the metric completely, as in dynamics where you know the motion if you know the force applied to a particle. In fact things are more subtle: there are things which can be completely recovered, but some which cannot, see more on this in §4.5. Even so-called *constant curvature* metrics hold some mysteries: see §6.6.

A important remark is in order: many people think that the curvature and its derivatives are the only Riemannian invariants. This is true and classical when looking for algebraic invariants which stem from the connection, see page 165 of Schouten 1954 [1111] and the references there. But things are dramatically different if one asks only for tensors which are invariant under isometries (called *natural*). Then there is no hope to get any kind of classification, as explained in Epstein 1975 [490]. For more see Muñoz & Valdés 1996 [951].

Formula 4.26 shows that R is exactly the infinitesimal defect of our ds^2 to be locally Euclidean. How about globally? We saw in equation 3.10 on page 118 that for surfaces, $K \equiv 0$ implies flatness (i.e. local isometry with $\mathbb{E}^2$). We will soon show that this extends easily to any dimension: $R \equiv 0$ implies flatness: see theorem 69 on page 252.

Note that the Riemann tensor will be recovered from other points of view: see the golden triangle in §15.3 for the parallel transport point of view, where the curvature tensor expresses exactly the infinitesimal defect along small loops of the parallel transport to be the identity, and see chapter 15, where the curvature tensor expresses the defect of higher order derivatives of functions to be symmetric.

Another conclusion is that Riemannian geometry cannot be very simple, especially in dimension 3 or more. This because quadrilinear forms are complicated objects, not much encountered in mathematical literature. But at least R is not a completely general quadrilinear form. From 4.26 it follows that R has the following symmetry properties: R is antisymmetric both in the two first entries and the two last, R is symmetric when one exchanges the first two with the last two entries and finally it satisfies the Bianchi identity, which says that the sum of the three first entries under circular permutations vanishes:

$$\begin{aligned} &R(x,y,z,t) = R(z,t,x,y) = -R(x,y,t,z) \\ &R(x,y,z,t) = -R(y,x,z,t) \\ &R(x,y,z,t) + R(y,z,x,t) + R(z,x,y,t) = 0 \quad \text{First Bianchi identity} \end{aligned} \tag{4.28}$$

the above relations being valid for any tangent vectors x, y, z, t at any point. The first "two by two" symmetry identity is deduced from the second and

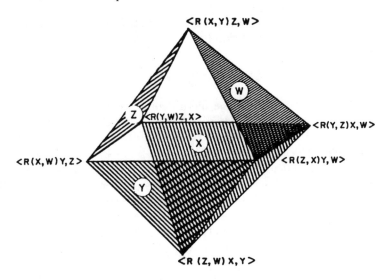

Fig. 4.46. Milnor's octahedron

third ones by elementary combinatorics. But a geometer will appreciate the trick called *Milnor's octahedron* in Milnor 1963 [921], drawn in figure 4.46.

The relations 4.28 are the only ones satisfied by R. To prove this, just take any set of real numbers R_{ijkh} satisfying these relations, and define (locally) a Riemannian metric by 4.26 stopping at the second order, i.e. take $o\left(x_1^2 + \cdots + x_d^2\right)$ to be identically zero.

Another question for an algebraic mind: in the real vector space of all quadrilinear forms on $\mathbb{R}^d$, what is the dimension of the subspace made up by the ones satisfying the relations 4.28? Call them *curvature type*.

Proposition 37 *The dimension of the vector space made up by the curvature type quadrilinear forms is $d^2(d^2-1)/12$.*

Do not despair if the curvature tensor does not appeal to you. It is frightening for everybody. We hope that after a while you will enjoy it a little.

A paradox: it might appear to you from formula 4.27 (as opposed to equation 37) that since the metric depends on $d(d+1)/2$ parameters and its second derivatives (which are symmetric) on also $d(d+1)/2$, the curvature should depend on $(d(d+1)/2)^2$ parameters, a number much larger than $d^2(d^2-1)/12$. However, choosing normal coordinates involves using the geodesics, and this forces extra relations between the $\partial_i \partial_j g_{kh}$. See more in §4.5.

4.4.2 The Sectional Curvature

To make a bilinear form more palatable, we plug in two equal vectors to obtain a quadratic form. Similarly, with the curvature tensor $R(x, y, z, t)$, we can plug in equal vectors. By the antisymmetry this can be done only at some places, or else we get zero. Moreover the symmetry relations show that, up to sign, there is only one possibility, namely to put $z = x$: $R(x, y, x, t)$. But we want a number, so again we set $y = t$ and consider $R(x, y, x, y)$. Since we have on $T_m M$ the Euclidean form $g(m)$ we can moreover normalize $R(x, y, x, y)$ as:

$$K(x, y) = -\frac{R(x, y, x, y)}{\|x\|^2 \|y\|^2 - \langle x, y\rangle^2} \qquad (4.29)$$

where the denominator can be replaced by $\|x \wedge y\|^2$ if you like exterior algebra.

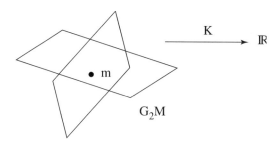

Fig. 4.47. The sectional curvature as a real valued function on Grass $(2, TM)$

Here we changed sign in order to ensure that K agrees when $d = 2$ with Gauß curvature. Formula 4.29 makes sense only when x and y are linearly independent i.e. define a tangent plane to M. Note also that we can avoid the normalization by asking the pair $\{x, y\}$ to be orthonormal. And finally the symmetry relations 4.28 imply immediately that in fact $K(x, y)$ depends only on the plane P spanned by x and y. It is called the *sectional curvature* of P.

To sum up, we came up with a real valued map K defined on the Grassmann manifold Grass $(2, TM)$. Recall that this is the manifold consisting of all the two dimensional tangent planes to M:

$$K : \operatorname{Grass}(2, TM) \to \mathbb{R}. \qquad (4.30)$$

When $d = 2$, K coincides with Gauß curvature, hence the coincidence of names. We have again a Bertrand–Puiseux formula as follows. Let $P \in \operatorname{Grass}(2, T_m M)$ be a 2-plane at a point $m \in M$, and draw all the geodesics starting from the point m with unit speed, and with velocity belonging to P, as in figure 4.48 on the following page. For a small ε, draw the circle $C(P, \varepsilon)$ made up by drawing these geodesics for time ε. Then (cf. theorem 17 on page 110 and equation 4.26 on page 202):

$$K(P) = \lim_{\varepsilon \to 0} \frac{3}{\pi} \frac{2\pi\varepsilon - \operatorname{length} C(P, \varepsilon)}{\varepsilon^3}. \qquad (4.31)$$

In coordinates (notations as above):

$$K(e_i, e_j) = -R_{ijij} \tag{4.32}$$

for a coordinate plane.

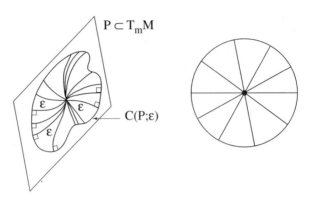

Fig. 4.48. Geodesic disks

Warning: you might have hoped that the quadrilinear form R could be diagonalized with respect to the Riemann positive quadratic form g as is the case for quadratic forms. Namely there would exist a nice basis $\{e_i\}_{i=1,\ldots,d}$ such that only the components R_{ijij} of R are nonzero. This is hopeless for a generic Riemannian metric except in dimensions 2 and 3. For 2 it is obvious, for 3 work it out yourself. In dimension $d = 4$ one can kill quite a few of the components of the type R_{1213} but there is no way to avoid the ones of R_{1234} type: see Singer & Thorpe 1969 [1145] and an application of this "killing technique" in the note 12.3.1.1 on page 579.

Now let us recall that a bilinear symmetric form P can be recovered from the quadratic form Q it generates by:

$$P(x,y) = \frac{1}{2}\left(Q(x+y) - Q(x) - Q(y)\right) \tag{4.33}$$

You will enjoy performing the same kind of computations with $K(x,y)$ and using the formulas 4.27. The explicit result is complicated, but one is soon convinced that:

Proposition 38 *From K one can recover R, so that the knowledge of K is equivalent to that of R.*

The relations in equation 4.28 on page 203 are necessary and sufficient, as we saw. In some sense one can say that R is well understood. Contrarily, as a function on the Grassmannian—say at one point m— the sectional curvature

$$K(m) : \mathrm{Grass}\,(2, T_m M) \to \mathbb{R}$$

is not well understood starting in dimension 4. For example, one does not understand the set made up by the planes which are critical for K, in particular where is the maximum and where the minimum. This is a drawback. For example, see §§11.3.1 for the functional called the *minimal volume*. Recall that a critical point of a smooth map is by definition a point where its differential vanishes. Critical points are a natural generalization for smooth functions of maxima and minima, we will meet them again for periodic geodesics in chapter 10. Returning to the curvature, to our knowledge only one case is understood: when $d = 4$ and moreover the Riemannian metric is *Einstein*. This condition (see Besse 1987 [183] and §§11.4.1 for the definition) implies immediately in dimension four that $K(P) = K(P^\perp)$ for every $P \in \mathrm{Grass}(2, TM)$, where $P^\perp$ denotes the orthogonal complement of P in the tangent space. Then the structure of the curvature is completely understood; see Singer & Thorpe 1969 [1145] for details. Historically, it seems that Churchill 1932 [380] was the first to consider the question, unsuccessfully.

There are other notions of curvature, e.g. Ricci curvature which will show up naturally in §6.2. For a hierarchy of these other notions of curvature, see §§12.1.2.

4.4.3 Curvature of Some Standard Examples

We will only give a list of basic examples, to provide the reader with a first feeling about curvature. Many more examples will appear all over the book.

Note 4.4.3.1 Before inspecting any examples, it helps to know how R and K behave under scaling: if one replace the Riemannian metric g by λg:

$$K_{\lambda g} = \lambda^{-1} K_g \qquad (4.34)$$
$$R_{\lambda g} = \lambda^{-1} R_g \qquad (4.35)$$

Beware that the above λ scaling replaces the metric $d(\cdot, \cdot)_g$ by

$$d(p, q)_{\lambda g} = \sqrt{\lambda}\, d(p, q)_g$$

For the sectional curvature, 4.34 and 4.35 agree with 4.31. ◆

4.4.3.1 Constant Sectional Curvature

After Euclidean space, for which $K \equiv 0$ and $R \equiv 0$, the next simplest cases are the spheres S^d and hyperbolic spaces Hyp^d. We claim that $K \equiv 1$ on spheres, and $K \equiv -1$ on hyperbolic spaces. The reason is that the group of isometries of any of those spaces acts transitively on $\mathrm{Grass}(2, TM)$ and by its very definition K is invariant under isometries. Changing the radius of the spheres and considering the spaces $\mathrm{Hyp}^d(K)$ (cf. §§4.3.2)

Theorem 39 *For every real number k there exists a complete simply connected Riemannian manifold with $K \equiv k$. It is denoted by $\mathbb{S}^d(k)$.*

The space referred to in theorem 39 is unique, of course up to isometries or, if you prefer, uniqueness of Riemannian structures. This will be easily proven in §§6.3.2.

Written in an orthonormal coordinate system, this is equivalent to: all the R_{ijkh} are zero except (up to obvious permutations) the $R_{ijij} = -k$ for every $i \neq j$. *Conformally flat* manifolds are those which are flat after rescaling the metric by a scalar function; for example, constant sectional curvature manifolds are conformally flat. Conformal flatness is equivalent to the vanishing of the Weyl conformal curvature tensor (in dimension 4 or more). This is an interesting subject, because there are many other examples of conformally flat manifolds besides those of constant curvature, and because of its applications to theoretical physics. The present state of the entire subject is treated in the book Matsumoto 1992 [903]; also see the more specialized Burstall, Ferus, Leschke, Pedit & Pinkall 2002 [289].

Theorem 40 (Symmetries of $\mathbb{S}^d(k)$) *There is a unique totally geodesic hypersurface of $\mathbb{S}^d(k)$ tangent to every hyperplane in every tangent space, as drawn in figure 4.49. (This hypersurface is naturally called a hyperplane. See §§6.1.4 for the definition of totally geodesic.) The geodesic reflection around that hypersurface is defined on the whole manifold and is an isometry.*

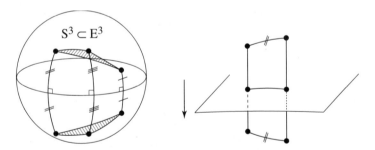

Fig. 4.49. A totally geodesic hypersurface

There are many characterizations of constant curvature (local and global); we will see one in §§6.3.2. In the spirit of the above, only constant curvature manifolds can have a totally geodesic hyperplane passing through each point and normal to each direction; this comes directly from §§6.1.4. This is the spirit of the chapters V and VI of Cartan 1988 [321], which can be interesting to read with a modern eye.

4.4.3.2 Projective Spaces $\mathbb{KP}^n$

The next examples are also important: the $\mathbb{KP}^n$ of §§4.1.3. We saw in §§4.3.4 that they have a canonical Riemannian metric up to scaling. In the literature one finds only two choices, up to a scaling factor of 2. Which one an author prefers can be seen precisely by looking at the value the author gives for K. We describe the sectional curvature K of $\mathbb{CP}^n$, leaving the reader to work out the sectional curvature of $\mathbb{HP}^n$ and $\mathbb{CaP}^2$. Fix some $m \in \mathbb{CP}^n$. The isometry group is transitive on unit tangent vectors: let us pick anyone v among them and look at $K(v, w)$ for w such that $\{v, w\}$ is orthonormal. This time the isometries which fix m and v do not act transitively on the choices of w. The tangent space $T_m M$ has the structure of a complex vector space, so we have a privileged choice of $w = iv$ (where $i = \sqrt{-1}$).

$$K(v, iv) = 1 \qquad (4.36)$$

$$K(v, w) = \frac{1}{4} \qquad \text{if } w \text{ is orthogonal to } iv.$$

In fact one has a complete determination:

$$K(v, w) = \frac{1 + 3\cos^2 \alpha}{4} \qquad (4.37)$$

where α is the angle between w and iv.

In the other scaling, K ranges from 1 to 4. According to the scaling, the diameter is π or $\pi/2$. For a detailed proof, see for example section II.6 of Sakai 1996 [1085] or 3.30 in Besse 1978 [182]. The geometric explanation is simple: our spaces contain two kinds of extreme totally geodesic submanifolds: first the $\mathbb{CP}^1$ (projective lines) which are spheres of diameter equal to $\pi/2$, hence of constant curvature equal to 4 and second the $\mathbb{RP}^n$ (real projective spaces) which are also of diameter equal to $\pi/2$ but hence of constant curvature equal to 1. See page 237 for more details.

4.4.3.3 Products

Even if trivial, the case of products (§§§4.3.3.1) is very important. The result is that $R(x, y, z, t)$ vanishes unless the four vectors are either all horizontal or all vertical. Then the sectional curvature is given by the formula:

$$K_{M \times N}(x, y) = K_M(x_M, y_M) + K_N(x_N, y_N) \qquad (4.38)$$

In particular this curvature vanishes for all of the mixed planes span $\{x, y\}$ where x is horizontal and y vertical. Here we have the obvious notations for the indices: M for example denotes the "M-part" in the product decomposition

$$T_{(m,n)}(M \times N) = T_m M \times T_n N.$$

Coverings and discrete quotients, since they are locally isometric, have identical curvature at corresponding points.

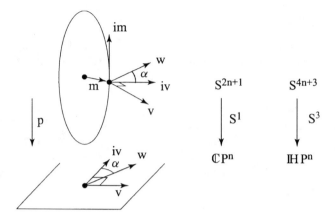

Fig. 4.50. Curvature of $\mathbb{C}\mathbb{P}^n$

4.4.3.4 Homogeneous Spaces

For Riemannian homogeneous spaces G/H (see §§4.3.4) the curvature is always computable in theory by Lie group and algebra techniques. It is enough to calculate curvature at any point of G/H, since G acts transitively on G/H by construction. In most cases such a computation is difficult. We refer to to §§15.8.1 and to chapter 7 of Besse 1987 [183] for a general reference. We will only mention a few special cases.

The first is a compact Lie group with a bi-invariant metric: then $K > 0$ and moreover the value zero can only be obtained for pairs of vectors tangent to an Abelian subgroup of G. Subsequent chapters will show that it is extremely hard to find Riemannian manifolds with $K > 0$, see §§12.3.1. For this reason compact Lie groups are important, as we will see. Consider some homogeneous space G/H with G compact and H a compact subgroup, and pick some bi-invariant metric g on G. The metric goes down to a metric g_0 on G/H. Such Riemannian homogeneous spaces are called *normal* (see also §§15.8.1). The sectional curvature of g_0 on G/H is also positive. The two above results were obtained by Samelson in 1958 via geometric reasoning. In this sense they are exceptional.[7] It is hard to resist sketching the proof of positivity of sectional curvature of G/H because it is beautifully geometric. First, the algebraic proof looks at formula 15.16 on page 719 which we give again here:

$$K(x,y) = \frac{1}{4}\|[x,y]_{\mathfrak{h}}\|^2 + \|[x,y]_{\mathfrak{m}}\|^2 \qquad (4.39)$$

Here the two tangent vectors x, y are considered in the tangent space to G at the identity element. This means that they are elements of the Lie algebra

---
[7] This work of Samelson was the first appearance of the concept of *Riemannian submersion* (see §§4.3.6).

of G. They have a bracket $[x,y]$ and the Euclidean structure on this Lie algebra, coming from our chosen Riemannian metric, yields an orthogonal decomposition of the Lie algebra associated to the subspace $\mathfrak{h}$ made up by the Lie algebra of H, and its orthogonal complement $\mathfrak{m}$. For irreducible symmetric spaces (see §§4.3.5) the Lie algebra formulas given there leave only one term, namely
$$\frac{1}{4}\|[x,y]\|^2 \ .$$
This is for the compact groups. For irreducible noncompact symmetric spaces, one has
$$-\frac{1}{4}\|[v,w]\|^2 \ .$$
In particular the curvature vanishes on the flat totally geodesic subspaces met in §§4.3.5. For nonnormal homogeneous spaces, 15.16 has to be replaced by the very complicated formula 15.15 on page 719. This explains why not everything is known concerning the curvature of homogeneous spaces; see for example §§§11.4.2.2.

Consider a Riemannian submersion
$$p:(M,g) \to (N,h) \ .$$
Nice geometric reasoning using the uniqueness of geodesics shows that a geodesic in (M,g) with initially horizontal velocity will always have horizontal velocity and moreover will project down to a geodesic in (N,h). By the above definition they will have the same length.

Let P be some horizontal tangent plane at $m \in M$ and P^* its projection in N. Samelson's claim is:
$$K(P^*) \geq K(P) \ . \tag{4.40}$$

The idea is to use formula 4.31 both for P and P^*. By the above, the small circle $C(P,\varepsilon)$ will project down to the small circle $C(P^*,\varepsilon)$. But in a Riemannian submersion, the differential of the projection can only diminish norms so that
$$\operatorname{length} C(P^*,\varepsilon) \leq \operatorname{length} C(P,\varepsilon) \tag{4.41}$$
and we are done.

Note 4.4.3.2 Today, we do not know which manifolds have Riemannian metrics with positive curvature or with nonnegative curvature. See §12.3 for that topic. ◆

4.4.3.5 Hypersurfaces in Euclidean Space

The curvature of submanifolds of Euclidean spaces can be easily computed. This can be achieved by nice formulas for the various presentations of submanifolds: chart-type parameterizations or solutions of equations, and of course the

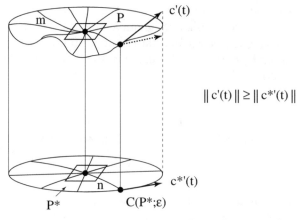

Fig. 4.51. $\|c'(t)\| \geq \left\|c^{*'}(t)\right\|$

case which is the intersection of these two: the graph presentation. See Dombrowski 1968 [454] for a unique (to our knowledge) complete formulary covering all possible ways of describing a submanifold: immersion, graph, equations. The general theory for submanifolds of $\mathbb{E}^n$, or more generally of submanifolds of Riemannian manifolds, i.e. the notions of the first and second fundamental forms and their relations (the so-called Gauß and Codazzi–Mainardi equations), is not treated in very many books. We know chapter VII of Kobayashi & Nomizu 1996 [827, 828], chapter XX (see section 20.14.8) of Dieudonné 1969 [446] and chapter 7 of Spivak 1979 [1155]. We will here content ourselves with the case of hypersurfaces in Euclidean space.

A very geometric way to see the sectional curvature of hypersurfaces in $\mathbb{E}^d$ is as follows. We just extend Gauß curvature by a simple trick. Given a plane P in the tangent space to our hypersurface, consider the three dimensional affine space generated by that plane and the normal to the hypersurface at the point of tangency. Locally, the section of the hypersurface by that three dimensional space is a surface. Its Gauß curvature at our point is nothing but the sectional curvature $K(P)$ (this was pointed out to us by Hermann Karcher).

Let $M^d \subset \mathbb{E}^{d+1}$ be a hypersurface. Like a surface in $\mathbb{E}^3$, it has a *second fundamental form* which is a quadratic form denoted by II and defined (up to ± 1) on its tangent spaces. Then it is not too hard to generalize Gauß's theorema egregium in the formula:

$$K(v,w) = II(v,v)II(w,w) - II(v,w)^2 \qquad (4.42)$$

In particular, if we diagonalize II with respect to the first fundamental form (which is $ds^2 = g$) and call its eigenvalues k_i the *principal curvatures* with $i = 1, \ldots, d$ we reveal:

$$K(e_i, e_j) = R_{ijij} = k_i k_j. \qquad (4.43)$$

4.5 Does the Curvature Determine the Metric?

If one compares this with the warning note 1.6.6.1 on page 61 one sees that Riemannian metrics on hypersurfaces of Euclidean spaces are very special. Moreover surfaces $M^2 \subset \mathbb{E}^3$ are also very special in the following sense. We saw on page 60 that to know how M^2 sits in $\mathbb{E}^3$, one needs to know more than the inner geometry. One must have at hand the first and the second fundamental forms. This is no longer true in dimension 3 or larger. Check as an exercise that formula 4.42 gives II as a function of K "in general". The hint is to look at equation 4.43. If you know the pairwise products of three numbers p, q, r

$$a = qr$$
$$b = rp$$
$$c = pq$$

then

$$p = \sqrt{\frac{bc}{a}}$$

etc.

This requires that no principal curvature vanishes, which explains the term "in general".

Theorem 41 (Rigidity of hypersurfaces) *In Euclidean spaces $\mathbb{E}^d$ with $d > 3$, generic hypersurfaces are determined by their inner geometry up to congruence.*

4.5 A Naive Question: Does the Curvature Determine the Metric?

We asked this question on page 203. We now treat it in detail. This question is rarely addressed. It might be because most people think the answer is always positive. In fact the answer is negative in general. This explains why there are difficulties in convergence problems for Riemannian metrics, problems to be met in §§12.4.2. Just one example: the normal coordinates, obtained by geodesics through a given point, are not the best ones for studying convergence. The best adapted coordinates will be the harmonic coordinates to be seen on page 267.

The negative answer is surprising, since a metric comprises $d(d+1)/2$ functions of d variables, while its curvature tensor comprises $d^2(d^2-1)/12$ functions of d variables, and moreover in view of formula 4.27 on page 203 it seems that we know all the second derivatives and can hope to rebuild the metric from them. Let us be more precise about the question of the curvature *determining* the metric. It turns out that this question splits into two completely different set of questions. This splitting is significant even for constant curvature manifolds. We have two natural questions:

1. Is every smooth map preserving the curvature tensor an isometry?
2. Is every smooth map preserving the sectional curvature an isometry?

We treat question 2 first. The answer is certainly negative for manifolds of constant curvature, since any map will preserve their curvature. The paradox here is that the metric in itself is known. This will naturally lead to another type of *determination* question below. Constant curvature manifolds are almost the sole exception as discovered in Kulkarni 1970 [839]. It is proven there that, when the sectional curvature is not constant and the dimension larger than 3, diffeomorphisms preserving the sectional curvature are isometries. This is not in contradiction with the above examples, because the definition of the sectional curvature involves not only the curvature tensor but also the metric. For dimension $d = 3$ there are examples of Yau and Kulkarni from 1974 of nonisometric diffeomorphisms preserving sectional curvature. For $d = 2$ this is trivial: any diffeomorphism preserving the level curves of the function $K : M \to \mathbb{R}$ does the trick, and of course need not be an isometry. For more on lower dimensions see Ruh 1985 [1075].

Enthusiasts of mathematical history will note that Riemann made a small mistake in his Habilitationsschrift of 1854, in characterizing flat manifolds; see a detailed historical account in Di Scala 2001 [443].

4.5.1 Surfaces

It is tempting to imagine that because of the theorema egregium, the Gauß curvature determines the metric. Of course, the converse is true. It is clear that the curvature (one function) is not enough to determine a metric (three functions). This question was considered in Darboux 1993 [433, 434] (or the other edition Darboux 1972 [429, 430, 431, 432]), and essentially finished in Cartan 1988 [321], pp. 322–323.

Proposition 42 (Cartan) *For the generic Riemannian metric on a surface, the differentials of K and $\|dK\|^2$ are linearly independent:*

$$dK \wedge d\|dK\|^2 \neq 0$$

on a dense open set.

Theorem 43 (Cartan) *Given two surfaces with Riemannian metrics, so that the functions K and $\|dK\|^2$ have everywhere independent differentials, a map between these surfaces is an isometry precisely when it preserves the four functions*

$$I_1 = K$$
$$I_2 = \|dK\|^2$$
$$I_3 = \langle dK, dI_2 \rangle$$
$$I_4 = \|dI_2\|^2.$$

See more in Cartan 1988 [321]; for example a characterization of surfaces of revolution by only $\|dK\|^2$ and the Laplacian of the curvature ΔK, under the assumption that there is a relation

$$F\left(K, \|dK\|^2\right) = 0 .$$

4.5.2 Any Dimension

To attack the subtle question 1 we need a formula. In general coordinates:

$$R_{ijkh} = \frac{1}{2}\left(\partial_i\partial_k g_{jh} + \partial_j\partial_h g_{ik} - \partial_i\partial_h g_{jk} - \partial_j\partial_k g_{ih}\right) + Q(g, \partial g, \partial g) \quad (4.44)$$

with the obvious notations for the second derivatives of g and where Q denotes a term which is quadratic in the first derivatives of g. We compare it with formula 4.27 on page 203. At the origin of normal coordinates:

$$\partial_i\partial_j g_{kh} = \frac{1}{6}\left(R_{ikhj} + R_{ihjk}\right) . \quad (4.45)$$

It is now clear from 4.44 that one cannot recover from the R_{ijkh}, which form a total of $d^2(d^2-1)/12$ numbers, all of the second derivatives $\partial_i\partial_j g_{kh}$, which form a total of $(d(d+1)/2)^2$ numbers. The apparent contradiction with 4.45 is explained by the fact that choosing normal coordinates involves extra information: one has integrated the geodesic equation. This forces extra relations between the curvature and the metric. This will agree completely with the philosophy of Élie Cartan to be seen on page 249.

This does not answer question 1: is there enough room between $(d(d+1)/2)^2$ and $d^2(d^2-1)/12$ to find a nonisometric map between two metrics which still preserves the whole curvature tensor? The subject was initiated quite recently: we know now many examples of nonisometric Riemannian manifolds admitting diffeomorphisms preserving their respective curvature tensors. Various authors managed to smartly squeeze into the room left over in the interval

$$\left[\left(\frac{d(d+1)}{2}\right)^2, \frac{d^2(d^2-1)}{12}\right] .$$

Complete mastery of this business has not been achieved today; see the books Tricerri & Vanhecke 1983 [1200], Berndt, Tricerri, & Vanhecke 1995 [180], and Boeckx, Kowalski, & Vanhecke 1996 [214] which can be used as surveys. See also Prüfer, Tricerri, & Vanhecke 1996 [1042] and the bibliographies of the references given.

A fundamental fact, to be seen in §6.4 is that bounds on sectional curvature, upper and lower, yield perfect geometric control in both directions. This does not contradict the above. We will come back to this philosophy in §§6.4.3, where one will see that the best adapted coordinates for some situations are not the normal ones but the so-called harmonic coordinates. This is a case where the analyst trumped the geometer.

A completely different meaning can be attributed to the question of the curvature determining the metric. Look for metrics whose curvature tensor satisfies at every point some purely algebraic condition. The first case is that of constant sectional curvature, where we know the answer is unique and perfect: theorem 69 on page 252. The next example: suppose that the curvature tensor is the same as that of a complex projective space. One would expect then in some cases isometry (local) and in general at least local homogeneity, as well as a description of the moduli. Since the founding Ambrose & Singer 1958 [30] and Singer 1960 [1144] the field today uses various definitions and enjoys many results. The books Tricerri & Vanhecke 1983 [1200], Boeckx, Kowalski & Vanhecke 1996 [214] can be used as surveys. See also Prüfer, Tricerri & Vanhecke 1996 [1042]. We only give the result in Tricerri & Vanhecke 1986 [1201], which is exemplary of simplicity both in statement and in significance:

Theorem 44 *If the curvature tensor of a Riemannian manifold is at every point the same as one of an irreducible symmetric space, then the manifold is locally symmetric and locally isometric to this model.*

The result is local. It uses a formula in Lichnerowicz 1952 [864] which expresses the Laplacian of the full square norm of the curvature tensor as

$$-\frac{1}{2}\Delta\left(\|R\|^2\right) = \|DR\|^2 + Univ(R,R,R) + Q(D\operatorname{Ricci}),$$

where $Univ(R, R, R)$ is a universal cubic form in the curvature tensor and $Q(D\operatorname{Ricci})$ is a quadratic form in the covariant derivative of the Ricci curvature. We have not yet defined the Ricci curvature, the covariant derivative D or the Laplacian Δ, but the impatient reader can turn to chapter 15. The hypothesis immediately implies that $D^2\operatorname{Ricci} = 0$ and that $\|R\|^2$ is constant as is $Univ(R, R, R)$, because it is obvious in the symmetric case which is characterized by $DR = 0$. It seems to us that this basic formula is still not used as much as one would have expected. However, it was used for studying manifolds with positive curvature operator, for example in Gallot & Meyer 1975 [543], Tachibana 1974 [1176], and Hamilton 1986 [679], and it was much used in Anderson 1989 [37].

A recent text on this subject is Ivanov & Petrova 1998 [760]. For the behavior of curvature mixed with parallel transport, see Ambrose's problem on page 704.

4.6 What are Abstract Riemannian Manifolds?

We saw in theorem 30 on page 161 that there are no more abstract smooth manifolds than submanifolds of $\mathbb{E}^n$. But submanifolds of $\mathbb{E}^n$ inherit a Riemannian metric from their embedding. For surfaces in $\mathbb{E}^3$ that was our initial motivation. So it is natural to ask the converse: let (M, g) be any Riemannian manifold. We know that, as a manifold, M can inherit its structure from an embedding $\phi : M \to \mathbb{E}^n$. But there is of course no reason for the induced Riemannian metric to be identical with g. If it is, we will say that ϕ is an *isometric embedding*. Does every (M, g) admit an isometric embedding into some $\mathbb{E}^n$?

Before answering this natural question, we should say that it is not a crucial one: we have made evident that it is most natural to "see" Riemannian manifolds abstractly.

4.6.1 Isometrically Embedding Surfaces in $\mathbb{E}^3$

Let us look first at the case of surfaces. Can we hope that every (M^2, g) is isometrically embeddable in $\mathbb{E}^3$? We remarked in theorem 5 on page 56 that any compact surface immersed in Euclidean space $\mathbb{E}^3$ has at least one point of positive Gauß curvature. So any compact abstract (M^2, g) with nonpositive Gauß curvature will never embed isometrically in $\mathbb{E}^3$. But there is a beautiful (and extremely hard to prove) theorem of Alexandrov, Weyl, Nirenberg and Pogorelov which says that an abstract surface (S^2, g) with positive Gauß curvature is always isometrically embeddable in $\mathbb{E}^3$ (see figure 4.52). Moreover the embedding is unique up to rigid motion, because of the rigidity result quoted in §§3.4.1. A recent survey of this problem, called Weyl's problem, is in I.2.1 of Burago & Zalgaller 1992 [284]. A recent reference is Guan & Li 1994 [662].

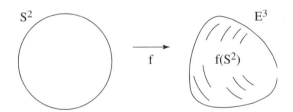

Fig. 4.52. Isometric embedding with some metric with $K > 0$

4.6.2 Local Isometric Embedding of Surfaces in $\mathbb{E}^3$

One may feel that there should be room to isometrically embed small pieces of a surface into $\mathbb{E}^3$, for instance some neighborhood of any point. There has been substantial progress on this question, but it is still not completely understood. Indeed, one can produce such local isometric embeddings near points of positive or negative Gauß curvature. In 1971, Pogorelov found a twice continuously differentiable metric (standard notation: C^2) on a topological disk in $\mathbb{E}^2$ with Lipschitz curvature, such that no neighborhood of the center of the disk can be isometrically embedded in $\mathbb{E}^3$. On the other hand, recently Lin has found partial positive results: if $K \geq 0$ and the metric has ten continuous derivatives, or if $dK(m) \neq 0$. The difficulty of isometrically embedding surfaces into $\mathbb{E}^3$ (even locally) could have been guessed from the paradox mentioned in note 3.1.1.2 on page 109: there is a forced loss of differentiability showing that the differential system to solve is quite intricate. See part III of Burago & Zalgaller 1992 [284] and also Gromov 1986 [620].

4.6.3 Isometric Embedding in Higher Dimensions

The first major event in embedding Riemannian manifolds was the Janet theorem.

Theorem 45 (Cartan–Janet, 1926) *Every real analytic Riemannian manifold of dimension n can be locally real analytically isometrically embedded into $\mathbb{E}^{n(n+1)/2}$.*

All was quiet until:

Theorem 46 (Nash, 1956) *Every smooth Riemannian manifold of dimension n can be smoothly isometrically embedded in $\mathbb{E}^N$ where*

$$N = \frac{(n+2)(n+3)}{2}.$$

The price to pay is in the dimension N. Various authors have since improved N; see again Gromov 1986 [620] for a recent reference. We now know that abstract Riemannian manifolds are no more general than submanifolds of the various $\mathbb{E}^N$.

In very low differentiability, Nash in 1954 and Kuiper in 1955 obtained surprising results:

Theorem 47 (Nash–Kuiper) *Any continuous embedding of a Riemannian manifold can be deformed into a C^1 isometric embedding. In particular, any n dimensional Riemannian manifold embeds C^1 isometrically into $\mathbb{E}^{2n+1}$.*

For example, this implies that the rigidity theorem in §§3.4.1 really needs a C^2 surface. See Bleecker's result in §§§3.4.1.2.

5 A One Page Panorama

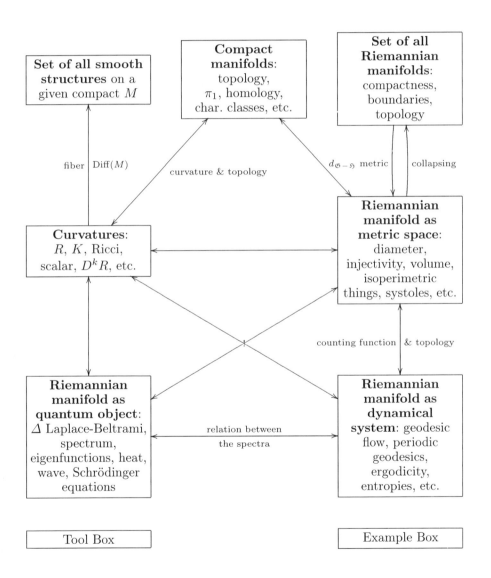

6 Riemannian Manifolds as Metric Spaces and the Geometric Meaning of Sectional and Ricci Curvature

Contents

6.1	**First Metric Properties**	**222**
	6.1.1 Local Properties	222
	6.1.2 Hopf–Rinow and de Rham Theorems	226
	6.1.3 Convexity and Small Balls	229
	6.1.4 Totally Geodesic Submanifolds	231
	6.1.5 Center of Mass	233
	6.1.6 Examples of Geodesics	235
	6.1.7 Transition	238
6.2	**First Technical Tools**	**239**
6.3	**Second Technical Tools**	**248**
	6.3.1 Exponential Map	248
	6.3.2 Space Forms	251
	6.3.3 Nonpositive Curvature	254
6.4	**Triangle Comparison Theorems**	**257**
	6.4.1 Bounded Sectional Curvature	257
	6.4.2 Ricci Lower Bound	262
	6.4.3 Philosophy Behind These Bounds	267
6.5	**Injectivity, Convexity Radius and Cut Locus**	**268**
	6.5.1 Definition of Cut Points and Injectivity Radius	268
	6.5.2 Klingenberg and Cheeger Theorems	272
	6.5.3 Convexity Radius	278
	6.5.4 Cut Locus	278
	6.5.5 Blaschke Manifolds	285
6.6	**Geometric Hierarchy**	**286**
	6.6.1 The Geometric Hierarchy	289
	6.6.2 Constant Sectional Curvature	290
	6.6.3 Rank 1 Symmetric Spaces	295
	6.6.4 Higher Rank Symmetric Spaces	296
	6.6.5 Homogeneous Spaces	296

6.1 First Metric Properties

6.1.1 Local Properties and the First Variation Formula

We want to study the metric of a Riemannian manifold. The first tasks to address are:

1. to compute the metric d as defined by equation 4.13 on page 174 (namely $d(p,q)$ is the infimum of the lengths of curves connecting p to q)
2. to determine if there are curves realizing this distance (called *segments* or *shortest paths* or *minimal geodesics* according to your taste) and
3. to study them.

It will become clear that these three questions are not independent nor is it wise to treat them in this order. Moreover except for extremely few Riemannian manifolds, explicit answers are not known. We will have to content ourselves with results of a general nature: bounds, inequalities, etc. We will try to have estimates or, if you prefer, some kind of control in terms of the curvature. We select curvature as our source of leverage because unlike distances and segments, curvature is relatively easy to compute, or at least to estimate. The deep reason for the near impossibility of computing distances is that they are given by an infimum definition. The curvature is given by computing derivatives of the metric tensor g_{ij}. Influence of curvature on global metric geometry will be the object of §6.4 and §6.5.

Most of the content of the present section belongs to the foundations of Riemannian geometry and is systematically treated in every one of the references we gave. When it is not, we will give more precise references.

We are going to follow the route of §§3.1.4 but in the general case and with more details. It is convenient to introduce some technical notions and notations. We set:

$$0_m \quad \text{is the zero vector of } T_m M$$
$$U_m M = \{v \in T_m M \mid \|v\| = 1\} \text{ is the } \textit{unit tangent sphere} \text{ to } M \text{ at } m$$
$$B(0_m, r) = \{w \in T_m M \mid \|w\| < r\} \text{ is a } \textit{metric ball} \text{ in } T_m M \tag{6.1}$$
$$B(m, r) = \{p \in M \mid d(m, p) < r\} \text{ is a } \textit{metric ball} \text{ in } M$$

From §§4.4.1 we know that segments can only be found among geodesics. We will always parameterize geodesics with unit speed unless otherwise stated. We know from §§4.4.1 that there is a unique geodesic γ_v with any initial velocity $v = c'(0) \in U_m M$. But in general this geodesic is only defined for small values of the parameter. Nonetheless this enables us to define a map:

$$\exp_m : W_m \subset T_m M \to M \tag{6.2}$$
$$w \mapsto \gamma_{w/\|w\|}(\|w\|)$$

called the *exponential map* at m.

6.1 First Metric Properties   223

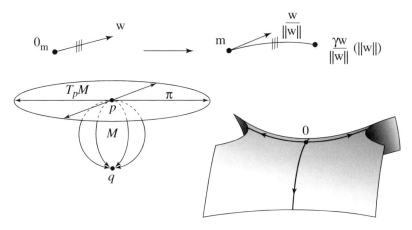

Fig. 6.1. The exponential map

Geometrically, one draws $\exp_m(w)$ by drawing the geodesic starting from m in the direction of w (and with unit speed) and traveling along it up to a length equal to $\|w\|$. The subset W_m is defined as the largest for which this process is well defined. From a simple compactness argument we know that for every $m \in M$ there exists some $r' > 0$ with $B(0_m, r') \subset W_m$. One is very eager to get some knowledge of the best such r' but this will need more work: see §6.5.

Through the exponential map, orthonormal coordinates in the Euclidean space $T_m M$ yield nice coordinates in M locally around m. These coordinates are the ones called *normal* in §§4.4.1. Not only do they have an elegant geometric definition, they are also very useful in resolving local geometric problems. But very recently it was discovered that there are better coordinates for purposes of analysis on a Riemannian manifold. They are the ones called *harmonic*, and will be discussed on page 267, on page 623 and on page 628.

The exponential map is a fundamental concept in Riemannian geometry. Geometrically it depicts how the geodesics issuing from m wrap or stretch around m. For the sphere they wrap, while in hyperbolic geometry they stretch, because they diverge exponentially. If we know the exact rate of variation of the geodesics issuing from m then we know the Riemannian metric—at least locally around m. Equivalently: the knowledge of g is equivalent to that of the measurement of the derivative of the exponential map. This derivative will be studied in §6.3. It is also essential to know when this derivative remains injective; if it is not then we have some kind of singularity. The control of that singularity will be considered in §6.5.

The differential of $\exp_m$ at the origin 0_m is the identity map, so we know, by the inverse mapping theorem, that by suitably restricting ourselves to $B(0_m, r)$ for some $r < r'$ on this last ball $\exp_m$ will be a diffeomorphism onto its image. As in §§3.1.4 we claim again that:

Fig. 6.2. The exponential map is in general defined only in a star-shaped part of $T_m M$: for a given ξ one stops when γ_ξ can not be prolonged

Proposition 48 *For any such r*

$$\exp_m B\left(0_m, r\right) = B\left(m, r\right)$$

In the proof, two major consequences are obtained:

Corollary 49 *The distance $d\left(p,q\right)$ defined in equation 4.13 on (M,g) is really a metric on M.*

Corollary 50 *For any point $q \in B\left(m,r\right)$ there is one and only one segment between m and q, namely the geodesic whose velocity is $\exp_m^{-1}(q)$.*

The spirit of the proof is that of §§3.1.4 and can be found in any of the books we quoted on Riemannian geometry. With one proviso: recall from page 174 and from the example in note 4.1.1.1 on page 145 one certainly requires the hypothesis that manifolds are Hausdorff if one wants to know that $d\left(p,q\right) > 0$ unless $p = q$. Hausdorffness is used in the custom passage theorem (see figure 3.11 on page 117), because the image of the compact ball which is the closure of $B\left(0_m, r\right)$ must be compact downstairs in M. Surprisingly enough, to the best of our knowledge, no book on Riemannian geometry explicitly mentions this subtlety, with one exception: see (iii) of §2.91 of Gallot, Hulin, & Lafontaine 1990 [542].

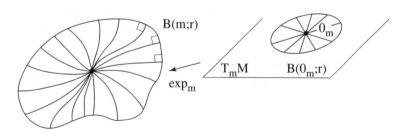

Fig. 6.3. Balls about 0_m go to balls about m under exp

6.1 First Metric Properties   225

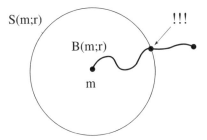

Fig. 6.4. The custom passage theorem: one cannot leave the ball $B(m,r)$ without crossing the boundary $\partial B(m,r)$

Two other main consequences: first the topology of the manifold M is identical with the metric topology. Second: for radii small enough, metric balls are diffeomorphic to $\mathbb{R}^d$.

As in §§3.1.4 to prove this we need the first variation formula:

$$\frac{d\delta}{dt} = d'(t) \left.\frac{\partial \gamma}{\partial s}\right|_{s=b(t)} - c'(t) \left.\frac{\partial \gamma}{\partial s}\right|_{s=a(t)} \tag{6.3}$$

Following with no real change the route of §§3.1.4, one proves proposition 48.

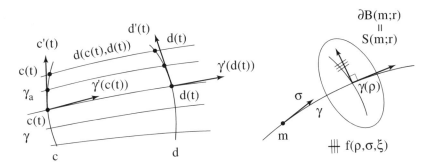

Fig. 6.5. First variation

A complete proof consists just in using the language of §15.2, in particular the fact that geodesics γ verify $D_{\gamma'}\gamma' = 0$. The end points of the geodesics emanating from a fixed point and drawn for a fixed length describe curves orthogonal to those geodesics (this is often called Gauß's lemma in books). There is a small change in equation 3.8 on page 116 describing g in polar coordinates:

$$g = d\rho^2 + f^2(\rho, \sigma, \xi)d\sigma^2 \tag{6.4}$$

where $d\sigma^2$ denotes the Riemannian metric of the unit sphere $U_mM \subset T_mM$ and ξ denotes a vector tangent to U_mM at σ.

226   6 Metric Geometry and Curvature

Recall that $T_m M$ has by definition a Euclidean structure. The function f is always positive because our restriction of $\exp_m$ is a diffeomorphism. Another way to interpret $f(\rho, \sigma, \xi)$ is to say that it measures the derivative of the exponential map.

We note in passing some consequences of the first variation. The first is the strict triangle inequality. If γ (resp. δ) is a segment from p to q (resp. from q to r) then

$$d(p,r) < d(p,q) + d(q,r) \quad \text{unless} \quad \gamma'(q) = \delta'(q) . \quad (6.5)$$

Consequently if γ and δ are two distinct segments from p to q :

Proposition 51 *After passing q, the geodesics γ and δ are no longer segments.*

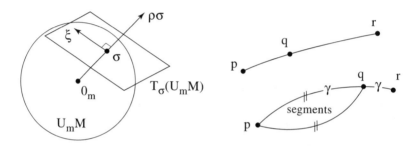

Fig. 6.6. γ is no longer a segment from p to r

Various examples of geodesics and segments will be given in §§6.1.6.

6.1.2 Hopf–Rinow and de Rham Theorems

Independently from asking them to be segments, it is important to know when geodesics are defined—or why they are not—for every t. When does one have $W_m = T_m M$? Surprisingly, the answer to this will also solve the basic question: when can one be sure that every pair of points is joined by at least one segment?

A clear answer appeared first in Hopf & Rinow 1931 [733]. Many people were aware of various results, but general topology was not very clear either in those times. The Hopf–Rinow paper proved the result below only in dimension two, and in great detail, but nevertheless for abstract Riemannian surfaces. Their proof was valid without any change in any dimension as remarked in Myers 1935 [954]. This applies to many other results in geometry before more recent times: mathematicians were principally interested in 2 and 3 dimensions, even while they were working in very abstract contexts. However in the case of Hopf, it was surprising, since he was often interested in general dimensions; see for example §12.1.

Recall the notion of *complete metric space*: a metric space is said to be *complete* when every Cauchy sequence converges to a point. Compact spaces are automatically complete (but the converse is of course not true).

Bear in mind the example of the plane with one point deleted. Two bad things happen: some geodesics cannot be extended to infinity, and some pairs of points cannot be connected by a geodesic. More generally, delete closed subsets of any "nice" manifold. This might make you feel that the two issues are connected. Moreover you can also guess that the hole we cut is an obstacle for extension to infinity of some geodesic emanating from each point. Conversely, after drawing a few pictures you can imagine that if geodesics emanating from a point are all extendable up to infinity, then this is valid for every other point. You might also consider these holes as some kind of "boundary" for the Riemannian manifold under consideration.

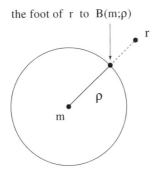

Fig. 6.7. The foot of r to $B(m, \rho)$

The relation to completeness is also foreseeable. A Cauchy sequence will occur entirely at a finite distance from a fixed point m, and then almost the entire sequence will be within a compact ball about m. Pick a Cauchy sequence along a geodesic. The completeness implies that you can keep going a little further along it.

Theorem 52 (Hopf–Rinow) *These four conditions on a connected Riemannian manifold M are equivalent:*

1. *For some $m \in M$, the map $\exp_m$ is defined on all of $T_m M$.*
2. *For every $m \in M$, $\exp_m$ is defined on all of $T_m M$.*
3. *M is complete for its Riemannian metric.*
4. *Closed and bounded sets of M are compact.*

Moreover, any one of these conditions implies that every pair of points is joined by at least one segment.

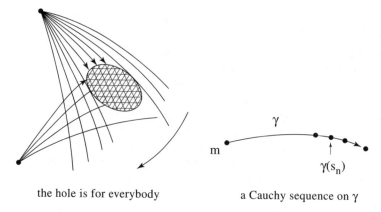

Fig. 6.8. (a) The hole is visible everywhere. (b) A Cauchy sequence on γ

The rigorous proof is a little lengthy, and is based on the foot trick seen in figure 6.7 on the previous page; see any reference on Riemannian geometry, and also see Gromov 1999 [633].

From now on, every Riemannian manifold we discuss will be assumed complete.

For any metric space M let us define the *diameter* of M to be

Definition 53
$$\operatorname{diam} M = \sup \{d(p, q) \mid p, q \in M\}.$$

Proposition 54 *A complete Riemannian manifold M is compact if and only*

$$\operatorname{diam} M < \infty.$$

Beware that $\exp_m$ might be defined on all of $T_m M$ even though large balls about m may not be diffeomorphic to Euclidean balls, and $\exp_m$ may not be a diffeomorphism between $T_m M$ and M.

A useful corollary of the Hopf–Rinow theorem:

Corollary 55 *Assume that M is a complete Riemannian manifold, and that at a point $m \in M$ the exponential map $\exp_m$ has full rank everywhere. Then it is a covering map. In particular, if M is simply connected, then $\exp_m$ is a diffeomorphism.*

The disaster of the exponential map only reaching a portion of the manifold cannot happen here because of the extendability of geodesics. The proof goes with the following trick: endow $T_m M$ with the inverse metric $\exp_m^* g$ (this makes sense because $\exp_m$ is of maximal rank). This Riemannian manifold $(T_m M, \exp_m^* g)$ is complete because radial geodesics through 0_m exist, extended to any length, since they are just straight lines covering the geodesics downstairs. Both $(T_m M, \exp_m^* g)$ and (M, g) being complete and the map being by construction a local isometry, the covering axioms are easily checked. We will use this corollary heavily later on in theorems 69 and 72. For a detailed proof see for example 2.106 in Gallot, Hulin & Lafontaine 1990 [542], or lemma 1.32, page 35, in Cheeger & Ebin 1975 [341].

6.1.2.1 Products

Remember the notion of Riemannian product

$$(M, g) \times (N, h) = (M \times N, g \times h)$$

in §§§4.3.3.1. Suppose now that some (M, g) is locally reducible. One imagines that a principle of analytic continuation will easily yield the:

Theorem 56 (de Rham) *A complete Riemannian manifold which is locally reducible and simply connected is a Riemannian product.*

In fact the de Rham theorem is quite subtle and lengthy to prove. To get a covering is quite easy; the difficulty lies in proving the injectivity. It is better phrased and understood in the language of holonomy; see theorem 394 on page 642. Most books omit the proof of the de Rham theorem, but not Sakai 1996 [1085]. The simple connectedness is of course necessary; look at a flat torus defined by a nonrectangular lattice; see figure 6.9.

Fig. 6.9. The quotient torus is not a Riemannian product

6.1.3 Convexity and Small Balls

Metric balls of small radius are diffeomorphic to Euclidean space, which is very interesting because it is a strong implication from the metric to the topology. But if you look at two such balls it is not true in general that their intersection is diffeomorphic (or even homeomorphic) to Euclidean space, as in figure 6.10 on the following page.

Balls intersect nicely in $\mathbb{E}^d$ (i.e. in balls, up to diffeomorphism) but on general Riemannian manifolds, balls can intersect poorly. You can guess from the picture that, unlike those in $\mathbb{E}^d$, balls in general Riemannian manifolds are not always convex.

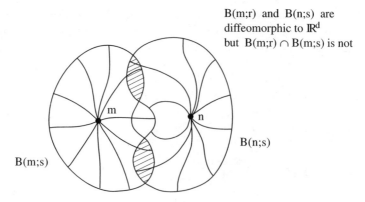

Fig. 6.10. $B(m,r)$ and $B(n,s)$ are diffeomorphic to $\mathbb{R}^d$, but $B(m,r) \cap B(n,s)$ is not

Let us say that a set in a Riemannian manifold is *(totally) convex* if for any pair of points in this set, *every* segment connecting these two points belongs to this set. The standard sphere is a typical example: metric balls with radius greater than $\pi/2$ are not convex. But you might guess:

Proposition 57 *Every point of a Riemannian manifold lives at the center of a convex ball such that any two points in that ball are joined by a unique segment contained in the ball.*

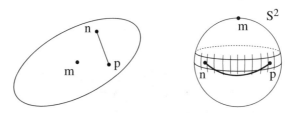

Fig. 6.11. Convexity and the lack of it

The proof is not too difficult; see for example Cheeger & Ebin 1975 [341] 5.14, Gallot, Hulin & Lafontaine 1990 [542] (the remark after 2.90), section 4 of chapter 3 of do Carmo 1992 [452]. In particular, any Riemannian manifold admits a covering by convex balls. This is particularly useful when studying the algebraic topology of a Riemannian manifold, for example under curvature conditions. But of course one will need to have a priori estimates from below for the radii of these convex balls: see §§6.5.3 below and the references there, and chapter 12 for applications. Let us just recall note 3.4.5.3 on page 139: the existence of a convex covering yields an easy proof of the existence of triangulations for any compact surface.

Convexity in Riemannian manifolds has been playing an increasingly important role since the 1960's. In the definition of convexity, one can ask for a weaker property: demand only that any two points of the set under consideration are joined by a segment belonging to that set. An example is a hemisphere of a standard sphere.

We will see in §§§12.3.1.3 and in §§12.3.4 that an essential observation concerning manifolds of nonnegative or of nonpositive sectional curvature is the convexity of suitable geometrically defined functions: standard distance functions or their extension, Busemann functions with a suitable sign.

Note 6.1.3.1 An extraordinarily simple question is still open (to the best of our knowledge). What is the convex envelope of three points in a 3 or higher dimensional Riemannian manifold? We look for the smallest possible set which contains these three points and which is convex. For example, it is unknown if this set is closed. The standing conjecture is that it is not closed, except in very special cases, the question starting typically in $\mathbb{CP}^2$. The only text we know of addressing this question is Bowditch 1994 [248]. ♦

Fig. 6.12. Some kind of lens

Let us develop a feeling for this problem. First in dimension two, and for three points close enough, the convex envelope is a triangle whose sides are segments. Now in higher dimensions: consider some point m and geodesics emanating from m and with velocity contained in some two dimensional subspace of $T_m M$. These build up at least locally a smooth surface $N \subset M$. But if you pick two points $q, r \in N$ (even as close as you like to m) the segment from q to r will not be contained in N. The segment will sit inside N when M has constant curvature (prove it), but not for generic metrics.

6.1.4 Totally Geodesic Submanifolds

This leads to the notion of *totally geodesic submanifold*: a submanifold N of a Riemannian manifold M is said to be *totally geodesic* if every geodesic starting from a point in N and tangent to N at that point remains always contained in N. Note that this is equivalent to the fact that, locally, the metric space

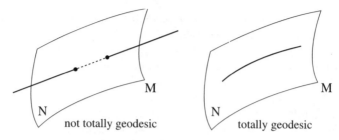

Fig. 6.13. (a) Not totally geodesic (b) Totally geodesic

structure induced in the sense of metric spaces coincides with the metric structure coming canonically from the induced Riemannian metric on N. Another equivalent condition is the fact that parallel transport in M, applied to curves lying in N, preserves the splitting $TN \oplus TN^\perp = TM$. A generic Riemannian manifold does not admit any such submanifold, except for curves (geodesics). Totally geodesic submanifolds appear only in special contexts; here are some. Space forms are the manifolds in which totally geodesic submanifolds are the most plentiful: see theorem 40 on page 208. To grasp the proof: the set of fixed points of a involutive isometry of (M, g) is always totally geodesic. The reader can prove this using the local uniqueness of segments.

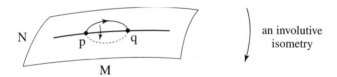

Fig. 6.14. An involutive isometry

Another circumstance where totally geodesic submanifolds appear is in Riemannian products: any two dimensional submanifold which is the product of an horizontal geodesic with a vertical geodesic is a totally geodesic surface. Therefore there are many of them. For example, the product of two periodic geodesics will be a flat two dimensional torus in the product. But in products, one also has completely different totally geodesic submanifolds: all of the horizontal and the vertical submanifolds: $M \times \{n\}$ and $\{m\} \times N$ in $M \times N$. Curiously similar is the case of a symmetric space of rank larger than 1: see §§4.3.5. But the difference between the two cases is not completely mastered today; see for example the simplicial volume on page 516.

The first variation formula implies that the elements of a one parameter family of totally geodesic submanifolds are all locally isometric. The isometries are obtained from the orthogonal trajectories. This remark implies the geometric characterization of space forms seen in note 4.3.2.2 on page 180. Totally geodesic submanifolds can be used in many instances to construct certain manifolds; see for example the Bazaikin manifolds of positive curvature on page 578 and the counterexamples of Gromov and Thurston in §§12.2.4.

The naive geometer will ask what in Riemannian geometry generalizes affine (Euclidean) or projective subspaces. The basic remark is that a general Riemannian manifold admits no submanifold which is stable under geodesy, i.e. so that geodesics in it are geodesics in the ambient space. Such submanifolds are exactly the totally geodesic ones. Manifolds with many totally geodesic submanifolds are extremely few and among them are the symmetric spaces of rank greater than one; see §§§4.3.5.2. We now know that having lots of "flats" (totally geodesic submanifolds) is very strong property.

Theorem 58 (Cartan) *A Riemannian manifold with a totally geodesic submanifold of dimension k passing through any point with any specified k dimensional plane as tangent space must be a space form.*

In fact the proof is easy, by simply taking a few derivatives, requiring only a little algebra to organize the repeated differentiation.

6.1.5 The Center of Mass

It appears as if there is no canonical method to fill up a triangle, or more general simplex, in a generic Riemannian manifold. But this is not true—the problem is solved by the notion of center of mass, modeled on Euclidean geometry. This center of mass exists only locally, except in special examples, for instance simply connected complete manifolds of negative curvature; see page 256 where it seems that its introduction by Élie Cartan was its first appearance in Riemannian geometry. The next appearance seems to have been in the unpublished proof by Eugenio Calabi of the differentiable pinching theorem; see page 555. Chapter 8 of Buser & Karcher 1981 [296] is a foundational text in this area. In Euclidean space one defines the center of mass of a finite set of points $\{x_i\}_{i=1,\ldots,n}$ as the sum

$$x = \frac{1}{n}(x_1 + \cdots + x_n) \ .$$

This is equivalent to asking the vector sum

$$\overrightarrow{xx_1} + \cdots + \overrightarrow{xx_n}$$

to vanish. But more important is the property (discovered by Appolonius of Perga) that x is in fact the (unique) point minimizing the function

$$y \mapsto \sum_{i=1}^{n} d(x, x_i)^2 \ .$$

This can be generalized by assigning to each x_i a weight $\lambda_i > 0$ with $\sum_i \lambda_i = 1$. The resulting points for all possible weights fill out the convex closure of the set $\{x_i\}$. Using measure and integration to replace the sums, this can be extended to any mass distribution on a compact subset of $\mathbb{E}^d$.

This extends without any problem to general Riemannian manifolds provided one stays within convex balls. Inside such a convex ball, we choose a compact subset $A \subset M$, and a mass distribution da on A of total mass 1.

Definition 59 *A function $f : M \to \mathbb{R}$ on a Riemannian manifold is (strictly) convex if its restriction to any geodesic is (strictly) convex as a function of one variable.*

Proposition 60 (Cartan) *The function*

$$f : m \in M \mapsto \frac{1}{2} \int_A d(m,a)^2 \, da$$

is strictly convex, achieves a unique minimum at a point called the center of mass *of A for the distribution da. Moreover this point is characterized by being the unique zero of the gradient vector field*[1]

$$\nabla f(x) = \int_A \exp_x^{-1}(a) \, da$$

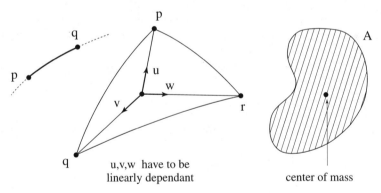

Fig. 6.15. (a) Center of mass of two points. (b) u, v, w have to be linearly independent. (c) center of mass

When A consists in two points, as in figure 6.15 (a), each mass distribution has center belonging to the segment whose extremities are these two points,

---

[1] The *gradient vector field* ∇f is defined to be the vector field dual to the differential df (which is a 1-form) in the sense that

$$g(\nabla f, v) = df(v)$$

for any tangent vector v.

and every point on that segment is the center of mass of some distribution. When we have three points, typically in dimension larger than 2, centers of mass fill in a unique way the three side segments by a "triangle". Points x inside the triangle are characterized by the fact that the tangent vectors at x to the three segments joining x to the vertices are linearly dependent. The center of mass technique will be used in various places, especially in §§12.4.1. The crucial point is that from pure metric information one gets implications on tangent vectors; this will be essential to get differentiable maps.

For the center of mass and some applications one can read Karcher 1989 [780], Buser & Karcher 1981 [296] chapter 8, Cheeger, Müller, & Schrader 1984 [350]. But the existence of a unique center of mass in the large for manifolds with nonpositive curvature was proven and used by Élie Cartan back in the 1920's. He used this to prove that maximal compact subgroups of Lie groups are always conjugate. The general case (but locally only of course, think of the standard sphere) was employed by Calabi in the unpublished result quoted above.

6.1.6 Explicit Calculation of Geodesics of Certain Riemannian Manifolds

In most Riemannian manifolds, geodesics are not computable nor are the distances between two points. We now list practically all Riemannian manifolds for which one can carry out such computations more or less explicitly.

In a Riemannian product manifold $(M \times N, g \times h)$ geodesics are exactly the curves which project to geodesics in both factors.

We recall the case of surfaces of revolution in $\mathbb{E}^3$, and also the complete description of geodesic behavior on ellipsoids, in §§1.6.2. For ellipsoids (as well as for the other hyperquadrics) in $\mathbb{E}^{d+1}$

$$\sum_{i=1}^{d+1} \frac{x_i^2}{a_i^2} = 1 \tag{6.6}$$

one knows how to integrate the geodesic equation and, in some sense, one knows their global behavior which is that of a completely integrable dynamical system; see for example Knörrer 1980 [824] (and the references there) and Paternain & Spatzier 1993 [1004]. There are relations between the geodesic flow on the ellipsoid and solitons.

Geodesics in domains of intersection with the confocal quadrics can be only either dense or periodic, with one exception: geodesics going through the umbilic points; see §§1.6.2, §§10.2.2 and §10.9. for an interesting use of these geodesics, namely to produce chaos on surfaces very close to an ellipsoid.

Some foolish people will say that everything is known about geodesics on ellipsoids of any dimension. But to our knowledge the cut locus is unknown (see section §§6.5.4 below for its definition) when $d \geq 3$. For $d = 2$, we saw this on page 39.

236   6 Metric Geometry and Curvature

Because Riemannian coverings are local isometries, geodesics upstairs project down to geodesics. Conversely, lifts of geodesics downstairs are geodesics in the covering. For example, in flat tori of every dimension, geodesics have a very simple behavior: they are either periodic or everywhere dense in a flat torus submanifold. In the case of the square tori (those glued together by unit translations along coordinate axes) the distinction is just whether the components of the initial velocity have rational or irrational ratios. How quickly does such a geodesic fill up the torus (or the square, the cube), or equivalently how quickly does your computer screen turn black? The answer in the plane is given by the continued fraction expansion of the slope. The *golden ratio* gives the fastest possible darkening. For the cube (three-dimensional case) it is an open problem to generalize the golden ratio. Equivalently: simultaneous approximation of two (or more) irrational numbers by rational ones is not understood. The dichotomy of geodesic deportments in the square torus is exceptional—in general one has more than two types of geodesics; see chapter 10. Another example is that all periodic geodesics in a Riemannian covering will project down to periodic geodesics, the converse being dependent on the homotopy class of the periodic geodesic in the covered (downstairs) manifold.

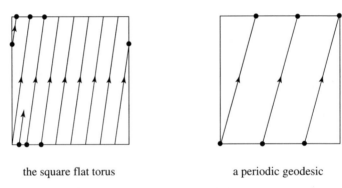

the square flat torus   a periodic geodesic

Fig. 6.16. (a) The square flat torus. (b) A periodic geodesic

In standard spheres geodesics are all periodic with the same period 2π and the ones emanating from a point m all meet again π units of time later at its antipode $-m$. This completely solves the problems of finding the distance between points, and finding all of the segments in S^d. To study segments of the real projective space $\mathbb{RP}^d$, one just projects down segments from S^d. The resulting geodesics are all periodic with periodic π. But this time those emanating from some given point never meet again except back at this same point. As a consequence, the diameter of $\mathbb{RP}^d$ is equal to $\pi/2$ and two points p, q are joined by a unique segment when $d(p,q) < \pi/2$ and by two when $d(p,q) = \pi/2$.

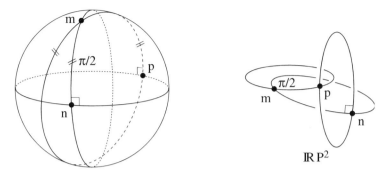

Fig. 6.17. Geodesics on S^2 and $\mathbb{RP}^2$

For a Lie group G with a bi-invariant metric, the local uniqueness of geodesics shows that they are exactly the one parameter subgroups and their translates. This is shown by a nice symmetry-uniqueness argument.

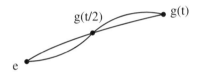

Fig. 6.18. The symmetry around $g(t/2)$ is an isometry. By uniqueness ...

For Riemannian submersions $p : (M,g) \to (N,h)$, by Samelson's results from §§4.4.3.4 one will know the geodesics in N if one knows the horizontal geodesics in M (and conversely). For example for homogeneous spaces G/H obtained from some bi-invariant metric on G the geodesics in G/H will be the projections of the horizontal one parameter subgroups.

Élie Cartan gave a complete description of the geodesic flow on symmetric spaces; we gave it in §§4.3.5. The standard reference is Helgason 1978 [701]; see also chapter 7 of Besse 1987 [183] and a few words on page 250. We recall that in compact symmetric spaces every geodesic is either periodic or everywhere dense (evenly) in a totally geodesic flat torus. This is not the complete description, especially if one wants to know when two geodesics emanating from one point meet again. The answer is completely known; see the reference given below in the cut locus story: §§6.5.4.

Back to the $\mathbb{KP}^n$. We saw in §§4.4.3.2 that for $\mathbb{K} = \mathbb{C}$ or $\mathbb{H}$ the projective spaces can be obtained from spheres by Riemannian submersion. Therefore we can completely describe their geodesic behaviour. But for $\mathbb{CaP}^2$, one cannot escape using all of Cartan's machinery from theorem 36 on page 190.

Let us treat the case $S^{2n+1} \to \mathbb{CP}^n$ and use the notation of §§4.4.3.2. All geodesics are periodic of period π. Let m be the starting point of a geodesic and

$v \in U_m M$ the initial velocity of the geodesic, call it γ_v. Then it is easy to see that γ_w will have the following behaviour: if w is in the tangent plane defined by v and iv (i.e. the complex tangent line containing v) then the geodesic γ_w will meet γ_v again at time $t = \pi/2$. Moreover the set of all these geodesics, for all w in this 2-plane, build up in $\mathbb{C}P^n$ a totally geodesic submanifold which is a standard round sphere S^2 of diameter equal to $\pi/2$ and of constant curvature equal to 4. These spheres are nothing but the complex projective lines of $\mathbb{C}P^n$. If w does not belong to the complex tangent line generated by v then γ_w and γ_v never meet before returning at time π to the original point m. Note that the set of geodesics γ_w where w turns in a tangent plane generated by v and some v' orthogonal to both v and iv (one could say a *real plane*) build up a totally geodesic submanifold which this time is a real projective plane $\mathbb{R}P^2$ of diameter π, hence of constant unit curvature.

For $\mathbb{H}P^n$, the situation is the same except that the projective lines are now four dimensional spheres S^4. For $\mathbb{C}aP^2$ one has 8 dimensional spheres S^8, but the proof is much more expensive, since we saw that there is no well behaved fibration $S^{23} \to \mathbb{C}aP^2$. For all of this geometry of $\mathbb{K}P^n$, some references are: section II.C of Gallot, Hulin & Lafontaine 1990 [542] and chapter 3 of Besse 1978 [182].

Note that for these manifolds the diameter is equal to the injectivity radius (see §6.5 for the definition of injectivity radius). It is not known today if this characterizes them; see §§6.5.5 for this baffling question.

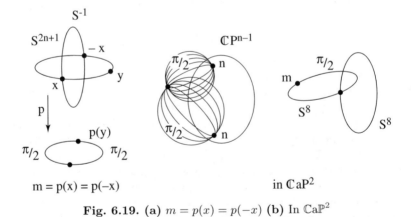

Fig. 6.19. (a) $m = p(x) = p(-x)$ (b) In $\mathbb{C}aP^2$

6.1.7 Transition

We have proven local existence and uniqueness of segments. In particular we know that short pieces of geodesics are segments. But we do not have any estimate of how short. For example, on a compact Riemannian manifold the length of a segment cannot exceed the manifold's diameter. Geodesics can also fail to be segments for curvature reasons: remember Bonnet's theorem 21 on page 125. Geodesic loops (a fortiori periodic geodesics) are obviously not segments (see figure 6.20). The picture shows that if the geodesic γ has

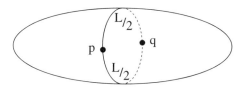

Fig. 6.20. A periodic geodesic

period L then a piece of γ is certainly not a segment if its length exceeds $L/2$. What will finally be seen is that, at least on compact manifolds, the above two situations are the only ones preventing geodesics from being segments: see 86. Note that flat tori can have periodic geodesics as small as desired. But to achieve this one needs some technical tools.

6.2 The First Technical Tools: Parallel Transport, Second Variation, and First Appearance of the Ricci Curvature

We follow the route of §§3.1.4: we have a geodesic $\gamma : [a, b] \to M$ and we look at a one parameter family $c_\alpha(t) = c(t, \alpha)$ of curves neighboring $\gamma(t) = c_0(t)$. We know that

$$\frac{\partial \text{length } c_\alpha}{\partial \alpha}\bigg|_{\alpha=0} = 0$$

because γ is a geodesic. We want to compute the second derivative

$$\frac{\partial^2 \text{length } c_\alpha}{\partial \alpha^2}\bigg|_{\alpha=0}$$

as a function of the *infinitesimal displacement*

$$Y(t) = \frac{\partial c}{\partial \alpha}\bigg|_{\alpha=0}.$$

One can always assume that this displacement is orthogonal to the geodesic. For surfaces ($d = 2$) this displacement was identified with a function f and we obtained the equation of second variation (equation 3.11 on page 119).

But as soon as $d > 2$ this displacement is now a vector field along γ taking values in the subspaces of the $T_{\gamma'(t)}M$ which are orthogonal to $\gamma'(t)$. Those spaces are of dimension greater than one and so there is no hope to work

240   6 Metric Geometry and Curvature

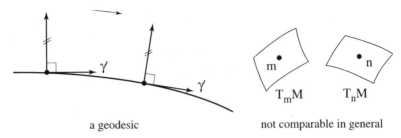

a geodesic

not comparable in general

Fig. 6.21. (a) A geodesic. (b) Not comparable in general

with plain functions. Moreover we do not know at this moment how to compare the various spaces $T_{\gamma'(t)}M$ when t varies in order to get an analogue of the derivative f' which figures in 3.11. For surfaces we succeeded to differentiate the infinitesimal displacement as follows: a good comparison between tangent spaces to a Riemannian manifold should be a Euclidean isometry. On a surface it is enough to know what to do with one nonzero vector plus an orientation. We determine any orientation of the surface near the initial point of the geodesic, and demand continuity in the choice thereafter, along the geodesic. We then identify tangent planes along the geodesic by identifying the velocities $\gamma'(t)$ and the orientations. Starting in three dimensions, this is no longer enough to produce a unique identification of tangent spaces. Geometrically, one can identify them with a trick using various geodesic families along γ as described in Arnold 1996 [66], appendix one and in figure 6.22.

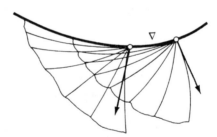

Fig. 6.22. Approximation of parallel transport with beams of geodesics, in dimension ≥ 3

We refer the reader to §15.4 for a rigorous proof (which is difficult), and assert the following:

Proposition 61 (Parallel transport) *On a Riemannian manifold M there is a canonical notion of parallel transport along an absolutely continuous curve $c : [A, B] \to M$ having the following properties: for any two values a, b of the parameter there is a map*

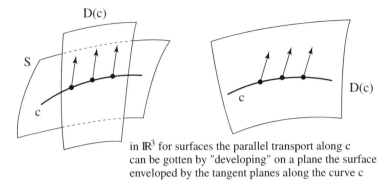

in ℝ³ for surfaces the parallel transport along c
can be gotten by "developing" on a plane the surface
enveloped by the tangent planes along the curve c

Fig. 6.23. For a surface S in $\mathbb{E}^3$, parallel transport along a curve c can be described by "developing" on a plane the surface S' enveloped by the tangent planes to S along c

$$c_{a \to b} : T_{c(a)}M \to T_{c(b)}M$$

which is an isomorphism of Euclidean spaces, and obeys the obvious composition rules. In particular, for a vector field Y along c there is a well defined derivative, denoted by Y', given by infinitesimal parallel transport:

$$Y'(t) = \lim_{h \to 0} \frac{Y(t) - c_{(t-h) \to t}Y(t-h)}{h}$$

and called the covariant derivative. Geodesics are precisely the curves γ whose velocity γ' is invariant under parallel transport:

$$(\gamma')' = 0 \ .$$

Other examples of parallel transport are: in a Riemannian product $M \times N$, the parallel transport along c is the product of the parallel transport along the two projections of c respectively on M and on N. The second example is a totally geodesic submanifold, see §§6.1.4.

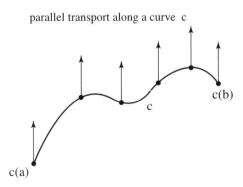

Fig. 6.24. Parallel transport along a curve c

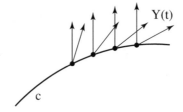

Fig. 6.25. $Y'(t) = \frac{dY}{dt}$ measures the defect of $Y(t)$ to be invariant under parallel transport

We now return to deriving a second variation formula; when the curves c_α have fixed ends $\gamma(a)$ and $\gamma(b)$ it reads:

$$\frac{\partial^2 \text{ length } c_\alpha}{\partial \alpha^2}\bigg|_{\alpha=0} = \int_a^b \left(\|Y'(t)\|^2 - K\left(\gamma'(t), Y(t)\right) \|Y(t)\|^2 \right) dt \qquad (6.7)$$

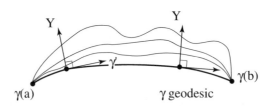

Fig. 6.26. Second variation—plucking a string

The second variation formula is an essential tool which will be used frequently. For its history, see section 4 of Petersen 1998 [1019]. Note that very often it is not the family of curves which is given but simply a vector field $Y(t)$ along γ. But it is trivial to built up a family of curves $c_\alpha(t) = c(t, \alpha)$ such that $\frac{\partial c}{\partial \alpha}\big|_{\alpha=0} = Y(t)$. For a complete proof one needs to extend the notion of covariant derivative, etc. to maps of surfaces into (M, g) more general than immersions. Even the definition on page 699 is not enough to get the proof of the formula. It is only easy when the variation of curves verifies $Y(t) \neq 0$. Otherwise it is quite technical, but necessary because Y never vanishing is too much to ask in most geometrical applications. The typical case is when the ends of γ are held fixed. Then Y vanishes at these ends. Technically in that case it is important to know the following interpretation for Y'. Assume that $c(0, \alpha)$ is a fixed point m and consider the velocities at m of the curves $c_\alpha(t)$, the vectors $\frac{d}{dt} c_\alpha(t)\big|_{t=0}$. They now belong to the fixed vector space $T_m M$ and so have an ordinary derivative vector with respect to variations of α. Then we have the following formula which is a sophisticated kind of symmetry rule for second derivatives:

$$\frac{\partial}{\partial \alpha}\left(\left.\frac{\partial c}{\partial t}\right|_{t=0}\right)\Bigg|_{\alpha=0} = Y'(0) \tag{6.8}$$

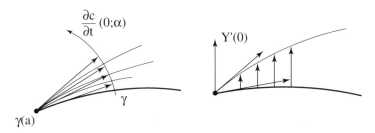

Fig. 6.27. Measuring the covariant derivative

For complete proofs, see various books on Riemannian geometry. It is interesting to compare the various proofs. We just note that Milnor 1963 [921] admits a technical point: he works with variations whose transversal vectors never vanish. To do this, one is forced to use the notion of induced connection or work with tricky computations in coordinates.

As in theorem 21 on page 125 we draw immediately from 6.7:

Theorem 62 (Bonnet–Schoenberg–Myers, 1935) *If a complete Riemannian manifold M has sectional curvature bounded below by a positive constant δ:*

$$K \geq \delta > 0$$

then it satisfies

$$\mathrm{diam}\, M \leq \frac{\pi}{\sqrt{\delta}}\,.$$

In particular M is compact.

Proof. Let p, q be any two points in M with $d(p,q) > \pi/\sqrt{\delta}$ and connect them by a segment $\gamma : [0, L] \to M$ using theorem 52. Pick any unit tangent vector $w \in T_p M$ which is orthogonal to $\gamma'(0)$ and define a vector field $Z(t)$ along γ by transporting w in parallel. Then $Z'(t) = 0$ by construction and definition of the derivative. Now inject into the second variation formula 6.7 the vector field

$$Y(t) = \sin\left(\frac{\pi}{L}t\right) Z(t)\,.$$

As in Bonnet's theorem 21 on page 125 one gets a negative second variation from which one can built up curves neighboring γ whose lengths are smaller than L.

A classical observation: consider a Riemannian manifold M with $K \geq \delta > 0$ and its universal Riemannian covering $\widetilde{M}$ (cf. §§§4.3.3.2). By construction

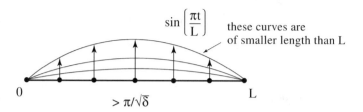

Fig. 6.28. These curves are of length smaller than L

it has the same curvature lower bound hence is also compact. This implies that the number of sheets of the covering is finite. Conclusion: a complete Riemannian manifold M whose sectional curvature has positive lower bound has finite fundamental group $\pi_1(M)$; but see the stronger theorem 63 on the next page.

Note 6.2.0.1 Today we know very little about which differentiable manifolds admit a complete Riemannian metric with positive sectional curvature: see §§12.3.1. ♦

In 1941, Myers made the following observation: in dimensions higher than two, there is a choice for w in the above proof, so that in some sense our proof does not used the full force of the assumption. Otherwise stated: can one get the same conclusion with a weaker assumption on the curvature? Indeed one can: we just have to take the mean value of the second variation formula for w running through the unit vectors in $T_m M$ which are orthogonal to $\gamma'(0)$. Practically this is done as follows: pick an orthonormal basis $\{e_i\}_{i=1,\ldots,d}$ of $T_m M$ so that $e_1 = \gamma'(0)$ and define $d-1$ vector fields $Y_i(t)$ for $i = 2,\ldots,d$ along γ starting with $Y_i(0) = e_i$. To aid in computing the sum of the associated second variation formulas, introduce the quantity:

$$\sum_{i=2}^{d} K\left(\gamma'(t), Z_i(t)\right)$$

Parallel transport being made up of isometries, one is led to introduce for any orthonormal basis $\{e_i\}_{i=1,\ldots,d}$ of $T_m M$ the quantity

$$\operatorname{Ricci}(e_1) = \sum_{i=2}^{d} K(e_1, e_i) \tag{6.9}$$

The above value is a trace (with respect to the Euclidean structure) of the quadratic form $-R(x, \cdot, x, \cdot)$. We define the *Ricci curvature* to be

$$\operatorname{Ricci}(x, y) = -\sum_{i=1}^{d} R(x, e_i, y, e_i) \tag{6.10}$$

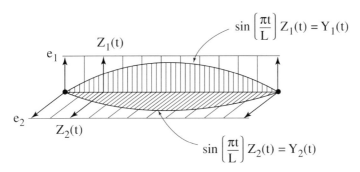

Fig. 6.29. Averaging variations in different directions

with $\{e_i\}_{i=1,\ldots,d}$ any orthonormal basis, and also set $\mathrm{Ricci}(x) = \mathrm{Ricci}(x,x)$.

Note that due to the symmetries of the curvature tensor (see equation 4.28 on page 203) this is the only nonvanishing quadratic form one can get from R. Historically, Ricci introduced it in order to find a generalization of the second fundamental form of surfaces in $\mathbb{E}^3$ and hoping that the integral curves of the directions given by diagonalizing this form with respect to the Riemannian metric tensor would yield nice curves in the manifold, analogous to the lines of curvature of a surface in $\mathbb{E}^3$. Curvature lines can play an important role in understanding even the inner geometry of surfaces. For example remember the curvature lines of the ellipsoid from on page 38: the geodesics of the ellipsoid oscillate between two curvature lines and are either periodic or everywhere dense in the annulus between these two curvature lines.

But for abstract Riemannian manifolds, Ricci was quite disappointed to find no geometric interpretation of the Ricci curvature lines. We know only of one instance where Ricci curvature lines have a geometric interpretation, to do with conformal changes of metric; see Ferrand 1982 [514]. However it will turn out in the sequel (more than once) that Ricci curvature is an extremely important invariant of Riemannian manifolds. Note the behavior under scaling:
$$\mathrm{Ricci}_{\lambda g} = \mathrm{Ricci}_g \ .$$
But the bounds of $\mathrm{Ricci}(x)$ on unit tangent vectors behave like K:
$$\inf \ \mathrm{Ricci}_{\lambda g} = \frac{1}{\lambda} \inf \ \mathrm{Ricci}_g$$
and the same for $\sup \mathrm{Ricci}$. For more philosophy on Ricci curvature, see §§6.4.3.

We return now to the above trace-like summation of second variations.

Theorem 63 (Myers–Cheng 1941,1975 [955, 362]) *If a complete Riemannian manifold M of dimension d satisfies*
$$Ricci(x) \geq \rho > 0$$

for every unit tangent vector x then

$$\operatorname{diam} M \leq \pi\sqrt{\frac{d-1}{\rho}}.$$

In particular, M is compact and $\pi_1(M)$ is finite. Moreover equality is realized only by the standard sphere.

Optimality is obvious by looking at the standard sphere. Only the sphere achieves equality; this is usually called *Cheng's theorem*, see Cheng 1975 [362]. The proof is surprisingly not so easy: looking directly at the equality case in the proof is not enough. One sees only that the sectional curvature of a 2-plane containing the velocity vector of a segment realizing the diameter must be one, but the problem is to propagate this all over the manifold. Various kinds of proofs exist today: see references in 3.6 of do Carmo 1992 [452] or in theorem 3.11 of Chavel 1993 [326]. But the fastest proof is to realize that it is a direct corollary of Bishop's theorem 107 on page 310.

This result of Myers can be seen as a global topological conclusion drawn from a curvature (infinitesimal) assumption. As such it belongs to the general type of questions relating curvature and topology; chapter 12 will be entirely devoted to these questions. You will see for example that the classification of differentiable manifolds which can carry a complete Riemannian metric of positive Ricci curvature is not known today. However we give right now a second simple and beautiful application of the second variation formula:

Theorem 64 (Synge 1936 [1172]) *A complete even dimensional Riemannian manifold with positive sectional curvature is simply connected if orientable. And if not orientable it is the quotient of a simply connected one by a single involutive isometry without fixed point.*

One cannot do better. Odd dimensional spheres have many quotients, called *lens spaces*. The simplest ones are obtained as follows: look at S^{2n-1} as embedded in $\mathbb{R}^{2n} = \mathbb{C}^n$ and quotient by the finite group generated by the maps

$$(z_1, \ldots, z_n) \mapsto (\alpha z_1, \ldots, \alpha z_n)$$

where α is some k^{th}-root of unity. The quotient is well defined, and the construction can employ various choices of α.

Assume that M is orientable and not simply connected; as we will see in chapter 10 compactness insures that M bears a periodic geodesic $\gamma : [0, L] \to M$ which is the shortest among all closed curves homotopic to it. Consider parallel transport $\gamma_{0 \to L}$ along γ from 0 to L. It preserves $\gamma'(t)$ for every t (cf. chapter 15) so it is an isometry

$$\gamma_{0 \to L} : T_{\gamma(0)}M \to T_{\gamma(L)}M = T_{\gamma(0)}M$$

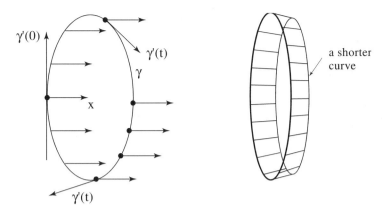

Fig. 6.30. Using positive curvature to make a shorter curve

which leaves invariant $\gamma'(0) = \gamma'(L)$. Since $T_{\gamma(0)}M$ is even dimensional, there is another unit vector $x \in T_{\gamma(0)}M$ which is invariant under $\gamma_{a \to b}$ and orthogonal to $\gamma'(0)$. We parallel transport that x along γ to produce a vector field $X(t)$ of norm one along γ and we apply the second variation formula 6.7 to it: $X'(t) = 0$ by construction of X so that the result is negative. This implies that there are shorter closed curves neighbouring γ, a contradiction. If M is not orientable, just consider its orientable two sheet covering.

The same trick will also be used in the proof of theorem 92 on page 276, and in spirit in the proof of theorem 93 on page 276, but of course the trick does not work with the weaker assumption of positive Ricci curvature. However, one can prove such a theorem under a extra mild assumption on the *systole* Sys (M) of the manifold, i.e. the length of the smallest noncontractible curve (see §§7.2.1):

Theorem 65 (Wilhelm 1995 [1262]) *If a complete Riemannian manifold M^d satisfies* Ricci $\geq d - 1$ *and*

$$\operatorname{Sys}(M) > \pi\sqrt{\frac{d-2}{d-1}}$$

then

1. *if d is even and M orientable, then M is simply connected and*
2. *if d is odd, then M is orientable.*

6.3 The Second Technical Tools: The Equation for Jacobi Vector Fields

6.3.1 The Exponential Map and its Derivative: The Philosophy of Élie Cartan

We still consider a given geodesic γ and a one parameter family of curves neighbouring γ but this time we ask that they all be geodesics $\gamma_\alpha(t) = \gamma(t,\alpha)$ and moreover all parameterized by arc length (unit speed). We again set

$$Y(t) = \left.\frac{\partial \gamma}{\partial \alpha}\right|_{\alpha=0}.$$

Using parallel transport along γ one can define not only $Y'(t)$ but also its second derivative $Y''(t)$. We find that Y satisfies the second order linear differential equation:

$$Y'' = R(\gamma', Y)\gamma' \tag{6.11}$$

called the *Jacobi field equation*, where the vector endomorphism

$$v \mapsto R(x,v)x$$

is defined by:

$$\langle R(x,v)x, w\rangle = R(x,v,x,w)$$

for any $w \in T_m M$. A solution of this equation is called a *Jacobi vector field*.

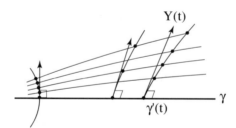

Fig. 6.31. A Jacobi vector field

As was the case for the second variation formula, the proof of equation 6.11 uses the results of the technical chapter (chapter 15) and again is much easier when Y never vanishes, but this asks too much for most geometrical applications, in particular for what we are going to do now.

Note: Jacobi vector fields $Y(t)$ along a geodesic γ are always orthogonal to γ' as soon as the initial value $Y(0)$ is orthogonal to $\gamma'(0)$. Indeed, the orthogonal projection on $\gamma'(t)$ has constant value, since $\gamma'(t)$ is a trivial solution of 6.11. For this reason, in general only such orthogonal Jacobi vector fields are considered. From standard results in ordinary differential equations we see:

Proposition 66 *Along a given geodesic γ there is exactly one Jacobi vector field Y with given values $Y(0)$ and $Y'(0)$, and it can always be realized by a one parameter family γ_α of geodesics.*

In some applications, it is useful to know the geometric interpretation of $Y(0)$ and $Y'(0)$. The value $Y(0)$ is easy to interpret—it is just the velocity of the curve described by the initial values of the γ_α. What $Y'(0)$ means geometrically is a little more subtle. One proves (using chapter 15) that it is the (absolute) derivative with respect to α of the initial velocity vectors of the γ_α.

The basic remark, which we call the *Élie Cartan philosophy* and which goes back to the 1928 first edition of his book Cartan 1988 [321], is that equation 6.11 on the preceding page gives the derivative of the exponential map $T_m M$ as soon as the various endomorphisms $R(\gamma', \cdot)\gamma'$ are known along the geodesics issuing from m. In fact, let x be some unit tangent vector to M at m and let us try to compute the derivative of $\exp_m$ at the point $tx \in T_m M$. By equation 6.4 on page 225 we have only to compute it for vectors y which are orthogonal to x (since in radial directions the exponential map preserves lengths). Then the very definition of the exponential map and the tricky formula 6.8 tell us:

Proposition 67
$$d\exp_m(y)(tx) = Y(t)$$
where Y is the solution of the equation
$$Y''(t) = R(g'(t), Y(t))g'(t)$$
with $Y(0) = 0$ and $Y'(0) = y$.

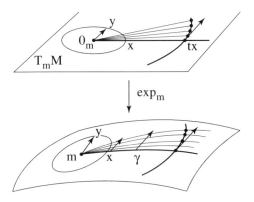

Fig. 6.32. How to compute the derivative of the exponential map with Jacobi fields

Standard theorems on differential equations tell us that such a solution is completely determined by such initial values. In the notation of equation 6.4 one has
$$f(\rho, \sigma, \xi) = \|Y|_{t=r}\|$$
plugging $x = \sigma$ and $y = \xi$ into proposition 67.

The exact content of Élie Cartan's philosophy is subtle. To know the endomorphisms $R(\gamma', \cdot)\gamma'$ we need to know the curvature tensor and the parallel transport operator along the various geodesics issued from m. Cartan used his philosophy in the spectacular case of the symmetric spaces described in theorem 36 on page 190. He was able to prove that those spaces are precisely the complete Riemannian manifolds for which the curvature tensor is invariant under parallel transport (see §15.4 for more on that). In that case proposition 66 shows that the Riemannian metric is known as soon as the curvature tensor at one point is known: then calculations become purely algebraic. More generally, on a real analytic Riemannian manifold, one need only know the curvature and all its covariant derivatives of any order (see §15.5) to know the metric, at least locally. The global problem of determining the metric from parallel transport and curvature is called the *Ambrose problem* (see on page 704) and is solved today in dimension 2.

Note 6.3.1.1 Beware that Élie Cartan's uniqueness philosophy of proposition 66 on the previous page does not say that the curvature determines the metric in general. This question has been treated at large in §4.5.

A contrario, upper and lower bounds on the sectional curvature completely control the geometry in both directions; see this, the main achievement of §6.4.
♦

6.3.1.1 Rank

Geometers recently introduced a notion of rank closely connected with the notion of Jacobi vector field: rank v for a (unit) tangent vector v in a Riemannian manifold. In Euclidean spaces, where the Jacobi vector fields verify $Y'' = 0$, and are consequently linear, there are certain Jacobi vector fields Y given by the translations. They verify $Y' = 0$. In a general Riemannian manifold we call a Jacobi vector field Y *trivial* when $Y' = 0$. Recall that we consider only Jacobi vector fields normal to γ'. This is equivalent to saying that $Y(t)$ is just parallel transported along the geodesic. This of course implies that $R(\gamma', Y(t))\gamma' = 0$ for every t.

Definition 68 *The* rank *of a nonzero vector v is the maximal dimension of the linear space made up by the trivial Jacobi vector fields along the geodesic generated by v, this time including the "super trivial" field $Y = \gamma'$. The* rank *of a Riemannian manifold is the minimum of the rank of all its tangent vectors.*

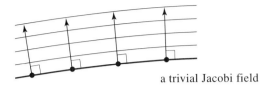

a trivial Jacobi field

Fig. 6.33. A trivial Jacobi vector field

Note that every vector is of rank at least one; only a rank larger than 1 is interesting. We met two completely different examples of manifolds of rank larger than one, namely the Riemannian products in §§§4.4.3.3 and the symmetric spaces in §§4.3.5. The notion of rank will be essential on page 612.

6.3.2 Spaces of Constant Sectional Curvature: Space Forms

In the special case of a constant sectional curvature metric, equation 66 completely determines the Riemannian metric without knowing anything a priori about the parallel transport operator. We saw in theorem 39 on page 208 that for every real number k there is at least one simply connected Riemannian manifold with constant sectional curvature equal to k. Conversely assume that is the case: from equation 4.29 on page 205 one deduces that for every orthonormal pair of vectors x, y:

$$R(x,y)x = -ky .\tag{6.12}$$

In particular the equation in proposition 66 is always the same:

$$Y''(t) = -kY(t)\tag{6.13}$$

and its solutions with initial conditions $Y(0) = 0$ and $\|Y'(0)\| = 1$ are

$$Y(t) = \begin{cases} \frac{\sin(\sqrt{k}t)}{\sqrt{k}} y & k > 0 \\ ty & k = 0 \\ -\frac{\sinh(\sqrt{-k}t)}{\sqrt{-k}} y & k < 0 \end{cases}\tag{6.14}$$

In the notation of equation 6.4 this implies:

$$g = d\rho^2 + \begin{cases} \frac{\sin^2(\sqrt{k}\rho)}{k} d\sigma^2 & k > 0 \\ \rho^2 d\sigma^2 & k = 0 \\ -\frac{\sinh^2(\sqrt{-k}\rho)}{k} d\sigma^2 & k < 0 \end{cases}\tag{6.15}$$

so that g is completely known. From this local uniqueness and because the models that we wrote down (the sphere, Euclidean space and hyperbolic space) when we proved theorem 39 are simply connected and complete, one deduces

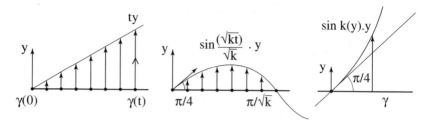

Fig. 6.34. Jacobi vector fields in space forms

Theorem 69 *For every real number k there is one and only one—up to isometry—simply connected complete Riemannian manifold of constant sectional curvature equal to k. We recall that it is denoted by $\mathbb{S}^d(k)$ (see theorem 39 on page 208).*

In this completely rigorous setting this result was available only in the late 1920s. Settling this uniqueness issue was one of the driving forces behind the Hopf–Rinow theorem.

Note 6.3.2.1 There are two little technical difficulties in the global proof of theorem 69. The simplest cases are $k = 0$ and $k < 0$. Then the formulas above show in particular that the derivative of the exponential map never vanishes, so the exponential map is locally one-to-one since the dimensions are the same. This is a necessary condition for a covering; sufficiency is achieved using completeness as stated in corollary 55 on page 228. For the sphere, $k > 0$, the derivative of the exponential map deteriorates at the preimage of the antipodal point. One wrangles the sphere by gluing two exponential charts at different points (antipodal if you wish). This gluing seems to us not explicit in many books. A reference: 3.82 in Gallot, Hulin & Lafontaine 1990 [542]. See another approach in part I of Vinberg 1993 [1221]. ♦

The above completely solves (at the first level, as we will see) the long-standing problem of existence and uniqueness of space forms. We present this problem as follows (see if needed the notes 1.6.1.1 on page 32 and 4.3.2.3 on page 182). We look for geometries for which once we know the distances from q to p and r to p, and the angle between the relevant segments, we can recover the distance from q to r by a universal formula involving only the angle and the two distances. This same formula should be valid for any triple of points, and the space should be complete. Such a space will present a nice generalization of Euclidean geometry.

The question can be asked locally or globally. Various considerations of axiomatic type led to the fact one has to look only among Riemannian manifolds: see part II of Reshetnyak 1993 [1055] for a recent survey on generalized Riemannian spaces. Considering smaller and smaller triangles, one sees easily that the sectional curvature has to be constant.

Let the side lengths be

$$a = d(q, r)$$
$$b = d(p, r)$$
$$c = d(p, q)$$

and take α, β, γ the angles opposite the sides of lengths a, b, c. We treat b, c, α as known, and want to know a. What was needed before the developments of §6.4 (material dating from the 1960's) was a Taylor expansion for a^2 as a function of b, c, α and the curvature at the vertex p, i.e. a more precise version of equation 4.31 on page 205:

$$a^2 = b^2 + c^2 - 2bc \cos\left(\alpha - \frac{KS}{3}\right) + o\left(b^2 + c^2\right) \qquad (6.16)$$

where K is the sectional curvature of the tangent plane generated by the two sides pq, pr and S is the area of the triangle. The error in the area S does not matter; one can estimate the area to be

$$S = \frac{1}{2} bc \sin \alpha \ .$$

See section V of chapter X of Cartan 1988 [319, 321] for a proof. It seems to us that more terms in the Taylor expansion for the distance between two points, close to a given one, are not to be found in books today. Since the 1960's our contention about spaces with a universal formula has become trivial by theorem 73 on page 258, which moreover yields the universal formula in one shot.

An equivalent axiom (called the *axiom of mobility*): for every pair of triangles having corresponding sides of equal lengths there is an isometry of the space sending one into the other. This appears stronger, but is in fact equivalent. The space forms are thus precisely the 3 point homogeneous spaces; see §6.6.

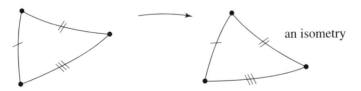

Fig. 6.35. An isometry

By theorem 69 this implies that the universal cover (see §§§4.3.3.2) of our space is a space form $\mathbb{S}^d(k)$ for some real k. Otherwise stated : a space form is the quotient by a group of isometries of one the $\mathbb{S}^d(k)$. Such a group should act without fixed point but also discretely, so that the quotient is a manifold.

This solves the problem of space forms at the first level. The next level is to find—up to isometries—all such quotients. We will shape this in a more general context in §6.6.

We kept for the end the most important fact:

Theorem 70 (Conformal representation theorem) *Let (M, g) be any compact Riemannian surface. Then there is one and only one metric g' conformal to g, i.e. $g' = fg$, such that g' has constant curvature and $f : M \to \mathbb{R}$ is a smooth function on M.*

This statement has a long history, and a huge importance because of the theory of one complex variable, number theory, algebraic geometry, etc; also see §§4.4.3. It was claimed by Riemann around 1850; he took it for granted. Many mathematicians thought that it did not need any formal proof, because it can be given an interpretation as an "obvious" result of physics, when reposed as a problem of finding a minimum. The first completely rigourous proof is due to Paul Koebe at the turn of the century. Note that it is a purely Riemannian geometry statement, and solves for compact surfaces the problem of finding the best metric on a given compact manifold. We will devote the entirety of chapter 11 to this theory of best metrics; see in particular a very modern proof on page 526 of the conformal representation theorem. This proof exhibits an explicit flow in the space of metrics (Ricci flow), attracting any metric toward one of constant curvature. The final proof is in Chow 1991 [378]; note that it uses sophisticated objects like the determinant of the Laplacian and a special entropy; see §§9.12.3.

Theorem 70 does not say that there is a unique metric on the surface (up to diffeomorphisms of course) of constant curvature. We now recall the classification of surfaces in §§§4.1.4.1. Uniqueness of constant curvature metric is easy to see for S^2 and $\mathbb{RP}^2$: by theorem 69 for the sphere and by a trivial game left to the reader for $\mathbb{RP}^2$. Flat structures on T^2, appropriately normalized, form the two dimensional modular domain drawn in figure 6.36 on the facing page; see the end of this story in §§6.6.2.

For the Klein bottle, one has only one parameter, because Klein bottles have to come from rectangles, not from arbitrary parallelograms.

6.3.3 Nonpositive Curvature: The von Mangoldt–Hadamard–Cartan Theorem

Recall proposition 66 which give the derivative of the exponential map in terms of solutions of Jacobi vector fields. It is interesting to know when this derivative is one-to-one. And if not, then try to control when the first bad point occurs. This will be done in §6.4 but first we remark that there is a simple case where we are sure never to get any kernel:

Proposition 71 *If $K < 0$ everywhere then the derivative of the exponential map is always one-to-one.*

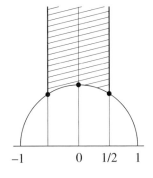
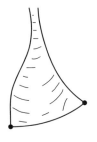

the modular domain is not compact, and not a manifold, having two singular points

Fig. 6.36. The modular domain is not compact, and not a manifold, having two singular points

From equation 6.11 on page 248 we get $\langle Y'', Y \rangle > 0$ which implies

$$\frac{d^2}{dt^2} \|Y\|^2 = 2\|Y'\|^2 + 2\langle Y, Y'' \rangle \tag{6.17}$$
$$\geq 0$$

so that $\|Y\|^2$ is a convex function. Vanishing at $t=0$, it cannot vanish elsewhere.

Using corollary 55, we find:

Theorem 72 (von Mangoldt–Hadamard–Cartan, 1881, 1928 [893, 321]) *If a complete Riemannian manifold M of dimension d has everywhere nonpositive curvature, then the exponential map at any point is a covering. In particular, as a manifold M is the quotient of $\mathbb{E}^d$ by a discrete group. If M is simply connected, the exponential is one-to-one, M is diffeomorphic to $\mathbb{E}^d$ and any two points are joined by a unique segment.*

Again this is a result passing from curvature to topology. But do not believe that this ends the classification of such manifolds. It is a extremely active field of research these days; see §§12.3.4 or (state of the art in 1985) Ballmann, Gromov & Schroeder 1985 [106], and thereafter Gromov 1991 [627]. Historically, von Mangoldt proved theorem 72 for surfaces. Hadamard gave two different proofs for surfaces again and embedded the issue in the more general study of the geodesic flow. He also mentioned an extension in three dimensions. Cartan put this theorem into its final form.

One can prove more results, which we leave as exercises:

1. In any homotopy class of a Riemannian manifold with nonpositive curvature there is one and only one geodesic loop.
2. In a simply connected complete Riemannian manifold with nonpositive curvature, any triangle obeys:

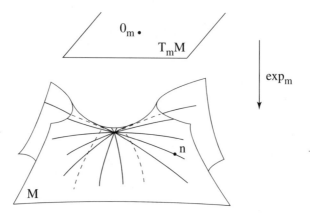

Fig. 6.37. The exponential map on a negatively curved surface

$$a^2 \geq b^2 + c^2 - 2bc \cos \alpha$$

(cf. formula 1.16 on page 54 and §6.4).

3. Consider an isometry $f : M \to M$ of a simply connected Riemannian manifold with nonpositive curvature. The associated displacement function $d(m, f(m))$ is a convex function (and strictly convex if the curvature is negative).
4. In a manifold of nonpositive curvature, the distance function

$$d : M \times M \to \mathbb{R}$$

is convex (see 60) on the Riemannian product $M \times M$. See figures 6.38 on the facing page and 6.39 on the next page to visualize this. In particular the distance functions $d(p, \cdot)$ are convex.
5. At the infinitesimal level, result 4 is related to the statement: for any Jacobi vector field Y in a manifold of nonpositive (resp. negative) curvature, the function

$$t \in \mathbb{R} \to \|Y(t)\| \in \mathbb{R}$$

is convex (resp. strictly convex).
6. In a simply connected Riemannian manifold with nonpositive curvature, there is a notion of *center of mass* (see proposition 60) for any compact subset. In particular, a group of isometries of M having a bounded orbit has always a fixed point.

Cartan 1929 [315] section 16 used this inequality to prove the existence and uniqueness of the center of mass in complete manifolds of negative curvature. From it he deduced that compact maximal subgroups of semisimple Lie groups are always conjugate. This holds because a symmetric space of nonpositive curvature is nothing but the quotient of a noncompact Lie group by one of its maximal compact subgroups as seen on page 192.

Besides Ballmann, Gromov & Schroeder 1985 [106], partial results can be found in chapter 12 of do Carmo 1992 [452], 3.110 of Gallot, Hulin & Lafontaine 1990 [542], chapter 9 of Cheeger & Ebin 1975 [341], section V.4 of Sakai 1996 [1085] and here in §§12.3.4.

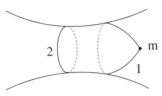

Fig. 6.38. Uniqueness: **(1)** loop based at m **(2)** A closed geodesic in its free homotopy class

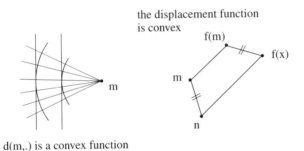

Fig. 6.39. (a) $d(m, \cdot)$ is a convex function. (b) The displacement function is convex

6.4 Triangle Comparison Theorems

6.4.1 Using Upper and Lower Bounds on Sectional Curvature

We now extend the results of §3.2 to any dimension and collect other interesting things on our way. What follows can be seen from various viewpoints. First we will try to control the global metric geometry of a Riemannian manifold, obtaining this control only via the sectional curvature, which is an infinitesimal (often explicitly calculable) invariant. This can be compared with controlling a complete motion by controlling the acceleration at each instant. One can also say this is a variation on the theme of the Euclidean equation

$$a^2 = b^2 + c^2 - 2bc \cos \alpha .$$

Other basic geometric consequences will be seen in the sequel, starting with §6.5, and of course in chapter 12. Control on the metric via curvature is important because, as remarked before, even in Riemannian manifolds described by elementary formulae, explicit calculations of distances are completely out of reach.

We state the results first. We assume in all this section that the sectional curvature of our Riemannian manifold obeys everywhere

$$\delta \leq K \leq \Delta \tag{6.18}$$

In referring to a *triangle* $\{p,q,r\}$ on a Riemannian manifold M we will mean three points $p,q,r \in M$ together with three segments joining them: note that in some cases those segments might be not unique.

Theorem 73 (Toponogov triangle comparison theorems) *Assume that the sectional curvature of a Riemannian manifold M of dimension d satisfies*

$$\delta \leq K \leq \Delta$$

and let $\{p,q,r\}$ be any triangle in M and in $\mathbb{S}^d(\delta)$ take some triangle $\{p',q',r'\}$ such that the corresponding sides from p and p' have equal lengths and equal angles at p and p'. Then

$$d_M(q,r) \leq d_{\mathbb{S}^d(\delta)}(q',r').$$

The reverse equality holds with $\mathbb{S}^d(\Delta)$ but under the proviso that—roughly speaking—the triangle $\{p,q,r\}$ can be filled by the exponential map at p; in particular this always works for small enough triangles.

As for surfaces (see §3.2) there is an equivalent way to state the theorem (particularly for the lower bound δ). One asks this time that the two triangles have corresponding sides of equal length. Then the conclusion is that the corresponding angles in M are never larger than those in $\mathbb{S}^d(\delta)$.

The above theorem can vaguely be compared with the "outer" comparison theorem 1 on page 15 given for plane curves, and theorem 6 on page 56 for rolling surfaces. Note that theorem 73 solves immediately the problem of the universal formula we alluded to on page 252.

The technique of the proof of theorem 73 is the same as for the formula 3.12 on page 121 for surfaces, except that two complications arise. The first: when cutting the triangles in small pieces. We will not comment on that. We refer to some good references for the topic: Karcher 1989 [780], Eschenburg 1994 [494], and chapter IV of Sakai 1996 [1085]. Historically the global theorem 73 was obtained by Élie Cartan for $K \leq 0$, while the global case with $\delta \leq K$ is due to Alexandrov for surfaces and to Toponogov for general dimensions.[2] The case of $K \leq \Delta$ comes from Rauch's inequality below. The most often used part of theorem 73 is the global result about $K \geq \delta$ which is most often called the *Toponogov comparison theorem*.

---

[2] Rauch had a statement for curves joining the extremities of the triangle.

6.4 Triangle Comparison Theorems

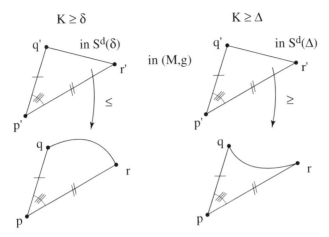

Fig. 6.40. Triangle comparison

The second difficulty is in getting bounds for Jacobi vector fields. On a surface this was easy because we had only to deal with real valued functions. But when $d > 2$ the functions in equation 6.11 are vector valued. The results of chapter 3 are still valid:

Proposition 74 (Rauch 1951 [1050]) *Any solution Y of 6.11 with $Y(0) = 0$ and $\|Y'(0)\| = 1$ satisfies*

$$\|Y(t)\| \leq \begin{cases} \frac{\sin(t\sqrt{\delta})}{\sqrt{\delta}} & \delta > 0 \\ t & \delta = 0 \\ \frac{\sinh(t\sqrt{-\delta})}{\sqrt{-\delta}} & \delta < 0 \end{cases}$$

under the proviso *that Y does not vanish between 0 and t.*

We recall (cf. §3.2) that the proviso is necessary; see figure 6.41 on the following page. The proof of proposition 74 is never simple.

Proposition 75 (Rauch 1951 [1050]) *Any solution Y of 6.11 with $Y(0) = 0$ and $\|Y'(0)\| = 1$ satisfies*

$$\|Y(t)\| \geq \begin{cases} \frac{\sin(t\sqrt{\Delta})}{\sqrt{\Delta}} & \Delta > 0 \\ t & \Delta = 0 \\ \frac{\sinh(t\sqrt{-\Delta})}{\sqrt{-\Delta}} & \Delta < 0 \end{cases}$$

for

$$0 \leq t \leq \begin{cases} \frac{\pi}{\sqrt{\Delta}} & \Delta > 0 \\ \infty & \Delta \leq 0. \end{cases}$$

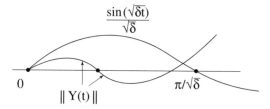

Fig. 6.41. The Rauch upper bound holds as long Y does not vanish between 0 and t

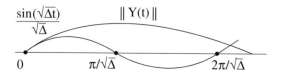

Fig. 6.42. The Rauch lower bound holds until $t = \pi/\sqrt{\Delta}$

Again the restriction on t is needed; see figure 6.42.

The proof of proposition 75 is not much more complicated than that of proposition 71 and is left as an exercise. Both of Rauch's theorems provide control with a second derivative—an acceleration. This is not surprising if one considers geodesics in a Riemannian manifold as free particle motions (no potential).

Applications of these Alexandrov–Toponogov theorems are innumerable. They are the corner stone of most of the results of chapter 12. We make an important remark: the similarity of the upper and lower bounds on the distance $d(q, r)$ seems to imply that if the curvature varies extremely slightly around zero, then the manifold is almost Euclidean. This is true locally (in a sense to be made precise) but globally we will see in §§12.2.3 that there are manifolds, called *nilmanifolds*, which have curvature as small as one wishes but without the topology of a flat manifold, i.e. they are never finite quotients of tori.

One can also expect that bounds on curvature will give bounds on volume. This is of course true but will be treated in chapter 7. The surprising and wonderful result is that for upper bounds on volumes, we only need to control the Ricci curvature.

Note 6.4.1.1 From formula 6.16 one gets the feeling that in some sense sectional curvature is the infinitesimal version of the metric geometry, and from the triangle comparison theorems 73 one feels that conversely the sectional curvature can be "integrated" and gives complete control on the metric. In other words one is tempted to say that sectional curvature is necessary and sufficient to control the metric. This is certainly true in one direction, the sufficiency, but the converse is extremely subtle and is discussed at large in §4.5. ♦

Note 6.4.1.2 The control theory we presented above was based on Jacobi vector fields. This technique is the one used in most books. But it is not the only one available. Another technique applies bounds on the second fundamental form of the metric balls centered at a fixed point. For those quadratic forms, the Jacobi vector field equation is replaced by a Ricatti type equation where the curvature enters. For a presentation in this spirit, good references are Gromov 1991 [627], Karcher 1989 [780] and Eschenburg 1994 [494]. This revolutionary technique was introduced by Gromov in 1979 (see Gromov 1999 [633]). In some sense one is playing with the distance function instead of the geodesics lines. This is no surprise since the gradient lines of a distance function are geodesics. Moreover, for the case of Ricci curvature in the next section, only distance functions can be used—Jacobi vector fields are not good enough.

The linear second order Jacobi vector field equation 6.11 on page 248 can be replaced equivalently by a first order nonlinear equation of square Ricatti type. It is the following equation, where the unknowns to control are the Hessian of the distance function f, denoted by $A = \text{Hess}\, f$. But one merely considers A as a linear symmetric map on tangent spaces. Then Jacobi's equation is equivalent to

$$A' + A^2 + R(\gamma', \cdot)\gamma' = 0 \qquad (6.19)$$

using the curvature tensor along the geodesic γ to form a symmetric linear map. See figure 6.43.

The geometer has a simple view of the Hessian of a distance function (to a point or more generally to some submanifold, most often a hypersurface). This Hessian is nothing but the second fundamental form of the level hypersurfaces $f^{-1}(t)$. This viewpoint was also initiated in Gromov 1999 [633]. One feels that if $K \geq \delta$, then these hypersurfaces will be less curved than the

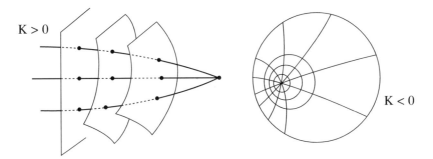

Fig. 6.43. R governs the evolution of the second fundamental form of the level hypersurfaces

corresponding ones in the comparison space form $\mathbb{S}^d(\delta)$. This in turn controls the distance functions, which are employed to control triangles. ♦

6.4.2 Using only a Lower Bound on Ricci Curvature

So far we have obtained optimal control of the geometry on both sides if we have bounds $\delta \leq K \leq \Delta$. Up to the end of the 1980s, Ricci curvature was believed to be only useful to control volumes, but in a very efficient way as we will see in §§7.1.2. This is not surprising heuristically. The philosophy of the second variation formula is that the second derivative of the metric is given by the sectional curvature, and requires knowledge of the sectional curvature. The Ricci curvature is a trace, and derivative of volumes, i.e. derivatives of determinants, are given by traces. It is important to know nevertheless that this only works when we have a lower bound

$$\text{Ricci} \geq (d-1)\delta$$

except in dimension 3. This is not surprising since in three dimensions the Ricci curvature and sectional curvature are essentially equivalent. As a striking departure from sectional curvature, we will see in §§12.3.5 that Ricci $\leq (d-1)\Delta$ is a condition which basically cannot have any consequence. Zhu [1305] is a systematic survey; also see Gallot 1998 [541].

Myers's theorem 63 is one metric exception to the rule that Ricci curvature can only control the measure. It yields a metric consequence through a mean value argument, thereby using only Ricci curvature. Very recently major breakthroughs appeared to the effect that a lower bound on Ricci curvature controls metric geometry of different types, and surprisingly enough, on both sides in some instances. Intermediate results were obtained by Calabi 1958 [300] and by Cheeger & Gromoll 1971 [343] in the particular case of noncompact manifolds with Ricci ≥ 0. We will now state the results. These results will be essential to chapter 12.

The first result is the Gromov precompactness theorem 382 on page 626 which involves only controls on the volumes of balls; see §§§7.1.1.5. The second is the excess theorem of Abresch & Gromoll 1990 [5]. The *excess* of a triangle $\{p,q,r\}$ is the number

$$e = d(p,q) + d(p,r) - d(p,\gamma)$$

where γ is a segment from q to r. This excess can be controlled when Ricci $\geq (d-1)\delta$ with δ of any sign. The most spectacular results are Colding's L^1 and L^2 triangle comparison theorems. We will briefly state some of them; more detailed formulas are in Colding 1996 [385, 386] and Colding 1997 [387].

Theorem 76 (Colding L^1) *In a Riemannian manifold with nonnegative Ricci curvature, for any $\varepsilon > 0$ there is some $\eta(d,\varepsilon) > 0$ so that for any $R > 0$, any $r < \eta R$, and any points p,q with $d(p,q) > 2R$ one has*

6.4 Triangle Comparison Theorems   263

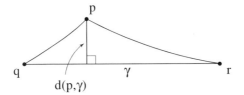

Fig. 6.44. A thin triangle

$$\frac{1}{\operatorname{Vol} UB(q,r)} \int_{v \in UB(q,R)} \|d(\gamma_v(r), p) - d(\gamma_v(0), p) - r\langle v, \nabla d(p, \cdot)\rangle\| \, dv \leq \varepsilon R$$

where $UB(q, R)$ denotes the unit tangent bundle over the ball $B(q, R)$.

There is a joint L^1 formula but for the derivative of the distance $d(\gamma_v(r), p)$, which yields an L^1 theorem for angles. Note that the above formula in the integrand is even weaker than the corresponding one in Euclidean space, but it is enough for Colding's applications, described in chapter 12. Colding also has a formula for manifolds with Ricci $\geq (d-1)\delta$ with δ of any sign.

One of the strongest of his formulas concerns the bound Ricci $\geq d-1$. He compares a manifold M satisfying this bound with the standard sphere S^d and he obtains L^2 theorems. We follow the pictures in figure 6.45. We compare triangles, for congruent initial conditions. Pick points $p, q \in M$ and $\underline{p}, \underline{q} \in S^d$, and unit vectors $v \in T_q M$ and $\underline{v} \in T_{\underline{q}} S^d$. Assume that $d(p,q) = d(\underline{p}, \underline{q})$ and that we have selected segments from p to q and from $\underline{p}$ to $\underline{q}$. Suppose that the angle between v and the tangent to the segment is the same as that between $\underline{v}$ and the corresponding segment. We are interested in the respective distances

$$d(t) = d(p, \gamma_v(t))$$
$$\underline{d}(t) = d(\underline{p}, \gamma_{\underline{v}}(t))$$

to the moving points $\gamma_v(t)$ and $\gamma_{\underline{v}}(t)$ along the geodesics with initial velocities v and $\underline{v}$ in the respective manifolds. The result is that we have, up to any

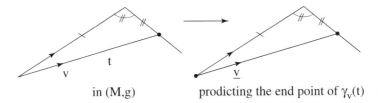

Fig. 6.45. (a) In M. (b) Predicting the end point of $\gamma_v(t)$

$\varepsilon > 0$ and for almost all triangles as in the figure, a two sided comparison theorem. This will hold provided that there exist points in the manifolds whose distance is close enough to π.

6 Metric Geometry and Curvature

Theorem 77 (Colding L^2 [386]) *Let M be a Riemannian manifold with Ricci $\geq d-1$. For any $\varepsilon > 0$ and $s_0 \in [\pi/2, \pi)$ there is a δ, depending only on ε, s_0 and d such that if there are two points p, q with $d(p,q) > \pi - \delta$ then for every $0 < s < s_0$ one has*

$$\frac{1}{s \operatorname{Vol} UM} \int_{v \in UM} \int_0^s |\cos d(t) - \cos \underline{d}(t)|^2 \, dv \, dt < \varepsilon \, .$$

The analogous inequality for the angles α_t and $\underline{\alpha}_t$ in figure 6.46 is also valid.

For the general case, Ricci $\geq (d-1)\delta$ with δ of any sign, there is still a result, but only for thin triangles:

Theorem 78 *Suppose that a Riemannian manifold M satisfies*

$$\operatorname{Ricci} \geq (d-1)\delta$$

and let d_t (resp. $\underline{d}_t$) and α_t (resp. $\underline{\alpha}_t$) be the distances and the angles in M (resp. in $\mathbb{S}^d(\delta)$) in the triangles as in figure 6.46. Then for any points p, q such that $d(p,q) > 2R$ and any $\varepsilon > 0$ there is a number $c = c(\delta, R, \varepsilon, d)$ such that for any $t < cR$ one has on the unit tangent bundles $UB(q, R)$ (resp. $UB(q, R)$) in M (resp. $\mathbb{S}^d(\delta)$) the bounds

$$\frac{1}{\operatorname{Vol} UB(q,R)} \int_{UB(q,R)} |d_t - \underline{d}_t| < \varepsilon$$

$$\frac{1}{\operatorname{Vol} UB(q,R)} \int_{UB(q,R)} |\alpha_t - \underline{\alpha}_t| < \varepsilon$$

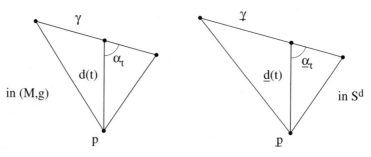

Fig. 6.46. (a) In (M, g). (b) In $\mathbb{S}^d(\delta)$

All these Colding results can be seen as statements of probability predicting the behavior of distance functions, or, say, where you land when starting from p; see a typical application in theorem 320 on page 572.

All preceding results rest finally on the following comparison theorem for distance functions. Assume Ricci $> (d-1)\delta$ and let f being any distance function
$$f = d(p, \cdot).$$
We have for the Laplacian Δf of f:

Proposition 79
$$\Delta f \leq \underline{\Delta f}$$
where $\underline{\Delta f}$ is the Laplacian in $\mathbb{S}^d(\delta)$ of a distance function $\underline{f}$ on $\mathbb{S}^d(\delta)$.

Of course this Laplacian is explicitly calculable in $\mathbb{S}^d(\delta)$, depending only on the dimension d and the lower bound δ. The proof is wonderfully simple and starts with the Bochner–Weitzenböck formula from equation 15.8 on page 707 applied to the square norm of the gradient of distance functions $f = d(p, \cdot)$:

$$-\frac{1}{2}\Delta\left(\|df\|^2\right) = \|\text{Hess } f\|^2 - \langle df, \Delta df \rangle + \text{Ricci}(df, df) \qquad (6.20)$$

For distance functions, $\|df\|$ is constant, so that the left hand side vanishes identically. Then one remarks that $\langle df, \Delta df \rangle$ is nothing but the derivative of Δf along the gradient lines of f, which are geodesics, say γ so that along γ one has, after applying Newton's inequality to $\|\text{Hess } f\|^2$ as in theorem 181 on page 408:

$$(\Delta f \circ \gamma)' + \frac{1}{d-1}(\Delta f \circ \gamma)^2 \leq -\text{Ricci}(\gamma', \gamma'). \qquad (6.21)$$

In $\mathbb{S}^d(\delta)$ one has equality with the constant d on the right hand side, so we are finished by comparing solutions of the Ricatti equation. If one compares this with what was done in §§6.4.1, it is nice to remark that here Δf is the mean curvature of the level hypersurfaces of f.

Note 6.4.2.1 (Busemann functions) If Ricci > 0 one has $\Delta f \leq \frac{d-1}{f}$. This was known to Calabi. In particular for Busemann functions (see definition 334 on page 585), which are distance functions based at infinity, then $\Delta f \leq 0$ and f is subharmonic. This will be the core of the proof of theorems of Cheeger & Gromoll 1971 [343] given in theorems 348 on page 595 and 349 on page 596.
♦

The above are the basic formulas which provide optimal control over any function (distance or not). It is difficult to complete the proof. One has to endure a lot of analysis, starting with Lichnerowicz's inequality (theorem 181 on page 408) on the first eigenvalue of the Laplacian when the Ricci curvature is positive. A second ingredient is the Cheng–Yau estimate of Cheng 1975 [362] for harmonic functions which says:

Theorem 80 *If* Ricci $> (d-1)k$ *(this k can be of any sign), and p is any point, and f is a harmonic function, then*

$$\sup_{B(p,R)} \|\nabla f\| \leq c(d,k,R) \sup_{B(p,2R)} f$$

for a constant depending only on the indicated data.

At first glance it is hard to believe Colding's result, if we are caught in the spirit of the two comparison theorems of §§6.4.1 with a lower and an upper bound for the sectional curvature. After all, the formula above says that, for small enough triangles, most of them are almost Euclidean in both directions since the difference in the integral is bounded in absolute value. But we are given only a lower bound for the Ricci curvature, so this seems odd. The explanation (heuristic, this is of course not the proof) is that the condition $d(p,q) > 2R$ implies that the sectional curvature along segments connecting p and q cannot be to large, thanks to the Bonnet–Schoenberg–Myers theorem 62 on page 243. So we have upper bounds in disguise.

Note 6.4.2.2 In all of the above we assumed the smoothness of the distance functions. They are smooth only outside of the cut locus (see §§ 6.5.4 on page 278). To overcome this, there are a lot of technical tricks to work out, for which we refer to the quoted references. ♦

Colding's comparison theorems are only probabilistic, not deterministic, but they provide control in both directions, and moreover they are optimal. Recently results like those of Rauch and Toponogov appeared with "only" lower Ricci curvature bounds, but not optimal. Still, they give upper estimates. For example:

Theorem 81 (Dai & Wei 1995 [423]) *In a Riemannian manifold M^d with*

$$\text{Ricci} \geq (d-1)\delta$$

let J be any Jacobi vector field defined on $[0, L]$ and in $\mathbb{S}^d(\delta)$ pick a ("the") Jacobi vector field J_0 with the same initial conditions. There is a universal constant $c = c(d, \delta, L)$ such that, if there is no conjugate point[3] before time L for J, then

$$\|J(t)\| \leq \exp\left(c\sqrt{t}\right) \|J_0(t)\|$$

for $t \in [0, L]$.

The proof is geometric, using a study of the geometry of the distance spheres, based on results of Brocks 1994 [259].

How about upper bounds for Ricci curvature? This question is today answered completely negatively. In §§12.3.5 we will meet Lohkamp's results, to the effect that with an upper bound on Ricci curvature one can still approximate any metric while obeying such a bound.

---
[3] See on page 268 for the definition of conjugate point.

6.4.3 Philosophy Behind These Bounds

We comment very briefly on the above triangle comparison theorems. The rough underlying idea for the case of two-sided bounds on sectional curvature is that the sectional curvature (alias the curvature tensor) is some twisted Hessian of the metric, say its *weird acceleration* or its *second derivative*. We saw this in detail in § 4.5 on page 213.

In the same rough spirit, Ricci curvature is the Laplacian of the metric, as can be guessed from formula 6.22, but there is a better answer. For this we need to introduce *harmonic coordinates*.

Definition 82 Harmonic coordinates *are sets of coordinates* $\{x_1, \ldots, x_d\}$ *such all the x_i are harmonic functions, i.e.* $\Delta x_i = 0$ *for all i (the Laplacian Δ is defined in chapter 9).*

It is easy to see that such coordinates always exist in any sufficiently small region. The precise optimal domains can be quite clearly specified. One chooses d harmonic functions which are linearly independent and satisfy ad hoc boundary conditions. This approach works only within the harmonic radius which was first systematically introduced and controlled in Anderson 1990 [39]. Such a control is basic in many results and we refer for this to the various references which we will meet in §12.4. With sectional curvature bounds it is easier than with Ricci ones. For the historian we note that harmonic coordinates were employed a long time ago by theoretical physicists such as Einstein 1916 [484] and Lanczos 1922 [844]. The founding paper is Jost & Karcher 1982 [769]; also see Hebey & Herzlich 1995 [693]. In §§11.4.1 harmonic coordinates are the basic tool to prove that Einstein manifolds are necessarily real analytic.

The underlying idea is briefly this: in normal coordinates, given control of the curvature (via for example Rauch comparison) one needs to integrate the equation of geodesics and thereafter to integrate the Jacobi vector field equation which is of second order, so that finally one loses at least one derivative in information on the metric. Harmonic coordinates enable us not to lose any derivative in the control. For Ricci curvature, which would be awful in general coordinates using equation 4.44 on page 215, the analysis is better in harmonic coordinates:

$$\Delta g_{ij} + Q(g, \partial g) = - \operatorname{Ricci}_{ij} \qquad (6.22)$$

where Q involves the first derivatives of g only quadratically.

This will turn out to be essential for convergence theorems under Ricci curvature lower bounds: see e.g. theorem 385 on page 629. We will sometimes play with the maximum principle, armed with Laplacian estimates: think of Liouville's theorem to the effect that any bounded harmonic function defined on the whole plane has to be constant. See also fact 281 on page 529 for some analogous ideas.

6.5 Injectivity, Convexity Radius and Cut Locus

6.5.1 Definition of Cut Points and Injectivity Radius

Pick some point $m \in M$ in a complete Riemannian manifold and look at the exponential map at m. We want to study the behavior of the various geodesics emanating from m. Our manifold being complete these geodesics can be extended indefinitely (this does not preclude their coming back and in some cases doing horrible things as their lengths become larger and larger). From §§6.1.1 we know that for a some small enough but positive number r the exponential map is a diffeomorphism: the geodesics from m do not meet again; they form a nice spray around m and they are segments between m and their various end points. We are then naturally forced to ask ourselves the following question:

Question 83 *What is the largest possible r for which the exponential map is a diffeomorphism? Try if possible to get some control on this r with various invariants—preferably with the curvature.*

One can ask first for less:

Question 84 *How long does a given geodesic γ emanating from a point m remain a segment from m to its extremity?*

The question was partially attacked in § 3.3 on page 125. We just remark that if one looks at spheres of various radii this forces the sectional curvature to enter into any result. But also the cylinder (or flat torus) shows that periodic geodesics cannot be avoided: curvature bounds are not enough. The fact that two completely different kind of conditions enter into the picture explains why the subject is difficult. This dichotomy can also be felt as follows: the exponential map can fail to be a local diffeomorphism either because it ceases being one-to-one or because its derivative does.

Optimal results for question 83 were uncovered by Klingenberg in 1959 and Cheeger in 1969. To present these results we first answer question 84. A convenient definition:

Definition 85 *Two points m and n on a geodesic γ are said to be* conjugate *on γ if there is a non-trivial Jacobi vector field along γ which vanishes both at m and at n. This is a symmetric relation. Moreover it is equivalent to say that the derivative of the exponential map $\exp_m$ is not one-to-one at the point $\exp_m^{-1}(n) \in T_m M$ obviously defined by n and γ.*

Now we sum up the elementary facts which completely solve question 84:

Theorem 86 *Let $m = \gamma(0)$ be the initial point of a geodesic γ. Then there is a number $t > 0$ (which can be infinite) such that γ is a segment from m to*

6.5 Injectivity, Convexity Radius and Cut Locus

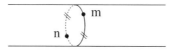

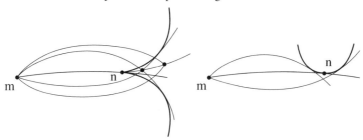

Fig. 6.47. (a) A geodesic on a flat torus or cylinder can cease to be a segment—curvature control is insufficient. (b) What happens (generically) when the derivative of the exponential map is no longer of maximal rank

$\gamma(s)$ for every $s \leq t$ and for $s > t$ thereafter never again a segment from m to any $\gamma(s)$ for $s > t$. This number t is called the cut value of γ, and denoted by $\mathrm{Cut}(\gamma)$. The point $\gamma(\mathrm{Cut}(\gamma))$ is called the cut point of γ. There are only two possible reasons (which can occur simultaneously) for n to be the cut point of γ:

1. n is conjugate on γ to m or
2. there is a segment from m to n different from γ.

Spheres are examples for which both forms of segment degeneration occur at the same time at the antipode. A more interesting one is a prolate ellipsoid of revolution. Look at the equator (which is a geodesic). The nature of the cut locus of an ellipsoid is still a matter of conjecture; see page 128. Please compute the exact cut value along the equator and look for where two different segments with common ends appear.

Theorem 86 is often used a contrario: before the cut value, the exponential has one-to-one derivative; in particular no Jacobi vector field vanishing at the origin can vanish between the origin and the cut value.

There are essentially two puzzles in the theorem; namely the two last assertions. The first is easy: look at the picture and apply the strict triangle inequality 6.5 on page 226 The second is less obvious. We sketch the proof, which is very interesting, and for details refer the reader to 2.2 of chapter 13 of do Carmo 1992 [452], 2.1.7 of Klingenberg 1995 [816], 5.2 of Cheeger &

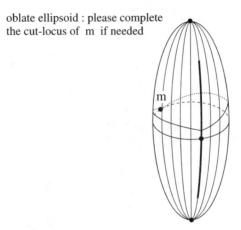

Fig. 6.48. Oblate ellipsoid: please complete the cut locus of m if needed

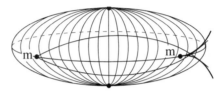

Fig. 6.49. Prolate ellipsoid: please complete the cut locus of m

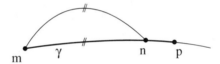

Fig. 6.50. On γ, from m up to p the derivative of $\exp_m$ is one-to-one

Ebin 1975 [341], 3.78 of Gallot, Hulin & Lafontaine 1990 [542], and III.4 of Sakai 1996 [1085].

We have to prove:

Lemma 87 *If $m = \gamma(0)$ and $n = \gamma(t)$ are conjugate along γ, then for every $s > t$ the geodesic γ is no longer a segment on the interval $[0, s]$.*

Let Y be the Jacobi vector field vanishing at 0 and t. It defines a one parameter family of curves with m and n as ends. The Jacobi equation 6.11 on page 248 implies that the second variation (equation 6.7 on page 242) vanishes for this variation. In fact the integrand reads now as an exact derivative:

6.5 Injectivity, Convexity Radius and Cut Locus   271

$$\int_0^t \left(\|Y'\|^2 + \langle Y'', Y \rangle \right) dt = \int_0^t \langle Y', Y \rangle' \, dt$$
$$= \langle Y'(t), Y(t) \rangle - \langle Y'(0), Y(0) \rangle$$
$$= 0$$

and gives zero at both ends. We have the same situation as with two different segments, but only infinitesimally. With a little gluing, it is not hard to make a family of curves out of Y from $\gamma(0)$ to $\gamma(s > t)$ with negative second variation (see figure 6.51).

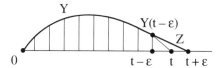

Fig. 6.51. Producing a negative second variation

Using continuity and compactness arguments one checks that the following definition and results make sense:

Proposition 88 *At a given point m in a Riemannian manifold M the infimum of the cut values of the various geodesics emanating from m is positive; it is called the* injectivity radius *of M at m and denoted by* $\mathrm{Inj}\,(m)$. *This radius* $\mathrm{Inj}\,(m)$ *is continuous in m. In particular if M is compact it has a positive minimum on M called the* injectivity radius *of M and denoted by* $\mathrm{Inj}\,(M)$.

The injectivity radius $\mathrm{Inj}\,(M)$ is the basic metric control one needs. Every piece of geodesic of length smaller than or equal to $\mathrm{Inj}\,(M)$ is a segment. Every open metric ball $B(m, r)$ of radius $r \leq \mathrm{Inj}\,(M)$ is diffeomorphic to $\mathbb{R}^d$. Note that without compactness the injectivity radius need not be positive, as in figure 6.52.

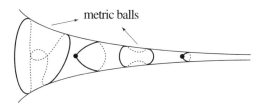

Fig. 6.52. Vanishing injectivity radius

6.5.2 Klingenberg and Cheeger Theorems

The first of these theorems says that Inj (M) can be controlled—for a compact manifold—with merely an upper bound for the sectional curvature and a lower bound for the length of the periodic geodesics:

Theorem 89 (Klingenberg 1959) *If M is a compact Riemannian manifold with sectional curvature $K \leq \Delta$ everywhere then the injectivity radius of M is not smaller than the lesser of the two numbers*

1. *$\pi/\sqrt{\Delta}$*
2. *half the length of the shortest periodic geodesic.*

A more precise statement comes from the proof itself. It says that, in the case where the second entry in the inf is used, we have equality. On the other hand, for the first entry, in most cases there is no periodic geodesic with length precisely $\pi/\sqrt{K}$. For example, for manifolds of nonpositive curvature, only periodic geodesics will matter here, since there are no conjugate points: see theorem 72 on page 255. The content of this section is covered by the following references: chapter 5 of Cheeger & Ebin 1975 [341], 3.2 of Chavel 1993 [326], 2.13 of do Carmo 1992 [452], and chapter V of Sakai 1996 [1085].

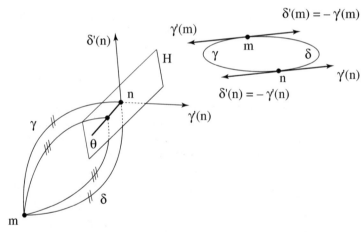

Fig. 6.53. $T_n H = \{\xi \mid \langle \xi, \delta'(n) \rangle = \langle \xi, \gamma'(n) \rangle\}$

The idea of the proof is too nice not to be presented. First, Rauch's inequality of proposition 74 on page 259 shows that there are never conjugate points before the fateful value $\pi/\sqrt{K}$. Assume that Inj $(M) < \pi/\sqrt{K}$. We apply the duality from theorem 86 on page 268 and consider any pair of points m, n which are joined by two distinct segments associated to geodesics γ and δ and with $d(m, n) < \pi/\sqrt{K}$. Look at the exponential map $\exp_m$ around

γ and δ : at both places its derivative is of maximal rank. This shows that locally around n there is some hypersurface H with every one of its points r situated equidistant from m on two different segments. Then by the first variation equation 6.3 on page 225 the tangent space $T_n H$ to the hypersurface H at n is precisely the bisector in $T_m M$ of the two tangent vectors $\gamma'(n)$ and $\delta'(n)$. If we are not in the case where $\gamma'(n) = -\delta'(n)$ then some points in H will be joined to m by two different segments of the same length, that length being smaller that $d(m, n)$. This implies that if you pick two points m and n such that
$$d(m, n) = \text{Inj}(M)$$
which is the absolute minimum (this is always possible by compactness) we necessarily have $\gamma'(n) = -\delta'(n)$. But replacing m by n, $\gamma'(m) = -\delta'(m)$. This says exactly that $\gamma \cup \delta$ is a periodic geodesic of length equal to $2d(m, n)$.

In general the length of periodic geodesics cannot be controlled with just the curvature: consider a flat rectangle two-dimensional torus. The curvature is the best possible, identically zero, but one can have periodic geodesics of length as small as desired. But assume that one knows that the area is not too small; then this will force the diameter to be large. See figure 6.56 on page 275 for the four required counterexamples. Conversely, Cheeger proved that no more is needed:

Theorem 90 (Cheeger 1970 [330]) *The length of periodic geodesics on a compact Riemannian manifold of given dimension can be controlled from below with the three following ingredients (besides the dimension, of course):*

1. *a lower bound δ for the sectional curvature*
2. *a lower bound for the volume* $\text{Vol}\, M$ *and*
3. *an upper bound for the diameter* $\text{diam}\, M$.

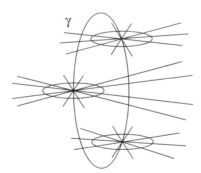

Fig. 6.54. Filling M completely by geodesics normal to γ

Cheeger employed beautiful "butterfly" arguments; see page 63 of Cheeger 1970 [330]. However, now we have a much simpler and more powerful argument, yielding better constants, which was given in Heintze & Karcher 1978 [699]. The idea is simple: start with a periodic geodesic γ of length L and draw all the geodesics orthogonal to it of length equal to the diameter diam M. This "tube" will certainly cover M completely; this is seen by considering, for any point $p \in M$ the point q of γ which is closest to p: γ is perpendicular to any segment from p to q by the first variation formula. But an extension to this situation of Rauch's comparison theorem (theorem 74 on page 259) will tell you that the volume of this tube (which in fact coincides with the volume of M) is bounded from above as a function of δ, diam M and of course L. So if the diameter and L are very small then Vol M will be too small.

Technically, one has to be a little more careful, as in the proof of the classical isoperimetric inequality given on page 65. For every point $m \in M$ take a point $p(m) \in \gamma$ as close as possible to m and pick a segment ξ from $p(m)$ to m. Of course

$$d(m, p(m)) \leq \operatorname{diam} g.$$

But when one applies Rauch's type of upper bound for computing the infinitesimal volume element along the segment from $p(m)$ to m, one has a proviso as on page 65. The proviso is that no Jacobi vector field along ξ starting with zero derivative at $p(m)$ can be allowed to vanish before m. This is technically proved mutatis mutandis exactly as in lemma 87 on page 270. We will use this technique again in §§§7.1.2.1.

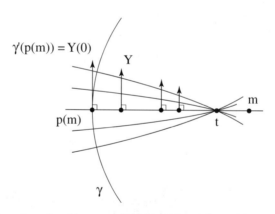

Fig. 6.55. If there is a Jacobi vector field Y with $Y(0) = \gamma'(p(m))$ and $Y(t) = 0$, with $t < d(m, p(m))$ then there is a negative second variation from $p(m)$ to m

Scholium 91 *The injectivity radius is either equal to half the length of the smallest periodic geodesic or equal to the smallest distance between two conjugate points (which is always bounded from below by $\pi/\sqrt{K}$ but not in general to equal to this bound).*

Note 6.5.2.1 Optimality, i.e. the need for these four ingredients entering Cheeger's theorem is evident in the pictures in figure 6.56. ♦

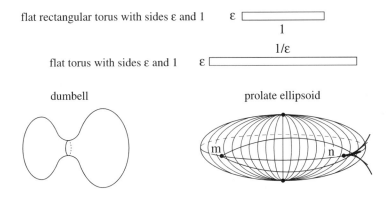

Fig. 6.56. (a) Flat rectangular torus with sides ε and 1. (b) Flat torus with sides ε and $1/\varepsilon$. (c) Dumbbell. (d) Prolate ellipsoid.

Note 6.5.2.2 In Heintze & Karcher 1978 [699] one finds a lower bound for the injectivity radius with weaker ingredients, namely: inf Ricci, sup K, diam, and Vol. Theorems 89 on page 272 and 90 on page 273 solve the injectivity radius problem for a "general" situation. ♦

Note 6.5.2.3 Using results from § 6.2 on page 239, on a manifold of positive curvature, with say $0 < \delta \leq K \leq \Delta$, a bound on the diameter is obtained for free. ♦

Note 6.5.2.4 On surfaces, with the Gauß–Bonnet theorem (theorem 28 on page 138) we can cancel the volume lower bound. More precisely let us assume, say, that $-1 \leq K \leq 1$. Then theorem 28 reads:

$$2\pi \|\chi(M)\| = \left\| \int_M K(m)\, dm \right\| \leq \int_M \|K(m)\|\, dm \leq \operatorname{Vol} M \qquad (6.23)$$

to the effect that we have a lower bound for the volume, employing only curvature bounds, provided that $\xi(M) \neq 0$. This excludes the torus and the Klein bottle.

This is one motivation for finding generalizations of the Gauß–Bonnet formula to any dimension. We will uncover such generalizations in § 15.7 on page 709. ♦

Note 6.5.2.5 We come back to positive curvature. The natural question is to find all possible differentiable manifolds which can carry some metric of positive curvature ("curvature and topology"). Among surfaces, only the sphere and the projective space can carry such metrics because of the Gauß–Bonnet formula. For higher dimensions, chapter 12 will show that the injectivity radius is important. We quote now what is known about the injectivity radius.

Theorem 92 (Klingenberg 1959 [811, 812]) *Let M be a compact oriented and simply connected manifold with $0 < K \leq \Delta$. If either*

1. *M has odd dimension and satisfies the "pinching" hypothesis*

$$\delta \leq K \leq \Delta$$

with $\delta/\Delta > 1/4$ or
2. *M has even dimension*

then the injectivity radius of M satisfies

$$\operatorname{Inj}(M) \geq \frac{\pi}{\sqrt{\Delta}}.$$

The even dimensional case is proven with a trick based on the proof of Synge's theorem (theorem 64 on page 246). The odd dimensional case is more elaborate—it uses Morse theory (see §§10.3.2 or 13.3 of do Carmo 1992 [452] or chapter 6 of Cheeger & Ebin 1975 [341]). The idea of Klingenberg is explained below, but is harder to work with when the ratio δ/Δ hits $1/4$, and was a little gappy there for a long time. The proof is very delicate and had to wait for Cheeger & Gromoll 1980 [345].

♦

What happens just below $1/4$ was a complete mystery until

Theorem 93 (Abresch & Meyer 1994 [8]) *If a compact manifold M has $\delta \leq K \leq \Delta$ with δ and Δ positive and*

$$\frac{\delta}{\Delta} > \frac{1}{4(1 + 10^{-6})}$$

then

$$\operatorname{Inj}(M) \geq \frac{\pi}{\sqrt{\Delta}}.$$

The proof is long and subtle. One can also look at the very recent survey Abresch & Meyer 1996 [7]. It starts with the Klingenberg–Cheeger–Gromoll "long homotopy" lemma: the initial idea is to take the smallest periodic geodesic and to consider a suitable deformation of it into a point (the manifold being simply connected). The key idea of Klingenberg is to use Morse theory to get the fact that, during the "long" homotopy, conjugate points will show up. Then one uses curvature assumptions to get a contradiction as follows. We assume that there is a small periodic geodesic γ of length smaller than $2\pi/\sqrt{\Delta}$ and pick any point $m \in \gamma$. Simple connectivity implies the existence of a homotopy between γ and the point m. If all the lengths of the curves in this homotopy are smaller than $2\pi/\sqrt{\Delta}$ then the entire homotopy can be lifted by the exponential map $\exp_m$ into T_mM, because the exponential map is a covering on the ball $B\left(0_m, \pi/\sqrt{\Delta}\right)$. This is a contradiction because the lift of γ will have two ends! In any homotopy, when one has a curve of length larger than $\pi/\sqrt{\delta}$ there is always a way to shrink all the lengths of the homotopy around it. Here, we require negative second variation on a two dimensional space of variations which is the worst possible case when our curve is a geodesic. Looking at the proof of Myer's theorem (theorem 62 on page 243) we see that there will always be a $(d-1)$ dimensional space of negative variations, and we are done since $d \geq 3$. It remains to be sure that our length is larger than $\pi/\sqrt{\delta}$, but our hypothesis says that one cannot have $\pi/\sqrt{\delta} > 2\pi/\sqrt{\Delta}$. This proceeds easily when $\delta/\Delta > 1/4$. A little more care is needed, as we have already said, when $\delta/\Delta = 1/4$.

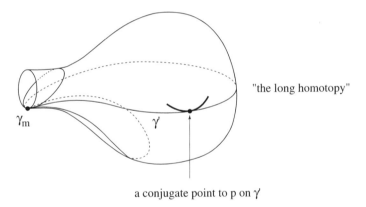

Fig. 6.57. The "long homotopy"

But when δ/Δ is actually below $1/4$, one has to work harder with the curvature assumption, and also with the various holonomy elements along the period geodesic. Holonomy will be discussed at large in chapter 13. In Synge's theorem, the holonomy element involved always had a fixed vector; here one has to put in more work and control the various angles showing up in the decomposition of the linear isometry into plane rotations given by the holonomy.

On the other hand, one cannot do better in a sense, because there are examples of compact simply connected Riemannian manifolds with positive curvature and injectivity radius $\mathrm{Inj}\,(g)$ as closed to zero as desired, with Δ

6.5.3 Convexity Radius

We come back to §§6.1.3 were it was proven that there are convex balls centered at any point of any Riemannian manifold. Again we ask the same question as for the metric balls: we want to control the radius of such balls.

Definition 94 *The* convexity radius *of a Riemannian manifold M is the infimum of positive numbers r such that the metric open ball $B(m,r)$ is convex for every $m \in M$.*

Compactness is really needed to ensure a positive convexity radius; see the picture in figure 6.52 on page 271. The following facts are not too difficult (see 7.9 of Chavel 1993 [326] or 5.14 of Cheeger & Ebin 1975 [341]).

Proposition 95 *If M is compact, then its convexity radius is positive. It is always less or than equal to half of the injectivity radius. The convexity radius $\mathrm{CvxRad}(M)$ satisfies*

$$\mathrm{CvxRad}(M) \geq \text{ the smaller of } \begin{cases} \frac{1}{2} \sup \frac{\pi}{\sqrt{K}} \\ \frac{1}{4} \text{length of shortest periodic geodesic.} \end{cases}$$

Apparently, there is no example in the literature with the convexity radius smaller than half of the injectivity radius. A natural conjecture is that such a bound should not be too difficult to prove.

To prove the above proposition, one can think of the sphere, for which $\pi/2$ is the convexity radius. On the other hand one, can look at the pictures in figure 6.58 on the next page of Jacobi vector fields; as soon as the norm decreases, convexity is lost.

An important application of the existence of a positive convexity radius on a compact Riemannian manifold was given above to triangulate surfaces; see the note 3.4.5.3 on page 139.

6.5.4 Cut Locus

We fix a point m of a Riemannian manifold M and use the name *cut locus* of m, denoted by $\mathrm{Cut\text{-}Locus}(m)$, the set of cut points of the various geodesics emanating from m. This is always a closed subset (possibly empty, or with branches going to infinity) of M. Just as the metric of a Riemannian manifold cannot in general be explicitly computed, the same is true for the cut locus. In fact, it is even harder to find its geometric structure. However it is of basic importance in control theory when the Riemannian manifold under consideration has any practical significance. Starting from m and arriving at

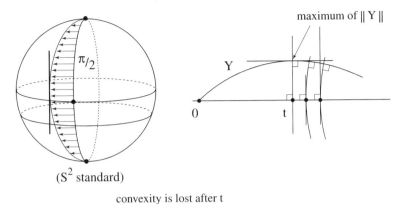

Fig. 6.58. Some Jacobi vector fields

the cut locus means some kind of catastrophe. Or, at least, it leaves us with a difficult decision as to the shortest path to follow; moreover the distance function will not be smooth there (see definition 305). We met the cut locus first on page 39 and then in more detail in the context of surface theory in §3.3 for surfaces. We are going to survey the state of knowledge about cut loci in general dimensions.

We are aware of only one survey on the cut locus: Kobayashi 1989 [825], and a great deal has been discovered since its date of publication. One can look at 3.81 in Gallot, Hulin & Lafontaine 1990 [542], at chapter 2 of Klingenberg 1995 [816] and at III.4 of Sakai 1996 [1085]. To make up for the absence of a more complete survey, we will present here a complement to Kobayashi's survey. The reader will find many references and much historical information in the first pages of Buchner 1978 [275]. For historians, it might be interesting to know that Élie Cartan used the cut locus in studying the topology of Lie groups in Cartan 1936 [316], precisely to prove that the second Betti number of a simple Lie group always vanishes, the reason being that the root structure forces the cut locus to be at least of codimension 3. Cartan also studied cut loci in Cartan 1988 [321], but it seems that he was unaware of the results of Poincaré and Myers.

By theorem 72 on page 255, the cut locus is always empty, for any point in any simply connected Riemannian manifold of nonpositive curvature. Recall that examples of surfaces were given in §3.3. We will now make a list of the positively curved compact manifolds where the cut locus is completely described. For spheres, the cut locus of every one of its points is a single point: its antipode. Taking the quotient by the antipodal map, one gets the standard real projective space; the cut locus of any point of the real projective space is the dual hyperplane, namely the image under the antipodal quotient of the associated equator in the sphere.

280   6 Metric Geometry and Curvature

More generally, the same applies to the various $\mathbb{KP}^n$ of §§§4.1.3.5: the cut locus of every point is the dual hyperplane which is a $\mathbb{KP}^{n-1}$; see §§6.1.5. So the cut locus of any point is a nice submanifold (located at constant distance π for a suitable scaling of the metric) of dimension

$$n-1, 2(n-1), 4(n-1), \text{ or } 8$$

if, respectively

$$\mathbb{K} = \mathbb{R}, \mathbb{C}, \mathbb{H}, \mathbb{C}\text{a} .$$

For more information, see 2.114 of Gallot, Hulin & Lafontaine 1990 [542], or 3.E of Besse 1978 [182]. Note that for these manifolds, the injectivity radius and the diameter are equal; see §§6.5.5 for more on that.

For a flat torus obtained as the quotient of $\mathbb{R}^d$ by a lattice Λ (see §§4.3.3) the cut locus is the image by the exponential map (which here is also the quotient map) of the boundary of the fundamental domain. We suggest to the reader that she make drawings, at least on a two-dimensional flat torus, to determine the structure of the cut locus of a point in a flat torus, paying attention to the different possible cases. For tori of revolution embedded in $\mathbb{R}^3$, see §3.3.

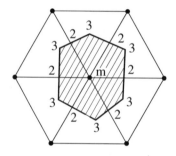

The cut-locus of this flat torus in $T_m M$ before $\exp_m$
The cut-locus of m in M is its image by $\exp_m$

Fig. 6.59. The cut locus of this flat torus in $T_m M$ before $\exp_m$. The cut locus of m in M is its image under $\exp_m$.

For other space forms (see §§6.3.2), cut loci are more or less workable. They were completely determined for (compact) symmetric spaces by Takeuchi 1979 [1179]. This illustrates the philosophy (see §§4.4.3) that symmetric spaces are completely accessible to explicit determination of their geometric invariants through algebra. An exception will be the isoperimetric profile; see §§§7.1.2.1.

We saw in §3.3 that the cut locus of the quadrics in $\mathbb{E}^3$ are conjecturally topological intervals. It seems to be an open problem to find the cut locus structure of quadrics in higher dimensions as well as to find, even to guess, a complete proof for the $\mathbb{E}^3$ quadrics.

There is only one other example that the author is aware of: namely the 3-dimensional sphere (or the 3-dimensional real projective space). We saw in §§4.3.3 that the motions of a rigid body around its center of gravity are given as the geodesics of the group $SO(3)$ for a left invariant metric depending on three parameters $\{a, b, c\}$. For simplicity, we will examine the universal covering which is (see §§4.1.2) the sphere S^3 considered as a Lie group. So S^3 has left invariant metrics depending on constants $\{a, b, c\}$ (of course, only different ratios will give different geometries, after rescaling). To our knowledge it is open to find the structure of the cut locus when $a < b < c$. When $a = b < c$, the cut locus was determined by Sakai in Sakai 1981 [1084]. It has the structure of a disk: inside the disk, two different segments meet; at the boundary only one segment reaches the disk. For the case $a < b = c$, it seems to us (to be checked) that the cut locus is a segment. Inside the segment, a one-parameter family of segments arrive, at both ends only one segment does. See figure 6.60.

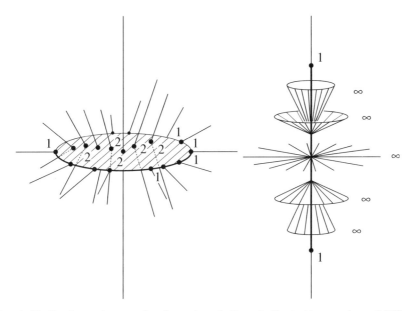

Fig. 6.60. In these pictures the characters 1, 2, ∞ indicate the number of different segments attaining the labeled point

A reasonable guess when $a < b < c$ might be that the cut locus has the same structure as the "cut locus" of a solid 3-dimensional ellipsoid with unequal axes. By this we mean the closure of the set of points inside the ellipsoid which are connected to the surface of the ellipsoid by more than one segment. The structure of this naive object was determined recently in

Degen 1997 [437]. It is always a topological disk located in the principal plane of coordinates. The proof is quite simple: such points are centers of spheres bitangent to the ellipsoid, and classically the line of their contact points is orthogonal to the principal plane.

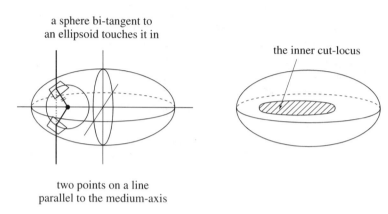

Fig. 6.61. (a) A sphere bitangent to an ellipsoid touches it in two points on a line parallel to the medium axis. (b) The inner cut locus

A positive answer for the homogeneous metric on S^3 (see §§4.3.8) will however not solve the following question: launch a body around its center of gravity and ask when the attained position can be obtained from a different initial position in the same time. This is because one has to work on $\mathbb{RP}^3$ instead of S^3. In exchange, a continuity argument shows that when the three moments of inertia are close enough, the cut locus is a smooth surface as it is in $\mathbb{RP}^3$. This is because accidents in smoothness could happen only when conjugate points show up. And they also depend continuously on the inertia parameters.

Note 6.5.4.1 (Poinsot motions) The motions of a body around its center of gravity have a marvelous geometric interpretation due to Poinsot 1842 [1035]; see figure 6.62 on the next page. Take the ellipsoid of inertia of the body. The motions of the body are the same as those obtained by rolling the ellipsoid of inertia on a plane without sliding. The distance of the plane to the center of the ellipsoid depends on the initial conditions. For proof, and more, see (for example) §29 of Arnold 1996 [66]. We cannot resist mentioning that the curves drawn out on the surface of the ellipsoid are the intersection of the ellipsoid with the various spheres centered at the center of the ellipsoid; this is pictured in figure 6.62. From this, you will deduce immediately the *Euler theorem*: around the longest and the shortest axes of inertia, the motions are stable, but the motion around the middle axis is unstable. Please check this

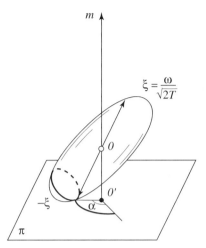

Fig. 6.62. The visual interpretation of the Poinsot motions

by launching a book (not a square book, a truly rectangular one) around the three possibilities. ♦

Note 6.5.4.2 Most often, a clear picture of the cut locus can only be attained from one of the first conjugate locus: this is the set made up by following the geodesics propagating from a given point, up to the first conjugate point. By theorem 86, the cut point always appear just at or before the first conjugate point. Conjugate points are apparently more accessible because they are given by the Jacobi differential equation for geodesics. For the ellipsoid, this requires only *hyperelliptic functions*; see the conjectured picture for an ellipsoid in figure 3.30 on page 130. ♦

We now present what is known about the cut locus (to our knowledge). First, as in §3.3 the cut locus is closed and is the closure of the set of points attained by two different segments; see 2.1.14 in Klingenberg 1995 [816] for a proof. Second, as already mentioned for surfaces in §3.3, the cut locus of a smooth manifold can be a very bad set, e.g. nontriangulable: see Gluck & Singer 1979 [570]. The cut locus of a real analytic metric is subanalytic, and in particular is triangulable: we saw in §3.3 that this was proven by Myers for surfaces in 1935. For higher dimensions, this result remains true but the proof is much much more expensive and due to Buchner 1977 [273]. In dimension two, this means that the cut locus is a nice graph. Myers showed that the order of a point in the graph is equal to the number of the segments connecting it to m.

For smooth manifolds, Buchner 1977,1978 [274, 275] and Wall 1977 [1228] proved:

Theorem 96 (Buchner & Wall) *The cut locus of a smooth compact manifold is triangulable for generic Riemannian metrics and moreover is stable under perturbations of the metric. The stability is smooth up to dimension 6, thereafter only continuous.*

In Buchner 1978 [275] the various possible local structures of the cut locus around a cut point are described for generic metrics up to dimension 6. We give the pictures for dimensions 2 and 3 in figure 6.63.

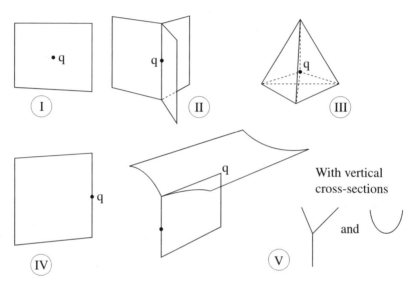

Fig. 6.63. Cut loci in 2 and 3 dimensional compact Riemannian manifolds

The difficulty for all these studies is an unavoidable dichotomy for cut points: the mixture of points with two different segments and conjugate points.

The structure of the generic conjugate locus was investigated back in 1970 by Weinstein 1968 [1247]. We only note here that, contrary to what most people think, the conjugate locus is not always closed in M: see Margerin 1992 [896].

Studies on the cut locus have been infrequent after Buchner. However, very recently there has been a strong revival. Firstly concerning surfaces, Itoh 1996 [758] showed that the cut locus of a surface always has finite 1-dimensional Hausdorff measure; this solves Ambrose's problem for surfaces: Ambrose asked (back in 1956) if curvature and parallel transport from one point determines completely the Riemannian metric. The Ambrose problem is discussed on page 250 and on page 704. Finally, cut loci are not incorrigible; see Itoh 1996 [758] for a relation between their Hausdorff dimension and the smoothness of the metric. See Itoh & Tanaka 2001 [759].

The reader versed in analysis will be pleased to note that we will meet the cut locus in studying heat diffusion in a Riemannian manifold in the discussion of theorem 168 on page 400.

It is interesting to contrast cut loci in Riemannian manifolds with those of a more general metric space. For example, on convex surfaces (defined as boundary of a convex body in $\mathbb{E}^3$) the set of points y which can be joined to a given point x by at least three different segments can be everywhere dense in the surface. In particular we are very far from having positive injectivity radius: see Zamfirescu 1996 [1301]. For Alexandrov spaces, see Shiohama & Tanaka 1992 [1135], for Carnot–Carathéodory metrics (see §§14.5.7) the injectivity radius is zero in general, see El Alaoui, Gauthier & Kupka 1996 [487].

6.5.5 Simple Question Scandalously Unsolved: Blaschke Manifolds

We start with this section with a trivial remark: for any Riemannian manifold, $\operatorname{Inj}(M) \leq \operatorname{diam} M$. A curious mind will ask which manifolds achieve equality:

Question 97 *What are the compact Riemannian manifolds M for which*

$$\operatorname{Inj}(M) = \operatorname{diam} M?$$

This is an extremely strong condition and the naive reader will suppose its solution easy, namely that only the standard $\mathbb{KP}^n$ and of course the spheres enjoy this property (see §6.5.4). Nevertheless the question is still open today. Indeed, no other examples are known than the $\mathbb{KP}^n$ and the spheres. It is even unknown if there are others. Here is the beginning of what we know. The basic reference, up to 1978, is Besse 1978 [182]. For more see §10.10.

The condition is in fact very strong. Let us normalize the manifold with

$$\operatorname{Inj}(M) = \operatorname{diam} M = \pi.$$

So the cut value of every geodesic is constant and equal to π. Constancy of the cut value for all the geodesics emanating from a given point is not a strong condition; this always happens in a surface of revolution for the north pole, the cut locus being reduced to the south pole. But we ask that this occurs for every point. Pick a point $m \in M$ and take n in its cut locus. Let γ be some segment from m to n: then looking at points of γ *after* n one sees that in fact γ is a periodic geodesic of total length 2π. So M is a manifold all of whose geodesics are closed. Such manifolds will be studied in §10.10. We turn our attention to digging out information that can be obtained with simple geometric means (essentially the first variation formula).

First, the set of all segments from m to n build up a smooth sphere which can be denoted by $S(m,n)$. All of these spheres have the same dimension. The cut locus of m is a smooth submanifold of M of dimension complementary to that of the $S(m,n)$.

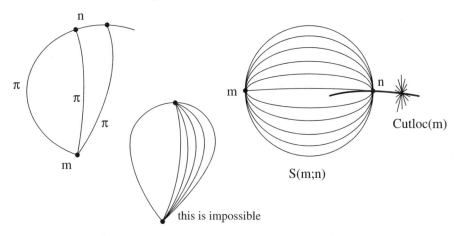

Fig. 6.64. Behaviour of the cut locus

With the use of expensive tools from algebraic topology, we will see in §§10.10.2 that the only possibilities for the dimension of the $S(m, n)$ are precisely the same as for the sphere and the $\mathbb{KP}^n$: if $k = \dim_{\mathbb{R}} \mathbb{K}$, then the possible dimensions are the multiples of k. And more: the topology of M is known to be almost that of a $\mathbb{KP}^n$. We will also see in theorem 257 on page 493 that the question is solved for topological spheres, and the answer is the expected one: our manifold has to be isometric to a standard sphere or a standard real projective space (where standard means of constant sectional curvature).

6.6 The Geometric Hierarchy and Generalized Space Forms

Except in §§6.6.1, we will only consider compact Riemannian manifolds, for simplicity. Noncompact cases, especially those of finite volume, are vitally important and under earnest study. The special manifolds of this chapter, and their quotients, arise naturally in number theory (as modular domains), geometry and dynamical systems. Some references to these neighbouring subjects will be given. References for space forms are Gromov and Pansu 1991 [637], Vinberg 1993 [1221] and, for the influence of Calabi, Berger 1996 [168].

Table 6.1. Geometric hierarchy of homogeneous Riemannian manifolds

Nature	Local geometry	Geometric hierarchy of homogeneous Riemannian manifolds $\pi_1 = 0$	Global existence	Global structure
Euclidean space: $K = 0$	Locally Euclidean: $a^2 = b^2 + c^2 - 2bc\cos\alpha$	$\mathbb{E}^d$	$\mathbb{E}^d/\Lambda$, and some finite quotients	Bieberbach 1911. Classification not completely extended.
K constant, $+1, 0, -1$; 3 point homogeneous, plane isotropic; universal $a = F(b, c, \alpha)$	Locally $\mathbb{S}^d(K)$	S^d $+1$ $\uparrow$ $\mathbb{S}^d(K)$ $\downarrow$ Hyp^d -1	$+1$: Vincent, lens spaces, etc. (d odd) $-1, d = 2$: Arithmetic, gluing. Riemann surfaces $-1, d > 2$: Arithmetic, any d. Reflections known, $d \leq 21$, impossible $d \geq 40$; (first Löbell 1931); $\exists$ nonarithmetic $\forall d$ (Gromov, Piatetski-Shapiro 1988) $d = 3, 4$ Markarov 1966	$+1$: almost known (d odd) $-1, d = 2$: Teichmüller $\infty^{6\gamma - 6}, (p, q, r)$ Klein $-1, d > 2$: Rigid π_1 (Mostow 1968). Reflections impossible (Vinberg) $d \geq 40$. Open: more arithmetic or more nonarithmetic

Table 6.1. (continued)

	Geometric hierarchy of homogeneous Riemannian manifolds			
Nature	Local geometry	$\pi_1 = 0$	Global existence	Global structure
$\mathbb{K}\mathbb{P}^n$, $\text{Hyp}_{\mathbb{K}}^n$, $\mathbb{K} = \mathbb{C}, \mathbb{H}, \mathbb{C}a$ ($n=2$ only). Symmetric space of rank 1. 2-point metric homogeneous. Isotropic. $\text{diam } M = \text{Inj } M$.	$K = \pm(1 + 3\cos\alpha^2)$	$\mathbb{K}\mathbb{P}^n$, $\text{Hyp}_{\mathbb{K}}^n$	None (up to $\mathbb{Z}_2$) if $K > 0$. For $K < 0$, Borel 1963: $\exists$ arithmetic. Mostow 1985: $\exists$ nonarithmetic $n = 2, 3$ (use reflection)	$\mathbb{C}$ & $d > 4$: unknown if $\exists$ nonarithmetic. $\mathbb{H}, \mathbb{C}a$ Gromov, Schoen 1992: all arithmetic. Mostow 1973: all rigid π_1.
Symmetric spaces. σ_m isometry $\forall$ points m. $DR \equiv 0$. G/H where $H = \text{Fixed}(\sigma)$, $\sigma: G \to G$ involution. They come in pairs: $G_\pm/H$, same H, $K \geq 0$ or $K \leq 0$.	$DR \equiv 0$, determined by their curvature tensor: Tricerri & Vanhecke 1986 [1201].	Completely classified (É. Cartan 1925): $\text{Grass}_{\mathbb{K}}(p,q)$ natural structures finite exceptions Lie groups themselves	Borel 1963: $K \leq 0$: existence in every case	$K \leq 0$, rank ≥ 2: Margulis superrigidity implies arithmeticity. Mostow 1977: at π_1 level.
Homogeneous spaces G/H (H necessarily compact)	*Not* determined by their curvature tensor.			

6.6.1 The Geometric Hierarchy

6.6.1.1 Space Forms

Let us consider Riemannian geometry as a generalization of Euclidean geometry. The first level of generalization is to look at metric spaces which are 3 point transitive (or satisfy the *mobility axiom*) as seen on page 253 We found that these spaces are precisely the simply connected, constant sectional curvature manifolds, namely (besides Euclidean spaces) the spheres and the hyperbolic spaces.

6.6.1.2 Rank 1 Symmetric Spaces

Pursuing our hierarchy, we look now for metric spaces which are "only" 2 point transitive i.e. there is always an isometry carrying one pair of points into another pair (provided of course that their respective distances are equal). Another formulation is: we are looking for spaces which are isotropic, in that all unit tangent directions are metrically equivalent. The simply connected ones are all known. Discarding the constant curvature ones discussed about, we are left with the $\mathbb{KP}^n$ and their negative curvature analogues, denoted here by $\text{Hyp}_{\mathbb{K}}^n$. They are the symmetric spaces of rank one, see §§4.3.5. Their curvature ranges from 1/4 to 1 (or from -1 to $-1/4$) after normalization. The $\mathbb{CP}^n$ together with the $\text{Hyp}_{\mathbb{C}}^n$ are the complex geometries corresponding to spheres and hyperbolic spaces. Their classification is in Wang 1952 [1233] for the compact ones and in Tits 1955 [1192] for the noncompact ones. It is still a long story to carry out the classification in detail; the best reference today to our knowledge is Karcher 1988 [779] (see page 120). The difficult part of the proof is to show that one is in a symmetric space, see Szabo 1991 [1173] for a short proof. Then the space is forced obviously to be of rank 1, see §§4.3.5.

6.6.1.3 Measure Isotropy

The search for spaces which are only *measure isotropic* (so called *harmonic*) started in the early 1940's. There are many equivalent definitions; the first says that at every point the solid angle (the infinitesimal measure along a geodesic starting from this point) depends only on the distance. The word *harmonic* was chosen because this condition is equivalent, at least locally, to the property that the value at every point p of any harmonic function f (i.e. a function satisfying $\Delta f = 0$) is equal to the mean value of f on every metric sphere centered at p. Lichnerowicz conjectured in 1944 that such spaces are, at least locally, isometric to a space form of rank one, so that measure-isotropy forces metric isotropy. After many intermediate results, the conjecture was settled positively for compact manifolds by Szabó in 1990 and then negatively for noncompact manifolds by Damek and Ricci. All of the references can be found in Berndt, Tricerri, & Vanhecke 1995 [180] and its bibliography. Conversely, harmonic compact manifolds of negative curvature are also proven in Besson, Courtois, & Gallot 1995 [189] to be symmetric spaces; this is a corollary of theorem 251 on page 484.

6.6.1.4 Symmetric Spaces

Next in the hierarchy are the symmetric spaces of §§4.3.5. In some sense, their geometry is completely computable; in particular answers to metric questions and the geodesic flow. This does not prevent certain features from being very hard to work out explicitly, e.g. the complete trigonometric formulas for triangles in Leuzinger 1992 [860].

6.6.1.5 Homogeneous Spaces

A further step down is to look for merely homogeneous spaces—namely we ask the isometry group to be transitive. Thus the geometry will be the same at every point, as in Euclidean spaces. It turns out that this last class is extremely large and then we will not look for their space forms. But we will meet many of them in the future, because their curvatures can be computed (to some extent, and sometimes awkwardly).

6.6.2 Space Forms of Type (i): Constant Sectional Curvature

Do not think that everything is known today about constant sectional curvature manifolds. Moreover the subject is a whole field in itself. Let us me add just one more motivation, among many others possible. To study dynamics on a plane polygonal billiard with angles rational multiples of π one attempts to build a space form by identification along the sides of the polygon. This very nearly yields a space form, of zero curvature (flat), with a finite number of exceptional points where the surface is not smooth but has a conical angle an integral multiple of π. However, to study such an object, one endows it with a metric of constant negative curvature. For various references and beautiful results on this topic, see the survey which is part II of Vinberg 1993 [1221]. In a different context, a very brief survey is given in Berger 1996 [168]. We now present another survey because of the importance of the subject.

Today our knowledge of space forms is as follows. Remark first that after normalization there are only three cases to consider: $K = 1, 0, -1$ i.e. one should look for quotients of $S^d, \mathbb{E}^d, \mathrm{Hyp}^d$. For the cases $K = 1$ and $K = 0$ the basic reference is Wolf 1984 [1276]. For $K = -1$, the main reference is Ratcliffe 1994 [1049].

The quotients of S^d and $\mathbb{E}^d$ are childishly simple for $d = 2$: namely one can only make out of S^2, besides S^2 itself, the projective plane $\mathbb{RP}^2$; as for $\mathbb{E}^2$, one can construct only the flat tori associated to various lattices and the

Klein bottles (which have to be rectangular ones). As to the higher dimensions, these quotient objects were completely classified in 1930 for S^3 and very nearly classified in 1948–1960 for general dimensions. But the complete classification is still not really finished. One has to look first for the possible discrete subgroups of the orthogonal group $O(d+1)$ of all isometries of S^d. This is treated in full detail in the book Wolf 1984 [1276] (try to get the latest edition). Contrary to what most people think, the classification is not completely finished but this is only an algorithmic problem. Moreover the rigidity problem for quotients of the sphere is almost virgin territory, in contrast with hyperbolic space forms, which we will turn to shortly. To get some feeling about the mysterious and hard component in spherical quotients, see Milnor 1966 [923].

Vanishing curvature occurs precisely on locally Euclidean Riemannian manifolds, called most often *flat*. The question of categorizing the complete flat manifolds was asked explicitly by Hilbert in 1900 in his famous address. In 1911, Bieberbach got the basic result (still not too easy to prove) that

Theorem 98 (Bieberbach 1911) *Every flat complete Riemannian manifold is a finite quotient of a product $\mathbb{E}^k \times T^{d-k}$ of some Euclidean space with a flat torus.*

If one looks only for compact flat manifolds, then there can be no Euclidean factor, and the compact flat manifolds are finite quotients of flat tori. Moreover Bieberbach proved that the number of possible quotients is finite in every dimension. But this is still not a complete classification: see again Wolf 1984 [1276]. We just note that, especially in dimension three, the problem was of great interest for crystallographers. And also that we will come back later on to flat tori: see chapters 9 and 10.

Manifolds which are quotients of hyperbolic space, called *hyperbolic manifolds* follow a completely different story, and are in some sense more important. The study of amazing relations between geometry (volumes of polytopes, cross-ratio and its generalizations) is currently in full swing and enjoys beautiful results. We just mention chapter 14 (by Kellerhals) of the book Lewin 1991 [862] and the informative Oesterlé 1993 [972].

We briefly comment on these two remarks and focus our attention on compact manifolds; also see §§6.3.2. Hyperbolic surfaces were quite clearly understood by the end of the 1930's. We discard the trivial case of positive and zero constant curvature surfaces (spheres and tori). It is easy to build up examples geometrically. For example, take suitable triangles or polygons in the hyperbolic plane and make tessellations with them. But then the group yielding the quotient is hard to visualize. Another way is to glue together pantaloon pieces along their boundaries, provided the boundaries are closed geodesics; we saw this in §§4.3.7 and in figure 4.10 on page 157. The pantaloon pieces are obtained by gluing two identical hyperbolic hexagons all of whose angles are equal to $\pi/2$. This latter method enables us to build up all compact hyperbolic surfaces, as seen by working back by dissection (this is perfectly exposed in Buser 1992 [292]). The other method is to use the conformal representation theorem for compact orientable topological surfaces of any genus; see theorem 70 on page 254 The complete classification of hyperbolic surfaces was worked out by Teichmüller in the late 1930's. On an orientable surface of genus γ, the constant curvature Riemannian structures build up a space with $6\gamma - 6$ parameters. Constructing all of these hyperbolic surfaces by group theory (i.e. quotienting the Poincaré disk, using the conformal representation theorem) is harder and consists largely in number theory; see for example Vigneras 1980 [1215]. The link between the algebra and the genus of the surface is already subtle. We note here for the curious reader that the unorientable hyperbolic surfaces (beyond the topological classification) have received less attention; we gave references in §§6.3.2.

Unorientable hyperbolic surfaces have been recently described in many places, because, although these surfaces are no longer objects generalizing a complex variable (describing, for example, complex algebraic curves) they do nonetheless appear in crystallography, and are important to geometry as representing "complexifications" of real algebraic curves. A surface, whether orientable or not, which is equipped with a family of complex coordinate charts which agree up to either holomorphic or conjugate holomorphic transformations is called a *Klein surface*. We give some references: Alling & Greenleaf 1971 [25], Bujalance, Etayo, Gamboa & Gromaszki 1990 [276], Seppälä 1990 [1123], Singerman 1974 [1146].

6.6.2.1 Negatively Curved Space Forms in Three and Higher Dimensions

Starting in three dimensions, the difficulties of classifying hyperbolic manifolds appear formidable. We will be very sketchy here. First: it is unbelievably hard to obtain examples. Our two dimensional techniques look easy to generalize to any dimension by considering suitable polytopes in Hyp^3. But researchers had to wait for Löbell in 1931 to give the first infinite family of nonisometric hyperbolic three manifolds. We remark that he used a geometric approach, and not at all a group theoretical one. It is far from easy to start from a pantaloon type gluing and find out what subgroup of $SO(3,1)$ (the group of all isometries of Hyp^3) generates the resulting three manifold that has been thus glued together. It is a dramatic fact that polyhedral (polytopal, for higher dimensions) techniques cannot work in every dimension. There is a good reason for that: in Vinberg 1984 [1220] it was discovered that geometric tessellation constructions do not exist in large dimensions. Left open today is the exact value of the limit dimension, see Vinberg 1993 [1221].

So we are now trying to build space forms in any number of dimensions. This is can be done only using number theory, and the spaces obtained (using suitable subgroups of $SO(d,1)$) are called *arithmetic*. There are also (see below) nonarithmetic space forms. The definition of arithmeticity, which is also valid for the more general space forms to be seen below, is as follows: we consider space forms of the type G/Γ, for a semi-simple Lie group G and a discrete subgroup Γ. Arithmeticity says, roughly, that Γ is given by matrices with integral entries, up to finite differences. More precisely, (G, Γ) is arithmetic if there exists a subgroup G' of the general linear group $GL(N, R)$, and a group epimorphism $\rho : G' \to G$ with compact kernel such that $\Gamma \cap \rho(G' \cap GL(N, \mathbb{Z}))$ has finite index both in Γ and in $\rho(G' \cap GL(N, \mathbb{Z}))$. In two dimensions, nonarithmetic examples arise easily from gluing. The possibility of finding such space forms was in the air in the 1960's, but they appeared explicitly for the first time in the founding paper Borel 1963 [223]. The difficulty is to get compact quotients; if not, then any $\mathbb{Z}$ valued discrete representation would do. It is not hard to realize the difficulty, since the first quotient is from Hyp^2 and is

$$SL(2, \mathbb{R}) / SL(2, \mathbb{Z}) \ .$$

This is the *modular domain* and is not compact (even though it has finite volume and has two singular points); see figure 6.36 on page 255.

6.6.2.2 Mostow Rigidity

The reader will want some kind of Teichmüller theory (deformations, number of parameters) for higher dimensions. On a given compact orientable surface of genus γ, there is a $6\gamma - 6$ parameter family of hyperbolic structures (metrics up to diffeomorphism). Teichmüller began to understand this family in 1939 (see however the historical notes 9.9 of Ratcliffe 1994 [1049]). We will come back to this in chapter 11. The big event:

Theorem 99 (Mostow 1968 [947]) *Two space forms of constant negative curvature and dimension larger than 2 whose fundamental groups are isomorphic (as groups) must be isometric.*

So important and conceptual is the rigidity of hyperbolic manifolds that a succession of simpler proofs have appeared. See chapter XI of Ratcliffe 1994 [1049], part II of Vinberg 1993 [1221], and the completely new Besson, Courtois & Gallot 1995 [189, 190]. Also see the survey Besson 1996 [191], the expository Pansu 1997 [999], Pansu 1995 [997], Gromov & Pansu 1991 [637] and the book Farrell 1996 [504].

Mostow's initial proof used the *sphere at infinity* (see §§§12.3.4.3) and the subtle notion of a quasiconformal map. Gromov's proof used ideal simplices in hyperbolic geometry (see this, and his notion of simplicial volume, in §§§11.3.5.2). We cannot resist the temptation to sketch the ideas behind the various steps of the complete proof, which is completely explained in the book Ratcliffe 1994 [1049]. From algebraic topology, the fact that our two space forms M and M' are covered by $\mathbb{R}^d$ implies that, if they have isomorphic fundamental groups, then they are in fact homotopic; let us write $f : M \to M'$ for some such homotopy. One lifts f to a map

$$\hat{f}: \text{Hyp}^d \to \text{Hyp}^d \ .$$

Then one uses the fundamental technique of §§§12.3.4.3: one proves first that $\hat{f}$ is a quasi-isometry of Hyp^d; then one can extend it to the sphere at infinity $S(\infty)$ to get a map

$$\check{f}: S(\infty) \to S(\infty) \ .$$

We will be done if we can prove that $\check{f}$ is a Möbius transformation, since this equivalent to saying that $\hat{f}$ is an isometry of Hyp^d. Now let us triangulate Hyp^d by regular simplices and let us prove that all of the images of simplices under $\check{f}$ are again regular. This is a volume argument, which after some work (recalling that the regular ideal simplices are characterized as those of maximal volume) boils down to proving that

$$\text{Vol}\, M = \text{Vol}\, M' \ .$$

But we can use theorem 273 on page 517 twice:

$$\|M\| = \text{Vol}\, M / \textit{VRS}\,(d)$$

and

$$\|M'\| = \text{Vol}\, M' / \textit{VRS}\,(d) \ .$$

But the simplicial volume of Gromov is an invariant depending only on the fundamental group; since the fundamental groups are isomorphic,

$$\|M\| = \|M'\|$$

so that

$$\text{Vol}\, M = \text{Vol}\, M'$$

as desired. We end by using a lemma to the effect that when two regular ideal simplices have a common face, they are deduced by a hyperbolic (hyperplane) symmetry. Look at the vertices of our two triangulations in $S(\infty)$, both of which are built of regular simplices. By the above, the restriction of $\check{f}$ to these sets is an isometry (a Möbius transformation). But these sets are everywhere dense in $S(\infty)$, so we are finished.

6.6.2.3 Classification of Arithmetic and Nonarithmetic Negatively Curved Space Forms

Mostow's rigidity, although very strong and wonderful, does not yield any classification. Today one still does not have a classification. On one side, we have the Borel type arithmetic examples. On the other hand, without employing polytopal tiling by reflexions on the polytopes' faces, one can still produce nonarithmetic examples by cutting an arithmetic form along some totally geodesic hyperplane and gluing the two parts again with a different identification. This was carried out in Gromov & Piatetski-Shapiro 1988 [638] and yielded examples which are definitely not arithmetic. The open main question is

Question 100 *Among compact hyperbolic manifolds, are the arithmetic or the nonarithmetic more numerous?*

But in higher dimensions, it is much more difficult to prove the existence of nonarithmetic examples. Such a proof was achieved only in 1988; see Gromov & Pansu 1991 [637] and part II of Vinberg 1993 [1221].

6.6.2.4 Volumes of Negatively Curved Space Forms

An important issue for negatively curved space forms are their volumes. The Gauß–Bonnet formula implies immediately that the set of volumes of hyperbolic surfaces is an arithmetic progression. The existence of an universal positive lower bound for any dimension is proven in Wang 1972 [1234] where it is proven that starting in dimension 4 and for space forms of any rank the set of volumes is discrete. In Prasad 1989 [1040], the volumes of all arithmetic space forms are explicitly computed. The case of dimension 3 was solved in Jørgensen 1977 [765] and Thurston 1997 [1189] and the result is fascinating: volumes are isolated from zero (however the best value is still unknown today), but the set of possible volumes has accumulation points located on a discrete scale. For more see Gromov 1981 [615] and Ratcliffe 1994 [1049], page 501. See the recent Goncharov 1999 [572] to realize the depth of the problem.

6.6.3 Space Forms of Type (ii): Rank 1 Symmetric Spaces

The rank 1 symmetric spaces are the first level of generalized space forms, covering different levels of generalization of the preceding spaces. We saw that they coincide (globally) with the 2-point transitive spaces, i.e. there is always an isometry carrying one pair of points into any other pair of of points (provided the distances are equal). The associated space forms will have that 2-point property only locally. We discard of course the constant curvature case, treated above. Now Synge's theorem 64 on page 246 shows that, in the positive curvature case, because the $\mathbb{KP}^n$ are even dimensional, only the simply connected spaces or the two sheet quotient $\mathbb{CP}^{2k+1}/\mathbb{Z}_2$ can appear. There is thus no classification problem for positively curved rank 1 symmetric spaces.

Let us consider the negatively curved rank 1 locally symmetric spaces. The complex case is particularly interesting because of the relations with complex analysis and algebraic geometry. General existence of lots of arithmetic examples was proven in Borel 1963 [223] (the same article mentioned previously for existence of arithmetic space forms). Nonarithmetic examples appeared in Mostow 1980 [949], built up by subtlely tessellating polyhedra in $\text{Hyp}_{\mathbb{C}}^2$ and in $\text{Hyp}_{\mathbb{C}}^3$. Also later on in Deligne & Mostow 1986 [439] by a different and expensive number theoretic technique. The question of deciding if there exist nonarithmetic manifolds covered by the $\text{Hyp}_{\mathbb{C}}^n$ is still open today for higher n.

For the quaternionic and the Cayley case, $\text{Hyp}_{\mathbb{H}}^n$ and Hyp_{Ca}^2, it was proven in Gromov & Schoen 1992 [639] that all compact quotients of these spaces are necessarily arithmetic. The proof uses hard analysis; more precisely it uses harmonic maps into manifolds with singularities (see §14.3). Their proof also uses Corlette 1992 [404], where there is an intermediate result, based on an essential tool which is a Bochner-type formula for manifolds with special holonomy.

The Mostow rigidity theorem 99 on page 293 has a companion rigidity theorem for the compact quotients of the $\text{Hyp}_{\mathbb{K}}^n$ in Mostow 1973 [948], but today it can be obtained by the general theorem 251 on page 484 of Besson, Courtois & Gallot [189] already mentioned.

6.6.4 Space Forms of Type (iii): Higher Rank Symmetric Spaces

The higher rank irreducible symmetric geometries (see §§4.4.3) split again into two classes: those of nonnegative curvature and those of nonpositive curvature. Finding all the quotients in the positive case is a finite job, which was completed back in Cartan 1927 [314]. For the negatively curved ones, it is again in Borel 1963 [223] where compact quotients were proven to exist for the first time, always by arithmetic considerations. But here, if there is a classification problem, it is for number theorists. Indeed, in Margulis 1975 [900], a superrigidity result was proven, which implies not only topological rigidity but also arithmeticity for ranks greater than one. Margulis's tools were expensive; today there are other ways to recover Margulis's result (see for example Jost & Yau 1990 [770]). Results like those of Besson, Courtois & Gallot 1995 [189] still do not yield superrigidity. Do not forget the survey Gromov & Pansu 1991 [637].

6.6.4.1 Superrigidity

We briefly explain the idea of *superrigidity*. Arithmeticity is a consequence of superrigidity, which concerns discrete subgroups Γ of isometries of a symmetric space, such the quotient by Γ is a compact manifold. On the universal cover, one can again define the *sphere at infinity* $S(\infty)$. One puts on it the *Tits metric* of §§§12.3.4.3. Now the fact that the rank is at least two makes the geometry of the covering space, and of its sphere at infinity, quite rigid. The flat spaces (flat totally geodesic submanifolds) are moreover very numerous, since they are all conjugate under the action of the group; see §§4.3.5. Finally one sees that nothing can be moved.

6.6.5 Homogeneous Spaces

These are the spaces whose isometry group is transitive. This implies that all points have the same geometry. Thus they are the natural (but weakest) Riemannian generalization of Euclidean spaces. Hence their properties can be understood only very weakly. We do not say much more here about them.

It is much more complicated to compute their curvature in the general case G/H that in the special case of a bi-invariant metric as seen in equation 4.39 on page 210. see formula 15.15 on page 719. But they will be of fundamental importance later on when building examples of Riemannian manifolds with various types of conditions imposed on the curvatures, e.g. §§§12.3.1.1

7 Volumes and Inequalities on Volumes of Cycles

Contents

7.1	Curvature Inequalities............................ 299	
	7.1.1 Bounds on Volume Elements and First Applications 299	
	7.1.2 Isoperimetric Profile 315	
7.2	**Curvature Free Inequalities on Volumes of Cycles** **325**	
	7.2.1 Curves in Surfaces........................ 325	
	7.2.2 Inequalities for Curves 340	
	7.2.3 Higher Dimensional Systoles: Systolic Freedom Almost Everywhere 348	
	7.2.4 Embolic Inequalities 353	

7.1 Curvature Inequalities

7.1.1 Bounds on Volume Elements and First Applications

7.1.1.1 The Canonical Measure and Computing it with Jacobi Fields

We start by stating, in case it is not obvious, that a Riemannian manifold M enjoys a canonical measure which can be denoted by various notations; we pick dV_M as our notation. On every tangent space we have the canonical Lebesgue measure of any Euclidean space. So the measure we are looking for is roughly the "integral" of those infinitesimal measures. If M is oriented (of dimension d) it will have a canonical *volume form*: see §§ 4.2.2 on page 166. The measure is the "absolute value" of that volume form. In any coordinates $\{x_i\}$ with the Riemannian metric represented by g_{ij}, the canonical measure is written

$$dV_M = \sqrt{\det(g_{ij})}\, dx_1 \ldots dx_d$$

where $dx_1 \ldots dx_d$ is the standard Lebesgue measure of $\mathbb{E}^d$. Similarly, the volume form of an oriented Riemannian manifold is

$$\omega = \sqrt{\det(g_{ij})}\, dx_1 \wedge \cdots \wedge dx_d\,.$$

Under scaling $g \to \lambda g$ the canonical measure undergoes

$$dV_{(M,\lambda g)} = \lambda^{d/2} dV_{(M,g)}\,.$$

For a domain $D \subset M$ we will use the notation $\mathrm{Vol}(D)$ for its measure. When M is oriented, $\mathrm{Vol}(M) = \int_M \omega$. Any submanifold of a Riemannian manifold inherits a Riemannian metric and from it a canonical measure. In particular, a compact submanifold will have a volume. In the sequel we will use the word *volume* whatever the dimension to avoid verbiage; but it should be clear that for curves volume is *length* and for surfaces it is *area*.

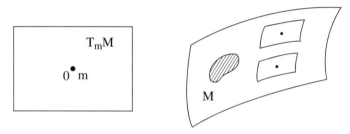

Fig. 7.1. The measure is the "integral" of the infinitesimal measures

Most often volumes in Riemannian manifolds are computed by the following scheme. One fills up a domain of the d dimensional manifold M by a $(d-1)$ parameter family geodesics, so that the dth dimension is nothing but the arc length s along the geodesics. By the Fubini theorem the volume of the domain under consideration will be the integral with respect to s of the integral with respect to the $d-1$ parameters of the infinitesimal $(d-1)$ dimensional volume element of the "tube". This is what we have to compute. It is reasonable to start at time $s = 0$ with initial point n running through some hypersurface H (our $d-1$ parameters) and the corresponding geodesics $\gamma = \gamma_n$ normal to H at n.

Since an infinitesimal variation of geodesics is a Jacobi field Y (see equation 6.11 on page 248), to answer our question we need to know the initial conditions $Y(0)$ and $Y'(0)$ for such Jacobi field. The value $Y(0)$ is nothing but some unit vector $e \in T_n H$ and by the same trick as in equation 6.8 on page 243 the value of $Y'(0)$ will be equal to the parallel transport derivative of the vector field along the curve $c_e(\alpha)$ of H (with $c'_e(0) = e$) given by the initial speed vector $c'_e(0)$ of the family of geodesics normal to H at $c_e(\alpha)$.

This is for a one parameter infinitesimal displacement of γ. Now we need $d-1$ of them. Pick up a orthonormal basis $\{e_i\}_{i=1,\ldots,d-1}$ of $T_n H$. To each e_i is associated a Jacobi field Y_i along γ with initial conditions $Y_i(0) = e_i$ and

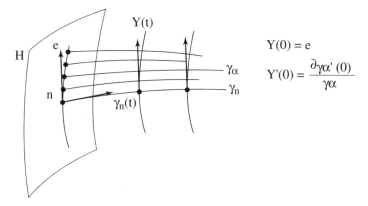

Fig. 7.2. Measuring volume by extending normal geodesics off of a hypersurface

$$Y'_i(0) = h_i = \left.\frac{\partial \gamma_\alpha}{\partial \alpha'}\right|_{\alpha=0}$$

where (as was said above) the derivative has to be taken using the notion of parallel transport (see § 15.4 on page 701) and where γ_α denotes the geodesic starting from $c_e(\alpha) \in H$ and proceeding orthogonally to H. Then the picture in figure 7.3 on the next page makes intuitive that we have:

Lemma 101

$$dV_M(\gamma(t)) = \det(Y_1(t), \ldots, Y_{d-1}(t))\, dt \wedge dV_H(n)$$

where the determinant is the canonical one associated to the $(d-1)$ dimensional Euclidean subspace of $T_{\gamma(t)} M$ orthogonal to $\gamma'(t)$.

This is in fact nothing but the change of variables formula.

Now suppose we want to compute the volume of a ball centered around a point $m \in M$. This time the natural $(d-1)$ parameter family of geodesics is that of all geodesics emanating from the center m, and its natural parameterization is the unit sphere $U_m M$. This time we have no hypersurface H, but we replace it by $U_m M$. Let γ be a geodesic starting from m with speed vector $\gamma'(0) = e \in U_m M$ and complete e into an orthonormal basis

$$\text{span}\{e_i\} \qquad i = 1, \ldots, d-1$$

of the tangent space $T_m M$. Denote by Y_i the Jacobi field along γ with initial conditions

$$Y_i(0) = 0 \quad \text{and} \quad Y'_i(0) = e_i\,.$$

Then the analogue of lemma 101 is:

$$dV_M(\gamma(t)) = \det(Y_1(t), \ldots, Y_{d-1}(t))\, dt\, d\sigma \tag{7.1}$$

where $d\sigma$ is the canonical measure on the sphere $U_m M$.

302   7 Volumes and Inequalities on Volumes of Cycles

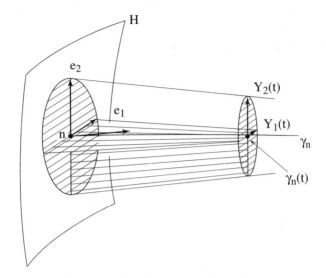

Fig. 7.3. Calculating the volume form

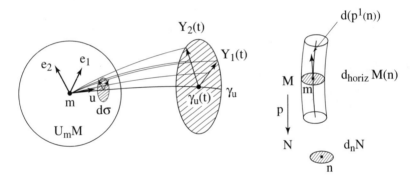

Fig. 7.4. Computing volumes of small balls

Note 7.1.1.1 There is at least one interesting type of volume computation not covered by the above technique, that of Riemannian submersions (see §§4.3.6). We leave as an exercise to prove that for a Riemannian submersion

$$p : (M, g) \to (N, h)$$

the volume elements are:

$$dV_M(m) = dV_{p^{-1}(n)}(m)\, dHm \tag{7.2}$$

where H is the hyperplane in $T_m M$ perpendicular to the tangent space of the fiber $p^{-1}(n)$.

In particular,

$$\mathrm{Vol}(M) = \int_N \mathrm{Vol}\, p^{-1}(n)\, dV_N(n)$$

In many cases, in particular when the fibers are all totally geodesic submanifolds, their volume is constant. Then

$$\mathrm{Vol}(M) = \mathrm{Vol}\,(\text{fiber}) \cdot \mathrm{Vol}(N) .$$

This is the case for example for the submersions from spheres to the $\mathbb{KP}^n$ seen in §§4.1.3. This is one way to compute the volume of the $\mathbb{KP}^n$ (except for $\mathbb{CaP}^2$) if one knows the volumes of spheres; we will find these next. ♦

7.1.1.2 Volumes of Standard Spaces

Formula 7.1 on page 301 enables us to compute the volume of balls of any radius in Euclidean spaces, spheres of various radii, hyperbolic spaces of any constant curvature and in the $\mathbb{KP}^n$. It is enough to remember formula 6.14 on page 251 to find volumes of Euclidean, spherical and hyperbolic balls. For balls in the $\mathbb{KP}^n$, one will need formula 4.36 on page 209 and need to pick an adapted basis. In particular, one can compute the (total) volume of those of the preceding spaces which are compact: spheres and $\mathbb{KP}^n$. Details are given, partly or completely, in III.H of Gallot, Hulin & Lafontaine 1990 [542], 3.3 of Chavel 1993 [326], Berger 1965 [151] and Sakai 1996 [1085]. With the notation

$$\sigma_d = \mathrm{Vol}\, S^d$$

the volumes of balls are

$$\mathrm{Vol}\, B\left(S^d, r\right) = \sigma_{d-1} \int_0^r \sin^{d-1}(t)\, dt$$

where since the sphere (as well all the spaces below) is homogeneous, we need not make precise which point is the center of the ball.

In particular, σ_d is obtained by induction using the value of the classical integral on the right. We then get:

$$\mathrm{Vol}\, S^{2n} = \frac{2^{n+1} \pi^n}{1 \cdot 3 \ldots (2n-1)}$$

$$\mathrm{Vol}\, S^{2n+1} = \frac{2\pi^{n+1}}{n!}$$

or in one shot, with the classical Γ function:

$$\mathrm{Vol}\, S^d = \frac{2\pi^{(d+1)/2}}{\Gamma\left(\frac{d+1}{2}\right)} .$$

In hyperbolic space,

$$\operatorname{Vol} B\left(\operatorname{Hyp}^d(-k^2), r\right) = \sigma_{d-1} \int_0^r \left(\frac{\sinh(kt)}{k}\right)^{d-1} dt$$

And for the $\mathbb{KP}^n$ of diameter $\pi/2$:

$$\operatorname{Vol} B\left(\mathbb{CP}^n, r\right) = \sigma_{2n-1} \int_0^r \frac{\sin(2t)}{2} \sin^{2n-2}(t)\, dt$$

and in particular

$$\operatorname{Vol} \mathbb{CP}^n = \frac{\pi^n}{n!}$$
$$= \frac{\sigma_{2n+1}}{\sigma_1}.$$

$$\operatorname{Vol} B\left(\mathbb{HP}^n, r\right) = \sigma_{4n-1} \int_0^r \left(\frac{\sin(2t)}{2}\right)^3 \sin^{4n-4}(t)\, dt$$

and in particular

$$\operatorname{Vol} \mathbb{HP}^n = \frac{\pi^{2n}}{(2n+1)!}$$
$$= \frac{\sigma_{4n+3}}{\sigma_3}.$$

$$\operatorname{Vol} B\left(\mathbb{CaP}^2, r\right) = \sigma_{15} \int_0^r \left(\frac{\sin(2t)}{2}\right)^7 \sin(8t)\, dt$$

and in particular

$$\operatorname{Vol} \mathbb{CaP}^2 = \frac{6\pi^8}{11!}$$

It is amusing to observe that this volume is equal to σ_{23}/σ_7 although there is no fibration $S^{23} \to \mathbb{CaP}^2$ as mentioned on page 155.

7.1.1.3 The Isoperimetric Inequality for Spheres

In the 1940's Schmidt proved the isoperimetric inequality for spheres of any dimension (endowed with the constant curvature metric) as well as for hyperbolic spaces. Great simplification comes from the fact that those spaces have as many as possible hyperplane symmetries[1] (see theorem 40 on page 208). The result is what one expects:

---

[1] An exercise for the reader: prove that this multitude of isometric involutions characterizes these spaces.

Theorem 102 (Spherical isoperimetric inequality) *Among all domains of a given volume living inside a sphere, the metric balls are precisely the domains whose boundary has the minimum volume.*

See an application to the concentration phenomenon in note 7.1.1.2. A detailed proof is given (for example) in 6.3 of Chavel 1993 [326]; see also IV.H of Gallot, Hulin & Lafontaine 1990 [542] and chapter VI of Sakai 1996 [1085]; but the basic reference today is Burago & Zalgaller 1988 [283].

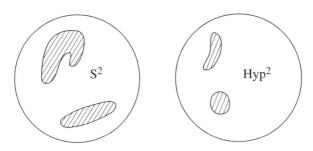

Fig. 7.5. The isoperimetric inequality on the sphere and in hyperbolic space

Note 7.1.1.2 (On the volumes of balls in spheres) Looking at the graph of the function $\sin^{d-1}(t)$ for large d we see that that the volume of a ball in S^d grows extremely slowly with the growth radius for small radii, and then changes drastically around the equator. This is called the *concentration phenomenon*: the sphere is concentrated around its equator in the sense that a very small tubular neighborhood of an equator has almost full measure. From this and from the spherical isoperimetric inequality (theorem 102) one can deduce this surprising phenomenon: on a sphere of very large dimension, any function is very close to its mean value on a subset of very large measure. To see this, one employs lemma 101 on page 301. This is basic in proving Dvoretzky's theorem for symmetric convex bodies in Euclidean spaces: for every ε a convex body admits ε-almost spherical sections of a suitable dimension (of logarithmic growth in d). References for this fascinating subject: Lindenstrauss 1992 [869], Milman 1992 [920], Pisier 1989 [1030], Lindenstrauss & Milman 1993 [870], Berger 2003 [173]. For a general setting for this story of "geometric probabilities", see Gromov's mm spaces in §14.6.

The values given above for the σ_d show also that σ_d goes to zero very quickly (use Stirling's formula to see that σ_d behaves like $d^{-d/2}$). An amusing exercise is to compute the first d for which $\sigma_d < 1$. ♦

7.1.1.4 Sectional Curvature Upper Bounds

We simply mix Rauch's inequality (proposition 74 on page 259) together with equation 7.1 on page 301 above to get a simple control result:

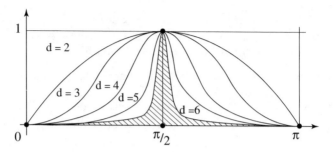

Fig. 7.6. The graph of $\sin^{d-1}(t)$ for various d

Theorem 103 *Assume $K \leq \Delta$ everywhere on a complete d dimensional Riemannian manifold M and that $\exp_m$ is a diffeomorphism on $B(m,r)$. Then*

$$\operatorname{Vol} B(m,r) \geq \operatorname{Vol} B\left(S^d(\Delta), r\right)$$

To get a complete proof one only needs the following trick: at a given t in equation 7.1 on page 301 replace $\{Y_i(t)\}$ by an *orthogonal* basis $\{Z_i\}$ obtained by taking an orthogonal transformation of the orthogonal complement of the velocity $\gamma'(0)$ in $T_m M$. Then the determinant

$$\det(Y_1(t), \ldots, Y_{d-1}(t))$$

is equal to the product

$$\det(Z_1(t), \ldots, Z_{d-1}(t)) = \Pi_{i=1}^{d-1} \|Z_i\|$$

and

$$\|Z_i\| \geq \frac{\sinh\left(\sqrt{\Delta} t\right)}{\sqrt{\Delta}}$$

by proposition 74. An analogous trick will be used in the proof of lemma 106 on page 309.

The inequality in theorem 103 was used in 1968 by Milnor (see also Švarc 1955 [1170] and Karcher 1989 [780]) to get:

Theorem 104 (Milnor 1968 [924]) *The fundamental group of compact manifold of negative curvature has exponential growth.*[2]

Definition 105 *The growth of a finitely generated group with fixed choice of generators is the function $\gamma(s)$ which tells us the number of elements of the group which can be expressed as words in those generators with length smaller than s. The growth is said to be* polynomial *if one has*

---

[2] It is easy to see that the fundamental group of a compact manifold is finitely generated.

$$\gamma(s) \leq s^k$$

for some integer k, and all sufficiently large s, while the growth is exponential *if one has*

$$\gamma(s) \geq a^s$$

for some number $a > 1$. Both of these notions make sense because they are invariant under changes of the choice of generators.

Fig. 7.7. A hyperbolic tiling

We sketch the proof, which is very geometric. We consider the universal Riemannian covering $\tilde{M} \to M$ (which is not compact, by the von Mangoldt–Hadamard–Cartan theorem 72 on page 255). Being simply connected and of negative curvature we know from that theorem that we can apply theorem 103 on the preceding page to it for any real number r and get that for any point $\tilde{m} \in \tilde{M}$:

$$\operatorname{Vol} B(\tilde{m}, r) > a \exp(br)$$

for some constants $a, b > 0$.

Fix $\tilde{m}$ in $\tilde{M}$ and set

$$N = B(\tilde{m}, \operatorname{diam} M) \ .$$

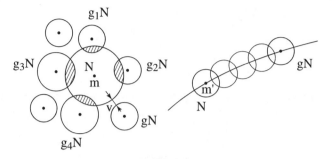

Fig. 7.8. Proof of theorem 104 on page 306

Recall (see §§4.3.3) that the elements of the fundamental group $\pi_1(M)$ act on $\tilde{M}$ by isometries. Introduce the subset of $\pi_1(M)$ made up of the g such that gN intersects N. These g build up a finite subset Γ of $\pi_1(M)$. The other gN are always at a positive minimum distance ν from N. If the distance from $\tilde{m}$ to some gN is such that

$$d(\tilde{m}, gN) < n\nu + \operatorname{diam} M$$

for some integer n then the picture below will show that g can be written as a product

$$g = g_1 \cdots g_n$$

of elements of Γ. This proves first that $\pi_1(M)$ is generated by Γ. Second, it proves that the set of the gN build up by the g of word length not larger than n will certainly cover the ball $B(\tilde{m}, n\nu + \operatorname{diam} M)$ so that:

$$\gamma(s)\operatorname{Vol} N \geq \operatorname{Vol} B(\tilde{m}, n\nu + \operatorname{diam} M)$$

and we are done via equation 7.1.1.4 on the previous page.

7.1.1.5 Ricci Curvature Lower Bounds

Without any trick, since we always have

$$\det(Y_i) \leq \Pi_i \|Y_i\|,$$

one would clearly bet that a lower bound $K > \delta$ would provide an upper bound on the volume of balls, through theorem 62 on page 243. But in 1963 Bishop made a striking discovery to the effect that a Ricci curvature lower bound is enough to get an optimal upper bound on the volume of balls. According to Gromov such a result is outlined already in Lévy 1951 [861] but it seems to have been unnoticed thereafter.

We will first state a more general inequality due to Heintze and Karcher in 1978 and then give two consequences of it (for reference see 7.5 of Chavel 1993 [326], and also 4.21 in Gallot, Hulin & Lafontaine 1990 [542]):

7.1 Curvature Inequalities

Lemma 106 (Heintze & Karcher) *As in lemma 101 on page 301 let*

$$\{Y_i(t)\}_{i=1,\ldots,d-1}$$

be Jacobi fields along some geodesic γ and set

$$F = \det(Y_1 \ldots, Y_{d-1})$$

and

$$f = F^{1/d-1}.$$

Then f satisfies:

$$f'' + \frac{\mathrm{Ricci}(\gamma')}{d-1} f \geq 0$$

under the proviso that f does not vanish (except possibly at the origin).

There is a straightforward computation plus a trick analogous to that of theorem 103 on page 306. Computation yields:

$$f'' = \frac{2-d}{(d-1)^2} f F^{-2} \left(\sum_i \det(Y_i') \right)^2$$

$$+ \frac{1}{d-1} f F^{-1} \left(\sum_i \det(Y_i', Y_j') + \sum_i \det(Y_i'') \right)$$

with the obvious notations for the determinants; for example Y_i' should appear at the ith place and at all the other places one puts Y_i etc. The last term is easy and classically by linear algebra yields the value

$$\frac{1}{d-1} f \, \mathrm{Ricci}(\gamma').$$

But the two other terms are a mess in the general expression for the $Y_i(t)$. To clear up things one must first fix some time t. By an orthogonal transformation of $T_{\gamma(t)}H$ one can change the sets

$$\{Y_i(t)\}, \{Y_i'(t)\}$$

into sets

$$\{A_i\}, \{B_i\}$$

such that

$$B_i = \lambda_i A_i.$$

This is possible because the bilinear form given by the scalar products

$$\langle Y{i}'(t), Y_j(t) \rangle$$

is symmetric. This is not obvious; it comes from the fact that the Jacobi field equation 6.11 on page 248 and the symmetries of the curvature tensor (see equation 4.28 on page 203) implies that

$$(\langle Y', Z \rangle - \langle Y, Z' \rangle)' = 0$$

so that we need only to insure symmetry at time $t = 0$ which can be done for example taking for the e_i the eigendirections of the second fundamental form.

Note that the determinant is not changed so that we are left with two terms:

$$\left(\sum_i \lambda_i\right)^2 \quad \text{and} \quad \sum_{i,j} \lambda_i \lambda_j .$$

The total result turns out to be

$$-\frac{1}{(d-1)^2} \sum_{i,j} (\lambda_i - \lambda_j)^2$$

proving the theorem.

Note 7.1.1.3 Beware that the above trick is only valid when the full action of $SO(d-1)$ is available. This is the case for the set of geodesics emanating from a given point or that of geodesics emanating orthogonal to a given hypersurface. It does not work, for example, for geodesics emanating orthogonally from a curve (as in theorem 90 on page 273 for example). In such a case one needs more assumptions that only controlling the Ricci curvature. ♦

The basic consequence of theorem 106 on the preceding page is:

Theorem 107 (Bishop–Gromov 1963,1999 [196, 633]) *Assume that*

$$\text{Ricci} \geq (d-1)\delta$$

where δ is any real number. Starting from any point $m \in M$ the function

$$r \to \frac{\operatorname{Vol} B(m, r)}{\operatorname{Vol} B\left(\mathbb{S}^d(\delta), r\right)}$$

is nowhere increasing. In particular for any $r \geq 0$ we have

$$\operatorname{Vol} B(m, r) \leq \operatorname{Vol} B\left(\mathbb{S}^d(\delta), r\right) .$$

We have only to integrate in polar coordinates centered at m and apply equation 7.1 on page 301: the determinant will be bounded by f^{d-1} where f satisfies

$$f'' + \delta f > 0 .$$

Looking at the limit when $r \to 0$, the initial condition yields then

$$f \leq f_\delta$$

where f_δ is the corresponding function in the space form $\mathbb{S}^d(\delta)$.

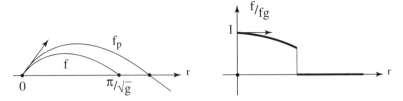

Fig. 7.9. Continue with 0 after the cut value

This was Bishop's result. Now one has to wonder about the proviso in lemma 106 on page 309 which concerns conjugate points. But it is irrelevant here because we are looking for lower bound. On every geodesic starting from m one has only to go no longer than the cut-value (see §§6.5.1) and one knows that before the cut value the function f will never vanish from theorem 86 on page 268. After the cut value there is only zero to integrate. This was the contribution of Gromov, which will turn out to be essential in the future, first for getting metric bounds from lower Ricci curvature without wondering about injectivity radius, and second for getting lower bounds for the volume of balls when the total volume

$$\operatorname{Vol} M = \operatorname{Vol} B(m, \operatorname{diam} M)$$

(for any point m) is bounded (this is because of the non-increasing property in theorem 107 on the facing page see equation 12.2 on page 626.

In fact, if this is not obvious at first glance, this is Gromov's philosophy as explained in 5.31 and in the Appendix B (see B.2) of Gromov 1999 [633]; we will see in the references quoted just above that what is at the root of the proofs is the:

Theorem 108 (The doubling property) *In a Riemannian manifold of dimension d with nonnegative Ricci curvature, any ball of radius $2R$ can be covered by a collection of 4^d balls of radius R. If we only know that Ricci $\geq (d-1)\delta$, then there is an analogous inequality, but only valid for "small balls" and with some constant depending only on d and δ.*

We have only to use the covering trick (lemma 125 on page 333): pack in the given ball of radius $2R$ as many balls of radius $R/2$ as possible. Then the balls of same centers and radius R must cover the initial ball. But now use the Bishop–Gromov theorem 107 on the preceding page and remember that the volume of a Euclidean ball of radius ρ is proportional to ρ^d. This shows that there cannot be more than 4^d disjoint balls of radius $R/2$ within a ball of radius $2R$. For the doubling property in a completely different context, see Toro 1997 [1197].

The natural analogue of theorem 104 on page 306 is:

Theorem 109 (Milnor 1968 [924]) *Let M be a complete Riemannian manifold with non-negative Ricci curvature (not necessarily compact). Then the growth function of any finitely generated subgroup of the fundamental group $\pi_1(M)$ satisfies*
$$\gamma(s) \leq as^d$$
for some positive constant a.

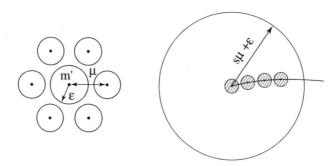

Fig. 7.10. Ball packing and growth of the fundamental group

Let us look once again at the universal Riemannian covering $\tilde{M}$ which has nonnegative Ricci curvature because M does. Pick some point $\tilde{m}$ in $\tilde{M}$. Recall that the fundamental group $\pi_1(M)$ acts as isometries on $\tilde{M}$. There is $\varepsilon > 0$ such that $B(\tilde{m}, \varepsilon)$ does not meet any other $B(g\tilde{m}, \varepsilon)$ for g running through $\pi_1(M)$ (except the identity). Let us also introduce the supremum μ of the distances $d(\tilde{m}, g_i\tilde{m}')$ when g_i runs through the generators of a finitely generated subgroup. Then the ball $B(\tilde{m}, s\mu + \varepsilon)$ will contain at least $\gamma(s)$ disjoint balls of radius ε to the effect that

$$\gamma(s)\operatorname{Vol} B(\tilde{m}, \varepsilon) \leq \operatorname{Vol} B(\tilde{m}, s\mu + \varepsilon)$$

and theorem 108 on the previous page (the doubling property) implies the desired assertion.

Note 7.1.1.4 (About Milnor's result) A deep theorem of Gromov says that a group of polynomial growth is a discrete subgroup of a nilpotent Lie group: see Gromov 1981 [614] or Gromov 1993 [629]. We will come back to this in theorem 352 on page 598.

Milnor conjectures in his article that the fundamental group is automatically finitely generated (for complete noncompact manifolds of course): to our knowledge it is still an open question.

Note that the doubling property theorem 108 is a refinement of Myers' theorem 63 on page 245. Also note that one cannot do better: take flat tori for example. Compare with Bochner's theorem 345 on page 594. ◆

7.1 Curvature Inequalities

Note 7.1.1.5 A priori one cannot expect a lower bound for the volumes of balls with only an upper bound on the Ricci curvature: in fact there are manifolds with identically zero Ricci curvature (see Besse 1987 [183]) and for which there is no lower bound in comparison with Euclidean space for the volume of balls. There is one exception (besides of course dimension two): this is dimension three. The reason is that in dimension three Ricci curvature and sectional curvature are equivalent (an exercise left to the reader): see Eschenburg & O'Sullivan 1980 [496]. This was *stricto sensu*. But as remarked above, if for example we know the total volume, then we can hope to control volumes of balls. ♦

We come now to the second application of lemma 106 on page 309. To express it we need to first extend to general Riemannian manifolds the notion of *second fundamental form* of a hypersurface. As in §§1.6.3 this is a quadratic form on the tangent space $T_n H$ at n of a hypersurface H of our Riemannian manifold M. It tells us how much H differs from the hypersurface generated by the various geodesics starting from n and tangent to H. With the notation of §§§7.1.1.1 it is nothing but the quadratic form given by the scalar products

$$\langle e_i, h_i \rangle .$$

The *mean curvature* of H at n is by definition equal to the trace of the second fundamental form of H at n divided by $d-1$ and is denoted by η.

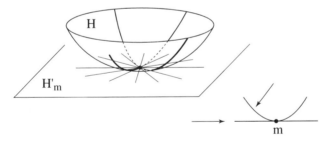

Fig. 7.11. Comparing a hypersurface with the hypersurface formed from ambient space geodesics emanating out of a point and tangent to our original hypersurface

The mean curvature of hypersurface in general Riemannian manifolds has the same property that it has in Euclidean spaces: formula 9 on page 59 extends without modification.[3]

Lemma 110 (First variation equation) *For a normal variation of length f of a piece H of hypersurface, the first derivative of the volume is equal to*

---

[3] Just recall that the mean curvature is only determined up to a sign but in theorem 111 on the next page below a preferred normal will be given by $\gamma'(0)$.

$$\int_H f\eta\, dV_H \ .$$

In particular, assume that D is a compact domain in M with boundary ∂D and that ∂D has minimal volume among all domains close to D having the same volume as D. Then ∂D has constant mean curvature.

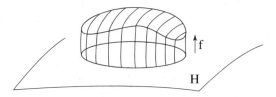

Fig. 7.12. The first variation of a hypersurface

We now come to the evaluation of the volume equation in lemma 101 on page 301:

Theorem 111 (Heintze–Karcher, 1978) *Assume* Ricci $> (d-1)\delta$ *and consider a pencil of geodesics normal to a hypersurface H neighboring some geodesic γ. Provided that the determinant in lemma 101 on page 301 does not vanish from 0 to t then*

$$dV_M(\gamma(t)) < f_\delta^{d-1}(t)$$

where η is the mean curvature of H at $\gamma(0)$ and f_δ is the standard function

$$f_\delta = \begin{cases} \cos(t\sqrt{\delta}) + \frac{\eta}{\sqrt{\delta}}\sin(t\sqrt{\delta}) & \text{when } \delta > 0 \\ 1 + \frac{\eta}{t} & \text{when } \delta = 0 \\ \cosh(t\sqrt{-\delta}) + \frac{\eta}{\sqrt{-\delta}}\sinh(t\sqrt{-\delta}) & \text{when } \delta < 0. \end{cases}$$

For the proof we first remark that f and f_δ have the same initial conditions

$$f(0) = f_\delta(0) = 1$$

and

$$f'(0) = f_\delta'(0) = \eta$$

by the very construction of the Y_i and the definition of the mean curvature. Then we proceed as in on page 121 by the Sturm-Liouville technique. As an exercise, compare the above with the proof in §§1.6.8.

7.1.2 Bounding the Isoperimetric Profile with the Diameter and Ricci Curvature

7.1.2.1 Definition and Examples

The isoperimetric inequality in $\mathbb{E}^d$ is a fundamental geometric notion and also has many applications. We gave one in detail: the Faber–Krahn inequality 1.22 on page 81. We saw also in §§7.1.1.3 that there is an isoperimetric inequality in the standard space forms, the spheres and hyperbolic spaces, so that it is natural to ask

Question 112 *Is there an isoperimetric inequality in a general Riemannian manifold?*

Let us say first that the state of affairs today is far from what we hope for. For example one would expect the domains optimal for isoperimetry in the symmetric spaces of rank one would be the metric balls. This is extremely reasonable since those spaces are precisely the isotropic (or two-point homogeneous) ones (see §§6.6.2). We will see below that is false for compact spaces, but that there is a very reasonable conjecture. But for the noncompact, simply connected ones, one would expect the balls to be optimal. However this is a completely open problem today, starting with the $\text{Hyp}_{\mathbb{C}}^n$.

To be more precise, we consider domains D in a Riemannian manifold M with a smooth boundary ∂D. As in the case of the sphere we cannot in general hope to have a universal bound for the dimensionless ratio

$$\frac{\text{Vol}(\partial D)^d}{\text{Vol}(D)^{d-1}}.$$

This was possible for Euclidean spaces essentially because they admit nonisometric similarities. We need a concept where $\text{Vol}(D)$ enters explicitly and is normalized in some way.

Definition 113 *Take M a compact Riemannian manifold. For every real number*

$$\beta \in [0, 1]$$

we define $h(\beta)$ as

$$h(\beta) = \inf_D \left\{ \frac{\text{Vol}(\partial D)}{\text{Vol}(M)} \mid \text{Vol}(D) = \beta \text{Vol}(M) \right\}.$$

The function $\beta \mapsto h(\beta)$ is called the *isoperimetric profile* of the Riemannian manifold M. The more crude Cheeger's constant h_c was introduced in Cheeger 1970 [329] and is defined as above, but in terms of domains D with

$$\text{Vol}(D) \leq \frac{1}{2} \text{Vol}(M).$$

It is then equal to
$$h_c = \inf_{\beta \leq \frac{1}{2}} h(\beta).$$

This profile can be quite wild, for example for the surface in figure 7.13.

Fig. 7.13. Study $h(\beta)$ for this surface

The isoperimetric inequality 102 on page 305 for spheres and the formula in §§§7.1.1.2 yields the isoperimetric profile for sphere. We suggest that you perform some computation and drawings, starting with two dimensions.

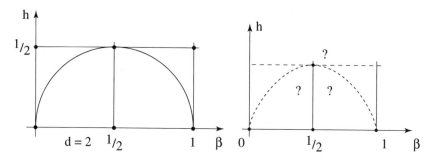

Fig. 7.14. Compute and draw isoperimetric profiles for dimensions $3, 4, \ldots$

Let us now try to compute the isoperimetric profile of the real projective plane $\mathbb{RP}^2$, equipping it with the standard metric of constant curvature (see §§§4.3.3.2). We follow Gromov's technique as in §§1.6.8. For a given β, we look for a domain D of area equal to
$$\beta \operatorname{Vol} \mathbb{RP}^2 = 2\pi\beta$$
and with smallest possible length for ∂D. We apply lemma 110 on page 313: the curve ∂D has constant curvature. In $\mathbb{RP}^2$ this implies that ∂D is a geodesic or a circle, i.e. the projection of a small circle of the sphere S^2. To see this, show that in S^2 the curves of constant curvature are small circles and use the fact that
$$S^2 \to \mathbb{RP}^2$$
is a local isometry.

7.1 Curvature Inequalities   317

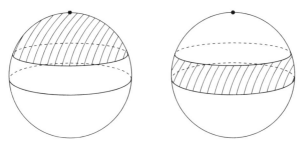

Fig. 7.15. Working in $\mathbb{RP}^2$

As expected, at least for small values of β, the optimal domains are those D whose boundaries have constant curvature, which could be a disk, or could very well be the tubular neighborhood of a geodesic. The picture in figure 7.15 represents the story on $\mathbb{RP}^2$ lifted up to the sphere. Down in $\mathbb{RP}^2$ such a domain is a Möbius band. But its complement is an ordinary disk and their boundaries have the same length. So finally the isoperimetric profile of $\mathbb{RP}^2$ is as indicated in figure 7.16.

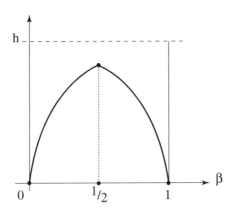

Fig. 7.16. The isoperimetric profile of the real projective plane

We now look at the square flat torus of two dimensions. Here again candidates for boundaries are ordinary plane circles or straight lines. But notice that when we look at a large disk, of radius $r > 1/\pi$, the band of width δ has a better isoperimetric ratio: the band has isoperimetric ratio $2/\delta$ while the disk has $2\pi r/\pi r^2$. Since disks and strips are the only ones with constant curvature boundary, the profile for the square T^2 is as pictured in figure 7.17 on the next page.

As an exercise, work out the case of a square Klein bottle and some other flat tori.

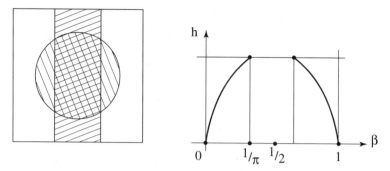

Fig. 7.17. The isoperimetric profile of the flat square torus

For compact hyperbolic space forms, the isoperimetric profile is complicated but theoretically answerable since again we know which boundary to look at.

The recent result of Ritoré & Ros 1992 [1059] solves the problem for real projective 3-space $\mathbb{RP}^3$ (but not for the three dimensional torus). The answer is the "reasonable guess" from figure 7.18 on the facing page.

Except for spheres of any dimension and $\mathbb{RP}^3$, to our knowledge there is not a single example with dimension at least three where the isoperimetric profile is known. See partial results for noncompact manifolds in Hsiang & Hsiang 1989 [740].

This is a surprise for us because we know everything about the metric of each of the $\mathbb{KP}^n$, in particular the geodesic behavior (see §§6.1.6). But the reason is simple. Boundaries will now, starting in dimension three, be hypersurfaces of constant mean curvature in the manifolds under consideration. Of course in flat tori or real projective spaces (or more generally in the $\mathbb{KP}^n$) the metric spheres (boundary of a metric ball) will be of constant mean curvature (because of homogeneity, for example) but there are certainly other constant mean curvature surfaces. For example the tubular neighborhoods of projective subspaces of any dimension in the $\mathbb{KP}^n$ or various cylinders in flat tori. But we do not know if there are other ones and if so who they are. There is no analogue to the result of §§3.4.2 for surfaces in $\mathbb{E}^3$.

Nevertheless, it seems reasonable to guess that the isoperimetric profile for a square flat torus is given by a succession of curves corresponding first to disks, second to tubular neighborhoods of straight lines, third to tubular neighborhoods of planes, etc. ending with tubular neighborhoods of hyperplanes. As we have seen, something like this is true for $\mathbb{RP}^3$.

For all of the $\mathbb{KP}^n$, it also seems reasonable that the isoperimetric profile will be that given by successive tubular neighborhoods of projective subspaces (starting with points). Now if we leave our favourite manifolds to look for theorems about general manifolds, it was discovered in Pansu 1997,1998 [998, 1000] that the isoperimetric profile can be quite wild; for example, not smooth at the origin. For more see the references in Pansu's articles and for some

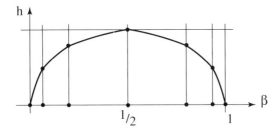

Fig. 7.18. The isoperimetric profile of real projective 3-space $\mathbb{RP}^3$

surfaces of revolution Morgan, Hitchings, & Howards 2000 [938]. In a different direction, the isoperimetric behavior at infinity of noncompact manifolds is basic in various contexts; for instance, group theory and geometry; see Burago & Ivanov 1994 [280].

7.1.2.2 The Gromov–Bérard–Besson–Gallot Bound

Thinking of the proof of Faber–Krahn inequality 1.22 on page 81, assume we want only some bound on the first eigenvalue of the Laplacian, λ_1 (which might not be an optimal bound). Then it is enough to have an isoperimetric bound of the type

Vol(∂D) is larger than some function of Vol(D).

This is what we are going to do for Riemannian manifolds. It is clear that optimal bounds for general Riemannian manifolds are hopeless. But our understanding of control is that we wish to get an estimate from curvature bounds. Recalling the Klingenberg and Cheeger theorems from §6.5.2, we might expect to need to know the diameter (the volume is already in the profile anyway). After intermediate results due to various authors, Bérard, Besson and Gallot in 1985 got a bound involving only the lower bound of the Ricci curvature and the diameter. The pictures above in §§7.1.2.1 show that one cannot do better. So their bound is optimal given the ingredients. Their explicit functions (which we are not going to give because they are very complicated and not yet having any clear interpretation) will probably be improved some day.

Theorem 114 (Gromov 1980 [633], Bérard–Besson–Gallot 1985[139])
There are three universal functions

$$A_\varepsilon(d, \alpha)$$

for $\varepsilon = 1, 0, -1$ such that for every compact Riemannian manifold M of dimension d with

$$\operatorname{diam}^2 M \cdot \inf \operatorname{Ricci} = (d-1)\varepsilon\alpha^2$$

one has for every β with $0 \leq \beta \leq 1$

$$h(\beta) \geq \frac{1}{\operatorname{diam} M} h_{S^d}(\beta) A_\varepsilon(d, \alpha)$$

where h_{S^d} denotes the isoperimetric profile of the sphere.

We have previously accumulated all of the material for the proof and are prepared for it. The general case is quite involved but the key ideas are in the following simpler case, initiated by Gromov (see Gromov 1999 [633]). The Ricci curvature is positive and we normalize it to have

$$\operatorname{Ricci} \geq d - 1 .$$

See also, if needed, IV.H of Gallot, Hulin & Lafontaine 1990 [542]. More precisely, we look only at the positive case and we forget the diameter. This is possible because of Myers' result 62 on page 243. Doing so, we are loosing an improvement of Gromov's inequality, and for the general case we again refer the reader to the original article Bérard, Besson & Gallot 1985 [139]. Before starting, you might like to reread the proof in §§1.6.8 from which we follow the scheme.

In the present case, we want to prove that

$$h_M(b) \geq h_{S^d}(\beta)$$

for every $\beta \in [0, 1]$ and that equality can occur only for the standard sphere, as the proof will show. We pick some β and we use deep results from geometric measure theory (see §§14.7.2) to get a domain $D \subset M$ with volume

$$\operatorname{Vol}(D) = \beta \operatorname{Vol}(M)$$

and with the following properties: (remember the idea on page 64)

1. its boundary ∂D has minimal volume among all domains with same volume as D,
2. the singular points where ∂D is not locally a manifold are of measure zero in ∂D and
3. when the tangent "object"[4] to ∂D at some point n is contained in a half space of $T_n M$ then n is a regular point.

These results of *geometric measure theory* are very hard to locate and to find in an handy form; we refer the reader to the various references given in the papers we quoted on the subject, and refer again to Morgan 2000 [937] for a wonderful first introduction to the subject.

We now compute the volume of D as a function of the volume of ∂D as we did in §§1.6.8. For every $m \in D$, let n be a point of ∂D as close as possible to m, and take γ a segment from n to m. Then the first variation formula 6.3 on

---
[4] generalizing the tangent plane

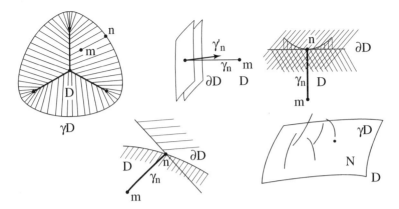

Fig. 7.19. Geometric measure theory in action: Isoperimetric inequalities

page 225 tells us that the tangent object to ∂D at n should stay in the half space of $T_n M$ determined by $\gamma'(n)$ so that n is regular. Denote by N the set of regular points of ∂D.

Starting in the opposite direction, this means that we can recover all of D by starting from all regular points n of ∂D and taking the geodesics γ_n normal to ∂D at n and heading along them inside D. On such a geodesic, we can stop when it ceases to be a shortest path in D. We call this stopping point the *focal value* of n and denote it by focal(n). As in §§1.6.8 and theorem 86 on page 268 we know that the Heintze–Karcher theorem 111 on page 314 is applicable.

By Milnor's theorem 109 on page 311 we know that ∂D has constant mean curvature, say η_0. We apply lemma 101 on page 301 and the Heintze–Karcher theorem 111 to get:

Lemma 115

$$\mathrm{Vol}\, D = \int_{\partial D} dV_{\partial D}(n) \int_0^{\mathrm{focal}(n)} dt\, (\cos(t) + \eta_0 \sin(t))^{d-1}$$

Denote by D^* a ball in S^d with

$$\mathrm{Vol}\, D^* = \beta \,\mathrm{Vol}\, S^d\ .$$

Let r^* be the radius of D^* and η^* be the mean curvature of its boundary. Assume first that
$$\eta_0 < \eta^*\ .$$

Check that
$$\mathrm{focal}(n) \leq r^*$$
for every $n \in \partial D$ so that the lemma 115 of Heintze and Karcher implies that

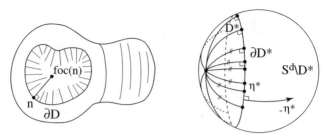

Fig. 7.20. Calculating volume by integrating up to the focal value

$$\mathrm{Vol}(D) \leq \mathrm{Vol}(\partial D) \int_0^{r^*} (\cos(t) + \eta^* \sin(t))^{d-1} \, dt \ .$$

When we carry out the same integral for D^*, the last integral will be exactly the same as for D except that it will take place in S^d and with equality everywhere. So the value of this integral is exactly

$$\frac{\mathrm{Vol}(D^*)}{\mathrm{Vol}(\partial D^*)}$$

Q.E.D. We have proven theorem 114 on page 319. If by bad luck $\eta_0 \geq \eta^*$, the amazing trick of Gromov is simply to perform the same computation but with the complementary sets

$$M \backslash D \text{ and } S^d \backslash D^* \ .$$

This has the effect of switching the orientation of the boundary, so switching the signs of the mean curvatures.

The fun is that is we have equality for only one manifold: the sphere. The above proof will imply knowledge of both D and $M \backslash D$. To prove isometry with the sphere, one just traces back the inequalities, and ends up with a metric given by

$$ds^2 = d\rho^2 + sin^2 \rho \, d\sigma^2 \ .$$

Note 7.1.2.1 Theorem 114 on page 319 will play a significant part in chapter 9 to get lower bounds for every eigenvalue in terms of only the infimum of the Ricci curvature and the diameter. ♦

7.1.2.3 Nonpositive Curvature on Noncompact Manifolds

An unanswered

Question 116 *Is the isoperimetric profile of a complete simply connected Riemannian manifold of nonpositive curvature always dominated by the isoperimetric profile of Euclidean space?*

Such manifolds are never compact, as we known by theorem 72 on page 255 (the von Mangoldt–Hadamard–Cartan theorem). We consider a simply connected manifold M of nonnegative curvature. By the von Mangoldt–Hadamard–Cartan theorem, we know that this manifold is diffeomorphic to $\mathbb{E}^d$. It is conjectured that

Conjecture 117 *For any domain D of M one has the same isoperimetric inequality as for Euclidean space:*

$$\frac{\operatorname{Vol}^d \partial D}{\operatorname{Vol}^{d-1} D} \geq \frac{\operatorname{Vol}^d S^{d-1}}{\operatorname{Vol}^{d-1} B\left(\mathbb{E}^d, 1\right)}$$

with equality if and only if the restriction of the metric of M to D is flat and for this flat metric D is a ball.

This was proved for surfaces by André Weil in 1926 using conformal representation and the theory of harmonic functions; it was in fact the first mathematical piece of work of André Weil, answering a question asked during or after a Hadamard seminar at the Collège de France. The lecturer was Paul Lévy, telling the audience about Carleman's result of 1921. What Carleman had obtained was the desired isoperimetric profile, but only for a minimal surface.[5]. In dimension 4, it was proven by Croke in 1984 using Santalo's equation 7.11 on page 364. For this demonstration the surprise is that Santalo's equation gave the right answer only in four dimensions; for all other dimensions it only gives an weaker inequality. For three dimensions, this isoperimetric inequality was recently proven by Kleiner in 1991 [808]; the proof is subtle and uses the fact that hypersurfaces in three dimensions are ordinary Riemannian surfaces.

Unlike two and three dimensional manifolds where one works, so to speak, downstairs (i.e. directly on the manifold itself), the proof for four dimensions requires one to work in the tangent bundle over the manifold, to use the geodesic flow, etc. See §§§7.2.4.2

The main difficulty in proving the above conjecture is that, contrary to the compact case in the beginning of this section, geometric measure theory cannot be applied directly thanks to noncompactness: the optimal domains can "go to infinity."

Remarkably, Cao & Escobar 2000 [309] proved the desired isoperimetric inequality of our conjecture for three dimensional piecewise linear manifolds. Here, nonpositive curvature means that at the singular points the sum of the measures of the spherical angles is never less than the total volume of the Euclidean unit sphere (of the given dimension). Unhappily, it is an open very interesting question to know if every Riemannian manifold of nonpositive curvature can be nicely approximated by a piecewise linear one (still of nonpositive curvature). The Cao–Escobar result is of course preserved under approximations.

---

[5] Recall that minimal surfaces have nonpositive curvature

The natural and fascinating, but very difficult, topic of piecewise flat approximations of Riemannian manifolds is treated in Cheeger, Müller, & Schrader 1984,1986 [350, 351]; see also the lecture Lafontaine 1987 [843].

Note 7.1.2.2 Another important control is that of λ_1 (the fist eigenvalue of the Laplacian) with the Cheeger's constant h_c:

$$h_c = \inf\left\{\operatorname{Vol} D \mid \operatorname{Vol} D \leq \frac{1}{2}\operatorname{Vol} M\right\}$$

which is obtained from the isoperimetric profile by

$$h_c = \inf\left\{\frac{h(t)}{t} \mid 0 \leq t \leq \frac{1}{2}\right\}.$$

In Cheeger 1970 [329] one finds the inequality

$$\lambda_1 > \frac{1}{4}h_c^2$$

which is optimal by Buser 1978 [291] but never attained by a smooth object. See §§ 9.10.2 on page 409 for more on this. ♦

The beautiful formula in Savo 1996 [1098], or in Savo 1998 [1099], studies the volume of various tubes and then encompasses many results of the above type with a very nice proof. The second derivative of volume with respect to the radius of the tube is computed and linked with the Laplacian. Precisely, let N be any submanifold and u any function. Let $M(r)$ denote the solid tube of radius r around N, and ρ the distance function to N. Then set

$$F(r) = \int_{M(r)} u\, dV_M$$

for the Riemannian measure dV_M. Then

$$-F''(r) = \int_{M(r)} \Delta u\, dV_M + \int_{\rho^{-1}(r)} u\, \Delta u\, dV_{\rho^{-1}(r)}$$

where the measure on the "sphere" $\rho^{-1}(r)$ is the one induced from its Riemannian geometry as a submanifold of M. This formula is valid within the injectivity radius of the tube. Beyond that radius, one has a distribution rather than a function.

Recently, in the spirit of many results in Riemannian geometry (some of them met above), the article of Gallot 1988 [540] succeeded in controlling the isoperimetric profile via integral bounds on the Ricci curvature; see §§§7.2.4.4 for precise statements. There are many corollaries, for example the finiteness theorems of §§12.4.1. and control on the spectrum. Moreover, Gallot also controls the volumes of tubes around hypersurfaces. A key tool as we will see in the proof of theorem 90 on page 273 or proposition 375 on page 620 would be to control the volume of tubes around geodesics. With only an integral Ricci bound, it is impossible to control the volume of tubes around geodesics; see the Eguchi–Hanson examples in Petersen, Shteingold, & Wei 1996 [1021] for more information in this domain.

The Sobolev inequalities are a basic tool in analysis. They are used in many results quoted in this text.

Theorem 118 (Sergei L. Sobolev) *Take a numerical function*

$$f : M \to \mathbb{R}$$

on a Riemannian manifold M of dimension d and take integers p and q with

$$\frac{1}{p} + \frac{1}{d} = \frac{1}{q}.$$

Between the function and its gradient one has

$$\|f\|_{L^p} \leq A\|df\|_{L^q} + B\|f\|_{L^q}.$$

In most cases the Riemannian geometer is interested not just in some A and B numbers; he wishes to control them with the curvature, etc. The problems of the optimal A and B are quite different. For the constant B, it is basically the work of Gallot; see the various references to this author above. For the constant A it is the work of Aubin, see for example Aubin 1998 [86]. See the recent references Hebey 1996 [690] and 1999 [691], Hebey & Vaugon 2001 [694] and Druet & Hebey 2000 [463] (where in some cases, sharp estimates of the A term characterize certain Riemannian manifolds). We just mention that in these works the control of the isoperimetric profile is crucial. The isoperimetric profile of a noncompact manifold is of course deeply linked with the geometry of the manifold at infinity. For this important perspective, as well as results, see Pittet 2000 [1032] and the references there.

7.2 Curvature Free Inequalities on Volumes of Cycles

The beauty of this section is its purity: the inequalities we will uncover are curvature independent, so that we completely lack local control—this is an entirely global game. General references are Chavel 1993 [326], Sakai 1996 [1085], two surveys Berger 1993,1996 [165, 168], and in Gromov 1999 [633] chapter 4 and appendix D.

7.2.1 Systolic Inequalities for Curves in Surfaces

7.2.1.1 Loewner, Pu and Blatter–Bavard Theorems

In 1949, Loewner made a wonderful discovery, apparently the very first of its kind. Consider a two dimensional torus T^2 with Riemannian metric g. It can be abstract (i.e. not embeddable in $\mathbb{E}^3$) but it is not forbidden to be an embedded one.

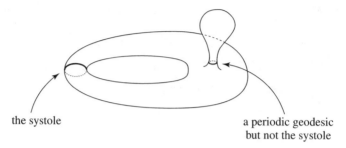

the systole

a periodic geodesic but not the systole

Fig. 7.21. Searching a torus for its systole

We see in the picture in figure 7.21 that there are closed curves on our torus which are not contractible, this is due to the fact that the torus is not simply connected. But beware that the set of closed curves up to homotopy is in general smaller that the fundamental group. This set is the set of conjugacy classes of the fundamental group; see VII.6 of Berger 1965 [151]; the name for a closed curve up to homotopy is "a class of free homotopy." The torus is an exception to the rule that the classes of free homotopy form a smaller set than the fundamental group, because its fundamental group is Abelian.

Now one might guess that there is, in each free homotopy class, at least one curve of smallest possible positive length. This is not hard to prove. The idea is that the minimum is achieved, even though the space of curves is infinite dimensional, in the present case for various possible reasons. One is the possibility of approximating curves in a compact Riemannian manifold by piecewise geodesics. This reduces the problem to a finite dimensional one (for this idea, basic in Morse theory, see Milnor 1963 [921], and §§10.3.2). A second reason is the so-called *semicontinuity* of the length of a curve. See for example Choquet 1966 [377], VI.3.11, where there is a proof for quite general metric spaces. Compactness is of course required, as the picture shows. Such a curve will then necessarily be a periodic geodesic, and not merely a geodesic loop. See also 4.12 of Chavel 1993 [326].

We refer to the *systole* of our torus to mean the smallest length of all possible noncontractible curves.[6] Denote the systole by $\text{Sys}\left(T^2, g\right)$. Looking at the picture, a reasonable guess is that if $\text{Sys}\left(T^2, g\right)$ is large, then the area of the torus in the metric g cannot be very small.

---

[6] The term *systole* comes from physiology (originally from the Greek word for contraction), and refers to the contraction of the heart that occurs when the heart pumps blood into the arteries. The analogy of this periodic behaviour with the periodic behaviour of the geodesics explains the origin of the term in geometry.

7.2 Curvature Free Inequalities on Volumes of Cycles 327

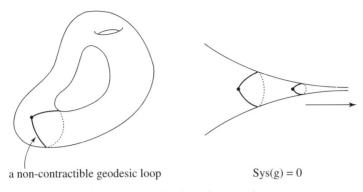

a non-contractible geodesic loop   Sys(g) = 0

Fig. 7.22. Looking for systoles

Theorem 119 (Loewner 1949) *Any Riemannian metric g on T^2 satisfies*

$$\mathrm{Area}_g\left(T^2\right) \geq \frac{\sqrt{3}}{2}\,\mathrm{Sys}\,(g)^2\ .$$

Moreover equality is achieved precisely when g is flat and corresponds to an equilateral torus (see §§4.3.3).

One can see that inequality as a kind of global isoperimetric inequality: there is no domain with a boundary here, the domain is the whole manifold and the systole plays the role of boundary. Of course we call such an inequality an *isosystolic inequality*. For such an inequality, the word co-isoperimetric could be used, in the sense that here we relate the area to the length of something looking like a boundary, but in the opposite way: instead of squared length bounding a constant multiple of area, we have area bounding a constant multiple of squared length.

The proof is beautiful to be concealed but a little expensive because there is no way to escape to use the conformal representation theorem (which we have already put in its general context in theorem 70 on page 254):

Theorem 120 *Let g be any Riemannian metric on T^2. Then there exists a flat metric g' on T^2 and a positive function $f : T^2 \to \mathbb{R}$ such that $g = fg'$.*

Proof. We now give two different proofs of Loewner's theorem. The first is Loewner's original. Being flat, (T^2, g') has a transitive group G of isometries ξ. (The group is identifiable with the torus itself, in some sense, but maintaining a distinction makes the proof clearer.) We calculate the mean value of $f^{1/2}$ under the action of G for its canonical measure $d\xi$ to get

$$(f^*)^{1/2} = \int_\xi f(\xi)^{1/2} d\xi\ .$$

We get a new metric f^*g'. It is easy to check that by the definitions one has

$$\mathrm{Sys}\,(f^*g') \geq \mathrm{Sys}\,(g)$$

and (using Schwarz's inequality) that

$$\mathrm{Area}\,(T^2, f^*g') \leq \mathrm{Area}\,(T^2, g)$$

so that

$$\frac{\mathrm{Area}\,(T^2,g)}{\mathrm{Sys}\,(g)^2} \geq \frac{\mathrm{Area}\,(T^2, f^*g')}{\mathrm{Sys}\,(f^*g')^2}\,.$$

But (if it didn't already occur to you) the function f^* is constant because G acts transitively. So the metric f^*g' is in fact flat. The problem is thus reduced to finding the minimum of the systolic quotient for flat tori. The picture below shows that is is attained exactly for the equilateral one.

The second proof consists in testing the systole on a nice set of curves, large enough to fill up the torus. We fix the flat torus to be given by the two vectors $(0,a)$ and (c,b) in an orthonormal coordinate system (x,y) in the plane, adding moreover that $(0,a)$ is the smallest non-zero vector in the lattice, and that (c,b) is the second smallest one. And, moreover that (c,b) is not a multiple of $(0,a)$ and that

$$c \in \left[-\frac{1}{2}, \frac{1}{2}\right]\,.$$

We compute the sum of the lengths—for the primitive metric fg'—of the horizontal curves and get, by the very definition of the systole:

$$\int_0^b \mathrm{length}\,(y_{-1}(t)) = \int_0^b \int_0^a f^{1/2}\,dt\,dx > b\,\mathrm{Sys}\,(g)\,.$$

But

$$\mathrm{Area}\,(T^2,g) = \int_0^a \int_0^b f\,dx\,dt\,.$$

So we conclude by the Schwarz inequality and looking at the picture in figure 7.23.

Now for some other surfaces. The real projective plane is the simplest. Take any Riemannian metric on it, and look at its Riemannian universal covering, which is a metric on S^2. Use the conformal representation theorem for the sphere. As an exercise, first compute the systole for the standard real projective plane. After a small trick involving the Möbius group of the standard sphere:

Theorem 121 (Pu 1952 [1043]) *For every Riemannian metric g on $\mathbb{RP}^2$ one has*

$$\mathrm{Area}\,(\mathbb{RP}^2, g) \geq \frac{2}{\pi}\,\mathrm{Sys}\,(\mathbb{RP}^2, g)^2$$

with equality if and only if the metric is the standard one (up to rescaling by a constant factor).

7.2 Curvature Free Inequalities on Volumes of Cycles

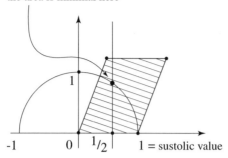

Fig. 7.23. Loewner's theorem

See VIII.12 of Berger 1965 [151] for details. Next comes the Klein bottle. Again we have a complete solution, but the result is not the one which we would naively expect. As an exercise, compute the systolic ratio for the flat Klein bottle. You will have first to know that the flat Klein bottle can only come about from rectangular flat tori.

Theorem 122 *For any metric g on the Klein bottle K,*

$$\text{Area}(K, g) \geq \frac{2\sqrt{2}}{\pi} \text{Sys}(K, g)^2$$

and if one wants to attain this lower bounded one should take a metric with a singular line obtained by gluing two copies of the Möbius band obtained from the standard sphere as indicated by the picture in figure 7.24 on the next page.

Note 7.2.1.1 The systolic inequality for Klein bottles is largely present in Blatter 1963 [206] and was rediscovered in Bavard 1988 [119]. Why is the optimal Klein bottle not a flat Klein bottle? Because the smallest geodesic in the homotopy class which is of order 2 is isolated. The neighboring curves are turning twice around it. So in the flat case one can reduces the area without changing the systole. The general idea is that there should be continuous families of curves having the systolic length; if not then one can reduce the volume. See below for more on this, and 4.D in Berger 1996 [168], and of course Calabi 1992 [304]. For example, for the equilateral torus there are three such families. The case of $\mathbb{RP}^2$ is exceptional: all of the geodesics are periodic. Look for yourself at these bands for Bavard's Klein bottle. ♦

7.2.1.2 Higher Genus Surfaces

We recall that compact surfaces are completely classified (see figure 3.36 on page 138). The orientable ones are the sphere, the torus and then surfaces with two or more holes; the number of holes γ is called the *genus* and the

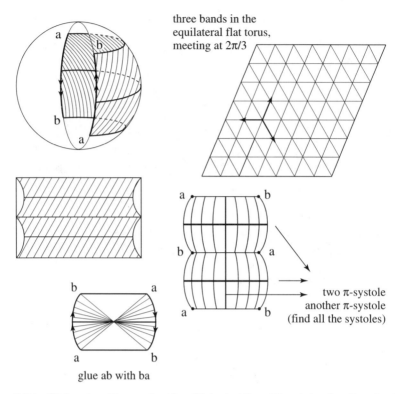

Fig. 7.24. Gluing together a singular Klein bottle with minimal ratio of area to squared systole

Euler characteristic χ (already met in equation 3.17 on page 138) is equal to $2(1-\gamma)$. The nonorientable ones are $\mathbb{RP}^2$, the Klein bottle and those of higher genus (obtained by add doughnut holes to $\mathbb{RP}^2$, for example).

Looking at the pictures, it is tempting to think that for surfaces M of genus higher than one, the systolic ratio

$$\frac{\text{Area}(M, g)}{\text{Sys}(M, g)^2}$$

can be given a lower bound and that moreover this bound will grow with the genus.

This was proven recently (in 1983) by Gromov, and has a very interesting history. In various fields of mathematics (algebraic curves, number theory, and the conformal representation theorem 70 on page 254) a very fruitful approach to the study of surfaces of any genus is via the theory of functions of a complex variable. Independently in Accola 1960 [11] and Blatter 1962 [206] complex function theory was used to attack the systolic problem for surfaces of genus higher than one. But counter to the intuition stemming from our pictures they

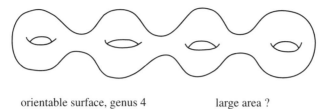

orientable surface, genus 4   large area ?

Fig. 7.25. An orientable surface of genus 4. Can a surface have large area with small systole, if it has large genus?

could only find a lower bound on the systolic quotient

$$\frac{\mathrm{Area}(M,g)}{\mathrm{Sys}\,(M,g)^2}$$

which tended toward zero for large genus.[7] The reason is that they used the conformal representation theorem, always a natural move when studying surfaces. They thereby reduced the question to the isosystolic inequality for flat tori (applied to the Jacobian of the surface M).[8] See the end of section §§§7.2.2.1 and the remarks at the end of section §§§7.2.2.3. This gave them a bound whose order of magnitude in γ is $1/g^{\gamma/2}$. This is shocking.

This is one of the rare cases where, for studying surfaces, conformal representation (alias complex analysis) is the wrong approach.

Theorem 123 (Hebda & Burago (independently) 1980)

$$\frac{\mathrm{Area}(M,g)}{\mathrm{Sys}\,(M,g)^2} \geq \frac{1}{2}$$

whatever the nonsimply connected surface and whatever metric g you place on it.

Proof. Let γ be a periodic geodesic of length equal to

$$L = \mathrm{Sys}\,(M,g)$$

and m any point on γ. We are going to see that

$$\mathrm{Vol}\,B\left(m, \frac{L}{2}\right) \geq \frac{L^2}{2}.$$

The subtlety is that we cannot say that the injectivity radius is larger than or equal to $L/2$, as one can see on the picture in figure 7.26 on the next page exhibiting "small fingers." So the simple argument used below in section §§§

---

[7] It was not the best possible lower bound, as we will see.
[8] For more on this, see e.g. section 4.B of Berger 1993 [165].

7.2.4.1 on page 353 is not available. Still we will retain part of it as follows. For any
$$r < L/2$$
we consider the closed metric disks
$$\bar{B}(m, r) \, .$$
By the very definition of the systole, $\bar{B}(m, r)$ is a topological disk, and in particular its boundary
$$S(m, r) = \partial \bar{B}(m, r)$$
is made up of two curves joining $\gamma(r)$ to $\gamma(-r)$. We cannot have both of these curves of length smaller than $2r$ because then one will get together with the remaining part of γ a closed curve homotopic to γ and smaller than γ, contradicting the fact that γ realizes the systole.

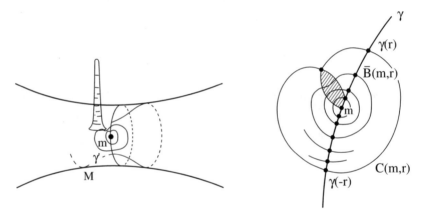

Fig. 7.26. Little fingers

Putting these two curves together, we know that
$$\text{length } S(m, r) \geq 4r$$
and by integration
$$\text{Area } \bar{B}\left(m, \frac{L}{2}\right) = \int_0^{L/2} \text{length } S(m, r) \, dr \geq \int_0^{L/2} 4r \, dr = \frac{L^2}{2} \, . \qquad (7.3)$$

Very soon after, Gromov used the same technique to prove what was expected, namely growth with the genus.

Theorem 124 (Gromov) *For any Riemannian metric on a compact surface M of genus γ*

$$\frac{\mathrm{Area}(M.g)}{\mathrm{Sys}\,(M,g)^2} \geq a\sqrt{\gamma}$$

for some positive constant a (see also note 7.2.1.2).

Proof. (Sketch) Build up a succession of periodic geodesics γ_i belonging to a basis of the homology of M.[9] We ask that each γ_i be the shortest in its homology class. The systole is still denoted by L. For any point m on any γ_i we still have

$$\mathrm{Area}\,B\left(m, \frac{L}{4}\right) \geq \frac{L^2}{8}.$$

Now consider a maximal set of pairwise disjoint balls

$$B\left(m_j, \frac{L}{4}\right)$$

each of which has center m_j belonging to one of the γ_i. We apply the following classical trick, which is no more than the triangle inequality in disguise, but which is incredibly useful in many contexts (see for example chapter 12):

Lemma 125 (The covering trick) *In a metric space, take a system of pairwise disjoint metric balls of a given radius r as large as possible (i.e. so that with larger r some of them will overlap). Then the system of balls obtained from the preceding one by keeping the centers the same, but doubling the radii, covers the whole space.*[10]

Remark now that any one of the γ_i needs at least two such balls, and that such a pair of balls cannot be used for any other loop γ_j. So the number N of our balls satisfies

$$\frac{N(N-1)}{2} \geq 2\gamma.$$

Thus

$$N = O(\sqrt{\gamma}).$$

Note 7.2.1.2 Gromov has a much better result on systolic inequalities for surfaces. By deeply refining the above proof, he got a lower bound of the type

$$\frac{\mathrm{Area}(M.g)}{\mathrm{Sys}\,(M,g)^2} \geq a\frac{\gamma}{\log^2 \gamma}$$

(the constant a being of course universal, i.e. independent of the metric g and the choice of surface M). For the proof, see 6.4.D' of Gromov 1983 [618] (where the wrong factor $\log \gamma$ appears) and 2.C of Gromov 1992 [632], or

---

[9] The homology of M is of dimension twice the genus on an orientable surface.
[10] Do not confuse this weak covering trick with the very strong doubling trick of theorem 108 on page 311, which does not apply to general metric spaces.

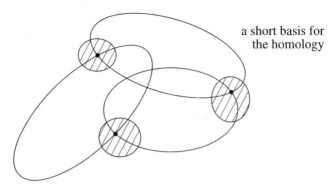

Fig. 7.27. A short basis for the homology

4.C of Berger 1993 [165] for very sketchy idea of the proof. This a very hard result, using Gromov's technique called *diffusion of cycles*. See also VIII. 9. B. 2 and IX. 4. 4. A detailed proof would most welcomed. Gromov can prove that asymptotically the ratio

$$a\frac{\gamma}{\log^2 \gamma}$$

is optimal (unpublished). ♦

Let us consider where the subject currently stands. There are many things to say. Here are the more important and natural. The first remark is that elementary geometry is solving a problem for surfaces that complex function theory was unable to solve. This is exceptional, to our knowledge. We turn now to two natural questions:

Question 126

1. *For a given surface, what is the optimal systolic ratio?*
2. *For which metrics is it attained?*

Above we saw that both questions are solved for the torus, the projective plane and the Klein bottle. As far as we know, both questions are open for any other compact surface. There are only partial answers.

The first is in section 5.6 of Gromov 1983 [618]

Theorem 127 *The infimum systolic ratio is attained by some metric, but for a Riemannian metric with singularities.*

The relevant kind of singularity is quite hard to grasp. This is an abstract statement. The value of the lowest systolic ratio and the metric achieving it are not known today. But very hard work was done by Calabi (see below). A result of Jenni 1984 [762] gives the best constant but within the restricted class of metrics of constant curvature (for example, curvature equal to -1) and for hyperelliptic surfaces of genus 2 and 5 only. Recall from §§6.3.2 that these space forms are very important mathematical objects. A good reference is the recent book Buser 1992 [292]. Also see Bavard 1993 [120] and of course Ratcliffe 1994 [1049].

Note 7.2.1.3 A metric of negative curvature will never achieve the best ratio. The philosophy for this was explained at the end of the preceding section (see figure 7.28). In fact, when the curvature is negative, periodic geodesics are isolated by the second variation formula, and then one can always reduce the volume without changing the systole. But if one decides to work within the world of Riemann surfaces, i.e. Riemannian surfaces of constant curvature, the game is different; see Buser 1992 [292]. ♦

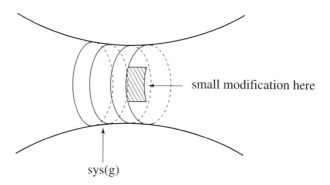

Fig. 7.28. Systoles on negatively curved surfaces

Note 7.2.1.4 We come back now to the general case. In general, when looking for the minimum of some quantity, the natural approach since the appearance of calculus is to look for minima among the cases where the first variation is zero. Then one studies the case at hand directly or computes the second variation, etc. We did that amply for geodesics and for the isoperimetric inequality. But systoles are not accessible to calculus, in some sense because of their nature, or perhaps because we lack the required tools. However Calabi 1992 [304] succeeded in (partly) reducing the problem to studying a partial differential equation. However, analysis of this equation is still in progress. Of course, the first case to look at is genus 2, but it does not seem to be simpler than, say, genus 4. ♦

Let us mention a simply stated plane geometry problem which, according to Calabi 1992 [304], is linked with the above. Consider the first Besicovitch theorem 144 on page 352 applied to a square in the plane: consider all possible Riemannian metrics on the square having the property that the Riemannian distance between any pair of points sitting in opposite sides of the square is always at least one. Then the Riemannian area of this Riemannian square is at least one, and equality can occur only when the square is the standard Euclidean square. One can compare this with the no-boundary torus case of Loewner.

Conjecture 128 (Calabi) *Suppose we take a hexagon and impose a metric on it so that the side lengths are at least one unit, and so that any pair of points on opposite edges are at least one unit apart. See figure 7.29. He has a conjecture for which we refer again to Calabi 1992 [304]: the extremal metric inside the hexagon looks like "a fried egg;" see figure 7.30 on the facing page. It will have regions with curvature of either sign, with singularities at the junctions.*

We have mentioned Besicovitch's result because it seems to have a very promising future in Riemannian geometry, since it was revived by Gromov in section 7 of Gromov 1983 [618]. This is essentially studying Riemannian manifolds with boundary; see §§ 14.5.1 on page 676 for more on this topic. One more remark: the story is really only about surfaces and one dimensional systoles, since Besicovitch 1952 [181] proves that for a cylinder, there are metrics with arbitrarily small volume even when both the infima of the length of the curves joining the two boundary disks and the area of disks whose boundary belong to the cylinder itself are larger than one.

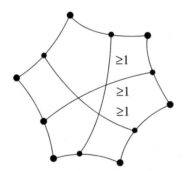

Fig. 7.29. What is the smallest possible area of our hexagon with at least unit distance between opposite sides, and which metric has that area?

7.2.1.3 The Sphere

The notion of systole on the sphere does not make sense at first glance since the systole is the least length of noncontractible curves, and all curves on a sphere are contractible. But the systolic value is achieved by periodic geodesics. So now forget about topology and retain only periodic geodesics. Consider a Riemannian metric g on S^2 and take the smallest periodic geodesic. It is not obvious that there is one. There is, but it is hard to prove and will be

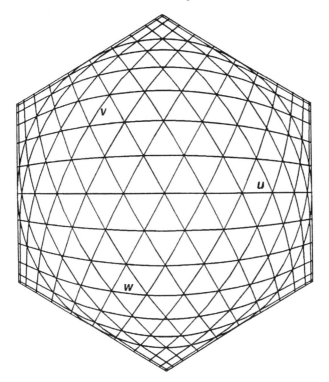

Fig. 7.30. Calabi's fried egg

amply treated in §§10.3.1. Let $\Lambda\left(S^{2}, g\right)$ be the length of that smallest periodic geodesic. You might feel that forces the area Area $\left(S^{2}, g\right)$ to be large. This question was asked by Gromov in 1980. The conjecture is

Conjecture 129 *For any Riemannian metric g on S^2*

$$\frac{\text{Area}\left(S^{2}, g\right)}{\Lambda\left(S^{2}, g\right)^{2}} \geq \frac{1}{2\sqrt{3}}$$

and equality can be achieved only by the singular metric corresponding to a doubly covered equilateral triangle.

Today there is only a partial answer by Croke 1988 [417] with the constant 1/961. For higher dimensional spheres, one needs to restrict to metrics with some condition on the curvature; see Wilhelm 1997 [1264], Rotman 2000 [1072].

Comparing the Birkhoff closed geodesic and the shortest one, see Nabutovsky & Rotman 2002 [960], Sabourau 2002 [1079, 1080, 1081] and Rotman 2000 [1072].

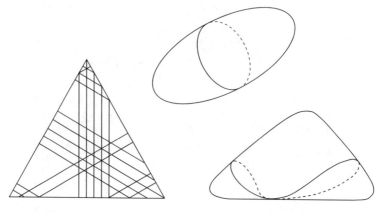

again three families and meeting at 2π/3

Fig. 7.31. The smallest area that a topological 2-sphere can have with fixed length for the smallest period geodesic occurs for the doubly covered equilateral triangle

7.2.1.4 Homological Systoles

We defined the systole of a surface in terms of the curves which are not homotopic to zero (noncontractible). But in topology, a closed curve defines a homology class, and even a noncontractible curve can have vanishing homology class.

Definition 130 *The* homological systole

$$\operatorname{Sys} H_1(M, g)$$

of a surface M with Riemannian metric g is the lower bound of the lengths of the closed curves whose homology is nonzero.

For the torus, the Klein bottle, or the real projective plane, the two notions coincide.[11]

But starting at genus two, the homological systole is radically different from the previously discussed free homotopy systole. Figure 7.32 on the facing page presents a famous picture of a noncontractible small closed geodesic which is homologous to zero.

The closed curve drawn there is homological to zero as is visibly obvious since it bounds a subsurface, but (also visibly obvious) it is noncontractible. In Gromov 1992 [632] it is proved, by a nice induction argument starting with the torus and and without too much difficulty, that for surfaces of any genus, theorem 124 on page 332 is still valid (asymptotically) for the homological

---

[11] Be careful with the nonorientable case, where you will have to decide whether you prefer to work with the integral or the $\mathbb{Z}_2$ homology.

7.2 Curvature Free Inequalities on Volumes of Cycles 339

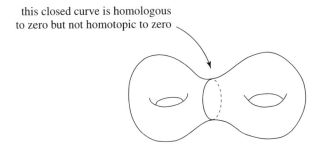

Fig. 7.32. The closed curve drawn is homologous to zero, but not freely homotopic to zero.

systole. The idea of the proof is to use induction, by cutting the surface along a periodic geodesic as in figure 7.33. Close the two pieces with a hemisphere. This will not change the systolic ratios. One piece is a torus and we apply to it Loewner's theorem 119 on page 326 and the induction hypothesis to the second piece. Finally, note that

$$\frac{\gamma}{\log^2 \gamma}$$

behaves essentially additively in γ.

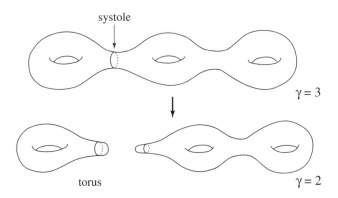

Fig. 7.33. Inductively obtaining asymptotics of homological systoles

This result is surprising since, starting with genus 2, the homology group is much smaller that the homotopy one. The homology of a compact surface is abelian and of finite dimension, while the homotopy group is infinite, and a little like a free group (so with exponential growth). In conclusion one would have been expecting a much smaller ratio for

$$\frac{\text{Area}(M)}{\text{Sys} \, H_1 \, (M)^2}$$

than for
$$\frac{\text{Area}(M)}{\text{Sys}\,(M)^2}$$
since many homotopically nontrivial curves are homologically trivial.

7.2.2 Systolic Inequalities for Curves in Higher Dimensional Manifolds

7.2.2.1 The Problem, and Standard Manifolds

Consider now compact Riemannian manifolds of dimension d larger than two and only nonsimply connected ones. For such a manifold M we still have the general notions of systole $\text{Sys}\,(M)$ and of course the volume $\text{Vol}(M)$. But, thinking of §§§7.2.1.4 above, we have to be careful to distinguish between homology and homotopy. We will make a distinction between $\text{Sys}\,H_1\,(M)$ and $\text{Sys}\,(M)$ only when needed and leave the reader in temporary confusion. He will soon discover that the distinction is insignificant (sadly enough).

There is no universal lower bound for
$$\frac{\text{Vol}(M)}{\text{Sys}\,(M)^{\dim M}}$$
(independent of the manifold M). Consider the manifold
$$M = S^1 \times S^{d-1}\ .$$

Taking product metrics for which the circle S^1 keeps length at least 1 and the volume of the S^{d-1} factor goes to zero, the systole will be equal to 1 and the volume will be as small as you like.

What was preventing us from employing this technique on a surface? Algebraic topology: on any surface other than the sphere S^2, the fundamental group (or the first homology group) generates the fundamental class. The systole represents the first homology and the volume the fundamental class. If they are dissociated as in the example above there is of course no hope to get an inequality. If you do not know what the fundamental class of a compact manifold M is, you can consider it as the top dimensional class in the de Rham cohomology group $H_{dR}^{\dim M}(M)$ as described in §§4.2.2.[12]

We have then to restrict the potential manifolds. The simplest examples which come to mind are tori T^d and the real projective spaces $\mathbb{RP}^d$. Also, we can look at space forms of negative curvature type. In those three cases, the first homology generates the fundamental class. No one could prove any systolic inequality for these spaces before Gromov cracked the nut in the very

---
[12] This is true for orientable manifolds. For unorientable manifolds, one has to work with $\mathbb{Z}_2$ homology.

important article Gromov 1983 [618] which is the basic reference for everything we will present here. See also Gromov 1992 [632].

Before presenting a sketch of Gromov's many new ideas, we look at a naive question which we almost skipped in the two dimensional case. Consider first a flat (locally Euclidean) torus and ask what is the systole and what is the volume? Denote by Λ the lattice yielding this flat torus as presented in §§4.4.3. Geodesics being projections of straight lines, the periodic ones are represented by segments joining the origin to some nonzero point of the lattice Λ and the volume is equal to the determinant of Λ. The systole is then equal to the smallest possible norm for a nonzero element of Λ. Our question is to find what is the best possible ratio for various lattices in $\mathbb{E}^d$ of the determinant to the dth power of the smallest nonzero element of the lattice.

In two dimensions, the best ratio was depicted in figure 7.22 on page 327. For a general dimension d this is a famous problem in the field of mathematics called *geometric number theory*. Compactness shows quite easily that there is a optimal lower bound and that it is achieved by some lattices, called *critical* lattices. They are completely known in vector spaces of all dimensions up to and including eight dimensions. After eight dimensions, neither the best ratio nor the lattices achieving it are explicitly known, but extremely good asymptotic values are known. We refer the reader to Berger 1993 [165], Gromov 1992 [632] and the various references there. We just want to note that the above bound is extremely small in d; its order of magnitude is

$$\frac{1}{d^{d/2}}.$$

The best references in our opinion are Gruber & Lekkerkerker 1987 [660] and Conway & Sloane 1999 [403]. Erdös, Gruber & Hammer 1989 [491] could be also useful.

An even more special case of flat tori is of great interest in algebraic geometry and number theory, namely we consider the so-called *Jacobian varieties* of Riemann surfaces. They are flat tori of (real) dimension 2γ if the genus of the surface is γ. Jacobians have a natural (flat) metric depending on the conformal structure on the given surface. It is a long standing problem to characterize Jacobians among flat tori (or equivalently, to characterize the lattices in $\mathbb{C}^\gamma$ which yield Jacobians). There are some very algebraic and complicated solutions of that problem; see Beauville 1987 [123]. Buser & Sarnak 1994 [297] proved that the order of magnitude of the systole is very different from the general case where the order of the systole which can go up to $\sqrt{\gamma}$, namely the systole has maximal order $\log \gamma$. See on page 352 for a geometric characterization of Jacobians using 2-systoles.

In our beloved $\mathbb{KP}^n$, the only case to consider is the real projective space (with its canonical metric) for which things are trivial: there is only one metric on projective space of constant curvature and since we know the geodesics and the total volume we are done.

7.2.2.2 Filling Volume and Filling Radius

Gromov proved the existence of a positive lower bound for the systolic ratio

$$\frac{\text{Vol}(M)}{\text{Sys}_d(M)}$$

for any compact Riemannian manifold M whenever M has sufficient 1-dimensional topology to generate its fundamental class. The exact condition will be expressed in §§§7.2.2.3. Before describing that condition, we will now introduce the concepts needed for the proof. These concepts have already much geometric interest and raise many open questions.

The difficulty in the topic—at least with current knowledge—is that one cannot avoid working in infinite dimensional vector spaces. The starting point is to construct the map

Definition 131

$$f : M \to L^\infty(M)$$

where $L^\infty(M)$ is the space of all bounded functions on M endowed with the sup norm and where the map f is given by the various distance functions

$$f : m \mapsto d(m,)$$

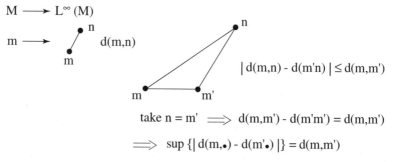

Fig. 7.34. The map taking a point $m \in M$ to the distance function $d(m,) \in L^\infty(M)$.

For the two metric spaces M and $L^\infty(M)$, the map f is an isometry.

Definition 132 *The* filling radius *of a compact Riemannian manifold M is the smallest positive number ε for which the image $f(M)$ in $L^\infty(M)$ bounds in its ε tubular neighborhood.*[13]

---

[13] The term "to bound" comes from algebraic topology and means that there is a cycle of one higher dimension whose boundary is the object under consideration.

7.2 Curvature Free Inequalities on Volumes of Cycles 343

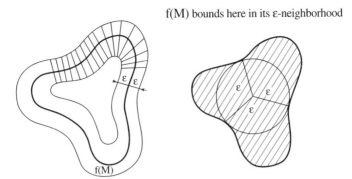

Fig. 7.35. (a) $f(M)$ does not bound in its ε neighborhood. (b) $f(M)$ bounds here in its ε neighborhood.

This filling radius is a geometric invariant which at first glance appears very simple. In fact its value is known only for the standard spheres S^d, the standard projective spaces $\mathbb{RP}^d$ and the complex projective plane $\mathbb{CP}^2$, but is still unknown for the remaining $\mathbb{KP}^n$ as well as for any other Riemannian manifolds (in particular, any of the above manifolds with nonstandard metrics); see Katz 1983,1991 [792, 793] and Wilhelm 1992 [1261]. For the filling radius of the sphere, see Sabourau 2002 [1079, 1080, 1081].

Definition 133 *The* filling volume *of a compact Riemannian manifold M of dimension d is the infimum of the volumes of the $d+1$ dimensional submanifolds in $L^\infty(M)$ whose boundary is $f(M)$.*

Today there is not a single manifold whose filling volume is known, not even the circle (for which Gromov conjectures the value is 2π). These two invariants are promised a future in Riemannian geometry. For example the filling radius is used in Greene & Petersen 1992 [592] to improve the recent finiteness theorems mentioned in §§ 12.4.1 on page 614. But some inequalities (due to Gromov of course) are valid for any compact manifold; namely

Theorem 134 *There are two positive constants $a(d)$ and $b(d)$ such that for every compact d dimensional Riemannian manifold M*

$$\mathrm{Vol}(M)^{d+1} > a(d)\,(\textit{Filling volume of } M)^d$$
$$\textit{Filling volume of } M > b(d)\,(\textit{Filling radius of } M)^{d+1}$$

The proofs are hard, especially since one is working in infinite dimension. But as seen in ordinary space they are quite reasonable. The first inequality is classical in the standard Euclidean setting, say of dimension N. You have a compact d dimensional submanifold M in $\mathbb{R}^N$ and you span it with a minimal submanifold H of dimension $d+1$ with the boundary $\partial H = M$ equal to your original manifold. Then there is an isoperimetric inequality in Michael & Simon 1973 [918] to the effect that

$$\frac{\operatorname{Vol}(M)^{d+1}}{\operatorname{Vol}(H)^d} > a(d)$$

for some universal constant $a(d)$ (independent of M and H).

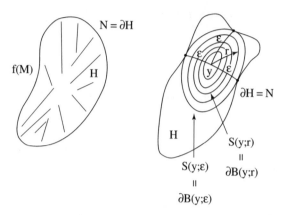

Fig. 7.36. There is a isoperimetric inequality of classical character for minimal submanifolds and their boundaries

We will continue to imagine this picture in a finite dimensional Euclidean space $\mathbb{R}^N$ of M as the boundary

$$\partial H = M$$

of a $d+1$ dimensional minimal submanifold H of $\mathbb{R}^N$. (Of course, things can be more subtle in infinite dimensions.) Now we will use the definition of the filling radius ε. This means exactly that there is some point $y \in H$ such that the metric ball $B(y, \varepsilon)$ is filled up nicely with the spheres $\partial B(y, r)$ for every $r < \varepsilon$. Compare this with the proof in theorem 123 on page 331. Using this and the minimality we see that volume should grow at least as quickly as in the flat case.

Note 7.2.2.1 Filling radius and filling volume have been introduced here, not only because they are basic ingredients in Gromov's proof, but because it seems highly probable that they will play an important role in the future of Riemannian geometry. ◆

7.2.2.3 Gromov's Theorem and Sketch of the Proof

First we state the technical formulation of the heuristic notion of the first dimensional homology generating the fundamental class.

7.2 Curvature Free Inequalities on Volumes of Cycles

Definition 135 *A topological space is said to be* aspherical *if except for dimension one all the homotopy groups vanish.*

Definition 136 *A compact Riemannian manifold M is said to be* essential *if there is some map*
$$g : M \to X$$
into some aspherical space X such that the induced map
$$g_* : H_*(M) \to H_*(X)$$
at the homology level maps the fundamental class of M into a nonzero element:
$$g_*[M] \neq 0 \ .$$

Theorem 137 (Gromov 1983 [618]) *There is a positive constant $c(d)$ such that for every essential compact d dimensional Riemannian manifold M one has*
$$\frac{\mathrm{Vol}(M)}{\mathrm{Sys}_1(M)^d} > c(d) \ .$$
(where $\mathrm{Sys}_1(M)$ is the homotopic 1-systole).

Using theorem 134 on page 343 the proof is finished with

Theorem 138 *For any essential compact Riemannian manifold M*
$$Filling\ radius(M) \geq \frac{1}{6} \mathrm{Sys}_1(M) \ .$$

This is the key point in the proof and it goes by contradiction assuming that
$$Filling\ radius = r < \frac{1}{6} \mathrm{Sys}_1(M).$$
We first identify M with its isometric image $f(M)$ inside $L^\infty(M)$. Then we know that M is the boundary of some cycle H which is contained in the r neighborhood of M, since r is the filling radius of M. This means that the inclusion map acts on homology
$$\iota_* : H_d(M) \to H_d(K)$$
trivially on the fundamental class of M:
$$\iota_*[M] = 0.$$
Let
$$\eta : M \to X$$
be a continuous map into an aspherical space X which is not trivial on the fundamental class of M at the dth dimensional homology:

$$\eta_* : H_d(M) \to H_d(X)$$
$$[M] \mapsto \eta_*[M] \neq 0.$$

We claim that under the assumption above we can extend η to a map
$$\eta : H \to X.$$
This will be the desired contradiction, since M is a boundary in H.

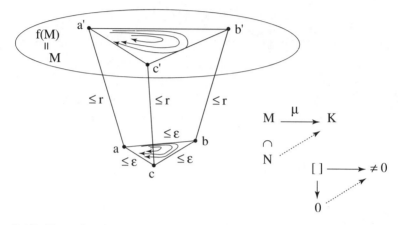

Fig. 7.37. Extending a continuous map from M to a higher dimensional cycle containing M

We note that we have only to carry out such an extension at the two dimensional level since X is aspherical. To do this we triangulate H (including M) in such a way that every edge is of length smaller than ε with
$$3\varepsilon + 6r < \mathrm{Sys}_1(M).$$
To extend η to H, we associate to each vertex $a \in H$ in the triangulation another vertex $a' \in M$ with distance
$$d(a, a') < r.$$
Let $\{a, b, c\}$ be any triangle of the triangulation and $\{a', b', c'\}$ the corresponding triangle in M. The segments which are the sides of $\{a', b', c'\}$ are geodesics in M since we have an isometric embedding. We draw the sides of $\{a, b, c\}$ and this gives us an extension from the original η up to the 1-skeleton. Now we remark that by our construction and the triangle inequality

perimeter of $\{a', b', c'\}$ $\leq$ perimeter of $\{a, b, c\} + 6r < \mathrm{Sys}_1(M).$

By the very definition of the systole, the perimeter of $\{a', b', c'\}$ is contractible in M. In order to extend η to the interior of $\{a, b, c\}$, we have only to make a point to point correspondence with a contraction in M of the perimeter of $\{a', b', c'\}$. This yields the desired extension of η to the 2-skeleton.

Note 7.2.2.2 The above theorem applies to every torus, to every negative curvature type space form (see §§6.3.2) and also to the projective spaces $\mathbb{RP}^d$. To see this you need to know that the infinite dimensional real projective space $\mathbb{RP}^\infty$ is aspherical. ♦

We are left with three questions. First:

Question 139 *What is the optimal value and if "extremal metrics" exist who are they?*

For this, besides the "Filling paper" of Gromov 1983 [618], see Calabi 1992 [304] and Gromov 1992 [632]. But in dimensions three and higher, there is to our knowledge no existence theorem of an optimal systolic metric (possibly with singularities) on any type of manifold. Nor is any optimal ratio known. It is not clear today if it is reasonable to conjecture that for the tori T^d and the projective spaces $\mathbb{RP}^d$ the best ratio is attained for flat and standard metrics and only for them (see §§§7.2.2.1). The first cases to look at would be T^3 and $\mathbb{RP}^3$. A classical idea is to sit at the standard metric g for $\mathbb{RP}^3$ or a flat one g for T^3 which is optimal among flat metrics and try to compute a variation of the systolic ratio

$$\frac{\mathrm{Vol}(M,g)}{\mathrm{Sys}_3(M,g)}.$$

This is difficult because the systole is not directly accessible to calculus: see question 126 on page 334. The volume is accessible to calculus. One still can show that the first variation is always zero but the second variation seems hard to work with. See Besse 1978 [182] 5.90 for a result in this spirit.

The second question:

Question 140 *What characterizes compact manifolds for which we have such a universal inequality for*

$$\frac{\mathrm{Vol}(M,g)}{\mathrm{Sys}_1(M,g)^{\dim M}}?$$

This is studied in Babenko 1992 [91] for the homotopical

$$\mathrm{Sys}_1(M)$$

associated to the fundamental group. Such a result can be seen as a "systolic" characterization of a topological property.

The third question concerns the homological systole:

Question 141 *What characterizes compact manifolds for which we have such an universal inequality for*

$$\frac{\mathrm{Vol}(M,g)}{\mathrm{Sys}\,H_1(M,g)^{\dim M}}?$$

Things are completely solved positively in Babenko 1992 [91] (but not everything figures explicitly in the text, and be careful about the nonorientable case).

7.2.3 Higher Dimensional Systoles: Systolic Freedom Almost Everywhere

Everything in the last section concerned closed curves in a Riemannian manifold. But there is no reason not to consider closed submanifolds of any dimension. For simplicity, we will stick from now on with homology (no longer homotopy). References are Berger 1993 [165], Gromov 1992 [632], chapter 4 and appendix D of Gromov 1999 [633] and references there. Interesting examples of systolic freedom occur in Freedman 2000 [521], where they are related to quantum computing; this relationship is still being developed. See Nabutovsky & Rotman 2002 [960], Katz, Kreck & Suciu 2002 [797], Katz & Suciu 2001 [799], Bangert & Katz 2002 [113], and Katz 2002 [796] for relations with calibrations (see §§§14.2.1.3).

We define the k dimensional systole

$$\mathrm{Sys}_k(M, g)$$

of a Riemannian manifold M with Riemannian metric g as the lower bound of the volume of the closed k dimensional submanifolds of M which are not homologous to zero. We will ask

Question 142 *Is there, at least for some compact d dimensional manifolds M, a constant $a(M)$ and an inequality*

$$\frac{\mathrm{Vol}(M, g)^k}{\mathrm{Sys}_k(M, g)^d} \geq a(M)$$

for any Riemannian metric g on M?

We might ask this for example for the $\mathbb{K}\mathbb{P}^n$ or various products like

$$S^k \times \cdots \times S^k$$

$$M \times N$$

for

$$\frac{\mathrm{Vol}(M \times N)}{\mathrm{Sys}_p(M) \mathrm{Sys}_q(N)}$$

etc. One will find in the two above references the state of affairs today. But we will now be extremely brief for the following reason. It might well be the answer to question 142 is *no* for every compact manifold and every $k \geq 2$.

For higher dimensional systoles, negative results started to appear in Gromov 1992 [632]. They were intermediate results. Gromov's example of $(1, 3)$

7.2 Curvature Free Inequalities on Volumes of Cycles

softness on $S^1 \times S^3$ is elementary and consists simply in gluing on a copy of $[0,1] \times S^3$ after twisting enough with the Hopf fibration. We have now a extremely large category of negative examples in Bérard Bergery & Katz 1994 [146], Babenko & Katz 1998 [93] and Babenko, Katz & Suciu 1998 [94]; also see Katz's appendix in Gromov 1999 [633] and Pittet 1997 [1031]. One says that a manifold M^d is *systolically* $(k, d - k)$ *soft* (or *systollically* $(k, d - k)$ *free*) if the infimum of the quotients

$$\frac{\mathrm{Vol}(M, g)}{\mathrm{Sys}_k(M, g)\,\mathrm{Sys}_{d-k}(M, g)}$$

among all metrics g on M^d is zero. The above authors proved softness in the following cases:

1. For any orientable $(k-1)$-connected d dimensional manifold with

$$d \geq 3, \quad k < \frac{d}{2}, \quad \text{and } k \text{ not a multiple of 4,}$$

   there is $(k, d - k)$ softness.
2. For simply connected manifolds of even dimension $M^{d=2n}$ with $d \geq 6$, provided $H_n(M)$ is torsion free, there is (n, n)-softness.

The $S^k \times S^k$ (for $k \geq 3$) and $\mathbb{HP}^2$ are startling examples. The $(4, 4)$-softness of $\mathbb{HP}^2$ is most surprising, since the projective lines fill up the whole space in the most geometric possible fashion and should have been "sniffing around enough" to prevent softness. The "freedom" can be very large, for example on $S^3 \times S^3$ one can even restrict attention to only homogeneous metrics and still see the freedom. It is not clear today if the various topological restrictions are really necessary as well as the dimension ones. In fact there is no known example of a hard inequality as soon as the involved systoles are in dimension 2 or more. The dimension 4 case was solved in Katz 1998 [795]. Even earlier, Katz had shown that $S^2 \times S^2$ and $\mathbb{CP}^2$ are systolically soft. Of course the case of $\mathbb{CP}^2$ is even more surprising, just as for all $\mathbb{KP}^n$ but especially because Gromov proved hardness for metrics close to the canonical metric, announced in Gromov 1985 [619]. The recent Katz & Suciu 1999 [798] and Babenko 2000 [92] completely finishes the proof of middle dimensional freedom on even dimensional manifolds.

All the above results are very geometric. We will give a flavor of the ideas and refer to the Appendix D of Gromov 1999 [633] for more details. We start with Gromov's initial example:

Proposition 143 *On the manifold $M = S^1 \times S^3$ there are Riemannian metrics g whose quotient*

$$\frac{\mathrm{Vol}(M, g)}{\mathrm{Sys}_1(M, g)\,\mathrm{Sys}_3(M, g)}$$

is arbitrarily small.

Proof. We will play with a real number R which will soon start to get larger and larger. Consider M the product of the sphere $S^3(R)$ of radius R with the interval $[0, 1/R]$ with the two boundary spheres identified

$$S^3(R) \times \{0\} \quad \text{and} \quad S^3(R) \times \{1/R\}$$

under the "Hopf rotation" map

$$S^3(R) \to S^3(R).$$

A Hopf rotation is the following map: we choose some Hopf fibration on $S^3(R)$ by great circles and then push any point by a unit distance along the fiber through it. Then the total volume of the resulting Riemannian manifold (a local product) is

$$2\pi^2 R^3 \cdot \frac{1}{R} = 2\pi^2 R^2.$$

Controlling the systoles is easy:

$$\mathrm{Sys}_3(M) = 2\pi^2 R^3$$

and

$$\mathrm{Sys}_1(M) = \sqrt{1 + \frac{1}{R^2}}.$$

$$\frac{\mathrm{Vol}(M)}{\mathrm{Sys}_3(M)\,\mathrm{Sys}_1(M)} \sim O\left(\frac{1}{R}\right) \to 0 \text{ as } R \to \infty$$

This counterexample metric is not very wild, since it is homogeneous.

To determine systoles on some more general manifolds, one performs surgery in such a manner that one can control the systole and the volume, employing quite a few subtle results of algebraic topology. Technically, one also uses calibrated geometry; see §§14.2.1. Using the concept of calibration, one can relate homological systoles to differential forms, which is not the case in general. When there are topological restrictions, it is because they are needed to be able to employ various algebraic topology results.

Note 7.2.3.1 (Stable systoles) There are results similar to those we have already studied, but for the so-called *stable systoles*. For details see chapter 4 of Gromov 1999 [633] or section 8 of Berger 1993 [165]. We briefly explain the theory of stable systoles. The story here is really in the realm of homology. As an introduction, let us consider the case of some Riemannian surface (you can think of a torus for example). Let γ be the smallest homologically nontrivial periodic geodesic. Travel along it twice; call this 2γ. Can you find some new curve σ which is homotopic to 2γ but of smaller length? An old theorem of Morse says that this is impossible, the reason being that the new curve will

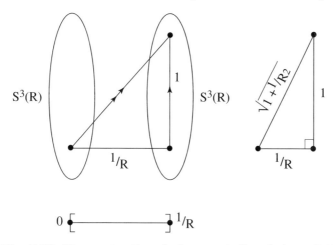

Fig. 7.38. The construction of a homotopically soft 4-manifold

necessarily intersect itself and then will not be a shortest curve by the first variation formula. But in higher dimension, there is room enough to avoid self-intersections. This was already mentioned in Hedlund 1932 [695]. But if we write

$$\text{length } \alpha$$

for the infimum of the lengths of all closed curves in homology class $\alpha \in H_1(M)$, then nonetheless one can prove that there is a positive limit for the real number

$$\text{Stab Sys } H_1(M) = \lim_{n \to \infty} \frac{\inf \{\text{length}(n\alpha) \mid \alpha \in H_1(M)\}}{n}$$

which number which we call the *stable 1-systole* of the compact Riemannian manifold M.

One can define the same way the stable systole of any dimension k in any compact manifold M, denoted $\text{Stab Sys } H_k(M)$.

It turns out that stable systoles enjoy the inequality one would naturally expect, e.g. for the $\mathbb{KP}^n$ and for products of manifolds. The reason at the heart of this is that the stable systole can be related to various norms of exterior differential forms on a Riemannian manifold, which is not true for standard systoles as is clear from the counterexamples above. One has at ones disposal the exterior product, Stokes' formula, etc. This is to be found only in chapter, 4, section D of Gromov 1999 [633]. In particular, for $\mathbb{CP}^n$ we see an inequality with the optimal constant

$$\frac{\text{Vol}(\mathbb{CP}^n, g)}{\text{Stab Sys } H_2(\mathbb{CP}^n, g)^n} \geq \frac{\text{Vol}(\mathbb{CP}^n, \text{Fubini–Study})}{\text{Vol}(\mathbb{CP}^1, \text{Fubini–Study})^n}$$

for any metric g on $\mathbb{CP}^n$. For products, one still finds inequalities but, at least today, with poor constants:

- For any Riemannian metric g on $\prod^n S^k$

$$\frac{\text{Vol}\left(\prod^n S^k, g\right)}{\text{Stab Sys } H_k \left(\prod^n S^k, g\right)^n} > c > 0$$

- For any Riemannian metric g on a product of compact manifolds $M^p \times N^q$

$$\frac{\text{Vol}(M \times N, g)}{\text{Stab Sys } H_p(M, g) \text{ Stab Sys } H_q(N, g)} > c > 0$$

◆

Note 7.2.3.2 k-systoles with $k > 1$ are interesting objects. In 2.E of Gromov 1992 [632], Gromov uses some 2-systoles to give a purely geometrical characterization of Jacobians of curves among flat tori, already met for the 1-systole in §§§7.2.2.1. ◆

Note 7.2.3.3 For simply connected manifolds, it makes sense as in §§§7.2.1.3 to consider the volume and the length of the smallest periodic geodesic. Today one can get results if one adds some curvature conditions, see Rotman 2000 [1072]. ◆

Note 7.2.3.4 (Besicovitch's results 1952 [181]) We imagine that the reader is surprised by the theory of systoles, which one can sketch as: for curves, one has very good positive results, but as soon as one works with submanifolds of dimension two or more, one has softness, i.e. rather weak results. In fact this is not surprising, in view of two results of Besicovitch for Riemannian manifolds with boundary:

Theorem 144 (Besicovitch 1952 [181]) *Take the standard topological cube*

$$C = [0, 1]^d$$

but put inside it any Riemannian metric. If L_i denotes the infimum of the length of any curve having its ends in the opposite faces

$$E_i = \{x_i = 0\} \cap C \qquad F_i = \{x_i = 1\} \cap C$$

for $i = 1, \ldots, d$ then the total volume $\text{Vol } C$ satisfies

$$\text{Vol } C \geq \prod_{i=1}^d L_i$$

with equality if only if the metric we have imposed on C is the standard Euclidean cube metric.

Theorem 145 (Besicovitch 1952 [181]) *Consider a cylinder*

$$C = D \times [0, 1]$$

where is D is a disk. Put any Riemannian metric inside and define L to be the infimum of the length of the curves with one end in the disk

$$D \times \{0\}$$

and the other in the disk

$$D \times \{1\}$$

and denote by S the infimum of the area of surfaces in C whose boundary lies in the surface

$$S^1 \times [0, 1].$$

Then the ratio

$$\frac{\operatorname{Vol} C}{LS}$$

can be made arbitrarily small (softness).

For more see Gromov 1983 [618]. ◆

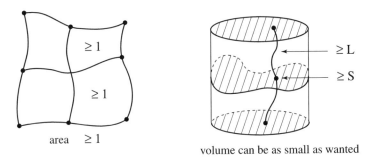

Fig. 7.39. (a) Area ≥ 1 (b) Volume can be as small as desired

7.2.4 Embolic Inequalities

7.2.4.1 Introduction, Questions and Answers

To our knowledge, the content of the present section is treated in book form only in chapters 6 and 7 of Chavel 1993 [326], and not completely even there. One might also consult Sakai 1996 [1085] chapter VI.

Looking at the above two pictures and also remembering the dichotomy of Klingenberg's theorem 89 on page 272, one is tempted to say that forcing the

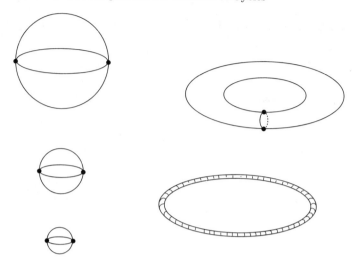

Fig. 7.40. Volume and injectivity radius

volume to be small will force the injectivity radius to be small. For example think of a flat torus or of a sphere.

This is easily seen on any surface; we apply the technique of equation 7.3 on page 332 as follows. Pick up any point $m \in M$ and look at the ball $B(m, \iota/2)$ where $\iota = \text{Inj}(M, g)$. That ball is nicely covered by the exponential map, by the definition of injectivity radius. For any $t < \iota/2$ and any geodesic γ through $m = \gamma(0)$ the two points $\gamma(r)$ and $\gamma(-r)$ cannot be joined by a curve of length smaller than $2r$, once again by the definition of the injectivity radius. In conclusion, the circle $C(m, r) = \partial B(m, r)$ is always made up of two pieces, each of length not smaller than $2r$. By integration one finds as in equation 7.3 on page 332

$$\text{Vol } B(m, \iota/2) \geq \frac{\iota^2}{8}.$$

Since one can cover M with two such balls we know that

$$\text{Vol}(M, g) \geq \frac{\text{Inj}(M, g)^2}{4} \tag{7.4}$$

This technique is very primitive. Indeed one never has equality (proof is left to the reader). Moreover it does not extend to higher dimension, since in higher dimensional manifolds, the boundary of a ball $B(m, r)$ is a sphere $S(m, r) = \partial B(m, r)$ which is no longer a curve and its area is almost impossible to control, at least by any simple mechanism.

We still can ask some natural questions. The first two stem directly from what we have just considered.

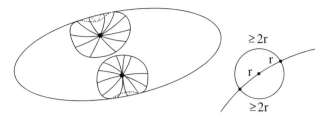

Fig. 7.41. Relating the injectivity radius of a surface to its volume

Question 146 *Is there a positive constant $a(d)$ such that for every compact Riemannian manifold M of dimension d, any point $m \in M$ and any $r \leq \mathrm{Inj}\,(M)/2$, the volumes of balls satisfy*

$$\mathrm{Vol}\,B\,(m,r) \geq a(d)r^d?$$

If so, what is the largest possible choice of $a(d)$?

Question 147 *Is there a positive constant $b(d)$ such that for every compact Riemannian manifold of dimension d the volume of the manifold satisfies*

$$\mathrm{Vol}(M) \geq b(d)\mathrm{Inj}\,(M)^d?$$

If so, what is the largest possible choice of $b(d)$?

Today there are some pleasing results, but not all of these questions answered. Here is what we know:

Theorem 148 (Berger 1980 [160]) *Question 147 has a positive answer. The largest possible constant is*

$$b(d) = \frac{\sigma(d)}{\pi^d}$$

(recall that $\sigma(d)$ denotes the volume of the standard sphere and that π is its injectivity radius). Moreover equality is attained only for standard round spheres.

Theorem 149 (Croke 1980 [411]) *Question 146 has a positive answer with the (not optimal) bound*

$$a(d) = \frac{2^{d-1}\sigma(d-1)^d}{d^d \sigma(d)^{d-1}}.$$

Theorem 150 (Croke 1984 [415]) *Question 146 has a positive answer with an optimal bound "on average." Namely, for every $r < \mathrm{Inj}\,(M)$,*

$$\frac{1}{\text{Vol}(M)} \int_M \text{Vol}\, B\,(m,r)\, dV_M\,(m) \geq \sigma(d) \frac{r^d}{\pi^d}$$

and equality occurs if and only if M is isometric to a standard sphere and $r = \text{Inj}\,(M)$.

The bound $a(d)$ in theorem 149 on the preceding page is certainly not optimal, in that equality can never be attained. It is not clear if it is reasonable to conjecture for the optimal bound the value obtained for standard hemisphere, namely

$$a(d) = \frac{\sigma(d)}{(\pi/2)^d}$$

with equality only for standard hemispheres.

Definition 151 *For a compact manifold M of dimension d its embolic constant[14] is*

$$\text{Emb}(M) = \inf_g \frac{\text{Vol}(M,g)}{\text{Inj}\,(M,g)^{\dim M}}$$

where the infimum is taken over all Riemannian metrics g on M.

By theorem 148 on the previous page we know that this is a positive constant which is not smaller than

$$\frac{\sigma(d)}{\pi^d}.$$

Imagine the subset of $\mathbb{R}^+$ made up by the different embolic constants of the various compact manifolds of a given dimension d. Notice the major result: theorem 267 on page 505 which is a direct corollary of theorem 377 on page 622.

Question 152 *Compute $\text{Emb}(M)$ for various standard compact manifolds e.g. the tori T^d, the various $\mathbb{KP}^n$, etc. In case one is able to compute the embolic constant, we may also wonder for which metric is it attained?*

Question 153 *Is the value*

$$\frac{\sigma(d)}{\pi^d}$$

isolated or not in the set of possible embolic constants for manifolds of a given dimension?

---

[14] The term *embolism* comes from the Greek for *insertion* and refers in physiology to a bubble or blood clot blocking an artery. We want to think of the injective image of a ball under the exponential map as our embolism, "blocking" the manifold in the sense that it strikes into itself, and this is the relevant analogy.

Question 152 is completely open today, even for projective spaces $\mathbb{RP}^d$ or tori. Except for Emb (S^d) not a single embolic constant Emb(M) is known when $d \geq 3$. As an exercise, the reader can find the embolic constants of $\mathbb{RP}^2$, the two dimensional torus and the Klein bottle (use Klingenberg's theorem 89 on page 272). If the embolic constant of $\mathbb{KP}^n$ turns out to be the one obtained for the standard metric then this will help to answer (and probably completely answer) the outrageously open question 97 on page 285 concerning Riemannian manifolds M with

$$\mathrm{diam}(M) = \mathrm{Inj}\,(M).$$

But there is a satisfactory answer to question 153:

Theorem 154 (Croke 1988 [418]) *Every manifold M of dimension d which is not a sphere satisfies*

$$\mathrm{Emb}(M) > \mathrm{Emb}\left(S^d\right) + c(d)$$

where $c(d)$ is an explicitly described positive constant depending only on the dimension d.

Note 7.2.4.1 (Best metrics) We have introduced a remarkable invariant of (smooth) compact manifolds: the embolic constant. This invariant comes from Riemannian geometry. In chapter 11 we will introduce two other ones:

- the minimal volume and
- the infimum of the integral

$$\int_M \|R\|^{d/2} dV_M.$$

Riemannian metrics achieving the lowest bound for a reasonable functional are, in some sense, the best possible on a given compact manifold. ♦

7.2.4.2 Starting the Proof and Introducing the Unit Tangent Bundle

We will give the ideas for the proofs of theorems 148 on page 355, 149 on page 355, 150 on page 355, and 154 when the manifold in question is a surface. We will just mention the technical difficulties for passing from surfaces to higher dimensional manifolds. But the main steps are already present in studying surfaces. In passing, we will now introduce the unit tangent bundle of a Riemannian manifold, which will also play various important roles in the future (for example, see chapter 10).

First, normalize the metric g in order that

$$\mathrm{Inj}\,(M,g) = \pi.$$

358   7 Volumes and Inequalities on Volumes of Cycles

We consider various balls $B(m,r)$ of radii $r < \pi$. We can be sure that these balls are nicely covered by the exponential map. In such a ball, the ds^2 is written as
$$ds^2 = dt^2 + f^2(u,t)du^2.$$
Consequently, we see
$$\text{Area } B(m,r) = \int_{u \in U_m M} \int_0^r f(u,t)\, dt\, dV_{U_m M}(u) \qquad (7.5)$$

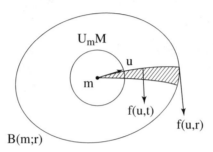

Fig. 7.42. Measuring the volumes of balls via the exponential map

Moreover, we can consider u as a unit tangent vector at m running through the unit sphere $U_m M$. If we denote by γ_u the geodesic with
$$\gamma_u'(0) = u$$
then the function f is given along γ_u by the Gauß–Jacobi equation
$$f'' + Kf = 0. \qquad (7.6)$$

The general idea is to take some kind of "mean value" of areas of all the balls when m runs through M and r ranges from 0 to π. This will involve four integrations:

1. in t from O to r
2. in r from 0 to π
3. in u running through $U_m M$ and finally
4. in m running through M.

The last two integrations are better packaged into one by introducing

Definition 155 *The set of all unit tangent vectors to a Riemannian manifold M is called its* **unit tangent bundle** *and is denoted by UM. It is a subbundle of the full tangent bundle TM. The canonical projection is denoted by*
$$p : UM \to M.$$

7.2 Curvature Free Inequalities on Volumes of Cycles   359

The bundle UM is automatically oriented, and endowed with a canonical Riemannian metric (see §15.2) which moreover makes $p : UM \to M$ a Riemannian submersion (see §§4.3.6). In particular, it has a canonical volume form which is denoted by ω. The geodesic flow on UM is the one parameter group G_t of diffeomorphisms of UM defined as follows:

$$G_t(u) = \gamma'_u(t).$$

Theorem 156 (Liouville) *The volume form ω is invariant under the geodesic flow (see chapter 10).*

In its canonical metric, the trajectories of the geodesic flow are themselves geodesics and (see on page 237) project down onto the geodesics of M.

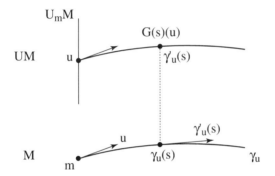

Fig. 7.43. The geodesic flow lines in UM project to the geodesics of M

A physicist might consider UM as the phase space of free particles on M. It is helpful to look at the geodesics as trajectories of a flow in UM because in UM they fill up the space and do not intersect each other. This will turn to be very important in chapter 10.

7.2.4.3 The Core of the Proof

The next idea in proving the theorems 148 on page 355, 149 on page 355, 150 on page 355, and 154 on page 357 is to apply Fubini's theorem many times. The starting point is to integrate things (i.e. the various functions figuring in equation 7.5 on the preceding page) first along a given geodesic segment of length π. To do this, we decorate the function f with a subscript as f_u when we want to describe its restriction to γ_u. We have to compute the double integral :

$$I(u) = \int_{s=0}^{s=\pi} ds \int_{t=s}^{t=\pi} dt\, f\left(G_s\left(u\right), t\right) \tag{7.7}$$

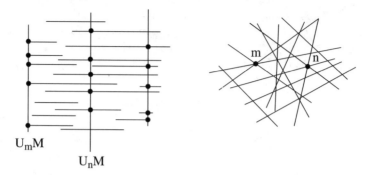

Fig. 7.44. (a) Geodesics in UM upstairs (b) Geodesics downstairs in M

The basic idea of the proof is that the various $f(G_s(u), t)$ (when s varies) can be explicitly computed as a function of $f(u,t)$. This is because they all are solutions of the Gauß equation

$$f'' + Kf = 0 \tag{7.8}$$

It is essential here that $f(u,t)$ does not vanish between 0 and π. This is because of the results on cut points which we found in §§6.5.1. Now elementary theory of ordinary differential equations yields

$$f(G_s(u), t) = f(s) f(t) \int_{w=s}^{w=t} \frac{dw}{f^2(w)} \tag{7.9}$$

where

$$f(s) = f(u, s).$$

We might guess that there will some kind of compensation of the following sort. Assume for example that f is very small on a large interval. If f were moreover a constant then this will imply $K = 0$. But then, on such an interval, $f(G_s(u), t)$ will be linear and as such will get large. This guess is correct

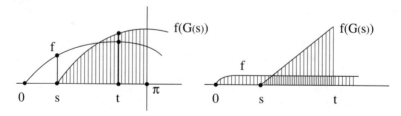

Fig. 7.45. Behaviour of the Jacobi field

and the sphere yields the minimal compensation since

Lemma 157 (Basic lemma)

$$I(u) \geq \pi$$

with equality only if f is proportional to $\sin t$, i.e. if

$$K \equiv 1.$$

The proof is very clever, but completely elementary; see Berger 1977 [156] if you are stuck. What you have to prove is that for any function ϕ vanishing at 0 and π

$$\int_{x=0}^{x=s} dx \int_{y=x} y = sdy \int_{t=x}^{t=y} dt\, \frac{\phi(x)\phi(y)}{\phi^2(t)} \geq \frac{s^3}{\pi^2}$$

and equality happens if and only if ϕ is proportional to the sine function.

The proof of theorem 148 on page 355 is now finished as follows. To every $u \in UM$ one attaches the picture in figure 7.46.

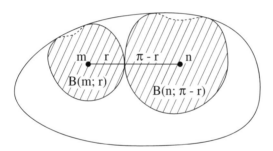

Fig. 7.46. Two balls sitting astride each other

By the definition of the injectivity radius, for every $u \in UM$ and every $r < \pi$, the two balls

$$B\left(m = p(u), r\right) \quad \text{and} \quad B\left(n = p\left(G_\pi(u)\right), \pi - r\right)$$

are disjoint. This show that

$$\mathrm{Area}(M) \geq \mathrm{Area}\,B\,(m, r) + \mathrm{Area}\,B\,(n, \pi - r). \tag{7.10}$$

We integrate this inequality on the whole of UM, and for r running from 0 to $\pi/2$, using equation 7.5 on page 358. Then we apply the Fubini theorem and on the way apply Liouville's theorem 156 on page 359. and finally lemma 157. This yields theorem 148 on page 355. It is easy to identify the manifold when we have equality in our inequality: we are forced to have equality in lemma 157 and this forces the curvature to be constant and equal to 1. We are then on a sphere: apply the results of §§6.3.2 and we remark moreover that M cannot be a nontrivial quotient of the standard sphere since its injectivity radius is

$$\mathrm{Inj}\,(M) = \pi.$$

When the dimension is higher than two, as we have seen in equation 6.11 on page 248 the Gauß ordinary differential equation

$$f'' + Kf = 0$$

has to be replaced by the vector Jacobi equation

$$Y'' + R(\gamma', Y)\gamma' = 0$$

and the function f by the determinant

$$\det(Y_1, \cdots, Y_{d-1})$$

of Jacobi fields. We replace the scalar equation 7.9 on page 360 above by the following one, which an endomorphism equation (or a matrix equation, if you prefer)[15]

$$A\left(G\left(s\right)\left(u\right),t\right) = A\left(s\right)\left(\int_s^t \left(A^t\left(r\right)A\left(r\right)\right)^{-1} dr\right) A^t\left(s\right)$$

For this relation, as well as for a detailed proof, one can consult for example section VI.2 of Sakai's book Sakai 1996 [1085], or appendix D of Besse 1978 [182]. With a clever trick, the classical inequality for integrals of convex functions, one has finally to replace the basic lemma 157 on the previous page by the following inequality:

Lemma 158 (Basic inequality) *If*

$$\rho : [0, \pi] \to \mathbb{R}^+$$

is any function satisfying

$$\rho(\pi - t) = \rho(t)$$

then for any nonnegative function ϕ

$$\int_0^\pi dr \int_r^\pi ds \int_r^t \frac{\phi(r)\phi(t)}{\phi^2(s)\rho(t-r)} \geq \text{``the value the same expression takes for } \phi = \sin\text{''}$$

and with equality if and only if ϕ is proportional to the sine function.

This very subtle inequality was proven by Kazdan in 1978: see section 5.3 of Chavel 1993 [326] or appendix E of Besse 1978 [182] or section VI.2 of Sakai 1996 [1085]. It is an open problem to find a proof of it with a clear underlying concept; one would expect interesting generalizations and applications of such a proof.

---

[15] Clearly, the order of matrix products is significant.

7.2.4.4 Croke's Three Results

We now sketch the proof of theorem 149 on page 355. The proof is a mixture of the basic inequality in lemma 158 on the preceding page with Santalo's equation 7.11 on the following page which was already used in §§7.1.2.3 by Croke to prove the conjecture for the isoperimetric profile of four dimensional manifolds of nonpositive curvature. Santalo's formula is proven in section 5.2 of Chavel 1993 [326].

The proof is concerned with balls of radius $r < \pi/2$. The sphere which is the boundary of this ball will be denoted

$$S(m,r) = \partial B(m,r).$$

This sphere is smooth because we are within the injectivity radius. Moreover, for any point $n \in B(m,r)$, the injectivity radius at n, $\text{Inj}(n)$, is not smaller than π, which has the effect that the geodesics through n will nicely sweep out over the whole ball $B(m,r)$. And there will be no conjugate points on these geodesics inside $B(m,r)$.

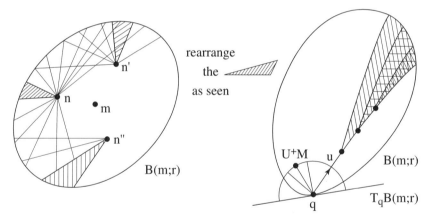

Fig. 7.47. (a) Sweeping out the whole ball about m from any point n inside it. (b) Sweeping out geodesics inside the ball from a point of the boundary.

This implies that we can compute

$$\text{Vol}\, B(m,r)$$

from the viewpoint of n, using an integral over $U_n M$ of suitable determinants of Jacobi fields. We now carry out these integrals over all n running through $B(m,r)$ and exchange the order of integration to integrate along the various geodesics γ_u where u runs through the unit tangent vectors $U_q^+ M$ where, for a point of the sphere $q \in S(m,r)$, the half tangent unit sphere $U_q^+ M$ is that made of the vectors pointing inside $B(m,r)$.

We use now Kazdan's inequality from lemma 158 on page 362 and Liouville's theorem just as in Santalo's formula below. Consider in some compact Riemannian manifold M a precompact piece of smooth hypersurface N having a side, so that we can talk about U^+N. Consider an arbitrary positive function $\psi(u)$ on U^+N, but small enough that

$$\psi(u) < \text{Cut}(u)$$

(smaller than the cut value) and any integrable function f defined on the set D made up by the geodesic flows $G(s)(u)$ where u runs through U^+N, and s runs through the interval $[0, \psi(u)]$. *Denote* by $\cos(u)$ the cosine of the angle between the vector u and the normal vector to N. Then one has *Santalo's formula*

$$\int_D f(v)\, dv = \int_{U^+N} \left(\int_0^{\psi(u)} f(G_s(u))\cos(u)\, ds \right) dV_{U^+N}(u) \qquad (7.11)$$

The proof is nothing but Fubini's theorem and the change of variables formula; for details see page 286 of Berger 1965 [151] or Santalo 1976 [1093].

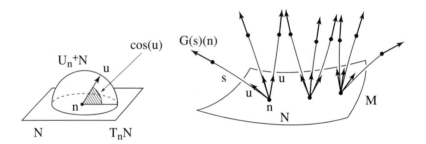

Fig. 7.48. Santalo's formula in pictures

All together this results in the following inequality between the volume of the ball $B(m,r)$ and that of its boundary $S(m,r)$:

$$\text{Vol}\, S(m,r)^d \geq a(d)\, \text{Vol}\, B(m,r)^{d-1} \qquad (7.12)$$

But

$$\text{Vol}\, S(m,r) = \frac{\partial}{\partial r} \text{Vol}\, B(m,r)$$

so that this last inequality is a differential inequality which by integration from 0 to $\pi/2$ furnishes theorem 149 on page 355. That this inequality is certainly not optimal is left to the reader. The proof of theorem 150 on page 355 is a refinement of that of theorem 149, still largely using only Santalo's formula and the basic inequality.

The proof of theorem 154 on page 357 now follows theorem 149 inequality with the following elegant modification:

Lemma 159 *Assume that M is not a sphere. Then for every $u \in UM$ and every $r \in [0, \pi/2]$ there is some $q \in M$ such that*

$$B\left(\gamma_u(0), r\right) \cap B\left(\gamma_u(\pi), \pi - r\right) \cap B\left(q, r\right) = \emptyset$$

and in particular

$$\operatorname{Vol} M \geq \operatorname{Vol} B\left(\gamma_u(0), r\right) + \operatorname{Vol} B\left(\gamma_u(\pi), \pi - r\right) + \operatorname{Vol} B\left(q, r\right).$$

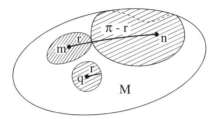

Fig. 7.49. Three balls that don't fill up all the whole manifold

Croke's result is obtained by contradiction, resting on the folk theorem that if a manifold can be covered by only two balls, then it is a sphere: see for example theorem 4.1 of Milnor 1963 [921]; this theorem is essential in the proof of theorem 297 on page 552. But assume there is no such point q. This means (by the triangle inequality) that M is covered by the two balls

$$B\left(\gamma_u(0), 2r\right) \text{ and } B\left(\gamma_u(\pi), \pi\right).$$

When one integrates the inequality in lemma 159 over all unit vectors u, as in the proof in §§§7.2.4.3, the two first terms yield exactly the volume of M but the third terms—thanks to theorem 149 on page 355—add up to a positive constant which is universal in the dimension d and this is exactly theorem 154 on page 357.

Such an inequality is certainly not optimal since the proof leaves a lot of room between the third ball and the other ones. Contrarily, for the standard spheres the two first balls never leave any room for a third.

Note 7.2.4.2 As Croke remarks in his work, the above proof admits with no modification the following generalization. The *category* of a compact manifold is the minimum number of topological balls needed to cover it. By the way, the category of a d dimensional manifold never exceeds $d + 1$. The desired result is

$$\operatorname{Emb}(M) \geq \operatorname{Emb}\left(S^d\right) + c(d)(\operatorname{category}(M) - 2).$$

See more on this in §§11.2.3. ♦

7.2.4.5 Infinite Injectivity Radius

From corollary 55 on page 228 and definition 85 on page 268 we know that the exponential map
$$\exp_m : T_m M \to M$$
is a covering for every $m \in M$ if and only if M has no conjugate points (for any interval and on any geodesic). For example, this happens on a manifold of nonpositive curvature; see §§ 6.3.3 on page 254. But there are Riemannian manifolds with no conjugate points but with parts having positive curvature (not too much).

For various reasons (some of them are explained in chapter 12), Riemannian manifolds without conjugate points are of great interest. Let M be such a manifold and consider its universal Riemannian covering $\tilde{M}$. It is simply $T_m M$ endowed with the pullback metric $\exp_m^* g$: see the proof of corollary 55 on page 228. Now of course $\tilde{M}$ is not compact, but is complete and its injectivity radius is infinite. In particular, for every positive r and every point $\tilde{m}$ the ball $B(\tilde{m}, r)$ is diffeomorphic to $\mathbb{R}^d$ and nicely covered by the geodesics emanating from $\tilde{m}$. Since the deck transformations are isometries, in particular two points $\tilde{m}, \tilde{n} \in \tilde{M}$ having the same projection $m \in M$ yield balls $B(\tilde{m}, r)$ and $B(\tilde{n}, r)$ which have the same volume. So one can refer without ambiguity to the volume
$$\operatorname{Vol} B\left(m, \tilde{M}, r\right)$$
as if it were attached to the point $m \in M$ downstairs.

When the curvature is nonpositive, we have seen in theorem 103 on page 306 that the volume of balls at any point grow with their radius r as fast as in Euclidean space $\mathbb{E}^d$. Can one expect the same conclusion, but with the weaker hypothesis that the manifold has no conjugate points? In 1991, Croke 1992 [421] discovered that this is the case. There are two results: a pointwise one and an averaging one:

Theorem 160 (Croke 1991 [421]) *Let M be a compact manifold without conjugate points. Then for every $r > 0$*
$$\frac{1}{\operatorname{Vol} M} \int \operatorname{Vol} B\left(m, \tilde{M}, r\right) \, dr \geq \omega_d r^d$$
with equality if and only if M is flat (where ω_d is the volume of the unit ball in d dimensional Euclidean space).

Theorem 161 *For every $r > 0$ and for every point $m \in M$*
$$\liminf_{r \to \infty} \frac{\operatorname{Vol} B\left(m, \tilde{M}, r\right)}{\omega_d r^d} \geq 1$$
with equality if and only if M is flat.

7.2 Curvature Free Inequalities on Volumes of Cycles   367

These results are applied in §10.11. The proofs of the above results use some of the same inequalities we used above but also much more sophisticated ones and subtle limit arguments. We refer the reader to Croke's paper.

7.2.4.6 Using Embolic Inequalities and Local Contractibility

Besides being very natural in the study of relations between curvature and topology, results of embolic nature are essential for some other reasons. We mention them here briefly before meeting them again later on.

Embolic results permit good, even optimal, constants in Sobolev inequalities; see theorem 118 on page 325. and the survey Hebey 1996 [690]. Various inequalities for the eigenvalues of the Laplace operator make essential use of Croke's local inequality from theorem 149 on page 355. One also finds embolic inequalities in the C^∞ compactness of isospectral moduli (see §§9.12.3). An optimal embolic inequality for $\mathbb{CP}^n$ will solve the question of §§10.4.2. The optimal inequality in theorem 148 on page 355 is used to solve the Wiedersehnmannigfalthigkeiten problem; see theorem 257 on page 493.

The essential property of (open) balls of radius smaller than or equal to the injectivity radius is that they are contractible. For example the inequality in theorem 149 on page 355 has the following immediate corollary (use the covering trick of lemma 125 on page 333):

Theorem 162 *A compact manifold of a given dimension d can be covered by a number of contractible sets, and this number is universally bounded in terms of the total volume and the manifold and its injectivity radius.*

We will see in §§§12.4.1.2 how important this corollary is. But luckily one can do surprisingly many things with a weaker notion of contractibility, which first appeared in Gromov 1981 [613]:

Definition 163 *A* contractibility function *for a metric space is a function*

$$\rho : [0, R] \to \mathbb{R}^+$$

with $R > 0$, so that

$$\rho(r) \to 0 \text{ as } r \to 0$$

and

$$\rho(r) \geq r$$

for all $r \in [0, R]$. A metric space is said to be locally geometrically contractible with contractibility function ρ, written symbolically LGC (ρ), if every metric ball $B(x, r)$ is contractible in the ball $B(x, \rho(r))$ for every point x and every radius $r \leq R$. The space is said to be geometrically contractible if the function ρ is actually defined on all $\mathbb{R}^+$ and the space is LGC (ρ) for ρ restricted to $[0, R]$ for any number $R > 0$.

Thinking along the lines of theorems 148 on page 355 and 149 on page 355, we would like to have analogies of these results in the realm of local or global contractible metric spaces. For (complete) noncompact Riemannian manifolds, on page 43 of Gromov 1983 [618] it is proven that geometric contractibility implies, via infinite filling volume, infinite volume. The local version of this sort of result for LGC(ρ) Riemannian manifolds was proven in Greene & Petersen 1992 [592]: in a complete noncompact LGC(ρ) manifold[16], the volume of balls grows at least linearly; but if moreover

$$\rho(r) \geq cr$$

then the volume of balls of radius r grows at least as fast as

$$\operatorname{Vol} B(m, r) \geq C r^d.$$

The concept of *filling radius* (defined on page 342) plays an important role. We will find the LGC(ρ) condition entering essentially in §§§12.4.1.2. This notion is becoming increasingly important in various spaces more general than Riemannian manifolds; for example see Semmes 1996 [1122].

---

[16] With R finite; otherwise it wouldn't be interesting.

8 Transition: The Next Two Chapters

Contents

8.1	Spectral Geometry and Geodesic Dynamics	... 369
8.2	Why are Riemannian Manifolds So Important?	372
8.3	Positive Versus Negative Curvature	 372

8.1 Spectral Geometry and Geodesic Dynamics

Toward the end of chapter 1, we met the heat and wave equations on compact regions of the plane. We also met billiard balls travelling in those regions, bouncing off of the sides. In §§1.8.6 we asked whether there might be links between these two completely different mathematical stories: elementary physics of fields in planar regions and long term geometry of billiard ball trajectories.

These two stories can be told on Riemannian manifolds just as easily as on planar regions. In the Euclidean plane, the geodesics are straight lines, giving us the planar billiard balls, while on Riemannian manifolds *with no boundary* we can consider the geodesics as the analogues of billiard ball paths. The periodic billiard ball paths have as analogue the periodic geodesics. Be aware that the phrase *closed geodesic* can be ambiguous, although it is frequently employed. It can either mean a mere geodesic loop, or a truly periodic geodesic. See figure 8.1 on the following page.

Thanks to the uniqueness of solutions of ordinary differential equations, in particular for the geodesic equations, periodicity of a geodesic occurs precisely when the geodesic is a loop with the same initial and final velocity. Using the point of view from the unit tangent bundle, and the notion of geodesic flow, the periodic geodesics are precisely the periodic flow lines of the geodesic flow. Note that a periodic geodesic is permitted self-intersections. Those without self-intersections are called *simple*.

The natural perspective to take in studying periodic geodesics is to think of a Riemannian manifold as a dynamical system. The viewpoint of dynamical system leads us to many natural questions:

- What can we say about a geodesic over large intervals of time?
- Are nonperiodic geodesics everywhere dense? Or somewhere dense?

8 Transition: The Next Two Chapters

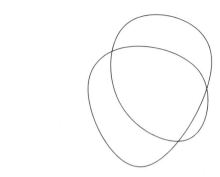

a A periodic geodesic

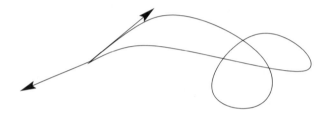

b A geodesic loop

Fig. 8.1. Geodesic loops may not always return with the same velocity, and so might not be periodic

- How many periodic geodesics are there?
- Where are the periodic geodesics situated?
- How many periodic geodesics have length less than a given number?

These geometric questions will be the subject of chapter 10.

We turn our attention to the physical fields of temperature and electrostatics, and to waves. These are less geometric questions, and so less natural from our perspective. We would like to have analogues of the heat, wave and Laplace equations on Riemannian manifolds. Then we want to relate the ideas we have formed already about eigenvalues and eigenfunctions to the resulting theory on Riemannian manifolds.

On any Riemannian manifold, there essentially a natural differential operator acting on functions on the manifold[1] called the *Laplace–Beltrami operator*, or *Laplace operator* or *Laplacian*, and denoted Δ. This operator enjoys all of the nice properties of the Laplace operator from Euclidean space. For the analyst, it is a positive definite second order elliptic operator. Moreover, one

---

[1] And, if we demand a few simple properties from it, then it is unique.

8.1 Spectral Geometry and Geodesic Dynamics

can prove the existence of eigenfunctions very cheaply, because they are the critical points of the Dirichlet quotient

$$f \mapsto \frac{\int_M \|df\|^2 \, dV_M}{\int_M f^2 \, dV_M}. \tag{8.1}$$

More details are given in chapter 9. The central idea is that on any Riemannian manifold, there is a complete orthonormal system of eigenfunctions, yielding a "Riemannian Fourier series" for any function. This brings the entire apparatus of classical analysis to hand. Moreover, we know very good (in some situations, optimal) asymptotics for the eigenvalues (which are the squares of the frequencies of vibration of our manifold) and the asymptotic behaviour of the eigenfunctions (the component vibrations themselves). Following the point of view of §§1.8.1, we can say that this is picture of a Riemannian manifold as a quantum mechanical world.[2]

In sharp contrast with the quantum mechanical point of view, the dynamical system point of view (the classical mechanics arrived at from the classical limit) is considerably more difficult. Perhaps this is surprising at first glance— soon it will be almost obvious. Indeed we hope the reader will be in awe of the tremendous difficulty of the results gathered in chapter 10. For example, it is known that every Riemannian metric on S^3 must have a periodic geodesic, but not known if it must have more than one. Needless to say, we know very little about the distribution of geodesics, or the number and location of periodic geodesics, in all but a handful of manifolds. Compact manifolds of negative curvature (especially those with constant negative curvature) are exceptionally well understood, which is paradoxical since they are are hard to construct (see §§6.6.2). We can summarize by saying that classical mechanics is more difficult than quantum mechanics, or that geometry is more difficult than analysis. It is for this reason that we treat analysis first, in chapter 9, before treating geometry in chapter 10.

The underlying reason for the simplicity of spectral geometry compared to geodesic dynamics is that the Dirichlet quotient in equation 8.1 is a quadratic function on the vector space of functions on our manifold M. This vector space is infinite dimensional, but we are still within the realm of linear algebra. On the other hand, one certainly can express the periodic geodesics as critical points of some functional (and we will proceed in this fashion), but this functional is defined on the space of closed curves on the manifold, which is not a vector space. The space of closed curves is in fact an infinite dimensional manifold. This will cost us dearly.

---

[2] Or thermodynamical, or fluid mechanical, or electrostatic. Indeed, analogies of many different physical theories can be implemented in Riemannian geometry.

8.2 Why are Riemannian Manifolds So Important?

One reason for the importance of Riemannian manifolds is that they are generalizations of Euclidean geometry—general enough but not too general. They are still close enough to Euclidean geometry to have a Laplace operator. This is the key to quantum mechanics, heat and waves. The various generalizations of Riemannian manifold which will be briefly mentioned in §14.5 do not have a simple natural unambiguous choice of such an operator.

Another reason for the prominence of Riemannian manifolds is that the maximal compact subgroup of the general linear group is the orthogonal group. So the least restriction we can make on any geometric structure[3] so that it "rigidifies" always adds a Riemannian geometry. Moreover, any geometric structure[4] will always permit such a "rigidification."

Similarly, if we were to pick out a submanifold of the tangent bundle of some manifold, distinguishing tangent vectors, in such a manner that in each tangent space, any two lines could be brought to one another, or any two planes, etc., then the maximal symmetry group we could come up with in a single tangent space which was not the whole general linear group would be the orthogonal group of a Riemannian metric. So Riemannian geometry is the "least" structure, or most symmetrical one, we can pick, at first order.

8.3 Positive Versus Negative Curvature

As the next two chapters proceed, we will often meet the dichotomy of positive and negative curvature. Currently there is no result in either geodesic dynamics or spectral geometry which presents a unified picture of both negatively curved and positively curved manifolds.

---

[3] In the sense of Cartan's theory of G structures.
[4] Again, in the sense of G structures.

9 Riemannian Manifolds as Quantum Mechanical Worlds: The Spectrum and Eigenfunctions of the Laplacian

Contents

9.1	History	374
9.2	Motivation	375
9.3	Setting Up	376
	9.3.1 Xdefinition	376
	9.3.2 The Hodge Star	378
	9.3.3 Facts	380
	9.3.4 Heat, Wave and Schrödinger Equations	381
9.4	Minimax	383
	9.4.1 The Principle	383
	9.4.2 An Application	385
9.5	Some Extreme Examples	387
	9.5.1 Square Tori, Alias Several Variable Fourier Series	387
	9.5.2 Other Flat Tori	388
	9.5.3 Spheres	390
	9.5.4 $\mathbb{KP}^n$	390
	9.5.5 Other Space Forms	391
9.6	Current Questions	392
	9.6.1 Direct Questions About the Spectrum	392
	9.6.2 Direct Problems About the Eigenfunctions	393
	9.6.3 Inverse Problems on the Spectrum	393
9.7	First Tools: The Heat Kernel and Heat Equation	393
	9.7.1 The Main Result	393
	9.7.2 Great Hopes	396
	9.7.3 The Heat Kernel and Ricci Curvature	401
9.8	The Wave Equation: The Gaps	402
9.9	The Wave Equation: Spectrum & Geodesic Flow	405
9.10	The First Eigenvalue	408
	9.10.1 λ_1 and Ricci Curvature	408
	9.10.2 Cheeger's Constant	409
	9.10.3 λ_1 and Volume; Surfaces and Multiplicity	410
	9.10.4 Kähler Manifolds	411
9.11	Results on Eigenfunctions	412
	9.11.1 Distribution of the Eigenfunctions	412

	9.11.2	Volume of the Nodal Hypersurfaces 413
	9.11.3	Distribution of the Nodal Hypersurfaces 414
9.12	**Inverse Problems** **414**
	9.12.1	The Nature of the Image 414
	9.12.2	Inverse Problems: Nonuniqueness 416
	9.12.3	Inverse Problems: Finiteness, Compactness .. 418
	9.12.4	Uniqueness and Rigidity Results 419
9.13	**Special Cases** **421**
	9.13.1	Riemann Surfaces 421
	9.13.2	Space Forms 424
9.14	**The Spectrum of Exterior Differential Forms .. 426**

9.1 History

While harmonic analysis on domains in Euclidean space is a long established field, as seen in §1.8, the study of the Laplace operator on Riemannian manifolds (together with the heat and wave equations, and the spectrum and eigenfunctions) seems to have begun only quite recently. Some of the earliest accomplishments were the computation of the spectrum of $\mathbb{CP}^n$ (see §§9.5.4) and Lichnerowicz's inequality for the first eigenvalue (see §§9.10.1). The first paper to address the Laplacian on general Riemannian manifolds was Minakshisundaram & Pleijel 1949 [930]. More narrowly, Maaß 1949 [889] investigated the Laplacian on Riemann surfaces. Also, one can turn to Avakumović 1956 [90]. But a spark was lit in the 1960's when Leon Green asked if a Riemannian manifold was determined by its spectrum (the complete set of eigenvalues of the Laplacian).

In the special case of Riemann surfaces, a deep study of the spectrum can be found as early as 1954 in Selberg 1954 [1120, 1121] and 1955 in Huber 1956,1959,1961 [744, 745, 747].

Green's isospectral question was answered in the negative in Milnor 1964 [922]. Kac 1966 [775] in 1966 was also very influential. But the two major events were the papers of McKean & Singer 1967 [910] and Hörmander 1968 [734]. We will meet them below; let us just say that the first paper pioneered the study of the heat kernel expansion in Riemannian geometry, and its consequences. The second was more general, treating the case of a general elliptic operator, without reference to any Riemannian structure on the manifold under consideration. But it introduced the wave equation technique, microlocal analysis and symplectic geometry. This technique is indispensable when studying the relations between the spectrum and the geodesic flow; see §9.9. Thereafter the subject became a vast field of inquiry.

Note 9.1.0.3 (On the bibliography) As we go on in this book, we will have to give less and less detail, in order to keep the book of reasonable size. Then the reader will want to ask for more and more references, especially those of general character, as opposed to research articles. There are now quite a few books which addressing the topic of the present chapter. Some people still like Berger, Gauduchon, & Mazet 1971 [174] for an introduction and basic facts. But on most of the more advanced topics that book is completely outdated. New texts are: Chavel 1984,1993 [325, 326], Buser 1992,1997 [292, 293] which discusses only the spectral geometry of surfaces, Sakai 1996 [1085] chapter VI, Gilkey 1995 [564] (try to get this second edition), Bérard 1986 [135] which is very expository but outdated on some advanced topics, Guillemin & Sternberg 1977 [670]. In particular, Gilkey 1995 [564] is important and contains an amazing collection of mathematics in a single book, e.g. the η invariant which is hard to find in books. The heat equation, in a very general context, is also analyzed in Berline, Getzler & Vergne 1992 [179] and in Gilkey, Leahy & Park 2000 [566].

There are few completely expository books on the wave equation technique and microlocal analysis. The bible of Hörmander 1983 [737, 738] is hard to read, but Trèves 1980 [1198, 1199] is very informative. With a view toward physics, Guillemin & Sternberg 1977 [670] is fascinating. Note also that Bérard 1986 [135] contains a very extensive bibliography, but up to date only to 1986.
♦

9.2 Motivation

Why should a geometer, whose principal concern is in measurements of distance, desire to engage in analysis on a Riemannian manifold? For example, pondering the Laplacian, its eigenvalues and eigenfunctions? Here are some reasons, chosen from among many others. We note also here that the existence of a canonical elliptic differential operator on any Riemannian manifold, one which is moreover easy to define and manipulate, is one of the motivations to consider Riemannian geometry as a basic field of investigation. For Laplacians on more general spaces, see §14.5 and §14.6.

Riemannian geometry is by its very essence differential, working on manifolds with a differentiable structure. This automatically leads to analysis. It is interesting to note here that, historically, many great contributions to the field of Riemannian geometry came from analysts. Let us present a few names (we do not pretend to be exhaustive). Hadamard's contribution in quotation 10.1 on page 434 goes back to 1901 and Poincaré's in §§10.3.1 to 1905. Élie Cartan was an analyst; see Chern & Chevalley 1952 [368]. More recently, let us mention Bochner, (see theorem 345 on page 594), Nirenberg (see §§4.6.1), Chern, Calabi, Aubin, Yau and Gromov.

We will see deep links between the spectrum (especially the first eigenvalue λ_1) and periodic geodesics in §9.9 as well as in theorem 205 on page 447. The proof of Colding's L^1 theorem 76 on page 262 rests essentially on analysis, as we have briefly seen there. Harmonic coordinates turn out to be a godsend when studying convergence of Riemannian manifolds: see §§6.4.3 and §§12.4.2.

The deformation of a Riemannian metric via a parabolic evolution equation, which is based on hard techniques from the theory of partial differential equations, is extremely useful. We will see this in more than one instance: see §§11.4.3, the smoothing techniques in §§12.4.2 and the new proof of the conformal representation theorem 70 on page 254. This is one of many evolution equations arising in Riemannian geometry. Another type of evolution equation is the heat equation which will turn out not only to be useful in establishing the existence and some of the first properties of the spectrum and of the eigenfunctions (see §1.8), but has become a basic tool in a large number of contexts; see §9.7.

Finally let us mention harmonic maps (see §14.3), minimal submanifolds (e.g. for the theorem on manifolds with positive curvature operator in §§§12.3.1.4), and the use of geometric measure theory. And do not forget harmonic coordinates.

From the point of view of theoretical physics, it is very natural to consider the *semiclassical limit*, which is the limiting behaviour of the solutions of the Schrödinger equation

$$i\hbar \frac{\partial \psi}{\partial t} = \hbar^2 \Delta f$$

as $\hbar \to 0$. In Euclidean space, this is equivalent to rescaling the spatial coordinates outward, looking at the large scale physics. The hope is that classical mechanics will emerge from this limit in some sense. This suggests looking at the asymptotic expansion of the eigenvalues λ_i as $i \to \infty$. This explains why we so often mention results such as theorems 164 on page 386, 172 on page 401, 174 on page 403, and 175 on page 404.

Every manifold is COMPACT and connected unless otherwise stated.

9.3 Setting Up

9.3.1 Xdefinition

Recalling §1.8 and §1.9, even before tackling the heat equation, the first thing to do is to define the Laplacian Δ on a Riemannian manifold M with metric g. It is a second order elliptic differential operator, attached intrinsically to M. It is not surprising that one can give many equivalent definitions of it. We start with the most natural, as soon as one knows that the Riemannian metric enables us to define an intrinsic second derivative (which is not the case for a manifold with "only" a smooth structure). To every smooth numerical function

$$f: M \to \mathbb{R}$$

we attach its Hessian

$$\text{Hess } f$$

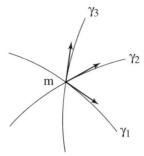

Fig. 9.1. Calculating the Laplacian by differentiating along an orthogonal system of geodesics and taking the sum

which is the bilinear symmetric quadratic differential form made up by the second derivatives of f. Namely, using the covariant derivative D, we set

$$\text{Hess}\, f = Ddf$$

(see §15.5 if needed). To get a numerical function from this Hessian, we need only take its trace with respect to the metric g. For technical reasons, we add a minus sign. Beware that this is a matter of convention, and the convention depends on the author. The negative sign insures us that the eigenvalues will be nonnegative (in fact, positive except the 0th whose eigenvalues are the constant functions). The *Laplacian* of f is then defined as

$$\Delta f = -trace_g \,\text{Hess}\, f \ .$$

Since along geodesics, the (covariant) second derivative coincides with the ordinary numerical second derivative, by the definition of the trace with respect to g, the geometer will define the Laplacian of f at a point $m \in M$ as

$$\Delta f(m) = -\sum_{i=1}^{d} \frac{d^2}{dt^2} f\left(\gamma_i(t)\right)\bigg|_{t=0}$$

where the γ_i are geodesics through m whose velocities at m form an orthonormal basis of $T_m M$. In particular, at the center m of a system of normal coordinates, this is written

$$\Delta f(m) = -\sum_{i=1}^{d} \frac{\partial^2 f}{\partial x_i^2}(m) \ .$$

This cannot be used as a definition directly, since one needs to show that such a description yields a well defined differential operator. Two other definitions can be given. The first uses the Hodge $*$ operation applied to differential forms, which will be defined in §§9.3.2. Then

$$\Delta f = - * d * d .$$

Using any definition, in general coordinate systems we find

$$\Delta f = -\frac{1}{\sqrt{\det g}} \sum_{j,k} \frac{\partial}{\partial x_j} \left(g^{jk} \sqrt{\det g} \frac{\partial f}{\partial x_k} \right) \tag{9.1}$$

where

$$\det g = \det (g_{ij})$$

and the g^{jk} are the matrix elements of the inverse matrix to g_{jk}. We won't need to use this complicated formula. From all of these definitions, one sees that this Δ extends to any Riemannian manifold the Δ of Euclidean space as defined in equation 1.27 on page 94 and the Δ of the sphere defined in equation 1.28 on page 95. In this formula, one sees that the Laplacian involves the metric g and its first derivatives; this makes it an invariant which is not C^0 robust, only C^1 robust. However, §9.4 will show that the spectrum is C^0 robust. This is the beginning of spectral analysis for more general geometries; see §14.5 and §14.6.

If you are familiar with the notion of *symbol* of a differential operator, then the best way to define and to see the uniqueness of the Laplacian is to say that Δ is the second order differential operator whose principal symbol is $-g$ (the quadratic form giving the metric g) and which has no term of order zero.

If we construct a function measuring distance from some point then, when written in polar geodesic coordinates centered at that point, the Ricci curvature comes into the formula giving the Laplacian of this distance function. We employed this fact when proving Colding's L^2 theorem 77 on page 264:

$$\frac{d}{ds}\Delta f \circ \gamma + \frac{1}{d-1} (\Delta f \circ \gamma)^2 \leq - \operatorname{Ricci}(\gamma', \gamma') .$$

9.3.2 The Hodge Star

To present many of the foundational facts in spectral geometry[1] we need the definition of the Laplace operator Δ on differential forms and the concept of adjoint operator. We first denote by $\Omega^p(M)$ the space of differential forms of degree p on the differentiable manifold M, which is defined on any differentiable manifold, without need for a Riemannian metric; see §§4.2.2. But if M is moreover equipped with a Riemannian metric and oriented, then there is an linear operator

$$* : \Omega^p(M) \to \Omega^{\dim(M)-p}(M)$$

called the Hodge star operator. Choosing a positive orthonormal basis

$$\{e_i\}_{i=1,\ldots,d}$$

---

[1] For example, theorems 338 on page 588 and 405 on page 665; also see §9.14.

for the tangent space $T_m M$ at a point $m \in M$, define
$$*\alpha\,(e_{p+1},\ldots,e_d) = \alpha\,(e_1,\ldots,e_p)\;.$$

This turns out to be independent of the choice of oriented orthonormal basis. The square of $*$ is plus or minus the identity on $\Omega^p(M)$:
$$*^2 = (-1)^{p(\dim(M)-p)}\;.$$

The differential operator d is transformed by $*$ into another first order operator, denoted by d^* (sometimes also by δ)
$$d^* = (-1)^{1+d(p+1)} * d *$$

which is not dependent on the choice of orientation, hence is intrinsic. The reason for the notation
$$\delta = d^*$$
is that it is the adjoint of d:
$$\int_M d\alpha \wedge \beta = \int_M \alpha \wedge d^*\beta \tag{9.2}$$

for any
$$\alpha,\beta \in \Omega^p(M)$$

and any $p = 0,\ldots,\dim(M)$. We can define a Laplacian for exterior forms of any degree by
$$\Delta = -\,(dd^* + d^*d) = -\,(d + d^*)^2\;. \tag{9.3}$$

For the moment, we will only use the Laplacian on functions, i.e. $p = 0$. This Δ is the same as the one previously defined in this chapter. A useful formula, valid for any pair of functions, is
$$\int_M g\Delta f = \int_M \langle df, dg\rangle$$
$$= \int_M f\Delta g \tag{9.4}$$

in particular
$$\int_M \Delta f = 0$$
for any function f.

When using integrals like the above on compact Riemannian manifolds, we will often omit the Riemannian canonical measure:
$$\int_M f = \int_M f\,dV_M\;.$$

9.3.3 Facts

The Laplacian on any compact Riemannian manifold provides us with all the tools of Fourier analysis on our Riemannian manifold. Let us call a function ϕ an *eigenfunction* with *eigenvalue* the number λ if

$$\Delta f = \lambda f .$$

The set of all eigenvalues of Δ is an infinite discrete subset of $\mathbb{R}^+$ called the *spectrum* of Δ

$$\operatorname{Spec}(M) = \{\lambda_k\} = \{0 < \lambda_1 < \lambda_2 < \ldots\} \qquad (9.5)$$

with λ_k tending to infinity with k.

For each eigenvalue λ_i, the vector space of eigenfunctions ϕ satisfying

$$\Delta f = \lambda_i f$$

is always finite dimensional and its dimension is called the *multiplicity* of λ_i. Once we have a basis of the eigenfunctions with this eigenvalue written out, it is trivial to find an orthonormal basis

$$\{\phi_k\}$$

(where k runs from 1 to the multiplicity) of eigenfunctions. Here the orthonormalcy is to be understood for the global scalar product

$$\langle f, g \rangle_{L^2(M)} = \int_M fg .$$

Note that equation 9.4 on the preceding page shows (a classical fact) that eigenfunctions with different eigenvalues are automatically orthogonal. Unlike the domains in Euclidean space which we treated in chapter 1, our compact Riemannian manifolds have no boundary. This explains why we get the "extra" eigenvalue

$$\lambda_0 = 0$$

whose eigenfunctions are the constant functions.[2]

Note 9.3.3.1 Beware now that there are two different ways of writing the eigenvalues and the eigenfunctions when making sums. In the first one, we understand that a sum over the spectrum sums each eigenvalue a number of times given by its multiplicity. In the other notation, the indices are not those used in equation 9.5, but instead the index moves up at each eigenvalue through the entire multiplicity. Which sort of summation is required will always be clear from the context, as in what follows for example. ◆

---
[2] Since the manifold M is assumed to be connected, the multiplicity of λ_0 is exactly one.

As for classical Fourier series, any reasonable function
$$f : M \to \mathbb{R}$$
has Fourier coefficients
$$a_i = \int_M f\phi_i$$
and f is recovered from these coefficients by the converging series
$$f = \sum_i a_i \phi_i .$$
In the same spirit, the scalar product of two functions is the sum of products of their coefficients:
$$\int_M fg = \sum_i a_i b_i$$
where
$$f = \sum_i a_i \phi_i$$
$$g = \sum_i b_i \phi_i .$$

9.3.4 Heat, Wave and Schrödinger Equations

We will follow the same steps that we did in §1.8: defining heat, wave and Schrödinger equations on Riemannian manifolds. The heat equation for the heat $f(m,t)$ at time t at a point m of the Riemannian manifold M is

$$\Delta f = -\frac{\partial f}{\partial t} . \tag{9.6}$$

The wave equation for the height $f(m,t)$ of the "water" after time t at a point m is

$$\Delta f = -\frac{\partial^2 f}{\partial t^2} . \tag{9.7}$$

where if M were a surface, you would consider M covered in a thin sheet of water, or for M of three dimensions, M is a place through which sound is propagating. The wave equation can also be considered as describing the manifold M as a vibrating membrane object. Finally the Schrödinger equation uses complex valued functions and is written

$$\hbar^2 \Delta f = i\hbar \frac{\partial f}{\partial t} \tag{9.8}$$

where $i = \sqrt{-1}$ and $\hbar$ is Planck's constant.

9 Spectrum of the Laplacian

To solve these equations, at least formally, one uses the same trick as in §§1.8.1. To solve such an equation depending both on time t and a point $m \in M$, the initial idea is to use the fact that, roughly by the Stone–Weierstraß approximation theorem, we need only to consider product functions

$$f(m,t) = g(m)h(t) .$$

One will subsequently consider series of them (as in the theory of Fourier series). Look for example at the heat equation. The function $f = gh$ satisfies the heat equation precisely when the functions g and h satisfy

$$\frac{\Delta g}{g} = -\frac{h'}{h} \tag{9.9}$$

where

$$h'(t) = \frac{dh}{dt}$$

is the usual derivative.

Since the first fraction depends only on the point $m \in M$ and the second only on the time t their common value has to be a constant, call it λ. Then the function

$$g : M \to \mathbb{R}$$

is an eigenfunction of Δ with eigenvalue λ, while h is an exponential decay at rate λ. If all eigenfunctions and eigenvalues of Δ are known, we can then solve the heat equation explicitly. Note that the time dependence $h(t)$ is

$$h(t) = \begin{cases} e^{-\lambda t} & \text{for the heat equation} \\ e^{i\lambda t} & \text{for the Schrödinger equation} \\ e^{i\sqrt{\lambda} t} & \text{for the wave equation.} \end{cases}$$

Physically, the product motions $g(m)h(t)$ are the stationary ones—they are the ones we can observe through some kind of "Riemannian stroboscopy."

As we did in Euclidean space, we will begin our analysis with the *fundamental solution of the heat equation*, denoted $K(m,n,t)$. One also calls it the *heat kernel*. It is a function

$$K : M \times M \times \mathbb{R}^+ \to \mathbb{R} .$$

It has the property that the solution $f(m,t)$ of the heat equation with initial temperature $f(m,0)$ at time zero is

$$f(m,t) = \int_M K(m,n,t) f(n,0) \, dn$$

and one can prove that the heat kernel is the sum of the convergent series

$$K(m,n,t) = \sum_{k=1}^{\infty} \phi_k(m)\phi_k(n) e^{-\lambda_k t}. \tag{9.10}$$

The reader can check this formally, ignoring convergence, by just plugging the series into the integral. The hard part, which required analysts' efforts, is to prove the convergence.

Another way to write the solution $f(m,t)$ with initial temperature $f(m,0)$ is to compute the Riemannian Fourier series

$$f(m,0) = \sum_{k=1}^{\infty} a_k \phi_k$$

and then

$$f(m,t) = \sum_{k=1}^{\infty} a_k \phi_k(m) e^{-\lambda_k t} .$$

For the wave equation, the fundamental solution similar to equation 9.10 on the facing page requires imaginary terms, i.e.

$$e^{i\sqrt{\lambda_k} t}$$

which are linear combinations of

$$\cos\left(\sqrt{\lambda_k} t\right) \text{ and } \sin\left(\sqrt{\lambda_k} t\right) .$$

But the dramatic difference between the heat equation and the wave equation is that waves demand not converging series, but distributions. Heat spreads out uniformly with time, while waves bounce up and down forever. This major difference explains why working with the wave equation (in Riemannian manifolds, but also in Euclidean spaces) is much more expensive mathematically. We refer to our bibliographical introduction for references. Note that the conservative nature of waves will provide an amazing source of information in §9.8. Another major difference between the heat equation and the wave equation is that for the waves one does really need to work in the tangent bundle and use the tools of microlocal analysis; a most informative book on the subject is Trèves 1980 [1198, 1199].

9.4 The Cheapest (But Most Robust) Method to Obtain Eigenfunctions: The Minimax Principle

9.4.1 The Principle

Analysis and convergence problems (which we will not attempt to explain) are very well exposed in Bérard 1986 [135]. We will begin as we did in §§1.8.3. One way to identify and then study the eigenfunctions is as follows. One pulls out the first one by the so-called *Dirichlet principle*. Among all functions, one looks for one minimizing the ratio

$$\text{Dirichlet}(f) = \frac{\int_M \|df\|^2}{\int_M f^2} \tag{9.11}$$

called the *Dirichlet quotient*.

The infimum value of zero is trivially attained for constant functions. So we look next to minimize this quotient among functions which are "not constant," more precisely among those functions orthogonal to constants, i.e. functions f with

$$\int_M f = 0.$$

Let us compute the derivative of this ratio with respect to a variation $f + \varepsilon g$ of the function f (assuming it exists) achieving such a minimum, and use formula 9.3 on page 379 together with the Lagrange multiplier technique. We find that f necessarily satisfies

$$\Delta f = \lambda_1 f$$

for the constant

$$\lambda_1 = \inf \left\{ \frac{\int_M \|df\|^2}{\int_M f^2} : \int_M f = 0 \right\}. \tag{9.12}$$

Rescale f to have unit norm

$$\int_M f^2 = 1.$$

This yields the first (nontrivial) eigenfunctions together with the first eigenvalue. Unlike a Euclidean domain, where there was only one first eigenfunction, here there may be a finite dimensional vector space of them; for example the sphere of dimension d has a $d+1$ dimensional space of eigenfunctions with the same eigenvalue λ_1.

To get the next eigenfunctions and values, one just applies the same trick, but restricting the set of functions f into consideration to the set of functions which are orthogonal to the first eigenfunctions

$$\int_M f\phi_i = 0$$

for $\phi_1, \ldots, \phi_h$ a basis for the eigenfunctions with eigenvalue λ_1. And keep going on in this way.

But this procedure necessitates calculating all of the eigenfunctions preceding the one that we might be looking for. To get around this obstacle, a wonderful trick was invented, the *minimax principle*. We first state the result, and then explain it geometrically on ordinary ellipsoids in $\mathbb{E}^3$. The eigenvalue λ_{k+1} is exactly

$$\lambda_k = \inf_V \sup_{f \in V} \text{Dirichlet}(f) \tag{9.13}$$

where V runs through all $k+1$ dimensional vector subspaces of the vector space of real valued functions on M.

The proof is detailed in the beginning of Bérard 1989 [135]. It involves a little linear algebra (geometrically pictured in figure 9.2) and of course some analysis, since we are working in an infinite dimensional space.

In Dirichlet(f) there are two positive definite quadratic forms. In $\mathbb{R}^3$, say that the first has as its unit level set an ellipsoid, and the second is the Euclidean structure (i.e. its unit level set is the unit sphere). Then the eigenfunctions correspond to the three principal axes of the ellipsoid, and the eigenvalues are their lengths. To find the length of the second principal axis, consider all of the ellipses obtained by cutting the ellipsoid by planes through the origin. The largest principal axis that occurs among all of the ellipses is the largest axis of the ellipsoid. The second largest axis of the ellipsoid is the largest number that occurs among all ellipses as the smaller of the two axes.[3]

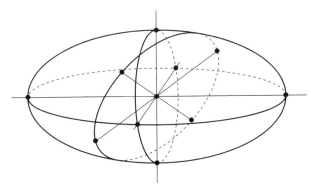

Fig. 9.2. The Dirichlet quotient is a quadratic function on the unit sphere in the infinite dimensional space of functions

The above method heuristically explains why every function is equal to a series of eigenfunctions and, since the space of functions is infinite dimensional, why the spectrum goes to infinity.

Note 9.4.1.1 A theoretical, but important, consequence of the minimax principle is that the spectrum is a robust invariant of the Riemannian metric; it depends only on the metric g, not on its derivatives (unlike the Laplacian itself); see equation 9.1 on page 378. Therefore the spectrum can be defined in a more general context; see §14.6. ♦

9.4.2 An Application

One of the main tasks when studying the spectrum of Riemannian manifolds is to relate the spectrum to the Riemannian invariants, for example the curva-

---
[3] It is harder to say than to see.

tures, the volume, the diameter, the injectivity radius, etc. This is the central objective of this chapter. So we start right away with an application of the minimax principle, given in Gromov 1999 [633]; for details, improvement and explicit constants we refer the reader to Bérard, Besson, & Gallot 1985 [139].

Theorem 164 *There is a universal constant*

$$\mathrm{univ}(d, r)$$

depending only on the dimension d of a compact Riemannian manifold M and the lower bound r of the Ricci curvature, such that for every k the eigenvalue λ_k of M obeys the upper bound

$$\lambda_k \leq \frac{\mathrm{univ}(d, r)}{\mathrm{Vol}(M)^{2/d}} k^{2/d} \ .$$

The asymptotic behavior in $k^{2/d}$ agrees with that which we will see in theorem 172 on page 401. Upper bounds are in general easier to get than lower ones. The reason is that the minimax principle, as we are going to see, shows that one can use upper bounds on the Dirichlet quotient for suitable functions to control the asymptotics of eigenvalues. For the proof, let us think of large indices k. The idea is to pack in M, as densely as possible, a set of metric balls

$$B_i = B(p_i, R) \ .$$

The number N of balls is controlled first by the usual metric trick of lemma 125 on page 333: if it is as dense as possible, then the balls

$$B(p_i, 2R)$$

will completely cover M. This enables us to estimate N with Ricci curvature thanks to Bishop's theorem 107 on page 310.

Now on every ball $B(p_i, R)$ we define a function f_i vanishing at the boundary of $B(p_i, R)$ and with a low Dirichlet quotient. This can be done by transferring (in polar coordinates on $B(p_i, R)$) the first eigenfunction g for the Dirichlet problem in the manifold with boundary

$$B\left(\mathbb{S}^d\left(\frac{r}{d-1}\right), R\right)$$

which is the metric ball of radius R in the comparison space

$$\mathbb{S}^d\left(\frac{r}{d-1}\right)$$

of constant curvature and whose Ricci curvature is our lower bound r. Knowledge of Ricci curvature permits us to control the Dirichlet quotient during the transfer (compare this with the geodesic transfer for Rauch–Toponogov theorems of chapter 6). This was done in Cheng 1975 [362]. There is a very nice proof today of Cheng's result, which is put in a very general context with a beautiful formula in Savo 1996 [1098]. On these balls in $\mathbb{S}^d\left(\frac{r}{d-1}\right)$ the first eigenvalue is known. This transplantation is similar, but not quite the same, as that of the Faber–Krahn inequality 1.22 on page 81. One finishes the estimate by applying the minimax principle to the N dimensional vector space of functions which is spanned by the f_i.

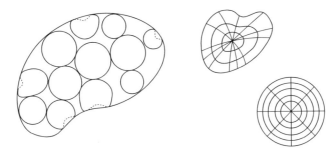

Fig. 9.3. Pack balls into your manifold, and transfer eigenfunctions into them from space forms $\mathbb{S}^d\left(\frac{r}{d-1}\right)$

9.5 Some Extreme Examples

Let us describe the spectral geometry of the most tractable Riemannian manifolds. We will follow more or less the geometric hierarchy of §6.6.

9.5.1 Square Tori, Alias Several Variable Fourier Series

The theory of eigenfunctions on tori, square or rectangular, is very much like that which we met in equation 1.21 on page 76 for the plane rectangle, except that now we use a periodic boundary condition, and of course we work in any dimension d. The variables $x_1, \ldots, x_d$ separate for the Laplacian and we still have the Stone–Weierstraß theorem enabling us to look only at product functions

$$f(x_1, \ldots, x_d) = f_1(x_1) \cdots f_d(x_d) .$$

Our torus is the quotient of $\mathbb{R}^d$ by the group $\mathbb{Z}^d$ of integral translations (this means that all sides of the box have unit length). The Riemannian structure we consider on it is of course the locally Euclidean one just obtained by the quotient operation. The functions $f_j(x_j)$ are linear combinations of

$$\cos(2\pi m_j x_j) \text{ and } \sin(2\pi m_j x_i)$$

with the m_j any integers; the resulting product function is an eigenfunction of Δ with eigenvalue
$$4\pi^2 \left(m_1^2 + \cdots + m_d^2\right) .$$
As in the classical theory of Fourier series, these functions are rich enough so that there are no other eigenfunctions except appropriate linear combinations of these ones. The set of eigenvalues is thus known. But their multiplicity is another story; it leads to many problems in number theory, far from being finished today. Indeed the question of multiplicity is the question as to how many ways an integer can be written as a sum of d squares: see the Gauß circle problem on page 76 which is an unsolved problem in number theory. Some references on the circle problem: Erdös, Gruber & Hammer 1989 [491], Gruber & Lekkerkerker 1987 [660] page 135, Gruber & Wills 1993 [661], Walfisz 1957 [1227] and Krätzel 1988 [833]. However the first order asymptotic estimate of
$$N(\lambda) = \text{number of eigenvalues (with multiplicity) smaller than } \lambda$$
is very easy geometrically. We look for the number of points with integral coordinates which are located inside the ball $B(0,r)$ (centered at the origin) of radius
$$r = \frac{\sqrt{\lambda}}{2\pi} ,$$
see figure 1.84 on page 76. This figure shows that, up to an error term which becomes negligible because it is "only" of order R^{d-1}, we find that $N(\lambda)$ is asymptotic to the volume of the ball of radius $2\pi R$, namely
$$\frac{\beta(d)}{(2\pi)^d} \lambda^{2/d} .$$
Hence the second term in the expansion is again connected to the circle problem, and so is unknown.

9.5.2 Other Flat Tori

This time we quotient our vector space $\mathbb{R}^d$ by any lattice Λ. A lattice is the set of all integral linear combinations of a basis of $\mathbb{R}^d$. Motivated by the preceding "cube" case, we look for functions which are eigenvalues of Δ and Λ periodic. We search for them among the imaginary exponentials of linear functions, which can be always written in the form
$$f(x) = e^{2\pi i \langle \xi, x \rangle}$$
where $i = \sqrt{-1}$ and ξ is a vector which we will try to find. We will have Λ periodicity exactly when the scalar product
$$\langle \xi, x \rangle$$

is an integer for each $x \in \Lambda$. Those ξ form a lattice, called the dual lattice of Λ and denoted by Λ^*. It is trivial to see that

$$\Lambda^{**} = \Lambda$$

and that

$$\operatorname{Vol}\left(\mathbb{R}^d/\Lambda^*\right) = \frac{1}{\operatorname{Vol}\left(\mathbb{R}^d/\Lambda\right)}$$

The eigenfunctions of Δ are

$$e^{2\pi i \langle \xi, x \rangle}$$

for $\xi \in \Lambda^*$ and the eigenvalue of this eigenfunction is

$$4\pi^2 |\xi|^2 \ .$$

But the precise description of the dual lattice is not so easy. It is only in dimension 2 that the dual lattice is always deduced from the original lattice by a similarity. Analysts know how to relate Λ and Λ^*, at least theoretically, with the *Poisson formula*:

$$\frac{1}{(4\pi t)^{d/2}} \operatorname{Vol}(\Lambda) \sum_{\lambda \in \Lambda} e^{-\|\lambda\|^2/4t} = \sum_{\xi \in \Lambda^*} e^{-4\pi^2 \|\xi\|^2 t}. \tag{9.14}$$

Stated another way, the set of eigenvalues of our torus is the set of $4\pi^2$ multiples of square norms (distance to the origin) of the points in the dual lattice Λ^*. It is important for future developments in this book that the distance to the origin from a point of Λ is precisely the length of a periodic geodesic of our torus. So the Poisson formula yields a relation between the spectrum and the length spectrum.

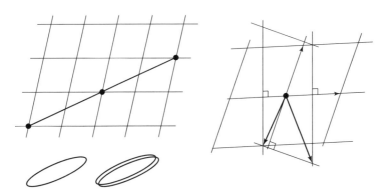

Fig. 9.4. (a) The same periodic geometric geodesic (b) Λ and Λ^* are similar in dimension 2 (only)

The proof of the Poisson formula is not very difficult. We can explicitly write down the heat kernel K^* of $\mathbb{R}^d$ (see §9.7). One then puts together the heat kernel $K(x, y)$ of our flat torus as a summation

$$\sum_i K^*(x, y+\lambda)$$

where λ runs through the lattice defining the torus. Using roughly the same idea, but with considerably more difficulty one can obtain Selberg's trace formula for space forms of negative curvature; see §§9.5.5.

9.5.3 Spheres

Harmonic analysis on spheres is a small miracle: we explained it in §§1.9.2 but the reader might like to see it again here. A polynomial p of degree k on $\mathbb{R}^{d+1}$ is said to be harmonic if

$$\Delta p = 0$$

for the Laplacian Δ on $\mathbb{R}^{d+1}$. The restriction

$$f = p|_{S^d}$$

to the sphere turns out to be an eigenfunction of the Laplacian on S^d with eigenvalue $k(k+d-1)$. Its multiplicity is just the dimension of the vector space of harmonic polynomials of degree d, namely

$$\binom{d+k}{k} - \binom{d+k-1}{k-1}.$$

Again as above, the Stone–Weierstraß theorem tells us that we have no other eigenfunctions and a complete orthonormal basis of eigenfunctions. This does not say that we know everything today about spherical harmonics, even if many people think we do. We turn now to the next objects in the hierarchy of §6.6.

9.5.4 $\mathbb{KP}^n$

Fourier analysis on $\mathbb{CP}^n$ goes back to Élie Cartan in his 1931 monograph Cartan 1992 [322]. The trick is the same as for the sphere, but here one starts with $\mathbb{C}^{n+1}$ and uses harmonic polynomials in the variables

$$z_j, \bar{z}_j .$$

Details can also be found in Berger, Gauduchon & Mazet 1971 [174].

Unhappily this trick does not work with the quaternions. This is linked with the following fact which we mention here because it is rarely known. It is impossible to define on $\mathbb{H}^n$ useful quaternionic derivatives analogous to the complex derivatives

$$\frac{\partial}{\partial z} \text{ and } \frac{\partial}{\partial \bar{z}} .$$

9.5 Some Extreme Examples 391

A related phenomenon: quaternionic structures on manifolds can be integrable only in the flat case. For all this, and a good notion of quaternionic functions, see Joyce 1997 [773] and the references there or the note 13.5.3.1 on page 653.

There are at least two ways to compute the spectra of the remaining $\mathbb{KP}^n$. One is to use a very general formula due to Hermann Weyl, and valid for all symmetric spaces. But the formula is explicit only in the sense that it is a summation over the roots of a certain Lie algebra. To get explicit expressions is hard. The other way is to use the general link between periodic geodesics and the spectrum, a quite deep result (unavoidably using the wave equation) which we will meet in §9.9.

The explicit result for all $\mathbb{KP}^n$ can be found on page 202 of Besse 1978 [182]. It is important to note the spectrum. Its square roots are in all cases included in intervals whose centers make up an arithmetic progression:

$$\operatorname{Spec}\left(\mathbb{KP}^n\right) \subset \bigcup_{k=0}^{\infty}\left[\frac{2\pi}{L}\left(k+\frac{\alpha}{4}\right)^2 - M, \frac{2\pi}{L}\left(k+\frac{\alpha}{4}\right)^2 + M\right] \tag{9.15}$$

where L is the common length of all of the geodesics (which are all periodic), M is some fixed constant and the "index" α is

$$\alpha\left(\mathbb{KP}^n\right) = \begin{cases} 0 & \mathbb{K} = \mathbb{R} \\ 1 & \mathbb{K} = \mathbb{C} \\ 3 & \mathbb{K} = \mathbb{H} \\ 7 & \mathbb{K} = \mathbb{C}\mathrm{a} \end{cases}.$$

As for the sphere, the multiplicities are very high but this is necessary to match the asymptotic behaviour of equation 9.20 on page 397. We will meet this special form of spectrum, as in equation 9.15, again in theorem 177 on page 406.

Fig. 9.5. The spectrum $\operatorname{Spec}\left(\mathbb{KP}^n\right)$

9.5.5 Other Space Forms

The spaces whose spectra we will look for include not only symmetric spaces of higher rank, but also space forms of negative curvature (any rank). For space forms of positive curvature and more generally for homogeneous spaces, the spectrum can be more or less handled in various cases, or only controlled in some instances. We do not give any details; they can be found in the various references which we will give later on.

The very hard but fascinating case is that of space forms of negative curvature. Then one needs to understand not only the Lie group and Lie algebra but also the discrete subgroup of isometries of the simply connected forms (of negative curvature) which yield compact quotients under study. The basic tool was discovered in 1956: it is the *Selberg trace formula*. This an entire subject in itself, intimately connected with number theory. We can only afford to give references on the subject. We choose to offer more or less expository references as opposed to partial results. We suggest for the Selberg trace formula on surfaces, which is quite special and exceptionally powerful: Buser 1992 [292] chapter 9, but the formula permeates a great deal of the book. Add of course the references given there. For higher dimensions, see Bunke & Olbrich 1995 [279]. For more about hyperbolic surfaces see §§9.13.2.

9.6 Current Questions

We can either concentrate on the eigenvalues or on the eigenfunctions. In each case, we can then ask how to derive information about the eigentheory from geometric information, and vice versa.

9.6.1 Direct Questions About the Spectrum

A typical result about eigenvalues is theorem 164 on page 386. It provides practically perfect upper control on the eigenvalues. It is optimal in the sense that none of the ingredients can be removed. Simple examples show that one needs a lower bound on the Ricci curvature and on the volume to obtain upper bounds on eigenvalues.

So the next natural question is to look for lower bounds. We will see below that lower bounds involve the diameter instead of the volume, and beyond that no more than a Ricci curvature lower bound; see §§9.7.3.

As explained in §9.2, the main question, vital for many physicists, is the asymptotic behavior of the spectrum. We will see that the first order term in the asymptotic expansion is easy to get. The next order term is another story, as we already saw for the flat torus case. The repartition of the spectrum about the asymptotic formula, the way the eigenvalues arrange themselves, is of equal significance in physics. Whatever a precise definition might be, one feels that the $\mathbb{KP}^n$ spectra given in equation 9.15 on the preceding page is an atypical distribution, with very high multiplicities, and poorly behaved if we want to tell different vibrations apart by hearing how they differ in frequencies. Looking at that equation, one might be led to wonder about *gaps* in the spectrum. We will meet some answers to this question, but some elementary questions of this sort are still completely open. Another important problem, also interesting for applications, is to have a lower bound for the first eigenvalue λ_1. It controls "resonances" and can prevent them. Control of all of the spectral data we have just discussed cannot be obtained only with a lower bound on Ricci curvature, volume and diameter. One will need to know more on the curvature, the injectivity radius, etc. For the behaviour of the spectrum when the metric varies, see Lott 2000 [881].

9.6.2 Direct Problems About the Eigenfunctions

There are very few results about eigenfunctions. It is natural to ask for control on the sup norm of the eigenfunctions, which amounts among other things to studying the asymptotic behavior of

$$\int_M \phi_i^2 \phi_j$$

for a fixed i with j going to infinity. The *nodal sets*, defined to be the zero sets of eigenfunctions, are of clear physical significance. Outside singularities, the nodal sets are hypersurfaces in the manifold. Do they have large measure (say $d-1$ dimensional Hausdorff measure)? How are they located? Think of the spreading out of nodal sets as a kind of even repartition in space. Today's harvest is quite meager: see §9.11.

9.6.3 Inverse Problems on the Spectrum

The literature on recovering Riemannian geometry from the spectrum is immense, this subject having excited people tremendously when it was triggered by Milnor 1964 [922]. There it was proven that two Riemannian manifolds which are not isometric can have the same spectrum. We will give below a brief account of the state of affairs today.

A completely different (still inverse) topic is to try to recover the Riemannian manifold from its geodesic flow. This can be asked in different ways. Suppose you know the lengths of all of the periodic geodesics (this is the so-called *length spectrum*); can you find the metric? But you might know even more, namely the complete structure of the flow on the unit tangent bundle (the phase space). See §9.12 and chapter 10 for the state of current knowledge.

9.7 First Tools: The Heat Kernel and Heat Equation

9.7.1 The Main Result

Theorem 165 (Minakshisundaram 1953 [929], McKean & Singer 1967 [910]) Let M be a compact Riemannian manifold. There is a function

$$K : M \times M \times \mathbb{R}_+^* \to \mathbb{R}$$

which is C^∞ and

1. Given any initial data $f: M \to \mathbb{R}$ the solution of the heat equation
$$-\frac{\partial F}{\partial t} = \Delta F$$
with
$$F(x, 0) = f(x)$$
is given by
$$F(x) = \int_M K(x, y, t) f(y) \, dy$$

2. K is given by the convergent series
$$K(x, y, t) = \sum_i \exp(-\lambda_i t) \phi_i(x) \phi_i(y)$$
(where the eigenfunctions ϕ_i of the Laplace operator Δ are chosen so that they form an orthonormal basis of the square integrable functions on M)

3. For every $x \in M$ there is an asymptotic expansion as $t \to 0$ of the form
$$K(x, x, t) \sim \frac{1}{(4\pi t)^{d/2}} \sum_{k=0}^{\infty} u_k(x) t^k$$
where the $u_k : M \to \mathbb{R}$ are functions given by universal formulae expressing $u_k(x)$ in terms of the curvature tensor of M and its covariant derivatives at the point x.

The three argument function K is called the *fundamental solution of the heat equation* on M, or the *heat kernel* of M.

If one assumes existence of the heat kernel, it is easy to check the properties 1 and 2. Note the surprising symmetry, which has no reason a priori to hold:
$$K(x, y, t) = K(y, x, t) .$$

We recall that the physical interpretation of the heat kernel is the following: $K(x, y, t)$ is the temperature at time t and at the point y when a unit of heat (a *Dirac δ function*) is placed at the point x.

To find the proof and to get a feeling for why property 3 is reasonable, we recall what we saw in equation 1.26 on page 88, namely that the fundamental solution of the heat equation for the Euclidean plane was explicitly determined as
$$K^*(m, n, t) = \frac{1}{4\pi t} e^{-\|m-n\|^2/4t}$$
For a Euclidean space of general dimension d it is also explicit and easy to find by formal computation, namely:
$$K^*(x, y, t) = \frac{1}{(4\pi t)^{d/2}} e^{-d(x,y)^2/4t} \tag{9.16}$$

9.7 First Tools: The Heat Kernel and Heat Equation

where we have replaced the square norm by the distance. To study heat on more general Riemannian manifolds, the idea is to get some function analogous to the above on a compact Riemannian manifold. It makes sense to consider equation 9.16 on the facing page in any Riemannian manifold, provided we cut it with a step function η. So we will set

$$H_0 = hS_0$$

for

$$S_0 = K^*$$

above and measure distance according to our Riemannian metric. This is a sort of first order approximation of the K that we are looking for. We have reason to hope that we can carry on in this direction, because the exponential decays very quickly with time t. In analysis, a function like S_0 (which approximates a kernel) is called a *parametrix*.

The sketch of the complete proof is as follows. We build up an exact solution in two steps. In the first step, one defines local parametrices with higher and higher orders of approximation by an induction formula and a sum as follows:

$$S_k = \frac{1}{(4\pi t)^{d/2}} e^{-d(x,y)^2/4t} \sum_{i=0}^{k} u_i(x,y) t^i$$

so that

$$\left(\Delta_x + \frac{\partial}{\partial t}\right) S_k = \frac{1}{(4\pi t)^{d/2}} e^{-d(x,y)^2/4t} \Delta_x u_k \qquad (9.17)$$

But these functions are only define locally. We now define global functions

$$H_k$$

on our manifold with the above and a step function η by setting

$$H_k = \eta S_k \ .$$

These functions are certainly not what we are looking, since for example they depend on the choice of η. The trick is to define K again as a series by a double convolution process which will "forget" the η function. The two variables in the convolution are the space and the time. We define the *convolution* $A * B$ of two functions of (x, y, t) by

$$(A * B)(x, y, t) = \int_0^t d\tau \int_M A(x, z, \tau) B(z, y, t - \tau) dV_M(z)$$

and the desired fundamental solution is

$$K = \sum_i \left(\Delta_x - \frac{\partial}{\partial t}\right)(Hk^*)^i \qquad (9.18)$$

which works as soon as k is large enough, namely

$$k > \frac{d}{2}.$$

For details of the proof we refer to III.E of Berger, Gauduchon & Mazet 1971 [174], Chavel 1984 [325] chapter VI, Gilkey 1995 [564] or chapter 2 of Berline, Getzler & Vergne 1992 [179]. Formal verification is trivial; the problems are principally in the convergence of the series and in the smoothness of the objects; smoothness is where we use the condition

$$k > \frac{d}{2}$$

which will not surprise readers used to Sobolev inequalities; see theorem 118 on page 325.

The universality of property 3 on page 394 is simply due to Élie Cartan's philosophy of normal coordinates. We saw one aspect of this philosophy when commenting on Jacobi's field equation 6.11 on page 248 in §§6.3.1. The second aspect is that Jacobi's equation can be differentiated as many times as we wish. In the result only the curvature tensor and its covariant derivatives of various orders will appear, and each of these in some universal polynomial expression. The Laplacian is also universal, involving only various derivatives of the Riemannian metric. It remains only to remark that, by construction of the kernel K in equation 9.18 on the preceding page, the u_k are the same as in property 3.

We mention here that the heat kernel is explicitly known for some special manifolds, as is the fundamental solution of the wave equation. Among these special manifolds are of course Euclidean spaces, spheres, and hyperbolic spaces. For example, in the case of the hyperbolic plane, these kernels are employed in Huber's result (theorem 192 on page 421). For the spheres, we can find the kernel in Cheeger & Taylor 1982 [355, 356]; see this text for previous results. For space forms, see the recent Bunke & Olbrich 1995 [279]. For symmetric spaces see Helgason 1992 [704].

9.7.2 Great Hopes

If in theorem 165 on page 393 we integrate over the manifold M and use property 2 we get the basic formula

$$\sum_k e^{-\lambda_k t} \sim \frac{1}{(4\pi t)^{d/2}} \operatorname{Vol}(M) \qquad \text{as } t \to \infty \qquad (9.19)$$

which gives us the first order term of the asymptotic behavior of the eigenvalues (counted with multiplicity). This is called the *Hermann Weyl estimate*, although Weyl was only interested in domains with boundary in Euclidean

9.7 First Tools: The Heat Kernel and Heat Equation

spaces as seen in §§1.8.5. If one is only interested in obtaining this estimate, it can be obtained less expensively with the minimax principle.

Regarding inverse problems, the knowledge of the spectrum gives you the dimension of the manifold and its volume.

As in §§1.8.5, the Hardy–Littlewood–Karamata theorem applies to yield what we are really interested in, namely

$$N(\lambda) = \#\{\lambda_i < \lambda\}$$
$$= \frac{\beta(d)}{(2\pi)^d} \operatorname{Vol}(M,g)\lambda^{d/2} + o\left(\lambda^{d/2}\right) \tag{9.20}$$

as $\lambda \to \infty$. From here, completely elementary calculus yields

$$\lambda_k \sim \left(\frac{(2\pi)^d}{\beta(d)\operatorname{Vol}(M,g)}\right)^{2/d} k^{2/d} \tag{9.21}$$

Note the perfect compatibility of this formula with theorem 164 on page 386.

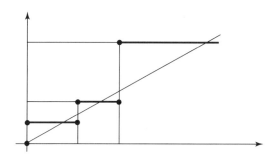

Fig. 9.6. The Weyl asymptotic for surfaces

The function $N(\lambda)$ is a step function. The next natural question on the spectrum is

Question 166 *How does the function $N(\lambda)$ distribute itself around the continuous function giving the asymptotic behaviour?*

Today we know very little about this question. But we will see in §9.9 that with the wave equation technique one can replace the little o by a capital

$$O\left(\lambda^{(d-1)/2}\right).$$

This will permit some rudimentary control, on the gaps for example.

There is also a heuristic principle to the effect that there is a deep relation between the jumps in the spectrum and the lengths of the periodic geodesics.

See §9.9, theorem 176 on page 405, note 9.9.0.1 on page 408, the proof of theorem 189 on page 420, and §§§ 9.13.2.1 on page 426.

The reader might wonder why we did not use the full asymptotic expansion obtained from parts 2 and 3 of theorem 165 on page 393. Let us look at it:

$$\sum_k e^{-\lambda_k t} \sim \frac{1}{(4\pi t)^{d/2}} \left(\mathrm{Vol}(M,g) + U_1 t + U_2 t^2 + \cdots \right) \quad \text{as } t \to \infty \quad (9.22)$$

where

$$U_k = \int_M u_k .$$

We did not do so first because the Hardy–Littlewood–Karamata theorem does not provide any information beyond the first order term. That is to say, the knowledge of the U_k is strictly useless for finding the higher order terms in $N(\lambda)$. We will need more than the above expansion—either a much more subtle analysis of the heat kernel or, better, the wave equation.

Still one can try to use theorem 165 on page 393 and see what one can extract from it. As expected the first u_k expressions should be simple. In fact various authors have computed the two first; if we write scalar for the scalar curvature of our manifold, and R for its Riemann curvature tensor, then

$$\begin{aligned} u_1(x) &= \frac{1}{6} \mathrm{scalar}(x) \\ u_2(x) &= \frac{1}{360} \left(2\|R\|^2 - 2\|\mathrm{Ricci}\|^2 + 5\,\mathrm{scalar}^2 \right) \end{aligned} \quad (9.23)$$

Beginning with the third term, the expressions become more and more complicated. For example, the third term involves the covariant derivative of the curvature tensor. We refer for those and their applications to: the end of this section for the uniqueness of the spectrum of low dimensional spheres, to theorem 188 on page 418 for the compactness of the sets of isospectral metrics on compact surfaces, to Gilkey 1995 [564] and Berline 1992 [179] for very general references.

If you integrate u_1 you get

$$\frac{1}{6} \int_M \mathrm{scalar} .$$

If M is a surface, the Blaschke–Gauß–Bonnet formula 28 on page 138 yields

$$\frac{1}{6} \int_M \mathrm{scalar} = \frac{\pi}{3} \chi(M) .$$

Although this is of no use for calculating $N(\lambda)$, it is very helpful for the inverse problem—it implies that the knowledge of the spectrum (1) tells you that you are on a surface (see above) but moreover (2) we now know its genus.

An important (but which will turn out eventually to be a "useless") remark: the fact we get a topological invariant, in particular an invariant of rescaling

9.7 First Tools: The Heat Kernel and Heat Equation

the metric, is not surprising because $U_1 t$ has to be divided by $t^{2/d} = t$ here, so it is dimensionless. So the natural question is: is U_2 a topological invariant of four dimensional manifolds? This question brought great excitement to spectral geometry in 1966, and was one of the reasons for serious study of the U_k. The answer is *no*. A simple reason is that the generalization of the Gauß–Bonnet theorem in four dimensions (see §§11.3.6 or §15.7) is

$$\chi(M) = \frac{1}{32\pi^2} \int_M \|R\|^2 - 4\|\text{Ricci}\|^2 + \text{scalar}^2 \qquad (9.24)$$

Asking U_2 to be invariant as well (even if not linked to the characteristic) is too much, as trivial examples show.

From the opposite point of view, it is easy to apply equations 9.23 on the facing page to a surface to prove that the round (constant curvature) sphere S^2 is characterized by its spectrum, as are flat tori. On the other hand, we will see in §§9.13.2 that there are isospectral nonisometric Riemannian surfaces of constant negative curvature. Still, using the higher U_k, it is proven in Tanno 1980 [1181] that round spheres of up to six dimensions are characterized by their spectra. The same question for higher dimensional spheres is still open today. This shows how far we are today toward understanding the spectra of Riemannian manifolds. Another nice application of the U_k is to be found in §§9.12.3.

Note 9.7.2.1 (Spectra of space forms) Let us reconsider that the knowledge of the spectrum yields the knowledge of all the U_k integrals. Look at the case of space forms (of constant sectional curvature). Then all the u_k are known and in particular the U_k are all known as soon as one knows the volume of the manifold. This does not yield the space form (up to isometry) except in one dimension. ♦

The spectral determination of the Euler characteristic χ above for surfaces is exceptional: today there is no known topological information in the spectrum in dimensions three and higher. Of course, so far we are discussing the spectra of the Laplace operator on functions. For the spectra of the Laplace operator on more general tensors, e.g. differential forms, see §9.14.

Note 9.7.2.2 (Futility of the U_k) Besides the theoretical interest of establishing a solid foundation for Fourier analysis on a Riemannian manifold, at this moment the heat equation technique seems to be of little use. It might seem that this is because the curvature appears in the asymptotic expansion in a too algebraically complicated manner. Except for the second term, the expansion involves not only the curvature but also its covariant derivatives; in particular geometric invariants (the volume excepted) like the diameter, the injectivity radius, the geodesic flow, do not enter into it. But one "explanation" for the impotence of the U_k is given by the following, which is a strong generalization of theorem 186 on page 415:

Theorem 167 (Lohkamp 1996 [874]) *Consider any compact manifold M of dimension larger than two, and any infinite sequence of positive numbers*

$$0 < \lambda_1 < \lambda_2 < \cdots .$$

Then there is a sequence of metrics g_m on M of fixed volume and fixed integral of scalar curvature such that not only does the spectrum of g_m coincide with the given sequence from 0 to λ_m, but all of the U_{2k} go to $+\infty$ and all of the U_{2k+1} go to $-\infty$. Under the same conditions, there is also another sequence of metrics with the same spectral condition but this time the volume is fixed and the Ricci curvature satisfies

$$\text{Ricci}\,(g_m) < -m^2 .$$

This explains the near inefficacy of the U_k and the poverty of the hypothesis of negativity of Ricci curvature (see §§12.3.5). For the nature of the proof, see §§9.12.1. ♦

There is also an important geometric formula which deserves to be mentioned, even if at the moment it has no geometric application:

Theorem 168 (Varadhan 1967 [1206])

$$\lim_{t \to 0} t \log K(x, y, t) = -\frac{d(x, y)^2}{2}$$

for any x, y close enough.

Varadhan's formula works within the injectivity radius. What happens when y moves to the cut locus of x is the subject of Malliavin & Stroock 1996 [890]; dramatic changes take place, for example on the standard sphere strange events occur at antipodal points. But theorem 168 is fundamental to modern probability theory, and in particular to the Malliavin stochastic calculus on infinite dimensional Riemannian manifolds (e.g. path spaces).

Exterior differential forms are canonically attached to a differentiable manifold and a Riemannian metric also provides a Laplace operator on them. But more generally there are other kind of bundles one can look at, as well as suitable differential operators. Some are canonical, as in the case of spinors, while others are built up with various techniques e.g. twisting canonical ones, etc. In this context the heat equation method works and yields important results. Some are of interest in themselves; these will be described briefly in §14.2. Some are basic tools for Riemannian geometry; we will meet such applications twice in §§12.3.3.

Still thinking about heat, we mention the recent notion of heat content of a domain in a Riemannian manifold. This notion has various applications, even in the Euclidean case, and probably some future: see Savo 1998 [1099].

9.7.3 The Heat Kernel and Ricci Curvature

In §§9.4.2, we used the minimax principle to get upper bounds on the spectrum. Lower bounds are more difficult. The case of the first eigenvalue λ_1 is treated separately in §9.11. We will now address the question of a lower bound for every eigenvalue. An optimal result can be found in Bérard, Besson & Gallot 1985 [139]; see the book Bérard 1986 [135] for a detailed exposition. To formulate their result we introduce some notation.

Definition 169
$$Z(t) = \sum_k e^{-\lambda_k t}$$

which we will write as
$$Z_{M,g}(t)$$

when we need to specify which Riemannian manifold M and metric g is being invoked. Similar notation is used to specify the manifold and metric when discussing the heat kernel:

Definition 170
$$K_{M,g}(x,y,t) = K(x,y,t)$$

Then we can state:

Theorem 171 *There is a universal constant*
$$c = \mathrm{univ}(\inf \mathrm{Ricci}, \dim, \mathrm{diam})$$

(where inf Ricci *is the lower bound of the Ricci curvature,* dim *the dimension and* diam *the diameter of a Riemannian manifold* M*) such that for any time* t

$$Z_M(t) \leq \mathrm{Vol}(M) \sup_{x,y \in M} K_M(x,y,t) \leq Z_{S^d}(ct)$$

This is a very strong result since it is a bound for the whole heat kernel. Since the spectrum of the standard sphere S^d is known, one gets immediately:

Theorem 172 *There is universal constant such that all eigenvalues satisfy the lower bound*
$$\lambda_k \geq \mathrm{univ}(\inf \mathrm{Ricci}, d, \mathrm{diam}) k^{2/d} .$$

The term $k^{2/d}$ agrees with Weyl's asymptotic 9.19 on page 396 for the power of k but not for the volume. Moreover, simple examples show that the diameter, not only the volume, is really needed. Examples also show that these results are optimal as far as the ingredients (see how they enter more explicitly in Bérard, Besson & Gallot 1985 [139]. Finding optimal explicit values is an open problem. The authors' values are explicit but not optimal, especially in the case of negative Ricci curvature. This will be seen from the proof.

The proof is very geometrical. Look again carefully at the proof of the Faber–Krahn inequality 1.22 on page 81 for the fundamental tone of a plane vibrating membrane. There we used function symmetrization—a transplantation, going from the membrane D under study to the circular membrane D^* having same area. From any function on D, a function on D^* was constructed. Then the key ingredient (besides Fubini's theorem and a change of variable) was the isoperimetric inequality for plane curves.

Bérard, Besson & Gallot 1985 [139] enact a double generalization of the same ideas. First we symmetrize the whole heat kernel as a function (which depends on three variables). Second we use the result on the isoperimetric profile obtained in theorem 114 on page 319 which needs precisely a lower bound on Ricci curvature and diameter. The transplantation here goes from M to a sphere whose radius is precisely defined as a function of inf Ricci, the dimension and the diameter. It is then clear that on a manifold of negative Ricci curvature, the comparison sphere cannot be optimal.

The proof is then concluded by expensive and technical details. In particular it uses the maximum principle for parabolic partial differential equations (because the heat equation is parabolic). Time is taken in account as follows. The heat equation for the symmetrized kernel becomes an ordinary differential equation and one then applies Sturm–Liouville theory, in some sense as for Jacobi fields in §3.2. One can find this technique in Bandle 1980 [109]. Details of the above results can be found in chapter V of Bérard 1986 [135] or in Berger 1985 [163].

Brownian motion on Riemannian manifolds is very closely related to the heat equation. The "propagation speed" of Brownian motion "is the Ricci curvature." The reader will enjoy Stroock 1996 [1164], Elworthy 1988 [489], and Pinsky 1990 [1029, 1028].

9.8 The Wave Equation: The Gaps

Put together, the bounds from theorems 172 on the preceding page and 164 on page 386 frame the λ_k between two asymptotic curves. This is reasonable control, but does not say much about how the eigenvalues are distributed. Questions can be asked about the "jumps," about the evenness of the distribution, and more simply about the gaps. The formulas 9.15 on page 391 for the spectrum of the spheres and the $\mathbb{KP}^n$ show spectra which are not evenly distributed, since they are concentrated in intervals. The heat equation is not a deep enough tool to get information on the gaps—we need to analyze the wave equation on our Riemannian manifold.

This is like climbing Jacob's ladder. To get information on the manifolds "downstairs" we have to travel to the unit tangent bundle UM and to work

Fig. 9.7. Spectral gaps

with distributions. Downstairs, a function only has a gradient, but a distribution on UM has a wavefront, which is nothing but the set of its directional singularities. The first result is that under time evolution the wave front evolves exactly by the action of the geodesic flow: "the waves (the light) travel along geodesics." We cannot say more about the wave equation, it will need an entire book. Today this topic is called *microlocal analysis*. It involves subtle notions such as *Fourier integral operators* and *canonical transformations*. To our knowledge, there are no "popular" expositions of microlocal analysis; the most picturesque, and closest to Riemannian geometry, is that of Guillemin & Sternberg 1977 [670]. The four volumes of Hörmander 1983 [735, 736, 737, 738] are complete and encyclopedic (get the second edition of volume I); Tréves 1980-82 [1198, 1199] is also very informative. The wave kernel

$$\sum_k \cos\left(\sqrt{\lambda_k}t\right)\phi_k(x)\phi_k(y)$$

is no longer a function (only a distribution) but in exchange it carries much more information. It can also be remarked that microlocal analysis involves a lot a symplectic geometry, which takes place in T^*M, the cotangent bundle. It is better to ignore the fact that (thanks to the Riemannian structure) T^*M is canonically isomorphic to TM (see §15.2). One also works with the canonical contact structure on the unit tangent bundle UM (which is *Sasakian*): see page 56 of Sakai 1996 [1085].

The following result is part of a very general theory which applies to any elliptic operator on a compact manifold; we employ it here only to the Laplace operator.

Theorem 173 (Hörmander 1968 [734]) *The number $N(\lambda)$ of eigenvalues smaller than λ obeys the asymptotic law*

$$N(\lambda) = \frac{\mathrm{Vol}(M)\beta(d)}{(2\pi)^d}\lambda^{d/2} + O\left(\lambda^{(d-1)/2}\right)$$

It should be mentioned that such a result had been obtained in Avakumović 1956 [90] in three dimensions using a technical study of the parametrix. The immediate corollary (by the very definition of a "capital O" and elementary calculus) is the one we are after:

Theorem 174 *For any Riemannian manifold M there is a constant C_M such that for any real numbers a and b with $b-a$ large enough, the set of eigenvalues λ of the Laplacian in the interval $[a, b]$ satisfies*

$$\#\left\{\sqrt{\lambda} \in [a, b]\right\} > C_M(b - a)a^d .$$

Note that such gap results cannot be too general; think of theorem 186 on page 415 to the effect that there is always a Riemannian manifold whose spectrum is any chosen finite subset of the real numbers.

For the geometer there is major drawback in Hörmander's result. The way the constant is found in Hörmander's proof is not constructive; the geometry of the Riemannian manifold does not come in. But we would like to be able to estimate $C(M, g)$ as a function of the geometric invariants of (M, g). At the moment there is no such result obtained by working with the wave equation on a Riemannian manifold. But the following recent result is to be found in section $6\frac{9}{10}$ of Gromov 1996 [631]. The proof is extremely intricate, and uses the *Kac–Feynman–Kato inequality*. This formula bounds the spectrum of any elliptic operator on any bundle on a Riemannian manifold with the spectrum downstairs of the manifold itself, and was always used the other way around. But Gromov looks at suitable bundles over a compact Riemannian manifold and uses various tools from Vafa–Witten, Bochner–Lichnerowicz and Atiyah–Singer. See chapter §14.2 for a brief survey of those tools. Using that incredibly high climb up Jacob's ladder one has:

Theorem 175 (Gromov 1996 [631]) *In any odd dimensional Riemannian manifold whose sectional curvature satisfies*

$$|K| \leq 1$$

and whose injectivity radius is larger than 1, the spectral gaps are controlled:

$$\#\left\{\sqrt{\lambda} \in [a, b]\right\} > C_d(b - a)^d \operatorname{Vol}(M)$$

for any positive real numbers a, b such that with

$$b > a + C'_d$$

where C_d and C'_d are universal constants in the dimension d.

We leave the reader to use appropriate scaling to replace C_d by a constant depending on $\sup |K|$ and $\operatorname{Inj}(M)$. Let us remark that some geometric control is required in view of theorem 186 on page 415. It seems to be an interesting question to prove the above result by working only with the wave equation "down" on the manifold itself. Note also that one knows more (but not everything) about the distribution of the spectrum on the real line for certain special manifolds; see §9.13.

9.9 The Wave Equation: Spectrum and Geodesic Flow

In the pioneering paper Balian & Bloch 1972 [99], which we have discussed in §§1.8.6, the authors suspected a relation between the spectrum of a plane domain and its length spectrum.[4] The fact that compact plane domains have a boundary rendered this study difficult. This is one reason why people turned first to compact Riemannian manifolds (without boundary of course). We speak now about general Riemannian manifolds; the special case of space forms will be taken care of in §9.13. Relations between the spectrum (of functions) and the length spectrum will be met again in §9.12. For flat tori, we met a perfect link between the spectra furnished by the Poisson formula 9.14 on page 389 So the problem is to find, if possible, various generalizations of this formula.

For the general case, the first result was Colin de Verdière 1973 [389], but the proof was very tricky, using the heat kernel and the stationary phase technique. Soon after it was realized that the wave equation is the more powerful and elegant technique: Chazarain 1974 [327] and Duistermaat & Guillemin 1975 [465]. This yielded:

Theorem 176 *For any Riemannian manifold M the series*

$$\sum_i \cos\left(\sqrt{\lambda_i}\,t\right)$$

defines a distribution whose singular support is contained (besides the value 0) in the set of the lengths L of the periodic geodesics of M. For a generic Riemannian manifold, this singular support is a sum of distributions T_L, with L ranging over lengths of periodic geodesics, and where each T_L has support located in a small neighborhood of L. Moreover each T_L can be expressed with the sole help of the Poincaré return map (see the definition 10.4.3.2 on page 469) associated to the periodic geodesics of length equal to L and the holonomy map (the effect of parallel transport) along these geodesics.

Here is a very primitive explanation for theorem 176; it is not even an idea of a proof but just help for the reader who needs to visualize things to get some grasp of them. We look at a surface and, like throwing a stone in a pond, look for the wave generated by this action. The picture in figure 9.8 on the following page shows what is happening at the beginning: no problem occurs at small distances, but as in §§9.7.2 we might expect trouble at the cut locus. Two waves meeting transversally generate only nice interferences—this has been known for a long time. But the wave interferences are different when the two waves come one against the other in exactly opposite directions; this will be the case for any periodic geodesic. If moreover their common frequency is

---

[4] Recall that the *length spectrum* of a plane domain is the set of lengths of its periodic (billiard or light) trajectories.

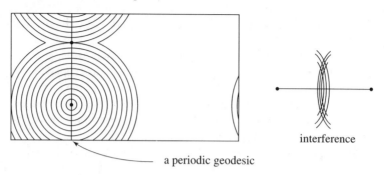

Fig. 9.8. (a) A periodic geodesic (b) Interference

of the form $2\pi nL$, where n is any integer and L is the length of the periodic geodesic under consideration, then we will have (probably) a resonance or, say, a tidal wave. This is the cause of singularities in the series above. Note that this does not happen for geodesic loops—they do not produce enough resonance.

We come back to more standard mathematical notation. First, there is a kind of reciprocal of the formula 9.15 on page 391:

Theorem 177 (Colin de Verdière 1979 [391] and Duistermaat & Guillemin 1975 [465]) If all the geodesics of a compact Riemannian manifold M are periodic with common length equal to L then for k large enough one has the inclusion

$$\operatorname{Spec}(M) \subset \bigcap_{k \in \mathbb{N}} \left[\frac{2\pi}{L}\left(k + \frac{\alpha}{4}\right)^2 - M, \frac{2\pi}{L}\left(k + \frac{\alpha}{4}\right)^2 + M \right]$$

and moreover the number of eigenvalues in every one of these intervals is polynomial in k.

We will see in §§10.10.2 the significance of the α which can only equal $0, 1, 3,$ or 7 (the reader can—and should—think of the $\mathbb{KP}^n$).

The end of theorem 176 on the preceding page was very imprecise about the Poincaré and holonomy maps and in particular was only passing from the singularity of T_L to the Poincaré map. Recall that T_L was defined in theorem 176 on the previous page. Further recall that this Poincaré map has to be viewed in the unit tangent bundle UM (at some starting point) and is the differential at the origin of the return map after going once around a periodic geodesic. After various partial answers, strong results became available only recently in Guillemin 1993 [665] and 1996 [666]. In those works, the singularity of T_L is completely determined by the Poincaré map, this being done in terms of the so-called *Birkhoff canonical form*. This is moreover carried out for a general elliptic linear differential operator, with the periodic geodesics being replaced by the *periodic bicharacteristics*.

9.9 The Wave Equation: Spectrum and Geodesic Flow

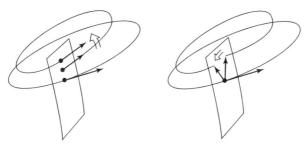

Fig. 9.9. (a) The Poincaré map works in $UM|_{\gamma^{\perp}}$ (b) The holonomy map works in $(\gamma')^{\perp}$

Our information about the gaps and relations with periodic geodesics (for general manifolds, see §9.13 for special manifolds) is still quite meager. Looking again at the picture in figure 9.6 on page 397, one can consider $N(\lambda)$ as a step function. Not only the repartition, but also the jumps are of interest. The common belief today is that those jumps are related in some way yet to be discovered to the length of the periodic geodesics (this set of lengths is called the *length spectrum*). This belief was initiated in Balian & Bloch 1972 [99]. But today we are still missing formal results. In exchange, there are many numerical computations, mainly done by theoretical physicists. This because they are extremely interested in the the semiclassical limit (see more about this on page 376). Recent numerical experiments and thoughts about them can be found in Sarnak 1995 [1095], Luo & Sarnak 1994 [885], Luo & Sarnak 1995 [886], Rudnick & Sarnak 1996 [1074] and the bibliographies of those articles.

The most baffling case will be seen in §§9.13.2; it is the case of negative curvature space forms. The idea is that we know that those forms are chaotic in the good sense: the geodesic flow is very ergodic, the behavior of periodic geodesics and of the geodesic flow are extremely well understood. Briefly speaking, the geodesic flow is extremely evenly distributed in the phase space UM. Because of theorem 176 on page 405, one would expect that the eigenvalues are evenly distributed as a subset of the reals. The answer should be that the spectrum looks like the eigenvalues of a random Gaussian symmetric matrix. This major question is almost completely open today; see §§9.13.2. There is a good result on the distribution of the eigenfunctions; see theorem 185 on page 412.

Question 178 *What is $N(\lambda)$ for a generic Riemannian manifold? Is it in*

$$o\left(\lambda^{(d-1)/2}\right)$$

instead of the extreme

$$O\left(\lambda^{(d-1)/2}\right) ?$$

We know only of the following intermediate result:

Theorem 179 (Bérard 1977 [132]) *If a compact Riemannian manifold has no conjugate points or has nonpositive sectional curvature then as $\lambda \to \infty$*

$$N(\lambda) = \frac{\mathrm{Vol}(M)\beta(d)}{(2\pi)^d}\lambda^{d/2} + O\left(\frac{\lambda^{(d-1)/2}}{\log \lambda}\right)$$

Note 9.9.0.1 (Quasimodes) An interesting link between the spectrum and the periodic geodesics is that of the *quasimodes*. The story started in Babich & Lazutkin 1967 [95] and is far from being finished today, remaining quite mysterious; see Colin de Verdière 1977 [390]. Briefly speaking, to one given periodic geodesic (satisfying certain conditions), one can associate a series of numbers which approach quite a few eigenvalues. The idea of the proof is to build up approximate solutions of the wave equation which will propagate along the geodesic.

Question 180 *Are they many cases for which one can obtain the whole spectrum in this fashion?*

The answer is that this possibility is exceptional and happens only when the geodesic flow is integrable. In general, the hyperbolic zones between the KAM tori will yield a contradiction. The entire book Lazutkin 1993 [852] is devoted to this topic. ♦

Note 9.9.0.2 For *scars*, see §§§ 9.13.2.1 on page 426. ♦

9.10 The First Eigenvalue

9.10.1 λ_1 and Ricci Curvature

The first nonzero eigenvalue λ_1 is of essential importance. It controls the Dirichlet quotient of functions of mean value zero, and it also controls resonances. Indirectly it controls even the pure metric geometry of the manifold—via the distance functions—as seen in Colding's formula 77 on page 264. Again lower bounds are the true prize; upper bounds can be useful but definitely are less useful and much easier to get. We now present results which are not simply a special case of theorems 164 on page 386 or 172 on page 401.

The first result on λ_1 to our knowledge is the following which is hidden on page 135 of Lichnerowicz 1958 [865] and used there to study transformation groups of Riemannian manifolds.

Theorem 181 (Lichnerowicz [865]) *If the Ricci curvature is larger than or equal to $d-1$ (that of the standard sphere of dimension d) then λ_1 is at least as large as the λ_1 of the sphere, namely d. Moreover equality happens only for manifolds isometric to the sphere.*

The proof is beautifully simple, based on Bochner's formula theorem 346 on page 595 (or equation 15.8 on page 707), applied to the 1-form which is the differential df of the first eigenfunction f. This df is not harmonic but

$$\Delta f = \lambda_1 f$$

is the trace of the Hessian

$$Ddf = \text{Hess}\, f \,.$$

Bochner's formula as applied to df becomes, after integration over the manifold and using Stokes' theorem:

$$0 = \int_M \|\text{Hess}\, f\|^2 - \lambda_1 \int_M \|df\|^2 + int_M \text{Ricci}(df, df)$$

The proof is concluded by using Newton's inequality

$$\|\text{Hess}\, f\|^2 \geq \frac{(\Delta f)^2}{d}$$

since after diagonalization at a point,

$$\|\text{Hess}\, f\|^2 = a_1^2 + \cdots + a_d^2$$

and

$$(Df)^2 = (a_1 + \cdots + a_d)^2 \,.$$

The equality is obtained quite easily tracing back each inequality, and appeared first in Obata 1962 [971] (also see Cheng 1975 [362]). This result should be compared with Myers' theorem 63 on page 245. We will come back to this in §§12.2.5. The general result of theorem 172 on page 401 as applied only to λ_1 is an improvement of theorem 181 on the facing page since it involves moreover the diameter (think for example of real projective space). But its main source of interest is that it can be applied when the Ricci curvature is nonnegative or negative.

For those who love Riemannian pinching, we mention Croke 1982 [413] for pinching λ_1, and the recent Petersen 1999 [1020].

9.10.2 Cheeger's Constant

A somewhat intermediate result between Lichnerowicz's theorem 181 on the preceding page and theorem 172 on page 401 is based on Cheeger's constant h_c introduced on page 315.

Theorem 182 (Cheeger 1970 [329]) *On any compact Riemannian manifold*

$$\lambda_1 > \frac{1}{4} h_c^2 \,.$$

It was proved in Buser 1978 [291] that this inequality is optimal, but equality never occurs for a smooth metric. For more on this and the role of λ_1, see §§9.13.1. There is a huge literature on λ_1 but still it seems that there has never been any practical application to various questions concerning "vibrations of great structures," or "nondestructive and noninvasive tests." There is a relation obviously, but vibration today is largely an experimental area of mechanical engineering. Bell casters have always used tests of the sound of a bell to check for possible cracks; see Bourguignon 1986 [238].

9.10.3 λ_1 and Volume; Surfaces and Multiplicity

Despite theorem 164 on page 386 (which used Ricci curvature and volume), there cannot exist an upper bound involving only the volume. This was proven in Dodziuk 1993 [453], by simply building up suitable examples (of course of larger and larger diameter, and this only for dimension larger than or equal to 3). The question was raised because the case of surfaces is exceptional. In fact:

Theorem 183 (Hersch, 1970) *The first three eigenvalues of any Riemannian metric on the sphere S^2 obey the inequality*

$$\frac{1}{\lambda_1} + \frac{1}{\lambda_2} + \frac{1}{\lambda_3} \geq \frac{3}{8\pi} \operatorname{Area}\left(S^2, g\right)$$

with equality only for the standard sphere. In particular

$$\lambda_1 < \frac{8\pi}{3} \frac{1}{\operatorname{Area}(S^2, g)} \ .$$

The proof is beautiful. It mixes three facts

1. the minimax principle of §§9.4.1,
2. the fact that the Dirichlet quotient of a surface is invariant under conformal change and finally
3. the fact that the conformal group of S^2 is large enough to transform any density on the sphere into a one whose center of mass is the origin.

For other compact surfaces, various authors found an upper bound involving only the area, with a constant depending on the genus. The optimal constant is still a pending problem. On this topic recent references can be found in the bibliographies of Dodziuk 1993 [453] and Nadirashvili 1996 [964].

The question of the highest possible multiplicity of λ_1 is also interesting for surfaces. Discard higher dimensions, thanks to the Colin de Verdière result 186 on page 415 to the effect that, starting in dimension three, any finite subset of the reals—including multiplicities—can always be realized as the beginning of the spectrum of a suitable Riemannian manifold. But for surfaces, the multiplicity of λ_1 is bounded with the genus of the surface. Results are optimal

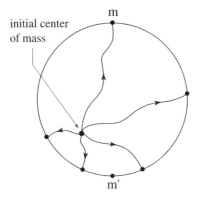

Fig. 9.10. Under radial conformal transformations sending all but one point into it, one can cover the whole ball with the initial center of mass

today for the sphere (triple) and the torus (sextuple). Optimal constants are still to be discovered for other surfaces. There is definitely a relation between this multiplicity and the chromatic number of the surface; see the definition on page 415.

To prove such an upper bound, one relies on the structure of the set of nodal lines, i.e. the set where an eigenfunction vanishes. Except at a finite set of singular points, the zero set is made up of regular curves. More important is that at singular points the curves meet with a set of tangents which are the directions of the diagonals of a regular polygon. Using this result of Bers the proof is concluded with arguments of algebraic topology; references are Besson 1980 [187], Yang & Yau 1980 [1289].

9.10.4 Kähler Manifolds

Mathematicians never stop asking questions. For example, can we have an upper bound on the first eigenvalue depending only on the volume when the manifold is "special"? Considering the geometric holonomy hierarchy introduced in chapter 13, the case to look at is that of Kähler manifolds (see §13.6). Indeed it is natural to wonder about the general spectrum of a Kähler manifold. One answer is the following extension to $\mathbb{CP}^n$ of Hersch's theorem 183 on the preceding page for S^2:

Theorem 184 (Bourguignon, Li & Yau 1994 [245]) *For any Riemannian metric g on $\mathbb{CP}^n$*

$$\lambda_1 \leq \frac{(n+1)\pi^n/n!}{\mathrm{Vol}(\mathbb{CP}^n, g)}.$$

We recall (see §§9.5.4 and §§§7.1.1.2) that the volume of the canonical metric of $\mathbb{CP}^n$ is

$$\text{Vol}\,(\mathbb{CP}^n, \text{Fubini–Study}) = \pi^n/n!$$

and its first eigenvalue is $n+1$. For the proof, the conformal group of the sphere is replaced here by the group of all biholomorphic transformations of $\mathbb{CP}^n$. In Bourguignon, Li & Yau 1994 [245] and Gromov 1992 [630] one will find generalizations of this result to various algebraic manifolds, and to the whole spectrum. For the big picture of the subject, it is important to remark that the spectrum is a robust invariant, while being Kähler is not: see note 9.4.1.1 on page 385 and §13.6 and §14.6.

The extremely important case of "Riemann surfaces", that is to say of constant curvature -1, whether or not compact, will be studied at large in §§9.13.1. In the spirit of §9.12, we are far from being able to recognize the spectra of Kähler manifolds.

9.11 Results on Eigenfunctions

9.11.1 Distribution of the Eigenfunctions

It seems hopeless to search for any general result valid for "any" Riemannian manifold. But one can hope for a regular distribution of the eigenfunctions when the manifold is generic (in any sense). A regular distribution would be one for which in any domain D of the manifold and for eigenfunctions with larger and larger eigenvalue, one finds the integral of the square of that function over that domain is in a proportion to the integral over the whole manifold which is closer and closer to the ratio of the volumes of D and M. To our knowledge there is not a single result in that direction; compare with the periodic geodesic result in §§10.3.5.

But if the manifold is "ergodic" (see §§10.5.1), then there are partial results. The conjecture is that ergodicity implies an even distribution of the eigenvalues and the eigenfunctions. Concerning the eigenfunctions one has only:

Theorem 185 *For an ergodic Riemannian manifold M, there is a sequence*

$$\{i(k)\}$$

of integers, of full density in the integers, such that for every $D \subset M$ with eigenfunctions $\phi_{i(k)}$ being normalized:

$$\lim_{k\to\infty} \int_M \phi^2_{i(k)} = \frac{\text{Vol}(D)}{\text{Vol}(M)}\ .$$

Measure theory aficionados would prefer to write this as

$$\lim_{k\to\infty} \phi^2_{i(k)} dV_M = dV_M$$

Full density means that the number of points in question in $[0, \lambda]$, compared to the whole spectrum, has a ratio closer and closer to one when λ goes to infinity. The latest general reference on this topic is Colin de Verdière 1985 [392] for our compact case, which completed the attempt of Shnirel'man 1973 [1137]. For the noncompact see Zelditch 1987 [1302] and Zelditch 1992 [1303]. The proofs involve a deep theorem of Yuri V. Egorov on Fourier integral operators and belong therefore to microlocal analysis. So again, the wave equation is used even if it disappears in the final statement. For the very special case of space forms of negative curvature, see further references in §9.13, but the results are still incomplete today.

9.11.2 Volume of the Nodal Hypersurfaces

Another way to look at regularity of eigenfunctions is to study their nodal hypersurfaces, namely the subsets of the manifold where they vanish. When the manifold is a surface, these subsets are curves. A reasonable behaviour to expect is that the volume of the $\phi_\lambda^{-1}(0)$ will grow as $\lambda \to \infty$, with some asymptotic order. The reader can check on examples (flat tori being the simplest) and also looking at spherical harmonics (see §§9.5.2 and §§9.5.3) that an eigenfunction with eigenvalue λ behaves like a polynomial of degree $\sqrt{\lambda}$. If this is more or less true for any compact Riemannian manifold, then one will have $\mathrm{Vol}\left(\phi_\lambda^{-1}(0)\right)$ roughly behaving like $\sqrt{\lambda}$. It was conjectured by Yau in 1982 that for every Riemannian manifold M with Riemannian metric g there are constants $c = c(g)$ and $c' = c'(g)$ such that

$$c\sqrt{\lambda} \leq \mathrm{Vol}\left(\phi_\lambda^{-1}(0)\right) \leq c'\sqrt{\lambda} \qquad (9.25)$$

for every eigenvalue λ. The intuitive idea behind Yau's conjecture was that eigenfunctions for λ behave roughly like polynomials of degree $\sqrt{\lambda}$, which is the case for the standard sphere for which the eigenfunctions are the restrictions to the sphere of the harmonic polynomials of Euclidean space. After the partial result of Brüning 1978 [266], this was proven in Donnelly & Fefferman 1988 [461]. The volume is to be understood as the $(d-1)$ dimensional Hausdorff measure to be sure to make sense. The proof is extremely hard, and involves various results from analysis. One needs to know the local behaviour of the eigenfunctions, their local sup norm and the distribution of their singular zeroes. Another basic fact is the analyticity of the eigenfunctions of an elliptic operator (here the Laplacian). And the proof tells us even more about the eigenfunctions.

The story does not end here for at least two reasons. The first in that the proof we need the analyticity of both the manifold and the metric. But for the geometer the major drawback is that the two constants $c(g)$ and $c'(g)$ are unknown. They come from an atlas and its coordinate changes. The geometer would like to be able to express $c(g)$ and $c'(g)$ as functions of Riemannian invariants of (M, g) (and of course the cheapest possible ones). We know of no work on this. Let us mention a recent paper addressing noncompact manifolds: Donnelly & Fefferman 1992 [462]. Also see Savo 2001 [1101].

9.11.3 Distribution of the Nodal Hypersurfaces

Figure 1.98 on page 91 shows the extraordinary regularity of a nodal line. There is some reason to believe that when the geodesic flow of a Riemannian manifold is ergodic, the nodal sets are evenly distributed. In saying that nodal lines are evenly distributed, we mean something like asking that given any domain $D \subset M$

$$\lim_{\lambda \to \infty} \frac{\mathrm{Vol}\left(D \cap \phi_\lambda^{-1}(0)\right)}{\mathrm{Vol}\left(\phi_\lambda^{-1}(0)\right)} = \frac{\mathrm{Vol}(D)}{\mathrm{Vol}(M)}.$$

Today there are only numerical experiments. Nodal sets might also be connected to periodic geodesics by some mysterious phenomenon called *scarring*; see figure 1.100 on page 93. For a discussion of scars, we refer to Sarnak 1995 [1095], also see §§§9.13.2.1.

9.12 Inverse Problems

The general scheme is to try to understand the map

$$(M, g) \mapsto \mathrm{Spec}\,(M, g)$$

from Riemannian structures on a manifold M to the set of all discrete subsets of the positive real line:

$$\mathrm{Spec} : \mathcal{RS}\,(M) \to \{\text{discrete subsets of } \mathbb{R}^+\}$$

By a Riemannian structure, recall that we mean a point in the quotient set of the set of Riemannian metrics by all possible diffeomorphisms. We do not want to distinguish between two isometric Riemannian manifolds (metrics). The first question is to determine the image of this map, the second is about its inverse: is it one-to-one, and if not what can be said about the preimages of various points in the image?

9.12.1 The Nature of the Image

We are far today from being able to guess a sufficient condition for a subset of the reals to be realizable as the spectrum of some Riemannian manifold; we know only of Omori 1983 [973]. Of course all of the results above can be viewed as necessary conditions, the most typical one being Weyl's asymptotic as made precise in Hörmander's result 173 on page 403, as well as its gap corollary. But for a finite set to be realized as the beginning of a spectrum (including imposed multiplicities) there is no obstruction:

9.12 Inverse Problems

Theorem 186 (Colin de Verdière 1987 [393]) *For any compact manifold M of dimension larger than or equal to 3 and for any finite subset of the positive real numbers, indexed with finite multiplicities, there exists some Riemannian structure on M whose spectrum begins with that subset.*

The proof is nice. It consists in putting points on the manifold, considering them as oscillators with the desired frequency and multiplicity. Then one joins them by nonintersecting curves, building up a tubular neighborhood of that structure and controlling everything to keep this finite spectrum. One can also see this result as first finding a (finite) graph whose spectrum for its standard graph Laplacian is the desired finite piece under consideration, and then playing some kind of "tunnel effect" along the edges. Technically the multiplicities give troubles, which can finally be controlled by a subtle transversality argument. But for infinite subsets of the reals, the question of sufficient conditions seems completely open; however see note 9.12.1.1 on the following page.

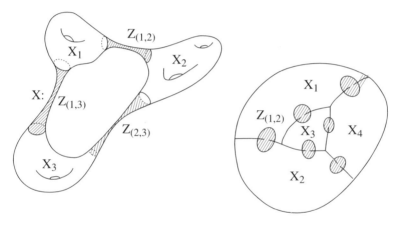

Fig. 9.11. Colin de Verdière's proof that one can choose any finite part of the spectrum of the Laplacian

It is when joining the points by nonintersecting curves that the condition on the dimension appears. This is of course impossible without extra conditions when the dimension is 2, since then some of those curves can be forced to meet. In fact this fits perfectly with the restriction on the multiplicity of λ_1 that we met in §§9.10.3. For the interested reader we mention here that pursuing this topic in the case of surfaces Colin de Verdière discovered recently a fascinating application to electrical circuits: see Colin de Verdière 1996 [396]. He was also led to make the following conjecture. For a compact surface M define its *chromatic number* $\mathrm{Chrom}\,(M)$ as the largest integer N so that there is an embedding into M of the complete graph with N vertices.

Conjecture 187 (Colin de Verdière) *For any surface M, the highest possible multiplicity of λ_1 of any Riemannian metric is equal to* $\operatorname{Chrom}(M) - 1$.

See page 601 of Colin de Verdière 1987 [393] for more on that.

Note 9.12.1.1 In theorem 167 on page 400 we saw a dramatic improvement of Colin de Verdière's results. The scheme for Lohkamp's proof is as follows: modify Colin de Verdière's construction by suitable "attachments of metrics". These constructions are hard and subtle—in particular the author uses Besicovitch's coverings and the technique of "crushed ice". ♦

9.12.2 Inverse Problems: Nonuniqueness

We have been studying direct problems: I know the manifold and some of its invariants. What can I say about the spectrum? Inverse problems have the form: I know various things about the spectrum, what can I recover of the metric? The first question is the uniqueness: are two isospectral manifolds necessarily isometric ?

The first time the author heard about this question was in letter written to him by Leon Green around 1960. In this letter, Green also remarked on an almost straightforward fact: if one knows not only the eigenvalues but also the eigenfunctions, then one knows the metric (two such manifolds can be called *homowave* or *homophonic*). This is because the completeness of the eigenfunctions (see §§9.3.3) implies knowledge of the Laplacian acting on functions, and then from the explicit formula of the Laplacian in coordinates, one recovers immediately the g_{ij}.

The isospectral question was a strong incentive in the sixties. In the case of Riemann surfaces, uniqueness was conjectured in Gel'fand 1962 [553]. For plane domains, we already met this question in §§1.8.4. The first counterexample came in Milnor 1964 [922]. It consists in two tori of dimension 16 with exactly the same spectra. By the results of §§9.5.2, we know the spectrum of a flat torus as soon as we know the lattice defining it. Then two lattices Λ and Λ' in $\mathbb{R}^d$ will yield isospectral tori if and only if the number N_m of points in them having a given norm m is always the same. The set of these numbers is completely encoded in the theta series of the lattices. Namely one defines the theta series of the lattice Λ by

$$\Theta_\Lambda(z) = \sum_{x \in \Lambda} q^{x \cdot x} = \sum_m N_m q^m \qquad (9.26)$$

(where $q = exp(\pi i z)$) defined for suitable values of the complex variable z. These functions have been exhaustively studied for purposes of number theory. An excellent presentation is 2.3 (pages 44–47) of Conway & Sloane 1999 [403]. There one will found out how to compute the theta series of various lattices, depending how they are defined.

Milnor's examples were the two lattices called $E_8 \times E_8$ and E_{16}. The lattice E_8 is the famous lattice attached to the exceptional Lie group denoted also by E_8. It can be defined as the set of tuples $(n_1, \ldots, n_8)$ where all n_i are integers or integers plus $1/2$ and with the extra condition that $\sum_i n_i$ is even. The lattice E_{16} is then simple to construct. What is subtle is to compute their theta series and to show that they are identical; a very good exposition of this is to be found in Serre 1973 [1125]. Checking that they are not isometric (congruent) is the trivial part.

It is a easy exercise to show that isospectral 2-dimensional lattices are congruent (i.e. they can be rotated into one another in the Euclidean sense). Various people found isospectral flat tori of various dimensions. We refer the reader to Conway & Sloane 1999 [403] for them. The dimension can go as low as four. For this dimension one will find on page xxi of the preface (of the second edition) of Conway & Sloane unbelievably simple examples, depending moreover on four parameters. The case of dimension 3 was finally solved positively in Schiemann [1103]; indeed one only needs to know that the eigenvalues are not too large.

Then people got more and more examples of different types. Using number theory (quaternionic number fields) Gel'fand's 1962 conjecture of uniqueness for Riemann surfaces was disproven in Vignéras 1980 [1216]. Thereafter the field blew up so much that we just give few references permitting the reader to go back to all of them. The landmark Sunada 1985 [1168] put things in the right context, at least when considering space forms obtained by quotienting by discrete subgroups. One then finds a sufficient algebraic condition between two such groups to yield isospectral quotients. Then Gordon 1993 [575] gives very geometric methods (using transplantation techniques, see Bérard 1989 [137] which can also be used as a survey) to construct isospectral Riemann surfaces. It is interesting to note that the plane isospectral domains, mentioned in §§1.8.4, were found using isospectral abstract Riemann surfaces with no boundary (compare figures 9.12 on the following page and 1.94 on page 86).

It is a natural instinct to search for more and more general examples; the preceding ones were all space forms. People found locally homogeneous spaces, then nonhomogeneous ones and even one parameter deformations; see Bérard & Webb 1995 [141], Gordon & Webb 1994 [578], Gordon 1994 [575], Gordon 2000 [576], Gordon & Mao 1994 [577], Gornet 1998 [582], and Schueth 1999 [1112].

In Szabo 2001 [1174], very interesting pairs of isospectral metrics are constructed on spheres; they can be made as close to the canonical metric as you like.

We will see in §9.14 that there is a natural Laplacian for exterior differential forms of any degree, hence associated spectra for each degree. We naturally meet the question of obtaining isospectral, nonisometric metrics at the level of differential forms. Today one has examples of different types. For instance, the counterexamples with flat tori are always isospectral for differential forms of any degree, since the eigenvalues for differential forms coincide trivially with

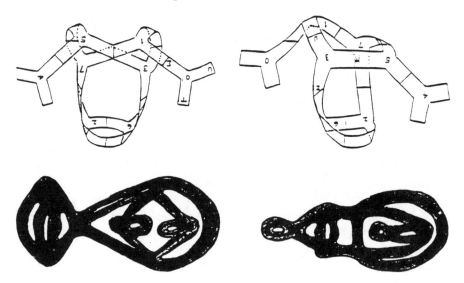

Fig. 9.12. Constructing isospectral plane domains out of isospectral surfaces of constant curvature

those for numerical functions, with only the multiplicities being multiplied by the fixed constant which is the binomial number $\binom{\dim M}{p}$. In the opposite direction, one will find in Gornet 1996 [581] examples distinguishing between isospectrality for functions and for differential forms.

9.12.3 Inverse Problems: Finiteness, Compactness

Since the geometry with a given spectrum is not unique, we can still try to have information on the possible geometries, i.e. sets of Riemannian structures having the same spectrum. How large can these sets be? Do they have any kind of structure, in particular are they "finite dimensional" in any reasonable sense or "compact"? To our knowledge, the finite dimensionality is a completely open problem (unlike the case of Einstein metrics as we will see in theorem 286 on page 531). The infinitesimal isospectral deformation equations in the space of metrics look hopeless; we will just meet a few exceptions.

But there is a nice result for surfaces:

Theorem 188 (Osgood, Phillips & Sarnak [979]) *For any choice of spectrum, the set of Riemannian structures (i.e. Riemannian metrics up to diffeomorphism) with that spectrum is compact.*

The proof is hard but two of its ingredients are of great importance for other purposes. The first is the collection of curvature terms which appear in the asymptotic expansion of the heat kernel; see theorem 165 on page 393. The second ingredient is new for us: it is the *determinant of the Laplacian*. Formally, it is defined as

$$\det \Delta = \prod_i \lambda_i \qquad (9.27)$$

Approached naively, as it is written, this product is not convergent, but one can define it anyway, using various regularization tricks. The most common trick is to take its logarithm: consider the ζ *function* of the spectrum:

$$\zeta(s) = \sum_i \frac{1}{\lambda_i^s}$$

and compute formally $\zeta'(0)$. You will find the determinant. All the effort now focuses on rendering this analysis rigorous. We refer to Osgood, Phillips & Sarnak [979] for details of the proof and also for references on this determinant. Recall that we mentioned on page 86 a compactness result that was obtained for isospectral plane domains. This determinant is also used for an extraordinary proof of the conformal representation theorem 70 on page 254 using Ricci flow; see references and the current state of affairs in Chow [378].

The use of the determinant cannot be avoided. The heat invariants are certainly not enough. This can be seen simply because all of these invariants coincide in the case of constant curvature metrics, as already remarked in note 9.7.2.1 on page 399. On the other hand the Teichmüller space of all Riemann surfaces of a given genus is not compact. In particular, (at least for higher genus) one cannot prove compactness inside the set of metrics conformal to a given metric, a case which is much simpler since it only involves scalar functions instead of metric tensors; namely they involve only the Gauss curvature K and its various iterated Laplacians $\Delta^m K$. But in two dimensions, extensive study shows that the heat invariants are simple enough when they are controlled by the determinant of Δ. For this determinant as a functional of Riemannian metrics, see Sarnak [1096].

The above proof suggests the conjecture that, in two dimensions, there can be only a finite number of metrics isospectral to a given metric. This is certainly false in higher dimensions, since there are one-parameter isospectral deformations.

The compactness of higher dimensional isospectral sets is open. One reason is that the proof above involves controlling the nature of the heat invariants, which are so much simpler for surfaces. However there are good partial results, in dimensions 3 and 4: Osgood, Phillips & Sarnak [980], Anderson [41] and Brooks, Perry & Petersen [264]. For topological finiteness of isospectral sets, see Brooks, Perry, & Petersen [263, 262]. Also see Gordon 2000 [576].

9.12.4 Uniqueness and Rigidity Results

We already saw on page 399 the uniqueness of the spectra of the standard spheres up to dimension 6. The analogous question is open for higher dimensions. But there are good results:

Theorem 189 (Guillemin & Kazhdan 1980 [667] and Croke & Sharafutdinov [422]) On compact manifolds of negative curvature, there are no isospectral deformations.

The proof for surfaces is beautiful and we explain it in some detail because it seems to us that this technique could be used more widely. It is a kind of double Fourier analysis leading to a contradiction. There are three steps. One looks at the derivative of a deformation of metrics on the unit tangent bundle UM. This bundle has fibers which are circles, which leads to Fourier analysis for functions $UM \to \mathbb{R}$. If the Fourier subspaces are called H_i then the deformation function

$$t : UM \to \mathbb{R}$$

belongs to the direct sum

$$H_{-2} \oplus H_0 \oplus H_2$$

because Riemannian metrics are quadratic forms. Now one invokes theorem 176 on page 405 to the effect that the lengths of periodic geodesics are preserved under our deformation since it is isospectral. A periodic geodesic when lifted up to UM is now a periodic trajectory of the geodesic flow. An easy computation, the "first variation formula for changes of metric," shows that the integral of the deformation function t is zero along any periodic geodesic. But a manifold of negative curvature has a lot of periodic geodesics, dense in the best possible sense (see §10.6). This explains (although it is not a proof) a result of Livtsic to the effect that there exists a new function $s : UM \to \mathbb{R}$ such that t is the derivative of s along the geodesic flow. It remains now to look at how the geodesic vector field behaves with respect to the Fourier analysis above. The negativity of the curvature implies that differentiating in G lowers the rank in Fourier analysis. In particular $s' = t$ implies that

$$s \in H_1 \oplus H_0 \oplus H_{-1} \ .$$

But s also should be like t in $H_2 \oplus H_0 \oplus H_{-2}$ and this finishes the proof: t has to be constant along the fibers. The proof of Croke & Sharafutdinov 1997 [422] for higher dimensions is somewhat different.

9.12.4.1 Vignéras Surfaces

The Vignéras examples of surfaces with the same spectrum appeared in Vignéras 1980 [1216] The recent basic uniqueness and rigidity result Besson, Courtois & Gallot 1995 [189], which will be addressed in detail in theorem 251 on page 484, has already had so many applications that its authors are conjecturing (see 9.20, page 780) a result which would be in some sense the best possible:

Conjecture 190 *Isospectral, compact, negatively curved manifolds of dimension larger than 2 are isometric.*

Question 191 *Is isospectrality a nongeneric phenomena? Otherwise stated: are generic Riemannian manifolds spectrally isolated (solitude)?*

A third remark concerns the length spectrum, i.e. the set of length of periodic geodesics. From theorem 176 on page 405 one is certain that isospectrality implies coincidence of the length spectra; but Vignéras counterexamples in §§§9.12.4.1 show that different Riemann surfaces can have the same length spectrum. In §10.11 we will see that is not the case for the marked length spectrum. This is true for example for negative curvature manifolds of dimension higher than 2 and supports the conjecture just presented: see 9.14 in Besson, Courtois & Gallot 1995 [189].

9.13 Special Cases

9.13.1 Riemann Surfaces

By a Riemann surface we understand a compact orientable surface of constant curvature -1. In our hierarchy they are the negative space forms of dimension 2. This means we exclude the sphere and the torus.

Riemann surfaces have been studied since Riemann in great detail, for their intrinsic interest. They appeared originally in complex variable theory, in algebraic geometry and in number theory. Recently they became a favourite object for theoretical physicists, in particular in string theory. It is then not surprising that we have many strong results for them, including for their spectra. The book Buser 1992 [292] is a very complete exposition of the subject at that date. A more recent survey is Buser 1997 [293]. We just note that in Buser 1992 [292] the question of the regularity (randomness) of the spectrum and that of the eigenfunctions (compare with §9.9 and theorem 185 on page 412) are still not well understood, we will discuss them in §§9.13.2.

The first basic fact is that for Riemann surfaces theorem 176 on page 405 can be inverted. What theorem 176 says is that the function spectrum of the Laplacian determines the length spectrum (the set of lengths of the periodic geodesics). But the converse is false in general; one needs much more that the length spectrum, namely essentially the Poincaré map and the parallel transport of periodic geodesics. But in the case of Riemann surfaces, the parallel transport is always the identity since the dimension is two and we have orientability. The Poincaré map is also known because the curvature is constant. This explains (but of course does not prove):

Theorem 192 (Huber 1959 [745, 747]) *On a Riemann surface, the spectrum of the Laplace operator on functions determines the length spectrum and vice versa.*

The proof is based on a formula for Riemann surfaces which is a generalization of the Poisson formula 9.14 on page 389 which was valid for flat tori. The formula computes the heat kernel by a suitable summation formula involving the length spectrum. It is possible simply because there is an explicit formula for the heat kernel K^* of the total (noncompact!) hyperbolic space Hyp^d, and in particular for Hyp^2. Our surface is a quotient of Hyp^2 by a discrete group of hyperbolic isometries. It is enough to know the primitive elements of this group. Being without fixed points, they have to consist in a *gliding* along an hyperbolic line (called the *axis* and denoted by γ). The length of the gliding corresponds exactly to the length of a periodic geodesic downstairs. As a matrix of the group $\text{Isom}\left(\text{Hyp}^2\right)$ that length is exactly the trace of this matrix. This explains the name "trace formula." This formula of Huber is a particular case of Selberg's trace formula which we will meet below. The proof is finished by remarking that the heat kernel downstairs is a suitable summation of the type

$$K = \sum_\gamma K^*\left(x, \gamma y\right)$$

for the axis γ above; details are to be found in chapter 9 of Buser 1992 [292].

We now present to the reader a choice of results that we find especially appealing; most of them are in the book Buser 1992 [292]. The heuristic possibility of these results comes from Huber's theorem, as explained in the preface of the book:

> *This theorem does not show only that the eigenvalues contain a great deal of geometric information, it also indicates that spectral problems may be approached by geometric methods....*
>
> Buser 1992 [292]

These geometric methods rest essentially on the fact that the set of all Riemannian surface structures on a given orientable surface of genus larger than 1 can be encoded in the lengths of the sides of the hexagonal pantaloon hyperbolic plane pieces and the twisting angles when one glues them together as was done in figure 4.10 on page 157. The study is still not too clear conceptually in Buser's book. But in Buser 1997 [293] the author made a decisive step. He succeeded, at least for a very large class of Riemann surfaces, to find the surface itself directly and explicitly from the spectrum. This means that the complete geometry is encoded in the spectrum. Those surfaces are called solitary because they don't have nonisometric isospectral companions.

We start with the eigenvalues called *small*. What is important for a Riemann surface is not only λ_1 and its position with respect to $1/4$, but also the set of λ's which are in $]0, 1/4]$ (called *small*). Why $1/4$ comes into the picture cannot be explained briefly; for details we refer the reader to Buser's book. From it we extract this. In writing the heat kernel as a summation, it is convenient to write the eigenvalues $\lambda = r^2 + 1/4$, so that the associated r are imaginary when λ is below $1/4$. A very heuristic reason is that in hyperbolic geometry, the modular domain is the one in figure 6.36 on page 255 and that

$$1/4 = (1/2)^2 .$$

Let us just recall that this modular domain is the quotient of the hyperbolic plane by the isometries whose matrix is integral. It might be the most important object of all mathematics, as its is connected with function analysis, complex variables, number theory, etc. Remember in this context the Riemann hypothesis for the zeros of the ζ function which "should" be all on the line $s = 1/2$.

Today the situation for small eigenvalues is satisfactory on one hand but on the other hand some conjectures are still open. Let us also mention that the small eigenvalues play a basic role in the refined version of the asymptotic expansion for the counting function of the length spectrum, as will be seen in theorem 205 on page 447. If we denote by $\mathcal{M}_\gamma$ the set of all Riemann surfaces of a given genus γ then

Theorem 193 (Buser 1992 [292] 8.1.1) *For any γ and any surface in $\mathcal{M}_\gamma$,*
$$\lambda_{4\gamma-2} > 1/4 \ .$$

Theorem 194 (Buser 1992 [292] 8.1.2) *For any γ and any integer n (think large) and for any $\varepsilon > 0$ (think of ε as small) there are elements of $\mathcal{M}_\gamma$ with*
$$\lambda_n \leq 1/4 + \varepsilon \ .$$

Together these two statements look surprising. There is a universal bound for the number of eigenvalues in $[0, 1/4]$ but not in any $[0, 1/4 + \varepsilon]$. A geometric reason is offered on page 211 of Buser's book; it mixes isoperimetric considerations for hyperbolic hexagons and the fact that $\mathcal{M}_\gamma$ is never compact—see just below.

Theorem 195 (Buser 1992 [292] 8.1.3) *For any $\varepsilon > 0$ there is a genus γ and a surface $\mathcal{M}_\gamma$ with*
$$\lambda_{2\gamma-3} < \varepsilon \ .$$

Theorem 196 (Buser 1992 [292] 8.1.4) *There is a universal constant $c > 0$ so that for any γ and any surface in $\mathcal{M}_\gamma$*
$$\lambda_{2\gamma-2} > c \ .$$

Although the conjectured value for c is in fact $1/4$, today the best known c is around 10^{-12}. There are many other results for small eigenvalues; see the Notes at the end of chapter 8 of Buser's book.

We turn now to the isospectral question. Recall that there are examples of isospectral but nonisometric Riemann surfaces: see §§9.12.2 and also that there is a general compactness result: see §§9.12.3. But in the present case we also have finiteness:

Theorem 197 (Buser 1992 [292] 13.1.1) *For a given genus γ there are at most $\exp\left(720\gamma^2\right)$ pairwise nonisometric isospectral Riemann surfaces.*

The last topic we will discuss in this section is *Wolpert's theorem* (1977-79). It says that for Riemann surfaces a certain finite part of the length spectrum determines the whole spectrum. In Buser's book the precise statement is theorem 10.1.4. Then Buser extends the theorem to the function spectrum as follows:

Theorem 198 (Buser 1992 [292] 14.10.1) *For any $\varepsilon > 0$ and any γ there is a universal constant $\mathrm{univ}(\varepsilon, \gamma)$ such that if two Riemann surfaces S and S' of the same genus γ both with injectivity radius larger than ε verify $\lambda_n(S) = \lambda_n(S')$ for every $n < \mathrm{univ}(\varepsilon, \gamma)$ then they are isospectral for their whole spectrum.*

Some remarks are now in order. First, the lower bound on the injectivity radius cannot be avoided. The noncompactness of $\mathcal{M}_\gamma$ is directly linked with the fact that the injectivity radius can go to zero. Conversely, compactness when there is lower bound on the injectivity radius is a very special case of the general compactness theorem which we will meet in complete detail in §§12.4.2 and also theorem 376 on page 621.

Second, the original proofs (both for the length and the function spectrum) were extremely expensive, using in particular the theory of real analytic varieties. Recently in Buser 1997 [293] the results on solitary surfaces (mentioned above) were used to give a much simpler proof of theorems like Wolpert's. Also see Schmutz 1996 [1106].

9.13.2 Space Forms

The preceding section concerned space forms of dimension two and of negative curvature. The case of zero or positive curvature was treated in section §§9.7.2 where we saw that the standard sphere and the standard $\mathbb{RP}^2$ are determined by their spectrum, as are flat tori. This was done using the asymptotic expansion of the heat kernel.

Looking now at higher dimensions, we saw in §§9.12.2 the state of affairs for flat tori and for spheres, circumstances being particularly unsatisfactory for spheres. Let us turn now to the compact manifolds of negative constant sectional curvature. This is very special case among manifolds of negative curvature. We saw at large in §9.11 that there are some results on the distribution of eigenvalues and of eigenfunctions for ergodic manifolds. But also that those results were very partial, the basic questions being completely open. Since negative curvature manifolds are ergodic in a very strong sense (see §10.6) and since we will see extremely satisfying results for them in §10.8 for the length spectrum with optimality for the space forms, it is then natural to expect for negative curvature space forms much stronger results than for the general ergodic or negatively curved ones. This was the case for Riemann surfaces as seen just above to some respect, in particular for the small eigenvalues.

There is a theoretical answer to every question concerning spectra of negatively curved Riemann surfaces, namely *Selberg's trace formula*, which in dimension 2 gives back part of Huber's theorem 192 on page 421. For higher dimensions, see Bunke & Olbrich [279].

These questions are under very intense study today. The hope is to use tools from number theory, since these space forms are mostly found by arithmetic means; see §§6.6.2. The tools are typically modular functions (for the flat tori in §§9.12.2 they were theta functions). Strong incentives come to this study from mathematical physics, in particular in what is called the semiclassical limit (see more on page 376) and in the present situation from *quantum chaos*.

There is no general picture arising from the various results obtained up to know. We already said in §9.9 that experts disagree, comparing mathematical results and numerical experiments (including dimension 2). We mention only references: Sarnak 1995 [1095], Luo & Sarnak 1994 [885], Luo & Sarnak 1995 [886], Rudnick & Sarnak 1996 [1074]. One should also of course look at the bibliographies of those. Today a conjecture is the following: there are numerical experiments from which it seems that the distribution of eigenvalues is not even for some arithmetic Riemann surfaces. That is, the distribution is not a *Gaussian orthogonal ensemble* (GOE), i.e. the set of the eigenvalues of a random $N \times N$ symmetric matrix as $N \to \infty$, with the whole business being rescaled to agree with Weyl's asymptotic. This negative statement was mathematically proven in Luo & Sarnak 1995 [886]. In figure 9.13 we see a picture taken from Sarnak 1995 [1095], comparing, for plane regions, arithmetic and the nonarithmetic spectra (see more on page 292 for the definition of arithmeticity in abstraction, but it is not really too much different for plane domains, and the plane domains are accessible to numerical computations).

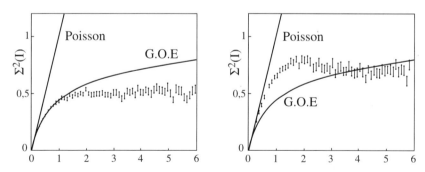

Fig. 9.13. (a) Σ^2 for a nonarithmetic triangle (b) $\Sigma^2(L)$ for an arithmetic triangle

However the geodesic flow is ergodic. Today it is believed that the distribution will be GOE for generic Riemann surfaces. And to explain the reason why arithmetic forms are exceptional, one should remember what was said in §9.9, namely that the jumps in the spectrum are linked with the structure of the length spectrum. But one knows that the length spectrum of an arithmetic form is very "degenerate" in the sense that the lengths are given by suitable integers—the reason for this is that we saw that the length shifts of gliding hyperbolic isometries are represented by the trace of an integral entry matrix. The asymptotic exponential behavior (see equation 9.19 on page 396) then forces all of these periodic geodesics to have very large multiplicities, hence huge jumps in the length spectrum.

In these results, precise descriptions of many quantities are studied for those space forms, not only the L^2 norms but also the sup norm. More: the behaviour of integrals like

$$\int_M P(\phi_{i_1}, \ldots, \phi_{i_k}) \, dV_M$$

for various polynomials P of degree k and their asymptotic behavior when one or more of the eigenvalues goes to infinity is related to possible scarring, which is the next problem we have to consider.

9.13.2.1 Scars

This is linked with the question of whether there are "scars." In some numerical experiments, people found that the nodal lines of some surfaces were, in some sense, accumulating along periodic geodesics. But in Sarnak 1995 [1095] it is proven that this can never happen for arithmetic space forms (for some suitable definition of what a scar is). A picture of a scar in a planar region is presented in figure 1.100 on page 93. This is an amusing paradox: the arithmetic case implies more regularity, and at the time it is a less common case (in the realm of space forms). The general state of affairs still divides experts, since scarring today is only purely experimental and because the definition of scars varies between authors; see Rudnick & Sarnak 1996 [1074], Shimizu & Shudo 1995 [1132] and the references there.

9.14 The Spectrum of Exterior Differential Forms

From equation 9.3 on page 379 we know that there is a sensible notion of Laplacian for exterior forms of any degree p from $p = 0$ (for functions) to the dimension $p = d = \dim M$. This time the kernel of Δ, i.e. the set of differential forms ω such that $\Delta \omega = 0$, is more subtle than for functions. From theorem 405 on page 665 we know that those forms, called *harmonic*, build up in degree p a real vector space isomorphic through the de Rham isomorphism 34 on page 171 to the cohomology space $H^p(M, \mathbb{R})$, hence of dimension equal to $b^p(M)$, the real p Betti number of M. This for the kernel of Δ. But Δ on

9.14 The Spectrum of Exterior Differential Forms

p-forms also has a spectrum, namely the set of its eigenvalues. We explain now how much information can be extracted, with our present state of knowledge, from the knowledge of the spectra for all degrees; we will denote by $\lambda_{p,k}$ the eigenvalues of exterior degree p.

There are today only two outcomes of spectral considerations for exterior forms which have a Riemannian geometry flavor; the Kähler case is richer and was briefly alluded to separately in §§9.10.4 above. First with McKean & Singer 1967 [910] a firework was ignited and in its brightness one could see with far greater clarity. We describe it briefly—a complete reference is Gilkey 1995 [564] (this second edition is very up to date). Roughly speaking what happens is the following. We look back at the asymptotic expansion in theorem 165 on page 393 for $\sum_k \exp(-\lambda_k t)$ with the U_k integrals, which are universal in the curvature tensor (this will mean always including its covariant derivatives). People were concerned that $U_{d/2}$ is not a topological invariant as soon as $d > 2$. Since differential forms also have a canonical Laplacian, we can do the same (it is not too much more expensive and appeared first in Gaffney 1958 [536]) with differential forms and get the pointwise invariants, denoted by $u_{p,k}(x)$ arising in the t^k term in the asymptotic expansions of the corresponding heat kernels. They are still universal in the curvature, but differ in general with various p. Their integrals over M will be denoted by capitals $U_{p,k}$ and the eigenvalues of the p-spectrum by $\{\lambda_{p,k}\}$. Now let us perform the alternate double sum

$$\sum_{p,k} (-1)^p \exp(-\lambda_{p,k} t) .$$

Because both of the operators d and d^* commute with the Laplacian Δ, they transform eigenfunctions into eigenfunctions. The Hodge decomposition theorem 406 on page 665 of any form into a harmonic part, a closed part and a coclosed part shows that into this alternate summation everything will disappear except at the harmonic level: there the zero eigenvalue $\lambda_{p,0}$ has a multiplicity equal to the pth Betti number $b^p(M)$. So in the alternating sum of the corresponding asymptotic expansions everything should also disappear for any k except when $k = d/2$. Hence the alternate pointwise sums

$$\sum (-1)^p u_{p,k}(x) ,$$

when integrated on M and adding after multiplication by t_k, will yield identically the constant

$$\sum (-1)^p b_p = \chi(M).$$

This explains McKean and Singer's dream: a fantastic pointwise cancellation might well take place in the pointwise $u_{p,k}$ functions to yield the forced integrated cancellation. This was indeed proven in Patodi 1971 [1005].

The rebound was taken first in Gilkey 1973 [562] and then in Atiyah, Bott & Patodi 1973 [76, 77]. One studies Patodi's cancellation result, but puts it in successively more general bundles equipped with suitable elliptic operators, including the Dirac operator on spinors and uses Gilkey's results. It then turns out that those structures are plentiful enough to yield all elliptic operators, giving a new proof of the index theorem in §§14.2.3. It is important to use the theory of invariants "à la Gilkey" and the functorial behaviour of indices. The harvest is large: Hirzebruch's signature theorem 417 on page 717 can be obtained this way and of course this new insight yields many results in differential topology. This domain is still blooming; see the two books already mentioned. One point in this philosophy is that "pointwise cancellation" shows that local index theorems can exist. But Riemannian geometry is quite far away. However here comes the second byproduct of the rebound: the η invariant.

The main trick in the founding papers Atiyah, Patodi & Singer 1975–1976 [81, 82, 83] is to obtain the characteristic $\chi(M)$, not as the alternating sum of the zero eigenvalues of the various Laplacians on the exterior forms of a given degree on (M,g), but in one shot as the index of the first order operator $B = d - d^*$ acting on the total set of exterior forms on M (one just has to be careful to put the right signs in front of B). The eigenvalues of Δ are of the form λ^2 where λ is an eigenvalue of B but different signs are possible here. Hence the function

$$\eta(s) = \sum_{\lambda \neq 0} \operatorname{sign}(\lambda)|\lambda|^s$$

makes sense for suitable s. In an strict sense (as usual for this kind of function) $\eta(0)$ is not defined, but with some extra work one can still make sense out of it. It is then called the η invariant of (M,g) and measures the "spectral asymmetry." This invariant is especially interesting for manifolds with boundary. For a $4k$ dimensional manifold M' with a $4k - 1$ dimensional boundary M (and provided that locally at the boundary the metric is a product) one can express the signature $\sigma(M')$ by the integral formula

$$\sigma(M') = \int_{M'} L(R) - \eta(M)$$

where L is the universal curvature integrand for the signature of Hirzebruch's theorem 417 on page 717. This invariant has many applications when looking at the subtle problem of the nonexistence of pointwise invariant integration formulas for the "signatures." Besides the original papers we refer the reader to Atiyah, Donnelly & Singer 1983 [78, 79] and Gilkey 1995 [564]. There are also relations with the secondary characteristic classes below, also with $\hat{A}$ genus when spinors are in view. The η invariant for 3-manifolds is applied in deriving the isolation result of Rong 1993 [1064] for the minimal volume in dimension 4 seen in equation 11.6 on page 518. The η invariant is also used in number theory: see Atiyah, Donnelly & Singer 1983 [78]. For η invariants of noncompact manifolds, see Hitchin 1996 [721]; for gluing and the η invariant see Bunke 1995 [278].

Another invariant based on the spectral analysis of differential forms is to be found in Ray & Singer 1971 [1052]. The result is that from the linear combination

9.14 The Spectrum of Exterior Differential Forms   429

$$\sum_{p=0}^{\dim M}(-1)^p p\zeta'_p(0)$$

of the $\zeta'_p(0)$ value of the ζ_p functions associated to the spectrum of the differential forms of all degrees p, one can recover a topological invariant. They conjectured that their invariant should coincide with the topological invariant called the *Reidemeister torsion* and gave some evidence for that. The conjecture was proven independently in Müller 1978 [952] and Cheeger 1979 [332]. The proof is very involved and was one of Cheeger's motivation for the study of the spectrum of certain singular manifolds, see Cheeger 1983 [333].

Do not hope that the knowledge of the differential form spectrum for all p from 0 to the dimension will determine the metric; in Milnor's examples discussed on page 417 all of those spectra coincide. For various questions concerning isospectrality of differential forms, see Gornet 1998 [582]. See Lott 2000 [881] for a subtle study of collapsing and the behaviour of differential forms.

10 Riemannian Manifolds as Dynamical Systems: the Geodesic Flow and Periodic Geodesics

Contents

- 10.1 Introduction 432
- 10.2 Some Well Understood Examples 436
  - 10.2.1 Surfaces of Revolution 436
  - 10.2.2 Ellipsoids and Morse Theory 440
  - 10.2.3 Flat and Other Tori: Influence of the Fundamental Group 442
  - 10.2.4 Space Forms 446
- 10.3 Geodesics Joining Two Points 449
  - 10.3.1 Birkhoff's Proof for the Sphere 449
  - 10.3.2 Morse Theory 453
  - 10.3.3 Discoveries of Morse and Serre 454
  - 10.3.4 Computing with Entropy 456
  - 10.3.5 Rational Homology and Gromov's Work 458
- 10.4 Periodic Geodesics 461
  - 10.4.1 The Difficulties 461
  - 10.4.2 General Results 463
  - 10.4.3 Surfaces 466
- 10.5 The Geodesic Flow 471
  - 10.5.1 Review of Ergodic Theory of Dynamical Systems 471
- 10.6 Negative Curvature 478
  - 10.6.1 Distribution of Geodesics 481
  - 10.6.2 Distribution of Periodic Geodesics 481
- 10.7 Nonpositive Curvature 482
- 10.8 Entropies on Various Space Forms 483
  - 10.8.1 Liouville Entropy 485
- 10.9 From Osserman to Lohkamp 485
- 10.10 Manifolds All of Whose Geodesics are Closed . 488
  - 10.10.1 Definitions and Caution 488
  - 10.10.2 Bott and Samelson Theorems 490
  - 10.10.3 The Structure on a Given S^d and $\mathbb{KP}^n$ 492
- 10.11 Inverse Problems: Conjugacy of Geodesic Flows 495

10.1 Introduction: Motivation, Problems and Structure of this Chapter

Following the transition presented in chapter 8, it is now quite natural to study the geodesic behaviour of a Riemannian manifold. For local metric geometry it is natural because geodesics are locally the shortest paths. This point of view was treated in §6.5. But geodesics of any length are of interest for the geometer. Another strong motivation for the study of geodesic dynamics comes from mechanics. Since Riemannian manifolds provide a very general setting for Hamiltonian mechanics, with their geodesics being the desired Hamiltonian trajectories, we are of course interested in their behaviour for any interval of time (any length). This perspective mixes dynamics and geometry and is extremely popular today. People always want to predict the future, more or less exactly. We will comment more on this below. Dynamics plays an ever larger role in geometry, even in very simple contexts such as the study of pentagons and Pappus theorems (see Schwartz 1993,1998 [1114, 1115] and the important work d'Ambra & Gromov 1991 [426]. Note also that Gromov 1987 [622] introduced dynamical systems into the study of discrete groups.

Let us have a panoply of problems. We note the obvious dichotomy: a geodesic can be periodic, and then we know exactly what happens when the time goes to infinity, or it can fail to be periodic, and then it can have many different possible behaviours: accumulating along some periodic geodesic, or becoming everywhere dense (in space, or in phase) everywhere or only in some subset of the manifold.

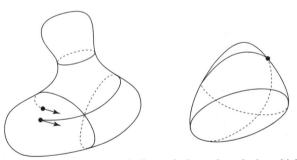

periodic geadesics and geodegics which get "lost"

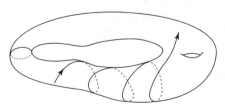

Fig. 10.1. Periodic geodesics and geodesics which "get lost"

10.1 Introduction

We think that the best motivation is to quote the masters. We begin with Henri Poincaré.

Dans mes Méthodes nouvelles de la Mécanique céleste *j'ai étudié les particularités des solutions du problème des trois corps et en particulier des solutions périodiques et asymptotiques. Il suffit de se reporter à ce que j'ai écrit à ce sujet pour comprendre l'extrême complexité de ce problème; à côté de la difficulté principale, de celle qui tient au fond même des choses, il y a une foule de difficultés secondaires qui viennent encore compliquer la tâche du chercheur. Il y aurait donc intérêt à étudier d'abord un problème où on rencontrerait cette difficulté principale, mais où on serait affranchi de toutes les difficultés secondaires. Ce problème est tout trouvé, c'est celui des lignes géodésiques d'une surface; c'est encore un problème de dynamique, de sorte que la difficulté principale subsiste; mais c'est le plus simple de tous les problèmes de dynamique; d'abord il n'y a que deux degrés de liberté, et puis si l'on prend une surface sans point singulier, on n'a rien de comparable avec la difficulté que l'on rencontre dans les problèmes de dynamique aux points où la vitesse est nulle; dans le problème des lignes géodésiques, en effet, la vitesse est constante et peut donc être regardée comme une des données de la question.*

M. HADAMARD *l'a bien compris, et c'est ce qui l'a déterminé à étudier les lignes géodésiques des surfaces à courbure opposées; il a donné une solution complète du problème dans un memoire du plus haut intérêt. Mais ce n'est pas aux géodésiques des surfaces à courbures opposées que les trajectoires du problème des trois corps sont comparables, c'est au contraire aux géodésiques des surfaces convexes.*

J'ai donc abordé l'étude des lignes géodésiques des surfaces convexes; malheureusement le problème est beaucoup plus difficile que celui qui a été résolu par M. HADAMARD. *J'ai donc dû me borner à quelques resultats partiels, relatifs surtout aux géodésiques fermées qui jouent ici le rôle des solutions périodiques du problème des trois corps.*

Poincaré 1905 [1034]

We also add

It seems at first that this fact [the existence of periodic solutions] could not be of any practical interest whatsoever... [however] what renders these periodic solutions so precious is that they are, so to speak, the only breach through which we may try to penetrate a stronghold previously reputed to be impregnable.

Henri Poincaré as quoted in Wayne 1997 [1244]

10 Geodesic Dynamics

Wayne adds few lines after

> However, contrary to what Poincaré's quotation might suggest, these periodic solutions are not only of theoretical interest but also have many practical applications.
>
> Wayne 1997 [1244]

We follow on with Jacques Hadamard:

> En second lieu, l'importance que ce géomètre a reconnue aux solutions périodiques, dans son Traité de Mécanique céleste, s'est manifestée également dans la question actuelle. Ici encore, elles se sont montrées « la seule brèche par laquelle nous puissions essayer de pénétrer dans une place jusqu'ici réputée inabordable. »
>
> D'une façon plus précise, elles ont joué pour nous le rôle d'une sorte de système de coordonnées auquel nous avons rapporté toutes les autres géodésiques.
>
> Hadamard 1898 [674]

Hadamard's indispensable text deserves comments. First let us recall that it is in this paper that the von Mangoldt–Cartan–Hadamard theorem 72 on page 255 was proven for surfaces. It is profitable to consider the result the author is alluding to when he says « une sorte de système de coordonnées auquel nous avons rapporté toutes les autres géodésiques. »

What he proves is that a typical aperiodic geodesic travels along beside some periodic one for quite a while, then travels through the manifold and again accumulates along another periodic geodesic, etc. This gives a coding for describing the geodesic flow; in fact Hadamard is the pioneer of the conception of coding in dynamical systems. Moreover, when the initial direction of the geodesic changes, the coding changes dramatically, as does the nature of the geodesic. Hadamard is a pioneer of the modern notion of chaos; moreover he realized that this picture remains accurate in on compact hyperbolic surfaces, as he explained in the article Hadamard 1898 [675] on hyperbolic billiards.

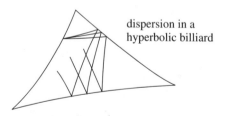

Fig. 10.2. Dispersion in a hyperbolic billiard table

The field of geodesic dynamics presents us with a host of natural questions. Periodic geodesics are the steady states of our mechanical object, and their lengths are the analogues of the frequencies of vibration: the square roots of the eigenvalues of the Laplacian. So one is led to ask:

1. the asymptotic behaviour of the lengths of the periodic geodesics and
2. the geometric distribution (in phase space UM or in configuration space M) of these geodesics.

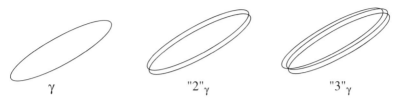

Fig. 10.3. (a) γ (b) "2γ" (c) "3γ"

We have implicitly assumed that there are infinitely many periodic geodesics; as we will see below, this is an open question. There is also the notion of geometrically different geodesics: turning more than once along a given periodic geodesic is not considered a different geodesic by a geometer, even if it might be in some sense different for a mechanics expert. The field of geodesic dynamics is dramatically different from that of spectrum geometry. The basic reason is that eigenfunctions are the critical points of the Dirichlet quotient on the (infinite dimensional) vector space of functions on the manifold. Periodic geodesics are the critical points of the length function on the space of all closed curves of the manifold. Sadly enough, this is not a vector space but an infinite dimensional manifold: one cannot play linear algebra with periodic geodesics.

Facing a Riemannian manifold with an infinite number of periodic geodesics, the basic question (comparing with the asymptotic expansion for the spectrum in chapter 9) is to understand the asymptotic behaviour, as the length L goes to infinity, of the counting function $CF(L)$ which is, by definition, the number of (geometrically distinct) periodic geodesics of length smaller than equal to L.

As for the spectrum, we naturally encounter both direct and indirect problems concerning the geodesic flow.[1] Given a Riemannian manifold, with some special conditions e.g. on the curvature, we would like to have some information on the geodesic flow. In particular, is the geodesic flow ergodic? Conversely, we will try to deduce geometric data from knowledge of the geodesic flow. In particular, the uniqueness question:

Question 199 *Are two manifolds with conjugate (identical) geodesic flows necessarily isometric?*

As before, we will restrict our enquiries to compact manifolds unless explicitly stated.

---

[1] See page 359 for a precise definition of the geodesic flow.

There are few books treating the geodesic flow as their main topic. One can look at the semi-popularization Berger 1994 [166] for a general treatment of geodesic flow on surfaces. For general dimensional manifolds, there is the very good survey Bangert 1985 [110]. There is also a book on the subject of periodic geodesics: Klingenberg 1978 [814]. But beware that the claim in it of the existence of at least three geometrically different periodic geodesics is still not proven, see the Russian translation Klingenberg 1982 [815] for a systematic detection of errors. The book Buser 1992 [292] is the essential reference for surfaces of constant negative curvature. The text book Klingenberg 1982 [815] discusses many topics related to geodesics, especially on ellipsoids, but should also be read with care. The geodesic flow appears as a special case of the results in Mañé 1987 [891] and also in the books of the Dynamical Series of the Russian/Springer Encyclopedia of Mathematical Sciences, for example Bunimovitch, Cornfeld, Dobrushin, Jakobson, Maslova, Pesin, Sinaĭ, Sukhov, & Vershik 1989 [277]. A more recent work, Katok & Hasselblatt 1995 [787], covers a large number of topics in differentiable dynamics. And there are two very new works in this field: Paternain 1999 [1002] and Knieper 1999 [822]. A special mention should be made of the very specialized Besse 1978 [182]. Concerning the bibliography and credits we will not be too detailed, in order to make the text lighter, but the origins of any material presented here can be found in the various references above.

Let us explain the table of contents of this chapter, which might look surprising. §10.2 is devoted to various manifolds where geodesic behaviour is quite well in hand, both for standard and nonstandard metrics. §10.3, starting with Birkhoff's proof of the existence of at least one periodic geodesic on any convex surface, will employ convex geometry as a typical example to introduce and recall the basics of Morse theory, which is the basic tool to obtain periodic geodesics. But we will see that this theory is better adapted to the study of geodesics joining two given points. The results on this problem will be explained and serve to point out the difficulties in extending Morse theory to handle periodic geodesics. §10.4 is devoted to what we know today about periodic geodesics: existence of an infinite number, and more: the asymptotic behaviour of the counting function. §10.5 will treat the geodesic flow.

10.2 Some Well Understood Examples

10.2.1 Surfaces of Revolution

10.2.1.1 Zoll Surfaces

From page 30, we know the geodesics of S^2: they are the great circles. Two questions come to mind: what are all the surfaces for which

1. all of the geodesics are simple curves, having the same length?

2. all of the geodesics emanating from a point focus at an antipodal point after the same interval of time?

There is a great deal known about these questions today. We will treat these geometric inverse problems in arbitrary dimensions in detail in §10.10. However, we will now treat these problems on surfaces in order to taste in advance some flavour of the topic, and also to gather some feeling for the difficulty of studying periodic geodesics.

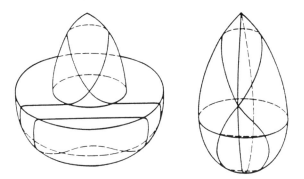

Fig. 10.4. Zoll's surface with singularities

Zoll 1903 [1310] found surfaces of revolution, even real analytic ones, which are not round spheres and yet all of their geodesics are periodic, simple and of the same length. We will call a surface enjoying these properties a *Zoll surface* (whether it is of revolution or not). In fact, the search for Zoll surfaces

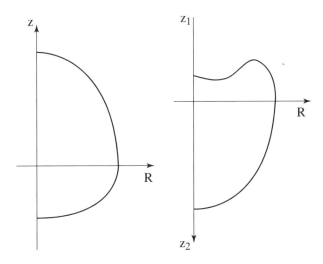

Fig. 10.5. A meridian of a smooth Zoll surface

438   10 Geodesic Dynamics

was already considered and their multitude partially demonstrated by Darboux around 1880. This is possible essentially because, as we saw in §§1.6.2, geodesics of surfaces of revolution can be obtained by integrating a numerical function given by the meridians. The proof still needs some clever change of variables, but no more. This was finally completely mastered by René Michel in 1978. He completely classified the Zoll surfaces of revolution. They are as many as any odd numerical functions from the interval $]-1,1[$ into itself. So Riemannian geometry has many "harmonic oscillators." Recall that the physicists use the term "harmonic oscillator" for the motion of a point under an attraction from a fixed point by a force proportional to the distance. All of the trajectories are ellipses (hence closed) and with the same period.

Why are Zoll surfaces of revolution so simple? Remember from §§1.6.2 that geodesics in a surface of revolution oscillate between two parallels. In fact, starting horizontally from the upper parallel and after meeting horizontally again the lower parallel, any geodesic comes back horizontally to the initial parallel, although not in general at the same point, but with a turn of some angle α (see figure 10.6), an angle which is the same for all geodesics with this same parallel (say of latitude λ). By the above integrability of the geodesic flow, the numerical function $\alpha(\lambda)$ can be explicitly computed by an integral involving the function defining the meridian (hence the surface). We just want for a Zoll surface to have $\alpha(\lambda)$ vanishing identically for all λ. This is an integral equation for the meridian, and Zoll solved it by a clever change of variables. In Darboux's time people could only manage to solve this for a partial range of λ, hence only "bands" of Zoll surfaces. Michel's trick is a cleverer change of variable. One difficulty was the singularity appearing at the north and the south pole.

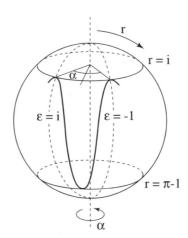

Fig. 10.6. Geodesics on surfaces of revolution

10.2 Some Well Understood Examples

And now for Zoll surfaces not of revolution. To attack this problem the idea is to look at deformations of the standard sphere, and linearize the problem by taking the derivative at the origin of a one parameter family of metrics. Using the conformal representation theorem 70 on page 254, and an easy formula for the derivative of the length, Funk 1913 [534] found that any Zoll deformation of the standard sphere is necessarily given by a function $S^2 \to \mathbb{R}$ all of whose integrals along all great circles have to vanish. This is certainly true if the function is odd, namely $f(-x) = -f(x)$. Funk proved the converse and it is now recognized that was the birth of the so-called *Radon transform*. Funk raised a conjecture which he could not prove: that any odd function gives rise to a one parameter family of Zoll surfaces whose derivative at the origin is the given function. This was achieved by Guillemin 1976 [664]. A basic tool is to prove that there is a Radon transform theory for every Zoll surface. For the Radon transform on the sphere, the basic references are Helgason 1980 [702] and Helgason 1984 [703].

So today we know something about the set of Zoll surfaces: its tangent space at the origin (where "the origin" means the standard sphere). But globally we are lost. It is even difficult to prove that complete periodicity implies a common period and the simplicity of geodesics (no self-intersections). This is true and proven in Gromoll & Grove 1981 [602] which is the only paper we know of on the subject. By the way: this paper uses the Lusternik–Schnirelman theorem 232 on page 466. The first question should be

Question 200 *Is the Zoll set connected?*

We have forgotten about the additional antipodal property of the sphere. It is trivial to see that an antipodal Zoll metric on S^2 (one for which the antipodal map is an isometry, and for which the geodesics are periodic) is equivalent to giving a metric on the real projective plane $\mathbb{RP}^2$ all of whose geodesics are periodic. This is the problem called the "Wiedersehenfläche Problem" by Blaschke in the first edition of Blaschke 1921 [203]. There is an interesting deadly false solution of it in the second edition Blaschke 1924 [204] and the error is pointed in the third edition Blaschke 1930 [205]. The real proof waited until 1963 to be found by Leon Green. It is elegant, mixing Santalo's integral geometry formula already seen in equation 7.11 on page 364 and §§§7.1.2.3 with the equation of Jacobi fields and the Gauß–Bonnet formula 28 on page 138. The conclusion then is twofold: an antipodal Zoll metric on S^2 has to be the standard one and a Zoll metric on $\mathbb{RP}^2$ must be the standard one (the elliptic geometry). Readers who like projective plane geometry can enjoy this. It is important to mention that this result is one of the few which are true in Riemannian geometry but false in Finsler geometry, as shown by Skorniakov in 1955. Blaschke's conjecture also exists in higher dimensions, and is proven as theorem 257 on page 493.

10.2.1.2 Weinstein Surfaces

Consider geodesics of a surface not as curves on the surface, but as flow lines of a vector field on the unit tangent bundle. To appreciate that such a viewpoint is possible, we can simply write out the geodesic equation, and think of it this way. Now we may ask what portion of the unit tangent bundle is taken up by periodic geodesics. For example, for geodesics of the two dimensional torus, the unit tangent bundle is a three dimensional torus, and the geodesic flow consists of straight lines. Those which are periodic are in fact dense. On the sphere, all geodesics are periodic, so again the periodic geodesics are dense in the unit tangent bundle. Moreover, in theorem 243 on page 478, we will see that all negatively curved compact Riemannian manifolds have periodic geodesics dense in their unit tangent bundle. So naturally, one conjectures that this is always the case. Weinstein 1970 [1248] found counterexamples as follows. From §§1.6.3 we have many surfaces of revolution of constant unit Gauß curvature, hence they are locally isometric with the standard sphere. In such a surface whose meridian looks like figure 1.61 on page 51 in the case of singular points of the axis of revolution, keep the portion S between two of these conical points p and q and deleting a very tiny part (of revolution) around p and around q. Replacing (smoothly) these two parts by caps of revolution, one finally creates a differentiable convex surface of revolution, say M. Now look at the equator and its length L. By results from §§1.6.3, the number L can be any number smaller than 2π and two cases are possible: either $L/2\pi$ is rational or it is irrational. From our discussion in §§1.6.2, we know the behaviour of the geodesics which oscillate inside S, namely those which start with a horizontal velocity contained in S and (by the results of §§1.6.2) stays forever inside S. From the geometry of the standard sphere, we know that if $L/2\pi$ is rational, then all of these geodesics are periodic with the same period L, while if $L/2\pi$ is irrational, they never close up. In this latter case, we have a large open set (of almost full measure) in UM such that any geodesic starting with velocity vector in it is not periodic, so disproving the former conjecture. Note that this does not prove that the periodic geodesics are not dense in space, i.e. in the surface M, since all of the meridians cover M. Note also the dramatic changes in the global picture of the flow when L changes continuously.

10.2.2 Ellipsoids and Morse Theory

We saw in §§1.6.2 how geodesics behave on an ellipsoid. They are very similar to the geodesics on surfaces of revolution. Geodesics oscillate between pairs of lines of curvature, except that they are of two different types. Those coming back on the same curvature line are in fact described by a numerical parameter α (an angle in the so-called *action-angle coordinates*). And the geodesics live in bands, of periodic geodesics if α is rational and everywhere dense in the band if α is irrational. This time α varies with the line of curvature, so that we get

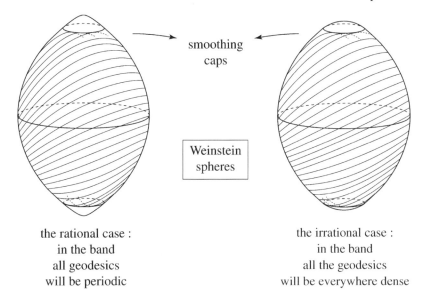

Fig. 10.7. Weinstein spheres: **(a)** The rational case: in the band, all geodesics will be periodic. **(b)** The irrational case: in the band, all of the geodesics will be everywhere dense

lots of periodic geodesics. The strict counting function is then always infinite. The counting function for the bands seems not to have been computed.

In the 1920's, Marston Morse proved a very interesting theorem; see the proof in Klingenberg 1982 [815], and if possible see the second edition, Klingenberg 1995 [816], section 3.4.7 (of course it is also found in Morse's work [see reference in Klingenberg's book]). One can guess Morse's result from the above behaviour. Assume we have an ellipsoid which is very close to a sphere. The three axis are very close to unit length. Then the above numbers α will be very close to zero. But a small nonzero rational number has to have a huge denominator. Hence the associated periodic geodesic will be very long. Together with Weinstein's examples this example shows that the behaviour of periodic geodesics, and of the geodesic flow, can be very weird and very sensitive.

Theorem 201 (Morse 1934) *Given any L (think of L as very large) there is an $\varepsilon > 0$ such that any three axis ellipsoid whose axis lengths are strictly between 1 and $1 + \varepsilon$ has all of its periodic geodesics of length larger than L except for the three sections by the coordinate planes.*

The proof of Morse's theorem is "elementary." When one writes the equation of a periodic geodesic in coordinates, the three projections are Sturm–Liouville equations:
$$y'' + fy = 0 \, .$$

which must be periodic equations. One uncovers a contradiction when one has three different axis. The theorem extends to any dimension d as applied to the $d(d+1)/2$ plane sections of the ellipsoids

$$\sum_{i=1}^{d+1} \frac{x_i^2}{a_i^2} = 1 .$$

One might call these very long geodesics *ghosts*; you see them less and less as your surfaces gets closer and closer to a sphere. Another surprising theorem concerning periodic geodesics will be found in theorem 203 on page 444.

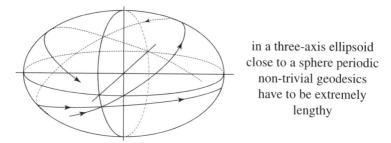

in a three-axis ellipsoid close to a sphere periodic non-trivial geodesics have to be extremely lengthy

Fig. 10.8. All but three of the periodic geodesics on a nearly spherical three-axis ellipsoid are very long

10.2.3 Flat and Other Tori: Influence of the Fundamental Group

10.2.3.1 Flat Tori

Flat tori were defined on page 185. We look first, for simplicity's sake, at the two dimensional tori. On the flat two-dimensional torus $\mathbb{R}^2/\Lambda$ defined by the lattice Λ, geodesics are the projections of the straight lines in $\mathbb{R}^2$. Their behaviour is completely understood. The periodic geodesics downstairs correspond exactly to the lines upstairs which go from the origin 0 to some point in the lattice Λ. One can suppose we are on a square torus $\mathbb{R}^2/\mathbb{Z}^2$ since affine transformations of $\mathbb{R}^2$ preserves lines. If the slope of the initial velocity vector is irrational then the geodesic in the torus is everywhere dense and moreover evenly distributed just as the mod 1 integral multiples of an irrational number. How fast this geodesic fills the torus is given by the continued fraction expansion of our irrational number. The fastest filling is given by the golden ratio. If the slope is rational then the geodesic is periodic. Note that then there is a continuous one parameter family of periodic geodesics associated to a given one by sliding it transversely (bands).

So at this stage one can say that the counting function $CF(L)$ (which is the number of periodic geodesics of length smaller than L) is infinite for any

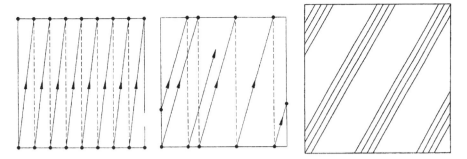

Fig. 10.9. (a) Rational slope = periodic geodesic **(b)** Irrational slope = everywhere dense geodesic **(c)** A band of geodesics

$L \geq 1$. This will be our point of view in this chapter: we will discard the estimation of $CF(L)$ as soon as one continuous band exists. In particular we then do not care about the difference between the various lattices Λ. However, for its own interest, we note here that the number of bands can be easily estimated: it is asymptotically quadratic in L and the first term is given by the area of Λ (its determinant) multiplied by some constant. This constant is different than the one for the spectrum, because here we want geometrically distinct bands. In dimension two, we have to divide the count of the number of points of Λ sitting inside the circle of radius L by the sum

$$\sum_n \frac{1}{n^2} = \frac{\pi^2}{6} = \zeta(2)$$

to take into account that only coprime pairs of integers will yield geometrically different bands. We saw in §§1.8.3 that the order of the next term is still an open question in number theory: the "circle problem."

For higher dimensional flat tori $\mathbb{R}^d/\Lambda$, the situation is basically the same: periodic geodesics come in bands, while nonperiodic ones are everywhere dense (evenly, moreover) in a subtorus of dimension ranging from 2 to d. For the counting problem one has only to replace

$$\frac{\pi^2}{6} = \zeta(2)$$

(the Riemann zeta-function) by

$$\zeta(d) = \sum_n \frac{1}{n^d} \ .$$

10.2.3.2 Manifolds Which are not Simply Connected

Let us consider now any Riemannian metric on the torus $T^d = \mathbb{R}^d/\mathbb{Z}^d$, and let us look again for periodic geodesics. In the case of a torus T^2 embedded in

$\mathbb{R}^3$ it looks plausible that, if it exists, the smallest curve C in the class of all curves which are obtained by deformations of the one in the picture should be a periodic geodesic. The proof is easy: for any two points of C close enough we know, by §§6.1.3 for example, that C is a piece of geodesic joining these two points. The deformations alluded to mean that we consider curves in a fixed free homotopy class of the fundamental group of T^2. We note here that a free homotopy class is a conjugacy class of the fundamental group $\pi_1(T^2)$.

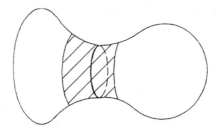

Fig. 10.10. Even on simply connected Riemannian manifolds, the minimum argument still yields a periodic geodesic

Now the difficulty is to prove the existence of a curve of minimal length. The space of curves is a space of infinite dimension, and compactness arguments are not available. Such a curve does exist however, and can be shown to exist quite cheaply for any compact manifold and by various techniques; see any modern book on Riemannian geometry, or prove it yourself (use coverings, deck transformations and the first variation formula 6.3 on page 225). So finally we have the simple but fundamental result, already noted in §§§7.2.1.1:

Theorem 202 *Any nontrivial free homotopy class γ on a compact Riemannian manifold contains at least one curve of minimal length, and it is a periodic geodesic.*

This result is often attributed to Hilbert, but it is already in Hadamard 1898 [674] for surfaces. For a proof in general metric spaces, see Choquet 1966 [377], VI.3.11.

When the set of free homotopy classes, denoted by $\pi_{free}(M)$, is large, or infinite, then we have many periodic geodesics. One has only to be careful about the geometric difference; we will see how to do that for tori just below. One might say: if you master the structure of π_{free}, you master the periodic geodesics. However it was discovered recently that things are in fact much more complicated. The following was conjectured by Gromov:

Theorem 203 (Nabutovsky 1996 [958]) *There are groups G so that if a compact manifold M has fundamental group $\pi_1(M) = G$, then for any Riemannian structure on M, the counting function $CF(L)$ (which counts num-*

bers of contractible periodic geodesics of length at most L) has exponential growth.

A closed curve is called *contractible* when it belongs to the trivial free homotopy class, so that in particular such periodic geodesics cannot be found by just taking those of minimum length. The idea of the proof is simple, but shows the intrusion of algorithmics in Riemannian geometry. The groups under consideration are groups given by generators and relations, such that it takes "an exponential number of steps to check that an element given by a word is in fact the neutral element." The complete proof is very interesting, as it needs (besides the injectivity radius) Cheeger's finiteness theorem 372 on page 616, Nash's isometric embedding theorem 46 on page 218, as well the Nash–Tognoli theorem 31 on page 161 that any submanifold of $\mathbb{R}^N$ can be approximated as closely as desired by connected components of semi-algebraic submanifolds.

An important remark, even if trivial:

Lemma 204 *If one has exponential growth of the counting function for periodic geodesics, then there is no need to worry about geometric distinctness since iteration of a periodic geodesic is linear with respect to its length.*

10.2.3.3 Tori, not Flat

Let us apply these ideas to a torus with any metric. The fundamental group is $\pi_1(T^d) = \mathbb{Z}^d$, hence
$$\pi_{free}(T^d) = \mathbb{Z}^d .$$
For simplicity's sake let us work first on T^2 and let us try to apply theorem 202 on the facing page to find an infinite number of periodic geodesics and if possible evaluate the counting function. A simple argument based on the compactness of T^2 yields a constant c such that a periodic geodesic whose homotopy class in $\mathbb{Z}^2$ is the couple (p,q) of integers has a length L smaller than $c(p^2 + q^2)$. By the argument in §§10.2.3.1 for the flat torus, we might think that the counting function $CF(L)$ is quadratic in L. But there is the problem of avoiding overcounting geodesics which are not geometrically different. Now here we are saved by the following remark: if p and q are coprime integers, then a periodic geodesic in the class of (p,q) cannot be an iterated one of some smaller periodic geodesic (one says that such a geodesic is *primitive*). And now, as for flat tori, the number of pairs (p,q) which are coprime and with $p^2 + q^2 < R^2$ behaves quadratically in R^2. In conclusion, for any metric on T^2, the counting function $CF(L)$ is at least quadratic in L as we saw in §§10.2.3.1. By the same token, for any dimension and any metric on T^d we have an infinite number of (geometrically distinct) periodic geodesics and $CF(L)$ grows at least like L^d.

But, even for T^2, today at least, there is not a single metric for which the exact asymptotic order of $CF(L)$ is known. Note that the flat case is disregarded because of the presence of continuous bands of geodesics. The exponential–generic result of theorem 227 on page 465 does not apply–the torus is a homogeneous manifold. It is not clear for tori, or many other manifolds not covered by theorem 227 if generically $CF(L)$ grows exponentially. The geodesic flow on tori can be extremely subtle; see Bangert 1988 [111].

10.2.4 Space Forms

10.2.4.1 Space Form Surfaces

Having treated the sphere and the torus, we look now at orientable surfaces of higher genus; we note here that the nonorientable case does not have a much different behaviour, since we can look at the oriented covering. The unorientable surfaces are never discussed in the literature since the primary interest of mathematicians in surfaces is in the study of one complex variable, number theory, algebraic geometry, etc. where all of the surfaces are oriented. We consider first the case of constant curvature, which we can take to be equal to -1. We expect many periodic geodesics because we know both that there is a periodic geodesic in any non-zero free homotopy class and that from algebraic topology the fundamental group is huge, even a free one. Even without this knowledge one can expect the counting function to be exponential by the following heuristic argument. We look at the number of geodesic loops based at some given point m. Lifted in the universal covering one has to count some number of points in a ball of radius L in the hyperbolic plane and this is exponential in L like the area of discs in Hyp^2. This is also in disguise the growth function of the fundamental group. But we are interested in free homotopy classes, for which one needs to divide the fundamental group by the equivalence relation associated to conjugation. The answer is still exponential and one has:

$$\lim_{L \to \infty} \frac{\log CF(L)}{L} = 1 . \tag{10.1}$$

This was first found independently by H. Huber and A. Selberg around 1955. It can be seen as a group problem or a geometric one. Both are well treated in Buser's book [292] and we refer the reader to it for an idea of the proof. Do not forget lemma 204 on the preceding page. However the above result should shock the reader for two reasons. There are many space forms of dimension 2. Topologically they are of different genera, and for a given genus they can have different constant curvature -1 structures (moduli, Teichmüller space).

The independence from the genus is the most striking, since pictures seem to show more periodic geodesic when there are more "holes," or larger π_1; or because the volume gets larger and larger with the genus, namely the volume is $2\pi(\gamma - 1)$, in view of theorem 28 on page 138. In fact we already mentioned in §§9.13.1 that the genus and the spectrum both appear (inextricably linked) when one looks at the next order in the asymptotic expansion of $CF(L)$. Here is the flavour of the promised formula for the next order of $CF(L)$:

10.2 Some Well Understood Examples

Theorem 205 (Huber 1959 [745], Buser 1992 [292] page 257) *Write x for*
$$x = e^L.$$
For any eigenvalue λ write
$$s = s(\lambda) = \frac{1}{2} + \sqrt{\frac{1}{4} - \lambda}.$$
Then
$$CF(L) = \text{li}(x) + \sum_{3/4 < s < 1} \text{li}(x^s) + O\left(\frac{x^{3/4}}{\log x}\right)$$
where the notation li *stands for*
$$\text{li}(x) = \int_2^x \frac{dt}{\log t}$$
and the summation is performed over the eigenvalues λ of the spectrum such that
$$\lambda < \frac{1}{4}$$
and
$$\frac{3}{4} < s(\lambda) < 1.$$

A glance at the formula explains why we cannot give any heuristic idea, and only refer to Buser's book. For the behaviour of nonperiodic geodesics, which is very well understood, as well as the manner in which the periodic geodesics are located in the surface, see §10.6 and §10.8. The behaviour of geodesics is completely understood in space forms. The geodesic flow is ergodic; see §10.5. In particular, almost all geodesics are everywhere dense. Moreover their paths in the phase space (the unit tangent bundle) are dense as well. Besides periodic geodesics and the preceding densely winding ones, a measure zero set of geodesics is made up by those which are asymptotic to one or more periodic geodesics.

How about a general Riemannian metric on a compact orientable surface of genus $\gamma \geq 2$? We have the best possible answer:

Theorem 206 (Katok 1988 [786]) *For any Riemannian metric on an orientable surface of genus higher than 1 and having the same area as when the curvature is constant and equal to -1, namely area $2\pi(\gamma - 1)$, the counting function satisfies*
$$\liminf_{L \to \infty} \frac{\log CF(L)}{L} \geq 1$$
and equality only if the curvature is constant.

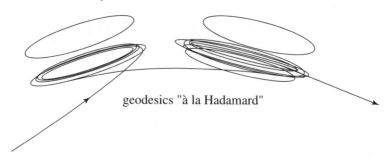

Fig. 10.11. Geodesics *"à la Hadamard"*

We already commented above on the fact that the genus does not appear in the formula for $CF(L)$. For notions of entropy and more results of Katok, see theorem 250 on page 483.

The proof makes elegant geometric use of the conformal representation theorem (see theorem 70 on page 254). Thereby the lengths of curves can be compared as measured by the metric under consideration and by the constant curvature one, by integrals of numerical functions. From this result one finds that a space form is an unbelievably sensitive creature: the smallest finger pressure at any place will make the manifold have a much larger number (a larger exponential factor) of periodic geodesics of given length, a sort of susceptibility to periodic geodesics. We will see in theorem 251 on page 484 that one has also an optimal result for the way the periodic geodesics are distributed inside the surface.

10.2.4.2 Higher Dimensional Space Forms

A generalization of Katok's theorem is probably true for all the space forms, strictly or of general type, see §6.6. For the moment one has a partial result, which is a direct corollary of theorems 251 on page 484 and 239 on page 476.

Corollary 207 *Consider any compact space form of negative curvature (M, g_0), i.e. a locally symmetric metric g_0 on a compact quotient of one of the standard $\mathrm{Hyp}_{\mathbb{K}}^n$, and consider any metric g on it of negative curvature and with the same volume as g_0. Then their respective counting functions satisfy:*

$$\limsup_{L \to \infty} \frac{\log CF(L|g)}{L} \geq \limsup_{L \to \infty} \frac{\log CF(L|g_0)}{L}$$

and moreover equality holds only in cases of isometry—not only must g be locally symmetric but globally isometric to g_0.

So our space forms are as sensitive in any dimension as in dimension two: any change, even very small, will immediately produce exponentially more periodic geodesics. It is open today to know if the negativity of the curvature is really necessary—remember it was unnecessary on constant curvature surfaces. Note also that the uniqueness fails on surfaces because of the moduli of Riemann surfaces. The negative curvature condition comes simply from the same condition in theorem 239 on page 476. To have a result for space forms of nonpositive curvature we want an extension of theorem 250 on page 483 to them, see below for this and more on the structure of the geodesic flow on these manifolds.

The symmetric metrics on negative curvature space forms have very subtle geodesic behaviour. For example, starting in dimension 3, one might wonder if it is possible to have all of the periodic geodesics simple (i.e. without self-intersection) or knotted. This is possible, but it requires subtle number theory; see Chinburg & Reid 1993 [373].

Do not forget to look at §10.6 for the theory of negatively curved manifolds.

10.3 Geodesics Joining Two Points

10.3.1 Birkhoff's Proof for the Sphere

The present section is intended to provide both motivation and an introduction. We look back into the past to the problem to find at least one periodic geodesic on a compact surface. This is trivial by theorem 202 on page 444 with the sole exception of the only simply connected surface, namely the sphere.

The sphere was tackled for the first time by Poincaré 1905 [1034]. This is a fascinating text, motivated by celestial mechanics because the sphere equipped with an arbitrary metric is the simplest object whose mathematics mimics the mathematics of the solar system. Poincaré tried to get existence of at least one periodic geodesic and this by three different approaches. See more about Poincaré's approaches in §§10.4.3.1. He got trapped in his three approaches by convergence problems in infinite dimensional spaces. The existence of a single periodic trajectory was first obtained in Birkhoff 1917 [195]. For many reasons the proof deserves to be presented to a larger audience. We consider the set of all smooth *tapestries* of the sphere as in figures 10.12 and 10.13 on the following page.

On this set of tapestries we hope that there is a minimax principle: we want to believe that there is at least one tapestry for which the length of the longest curve of the tapestry is as small as it is for any other tapestry. This is plausible because our tapestries can never be torn open since they are smooth. Birkhoff had two tricks to overcome the infinite dimensional nature of the space of tapestries. The first is to deform any curve into a curve made up of small geodesic pieces. Just divide the curve into pieces of length smaller that the injectivity radius.

Then have the following strictly decreasing operation on the set of these "broken geodesics:" take the middles of the pieces and connect them again by a geodesic segment. The strict triangle inequality (see equation 6.5 on page 226)

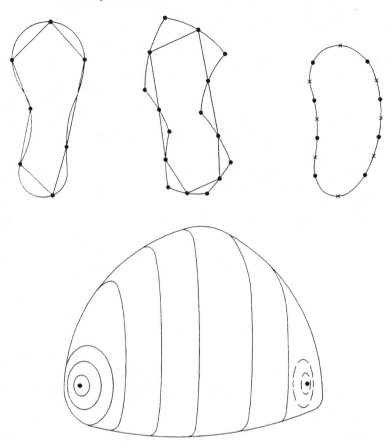

Fig. 10.12. A tapestry and some broken geodesics

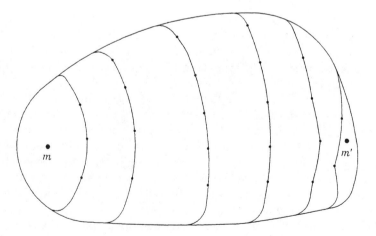

Fig. 10.13. A tapestry

10.3 Geodesics Joining Two Points

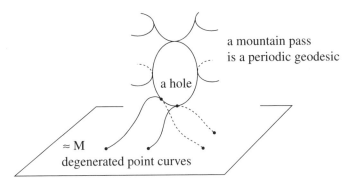

Fig. 10.14. A mountain pass is a periodic geodesic

tells you that this operation strictly decreases the length of a curve unless the curve is already a periodic geodesic. The existence is now finished because the set of broken geodesics is finite-dimensional. One never ends with a degenerate point geodesic because tapestries are not trivial topologically. This can be seen as follows: a tapestry is a curve in the set of all closed curves (broken geodesics or smooth curves, these sets are the same because of the deformation operation above) and it starts and ends at individual points. Moreover a tapestry is not homotopic to zero among curves joining two points in the set of degenerate curves made of points of the surface. Then the minimax result is just proving that somewhere we have a "mountain pass" as in figure 10.13 on the facing page.

By the way, Birkhoff's geodesic need not nicely behaved, say simple, i.e. without self- intersection. It might be neither simple nor the smallest periodic one. Use *strangulation* and a *three-legged starfish* as in figure 10.15 on the next page

Calabi & Cao 1992 [305] proved that Birkhoff's geodesic is indeed both simple and the smallest periodic geodesic when the curvature is positive. The proof is beautiful geometry and uses the second variation formula 6.7 on page 242. However we will see below that a simple geodesic always exists for any metric.

There cannot be some cheaper proof which works for example in more general metric spaces as can be seen from the following two examples.

Theorem 208 (Gal'perin 1995 [544]) *Most tetrahedra do not admit any simple periodic geodesic (see figure 10.16 on the next page).*

The argument can be recovered by the reader. Look at the holonomy around a periodic geodesic (here made up of line segments). Because we are on a surface with flat sides, a simple geodesic will yield a relation between the vertex angles—just deform a simple geodesic by parallelism and remark that the holonomy does not change.

The second example:

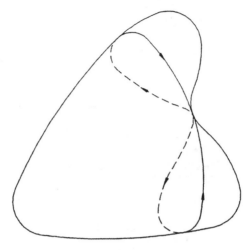

Fig. 10.15. A three-legged starfish

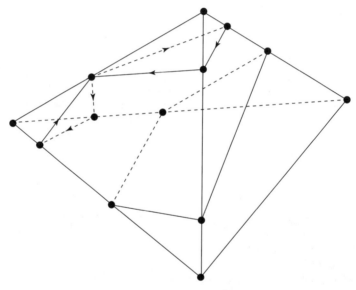

Fig. 10.16. Tetrahedra do not usually bear a closed simple geodesic

Theorem 209 (Gruber 1991 [659]) *Generic convex bodies do not admit periodic geodesics.*

Note that our two counterexamples are convex, which is quite a strong condition.

10.3.2 Morse Theory

In brief, Morse theory as a starting point is the essence of Birkhoff's proof. To obtain one periodic geodesic the fundamental fact is that the space of all closed curves of the sphere has a "hole in it," i.e. one can find in it a loop which is not contractible to a point, so that we have mountain pass points, which are periodic geodesics. Morse put this in a very general frame, in order in particular to get, not only one, but a lot of periodic geodesics. His central contribution to the subject is the book Morse 1934 [944], but a modern classic is Milnor 1963 [921]. For the huge realm of applications of the theory see Bott 1988 [228].

The setting for Morse theory is the couple made up by a compact differentiable manifold M and a numerical smooth function f on it. We will see later on how to apply Morse theory to geodesics. A critical point $m \in M$ is a point where the differential of f is zero : $df(m) = 0$. When M is a surface in $\mathbb{R}^3$ and f is the latitude, the critical points are the points where the tangent plane to M is horizontal. Note that we are working in a general manifold, and no metric is needed

At a critical point m, the second differential d^2f of f makes sense and is a quadratic form because $df(m) = 0$ and the chain rule is then applicable to change coordinates. One says that a critical point m is *nondegenerate* if $d^2f(m)$ is not degenerate as a quadratic form.[2] The *index* of the critical point m is the maximal dimension of vector subspaces of T_mM on which the restriction of d^2f is negative definite. Morse's fundamental theorem is:

Theorem 210 (Morse) *If f has only nondegenerate critical points, then the number of critical points of index equal to the integer i is larger than equal to the Betti number $b_i(M, \mathbb{K})$ over any field $\mathbb{K}$.*

The idea of the proof is this: first pick a metric, and when there is no critical point between two level sets, flow the manifold along the gradient of the function. This will not change the topology of the part of the manifold located below those levels. If one has to pass a critical point, the effect in the topology is to "add a handle" of a degree equal to the index of the critical point. The proof is pictured in figure 10.17 on the following page, the figures being taken from Milnor 1963 [921]. This makes very precise the intuition of the picture: many holes imply many horizontal points. We note here that the values $f(m)$ at the critical points have no relation with their indices; this will be dramatic in the future. The proof, besides a little play with algebraic topology, consists in two steps: the first is to prove that the topology of the "low" level parts $f^{-1}(]-\infty, L])$ of the manifold do not change topology when L runs into in interval where there is no critical point. We mention this, because

---

[2] Nondegeneracy of a quadratic form Q means that if we write $Q(v+w) = Q(v) + Q(w) + P(v,w)$ with $P(v,w) = P(w,v)$, then there is no v for which every w satisfies $P(v,w) = 0$,

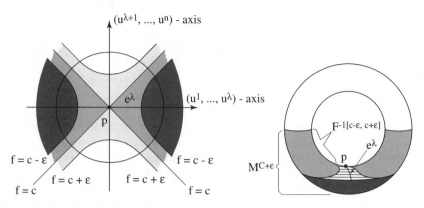

Fig. 10.17. (a) Nondegenerate critical point of a function on the plane, index equals 1 (b) Sublevel sets of a nondegenerate function

a basic Riemannian geometry extension of this result will be introduced in §§12.2.2. The second step is to study what happens when L passes through a critical point: the effect on the topology is to add one cell whose dimension is equal to the index of this critical point.

10.3.3 Discoveries of Morse and Serre

We want now to apply theorem 210 on the previous page to obtain "many" periodic geodesics. We will begin with geodesics joining two points. It will be soon clear why that case is much easier to master. We want now to apply the above Morse theory to get geodesics joining two given points p, q in some compact Riemannian manifold M. The idea is to take for our "space" $\Lambda'_{p,q} = \Lambda'_{p,q}(M)$ the set of all smooth curves from p to q and for our function f we take the length of such a curve. Three things have to be done. First we need to get off of $\Lambda'_{p,q}$ and on to a finite dimensional manifold. This is done as in Birkhoff's proof by considering only curves made of successive geodesic segments. This is no problem because we know that our manifold always has a positive injectivity radius, by proposition 88 on page 271. And an easy geometric deformation shows that this new $\Lambda_{p,q}$ has the same topology as the initial $\Lambda'_{p,q}$. To ensure that we are always on a compact manifold. we restrict to a compact subset of $\Lambda_{p,q}$, by working temporarily only with curves of length shorter than some fixed bound. We remark that critical points of length are "pure" geodesics: the strict triangle inequality shows that the angles at the breaks have to be all equal to π.

The third thing to do is to understand nondegeneracy and compute the index. This was also achieved by Morse:

Theorem 211 (Morse's index theorem) *A geodesic γ from p to q is nondegenerate if and only if the point q is not conjugate to p on γ (see defini-*

replacing curves by
broken geodesics
as in Birkhoff
for closed curves

Fig. 10.18. Replacing curves from p to q by broken geodesics as Birkhoff did for closed curves

tion 85 on page 268). And its index is equal to the weighted sum of the number of points q_i conjugate to p on γ, the weight being the dimension of the space of Jacobi fields along γ which vanish at p and at that point q_i.

Ignoring for the moment the nondegeneracy condition, to get many geodesics from p to q we need to have information on the Betti numbers of the loop space $\Lambda_{p,q}$. Note that all $\Lambda_{p,q}$ have the same topology, namely that of the set of loops $\Lambda = \Lambda_{p,p}$ (for any p, they are all the same). Morse could only get very partial results. One of the basic results of Serre 1951 [1124] (which made him famous overnight) is

Theorem 212 (Serre 1951 [1124]) *For any compact manifold M, there is an infinite number of integers k for which $b_k(\Lambda(M)) \neq 0$.*

For a geometer the proof is too nice not to be quoted. Fix any $p \in M$ and use the fibration
$$\Lambda_{p,p} \to \Lambda_{p,\cdot} \to M \tag{10.2}$$
which maps into M the end of any curve starting from p. Since $\Lambda_{p,\cdot}$ retracts trivially on the point p and hence has trivial topology, the fundamental class of the compact manifold M by the spectral sequence relating the topologies of the base and of the fiber should "transgress" an infinite times to be finally completely killed. This forces the result.

The point p being given in M, the set of points q which can be conjugate to p along some geodesic is of measure zero: this comes from the fact that in T_mM we have zero measure for them and then now also downstairs since the exponential map is differentiable. If one remarks moreover that one has compactness for geodesics of bounded length, one finally obtains

Theorem 213 *There is an infinite number of geodesics joining any pair of points in any compact Riemannian manifold.*

Any reader will be shocked if she looks at the standard sphere. The obtained geodesics cover the same great circle but of course have different indices. We eventually want to have geometrically different geodesics, i.e. covering different ground. It seems to us that the following naive question is open today:

Question 214 *What are the manifolds for which the set of pairs of points not joined by infinitely many geometrically distinct geodesics is of full measure?*

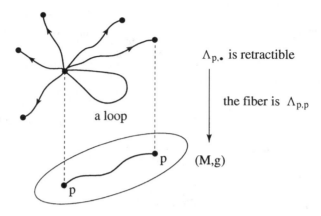

Fig. 10.19. Transgression of the loop space

Certainly the spheres and the $\mathbb{KP}^n$ are counterexamples, but are there other ones?

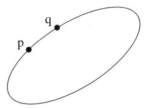

Fig. 10.20. Can this happen for many pairs of points?

10.3.4 Computing with Entropy

We will now consider a very nice recent result which says that the topological entropy (of the geodesic flow, see §§§10.5.1.2) is the mean value over all pairs of points of the exponential factor in the growth of the counting function $CF(L|p,q)$.

Theorem 215 (Mañé 1994 [892], Paternain & Paternain 1994 [1003])

$$h_{top}(M,g) = \lim_{L \to \infty} \frac{1}{L} \int_{M \times M} \log CF(L|p,q) \, dV_M(p) \, dV_M(q)$$

for every compact Riemannian manifold.

The proof is very geometric, working on the product manifold $M \times M$. There the set of geodesics joining two points becomes the set of geodesics

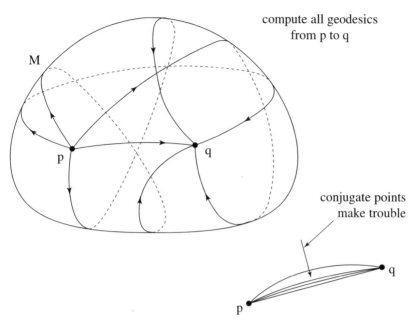

Fig. 10.21. (a) Compute all geodesics from p to q (b) Conjugate points make trouble

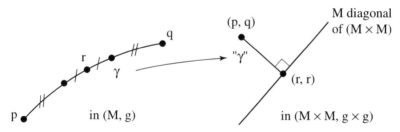

Fig. 10.22. M sits in $M \times M$ as the diagonal. A geodesic in M from p to q with midpoint r is represented as a geodesic in $M \times M$ from (p,q) to (r,r)

initiating from the diagonal. To obtain this start from the midpoint of the geodesic on M and travel an equal distance in both directions. This idea is basic already in Grove & Petersen 1988 [648]. Then $CF(L|p,q)$ is nothing but the volume of the cylinder of radius $L/2$ around the diagonal, this volume being taken with the measure on $M \times M$ which is the product of the canonical measure on the diagonal by the orthogonal measure obtained by taking the inverse measure under the exponential map of the measure on M. To get the entropy at the end ones uses a hard theorem of Yomdin 1987 [1299] (see also Gromov 1987 [621]) where the computation of the entropy uses differentiability. Yomdin's result stands as an exception in dynamical system theory in the sense that it uses essentially smoothness (the proof involves algebraic geometry notions).

For the Riemannian geometer, this is a simple definition of entropy. A corollary of theorem 215 on page 456 is that for many pairs of points p and q the counting function $CF(L|p,q)$ is exponential when h_{top} is positive. We will see in §10.9 that many metrics on many compact manifolds enjoy positive topological entropy.

Open questions today:

Question 216 *On a negatively curved compact manifold, can one still have the same formula with only one pair (p,q) (say for almost every pair)?*

Question 217 *The same question but for generic metrics.*

One need not wait for an answer for the same question applied to arbitrary metrics: the very inspiring Burns & Paternain 1996 [287] exhibits a metric on the two dimensional sphere for which

$$\limsup_{L\to\infty} \frac{1}{L} \log CF(L|p,q) < h_{top}$$

for (p,q) in a set of positive measure.

Question 218 *Is it true that for any pair of points p and q*

$$\limsup_{L\to\infty} \frac{1}{L} \log CF(L|p,q) \leq h_{top} ?$$

The answer is also no here—Burns & Paternain 1997 [288] exhibit metrics on S^2, arbitrarily close to the standard metric, with a point p never conjugate to itself along any geodesic loop from p to p and such that $CF(L|p,p)$ grows as quickly as desired. Both constructions are very geometric.

Final but important remarks: conjugate points cause no problem in the integral in theorem 215 on page 456 because points q conjugate to a given point p are of measure zero. Secondly, the iterates of a given geodesic have the same geometric support but their length grows only linearly, so they do not contribute after taking the logarithm (see lemma 204 on page 445).

10.3.5 Rational Homology and Gromov's Work

We come back to the problem of estimating the counting function of a manifold using algebraic topology. The infinity of Serre's theorem was not enough. Moreover the dramatic point is that Morse theory is useless as it stands for estimating the counting function, since it gives information on critical points of a function in terms of their indices but not in terms of the value of the function. So there is apparently no way to get from the knowledge of the

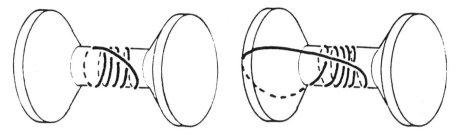

Fig. 10.23. The skein technique

Betti numbers of $\Lambda(M)$ any information on the counting function $CF(L|p,q)$ which counts the number of geodesics joining two points p and q and of length smaller than L. The situation was rescued by the pioneering result which can be found in Gromov 1978 [612], Gromov 1981 [616], Gromov 1999 [633]:

Theorem 219 (Gromov) *For every simply connected Riemannian manifold M there are two positive constants a and b such that, for every point p and q one has*

$$CF(L|p,q) \geq \frac{a}{L} \sum_{i \leq bL} b_i(\Lambda(M), \mathbb{K})$$

with $\mathbb{K}$ any field.

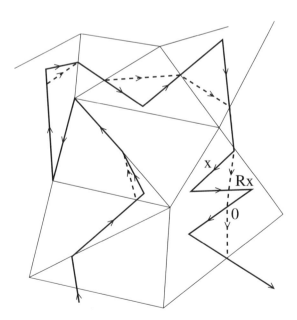

Fig. 10.24. Gromov's approach to geodesics via convex triangulation

The proof is illustrated by the two pictures in figure 10.23 on the previous page. One should attempt to retract every geodesic to some fixed segment from p to q. This is certainly possible by the simple connectedness. But the retraction we have to perform should control, for a given length bound, the i topology of $\Lambda(M)$ at the same time. The first idea is pictured on the left: unwrap a geodesic twisted many times very carefully, by contracting successively turn after turn, not trying to contract all of the turns together; if not carefully handled, the length and the index will grow in an uncontrolled fashion. This "skein" technique will be meet again in §§§10.4.3.2. But Gromov has a more powerful technique. One starts with a given convex triangulation of M and, as in classical Morse theory, one works only with broken geodesics (using the positivity of the injectivity radius). Now, to any broken geodesic from p to q, one associates the sequence of simplexes it enters into, as well as the points of entrance and exit, the exit coinciding with the next entrance. This yields a triangulation of $\Lambda(M)$ as follows: the simplices are made up by the broken geodesics which, in the preceding coding, have entrance (and hence exit) points varying in the same faces of a succession of initial simplices of M. The resulting dimension of such a simplex is just d times the number of simplices met by the geodesics. And clearly this number is proportional, with a fixed constant b, to the length L. One point was skipped in the above reasoning, namely that the geodesics traveling along the edges of the simplices of M will give trouble, since they yield only dimension 0. The way out of this problem is to contract this 1-skeleton of the given triangulation of M into a point. This is where the simply connectedness of M is used. Then one has to control everything in $\Lambda(M)$ during the contraction, which is not too hard. Finally note that the factor of $1/L$ in the formula is put there in order to get geometrically distinct geodesics. For example, in the case of the standard sphere, where the Betti numbers of $\Lambda\left(S^d\right)$ are bounded, the formula will yield only one geodesic, as we saw previously.

It remains now to ask algebraic topologists what they know about the $b_k\left(\Lambda(M)\right)$ and the results are wonderful, even if not completely finished. We will see in theorem 343 on page 592 that compact differentiable manifolds are divided, by the behaviour of their rational homology $H_*\left(M, \mathbb{Q}\right)$, into exactly two classes—the first are called *rationally hyperbolic*, the other *rationally elliptic*. The Betti numbers of the rationally hyperbolic manifolds grow exponentially, so that one has the strong statement:

Theorem 220 *For any metric on any rationally hyperbolic manifold and all pairs of points p and q the counting function $CF(L|p,q)$ grows exponentially, with exponent moreover which can be estimated from below by the Betti numbers of the loop space.*

We recall here again that one is sure to count geometrically distinct geodesics, since the iterates have length growing linearly—one does need the bumpy metric hypothesis here. Since rationally hyperbolic manifolds are the majority, and a generic manifold is rationally hyperbolic, one sees that, generically in the realm of manifolds, the counting functions $CF(L|p,q)$ grow exponentially for any pair of points.

For not necessarily rationally hyperbolic manifolds, one knows from Ziller 1977 [1307] for symmetric spaces and McCleary & Ziller 1987 [906] for general homogeneous spaces that, for any metric, the counting function grows polynomially, but if one wants to be sure of geometrically distinct geodesics, one should subtract 1 from the degree of the polynomial, since the coverings grow linearly. For symmetric spaces the degree of the polynomial growth of the Betti numbers of the loop space is precisely equal to the rank r, hence $CF(L|p,q)$ has growth at least polynomial of degree $r-1$.

Note 10.3.5.1 (Topology of positive Ricci curvature manifolds)
Sadly this point of view cannot be use to study the topology of manifolds with positive Ricci curvature, one of the main open problems in Riemannian geometry (see §§12.3.2). This precisely because the fact that Ricci curvature implies a growth of the indices of geodesics at least linear in length is superseded by theorem 219 on page 459. One can consider theorem 219 as a successful step in Gromov's program to quantify algebraic topology—another example will be seen in §§12.3.3. ♦

10.4 Periodic Geodesics

10.4.1 The Difficulties

As already mentioned, encouraged by Fourier analysis on compact Riemannian manifolds which is provided by the Laplacian and its eigenfunctions, say in quantum mechanics, we would like to do the same for Hamiltonian/classical mechanics. The parallel being that to eigenfunctions, the stationary states, correspond periodic geodesics. A Fourier analysis based on periodic geodesics would need first an infinite number of such objects. And to Weyl's asymptotic equation 9.20 on page 397 would correspond an estimate of the counting function $CF(L)$ for periodic geodesics, where

$$CF(L) = \#\{\text{periodic geodesics of length smaller than or equal to } L\}.$$

Moreover the $\lambda^{d/2}$ should naturally correspond to an exponential type, the exponent to be computed we would imagine via some Riemannian invariants as it was with only the volume for Weyl's asymptotic. We emphasize here that one is still missing the existence of an infinite number of geometrically different periodic geodesics for any Riemannian manifold, the most spectacular first case being that of S^3. We will now explain the difficulties and then what is known today, beside of course the cases already examined.

The naive idea, absolutely natural, is to imitate the successes of Morse–Serre theory but to replace the $\Lambda_{p,q}$ by the set $\Omega(M)$ of loops in M, i.e. the set of smooth maps $S^1 \to M$ from the circle into M. On the loop space we have the numerical function given by the length of the image curve and periodic geodesics are nothing but once again the critical points of this function. There is no problem, within the class of closed curves of bounded length to replace this infinite dimensional space by a finite dimensional one, again using broken geodesic polygons (whose edge lengths are smaller than the injectivity radius). And we hope that the topologist will be able to compute the Betti numbers of $\Omega(M)$ "à la Serre." The first difficulty: the computation works well for the set $\Omega_*(M)$ of pointed maps from S^1 into M, using the obvious fibration associated fibration

$$\Lambda(M) \to \Omega_*(M) \to M \ .$$

For a compact manifold this will yield an infinite number of nonzero Betti numbers for $\Omega_*(M)$. We have first to take care of few things which are not too difficult. But to go from $\Omega_*(M)$ to $\Omega(M)$ one has to take the quotient by the orthogonal group $SO(2)$ and the quotient has singularities. This is a trap in which many mathematicians fell and obtained incomplete or false results. This is the first reason for the lack of decisive results for periodic geodesics. We will not go into details of the various ways people got around that trap; see the literature for that.

The second problem is even more important and can be suspected from the Morse–Serre result. One can still prove that, in some sense, $\Omega(M)$ has infinite topology for any compact manifold by working with the spectral sequence of the fibration

$$\Lambda(M) \to \Omega_*(M) \to M \ .$$

Thus we always have an infinite number of periodic geodesics in any compact Riemannian manifold, but they can all be the iterates of a single one and hence geometrically identical.

The third trouble is the nondegeneracy hypothesis in Morse's main theorem 210 on page 453. First the notions of index and of degeneracy for periodic geodesics are more delicate to handle than for geodesics joining two points; we will not enter into these fine points, because it cannot be defined simply with conjugate points—one has also to consider the second variation formula for nonpointed curves. Second, when the metric is given the periodic geodesics are given, "they stay at their place" and one cannot, as for the (p,q) case, play the game of looking at nondegenerate ones and from them get every pair by a limit argument. We will see below that one way to bypass this difficulty is to restrict statements to generic metrics.

And here again, for the counting function (if there is one) we have the fact that Morse theory gives no relation between the lengths and the indices. On the other hand the difficulty is practically seen on the Weinstein example in §§§10.2.1.2 and Morse's "ghost" geodesics in theorem 201 on page 441. In fact there is a much worse example, due to Katok, which shows that something quite specific to Riemannian geometry must be used—the conjunction of Morse theory and algebraic topology on the loop space $\Omega(M)$ is definitely not enough; see Ziller 1983 [1308] for even more examples and §§14.5.8 for Finsler metrics:

Theorem 221 (Katok) *There are Finsler metrics on S^3 which admit no more than two periodic geodesics.*

Katok's examples are frightening, because general Morse theory applies as well to Finsler spaces. So even when the Betti numbers of the loop space are huge, one has really to play with some geometry specific to the Riemannian case to get periodic geodesics. The above difficulties explain why most results are quite recent. The first proof of the existence of at least one periodic geodesic is attributed to Lusternik & Fet 1951 [887], though the technique was essentially Birkhoff's, where the notion of degeneracy is not needed as it was for Morse's approach. We will now describe the state of affairs today. General references for this topic were given in §10.1.

10.4.2 General Results

10.4.2.1 Gromoll and Meyer

We now look more carefully at the same time at Morse theory and at the geometry of the manifold. Indices here matter a lot. One should decouple if possible indices and homology classes, looking at the behaviour of the indices of the various iterates of periodic geodesics. Here is the current state of the art in analyzing the behaviour of the index of the iterates:

Theorem 222 (Gromoll & Meyer 1969 [606]) *Given a periodic geodesic γ, write its iterates as γ^k. There are constants $a(\gamma), b(\gamma), c(\gamma)$ so that*

$$a(\gamma) - kb(\gamma) \leq \text{index}\left(\gamma^k\right) \leq a(\gamma) + kb(\gamma)$$

for every integer k.

We already said above that the difficulty in proving such a theorem is that the index of a periodic geodesic is a subtle object. In the case of curves with both ends fixed the counting of conjugate points immediately would yield an estimate as above. An initial partial result was found in Bott 1956 [227], explained in the book Klingenberg 1978 [814] section 3.2.15.

Using Bott's result one obtains the strongest result available today for general metrics

Theorem 223 (Gromoll & Meyer 1969 [606]) *Let M be a compact manifold with finite fundamental group. Assume there is at least one field $\mathbb{K}$ such*

that the Betti numbers $b_k\left(\Omega\left(M\right),\mathbb{K}\right)$ are not bounded as a function of k. Then any Riemannian metric on M carries infinitely many geometrically distinct periodic geodesics.

The idea is to use theorem 222 on the previous page to the effect that the index of the iterates of a periodic geodesic grows only very close to linearly when iterating it. A trivial counting argument will show that if we have only a finite number of geometrically different periodic geodesics then the number of periodic geodesics of a given index k will be bounded in k. The nondegeneracy restriction in Morse's theorem is bypassed by the authors by a "technical tour de force" and is the essential difficulty of the paper.

We are left with the pure algebraic topology question:

Question 224 *Which compact smooth manifolds M have all Betti numbers of their loop space $\Omega\left(M\right)$ bounded, and this for every field?*

This is a nice question for modern topologists. It is not completely settled today, but we have a good partial answer (compare with theorem 343 on page 592).

Theorem 225 (Sullivan and Vigué-Poirrier 1975 [1167]) *The condition in question 224 implies that the rational cohomology ring $H^*\left(M,\mathbb{Q}\right)$ has only one generator.*

And so we are back to our favourite manifolds: spheres S^d and $\mathbb{KP}^n$ because they enjoy that condition. And look at the paradox: for their canonical metrics all of their geodesics are periodic. However there are sadly enough other candidates for possible or eventual bad behaviour of periodic geodesics. First there are plenty of $\mathbb{Q}$-spheres, and second there are manifolds having the same homotopy type as the $\mathbb{CP}^n$ but not homeomorphic to them, and third there are exotic quaternionic projective planes $\mathbb{HP}^2$. We will meet these manifolds again in §10.10.

Note that theorem 223 on the preceding page yields an infinite number of periodic geodesics, but does not say anything about the counting function for them. In fact, we know that nothing can be said about the counting function, unless some more explicit geometric relations are brought into Morse's theorem.

At the risk of being a little technical about periodic geodesics, the situation is in fact more dramatic. On the one hand as soon as a periodic geodesic is of *twist type*, in any tubular neighborhood of that geodesic there is an infinity of other periodic geodesics, as follows from the KAM theorem (see Moser 1977 [946]). On the other hand one does not know a single example of a manifold with no periodic geodesics of twist type! See Rademacher 1994 [1046] for more on this as well Rademacher 1994 [1045]. Note that for the twist type geodesics, the nearby periodic geodesics have a counting function growing at least like $L/\log L$. The twist type condition is obtained when looking at the Poincaré return map seen on page 469: its eigenvalues should be of modulus equal to 1 and satisfy an extra technical but mild condition (of a generic type in the category of Poincaré maps).

10.4.2.2 Results for the Generic ("Bumpy") Case

Lacking a general statement, we look for generic behaviour (on any manifold, having to include in particular the spheres and the $\mathbb{KP}^n$). The assumption that a metric is generic allows us to apply Morse theory. The notion of genericity for Riemannian manifolds is usually denoted today by the word *bumpy*; there is no unique precise meaning of this word and it is always difficult to define carefully. We will state the "vague"

Theorem 226 *On any compact manifold, "almost every" Riemannian metric is bumpy, and in particular all of its periodic geodesics are nondegenerate.*

We have "decoupled" things as much as possible. We have to be more precise about what "almost every" means and which sort of bumpyness we will ask for. For precise definitions, one can look at Rademacher 1994 [1046]. Historically the first results were in Klingenberg & Takens 1972 [817], Anosov 1982 [54], Klingenberg 1978 [814]. As for which kind of bumpyness, it may be in the various C^r topologies for the set of all C^r metrics on the manifold, with $r \geq 2$. Because there is no nice measure on the set of all C^r metrics, the meaning of "almost every" has to be phrased in the language of residual versus meager sets: a set is said to be *residual* inside a given topological space if it can be expressed as a countable intersection of open dense subsets.

There are two results for generic metrics; the first is

Theorem 227 (Gromov 1978 [610]) *For any bumpy metric on a compact simply connected manifold M, there are constants a and b such that the counting function of periodic geodesics enjoys the inequality*

$$CF(L) \geq \frac{a}{L} \sum_{i \leq bL} b_i\left(\Omega\left(M\right)\right) .$$

In particular, if M is rationally hyperbolic then the counting function $CF(L)$ grows exponentially.

Here again the factor $1/L$ is needed to get geometrically distinct geodesics. Also, as on page 461 one finds explicit results for homogeneous spaces and in particular symmetric spaces. Since most manifolds are hyperbolic, the above can be summed up by saying that, with a double genericity (for the manifold and for its metric), one has an infinite number of periodic geodesics and moreover exponential growth of the counting function. The exponential rate can be estimated from the $b_i\left(\Omega\left(M\right)\right)$. Gromov's result was sharpened:

Theorem 228 (Ballmann & Ziller 1982 [108])

$$CF(L) \geq a \sup_{i \leq bL} b_i(\Omega(M)).$$

For the proof of theorem 227 on the previous page one has only to remark first that Gromov's geometric contraction seen on page 460 works as well in $\Omega(M)$ as in $\Lambda(M)$. Algebraic topologists know that the exponential growth of $b_i(\Lambda(M))$ for rationally hyperbolic manifolds also holds true for $\Omega(M)$. Finally the bumpyness condition enables us to apply Morse theory.

The second result is

Theorem 229 (Rademacher 1994 [1046]) *On any compact manifold every bumpy metric has infinitely many periodic geodesics.*

The proof here uses the full "decoupling" implications of bumpyness. Moreover the proof will probably yield more, namely: the counting function grows at least as a constant times $L/\log L$. But some cases are still to be checked. This $L/\log L$ which we saw above and will see below for surfaces is not mysterious. It comes from the fact that one has to look at prime numbers in a given arithmetic progression. The famous Dirichlet theorem gives precisely such an order of magnitude.

The open questions today are

Question 230 *Is the bumpyness really needed in Gromov's theorem?*

Question 231 *Is the counting function exponential for elliptic manifolds?*

Do not forget to look at §10.6 for the geodesics of negatively curved manifolds.

10.4.3 Surfaces

10.4.3.1 The Lusternik–Schnirelmann Theorem

Theorem 232 (Lusternik & Schnirelmann 1929 [888]) *Every Riemannian metric on S^2 has at least three geometrically different simple periodic geodesics.*

This was claimed in Lusternik & Schnirelmann 1929 [888]. The authors' original idea was to apply Birkhoff's shortening tapestry technique to tapestries of the manifolds using the totality of the set of circles (every circle, not only equators, including points) of S^2, this set being three dimensional.

The approach of Lusternik and Schnirelmann is very treacherous, since again here the problem is not to get three periodic geodesics, but to prove that they are geometrically distinct. The relevant technical questions seem to be settled in favour of this approach, after many incomplete proofs, see

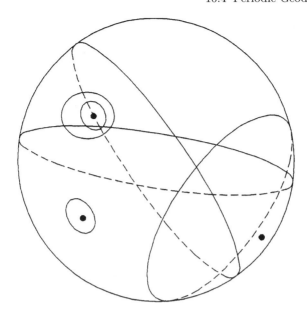

Fig. 10.25. The circles on a sphere

Taimanov 1992 [1177] and the analysis there. It is fair to say that the "intermediate" results brought about important observations. Note that in recent works some authors have employed a deformation technique involving heat equations; see references in Jost 2002 [768]. The idea was first used in the plane in Gage & Hamilton 1986 [537], see §§1.4.5. Deforming a plane curve by normal variations which are proportional to the curvature makes the curve ever closer to a circle (i.e. a curve of constant curvature). In a Riemannian manifold, with suitable modifications to the definition of the flow, the curve will have as a limit a curve of zero geodesic curvature, i.e. a geodesic. The proof is difficult, requiring significant analysis. The equation is a parabolic partial differential equation for which the existence of solutions for small time is quite easy; most of the effort is to show first existence of solutions for any time and second that the limit curve is smooth; see Grayson 1989 [585].

One of the plans of attack proposed by Poincaré to prove the existence of a periodic geodesic on a convex surface was to take the shortest curve dividing the surface into two pieces of total curvature 2π each. Should such a shortest curve exist, it would indeed be a periodic geodesic. After various attempts, in particular by Berger & Bombieri 1981 [177], the existence of such a curve was finally established by Croke 1982 [414]. Hass & Morgan 1996 [683] found a wonderfully simple proof.

10.4.3.2 The Bangert–Franks–Hingston Results

Let us give further consideration to a topological sphere with any Riemannian metric, but look for more that the three periodic geodesics we have found so far, and hope for an infinity of them. When we find such an infinity, then we will want an estimate on the counting function for periodic geodesics of at most a given length. This infinity was obtained only very recently by the conjunction of Bangert 1993 [112] and Franks 1992 [519]. The result was received with enthusiasm by the popular mathematical press, referred to as "a zillion rubber bands around a potato," etc.

Theorem 233 (Bangert–Franks–Hingston) *Every Riemannian metric on the two dimensional sphere has infinitely many geometrically distinct periodic geodesics. Moreover the number of periodic geodesics of length at most L, $CF(L)$, grows at least as fast as*

$$c \frac{L}{\log L}$$

where c is a positive constant.

We now briefly present the ideas behind the proof. The problem naturally splits into three problems. The difficulty of finding a fourth geodesic, let alone an infinite number, is not too much of a surprise considering Morse's theorem 201 on page 441 which shows that a fourth geodesic can be very long and tangled. Birkhoff attacked the problem as follows.

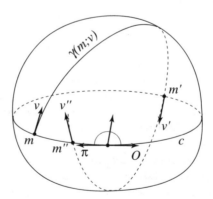

Fig. 10.26. Birkhoff's picture: shooting geodesic rays across a periodic geodesic, following them back to the same periodic geodesic.

We start with a fixed periodic geodesic c and pick a unit tangent vector v at $m \in c$ pointing toward a particular side of c (fix one of the two possible choices). This yields a geodesic entering the inside and getting out again at some point m' (with speed v'). Keep going now on the other side and finally come back to the curve c at a point m'' with speed v'' again pointing inside. This gives a map which we will call the *Poincaré return map*. One gets a new periodic geodesic just exactly when $v'' = v$, i.e. whenever one has a fixed point for the map $v \to v''$. For people familiar with dynamical systems this situation is familiar. The set of possible unit vectors v pointing across c toward the selected side of c is the annulus $c \times]-\pi, \pi[$ and the map $v \to v''$ preserves the measure $ds \wedge \cos\theta\, d\theta$ just by the first variation formula. Since the pioneering work of W. Neumann it is well known that such a map has a infinite number of periodic points (see Arnol'd 1996 [66], appendix 9).

What is faulty with the above reasoning? Two things: (1) the geodesic starting with speed v might never come back to c, it can be *trapped* on one side of c, and (2) the Neumann result applies only to the so-called *rotating* maps. This means that when one looks at the extended map along the boundary of the annulus that this map moves the boundary circles in opposite directions. It is not surprising that this, together with the preservation of measure implies existence of fixed points. The behaviour of this map on the edges of the cylinder is given by the location of the conjugate points along c.

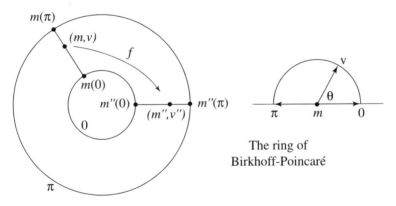

Fig. 10.27. The ring of Poincaré–Birkhoff

The trichotomy: (1) rotating in opposite directions without trapping, (2) not rotating in opposite directions but without trapping, and (3) trapping. Franks 1992 [519] took care of the first case by a technique coming from the theory of Riemann surfaces. An interesting point in the proof is that what Franks proved exactly is:

Theorem 234 (Franks 1992 [519]) *Every area preserving map of an annulus has either no fixed points or infinitely many.*

But here there must be at least two fixed points, thanks to the Lusternik & Schnirelmann theorem 232 on page 466.

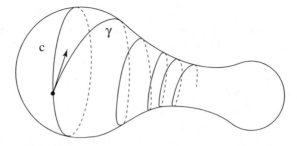

Fig. 10.28. A trapped geodesic crossing a periodic geodesic

The third (trapping) case is solved in Bangert 1993 [112]. It is nice geometry, using tapestries of only one side of a periodic geodesic which is an accumulation set of a nonperiodic geodesic. To use the minimax principle to find geometrically distinct periodic geodesics, he used a "skein" technique as in figure 10.29 (compare with figure 10.23 on page 459).

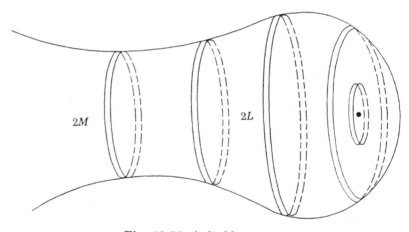

Fig. 10.29. A double tapestry

The estimation of $CF(L)$ as being like $L/\log L$ is in Hingston 1993 [714]. This type of estimate comes from the Dirichlet theorem counting prime numbers in arithmetical progression, these arithmetical progressions coming of course from the lengths of the iterates of periodic geodesics. This has to be done for every case in the trichotomy.

In all the above we forgot the poor projective plane $\mathbb{RP}^2$. But taking its spherical covering, it is clear that he will admit infinitely many periodic geodesics, with a growth at least like $L/\log L$. It still remains an open question to see if this is also a reliable estimate of the number of noncontractible periodic geodesics of length at most L. This question is open, but the answer might be settled affirmatively by working more carefully through the steps of

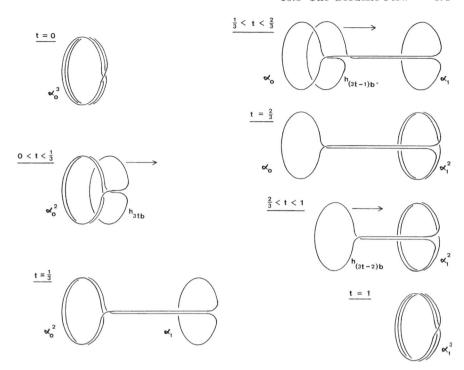

Fig. 10.30. The skein method

Bangert and Franks' work, in particular using the analogue for $\mathbb{RP}^2$ of the Lusternik–Schnirelmann theorem found in Ballmann 1978 [102].

There is not a single explicit surface of the type of the sphere, the projective plane, the torus or the Klein bottle for which the asymptotic order of the counting function is exactly known. No expert seems to even have a guess like: is it always polynomial, exponential, or exponential generically? Note that you can discard the question when a manifold, e.g. surfaces of revolution and ellipsoids, possesses a one parameter continuous family of periodic geodesics (which are then necessarily locally of the same length).

10.5 The Geodesic Flow

10.5.1 Review of Ergodic Theory of Dynamical Systems

10.5.1.1 Ergodicity and Mixing

A very pleasant (but not complete) introduction to ergodic theory is Walters 1982 [1232]. More complete are Arnol'd & Avez 1968 [68], and Mané 1987 [891] (quite dense but more complete for our purposes); also see Katok & Hasselblatt 1995 [787], Anosov, Aranson, Arnol'd, Bronshtein, Grines & Il'yashenko

1997 [55], Sinai 1976 [1140]. To invest today in learning about dynamical systems is very rewarding, considering the pervasive nature of the word *chaos* throughout mathematics and physics. We will skip most references for well known results and concepts.

A discrete dynamical system is a measure space (X,μ) together with a continuous map $f: X \to X$ which preserves the measure μ. In some cases one does not require continuity of the map nor for it to be one to one. What one is really interested in is the behaviour of the iterates f^k when k goes to infinity. Let us recall (see definition 155 on page 358) that the geodesic flow is the set made up by the collection of the maps G^t (t a real number) which act on the unit tangent bundle UM as follows: $G^t(v)$ is the speed vector at the point $\gamma_v(t)$ of the geodesic γ_v whose initial speed vector is precisely v. In all this study a basic fact is the *Liouville theorem* that the flow respects the canonical symplectic structure of UM and hence the attached measure μ. So we have a natural and geometrical dynamical system in Riemannian geometry. We will consider the geodesic flow

$$\{G^t \,|\, t \in \mathbb{R}\}$$

of the unit tangent bundle UM of a Riemannian manifold; this is a continuous dynamical system. We will use both languages, discrete and continuous, to simplify the writing. For the geodesic flow, the link is simply to take the discrete system $f = G^1$ on $X = UM$. It does not matter if the "length" (or time) 1 looks special because in fact we are interested only in what is happening when $t \to \infty$. Note also that we make no explicit distinction between a geodesic in the manifold M and its "image" or "lift" in UM, namely the set of its speed vectors. When speaking of various ergodic theory notions about Riemannian manifolds, we will always understand the dynamical system to be the geodesic flow and work with the Liouville measure on UM.

The most special dynamical systems are the *Bernoulli shifts*. They are defined as the transformations $x_n \to x_{n+1}$ on an infinite product $\prod_{n=-\infty}^{\infty}(X,\mu)$ of copies of a single measure space (X,μ) so that they are extremely simple to describe and work on, since the dynamics is given by a coding. Moreover classes of isomorphic Bernoulli shifts are classified exactly by their metric entropy.

From a physical point of view the basic interest is in the spatial mean values and the time mean values of any physically measurable quantity. We consider a numerical function $F: UM \to \mathbb{R}$ and a point $v \in UM$.

the *time mean value* at v is $F^*(v) = \lim_{L \to \infty} \dfrac{1}{L} \int_0^L F\left(G^t(v)\right) dt$

the *space mean value* is $\langle F \rangle = \displaystyle\int_{UM} F\mu$

Their existence is provided by Birkhoff's theorem:

Theorem 235 (Birkhoff) *The time mean value $F^*(v)$ of a measureable function $F : UM \to \mathbb{R}$ exists almost everywhere (i.e. except possibly on a set of measure zero) and*

$$\int_{UM} F^*(v)\,\mu = \int_{UM} F\,\mu \ .$$

For discrete systems one just replaces the time mean value by a finite sum

$$F^*(v) = \lim_{n \to \infty} \frac{1}{n} \sum_{i=1}^{n} F\left(f^n(x)\right) \ .$$

It is essential in physics to have some idea when an measurement of the time mean value (or an approximation, over a long time) will yield the spatial mean. This is precisely insured by the notion of ergodicity:

Definition 236 *A dynamical system (discrete or continuous) on a space X is said to be* ergodic *if the time mean values $F^*(v)$ are equal to the space mean value $\langle F \rangle$ for almost every v. This is equivalent to asking that under the flow of the dynamical system, the only invariant subsets of X are of measure zero or full.*

Moreover almost all trajectories are everywhere dense; see corollary 245 on page 481. In a demagogic language an ergodic flow presents "total chaos." There is still however a stronger notion, that of *mixing:* we will say that the dynamical system is mixing if for every pair of subsets A, B of X one has

$$\lim_{n \to \infty} \mu\left(f^n(A) \cap B\right) = \mu(A)\mu(B) \ .$$

This implies not only chaos but an everywhere evenly chaotic nature. The definition is very appealing: think of X as a glass of tonic and f as a mixing operation with a spoon. The set A is some gin you put into X and the result is that after a long time the gin, i.e. $f^n(A)$, will be distributed with a regular density everywhere. The subset B serves to measure the amount of gin found at various places. Here have complete and evenly dispersed chaos. Bernoulli shifts are mixing, hence ergodic.

10.5.1.2 Notions of Entropy

Entropy is a difficult notion. It was discovered only in 1958 by Kolmogorov. The author of the present book believes that when a notion came about only recently it is almost always because it was difficult. Think of abstract groups, vector spaces, etc. Heuristically entropy is not difficult—entropy is nothing but the measurement of the exponential factor of dispersion in a dynamical system of anything you want to compute. For example, the loss of information as time goes on, or the dispersion of trajectories, or the rate of mixing,

474   10 Geodesic Dynamics

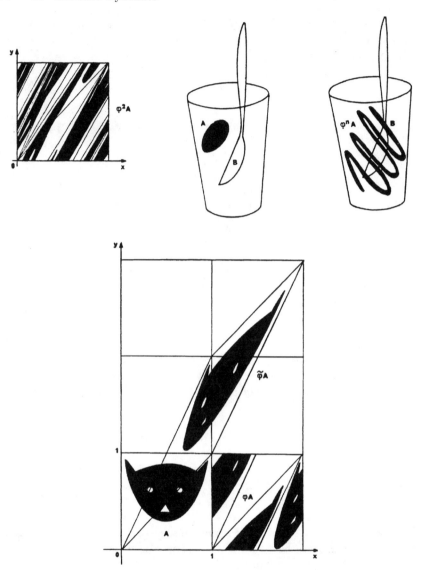

Fig. 10.31. Some dynamical systems

etc. Strict definitions are less simple. Moreover there is more than one entropy. Two are standard: the metric (or measure) entropy and the topological entropy. The labeling is ridiculous: metric entropy needs only a measure to be defined and is really a measure notion, while topological entropy needs a metric (even if finally it turns out to be independent of it) and is really a metric notion. The third entropy, the volume entropy, makes sense only for manifolds whose universal covering is "huge." Metric entropy was the first to arise historically. General references: Walters 1982 [1232] is the standard and a very good book, Sinai 1976 [1140] for its very informative style, Mané 1987 [891] cannot be avoided for the study of Riemannian manifolds, but there is now the bible Katok & Hasselblatt 1995 [787]. For references we refer mainly to the bibliographies there.

The simplest notion of entropy to define is the *volume entropy* h_{vol} in a compact Riemannian manifold M. Let $\tilde{M}$ be its universal cover. We look at the metric balls $B(p,R)$ in $\tilde{M}$ and set

$$h_{vol}(M) = \lim_{R \to \infty} \frac{1}{R} \log \mathrm{Vol}\,(B(p,R))$$

which is easily seen to be independent of the base point $p \in \tilde{M}$. Unhappily this entropy is of no interest unless $\tilde{M}$ is "huge" (a little more precisely, when $\pi_1(M)$ is "huge"). We remark that the geodesic flow is quite hidden here.

For a general dynamical system[3]

$$f: X \to X$$

on a space X, put any metric on X and look at the iterates f^k of f. One gets new metrics for every integer n by setting

$$d_n(x,y) = \sup_{0 < k < n} d\left(f^k(x), f^k(y)\right) .$$

Let $N_n(\varepsilon)$ be the minimum number of ε radius balls in the d_n metric needed to cover X and define the *topological entropy* to be

$$h_{top}(X,f) = \lim_{\varepsilon \to 0} \lim_{n \to \infty} \frac{1}{n} \log N_n(\varepsilon)$$

It is easy to check that this does not depend on the metric chosen on the compact manifold under consideration.

Finally the *measure (or metric) entropy* was the first to be defined historically and by a mixing definition but a very complicated definition, quite lengthy and moreover almost intractable in many situations. Moreover it was based on the function $x \to x \log x$ which needs some consideration before one realizes the final simplicity of the notion. We will still make use of an auxiliary metric which can be forgotten at the end. This time we define $N_n(\varepsilon, \delta)$

---

[3] For a change, and simplification of notation, we will now use the discrete language, leaving to the reader to translate it into the continuous one.

to be the minimum number of ε radius balls needed to cover some subset of X whose complement has measure less than δ.

$$h_{meas}(X,f) = h_{met}(X,f) = \lim_{\delta \to 0} \lim_{\varepsilon \to 0} \lim_{n \to \infty} \frac{1}{n} \log N_n(\varepsilon, \delta) \ .$$

In the case of a geodesic flow, we will also denote measure entropy by $h_{Liouville}$.

These different entropies enjoy many inequalities and properties on special manifolds. We note first that all of these entropies coincide on space forms (locally symmetric of rank one, defined in §6.6). This is not obvious; see the various references already given for proof. Their common value is trivial to compute with the volume entropy; it is enough in the formulas of §§§7.1.1.2 giving the volume of the $\mathbb{KP}^n$ to replace the sine function by the hyperbolic sine.

Theorem 237 *Every compact space form M locally isometric to $\mathrm{Hyp}_\mathbb{K}^n$ has entropy*

$$h(M) = n + k - 2$$

where

$$k = dim_\mathbb{R} \mathbb{K} \ .$$

But the various possible converses of these equalities are not completely understood today, see §10.8.

Theorem 238 (The variational principle) *Every dynamical system satisfies $h_{top} \geq h_{meas}$; in fact $h_{top} = \sup h_{meas}$ where the supremum is over all possible invariant measures.*

The problem is often to discover which measure has measure entropy precisely realizing h_{top}; we will see some examples below.

Theorem 239 (Manning 1979 [894]) *Every Riemannian manifold of nonpositive curvature satisfies*

$$h_{vol} \leq h_{top} \ .$$

There is no very general link with periodic geodesics, but

Theorem 240 (Margulis 1969 [899]) *On a compact Riemannian manifold of negative curvature,*

$$h_{top} = \lim_{L \to \infty} \frac{\log CF(L)}{L}$$

Do not think that positive measure entropy implies ergodicity even locally—it only implies some kind of "local chaos". Measure entropy is the strongest invariant. For example, for a Riemannian manifold, positive measure entropy implies the existence of a set of positive measure where a geodesic is everywhere dense. This is definitely not the case for topological entropy.

There is one implication from the topology to the entropy, namely

Theorem 241 (Paternain 1992)
$$h_{top} = 0$$
for any rationally elliptic manifold.

The result is not surprising if one considers theorem 215 on page 456 and §§10.3.5. The inverse question will be treated in §10.9. We end with a formula giving explicitly a Riemannian quantitative description of divergence.

Theorem 242 (Ballmann & Wojtlowski 1989 [107]) *The geodesic flow of a compact manifold of nonpositive curvature satisfies*

$$h_{Liouville} \geq \int_{UM} \mathrm{tr}\left(\sqrt{-R_v}\right) dv$$

where the integral is on unit vectors v and where R_v denotes the linear map

$$u \to R(v,u)v$$

constructed from the curvature tensor.

The trace is taken with respect to the Riemannian metric. Moreover equality holds only for locally symmetric spaces. This formula was preceded by weaker ones; see references in (7.3) of Eberlein, Hamenstädt & Schroeder 1990 [472]. In Foulon 1997 [518] there is a generalization of the above formula for Finsler spaces (see §§14.5.8).

It is interesting to compare the formula above with the special "entropy" introduced by Hamilton for proving the standard conformal representation theorem 70 on page 254 for surfaces. He used the Ricci flow on metrics to carry out the proof (see §§11.4.3 for the definition of Ricci flow). Namely this entropy is defined as

$$\int_M K \log K \, dV_M$$

where K is the Gauss curvature: see Chow 1991 [378].

10.6 Negative Curvature

Manifolds of negative curvature[4] will be treated at large, from the geometric point of view, in §§12.3.4. But the dynamic properties of the geodesic flow are closely interwoven with the global geometry of negatively curved manifolds. A partial survey of the theory of negatively curved manifolds is found in section 3 of Besse 1994 [184] and a very detailed discussion is presented in Eberlein, Hamenstädt & Schroeder 1990 [472].

The ergodicity of surfaces of negative curvature was emerging in Hadamard 1898 [675]. For constant curvature (Riemann surfaces) one found it explicitly in Hedlund 1934 [696]. The article Hopf 1939 [725] was extremely important; it proves that the geodesic flow on compact surfaces of negative curvature is ergodic (note that this is Eberhard Hopf, not Heinz Hopf whom we met and will meet many times again in considering the relations between curvature and topology). Hopf's result was extended to any dimension in

Theorem 243 (Anosov 1967 [53]) *The geodesic flow of any compact Riemannian manifold of negative curvature is a Bernoulli shift; in particular it is ergodic and mixing.*

Even for surfaces the proof is never simple; the best exposition is in the appendix by Brin of Ballmann 1995 [103], where one can see why the proof is much harder for general dimensions than for surfaces. Historically in fact Hopf's result had a precursor in Morse 1921 [942],[943] where first topological transitivity[5] was obtained and, even better, a coding of the trajectories was uncovered. Coding is extremely important—it enables people to study the geodesic flow by looking only at a discrete "shift". Let us recall Hadamard's coding in Hadamard 1898 [674] and see Katok 1996 [789] as well as Katok & Hasselblatt 1995 [787].

The basic starting idea for proving Anosov's theorem 243 is to look at the geodesic behaviour when the time (the length) goes to infinity in both senses (directions). We fix a geodesic γ and look at Jacobi vector fields along it.[6] The negative curvature assumption, in the spirit of §§6.3.1, shows us something remarkable. Given any unit tangent vector v, orthogonal to $\gamma'(0)$ in $T_{\gamma(0)}M$, and any time t, there is a unique Jacobi field Y_t along γ such that

$$Y_t(0) = v$$

and

$$Y_t(t) = 0 \ .$$

---

[4] Curvature is always understood here to mean sectional curvature, unless otherwise noted.

[5] *Topological transitivity* means the existence of at least one trajectory which is everywhere dense

[6] See page 248 for the definition of *Jacobi field*.

It is not hard to see, again using the negative curvature property, that

Theorem 244 *When t tends to $+\infty$ (resp. $-\infty$), the Jacobi vector field Y_t has a limit Jacobi vector field denoted by $Y_{+\infty}$ (resp. $Y_{-\infty}$).*

It is important for future considerations to realize that this is the infinitesimal version of the Busemann function to be met in §§12.3.2 and §§12.3.4.[7] More precisely, in UM the derivatives $Y'(v)(0)$ are tangent vectors in $T_v UM$ and the distribution they form, locally in UM and globally in $U\tilde{M}$ are integrable distributions,[8] called respectively *stable* (for $+\infty$) and unstable (for $-\infty$). Integrated, they are the level lines of the two Busemann functions along γ, namely the spheres centered at the two points at infinity of γ, respectively $\gamma(+\infty)$ and $\gamma(-\infty)$. In the canonical hyperbolic spaces, one gets nothing but the horocycles.

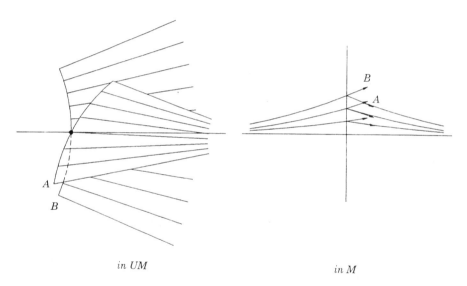

Fig. 10.32. (a) In UM (b) In M

The Rauch comparison theorems (propositions 74 and 75) show that, due again to the negative curvature condition, the norms $\|Y_{+\infty}(v)(t)\|$ (respectively $\|Y_{-\infty}(v)(t)\|$) decrease (resp. increase) exponentially with the time (a.k.a. length) t. A sketch of a proof consists in realizing that the geodesic flow, having both exponential decay and exponential growth, can permit only invariant sets of full measure or measure zero. The details consist in checking that, like in a jigsaw puzzle, one can go everywhere by traveling alternately

---

[7] See the definition of Busemann function in definition 334 on page 585.
[8] This $U\tilde{M}$ is the unit tangent bundle of the universal covering $\tilde{M}$ of M.

along stable and unstable distributions. This simultaneous exponential convergence and divergence insures this. Working out the details is difficult, because in fact the distributions above are not well behaved; except in the locally symmetric case they are not differentiable. Indeed, these distributions are not even Lipschitz for generic negatively curved metrics; but they are at least absolutely continuous. Hopf knew of the absolute continuity and the poor behaviour of the distributions for surfaces, and Anosov extended these results to any dimension. The reason for this savagery is simply that the Jacobi fields are given by a limiting process at infinity, and such limits need not even be continuous. Historically, these foliations appeared first in Hadamard 1901 [676]. Remarkably, smoothness of the distributions forces local symmetry; see §10.11.

The negativity of the curvature is absolutely necessary. For example do not hope for some kind of generic result; see however Lohkamp's results in §10.9. In figure 10.33 we see bad behaviour for the geodesic flow of a Riemannian manifold. Put a piece of a Weinstein surface of revolution (see §§§10.2.1.2) of rational type into the mushroom. In this band, all the geodesics are periodic. So we certainly do not have ergodicity. But we are looking for genericity. So we take any small perturbation of this mushroomed surface. Then we appeal to the KAM theorem which yields the existence of many invariant tori in the phase space close to the Weinstein piece. Because we are in dimension 2, the phase space is three-dimensional and the KAM tori disconnect it.

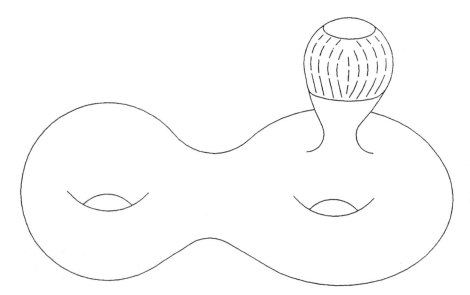

Fig. 10.33. A Weinstein piece attached to a Riemannian manifold

Note 10.6.0.1 We note two interesting problems. First, the behaviour of the geodesic flow on a torus is not understood. Second the above genericity argument fails in higher dimension because the KAM tori will no longer be disconnected. The genericity question in higher dimension is then much more difficult and we know of no result, save Boucetta 1998 [231]. ♦

10.6.1 Distribution of Geodesics

We now look at applications of the main result above. For the geometer seeing things downstairs on the manifold, we note first:

Corollary 245 *In a manifold of negative curvature, almost all geodesics are evenly distributed in the manifold.*

The precise definition of "evenly distributed" is the most natural view for a geometer: we say that a geodesic γ is evenly distributed if, for every domain $D \subset M$ the time that this geodesic spends inside D is proportional to the relative volume of D, i.e. $\operatorname{Vol} D / \operatorname{Vol} M$. Precisely:

$$\lim_{L \to \infty} \frac{1}{L} \operatorname{length}(D \cap \gamma[0, L]) = \frac{\operatorname{Vol} D}{\operatorname{Vol} M}$$

The proof is an application of the definition of ergodicity 236 on page 473. We apply it to the function $F(v) = 1_D(p(v))$ which is the composition of the characteristic function of D with the projection $p : UM \to M$. We apply it to a unit tangent vector v and for most of them we have that the time mean value is exactly what we want in the left hand side, while the right hand side is the space mean value. This proves even distribution in the *configuration space* M. By the same token, applied to the characteristic function of a subset of the *phase space* UM, we see that geodesics are evenly distributed in the phase space.

10.6.2 Distribution of Periodic Geodesics

We will continue to consider compact negatively curved manifolds. Since we know that we have an exponential number of periodic geodesics (see e.g. corollary 207 on page 448) we expect them to be evenly distributed in the same way almost all of the nonperiodic geodesics are evenly distributed, but this is false in general. Let us make precise the notion of even distribution for periodic geodesics.[9]

Theorem 246 *The unit bundle of a compact manifold of negative curvature carries a canonical measure, called the* Bowen–Margulis measure *(or BM*

---

[9] References are the same as the ones at the beginning of §§§10.5.1.1 and add Pansu 1991 [996] for a fast informative survey, and Knieper 1998 [821].

measure). One can define it as the limit, as $L \to \infty$, of the mean value of the Dirac distributions defined by the periodic geodesics of length smaller than L. Or more conceptually (once one knows its existence and uniqueness employing the previous definition) as the measure invariant by the geodesic flow which has maximal metric-measure entropy.

This measure is of course invariant under the geodesic flow and can be defined also as the product of three measures:

1. dt along the flow and two measures
2. $d\mu_{\text{stable}}$ and
3. $d\mu_{\text{unstable}}$

given on the distributions introduced above. Of course by the very definition, the periodic geodesics are evenly distributed for the BM measure. For locally symmetric metrics the BM and Liouville measures coincide. Katok proved that the Liouville and the BM measures on the tangent bundle of a surface coincide if and only if the curvature is constant. The analogous result for higher dimension remains an open question today.

We will come back to Anosov's result in detail later. We first address two questions stemming out of Hopf's and Anosov's results.

Question 247 *Is negative curvature needed to get ergodicity?*

Question 248 *Does one really need a manifold with a large fundamental group to have ergodicity?*

This question will be the topic of §10.9.

10.7 Nonpositive Curvature

We will see in §§12.3.4 that, at least today, hyperbolic groups are the "final answer" for the fundamental groups of negatively curved compact manifolds. But we will also see nice results, even if still not complete, to expose the very subtle difference between negative and nonpositive curvature. We will address here the splitting question between negative and nonpositive at the level of the geodesic flow. Today there are extremely strong results which show in essence that the two cases can be dramatically separated. Otherwise put, to go from one to the other you have to make quite a jump.

We give the latest result and refer to Eberlein, Hamenstädt & Schroeder 1990 [472] for a survey and to Ballmann 1995 [103] for the proof. The present text will intersect a lot with §§12.3.4. The contemporary theory of negative or nonpositive curvature manifolds is full of results. This is due to the rich interplay between geometry, structures at infinity, measure theory, dynamical system theory, symplectic geometry and group theory. See §§12.3.4 for more on this. A contrario the reason why we have so few results on the spheres and the $\mathbb{KP}^n$ might well be because of the lack of available tools. We now just extract one striking result here.

To understand the result we recall briefly the notion of rank in a Riemannian manifold (see the definition 68 on page 250) met also for symmetric spaces (see §§4.3.5 and §§§12.3.4.5). To say that the rank is at least two implies that, given any unit vector, along the geodesic it generates there is a nontrivial parallel Jacobi field (orthogonal to the geodesic of course, and not zero), something exceptional in a generic manifolds. The strongest result today is the following "all or nothing" result for compact Riemannian manifolds of nonpositive curvature. We assume any reasonable irreducibility condition.

Theorem 249 (Ballmann & Eberlein 1987 [105]) *Either the rank of the space is greater than or equal to 2, and then we are on a space form with its symmetric space metric, or the rank is 1 and then the geodesic flow is ergodic on the subset of the unit tangent bundle made up of the vectors of rank 1.*

Today it is still unknown if this subset of rank 1 vectors is of full measure or not. The above dichotomy admits a lot of extensions with weaker hypotheses, like completeness with finite volume to replace compactness, or a so-called "group duality condition," and also to metric spaces mimicking Riemannian manifolds of nonpositive curvature. For all this the reader can look at the various references already given.

For the proof, the idea is to be able to integrate in some sense the infinitesimal condition given by having a rank at least two, and to reconstruct all of the totally geodesic flat submanifolds of a locally symmetric space. One trick is to relate the rank to the nontransitivity of the holonomy group, which plays a basic role in Simons' proof of the classification theorem 397 on page 643. Even if the rank is one, there could exist vectors of larger rank; the corresponding periodic geodesics are called singular and counting them among the regular ones is linked with the Bowen–Margulis measure. For a subtle result on this see Knieper 1998 [821].

10.8 Entropies on Various Space Forms

We already said that for the space forms of negative curvature (see §6.6 and §§10.5.1) the four possible entropies are equal. One expects a converse. The first result in this direction is that of Katok, which is extremely complete.

Theorem 250 (Katok 1988 [786]) *Consider a compact surface which admits a metric g_0 of constant negative curvature. Let $g = f^2 g_0$ be any metric conformal to g and of same area, i.e.*

$$\int_M f^2 \, dV_{g_0} = 1 \, .$$

Then one has

$$h_{Liouville}(g) \leq \rho h_{top}(g_0) \leq \rho^{-1} h_{top}(g_0) \leq h_{top}(g)$$

where

$$\rho = \int_M f \, dV_{g_0} .$$

Plug this into the conformal representation theorem 70 on page 254 and the common value $h_{top}(g_0) = 1$. Then we see that on any compact Riemannian surface of genus larger than one and normalized, the topological entropy is at a minimum if and only if the curvature is constant, and the reverse for the Liouville entropy. Katok conjectured the same result for higher dimension and for the space forms locally isometric to $\text{Hyp}^n_{\mathbb{K}}$. Gromov suggested to use the volume entropy. Today we have the optimal result for the topological entropy:

Theorem 251 (Besson, Courtois, Gallot 1995 [189]) *Let (M, g_0) denote a compact locally symmetric space of negative curvature (hence of $\text{Hyp}^n_{\mathbb{K}}$ type). Then for any other metric g on M one has an inequality on volume entropy:*

$$h_{vol}(g) \geq h_{vol}(g_0) .$$

Moreover when the dimension is larger than two, equality implies isometry: $g = g_0$.

Using the hyperbolic analogues of the formulas in 7.1.1.2 for the volumes of the $\mathbb{KP}^n$ one gets the immediate

Corollary 252 *For space forms of type $\text{Hyp}^n_{\mathbb{K}}$, the common value of the three entropies is equal to $d + k - 2$ where $k = dim_{\mathbb{R}} \mathbb{K}$.*

From the union of theorems 237 and 238 one sees immediately corollary 207 for counting periodic geodesics, with equality only in cases of isometry. Another corollary of theorem 251 is the Mostow rigidity theorem 99 and in §§6.6.3; see Besson, Courtois & Gallot 1995 [189] for the proof. This turns out to be the simplest available proof of Mostow's rigidity. In theorem 285 and theorem 261 we will meet two other corollaries of the strong result above.

A few words about the proof of Besson, Courtois & Gallot 1995 [189], with Besson, Courtois, & Gallot 1995 [190] providing a simpler one. The more natural and conceptual tack would be to prove that the entropy is a strictly convex functional on the space of all Riemannian metrics and that the locally symmetric metric is a critical point. Unhappily this program does not work as such except in a conformal class (see Robert 1994 [1062]). The original proof was quite involved but used wonderful techniques, like the center of mass (see §§6.1.5) for the structure of the sphere at infinity (see the sphere at infinity in definition 362 on page 608) and the technique of calibration (see the definition of calibration on page 667). In Besson, Courtois & Gallot 1995 [189] the proof is greatly simplified by introducing the Patterson–Sullivan measure on the sphere at infinity in order to get a better notion of a center of mass.

10.8.1 Liouville Entropy

The Liouville entropy cannot be as simple to analyze in higher dimensions as it was in dimension 2, since in Flaminio 1995 [517] there are examples of compact space forms locally isometric to Hyp^3 whose Liouville entropy can be made larger by switching to a different metric. However one has to go far away from the locally symmetric metric, since by Knieper 1997 [820], the Liouville entropy has a local maximum at any constant curvature metric (in any dimension). Today the total picture is not clear.

10.9 From Osserman to Lohkamp

Ergodicity discovered by E. Hopf on surfaces of negative curvature seems to be linked to divergence of geodesics, and negative curvature is linked to higher genus by the Gauß–Bonnet theorem 28 on page 138. One might still wonder if ergodicity or partial chaotic behaviour can be obtained

- on topological spheres
- or on manifolds of positive curvature.

Recall also Anosov's theorem 243 on page 478. These questions have very clear answers. We will start with two examples which are beautiful geometric explicit constructions, and finally describe Lohkamp's recent and completely different approach to the question addressed.

The fact that ergodicity is possible on the topological sphere S^2 comes from an idea of Osserman: take a pair of pants of constant negative curvature (see the description of pairs of pants on page 197) such that the three ends are periodic geodesics of length π and complete it by standard hemispheres; hence equators are also of length π. Geodesics from inside the pants arriving at the boundary reenter the pants antipodally because of the hemisphere property. And so on. Therefore the geodesic behaviour can be obtained by that of the geodesic flow on the surface obtained from the pants after identification by antipody on the three boundaries. This is a nonorientable surface but of (constant) negative curvature and E. Hopf's result (see theorem 243 on page 478) still applies in the nonorientable case. So we have ergodicity on the quotient and this implies immediately ergodicity on the sphere.

Osserman's example is not very smooth, only continuously differentiable, the curvature having to jump from $+1$ to -1 along the gluing. A smoothing was obtained in Burns & Gerber 1989 [286]. Their proof is very geometric. One replaces the hemispheres by pieces of surfaces of revolution. If the meridian has a convex varying curvature then the study of the Jacobi fields shows that divergence of geodesics is still preserved. But this was an abstract surface (compare with Hilbert's impossibility in theorem 27 on page 135). Embedded ergodic surfaces were obtained in Burns & Donnay 1997 [285]. The proof is based on the existence of triply periodic minimal surfaces in $\mathbb{E}^3$ instead

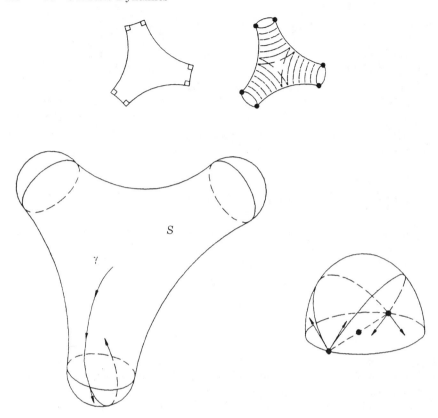

Fig. 10.34. (a) In Hyp^2, right angled hexagons (b) The antipodal map

of hyperbolic pants. One still adds caps to ends so that diverging geodesics entering the minimal surface still diverge again after the cap focusing. By the same approach, one can obtain such surfaces of any genus, including tori. In Burns & Gerber 1989 [286], still working "à la Osserman," one builds on S^2 real analytic metrics which are more than ergodic, that is to say are a Bernoulli shift, in particular the geodesic flow is ergodic and mixing (see the definition of Bernoulli shift on page 472).

To get positive topological entropy on a sphere of positive curvature one creates local chaos by deforming a three-axis ellipsoid as follows. One begins by trying to understand the behaviour of the geodesics going through two antipodal umbilics. Since Birkhoff, some people like to get intuition by flattening our ellipsoid onto a double-faced ellipse. It is similar to that of a light ray traveling through the two foci. When it comes back to the first umbilic it is never with the same direction so that it is never periodic but it is asymptotically closer and closer to the periodic geodesic going through the four umbilics. This holds in both $+\infty$ and $-\infty$ direction, because in the intermediate stage it "jiggles" in the middle of the ellipsoid. A small perturbation of the ellipsoid

10.9 From Osserman to Lohkamp   487

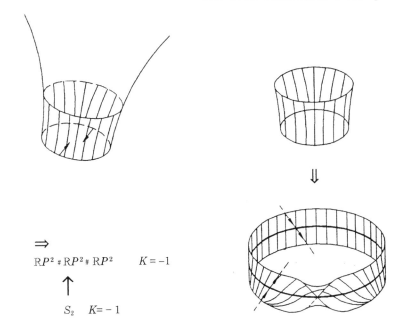

$\Rightarrow$
$RP^2 \# RP^2 \# RP^2$ $K = -1$
$\uparrow$
S_2 $K = -1$

Fig. 10.35. A double covering

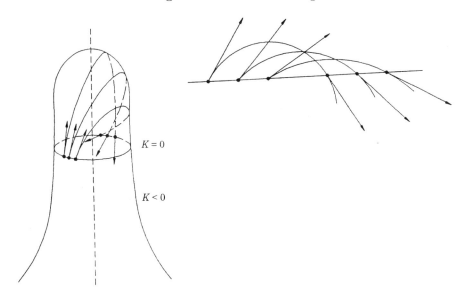

Fig. 10.36. If diverging when entering, geodesics will diverge still when leaving provided $dK/dl \leq 0$

in two small open sets in the jiggling part produces a horseshoe, and hence local chaos and positive topological entropy (and then moreover an exponential $CF(L)$ counting function). The complete proof is delicate and figures in Knieper & Weiss 1994 [823]. One can even get positive topological entropy with a real analytic metric, thanks to general approximation theorems. See Bolsinov & Taimanov 2000 [218].

Recently Lohkamp revolutionized the subject. Using a hard technique, similar to his work (see §§12.3.5) on negative Ricci curvature, there is a game to push enough pieces with ergodic behaviour and larger and larger entropy, but at the same time ensure that the positive curvature pieces do not destroy the ergodicity.

Theorem 253 (Lohkamp) *Within the set of Riemannian metrics of any compact manifold, there is a Hausdorff dense set of Riemannian metrics whose geodesic flow is ergodic.*

As example, Lohkamp can use his flexibility to obtain ergodic (Bernoulli) spheres embedded in $\mathbb{E}^3$. But at the moment none of them is convex. The major question left:

Question 254 *Is ergodicity impossible or not on positive curvature manifolds?*

10.10 Inverse Problems: Manifolds All of Whose Geodesics are Closed

We turn now to an inverse problem announced in the introduction. Motivated by Zoll's surfaces in §§§10.2.1.1, we address the problem of classification of the manifolds all of whose geodesics are closed. There is an entire book on the problem, namely Besse 1978 [182], and sadly enough there are basically very few new results in the subject. To our knowledge there are only Gromoll & Grove 1981 [602], Kiyohara 1984 [806], Tsukamoto 1981 [1202], and Tsukamoto 1984 [1203]. We met the problem and some answers for surfaces in §§§10.2.1.1. One must remember §§4.4.3 and §§6.6.1: the spheres and the $\mathbb{KP}^n$ have a beautiful geodesic flow, since all of the geodesics are periodic, simple and of the same length (for their canonical metric). But from Kiyohara 1984 [806] we are going to see that the inverse problem is still partly open. The spheres and the $\mathbb{KP}^n$ behave very differently, at least as far as we know.

In all of this section, lacking references can be looked up in the book Besse 1987 [182].

10.10.1 Definitions and Caution

For the reader who might underestimate the difficulty of the subject, we mention a few facts; for more, and for references, see Besse's book. Remark first

10.10 Manifolds All of Whose Geodesics are Closed

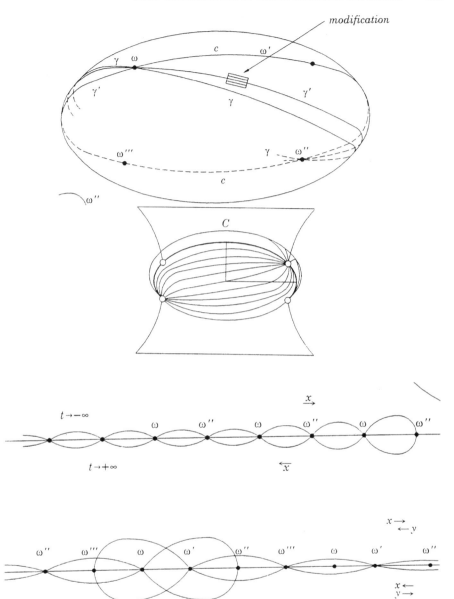

Fig. 10.37. An umbilical geodesic accumulating on C

that any quotient of a manifold whose geodesics are periodic has only periodic geodesics, but they need not all have the same length. This occurs for the lens spaces (the quotients of spheres by finite groups of isometries). Remember that the $\mathbb{KP}^n$ of even dimension can admit only $\mathbb{Z}^2$ quotients thanks to Synge's theorem 64 on page 246. The simplest example to have in mind is the quotient of $S^3 \subset \mathbb{R}^4$ by $\mathbb{Z}^2$ on (x_1, x_2) and $\mathbb{Z}^4$ on (x_3, x_4). All geodesics have length π except for one which has length $\pi/2$. Getting close to this $\pi/2$ geodesic, the others accumulate on it in double coverings. Some people might prefer to use the fact that $\mathbb{RP}^3$ is the same as $SO(3)$ and take the quotient of this rotation group by the symmetry around a line in $\mathbb{R}^3$.

Let us assume no more than that all geodesics are periodic. It is trivial to show that the lengths are all multiples of a smallest one (in the spirit of the first variation formula). It is true, but hard to prove, that these lengths are bounded; see Wadsley 1975 [1225]. An interesting fact about this theorem is that it is false in affine geometry where there are compact manifolds foliated by circles but the circle lengths (in any metric) are not bounded—they can turn more and more. See references in Besse's book for the examples of Sullivan in dimension 5 and D. Epstein in dimension 4. Worse: if there are plenty of examples for which the lengths are different, e.g. lens spaces, it is unknown if on simply connected manifolds with all geodesics periodic, all the lengths are equal. There is also the question of simple or self-intersecting periodic geodesics. The only result we know of is on S^2: Gromoll & Grove 1981 [602] mentioned already on page 439.

The very intricate relations between various definitions concerning periodic geodesics, geodesic loops, and whether one asks for periodicity at one point or and all points are addressed in chapter 7 of Besse's book. It seems that nobody has recently taken up the study of these relations and various possibilities.

10.10.2 Bott and Samelson Theorems

The greatest imaginable accomplishment concerning manifolds with only periodic geodesics would be to prove that, as manifolds, only the spheres and the $\mathbb{KP}^n$ can admit a Riemannian metric all of whose geodesics are periodic and of the same length. There are no other ones known today, but there are some results pointing in this direction. The most irritating fact is that the strong topological statement concerning the topology of the possible manifolds is obtained by assuming just that all geodesics through some point are periodic and of the same length. Nobody could deduce anything more even by assuming this is true for every point. Thus far the spectral result theorem 177 on page 406 has been useless.

To state the results one is best to look at the beginning of the proofs which follow two completely different approaches, the one of Bott and the one of Samelson. But they have the same starting point. Let m be our point with the above property. By compactness and existence of an injectivity radius is it easy to see that all points q close enough to m have the property that

10.10 Manifolds All of Whose Geodesics are Closed   491

all geodesics joining m to q are the covering of one periodic geodesic γ and moreover one of them is a segment σ with ends m and q not conjugate. It is now necessary to look at all of these coverings, which have two possible directions, as in figure 10.38.

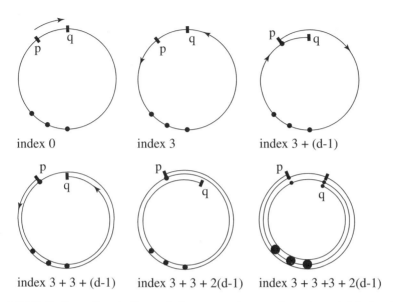

Fig. 10.38. Indices of periodic geodesics connecting nearby points: **(a)** index 0 **(b)** index 3 **(c)** index $3 + (d-1)$ **(d)** index $3 + 3 + (d-1)$ **(e)** index $3 + 3 + 2(d-1)$ **(f)** index $3 + 3 + 3 + 2(d-1)$

A well studied integer k comes now into the picture, namely the *index* (in the sense of Morse, as above) of the complement $\gamma - \sigma$ of the segment σ inside γ.

Theorem 255 (Bott 1954 [226], Samelson 1963 [1091]) *The integer k does not depend on the point q and if $k > 0$ then the manifold M^d is simply connected and its integral cohomology ring has exactly one generator. More precisely there are only the following possibilities:*

Index	Dimension	Topology
$k = 1$	$d = 2n$	M has the homotopy type of $\mathbb{CP}^n$
$k = 3$	$d = 4n$	M has the homology ring of $\mathbb{HP}^n$
$k = 7$	$d = 16$	M has the integral cohomology ring of $\mathbb{CaP}^2$
$k = d - 1$	d arbitrary	M has the homotopy type of S^d and hence is homeomorphic to S^d if $d \neq 3, 4$
$k = 0$	d arbitrary	M is diffeomorphic to $\mathbb{RP}^d$

The proof is a mixture of two ideas. Samelson's starting remark is that the geometric structure of the geodesics through m enables us to build, in an obvious and canonical way, a map $\mathbb{RP}^d \to M$, as in figure 10.39.

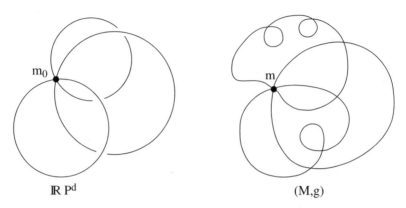

Fig. 10.39. Samelson's map $\mathbb{RP}^d \to M$

A little more geometry shows that if $k = 0$ then the map is one to one. Bott now comes into the picture when $k > 0$. This situation can be approached via Serre's theory (described on page 455). The sequence of indices of all geodesics joining m to q is exactly the set of integers $n(d + k - 1)$ and $n(d + k - 1) + k$ with all integers $n \geq 0$. The reader has only to notice that m is conjugate to itself with index $d - 1$. Such a sequence is very special, and very gappy since the indices jump by $d - 1$. This observation implies that the spectral sequence of the fibration

$$\Lambda \to \Lambda_{p,.} \to M$$

(already met in the fiber map 10.2 on page 455) degenerates (only its d_2 differential is not zero) and a nice algebraic calculation shows that this forces the cohomology ring of M to be generated by a single element. The conclusion in the theorem follows from "classical" results in algebraic topology about such manifolds.

One cannot do better: there are exotic "$\mathbb{KP}^n$"; we mentioned them on page 464 when talking about spaces for which we still do not know whether they admit infinitely many periodic geodesics for any metric. These exotic $\mathbb{KP}^n$ seem for the moment to be "cursed."

10.10.3 The Structure on a Given S^d and $\mathbb{KP}^n$

We already mentioned Weinstein's remark:

Theorem 256 (Weinstein) *In every dimension, there are "exotic" metrics all of whose geodesics are simple, periodic and of the same length.*

But starting in three dimensions, nobody seems to have the foggiest idea "how many" they are, in particular what is the tangent space at the canonical metric of Zoll's space of purely periodic S^d metrics. We just have from Kiyohara 1984 [806] some restrictions on the potential tangent vectors of one parameter deformations.

The only case in which our understanding is satisfactory is that of $\mathbb{RP}^d$, which is a corollary of the "Wiedersehenmannigfaltigkeit" theorem. We look on the sphere S^d at a metric such the cut locus of any point is reduced to a single point, its *antipode*. This is particular case of the situation where the diameter is equal to the injectivity radius; see §§6.5.5 for more on this very special, but baffling, property. On spheres, theorem 257 solves the question: only the standard spheres are "Wiedersehen". Another language is: the standard sphere is everywhere a perfect (stigmatic) optical instrument—we have everywhere perfect focusing with any angle up to 2π. There is no way to deform the standard sphere and preserve this property. The problem in geometrical optics to look for local perfect optics (Maxwell's fisheye, see Born & Wolf 1965 [225]) is treated in Deschamps 1982 [441].

Theorem 257 ("Wiedersehenmannigfaltigkeit") *On a sphere S^d only the metric of constant curvature can be "Wiedersehen".*

Theorem 258 *The only metric on $\mathbb{RP}^d$ such that all geodesics are periodic, simple and of same length is the standard one.*

If we lift a metric from $\mathbb{RP}^2$ up to S^2, then the latter one is "Wiedersehen" (i.e. a metric such the cut locus of any point is reduced to a single point, its antipode), hence the corollary.

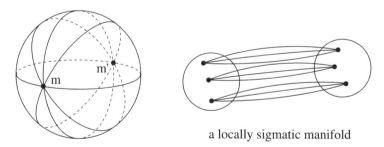

a locally sigmatic manifold

Fig. 10.40. (a) Light rays (i.e. geodesics) on the standard sphere (b) A locally stigmatic manifold

In the study of surfaces, theorem 257 (as applied to S^2) was a long standing conjecture of Blaschke in the 1930's, with even a deadly wrong proof figuring in the first edition of his book Blaschke 1921 [203]. The conjecture was settled in Green 1963 [586] as follows. We will normalize the diameter to π. One needs two tricks: first apply Santalo's formula 7.11 on page 364 to the surface S^+ made up by one side of a fixed periodic geodesic γ. This S^+ is exactly covered by the segments of length π issued from the points of γ, so that Santalo's formula yields 2π for the area of our S^+ and by the same token the total area Area (S^2, g) of a our surface is equal to 4π. The second trick is to integrate over the unit tangent bundle US^2 the family of all second variations, from formula 6.7 on page 242 (see theorem 62 on page 243) for every $v \in US^2$ and the sine function along it:

$$\int_0^\pi \left(\cos^2 t - K(\gamma(t)) \sin^2 t \right) dt \ .$$

These integrals all have to be nonnegative, since the injectivity radius is π. From Fubini's theorem,

$$\int_{S^2} K \, dV_{S^2} \leq \operatorname{Area}(S^2) \ .$$

But $\int_{S^2} K \, dV_{S^2} = 4\pi$ by the Gauß–Bonnet theorem 28 on page 138. So we have equality everywhere, and in particular for all of the integrated second variation. Because the first conjugate point is a distance exactly π, there is only (up to a constant) one Jacobi field vanishing on $[0, \pi]$. Vanishing of the second variation for the sine field along any geodesic forces this sine field to be a Jacobi field (see §6.3), and therefore the Gauß curvature K has to be everywhere equal to 1.

Now for higher dimensions, we are doubly vexed. First, the trick of integrating the sine-square second variation all over UM will only yield the scalar curvature, and second Santalo's formula would work only if we had a totally geodesic "equator." New approaches were required. The first idea came in Weinstein 1974 [1249]: he introduced the smooth manifold Σ which is the totality of all periodic geodesics on M. This yields a circle fibration

$$S^1 \to UM \to \Sigma \ .$$

Consideration of the Chern class of this circle bundle shows that the volume of M is an integral multiple of the volume of the standard sphere. Weinstein proved that (for topological reasons) this integer is equal to 1 when the dimension is even, and this was extended into the odd dimensional case in Yang 1980 [1284]. So that finally the volume of our manifold is equal to that of the standard sphere, but since the injectivity radius is equal to π, the proof is concluded with the embolic inequality from theorem 148 on page 355.

In the same vein, there are no known metrics on any $\mathbb{K}\mathbb{P}^n$ all of whose geodesics are periodic (except of course the standard ones) when $\mathbb{K} = \mathbb{C}, \mathbb{H}, \mathbb{C}\mathrm{a}$. But a proof of the uniqueness (if true) seems far away. We have only a partial result:

Theorem 259 (Tsukamoto 1981 [1202]) *On any* $\mathbb{KP}^n$, *there is no metric all of whose geodesics are periodic close to the standard metric, except the standard metric itself.*

This is used in Kiyohara 1987 [807] to prove the spectral solitude of the $\mathbb{KP}^n$. The link with the spectrum is insured by the results quoted above relating periodic geodesics and the spectrum. Other results of this type are in Gasqui & Goldschmidt 1994 [550] and various texts by the same authors, past and future.

10.11 Inverse Problems: Conjugacy of Geodesic Flows

We address the inverse (recognition) problem in the spirit of the geodesic flow: does the geodesic flow determine the metric up to isometry? The precise formulation is this. The geodesic flows of two Riemannian manifolds (M, g) and (N, h) are said to be *conjugate* if there exists a smooth map $f : UM \to UN$ which commutes with the geodesic flows G_M^t and G_N^t of M and N, namely we have for every time (length) t a commutative diagram:

$$\begin{array}{ccc} UM & \xrightarrow{f} & UN \\ \downarrow{G_M^t} & & \downarrow{G_N^t} \\ UM & \xrightarrow{f} & UN \end{array}$$

valid for every time t.

The existence of such a conjugacy can be a more or less strong assertion; we will see below that sometimes one needs not only a continuous conjugacy f, but an f which is continuously differentiable. It is important to note that the map is not asked to commute with the projections $UM \to M$ and $UN \to N$. It is like comparing trajectories of a plane but never looking through the window down at the earth. Now the precise question is:

Question 260 *Are two Riemannian manifolds with conjugate geodesic flows necessarily isometric?*

In this active field, besides research publications, there is the survey Eberlein, Hamenstädt & Schroeder 1990 [472].

From our previous experience, two families of possible counterexamples come to mind. The first are Zoll surfaces. It was proven by Weinstein that all Zoll surfaces of revolution have conjugate geodesic flow (see section 4.F of Besse 1978 [182]) and in particular the same geodesic flow as that of the standard sphere. The fact that geodesics are all periodic and of the same length is not enough; one has to built up explicitly some map f. Note also that the antipodal property is not preserved, since the conjugacy is not asked to commute with the canonical projections (see theorem 257 on page 493).

The second example needs first a remark: if the geodesic flows are conjugate then the length spectra (i.e. the set of the lengths of the periodic geodesics) are the same. Then the Vignéras examples (see §§§9.12.4.1) of Riemann surfaces having the same Laplace spectrum and hence the same length spectrum (by Huber's theorem 192 on page 421) are different Riemannian manifolds but with the same length spectra. Of course this does not prove that their flows are conjugate.

The situation is sometimes different for the marked length spectrum. This means one "remembers" not only length but also which free homotopy class a periodic geodesic comes from. Of course geodesic conjugacy implies the same marked length spectrum. We now have three notions:

1. same length spectrum
2. same marked length spectrum
3. conjugacy of geodesic flow.

They have a rich interplay in the realm of nonpositive and negative curvature. The basic survey is Eberlein, Hamenstädt & Schroeder 1990 [472]. In particular for manifolds of nonpositive curvature, equal marked length spectrum is equivalent to conjugacy of geodesic flow. Many partial but strong results are available. As already said they share many considerations with §§12.3.4. We just quote some results in this very active topic.

For manifolds of nonpositive curvature, conjugacy of geodesic flows implies isomorphy of fundamental groups. Then apply Farrell & Jones 1989 [507] to get homeomorphism.

The strongest result known about conjugacy of geodesic flows is a corollary of theorem 250 on page 483:

Theorem 261 *If M is any Riemannian manifold of dimension at least 3 and N is a locally symmetric manifold of nonpositive curvature, and M and N have C^1 conjugate geodesic flows, then M and N are isometric.*

This very strong statement uses various previous results; for details we refer to Besson, Courtois & Gallot 1995 [189].

Theorem 262 *If M and N have identical marked length spectra between, and dimensions greater than two, and N is of constant negative curvature, then M and N are isometric.*

Here one also needs dimensions at least three, because of the moduli spaces of constant curvature surfaces.

One may ask the question of extending the results above to the case of locally symmetric manifolds of only nonpositive curvature. This research domain is very active today.

Theorem 263 (Croke 1990 [419] & Otal 1990 [985]) *For surfaces of nonpositive curvature, conjugacy implies isometry.*

10.11 Inverse Problems: Conjugacy of Geodesic Flows

The main question today is if this is still valid in any dimension.

To end the marked length spectra story, we note that Gornet 1996 [580] presents nonisometric manifolds with identical marked length spectra.

Space forms can be recognized by an apparently very weak property: the continuity (or the smoothness) of the stable and unstable distributions introduced in §10.6.

Theorem 264 *If the stable distribution of a manifold of negative curvature and dimension at least three is C^∞ smooth, then it is isometric to a locally symmetric manifold of negative curvature.*

The proof uses hard results obtained by various authors, e.g. Benoist, Foulon & Labourie 1992 [131] and theorem 250 on page 483; see Besson, Courtois & Gallot 1995 [189] for more references.

The geodesic flow on flat tori can be characterized by the fact that there are no conjugate points. Namely, a metric on a torus with no conjugate points has to be flat. This was proven in Hopf 1948 [726] for the two-dimensional tori, where the Gauß–Bonnet theorem 28 on page 138 is used, similarly to its use in the proof above of theorem 257 on page 493, using the second variation. For higher dimensional manifolds, it was a long standing conjecture, but hard to solve since the second variation yields only a very weak invariant, namely the integral of the scalar curvature. The conjecture was finally solved:

Theorem 265 (Burago & Ivanov 1994 [280]) *The only metrics on tori which have no conjugate points are the flat ones.*

The proof makes a subtle analysis of the behaviour at infinity of the manifold and there employs approximation of symmetric convex bodies by ellipsoids. For the strict conjugacy of geodesic flows on tori, there are partial results in Croke 1992 [421].

In all of the above results, the proofs mix more or less classical results of ergodic theory with various geometric techniques; see §§12.3.2 and §§12.3.4. The notion of Busemann function will also be met there. Variations of these proofs can be caricaturally described as follows: one looks at the universal covering of the manifolds of nonpositive curvature under consideration. They look topologically like $\mathbb{E}^d$. One defines on them various kinds of structures "at infinity" and looks at the metric "from infinity." In particular the isoperimetric profile plays a role.

Note 10.11.0.1 (Different metrics with the same geodesics) An old local question, to find Riemannian manifolds with the same geodesics (i.e. nonisometric maps preserving geodesics) has been reconsidered recently in Matveev & Topalov [904]. These authors have discovered that such maps allow one to integrate the geodesic equation, and the Schrödinger equation.

♦

11 What is the Best Riemannian Metric on a Compact Manifold?

Contents

11.1	Introduction and a Possible Approach 499	
	11.1.1 An Approach 501	
11.2	**Purely Geometric Functionals**................ **503**	
	11.2.1 Systolic Inequalities 503	
	11.2.2 Counting Periodic Geodesics 504	
	11.2.3 The Embolic Constant 504	
	11.2.4 Diameter and Injectivity 505	
11.3	**Least curved**................................ **506**	
	11.3.1 Definitions 506	
	11.3.2 The Case of Surfaces...................... 508	
	11.3.3 Generalities, Compactness, Finiteness and Equivalence......................... 509	
	11.3.4 Manifolds with inf Vol (resp. inf $\|R\|_{L^{d/2}}$, inf diam) = 0 511	
	11.3.5 Some Manifolds with inf Vol > 0 and inf $\|R\|_{L^{d/2}} > 0$ 515	
	11.3.6 inf $\|R\|_{L^{d/2}}$ in Four Dimensions 518	
	11.3.7 Summing up Questions on inf Vol, inf $\|R\|_{L^{d/2}}$ 519	
11.4	**Einstein Manifolds** **520**	
	11.4.1 Hilbert's Variational Principle and Great Hopes.................................... 520	
	11.4.2 The Examples from the Geometric Hierarchy 524	
	11.4.3 Examples from Analysis: Evolution by Ricci Flow..................................... 525	
	11.4.4 Examples from Analysis: Kähler Manifolds .. 526	
	11.4.5 The Sporadic Examples 528	
	11.4.6 Around Existence and Uniqueness 529	
	11.4.7 The Yamabe Problem 533	
11.5	**The Bewildering Fractal Landscape of $\mathcal{RS}(M)$ According to Nabutovsky**................... **534**	

11.1 Introduction and a Possible Approach

What is the best Riemannian structure on a given compact manifold? René Thom asked the author this question in the Strasbourg mathematics department library around 1960. I should say not only that I liked it, but also that I found it very motivating and frequently advertised it. Moreover, the question is the first problem in the problem list Yau [1296]. It is only recently that I discovered that the question of best metric was posed much earlier by Hopf in Hopf 1932 [730], page 220.

There are obvious motivations; a startling one is to endow manifolds with a nice privileged geometrical structure to obtain purely topological or other types of results using metric methods: think of the conformal representation for surfaces. The problem was completely solved for surfaces in the mid 1930's, as we saw in more than on place, e.g. in note 1.6.1.1, §§4.3.2, §§6.3.2 and §§11.4.3: every compact surface (i.e. two dimensional differentiable manifold) admits best metrics (one only for the sphere and the projective plane), namely those of constant curvature, and (up to diffeomorphism) they are completely classified. The sets of such best metrics are referred to (following Riemann) as *moduli spaces*, of dimension $6g - 6$ for an orientable surface of genus g.

I have put in this chapter some of the possible ways to formulate the question more precisely and what is known in these various cases. I agree that this might seem artificial. Notwithstanding I am offering this chapter as I see things, leaving my readers to appreciate it or not.

The present question is a particular case of the study of the total landscape of the set of Riemannian structures on a given compact manifold. For more general questions about that landscape, a reference of fundamental importance is Nabutovsky & Weinberger 2000 [962] and also see Nabutovsky & Weinberger 1997 [961], Nabutovsky 1996 [958], and §11.5.

For higher dimensions we are going to see that the problem is mostly open. In dimensions three and four, it is full of activity. One of the sources of interest in low dimensional Riemannian geometry is the hope that we might use Riemannian tools to solve topology and differential topology problems which are still haunting topologists. In dimension three, the "best metric" approach based on functionals and their critical points, and the link with the geometrization program of Thurston, is discussed in the various references to Anderson below, in particular Anderson [47, 46, 48]. For dimension four, besides §§11.3.6, see the recent reports Donaldson [456], the books Donaldson & Kronheimer [457] and Morgan [939]. They present in particular the basic tools: Yang–Mills theory, twistors, anti-self-duality in dimension 4 and, from Seiberg & Witten [1118], the Spin^c theory; see the end of §§14.2.2.

For general dimensions, one case however is well understood: that of space forms of negative curvature, thanks to the results of Katok, discussed in theorem 250 on page 483 and those of Besson–Courtois–Gallot discussed in §§10.8 which say that the best metric on compact space forms of negative curvature is (up to trivial scaling) the (unique when the dimension is at least three) locally symmetric one which in fact defines it. This is true for two functionals: the counting function for periodic geodesics (with the restriction today of negative curvature) and the three entropies. Another direction was the Loewner theorem on the torus (theorem 119 on page 326): the flat hexagonal torus minimizes the ratio between the area and the square of systole (the length of the smallest non contractible closed curve).

In a different realm, Hersch's theorem (theorem 183 on page 410) says that the area of a metric on S^2 and the first eigenvalue λ_1 of its Laplacian satisfy

$$\lambda_1 \text{ Area} \leq \frac{8\pi}{3}$$

with equality only for the standard metric. This is particularly remarkable, as we will see that for spheres and the $\mathbb{KP}^n$ there are basically no results justifying our intuition that the standard metrics are the best, while many results assert this for the negative curvature space forms.

11.1.1 An Approach

The scheme of attack is clear from the above examples, and one can find it in a systematic exposition in chapter 4 of Besse 1987 [183], in Anderson 1990 [44], in Blair 2000 [201] and also in Sarnak 1997 [1096]. Sarnak takes as his functional the determinant of the Laplacian; see equation 9.27 on page 419.

The compact manifold being M, let us denote by

$$\mathcal{RM}(M) = \{g \,|\, g \text{ is a Riemannian metric on } M\}$$

the space of all Riemannian metrics on M. The set of all Riemannian structures $\mathcal{RS}(M)$ is its quotient by the group of all diffeomorphisms of M (see equation 4.14b on page 175)

$$\mathcal{RS}(M) = \mathcal{RM}(M) / \text{Diff}(M). \tag{11.1}$$

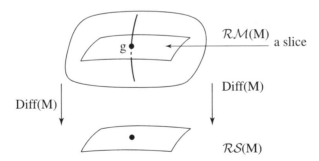

Fig. 11.1. Riemannian metrics modulo diffeomorphism are Riemannian structures

The idea is to consider some map (called a *functional*)

$$F: \mathcal{RS}(M) \to \mathbb{R}$$

defined on the Riemannian structures of any compact Riemannian manifold M, and to look for its infimum $\inf_{\mathcal{RS}(M)} F$, where the infimum is taken over the metrics on M. Usually, the functionals under consideration are positive, or at least nonnegative, but this is not always the case, as we will see with the Hilbert functional in the theory of Einstein metrics in §11.4. Some normalization might be needed for there to be any hope of that some metric will actually minimize the functional. All of the functionals F we will consider are defined by universal formulae defined for any compact Riemannian manifold M, so we can compare F on different manifolds. We can ask some sensible questions:

Question 266 1. *Is $\inf_{\mathcal{RS}(M)} F$ zero or positive?*
2. *If positive, then is it attained by some Riemannian metric g on M? Such a g will be of course called a best metric.*
3. *When these metrics exist, how can we classify best metrics? This is the moduli problem.*
4. *Compute $\inf_{\mathcal{RS}(M)} F$ for various F and frequently encountered manifolds M.*
5. *(This question might seem less natural, but it will appear soon below.) Classify the manifolds M for which $\inf_{\mathcal{RS}(M)} F$ is positive and those for which it is zero.*
6. *Find the possible values $\inf_{\mathcal{RS}(M)} F$ when M runs through all compact manifolds of a given dimension. Or at least, we can ask if the set of these values is discrete, or if zero is an isolated point of this set.*

The content of the chapter is clear. We will study the above problems for various functionals F. But there is one important exception: in §11.4 the functional under consideration (the integral of the scalar curvature, appropriately normalized, known as the Hilbert functional) never has a minimum. So we will look for its critical metrics g; a critical metric means simply a metric for which all of the directional derivatives

$$\frac{d}{dt} F(g + th)$$

vanish, for every symmetric 2-tensor h.

This will turn in fact to be the most satisfying case. The order we follow for the functionals suits the author's taste, starting with the most geometrical ones. Note that the derivative condition above is of course satisfied for minima. But one essential difficulty is that, except for the Hilbert functional, most geometric functionals have no "accessible" derivative.

A contrario, the functional encountered in §11.3 will turn out to be of great interest, defining deep invariants arising from Riemannian geometry for compact smooth manifolds. The infimum being zero or not will be the most important question, along with the question of isolation of the zero value. Except in one case, those invariants are so deep (at least as they appear today) that their explicit value is unknown for almost all manifolds—even space forms.

Some sections will consist in merely summarizing previous presentations of results and adding few comments, questions, etc. We hope that redundancy will not be counted a vice in our encyclopaedic book.

11.2 Purely Geometric Functionals

11.2.1 Systolic Inequalities

We will summarize the content of sections §§7.2.1, 7.2.2, and 7.2.3. In those sections, the game was essentially to have inequalities for volumes independent from the metric, depending only on the topological structure of the manifold. We saw that the results are basically complete, all major questions have been solved. From the systolic point of view, there only remain open a few points (which might be very hard, however). In the present perspective of "the best metric," the open points are precisely the ones of interest. We are looking for the best constants in the systolic inequalities (when they exist); few of these are known. Let us be more precise now.

We look first at manifolds M which are not simply connected, and take as functional F the systolic quotient MinSys(M) of a compact Riemannian manifold M of dimension d to be

$$\text{MinSys}\left(M^d\right) = \text{MinSys}_1\left(M^d\right)$$
$$= \inf\left\{\left.\frac{\text{Vol}(g)}{\text{Sys}(g)^d}\;\right|\; g \text{ a Riemannian metric on } M\right\}$$

Recall that the *systole* Sys(g) (also called the *1-systole*) of a Riemannian manifold was the infimum of the length of closed noncontractible curves. The results of chapter 7 are summarized exactly by saying that:

- The questions (1) to (4) on the preceding page are completely solved for only two manifolds, namely the torus T^2 and the real projective plane $\mathbb{RP}^2$, and they yield the flat hexagonal torus and the constant curvature projective plane as best metrics. For all other surfaces (which are not simply connected) one has a positive MinSys but its value is always unknown—with the exception of the Klein bottle, where the minimum is attained for a unique metric, but the metric has singularities.
- In higher dimensions, Gromov's main result is that MinSys is positive for all reasonable manifolds which are not simply connected (for instance, if that the first dimensional topology is rich enough to generate the fundamental class [in a suitable sense], which finally yields the volume). In not a single case is the exact value of MinSys known. It would be particularly desirable for the projective spaces $\mathbb{RP}^d$ and the tori T^d.

There is an obvious homological or homotopical notion of higher dimensional systole $\text{Sys}_k(M)$ and an associated $\text{MinSys}_k(M)$. Results today, described in §§7.2.3, indicate that this is never positive except possibly in some very, very special cases. In particular there is not a single nonzero $\text{MinSys}_k(M)$ known today. Even for the $\mathbb{KP}^n$, the systole associated to the generating homology class of

$$\mathbb{KP}^1 \subset \mathbb{KP}^n$$

and

$$\dim_\mathbb{R} \mathbb{K} = 2, 4, 8\,,$$

appears at first glance not incalculable and likely positive, because this is the only situation (by the very axioms of projective geometry) where the "systolic submanifolds," namely the projective lines, do really "sneak" all around the space without ignoring anywhere. But already $\mathbb{CP}^2$ and $\mathbb{HP}^2$ are "systolically soft!"

11.2.2 Counting Periodic Geodesics

We saw in §§§10.2.4.1 that on surfaces of higher genus, with the volume normalized, the rate of exponential growth of the counting function for periodic geodesics is the least exactly on constant curvature manifolds. This solved completely question 266 on page 502, parts 1 to 4. For higher dimensions this is also known; but, at least today, only if the curvature is assumed to be negative.

Recall that the *counting function* $CF(L)$ for periodic geodesics denotes the number of geometrically distinct periodic geodesics of length smaller than or equal to L. If one excepts the two cases of surfaces of constant curvature, and arbitrary compact manifolds of constant negative curvature, then one can summarize a good part of the second part of chapter 10 by saying that the functional

$$\inf \ CF(M) = \inf_g \frac{\text{Vol}(M,g)}{\limsup_{L\to\infty} \frac{1}{L} \log CF(L|g)}$$

(where the infimum is carried out over all Riemannian metrics g on M) is completely mysterious today.

11.2.3 The Embolic Constant

Summarizing §§7.2.4 is more pleasant. We recall that the embolic constant $\text{Emb}(M)$ of a compact smooth manifold is the minimum of the *embolic functional*

$$g \mapsto \frac{\text{Vol}(M,g)}{\text{Inj}\,(M,g)^d}$$

over all metrics g on M, where $\text{Inj}\,(M,g)$ means the injectivity radius of M in the metric g. Then we know from theorems 148 on page 355 and 154 on page 357 that:

$$\mathrm{Emb}(M) \geq \mathrm{Emb}\left(S^d\right) = \frac{\sigma(d)}{\pi^d}$$

and equality holds only if M is the sphere. Moreover the standard metric is the only one yielding that minimum (in particular, Emb is positive for every compact manifold). We also know that the value $\sigma(d)/\pi^d$ is isolated.

What we do not know, among other things, is the value of $\mathrm{Emb}(M)$ for any manifold other than the sphere. It would be helpful for various applications to have the exact value for the $\mathbb{KP}^n$ and the tori T^d, see §§7.2.4.6 and §§10.4.2. What we did not say in chapter 7 is that question 266 on page 502 part 6 is solved by

Theorem 267 (Grove, Petersen & Wu 1990 [651]) *The set of embolic constants* $\mathrm{Emb}\left(M^d\right)$ *is discrete when M runs through all manifolds of a given dimension d.*

This is a direct consequence of the finiteness theorem 377 on page 622 of to which we add the above isolation of zero. So that finally $\mathrm{Emb}(M)$ is quite a good functional for classifying compact manifolds. However it seems to us that there is no more than this general statement to be had. The embolic volume appears as if it is measuring the complexity of M. But if one excepts Croke's embolic theorem 154 on page 357 which concerns the category of manifolds, there is today no result relating it to various topological invariants as e.g. the Betti numbers, etc. The category is in fact only an extremely weak invariant since every manifold is of category at most $d+1$.

As far as question 266 on page 502 part 2 is concerned, the existence of extremal metrics is open, but one would expect anyhow only metrics with singularities. The reason for the absence of any existence theorem so far is that all compactness results available, as in §§12.4.2, need some sort of curvature condition.

11.2.4 Diameter and Injectivity

It seems to us that the invariant

$$\inf \frac{\mathrm{diam}}{\mathrm{Inj}}(M) = \inf_g \frac{\mathrm{diam}(M,g)}{\mathrm{Inj}\,(M,g)}$$

(where g runs through all Riemannian metric on M) is a very natural one. Still it is a very puzzling one. The only known result, quite a strong one, is for $M = S^d$: theorem 148 on page 355 says that

$$\inf \frac{\mathrm{diam}}{\mathrm{Inj}}\left(S^d\right) = 1$$

and attained (which is obvious), but only for the standard metric.

But now it is still unknown if spheres and the $\mathbb{KP}^n$ (with their standard metrics) are the only compact Riemannian manifolds for which inf $\frac{\text{diam}}{\text{Inj}} = 1$ i.e.

$$\text{diameter} = \text{injectivity radius},$$

see §§6.5.5.

Answering other question concerning inf $\frac{\text{diam}}{\text{Inj}}$: in each given dimension, the value 1 is not isolated. The discreteness outside 1 is open. It is not clear if there are any relations between inf $\frac{\text{diam}}{\text{Inj}}$ and the topological complexity of M.

11.3 Which Metric is Less Curved?

11.3.1 Definitions

It seems to us that the most natural definition of an optimal metric is that it is the least curved one.

11.3.1.1 inf $\|R\|_{L^{d/2}}$

The first obvious dimensionless functional is

$$\|R\|_{L^{d/2}} = \int_M |R|^{d/2} dV_M .$$

When one scales a metric g into λg, the curvature tensor R_g becomes

$$R_{\lambda g} = \lambda^{-1} R_g$$

and its norm

$$|R_{\lambda g}| = \lambda^{d/2} |R_g|$$

while the volume becomes

$$\text{Vol}(M, \lambda g) = \lambda^{d/2} \text{Vol}(M, g) .$$

For a compact smooth manifold M we introduce the number

$$\inf \|R\|_{L^{d/2}}(M) = \inf_g \int |R_g|^{d/2} dV_M \qquad (11.2)$$

We note in passing that the Gauß–Bonnet–Blaschke theorem 28 on page 138 shows that this functional is certainly not of any use when M is a surface.[1] But we are mainly interested in dimensions higher than two; see §11.1 and §11.3.2 to see why.

---

[1] Surprisingly, there is an elaborate quantum field theory associated to this functional, known as *topological gravity*.

11.3.1.2 Minimal Volume

The next natural functional to evaluate when one wants to get the "least curved" metric is the minimal volume of Gromov. The *minimal volume* of a compact manifold M is defined as the minimum of the volumes of all Riemannian metrics on M such that the sectional curvature satisfies

$$-1 \leq K \leq 1,$$

i.e.

$$\inf \operatorname{Vol}(M) = \inf_g \operatorname{Vol}(M, g) \tag{11.3}$$

where g runs through all Riemannian metrics on M such that $-1 \leq K \leq 1$.

We remark again that when one scales a metric g into λg the curvature K_g becomes

$$K_{\lambda g} = \lambda^{-1} K_g$$

while the volume becomes

$$\operatorname{Vol}(M, \lambda g) = \lambda^{d/2} \operatorname{Vol}(M, g) .$$

After having the curvature of g squeezed into $[-1, 1]$ by a suitable scaling, one looks for the smallest possible volume (not by scaling now). Equivalently, one can look at the infimum over all metrics of $\sup_M |K|$ under a volume normalized to 1. Putting everything together, we see that equivalently we could study

$$\inf \operatorname{Vol}(M, g)^{d/2} \sup |K| .$$

11.3.1.3 Minimal Diameter

A third functional is the *minimal diameter* $\inf \operatorname{diam}(M)$, that is to say the minimum of the diameter of M in the various metrics g under again the constraint

$$-1 \leq K \leq 1$$

on the sectional curvature, i.e.

$$\inf \operatorname{diam}(M) = \inf \operatorname{diam}(M, g) \tag{11.4}$$

where g runs through all Riemannian metrics on M such that

$$-1 \leq K \leq 1 .$$

Having the minimal diameter is also a notion of being "less curved," since

$$\operatorname{diam}(M, \lambda g) = \lambda^{1/2} \operatorname{diam}(M, g) .$$

The functional $\inf \|R\|_{L^{d/2}}$ is the only one which seems to be accessible to taking derivatives, but in fact we will see in §§11.3.6 that this is not really the case, at least today and with the exception of dimension 4. Of course the volume is accessible to differentiation by the formula

$$\frac{d}{dt} \operatorname{Vol}(M, g + th) = \int_M \operatorname{tr}_g h \, dV_M \tag{11.5}$$

but it is the condition

$$-1 \leq K \leq 1$$

which is not, nor is the diameter. For the diameter, contemplation of the cut-locus from §§6.5.4 will convince you why. For $\sup |K|$ the reason why this functional is easily differentiated can be seen from §§4.4.2, and this difficulty will be met again on page 579 when trying to understand manifolds of positive curvature, and in note 11.3.1.1 just below for the "best pinching" problem.

It will turn out that our three functionals have profound relations to other topics of Riemannian geometry, in particular in §12.4 where we study the set of all Riemannian metrics on a given compact manifold. These are the real reasons which obstruct proofs of many results concerning these functionals, since a contrario they are still (for the moment) useless as tools to find a best metric on all but a handful of manifolds, and on surfaces.

Nabutovsky & Weinberger 1997 [961] took a new look at the search for minimal diameter; we will examine their results in §11.5. Their text is very interesting—it shows examples of manifolds where the space of all Riemannian structures has a very complicated structure, namely infinitely many "very deep" local minima for the diameter under the curvature condition $|K| < 1$, but the metrics yielding these local minima cannot be connected within the space of metrics satisfying $|K| < 1$. Note that the theory of algorithmic complexity is heavily used, as it was in theorem 203 on page 444.

Note 11.3.1.1 (The best pinching problem) Some readers might want to define the least curved metric as follows. In the case where our manifold M admits some metric whose curvature is of a constant sign, then the least curved metric would be one for which the ratio

$$\left| \frac{\sup K}{\inf K} \right|$$

is as small as possible. This kind of question belongs to §§12.2.2. We will see there that is a theoretical answer, thanks to compactness theorems, but the difficult structure of the curvature tensor (see §§4.4.2) is precisely what will prevent us, at least today, from drawing any significant conclusions. ♦

11.3.2 The Case of Surfaces

As an introduction, and for simplicity's sake, we first treat best metrics for surfaces. The complete description of surfaces of minimal volume[2] is in fact understood thanks to the Gauß–Bonnet formula 28 on page 138:

---

[2] *Minimal area* would be a better name.

$$\int_M K\, dA = 2\pi \chi(M).$$

Taking the absolute value yields

$$\inf \mathrm{Vol}(M) \geq 2\pi |\chi(M)|$$

and equality holds if and only if the curvature is constant. If M is the torus T^2 or the Klein bottle K then

$$\inf \mathrm{Vol} = 0$$

and is of course never attained. We are using here theorem 69 on page 252: every compact surface admits a metric of constant curvature.

This result has solved all of the questions above since the metrics of constant curvature (space forms) are completely understood in dimension 2, except for T^2 and K. For these two surfaces, inf Vol is not good enough to find the best metric. But note that you can find the best metrics even here by taking as functional the integral

$$\int_M K^2\, dA$$

under the normalization $\mathrm{Vol}(M, g) = 1$. Then you get the expected answer for any compact surface. Finally note that the set of all numbers $\inf \mathrm{Vol}(M^2)$ as a subset of $\mathbb{R}$ is the discrete set $2\pi\mathbb{N}$.

We recall that the case of surfaces enjoys even a stronger property, that of conformal representation seen in theorem 70 on page 254; this implies that the set of all Riemannian metrics on a compact surface is structured (fibered) above the moduli space made up by the set of constant curvature metrics.

11.3.3 Generalities, Compactness, Finiteness and Equivalence

If one excepts the case of dimension 4 below, $\inf \|R\|_{L^{d/2}}$ is not accessible to differentiation in the sense that when one computes the equation of critical points, one finds a system of partial differential equations of order 4 which one cannot simplify today (even, by the way, in dimension 4). For more details, see §§11.3.5. See Besse 1987 [183] section 4.H for these equations as well as other ones given by quadratic functionals, and Anderson 1990 [44] and the deep study Anderson 1992 [43] for dimension 4. For general dimensions, a naive idea is to use

$$\int_M |R|^2\, dV_M$$

under the normalization $\mathrm{Vol}(M, g) = 1$, since this quadratic functional differentiates nicely in taking the first variation. But Gromov remarked that if the dimension of M is at least 5, then the infimum is zero for every manifold

M. To see this, embed in M a torus T^{d-1} and enlarge it as $[-\alpha, \alpha] \times T^{d-1}$. Playing with large α and a flat metric on this product, let the ambient region in M away from the torus carry a fixed metric; then

$$\int_M |R|^2 \, dV_M$$

remains fixed while the volume tends to infinity. But when one normalizes the metric by a factor λ, the integral

$$\int_M |R|^2 \, dV_M$$

will behave like $\lambda^{2-d/2}$ which goes to zero because $d \geq 5$. This counterexample was one of the reasons for Gromov to introduce the a priori less natural minimal volume of §§§11.3.1.2.

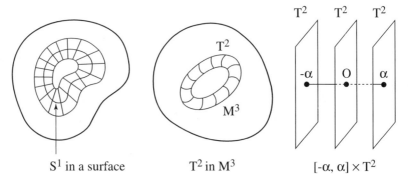

Fig. 11.2. (a) A circle in a surface (b) A two dimensional torus in a three dimensional manifold (c) $[-\alpha, \alpha] \times T^2$

So we are back to inf Vol and inf $\|R\|_{L^{d/2}}$, and we ask first if they are related. But remember proposition 38 on page 206: the sectional curvature and the curvature tensor are universally related, so that from $-1 \leq K \leq 1$ we deduce immediately that

$$\|R\|^{d/2} < c(d)$$

for some constant depending only on the dimension. Consequently

Lemma 268 *For any manifold M*

$$\inf \|R\|_{L^{d/2}}(M) < c(d) \inf \, \mathrm{Vol}(M).$$

In particular inf Vol $= 0$ *implies* inf $\|R\|_{L^{d/2}} = 0$.

The converse is still an open problem today, despite many efforts. Here is why: assume that $\inf \|R\|_{L^{d/2}}(M) = 0$. Then we have a sequence $\{g_k\}$ of metrics for which the limit of the integral of the curvature is zero. Since we use the dimensionless power $d/2$ we can assume that $-1 \leq K \leq 1$ for all of these metrics. Assume by contradiction that the volume as measured in these metrics is bounded from below and think of the compactness theorem 384 on page 627. We almost fulfill the conditions for convergence except that we do not know if the diameter stays bounded when k goes to infinity. If it where we would be essentially finished: the limit would be a flat metric, a contradiction. Controlling the diameters of the g_k is the principal concern today; however is still conjectured that $\inf \mathrm{Vol} = 0$ and $\inf \|R\|_{L^{d/2}} = 0$ are equivalent.

We are far today from knowing how to solve question 266 on page 502 parts 1 to 6 above for these two functionals. The first thing to do is to decide if such a functional is zero or not. Independently of its own interest, the question about whether $\inf \mathrm{Vol} > 0$ enters naturally as we have just seen in the compactness theorem. It is also important to remark that the starting key to that compactness theorem 384 on page 627 is Cheeger's lower bound for the injectivity radius in theorem 90 on page 273. To be sure of this, just note that by rescaling there is not much difference between the condition $-1 \leq K \leq 1$ and the condition $a \leq K \leq b$ with any a and b. The interest of inf diam will appeared below in theorem 312 on page 568.

11.3.4 Manifolds with inf Vol (resp. inf $\|R\|_{L^{d/2}}$, inf diam) $= 0$

11.3.4.1 Circle Fibrations and Other Examples

Of course flat manifolds[3] have identically zero curvature and hence $\inf \mathrm{Vol} = 0$. The main point is that there are many other compact manifolds with $\inf \mathrm{Vol} = 0$. The basic example is when M admits a free action of the circle S^1 i.e. is an S^1 fibered manifold over some manifold N. The first non trivial example is Hopf's fibration

$$S^3 \to S^2.$$

(See §§§4.1.3.5 if needed.) Then pick any Riemannian metric g on M. This yields at every point a Euclidean product decomposition for the tangent space, and we write this as $g = g_{vert} + g_{hor}$ for the vertical and horizontal parts. Pick now any $\varepsilon > 0$ and introduce the new metrics

$$g(\varepsilon) = \varepsilon g_{vert} + g_{hor}.$$

One then verifies easily that the sectional curvature stays bounded, for example by using the O'Neill formulas in equation 15.17 on page 721. Heuristically one can say that when shrinking the fibers, they do not generate large curvature because they are one dimensional. Curves have zero sectional curvature.

---

[3] We know that they are finite quotients of tori by Bieberbach's theorem 98 on page 291.

And the fact that the fibers are twisted contributes to a fixed amount, independent of the length of the fibers.

Hence inf Vol $= 0$, inf $\|R\|_{L^{d/2}} = 0$ and inf diam $= 0$ by construction. This was first discovered in Gromov 1983 [617] along with various generalizations. Note that in view of equation 15.17 on page 721 we will automatically get inf $\|R\|_{L^{d/2}} = 0$.

There are two ways to generalize the above example. The first will be treated in §§§11.3.4.3. The second consists in putting the following example into a very general setting. A typical example is as follows. Take two surfaces N and N' each with a circle as boundary, and consider the products $M = S^1 \times N$ and $M' = S^1 \times N'$. They are manifolds with boundary, and in both manifolds the boundary has the topological type of a torus T^2. Then glue M and M' along that T^2 not in the trivial way but interchange the two circles in T^2 (exchange the parallels and the meridians); see figure 11.3. The resulting manifold does not in general admit an S^1 action. It does admit such an action locally on both parts but these two actions agree in the common parts since we have a torus action there.

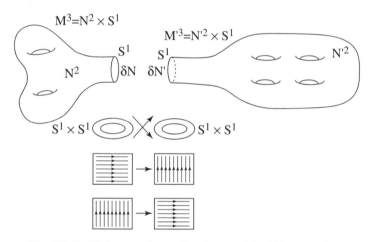

Fig. 11.3. Gluing products of surfaces with circles together.

We will see more in detail in §§§12.4.3.1 that this example has been put in a very general context in Cheeger & Gromov 1986-1990 [348], [347] and Cheeger & Rong 1995 [352], where the notions of F and T structures are defined. Important refined notions are that of polarized and pure polarized F structures. Detailed definitions are also to be found at the end of the survey Fukaya 1990 [531]. Existence of various structures of those types is linked with the topology of the manifold, but how exactly is still unclear.

For example for a general F structure one can build up only Riemannian metrics with $-1 \leq K \leq 1$ and injectivity radius going to zero. If the structure is moreover polarized then one can get the volume going to zero, hence inf Vol = 0. We will see in §§§11.3.4.4 below results showing that probably "most" manifolds have inf Vol = 0. This makes it an important goal to find ones with positive inf Vol.

11.3.4.2 Allof–Wallach's Type of Examples

These Riemannian manifolds deserve special consideration because we will meet them again in more than one topic—namely in the study of manifolds of positive sectional curvature (see §§§12.3.1.1) and just below in §§§11.4.2.2. They were found in Wallach [1231] originally as new manifolds of positive curvature. They are the homogeneous spaces of Lie groups defined as

$$\mathfrak{W}_{p,q} = SU(3)/S^1(p,q)$$

where (p,q) runs through the lattice $\mathbb{Z}^2$. The compact simple Lie group $SU(3)$ is of rank 2. This means that it admits subgroups isomorphic to the torus T^2. These maximal abelian subgroups are all conjugate (from general results of Élie Cartan) and their Riemannian structure is that of $\mathbb{R}^2/\Lambda$ where Λ is the regular hexagonal lattice. The notation

$$S^1(p,q)$$

means the circle subgroup of the torus T^2 which is the quotient by Λ of the line in $\mathbb{R}^2$ passing through $(0,0)$ and (p,q). Thus any $\mathfrak{W}_{p,q}$ is fibered by circles over the quotient manifold which is $SU(3)/T^2$ where T^2 is any maximal torus in $SU(3)$. In particular, inf Vol = 0 for all of them.

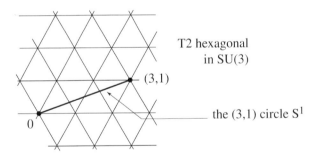

Fig. 11.4. T^2 hexagonal in $SU(3)$. The $(3,1)$ circle S^1

On any $\mathfrak{W}_{p,q}$ there are many homogeneous Riemannian metrics because the isotropy group is reducible (see §§§11.4.2.2 for the spirit) but for a suitable choice the sectional curvature is positive; see section §§§11.4.2.2 and §§§12.3.1.1.

11.3.4.3 Nilmanifolds and the Converse: Almost Flat Manifolds

Assume that we have a manifold M^d which admits d successive fibrations by S^1. Then not only the minimal volume but the minimal diameter will be zero since we will be able to shrink an initial metric in d linearly independent directions while keeping the curvature bounded on both sides. This is by a systematic use of formulas for Riemannian submersions (see equation 15.17 on page 721). Such manifolds need not be tori. So there are manifolds with inf diam $= 0$ which are not tori.

More precisely they are easily seen to be the so-called *infranilmanifolds*, see more on them in theorem 312 on page 568. They are quotients by suitable discrete subgroups of nilpotent Lie groups. Another way to look at inf diam $= 0$ is to scale things to work within manifolds of diameter bounded by 1. Then the above examples admit metrics with $-\varepsilon < K < \varepsilon$ for any $\varepsilon > 0$. In §§12.2.3 we will look at this as a pinching problem. A great event was when the converse was proven in 1978 by Gromov, solving then the 0-pinching problem: see theorem 312 on page 568.

11.3.4.4 The Examples of Cheeger and Rong

Fixing a dimension d, let us look at the subset of $\mathbb{R}^+$ consisting of the minimal diameters of all compact smooth d dimensional manifolds. We are addressing question 6 on page 502. The above theorem is a gap theorem: it shows that in this set, zero is isolated. Put together, two very recent works show, that except at zero, this set is far from discrete:

Theorem 269 (Cheeger & Rong 1995 [352]) *For every dimension there is a scale $\{D_i\}$ of diameters (with $D_i \to \infty$) such that in every interval $[D_i, D_{i+1}]$ there are infinitely many diffeomorphism types of compact manifolds with*

$$\text{inf diam} \in [D_i, D_{i+1}].$$

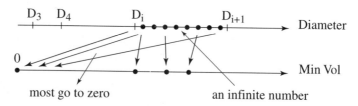

Fig. 11.5. Minimal diameters and minimal volumes of compact manifolds of given dimension

These D_i are obtained by building up geometrically suitable examples with various techniques: T^2 actions, gluing, collapsing, F structures which are not polarizable, etc. For inf Vol there is only a partial gap result:

Theorem 270 (Cheeger & Rong 1996 [353]) *There is a constant $c(d,D)$ depending only on the dimension d and a positive number D so that every manifold M of dimension d which has*

$$\inf\ \mathrm{diam}(M) < D \text{ and } \inf\ \mathrm{Vol}(M) < c(d,D)$$

has
$$\inf\ \mathrm{Vol}(M) = 0.$$

Now from the finiteness theorem 372 on page 616 one deduces immediately

Theorem 271 *For any dimension d and diameter D there are only a finite number of manifolds of dimension d with $\inf\ \mathrm{diam} < D$ and $\inf\ \mathrm{Vol} > 0$.*

Let us look now at $\mathbb{R}^+$ and the subset of possible minimal diameters for a given dimension as sitting in the scale made up by the D_i. In every interval $[D_i, D_{i+1}]$ we have an infinite number of different manifolds (differentiable). But only a finite number of them can have a positive minimal volume. So that in a sense most manifolds of a given dimension have zero minimal volume.

In our search for a best metric, this shows that the minimal volume is not a good candidate for "all" manifolds. Perhaps it will be for some special manifolds, though this has not yet become clear. We will have to look to other functionals and will come back to the minimal volume in §§11.3.7.

11.3.5 Some Manifolds with $\inf\ \mathrm{Vol} > 0$ and $\inf\ \|R\|_{L^{d/2}} > 0$

11.3.5.1 Using Integral Formulas

From §15.7 we know that there are integral formulas which are universal in the curvature and which yield the Euler–Poincaré characteristic and all of the characteristic numbers. Let Ω be any such invariant of a smooth manifold M. This means (see also equation 15.14 on page 717) that:

Theorem 272
$$\Omega = \int_M \mathrm{univ}_\Omega(R) dV_M.$$

Then on one hand knowing the full norm of the curvature tensor we certainly have
$$|\mathrm{univ}_\Omega(R)| \leq c(\Omega)|R|^{d/2}.$$

Taking absolute values in the above formula one gets immediately

$$\inf \|R\|_{L^{d/2}}(M) \geq \frac{1}{c(\Omega)}.$$

If we know that $-1 \leq K \leq 1$ then by the fact that R and K are "equivalent" we also get

hence:
$$|univ_\Omega(R)| \leq c'(\Omega)$$

$$\inf \text{Vol}(M) \geq \frac{|\Omega|}{c'(\Omega)}.$$

Then we know that both inf Vol and $\inf \|R\|_{L^{d/2}}$ are positive for every manifold having a nonzero characteristic number. This was remarked by Cheeger back in 1970. This can at best take care of manifolds whose dimension is even and a multiple of 4 to employ characteristic numbers. It is also important to remember from §§15.7.7 that there are no other integral formulas universal in the curvature. So we have to find a new tool in order to get more manifolds with

$$\inf \text{Vol} > 0.$$

11.3.5.2 The Simplicial Volume of Gromov

In the founding text, Gromov 1983 [617], proved that the minimal volume is nonzero for space forms of constant negative curvature of any dimension. From there, Gromov finds the same result for many other manifolds using various functorial properties of the simplicial volume. Almost no new result has appeared since that one.

The proof uses essentially as written above the notion of simplicial volume for topological manifolds, which is a fundamental invariant for compact smooth manifolds, but unbelievably deep and baffling: whether or not it is zero, not to speak of its exact value when it is positive, is unknown up to very very few exceptions to be met below. The simplicial volume was used for Mostow rigidity in §§6.6.2.

Assume that M is oriented and write the fundamental class $[M]$ in terms of some triangulation as

$$[M] = \sum_i a_i \sigma_i$$

with the real numbers a_i not necessarily integers. Then the *simplicial volume* is:

$$\|M\| = \inf \left\{ \sum_i |a_i| \ : \ [M] = \sum_i a_i \sigma_i \right\}.$$

For example it is zero for spheres: cover the sphere by turning more and more times around the equator. The volume of the simplices gets larger and larger, forcing $\sum_i |a_i|$ to be smaller and smaller. More generally, the simplicial volume of a manifold is zero if the manifold admits self-maps of degree larger than 1. In fact, if

$$f : M \to M'$$

is any map of degree d, then

$$\|M\| \geq d \, \|M'\|.$$

But much more generically, $\|M\| = 0$ not only for any simply connected manifold, but as soon the fundamental group is not too large. Here, "not too large" means amenable. In fact the simplicial volume is technically determined by $\pi_1(M)$ and the classifying map

$$M \to K\left(\pi_1(M), 1\right),$$

see Gromov's text for details. However we are going to see that the story is far from being finished by this theoretical answer.

The simplicial volume enjoys functorial inequalities for the product and the connected sum of manifolds:

$$\|M \# N\| = \|M\| + \|N\|$$
$$a(d)\|M\|\|N\| \leq \|M \times N\| \leq b(d)\|M\|\|N\|$$

where a and b are universal in the dimension.

Gromov proved two results in the above text. The first is:

Theorem 273 (Gromov 1983 [617]) *For a compact space form M of constant negative sectional curvature the simplicial volume is positive; better its exact value is*

$$\|M\| = \operatorname{Vol} M / VRS(d) .$$

Here $VRS(d)$, introduced in note 4.3.2.2 on page 180, denotes the volume of the regular simplex (simplices) in d dimensional hyperbolic space Hyp^d. The very geometrical proof is basic in contemporary negative curvature techniques and was used by Gromov to get a proof of Mostow's rigidity theorem (see theorem 99 on page 293 and §§6.6.3). This is extremely well written in Benedetti & Petronio 1992 [129]. It uses the sphere at infinity on the universal covering (which is the hyperbolic space Hyp^d; see §§4.3.2), ideal simplices, and the basic fact that in hyperbolic geometry, the volume of any simplex is smaller than or equal to that of an ideal simplex. Finally, the volume of ideal simplices is universally bounded by the volume of the "regular" ones, namely with the maximum number of symmetries. In dimension 2, they are all regular. In short: one can compute the volume of M as if M were triangulated exactly by regular ideal simplices. This is of course the case for the whole of Hyp^d but for a compact object "downstairs" it needs a mean value process.

Heuristically: when you stretch a simplex in hyperbolic geometry its volume is bounded. Oppositely, in spherical geometry or Euclidean geometry, stretching and covering more and more yield infinite volume.

A typical mystery: from the functorial properties above and theorem 273 we know that any product of compact manifolds of negative curvature has a nonzero simplicial volume. Such a product has a lot of zero curvature, namely for the planes corresponding to the product decomposition. So some (a lot of) zero curvature still permits $\|M\| > 0$. Besides products of negatively curved manifolds, the simplest example of strongly structured manifolds of nonpositive curvature are the locally symmetric space forms of §§6.6.4 of rank at least two. It is natural to conjecture that the simplicial volume of these forms is still positive. But the question remains open today. Manifolds with positive simplicial volume are extremely difficult to construct. In Nabutovsky & Weinberger 1997 [961], quite a few such manifolds are constructed, but this is extraordinarily difficult. We will meet this important text in §11.5.

From the above one might expect some relation between the simplicial volume and the minimal volume. This is in fact precisely the second result of Gromov:

Theorem 274 (Gromov 1983 [617]) *If on the Riemannian manifold M one has inequality*
$$\mathrm{Ricci}(g) \geq -(d-1)g$$
then
$$\mathrm{Vol}(g) \geq c(d)\|M\|$$
for some universal constant $c(d)$.

So we have a positive minimal volume as soon as the simplicial volume is positive since already $K \geq -1$ implies
$$\mathrm{Ricci}(g) \geq -(d-1).$$

Note that it does not say anything about $\inf \|R\|_{L^{d/2}}$. The proof of theorem 274 is extremely hard and to our knowledge does not figure in any other text. The technique is that of the "diffusion of cycles," also used in note 7.2.1.2 on page 333.

But note that those manifolds are never simply connected. Today no one knows of any odd dimensional simply connected compact manifold for which one can prove that its minimal volume is positive. For the simplicial volume, results are also very difficult. Ville 1996 [1219] studies the case of compact complex manifolds.

11.3.6 $\inf \|R\|_{L^{d/2}}$ in Four Dimensions

Besides the study of surfaces, there is one other exceptional dimension, namely $d = 4$. In that dimension, the generalization of the Gauß–Bonnet theorem found in equation 15.11 on page 712 reads

$$8\pi^2 \chi(M) = \int_M \left(|R|^2 - \left|\mathrm{Ricci} - \frac{\mathrm{scalar}}{4}g\right|^2 \right) dV_M. \tag{11.6}$$

On a given compact Riemannian manifold M with $\chi(M) \geq 0$, our functional has a lower bound of $8\pi^2 \chi(M)$. This potential minimum is attained if and only if M has a metric with

$$\mathrm{Ricci}_g = \frac{\mathrm{scalar}_g}{4} g.$$

But this condition is nothing but what we (in §11.4) are going to call an *Einstein metric*. In the context of various questions of convergence, etc. for manifolds under various conditions, equation 11.6 on the preceding page is essential. For example see theorem 379 on page 623 and we will find it again in §§§11.4.6.1.

11.3.7 Summing up Questions on inf Vol, inf $\|R\|_{L^{d/2}}$

Let us finish the inf Vol story by addressing some of the parts 1 to 6 of question 266 on page 502. The first is: how does the subset of the reals made up by the minimal volumes of all manifolds of a given dimension d look? Is it a discrete set? For surfaces we saw that the Gauß–Bonnet–Blaschke formula from theorem 28 on page 138 tells us that our subset is the arithmetic progression $2\pi\mathbb{Z}$ and in particular of course 0 is isolated.

Theorem 275 (Cheeger & Gromov 1986-1990, Rong 1993 [1064])
There is a gap for inf Vol *in dimensions 3 and 4, namely there are constants c_3 and c_4 such that, if*

$$\inf \mathrm{Vol}\left(M^3\right) < c_3 \qquad \textit{(resp. } \inf \mathrm{Vol}\left(M^4\right) < c_4 \textit{)}$$

then

$$\inf \mathrm{Vol}\left(M^3\right) = 0 \qquad \textit{(resp. } \inf \mathrm{Vol}\left(M^4\right) = 0 \textit{)}.$$

In dimension 3, the proof is an implicit corollary of the first authors results on the theory of collapsing, which we will meet in §§§12.4.3.1: small volume implies uniformly small injectivity radius by the results in §§§7.2.4.1. As a consequence, one has an F structure. One needs only to prove it admits also a polarized F structure. Then inf Vol has to vanish. Contrarily, the proof of the isolation of 0 in dimension 4 obtained in Rong 1993 [1064] is extremely involved and not completely geometric. It uses the η invariant (see §9.14 and §14.2.1) in a basic way. Remember also theorem 270 on page 515 which yields a gap when one restricts the problem to simply connected manifolds of a given bounded diameter.

A gap theorem is not be expected, already in dimension 4, if one adds the condition of bounded diameter when the volume goes to zero. This is a consequence of the example 6.4 in Cheeger & Rong 1995 [352].

The second question is to compute the minimal volume of "standard" manifolds, e.g. space forms of different types. Minimal volumes are dramatically different according the sign of the curvature. For positive curvature or non-negative curvature not a single minimal volume (when nonzero of course) is known. The problem starts with S^4. Equation 11.6 on the facing page will not help: we will see in §§11.4.5 that we still do not know if the standard metric on S^4 is the only Einstein one. The only known fact is a local result for even dimensional spheres in Ville 1987 [1218].

One reason why we know so little about explicit values of minimal volume when we know that it is positive is that the following scheme of attack is blocked for the moment. One could hope to get an best metric by a convergence theorem like theorem 384 on page 627: we work in the realm of manifolds with $-1 \leq K \leq 1$ and with a lower bound for the volume. But unhappily the convergence theorem 384 needs an upper bound for the diameter. There is no way today in general to get around this difficulty; see however Anderson 1990 [44] and remark that in pages 74-75 of Gromov 1983 [617] there is an existence result for an "extremal" metric, but it is not too detailed.

There is one exception: the very special case of negative curvature space forms. One has an unbelievably strong result. A corollary of the main theorem 251 on page 484 is:

Theorem 276 (Besson, Courtois & Gallot 1995 [189]) *Under the sole condition*
$$\text{Ricci} \geq -(d-1),$$
any metric on a compact hyperbolic space form has a volume larger than or equal to that of the hyperbolic metric and equality if and only if the metric is of constant curvature, and hence equal to the original one if $d \geq 3$ by Mostow rigidity.

Once again we know that the hyperbolic metric on a negative space form is the best one under the minimal volume criterion. The proof is fast. The result 213 on page 455 is a minimum for the volume entropy. But this entropy is given by the volume of balls in the universal covering. Then one just needs to apply Bishop's theorem for the volume of ball as in theorem 107 on page 310. The recent Bessières 1998 [186] is important: it shows in particular that the minimal volume is not preserved by connected sum.

11.4 Einstein Manifolds

11.4.1 Hilbert's Variational Principle and Great Hopes

We turn now to Einstein manifolds. A large book is dedicated to this sole topic: Besse 1987 [183] and was very complete in 1987. But since then the field has grown explosively, so that we are happy to have the book LeBrun & Wang 1999 [858]. In Besse's book, the chapter 0 is a very informative introduction. One should look at Besse's book for the references missing here. The topic being very large and well studied, we are only trying now to draw some kind of a map for helping the reader to find his way in it. But we find still the topic very hard to organize.

11.4 Einstein Manifolds

In a sense the notion of Einstein manifold can be traced back to Hilbert 1915 [711] who computed (motivated by theoretical physics) the derivative in the space of all Riemannian metrics of the simplest possible curvature functional: namely the "total" scalar curvature

$$F(g) = \int_M \text{scalar } dV_M.$$

The formula is obtained by a straightforward computation and is:

$$dF(g) \cdot h = \int_M \left\langle \frac{\text{scalar}_g}{2} g - \text{Ricci}_g, h \right\rangle dV_M \qquad (11.7)$$

In dimension 2 one gets identically zero, a phenomenon explaining in some sense the Gauß–Bonnet–Blaschke formula 28 on page 138: the integral of Gauss curvature is constant, but this of course does not tell us what this constant is. When the dimension is $d \geq 3$, criticality of this functional implies that $\text{Ricci}_g = 0$, which is too restrictive in general (see §§11.4.6.4 if you are wondering why) and should not be surprising, because our functional is not dimensionless. So we need to normalize the metric, and this in the simplest (computational) way: namely ask the volume to be some given constant (which does not matter). Then we get the condition

$$\text{Ricci}_g \text{ is proportional to } g$$

or

$$\text{Ricci}_g = \frac{\text{scalar}_g}{d} g. \qquad (11.8)$$

Such metrics will be called from now on *Einstein metrics*.

This condition is obtained by the Lagrange multiplier technique, the constant having to be scalar/d by just taking the trace with respect to g of both sides.

Note that in dimension 2, any metric is Einstein. But taking the trace of the (covariant) derivative of both sides of equation 11.8 and applying the Bianchi identity 4.28 on page 203 (or 15.2 on page 700) yields:

Theorem 277 *For any Einstein metric on a manifold of dimension $d \geq 3$, the scalar curvature is constant. The sign of the scalar curvature is called the sign of the Einstein metric and can be written as 1, 0, or -1 (or positive, zero, negative). An Einstein metric with zero sign, i.e. vanishing Ricci tensor, is also called* Ricci flat.

In three dimensions, the Einstein condition is "too strong." It implies constant sectional curvature (and conversely of course), hence a space form. so that we mainly discard this dimension from now on. In dimension 4, and when the characteristic is nonnegative, results from §§11.3.6 tell us that this is equivalent to minimizing the total norm of the full curvature tensor.

What is known today about existence, uniqueness or moduli of Einstein metrics? But first discard immediately looking for an extremum. It will follow from facts just below that the present functional has no local minima or maxima. However the equation 11.8 on the previous page creates great hopes to find a best metric, for many apparently good reasons. Here are some.

First we have an equation (system of partial differential equations) for the metric g. So we can hope to use more or less classical results about partial differential equations. This will work perfectly in dimension 3, using an evolution equation; see §§11.4.3 For higher dimensions, the prospects are rather poor.

A very nice result of DeTurck & Kazdan 1981 [442] says that Einstein metrics are forced to be real analytic. A sketch of the proof is in section 5.E of Besse's book; it is another nice application of the notion of harmonic coordinates which we saw on page 267.

A third fact is that we will see many examples of Einstein manifolds, both in the geometric hierarchy and in the Kähler case, but also a lot of more sophisticated ones of various types.

But a great hope would be this: critical points make us think of Morse theory as described in §§10.3.2. Before going on, we suggest to the reader who is not familiar with Morse theory to have another look at §§10.3.2. He will see that we have various things to do to employ Morse theory. We have in the first place to set a frame: here the infinite-dimensional realm cannot be avoided as it was for geodesics by taking a finite-dimensional approximation using broken geodesics, since we are in fact in the set of all Riemannian metrics $\mathcal{RM}(M)$. But there is a Morse theory for infinite dimensional manifolds called Palais–Smale theory: Palais & Smale 1964 [992], or 5.40 of Besse 1987 [183] or Jost 2002 [768]. Besides classical finite dimensional Morse theory, one needs a convergence condition for critical points, called "condition C of Palais–Smale;" we will come back to that below.

Moreover one cannot work in $\mathcal{RM}(M)$; one has to take its quotient

$$\mathcal{RS}(M) = \mathcal{RM}(M) / \text{Diff}(M)$$

(see equation 11.1 on page 501) by the diffeomorphisms, this for at least two reasons. The first is that we hope to get some nontrivial topology for this quotient, since $\mathcal{RM}(M)$ is a convex cone and hence topologically trivial and of no interest when one wants to use Morse theory. The second point is this: the diffeomorphisms leave invariant any "reasonable" functional since they are invariant by construction. We cannot hope to have here the nondegeneracy condition of theorem 210 on page 453. But we can hope to have this condition satisfied for the quotient $\mathcal{RS}(M)$. The second thing to do is to compute the second derivative $d^2F(g)$ to see if we have a finite index at critical points in $\mathcal{RS}(M)$. The computation of $d^2F(g)$ in $\mathcal{RM}(M)$ is quite straightforward. It can be found explicitly in 4.G of Besse 1987 [183] but the result is awkward, with exception in dimension 4 where it is used in Anderson 1992 [43], see §§§11.4.6.3. But there is the problem of working in the quotient

$$\mathcal{RS}(M) = \mathcal{RM}(M) / \text{Diff}(M).$$

This quotient is not a manifold in any reasonable (even infinite dimensional) sense, but one can use the notion of slice (see 12.22 in Besse's book) to work in conditions similar to those of manifolds.

What have these hopes yielded? Here are the main conclusions. The positive points are this: in the space $\mathcal{RS}(M)$, the index of a critical point (an Einstein structure) is never finite, but there is still hope because this index is in fact always finite if one restricts oneself to the subspace of manifolds of constant scalar curvature. And this is not a problem, since the set $\mathcal{RS}(M)$ retracts nicely onto such a subset (but beware: not necessarily conformally as we will see in §§11.4.7: see section 4.G of Besse 1987 [183].

The negative points are the following. First, the Palais–Smale condition C is not satisfied even if the indices are finite. In the subspace of constant scalar curvature, counter-examples will quoted below in §§11.4.5. Moreover today, believe it or not, the topology of $\mathcal{RS}(M)$ is almost totally unknown. It is of course a basic topic, as it is so natural, but with essentially no result; we know only of Bourguignon 1975 [237].

Experts are divided today about the eventual possibility to find Einstein metrics: some think it cannot work by essence. Some other think a tool like Floer homology (see Hofer, Taubes, Weinstein & Zehnder 1995 [723]) could help. See Besse's book and below for more comments. Note also that such a general scheme cannot always work since in dimension 4 there are necessary topological conditions for the existence of Einstein metrics on a given manifold: see §§§11.4.6.1. In Nabutovsky 1995 [956] a new approach is taken to the research of Einstein metrics: it is linked to algorithmic complexity questions. We will discuss this further in §11.5.

Note 11.4.1.1 (Connes' description of the Hilbert functional) This Hilbert functional received an interesting new direction in Connes 1995-6 [401] where it appears as the $(d-2)$ volume of the manifold, therefore an "area" in dimension 4 "á la Morse." It might be useful for studying Poincaré's conjecture.
♦

There are good moduli results for Einstein metrics; see §§§11.4.6.2. Of special interest and completely mysterious is the case of Ricci flat metrics; see §§11.4.5.

We sum up now what will follow in the next sections. First there are plenty of examples of Einstein metrics, for homogeneous spaces and Kähler manifolds. Second the general existence theorem is in the worst possible condition: it is possible that there is an Einstein metric on every manifold of dimension larger than 4. However it might be that an existence theorem would require that we admit manifolds with reasonable singularities; see Anderson 1994 [45]. The scheme here is as follows: the Ricci flow §§11.4.3 diminishes the defect to be Einstein, so that one can hope for an Einstein limit. Compare also with the Ricci curvature pinching situation in §§12.2.5.

For dimension 4 there are strong necessary topological restrictions and nobody knows if they are sufficient; see §§§11.4.6.1. There is a long survey of the situation in four dimensions in Derdzinski 2000 [440].

11.4.2 The Examples from the Geometric Hierarchy

11.4.2.1 Symmetric Spaces

First there are the locally irreducible symmetric spaces (§§4.3.5, in particularly generalized space forms in §6.6). The reason is that the linear action of the isotropy group respects both the metric and the Ricci curvature, but it is irreducible and then cannot leave invariant, up to a scalar, more that one quadratic form. Simply use diagonalization to see this. And now how about reducible ones? The conclusion is left to the reader. It follows from the following remark (see also §§§11.4.6.4). If in the product $M \times N$ of two manifolds, both manifolds admit an Einstein metric of the same sign (see theorem 277 on page 521), then the product also does: one has only to make a suitable normalization and keep a Riemannian product, but this procedure does not work when the signs are different (zero included). Do not forget the following: the impossibility of this product construction does not tell us that a product of symmetric spaces of different signs cannot bear some other "wild" Einstein metric. See the interesting Rollin 2002 [1063].

11.4.2.2 Homogeneous Spaces and Others

The situation is detailed in section 7.E of Besse's book. We extract only a few examples to give a general picture of the state of affairs. First, for homogeneous but nonsymmetric spaces, the above remarks show that they are Einstein for an invariant metric as soon as the linear action of the isotropy group on the metric is irreducible. Sadly enough it turns out that such homogeneous spaces are very scarce. There are not many more than symmetric spaces. They were classified independently by O. Manturov in 1961 and J. Wolf in 1968; see an accurate and simpler classification, together with more study of those spaces, in Wang & Ziller 1991 [1241].

For the reducible case, it seems however that we have to really look directly at the Nomizu curvature formulas 15.15 on page 719 to reduce the problem to an algebraic one, which is probably quite easy using Dynkin's classification of Lie subgroups of simple Lie groups (see §§4.3.5). But it turns out in fact that these equations, theoretically algebraically computable, are very subtle when one searches for Einstein metrics. First came:

Theorem 278 (Jensen 1973 [763]) *On the spheres S^{4n+3} there are Einstein metrics different from the standard one. The same is true for $\mathbb{CP}^{2n+1}$.*

In this case, one simply uses the fact that on odd dimensional spheres S^d one has transitive actions of smaller groups that the whole orthogonal group $O(d+1)$. The sphere S^{15} admits three Einstein metrics different from the standard one, because of the fibration

$$S^7 \to S^{15} \to S^8.$$

Then came a series of papers by Wang and Ziller (they are quoted in Besse 1987 [183] and Wang 1992 [1237]) which yield many examples, and counterexamples, but we are still far from a classification of homogeneous Einstein manifolds. The difficulty is well illustrated by the fact that it is very involved to give a complete classification of Einstein homogeneous metrics on the Aloff–Wallach spaces which we met in §§11.3.4.2: Kowalski & Vlasek 1993 [832]. Add also the recent Lanzendorf 1997 [849], Boyer, Galicki & Mann 1996 [254], and Boyer & Galicki 2000 [251]. Dancer & Wang 1999 [428] give some inhomogeneous examples, and some very subtle five dimensional inhomogeneous examples appeared in Boyer & Galicki 2000 [251].

11.4.3 Examples from Analysis: Evolution by Ricci Flow

Remember the deformation-evolution approach to get periodic geodesics in §§§10.2.3.2, and even to obtain circles in §1.4. It consisted in deforming "any" curve in directions orthogonal to the curve and at a rate proportional to its geodesic curvature. Then the geodesic curvature gets ever smaller, as long as one can prove local existence. Moreover this process will yield a periodic geodesic if one can prove existence up to infinite time and the existence of a smooth limit. In book form, the Ricci flow is the object of the last chapter of Hebey 1999 [691].

The "naive" scheme to find an Einstein metric on any compact manifold is to look for a suitable evolution equation in the space of Riemannian metrics instead of inside the space of closed curves. As it stands, it does not work in general, even with topological restrictions (see the nonexistence results below). But it works wonderfully in dimension 3 under the added condition that we start with a manifold of positive Ricci curvature:

Theorem 279 (Hamilton 1982 [678]) *If M is a three dimensional compact manifold with a positive Ricci curvature metric, then M admits a metric of constant positive sectional curvature.*

The most natural "Ricci flow" in any dimension d would be given by the equation:

$$\frac{dg}{dt} = \frac{2}{d} \operatorname{scalar}_g g - 2 \operatorname{Ricci}_g \tag{11.9}$$

In fact there is then no local solution of this system of partial differential equations. Hamilton had to replace $2/d$ and -2 by other suitable constants; see page 147 of Besse's book for more.

Hamilton's text was the first to use evolutions equations successfully in the space of all Riemannian metrics. We will come back to this technique later in more than one place. It is now of basic use in Riemannian geometry: see §§§12.3.1.4 for its use in studying positive curvature operators and also see this approach used in smoothing in various places, e.g. in §§12.4.2. For the moment it does not yield Einstein metrics in higher dimensions, where they would be extremely desirable. For more on the Ricci flow approach, see Lu 2001 [882].[4]

Here is the place to mention a very recent interesting proof of the conformal representation theorem 70 on page 254: on any compact surface and for any metric g there is a positive function f such that fg has constant Gauß curvature. Starting with Hamilton in 1988, and others, finishing with Chow 1991 [378], there is now a proof using a kind of Ricci flow. Since in dimension 2 every metric is Einstein or, equivalently, an Einstein metric need not be of constant scalar (=Gauß) curvature, one has to replace equation 11.9 on the preceding page by a suitable equation where dg/dt is proportional to g and where one tries to reduce the defect of the curvature to be constant. This equation is

$$\frac{dg}{dt} = (K_0 - K)\,g$$

where K_0 is the average of the Gauss curvature. This result is difficult to prove in the realm of partial differential equations, here of nonlinear parabolic type, especially in the case of the sphere when the curvature is not initially everywhere positive. An important ingredient is a kind of suitable "entropy" (compare with page 477 at the end of §§§10.5.1.2), namely $\int K \log K$ which one can show decreases over time under this flow.

11.4.4 Examples from Analysis: Kähler Manifolds

We turn now to the only domain where the search for Einstein metrics is satisfactory, namely the case of Kähler manifolds. We refer to chapter 13 for Kähler definitions and notation. Moreover this topic turned out to be important in mathematical physics; see references in §13.6. As we will see, the results are almost complete in a certain sense.

Here the basic remark is made in §13.6: using the complex structure one can transform the symmetric differential form which is the Ricci curvature into a exterior form ρ (of degree 2, more precisely of type $(1,1)$) by setting

$$\rho(x,y) = \text{Ricci}(x, Jy)$$

where J is the complex structure. By Chern's formulas in §15.7 and via de Rham's theorem 32 on page 168, this form (up to $1/2\pi$) belongs to the first Chern class $c_1(M)$ of the Kähler manifold M under consideration. So we have an immediate necessary condition for a Kähler manifold to admit any

---

[4] Very recently, using Hamilton's technique with various sophisticated improvements, G. Perelman succeeded to prove Poincaré's conjecture, see further details in section 14.4 and chapter 16, page 723.

Einstein metric: the first Chern class should have some de Rham representative 2-form which (via the complex structure) is either positive definite, identically zero, or negative definite. This condition is not that strong. Many algebraic manifolds satisfy it and this can be decided by using standard techniques to compute their Chern classes; see for example Besse's book section 11.10.

We briefly explain now why things are "workable" for Kähler manifolds; see chapter 11 of Besse's book for an informative sketch of the proofs. In fact the Kähler structure g yields a 2-form of complex type $(1,1)$, its *Kähler form* ω and Kähler variations in the same cohomology class are easily seen to be necessarily of the form

$$\omega + \sqrt{-1}\partial\bar{\partial}f$$

where f is a numerical function on the manifold. The Einstein condition is then the partial differential equation

$$\rho_\omega = \frac{\text{scalar}_\omega}{d}\omega$$

and when written explicitly in terms of the function f it turns out to be of Monge-Ampère type and then hopefully solvable.

This led Calabi in the insightful paper Calabi 1954 [299] to conjecture that as soon as the zero or the definiteness of the Chern class is ensured, one can change the initial Kähler metric into a new one which is Einstein. More: in the negative case one has uniqueness. It was only in Aubin 1970 [85] that any progress was obtained and then next in Yau 1978 [1294]. Aubin proved the existence of such a metric in the negative case, while Yau proved it in the zero and the negative case, so that now

Theorem 280 (Aubin 1970 [85], Yau 1978 [1294]) *If the first Chern class of a Kähler manifold admits a zero or a negative definite representative, then it admits in the same Kähler class an Einstein metric. Moreover in the negative definite case, the Einstein metric is unique.*

This definitely does not work for the positive definite case: an obstruction was discovered in Futaki 1983 [535]. In Tian 1990 [1190] one has a definite answer for complex surfaces, i.e. real dimension four, giving necessary and sufficient conditions to get Einstein Kähler metrics of the positive type. For larger dimensions, the latest result is Real 1996 [1054], and a complete conjecture is formulated in Tian 1997 [1191]. An expository text is Bourguignon 1997 [242].

The Calabi–Yau existence results are extremely useful. First the uniqueness of the Kähler structure on $\mathbb{CP}^n$ (for any n) was uncovered in Yau 1977 [1293]. Besides Yau's existence result one needs the deep Hirzebruch & Kodaira 1957 [718]. Another application is the inequality

$$c_1^2 \leq 3c_2$$

between the first two Chern classes of any Kähler manifold. For this and more see 11.B in Besse 1987 [183]. A still mysterious application is the fact that, for a Kähler manifold, the vanishing of the two first Chern classes c_1 and c_2 implies that the manifold can be made flat and in particular is a finite quotient of a torus: see references and comments in page 67 of Bourguignon 1996 [241]. The mystery lies in the fact that nobody knows a direct proof.

To stay in the Kähler–Einstein domain, we mention Hulin 1996 [749]. The author uses the very interesting notion of *diastasis* introduced by Calabi in 1953; see the expository Berger 1996 [168] and §13.6. And then Hulin proved very strong results for the complex submanifolds of $\mathbb{CP}^n$ which are Einstein. The diastasis has been rarely used up to now but it might still have a rich future; see Herrera & Herrera 2001 [708].

11.4.5 The Sporadic Examples

The first examples were obtained using almost homogeneous spaces, so-called "low cohomogeneity spaces," mixed with Riemannian submersion techniques (see §§15.8.2), a technique used in another context in paragraph 12.3.1.1.2 on page 578. Details of results prior to 1987 and proofs are at large in the chapter 9 of Besse's book. Those examples were quite scarce until recently; since Page 1978 [990] and Bérard Bergery 1982 [143]: there is an Einstein metric on the connected sum $\mathbb{CP}^2 \# \overline{\mathbb{CP}^2}$. For the proof one considers $\mathbb{CP}^2 \# \overline{\mathbb{CP}^2}$ as acted by $SO(3)$; then the orbits are a one parameter family of hypersurfaces. The Riemannian submersion formulas enable us to reduce the Einstein condition to an ordinary differential equation on $[0, 1]$; the remaining difficulty being to take care of the singularities at 0 and 1.

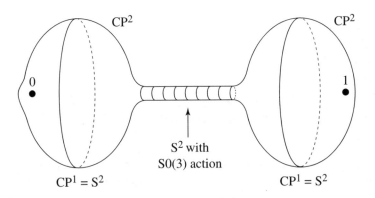

Fig. 11.6. Construction of an Einstein metric on $\mathbb{CP}^2 \# \overline{\mathbb{CP}^2}$

Then came quite a large variety of inhomogeneous examples, obtained by different techniques: fibre bundles of various types, twistor theory (see §§14.2.3), 3-Sasakian manifolds (see figure 13.6 on page 652). Today examples are getting more and more numerous. We only give some references posterior to Besse's book. The very recent Böhm 1998 [216] is important in two respects: first it gives among other examples some nonstandard Einstein metrics on low even dimensional spheres. The technique uses cohomogeneity one manifolds and the equations obtained in Bérard Bergery 1982 [143] for obtaining an Einstein metric. The cohomogeneity one technique is put in a systematic frame in Eschenburg & Wang 1997 [497]. The second point in Böhm 1997 [216] is that it presents examples of infinite sequences of Einstein metrics on given compact manifolds with fixed volume which do not converge to any metric, and this shows that the "condition C" of Palais & Smale 1964 [992] is not satisfied in general for Riemannian metrics. Contrast this with the false hopes in §§11.3.7. Earlier such "counterexamples" had been given by Wang & Ziller.

Quite recently Einstein manifolds appeared as a byproduct of the theory of 3-Sasakian manifolds; see §§13.5.2. This last sporadic category comes in some way from the rare holonomy groups; see §13.5. There is also a link with Kähler manifolds from §§11.4.4. We recall from chapter 13 that manifolds with holonomy group $SU(n)$ (of dimension $2n$), $Sp(1) \times Sp(n)$ (of dimension $4n$), G_2 (of dimension 7) and Spin(7) (of dimension 8) are automatically Ricci flat. If one excepts $SU(n)$, we have today very few examples of such manifolds. Moreover

Fact 281 *There is no known compact simply connected irreducible Ricci flat manifold whose holonomy group is not one of these.*

We say again that if Einstein manifolds are mysterious, Ricci flat ones are even more mysterious; cf. also the trichotomy in §§12.3.3. It is important to realize that the fact 281 raises a global question. Locally, analysts are able to produce "a lot" of Ricci flat metrics without these special holonomy groups, just by deformation: see Gasqui 1982 [549] and Bokan, Gilkey & Zivaljevic 1993 [217]. Gasqui's metrics are local, i.e. not complete. But in LeBrun 1991 [853] there are examples of Ricci flat complete metrics on $\mathbb{C}^n$ which are moreover Kähler.

11.4.6 Around Existence and Uniqueness

11.4.6.1 Existence

We again insist on the fact that nobody knows today the answer to:

Question 282 *Does every compact manifold of dimension ≥ 5 admit at least one Einstein metric? Otherwise stated, there is not a single topological condition known for the manifold M derivable from the fact that it bears some Einstein metric.*

We saw above that the question is of no interest when the dimension is 2 or 3. Now for dimension 4, the question 282 has a negative answer. There are various necessary conditions known to exist. This might be guessed from the formula 11.6 on page 518:

$$8\pi^2 \chi(M) = \int_M \left(|R|^2 - \left| Ricci - \frac{scal}{4} g \right|^2 \right) dV_M$$

which trivially shows that to be Einstein implies $\chi(M) \geq 0$ (and 0 only in the flat case as already remarked above). A more careful study of the integrand led to

Theorem 283 (Thorpe 1969 [1187]) *On any 4-dimensional Einstein manifold M*

$$\chi(M) \geq \frac{3}{2} |\tau(M)|$$

between its characteristic $\chi(M)$ and its signature $\tau(M)$.

The idea is to use simultaneously the integral formula for the characteristic and for the signature of the manifold; see the details at the very end of §15.7. See chapter 6 of Besse 1987 [183] for more. The above inequality is the only one which one can extract from these two integral formulas. A completely different direction was initiated by Gromov, using his inequality from theorem 274 on page 518, from which one deduces immediately

Theorem 284 (Gromov 1983 [617]) *For an Einstein manifold M of dimension 4 one has for its minimal volume the necessary condition:*

$$\|M\| \leq 2592\pi^2 \chi(M).$$

Mixing both techniques and results from the topology of 4-manifolds, and adding the Seiberg–Witten equations (see §11.1), one can see that in some sense "many 4-dimensional manifolds do not admit any Einstein metric": see Sambusetti 1996 [1089], which uses techniques from Besson, Courtois & Gallot 1995 [189], and also see LeBrun 1995 [855].

In dimension 4 there is notion weaker than that of Einstein, namely anti-self-duality. See the entire chapter 13 of Besse 1987 [183] and complete it with Donaldson & Kronheimer 1990 [457], and the expository Gauduchon 1992/93 [551] on Taubes 1992 [1183]. For the need of the proof of Taubes's result, special norms are introduced which are close to the various global norms we mentioned in theorem 118 on page 325.

The theory for noncompact manifolds is also very interesting; we just mention among others Heber 1997 [687], Hitchin 1995 [720] and Hitchin 1996 [721].

11.4.6.2 Uniqueness

The examples of §§11.4.5 show the complexity of the topic. We will meet here our favourite paradox: one knows of the uniqueness of Einstein metrics on some negative curvature space forms (those which are quotients of Hyp^4 and $\text{Hyp}_\mathbb{C}^2$).

Theorem 285 (LeBrun 1995 [855]) *On a space form of type Hyp^4 or of type $\text{Hyp}_\mathbb{C}^2$, only the locally symmetric metric can be Einstein.*

For $\text{Hyp}_\mathbb{C}^2$ the proof is in LeBrun 1995 [855], where the tools are the new Seiberg–Witten invariants (see §11.1). For Hyp^4, the result is a byproduct of the strong result of Besson, Courtois & Gallot 1995 [189] which is quoted as theorem 251 on page 484.

This is an irritating question. For standard manifolds, in particular space forms, uniqueness is not understood. The difficulty starts with S^4, $\mathbb{CP}^2$ and many other one manifolds. Note that in dimensions just above that there is no uniqueness for a lot of spheres. But this is far from a classification of Einstein metrics on spheres, or more generally for the $\mathbb{KP}^n$. We meet here again the "paradox" for space forms mentioned in §6.6: in the negatively curved space forms, existence of certainly structures on those manifolds is hard, but then they enjoyed many uniqueness results, while in the positive case examples come very easily but there are almost no results for them. Also see Yang 2000 [1288].

11.4.6.3 Moduli

The general situation is not too bad, in the sense that

Theorem 286 (Koiso, see chapter 12 of Besse [183]) *On a given compact manifold, the set of all Einstein metrics up to diffeomorphism (Einstein structures, in our terminology) is a finite dimensional, real analytic stratified space.*

A starting point is to compute the derivative for the defining equation 11.8 on page 521 of a one parameter family of Einstein metrics. This derivative satisfies an elliptic equation, and hence see that deformations form finite dimension family by standard elliptic theory, which remains of course to be organize globally. The tricky part is that it is only elliptic modulo diffeomorphism, so not elliptic in quite the usual sense.

But those moduli spaces are explicitly known in extremely few cases, the most spectacular being that of $K3$ surfaces seen just above. See Besse's book, section 12.K. The final result is a complete description of the moduli space of Ricci flat metrics on $K3$ surfaces; namely it is an open subset of the symmetric space
$$SO\,(3,19)\,/\,SO\,(3) \times SO\,(19)\,.$$

The complete proof is extremely involved, mixing a lot of mathematics running from algebraic geometry to analysis.

In Anderson 1992 [43] there is a deep study of the moduli space of Einstein metrics for any compact manifold of dimension 4, and this for the three possible signs. Special attention is given to the closure, here orbifolds (see §§14.5.2) appear and in particular the above results for $K3$ surfaces are made precise and algebraic surfaces with singularities come into the picture. However we recall from §§11.4.1 and §§11.4.5 that counter-examples show that in general there is no convergence for sequences of Einstein metrics on a given manifold to any smooth metric.

An easy case is obvious from formula 11.6 on page 518 when applied to the four-dimensional torus T^4. Since $\chi\left(T^4\right) = 0$, every Einstein metric on T^4 has to be flat and the problem is reduced to the classification of flat tori and their quotients; see §§6.6.1.

The finiteness result of theorem 286 on the preceding page really needs compactness; for example in LeBrun 1991 [854] one finds an infinite dimensional family of quaternionic-Kähler structures (hence automatically Einstein) on $\mathbb{R}^{4n}$ obtained by deformation of the symmetric spaces $\text{Hyp}_\mathbb{H}^n$.

11.4.6.4 The Set of Constants, Ricci Flat Metrics

A puzzling question is that of the signs of Einstein manifolds (see theorem 277 on page 521): the sign is the sign of the proportionality factor, i.e. of the scalar curvature. For example it is positive (resp. zero, negative) in the case of space forms of positive (resp. zero, negative) curvature. However do not hope for too much:

Theorem 287 (Catanese & LeBrun 1997 [323]) *In any dimension $4k$ (with $k \geq 2$) there are manifolds which carry two Einstein metrics, one with positive sign and one with negative sign.*

The construction is a follows: in dimension 4, one considers two Kähler surfaces M and M', the first being a deformation of the classical Barlow surface and the other having as underlying manifold the connected sum $\mathbb{CP}^2 \# 8\overline{\mathbb{CP}^2}$, where the $\overline{\mathbb{CP}^2}$ denotes just a change of orientation. Now one can choose these Kähler structures in order that the Chern class is positive (resp. negative) definite on M' (resp. M). Then results of Aubin–Yau and Tian mentioned in §§11.4.4 show that M (resp. M') admits an Einstein metric with positive sign (resp. negative sign). Now deep results of algebraic topology imply that those two manifolds, apparently coming from two different worlds, are in fact homeomorphic. But more: the products $M \times M$ and $M' \times M'$ are diffeomorphic, and by the above, carry Einstein metrics of different signs.

Koiso's result in theorem 286 on the preceding page implies by standard theorems that the moduli space is locally arcwise connected, so that the Einstein constant is constant on such components. Hence, by standard results on real analytic sets, the total set of possible Einstein constants on a given compact manifold is then always countable. This does not say how this set sits in $\mathbb{R}$. Is it discrete? finite? Certainly not finite: there are examples in Wang & Ziller 1990 [1240] of manifolds as simple as the products $S^2 \times S^{2m+1}$.

One result is easy: there is an upper bound for the set of Einstein constants of a given compact manifold (with volume normalized). This a trivial consequence of Bishop's inequality in theorem 107 on page 310 for the volume of balls under a lower bound for Ricci curvature.

We insist again on the fact that Ricci flat metrics are extremely mysterious. In some sense it is not clear at all today if this condition is very strong or not. For the moment we have only the fact 281 on page 529. We will meet these manifolds again, e.g. in §§12.3.2. A typical ridiculous open question, the first to come to mind: does S^4 admit a Ricci flat metric? If the denomination were not already used in §§6.6.1.3 for isotropic-measure manifolds, one could have suggested to call Ricci flat manifolds harmonic, for example in view of formula

$$\Delta g_{ij} = \text{Ricci}_{ij} + Q$$

presented in equation 6.22 on page 267.

11.4.7 The Yamabe Problem

We present here an interlude with a special topic, namely the Yamabe problem. In Yamabe 1960 [1280], the author tried, starting with any Riemannian metric g, to deform it conformally into a metric fg of constant scalar curvature, where f is a positive function. Conformal deformations are basic for surfaces. One suspects that the secret hope of the author was in dimension 3, to attack the Poincaré conjecture (compare with Hamilton's result of theorem 279 on page 525). To go further in dimension 3 is the content of Anderson 1997 [47],[46] and [48].

The problem to solve was a nonlinear elliptic equation in the unknown function f. This question in analysis is extremely hard and was carefully studied. This because the equation is nonlinear:

$$\Delta f + \frac{d-2}{4(d-1)} \text{scalar} = f^{(d+2)/(d-2)}$$

where the unknown function f comes with the "limiting" exponent for the Sobolev embedding theorem. For larger or smaller exponents, standard results are available but are of no use here. Today the Yamabe problem is almost completely understood after basic contributions of Aubin and Schoen. We refer to the surveys Besse 1987 [183] 4.D, Hebey 1993 [689], Hebey 1996 [690]; also see Hebey 2000 [692] and the references there. We just note that the set of metrics of constant scalar curvature, in any dimension above 2, is always an infinite dimensional space and is definitely not what we want when looking for a "best" metric.

Except in discussing surfaces, in theorem 70 on page 254, we never touched on conformal Riemannian manifolds, (i.e. manifolds whose Riemannian metric is only defined up to multiplying by a positive function). This is an important topic, and a very good reference is Matsumoto [903]; for locally conformally flat manifolds, see §§§4.4.3.1.

11.5 The Bewildering Fractal Landscape of $\mathcal{RS}(M)$ According to Nabutovsky

Since 1995, Nabutovsky has been unveiling "awful" properties of $\mathcal{RS}(M)$ for compact manifolds M of dimension five or more. We will sketch here some results of this kind, which for us are fascinating both in themselves and in his method of proof. Posterity will judge their value to mathematics, but the least one can say is that they are *disturbing*. Some references: Nabutovsky 1996 [957], Nabutovsky & Weinberger 1997 [961], and the survey Nabutovsky & Weinberger 2000 [962]. For applications of Nabutovsky's results to the physics of Euclidean quantum gravity, see Ben-Av & Nabutovsky 2002 [128].

On a compact surface M, thanks to the conformal representation theorem, and the results at the end of §§11.4.3, we have a beautiful picture of $\mathcal{RS}(M)$. There is a flow on $\mathcal{RS}(M)$ taking any Riemannian metric to a metric of constant curvature. But the space of constant curvature metrics is the Teichmüller space described in §§6.6.2. Thus one sees a picture of $\mathcal{RS}(M)$ as a fiber bundle whose base (Teichmüller space) is finite dimensional, and whose fibers are infinite dimensional and contractible. We will briefly discuss dimensions three and four, but first let us see how terrible $\mathcal{RS}(M)$ is for all manifolds of dimension five or more.

We want to survey the topography of $\mathcal{RS}(M)$, where the notion of height of a mountain or depth of a valley is measured using one of the functionals

$$F : \mathcal{RS}(M) \to \mathbb{R}$$

which we encountered previously in §11.1. In particular, we are interested in the critical points of that functional, concerning ourselves with where these points appear on our relief map of $\mathcal{RS}(M)$. We first have to explain what sort of geometry and topology we are employing on $\mathcal{RS}(M)$. We will not use the results of Ebin 1970 [474], even as augmented by those of Bourguignon 1975 [237] to the effect that $\mathcal{RS}(M)$ is an infinite dimensional generalized orbifold, in the sense that it is locally the quotient of an infinite dimensional manifold by a Lie group—this Lie group being at each point the isometry group of the corresponding Riemannian metric. Instead we shall use the topology defined on $\mathcal{RS}(M)$ by the Gromov–Hausdorff metric $d_{\mathfrak{G}-\mathfrak{H}}$ (see theorem 380 on page 626).

We will need to refine our comments from §§11.1.1: it is a very natural notion to exploit Morse theory to describe the topology of $\mathcal{RS}(M)$ by using the fibration

11.5 The Bewildering Fractal Landscape of $\mathcal{RS}(M)$

$$\text{Diff}(M) \longrightarrow \mathcal{RM}(M)$$
$$\downarrow$$
$$\mathcal{RS}(M) = \mathcal{RM}(M) / \text{Diff}(M)$$

as in Serre's theory which we saw in §§10.3.3. Once again we find that the total space $\mathcal{RM}(M)$ is topologically trivial, since it is convex, and then we can use a spectral sequence to uncover a rich topology on $\mathcal{RS}(M)$ from the rich topology of $\text{Diff}(M)$. Sadly, unlike the theory of compact manifolds from §§10.3.3, here we are at sea because the topology of $\text{Diff}(M)$, one of the most natural objects of study in differential topology, is almost completely a mystery. It is very hard to prove that for some integer n, the fundamental group

$$\pi_1(\text{Diff}(S^n)) \neq \{1\} .$$

Before stating a few of Nabutovsky's results, we recall the few geometric discoveries that were made before his work concerning the metric space $\mathcal{RS}(M)$ with metric $d_{\mathfrak{G}-\mathfrak{H}}$. In §§12.3.1 we saw that for spheres of certain dimensions, the space of metrics of positive scalar curvature is not connected. There are also manifolds for which the space of metrics with positive sectional curvature is not connected. One of Nabutovsky's theorems will show even more: it will show that there are infinitely many connected components, and that there are infinitely many connected components of every level set of many functionals. Let us describe one such result of Nabutovsky in more detail.

From now on in this section, all manifolds are assumed *compact* and *simply connected*.

We first need to describe the functional

$$F : \mathcal{RS}(M) \to \mathbb{R} .$$

Let $n = \dim(M)$. We will take the *embolic functional*

$$F(g) = \frac{\text{Vol}(M, g)}{\text{Inj}(M, g)^n}$$

which is dimensionless. We will now look at the corresponding landscape, for metrics of unit total volume, and with a small positive lower bound for injectivity radius:

$$\text{Vol}(M, g) = 1$$
$$\text{Inj}(M, g) \geq \varepsilon$$

We will write the set of such Riemannian metrics as

$$\mathcal{RM}^{\text{Vol}=1}_{\text{Inj}\geq\varepsilon}(M) = \{g \in \mathcal{RM}(M) \mid \text{Vol}(M, g) = 1 \ \& \ \text{Inj}(M, g) \geq \varepsilon\} .$$

We need the concept of *computable function*. We will not give a precise definition, but only roughly describe such a function as one which grows slowly enough that it can be described by a computer program; for example any power function
$$n \mapsto n^k$$
or even a finitely iterated exponential function
$$n \mapsto a_1^{a_2^{\cdot^{\cdot^{\cdot^n}}}}$$
with $a_1, a_2, \ldots \in \mathbb{N}$. Of course, we can not expect to write down an example of an uncomputable function.

Theorem 288 *Let M be any compact manifold of dimension $n \geq 5$, and*
$$\alpha : \mathbb{N} \to \mathbb{N}$$
and computable function satisfying
$$\alpha(n) \geq n \ .$$
There is a positive number ε_0 depending on α and M so that for any positive number $\varepsilon \leq \varepsilon_0$, there are metrics
$$g, h \in \mathcal{RM}_{\mathrm{Inj} \geq \varepsilon}^{\mathrm{Vol}=1}(M)$$
so that there is no finite sequence of metrics g_i with each
$$g_i \in \mathcal{RM}_{\mathrm{Inj} \geq \varepsilon'}^{\mathrm{Vol}=1}(M)$$
with
$$\epsilon' = \frac{1}{\alpha\left(\lfloor \frac{1}{\epsilon} \rfloor\right)}$$
and
$$g = g_1, \ldots, g_N = h$$
and such that
$$d_{\mathfrak{G}-\mathfrak{H}}(g_i, g_{i+1}) \leq \frac{1}{9} \min \left(\mathrm{Inj}(g_i), \mathrm{Inj}(g_{i+1})\right) \ .$$
In particular, $\mathcal{RM}_{\mathrm{Inj} \geq \varepsilon}^{\mathrm{Vol}=1}(M)$ is not connected for all such ε.

Note 11.5.0.1 Before discussing the proof, let us recall the properties we have already encountered of the embolic constant
$$F(g) = \frac{\mathrm{Vol}(M, g)}{\mathrm{Inj}(M, g)^n}$$

11.5 The Bewildering Fractal Landscape of $\mathcal{RS}(M)$

which demonstrate its significance, even though it does not depend (directly) on curvature. In theorem 148 on page 355 we saw that the embolic constant of the sphere is the smallest possible:

$$F(M,g) \geq F(S^n, \text{canonical})$$

among all compact n dimensional manifolds, with equality occurring only for the canonical metric on the sphere, up to rescaling. Secondly, by theorem 149 on page 355, for any metric g on any compact manifold M, small balls, of radius $r \leq \text{Inj}(M,g)$, have lower bounded volume

$$\text{Vol}\, B(m, r) \geq c_n r^n$$

where the constant c_n is independent of the choice of manifold M (but the largest possible c_n is still unknown). In theorem 377 on page 622 of Grove, Petersen & Wu, one sees that the number of possible manifolds (up to diffeomorphism) of dimension n bearing some metric g with $F(g) \geq k$ is finite for each k, and one even has an explicit bound universal in n and k. ♦

Proof. The key to the proof is to introduce the following invariants, arising out of Gromov's work.

Definition 289 *For a given closed curve c on M, denote by* $\text{FillVal}(c)$ *the filling value of c, which is the infimum of the numbers k such that there is a homotopy c_s from $c = c_0$ to a point $* = c_1$ such that for every value of s*

$$\text{length}\, c_s \leq k\, \text{length}\, c\,.$$

Now for a compact, simply connected manifold M with Riemannian metric g, define

$$\text{FillVal}(M, g) = \sup\{\text{FillVal}(c)\,|\, c \text{ is a closed curve in } M\}\,. \qquad (11.10)$$

A priori there is no reason for $\text{FillVal}(M,g)$ to be finite; indeed its finiteness was proven by Gromov in [610] and is one of the key results for proving his theorem 227 on page 465. The invariant that we must focus on now is

$$F_\varepsilon(M) = \sup\left\{\text{FillVal}(M,g)\,|\, g \in \mathcal{RM}_{\text{Inj} \geq \varepsilon}^{\text{Vol}=1}(M)\right\}\,.$$

We do not need to prove finiteness of this invariant, since the proof of theorem 288 on the preceding page proceeds by contradiction. Heuristically, the aim of the theorem is to construct on M *incorrigible* metrics. It is enough to find such metrics on spheres S^n; then one sees easily that every compact manifold carries an incorrigible metric by performing connect sums. The contradiction to prove is

Lemma 290 *For any computable α one has*

$$F_\varepsilon(S^n) > \alpha\left(\left\lfloor\frac{1}{\varepsilon}\right\rfloor\right)$$

for any ε.

Lemma 291 *If theorem 288 is false, then there is a computable α such that*

$$F_\varepsilon(S^n) < \alpha\left(\left\lfloor\frac{1}{\varepsilon}\right\rfloor\right)$$

for some ε.

The second step is pure Riemannian geometry. We have to prove some kind of precompactness theorem in $\mathcal{RM}_{\text{Inj}\geq\varepsilon}^{\text{Vol}=1}(S^n)$, and this is achieved in the spirit of theorem 384 on page 627. The crux of the argument is a lower bound for the volume of balls; in that theorem Bishop's inequality provided such a bound, while here it comes from Croke's inequality quoted above. After this, one proves that any pair of metrics g, h satisfy

$$\text{FillVal}(M, g) \leq \text{univ}\left(\varepsilon, d_{\mathfrak{G}-\mathfrak{H}}(g, h)\right) \text{FillVal}(M, h) .$$

The contradiction follows using arguments coming from Gromov's technique for proving the finiteness of FillVal in equation 11.10 on the previous page. This same technique also proves that there is some uncomputable function α for which the analogous result does *not* hold.

We now enter the realm of algorithmic computability theory, and the proof is a deep Riemannian refinement of Novikov's proof of the algorithmic unsolvability of the problem of recognizing a sphere S^n of dimension $n \geq 5$. The trick is to associate to every Turing machine[5] T a polynomial with rational coefficients

$$p_T(X_1, \ldots, X_{n+1}) \in \mathbb{Q}[X_1, \ldots, X_{n+1}]$$

such that the hypersurface

$$\{X \mid p_T(X) = 0\} \subset \mathbb{R}^{n+1}$$

is a smooth manifold and a homology sphere. Then it will be a sphere precisely when its fundamental group is trivial. Now suppose that there is some computable function $\alpha : \mathbb{N} \to \mathbb{N}$ with

$$F_\varepsilon(S^n) < \alpha\left(\left\lfloor\frac{1}{\varepsilon}\right\rfloor\right)$$

for some ε. Following a refinement of a classical theorem in algorithmic computability, one can construct a contradiction in the world of Turing machines

---

[5] For those unaccustomed to Turing machines, just imagine that they are computers–the notion of Turing machine is essentially the notion of computer.

11.5 The Bewildering Fractal Landscape of $\mathcal{RS}(M)$

as follows. Because our $p_T = 0$ hypersurface is algebraic, we can algebraically compute its volume, diameter, and sectional curvature. But putting together the theorems of Cheeger and Klingenberg from §§6.5.2, we obtain a lower bound on the injectivity radius. Using Croke's inequality again, one can design an algorithm to check if the fundamental group of our hypersurface is trivial. Take generators for the fundamental group of the hypersurface, and approximate them (as in Morse theory) by broken geodesics. We can do this with a maximum number of "breaks," because the injectivity radius is bounded from below. This allows us to work in a finite dimensional space, and control all of our geometric data algebraically, thanks to the Tarski–Seidenberg theorem and the essential fact that—by hypothesis—we have computable control on the lengths of geodesics when we try to deform them into a point. Note that the very deep finiteness of the number of manifolds (up to diffeomorphism) satisfying a bound on the functional F, which comes from theorem 377 on page 622, is required in this proof.

This theorem does not supersede the natural approach to picking best metrics as critical points of Riemannian functionals. We know that the various $\mathcal{RM}_{\text{Inj} \geq \varepsilon}^{\text{Vol}=1}(M)$ spaces have more than one component, but are the components numerous? More importantly, there is no theorem to prove the existence of a metric for which the infimum of

$$F(g) = \frac{\text{Vol}(M, g)}{\text{Inj}(M, g)}$$

is attained. Indeed, there likely is no such metric on some manifolds. In the spirit of §§12.4.2, one would like connected components which are compact modulo diffeomorphism, and then minima of our functional would exist. This is achieved in Nabutovsky & Weinberger 1997 [961] by the following theorem. Here the functional $F(g)$ is the diameter, but on the subset of $\mathcal{RS}(M)$ called

$$\mathcal{RS}_{\text{Vol}=1}^{|K| \leq 1}(M)$$

and defined as the set of Riemannian metrics on M for which M has unit volume and sectional curvature between -1 and 1, but with any two such metrics identified if they differ by diffeomorphism.

Theorem 292 *(Informally stated:) For any manifold M of dimension $n \geq 5$ and any computable function α there is some number $x_0 = x_0(\alpha, M)$ so that for any $x \geq x_0$, the set of Riemannian structures in $\mathcal{RS}_{\text{Vol}=1}^{|K| \leq 1}(M)$ with diameter less than x*

$$\{\text{diam} \leq x\} \cap \mathcal{RS}_{\text{Vol}=1}^{|K| \leq 1}(M)$$

has an exponential number

$$e^{c(n)x^n}$$

of basins (i.e. connected components) inside the set

$$\{\text{diam} \leq \alpha(x)\} \cap \mathcal{RS}^{|K|\leq 1}_{\text{Vol}=1}(M) \ .$$

Moreover, on every basin there is a positive lower bound for the volume $\text{Vol}(M,g)$.

The precise statement is even stronger, with finite chains for $d_{\mathfrak{G}-\mathfrak{H}}$ as in theorem 288 on page 536. The universal lower bound for the volume, applying the compactness theorem 377 on page 622, gives a $C^{1,\alpha}$ metric on each basin which is a local minimum for the functional

$$F(g) = \text{diam}(M,g)$$

on $\mathcal{RS}^{|K|\leq 1}_{\text{Vol}=1}(M)$.

Corollary 293 *The functional*

$$F(g) = \text{diam}(M,g)$$

on $\mathcal{RS}^{|K|\leq 1}_{\text{Vol}=1}(M)$ *has an awesomely complicated fractal panorama, since there are so many local minima for it. Namely, at a given level for the diameter, there is an exponentially growing number of local minima. Moreover these minima cannot be connected at a larger level of diameter even if we allow the larger diameter to be any computable function of the initial diameter.*

For the proof, one needs more sophisticated homology spheres than the very general ones used previously. The main difficulty is to find homology spheres which can be endowed with metrics in $\mathcal{RS}^{|K|\leq 1}_{\text{Vol}=1}(M)$, with a universal lower bound for their volume. This is very difficult, and is the main result in Nabutovsky & Weinberger 1997 [961]. One needs to find suitable fundamental groups, i.e. nice finitely presented groups. In terms of group theory, the proof uses many very deep and recent discoveries. In terms of Riemannian geometry, the lower bound on volume is immediate on manifolds with positive simplicial volume; see §§§11.3.5.2 where simplicial volume is explained. So Nabutovsky & Weinberger bring to topologists and geometers new examples of manifolds with positive simplicial volume.

In Nabutovsky & Weinberger [962], one will find analogous stories told about quite a number of natural functionals. For the moment, Nabutovsky has only achieved partial results for the Hilbert functional (the integral of the scalar curvature), which ones imagines will yield Einstein structures. It seems that the common moral of these stories is that (at least for standard functionals) looking for critical points, even for local minima, will not yield natural "best" metrics—when one has too many minima (uncomputably growing numbers) looking for the absolute minimum is not a tractable job. This also explains why the problem of distributing points spaced apart on a sphere,

11.5 The Bewildering Fractal Landscape of $\mathcal{RS}(M)$ 541

which is problem 7 in Smale's 1998 list [1149], is today apparently a hopeless task: see Saff & Kuijlaars 1997 [1083], and section III.3 of Berger 2000 [172]. In a similar vein, Nabutovsky shows that there is no "reasonable" flow in general; this can happen in some special cases, and close to good metrics, as we saw in §§11.4.3.

In conclusion, to get best metrics we need a more refined approach, correcting suitable functionals in an appropriate manner; for this spirit see Margerin 1991 [895] and 1993 [897].

Note 11.5.0.2 (Low dimensions) We have already examined Riemannian structures on surfaces closely. On three dimensional manifolds, the above theorems as stated are certainly false, since Thurston has proven that there is an algorithm to recognize the three dimensional sphere. But this does not solve his geometrization programme (see §14.4). On four dimensional manifolds, Nabutovsky thinks that most of his results may be valid and approachable by the same avenues. ♦

12 From Curvature to Topology

Contents

- 12.1 **Some History, and Structure of the Chapter** .. 543
  - 12.1.1 Hopf's Inspiration 543
  - 12.1.2 Hierarchy of Curvatures 546
- 12.2 **Pinching Problems** **549**
  - 12.2.1 Introduction 549
  - 12.2.2 Positive Pinching 552
  - 12.2.3 Pinching Near Zero 568
  - 12.2.4 Negative Pinching 569
  - 12.2.5 Ricci Curvature Pinching 571
- 12.3 **Curvature of Fixed Sign** **576**
  - 12.3.1 The Positive Side: Sectional Curvature 576
  - 12.3.2 Ricci Curvature: Positive, Negative and Just Below 593
  - 12.3.3 The Positive Side: Scalar Curvature 599
  - 12.3.4 The Negative Side: Sectional Curvature 605
  - 12.3.5 The Negative Side: Ricci Curvature 613
- 12.4 **Finiteness and Collapsing** **614**
  - 12.4.1 Finiteness 614
  - 12.4.2 Compactness and Convergence 624
  - 12.4.3 Collapsing and the Space of Riemannian Metrics 630

12.1 Some History, and Structure of the Chapter

12.1.1 Hopf's Inspiration

We have frequently encountered relations between curvature and topology. Among others we can mention the Gauss–Bonnet–Blaschke theorem 28, the von Mangoldt–Hadamard–Cartan theorem 72, Myers' theorem 63, and Synge's theorem 64. The topic of curvature and topology has been for some time the most popular and highly developed topic in Riemannian geometry.

The first to investigate relations between curvature and topology in a general and systematic context was Heinz Hopf:

12 From Curvature to Topology

> *The problem of determining the global structure of a space form from its local metric properties and the connected one of metrizing—in the sense of differential geometry—a given topological space, may be worthy of interest for physical reasons.*
>
> Heinz Hopf 1932 [730]

This implies two opposite problems. The second one was the object of chapter 11. The first one is the object of the present chapter. In fact Hopf had already started to work on the subject and was especially motivated by the two following questions:

Question 1. When the sectional curvature of an even dimensional compact manifold is of a constant sign $\varepsilon = \pm 1$, is the sign of its Euler characteristic equal to $\varepsilon^{d/2}$?

Question 2. Extend to any even dimension the Gauss–Bonnet theorem 28.

It seems quite clear that he was thinking (or perhaps only hoping) that the solution of question 1 would follow from that of question 2. In the papers Hopf 1925–1927 [727, 728, 729] he succeeded in proving the sign conjecture for space forms (in the sense of question 1, not of his quotation above) and to prove that the Euler characteristic of a hypersurface M^d in $\mathbb{E}^{d+1}$ is, up to a universal constant, the integral of $\sqrt[d]{\det II}$, the $(d-1)$-th root of the determinant of the second fundamental form, i.e. the product of the eigenvalues of that form (the principal curvatures; see equation 4.43 on page 212 and §§14.7.1). Note that this is exactly the Gauß–Bonnet theorem when $d=2$, since $\det II = K$. To prove this formula he first extended the Rodrigues–Gauß map to hypersurfaces and proved theorem 415 on page 712. But he was facing a topological problem, which he solved in the third paper, where he invented and proved his famous and beautiful formula giving the Euler characteristic as the sum of the indices of any vector field on the manifold: equation 15.12 on page 713. For hypersurfaces, his definition of the index of a singularity of a vector field is directly connected with the Rodrigues–Gauß map.

This $\det II$ is an invariant of the metric, for any dimension d (properly speaking, its absolute value is invariant, in odd dimensions). So that his formula was solving questions 1 and 2 only in the very special case of hypersurfaces in $\mathbb{R}^{d+1}$. This intrinsic nature comes from the fact that for hypersurfaces the curvature tensor is very special (see equation 4.43 on page 212): for an orthonormal basis diagonalizing the second fundamental form, the only nonzero components of the curvature tensor are the R_{ijij} and moreover their values are $R_{ijij} = \lambda_i \lambda_j$ where the λ_i are the principal curvatures. This is nothing but the generalization of Gauß's theorema egregium 16 on page 105, see section 14.7 on page 690. This means that the curvature tensor of a hypersurface is extremely special. The difficulty was to find the expression in the curvature tensor which generalizes this. This did not stop Hopf from advertising the topic. But note that Myers' first results were those of his dissertation under Morse. There is also of course the Hopf–Rinow theorem in section 6.1.2.

12.1 Some History, and Structure of the Chapter

We will see in some detail in §15.7 that the generalization of Gauß–Bonnet theorem was obtained by Allendoerfer and Weil in 1943 (and guessed by Allendoerfer and Fenchel before). But we will see in the note 12.3.1.1 on page 579 that this formula solves question 1 on the facing page in dimension 4 but is not enough to solve question 1 on the preceding page when the dimension is at least 6. Then, despite the huge harvest of the present chapter, question 1 remains open today for any (even) dimension starting with 6, with no guess from experts. See however the very end of §14.1 for the special case of Kähler manifolds.

Besides generalizing the Gauß–Bonnet theorem and the sign conjecture, Hopf had two other favourite questions concerning "curvature and topology."

Question 294 *Does $S^2 \times S^2$ admit a metric of positive sectional curvature?*

Note that it obviously admits metrics of nonnegative sectional curvature, namely Riemannian products of any positive curvature metrics on both S^2. We will see again below that today this question, along with its natural generalizations, is completely open with no guess from the experts. This is also surprising; see Yau's fact 325 on page 579.

So much for positive curvature. A third favourite question of Hopf was the pinching problem for the sphere. We saw that complete simply connected manifolds of positive sectional curvature are known: they are the standard spheres. A heuristic sense of continuity produces the following question: assume that a simply connected manifold has its sectional curvature between 1 and $1 - \varepsilon$. Can you infer that this manifold is a sphere?

Harry Rauch was visiting Zürich in 1948-1949. He was a specialist in Riemann surfaces. But he was so enthusiastic about Hopf's pinching that, back at the Institute for Advanced Study in Princeton, he finally cracked the nut and proved Hopf's conjecture in the pioneering paper Rauch 1951 [1050] with a $1 - \varepsilon$ approximately around $3/4$. We will come back soon to this seminal paper in §§12.2.3. Let us just say here that it really triggered "geometric global Riemannian geometry," in the form it has today.

But it would be unfair not to mention Bochner's results and those of his followers who used the so-called "Bochner technique" via the Weitzenböck formula. The results were from curvature to topology but using analysis via the Hodge theory of harmonic forms. We will meet this topic below in an appropriate place: §§§12.3.1.4 and theorem 345 on page 594 and explain it at large in §15.6.

We now comment briefly on the content of this chapter. The author had a very hard time organizing it. There are so many very different yet relevant results that to categorize them in a completely rational way is a difficult job, say impossible. We hope that our presentation will help the reader to find his way in the present blossoming period of the topic, and find our structure not too artificial. Berger 1960 [149], 1998 [171], 2000 [172], and Petersen 1998 [1019] provide surveys. Also see Yau 2000 [1297].

The section §12.2 is about pinching problems. Hopf's question was a kind of comparison theorem for the simply connected compact space forms. We can ask the same question about any space forms, simply connected or not and with curvature of any sign. They are called the positive, the zero and the negative pinching problem. Now that the pinching problem is solved for some ε, the story is far from being finished. The mathematician's natural imperative is to obtain the best result. In the positive case there is good news: see theorem 297 on page 552: $1/4$ is the best value and one cannot do better than a strict $1/4$ since the $\mathbb{K}\mathbb{P}^n$ do have curvature ranging between 1 and $1/4$ (see equation 4.37 on page 209). The second type of question is the following: assume that the curvature ranges from 1 to $1/4 - \varepsilon$ with a small enough ε for a simply connected manifold. Having again in mind some heuristic continuity principle, can one infer that the manifold is topologically a sphere or a $\mathbb{K}\mathbb{P}^n$? A third type of question is to replace the hypothesis of the type $a < K < b$ by some weaker condition on one of the sides. We will see below various types of geometric invariants coming into the picture, but the main discovery was that most often one can replace the strong condition $K < b$ by a bound involving only an inequality for the volume or the diameter: the moral is that large positive curvature does not affect the topology.

12.1.2 Hierarchy of Curvatures

12.1.2.1 Control via Curvature

Let us formulate another type of question. The sectional curvature (or the curvature tensor) is very "parameter-redundant" for a metric: see equation 4.28 on page 203. The number of parameters in the curvature tensor is

$$\frac{d^2 \left(d^2 - 1\right)}{12}$$

but there are only

$$\frac{d(d+1)}{2}$$

parameters in the metric. The curvature invariant which has as many parameters as a Riemannian metric is the Ricci curvature. The Ricci curvature (see equation 6.10 on page 244) is a differential quadratic form on the manifold, so its number of parameters is $d(d+1)/2$. So the urge is to use only Ricci curvature bounds (as for example in Myers' theorem 63 on page 245). Today there is a bountiful harvest of results and also of counterexamples to prove more or less some optimality of the results. We explained in §§6.4.3 why such results were possible with only a Ricci lower bound instead of a sectional curvature lower bound. We will report on these again in §§12.3.1. The harvest very recently became amazingly plentiful; see for example theorem 318 on page 572, theorem 319 on page 572 and §§12.3.2. An expository text is Gallot 1998 [541].

Finally one can ask if the scalar curvature (although such a weak invariant, namely a numerical function) can control something. The case of dimension 2 should be mainly discarded.

Thinking for example of Hopf's question on $S^2 \times S^2$ that we have met just above and of the von Mangoldt–Hadamard–Cartan theorem 72 on page 255, one might try to classify Riemannian manifolds whose curvature has a given sign, strictly or not: this will be the purpose of §12.3. Here the harvest is not poor, but for example the classification of positive or nonnegative sectional curvature manifolds is practically completely open. The harvest on the negative sign is much better.

12.1.2.2 Other Curvatures

In the above a hierarchy for curvatures was obvious: the strongest one is the sectional, then comes the Ricci and finally the scalar. For example the positivity of the sectional curvature implies that of the Ricci curvature, and the latter implies the positivity of the scalar curvature. And there is of course no inverse implication.

Quite recently other curvature invariants and notions came into this too simple three level hierarchy. We will meet them only occasionally, but one cannot ignore them totally. First comes the curvature operator: the two first lines of the curvature tensor identities in equation 4.28 on page 203 show that R defines a symmetric bilinear form on the exterior product $\Lambda^2(TM)$ of the tangent space. This bilinear form is called the *curvature operator* and denoted (if needed) by R^*. The word *operator* refers to the fact that, under the Riemannian structure, one can identify bilinear forms with endomorphisms. Beware that knowing R^* does not tell us more than knowing R. However knowing the positivity of R^* is much stronger that the positivity of the sectional curvature K; see §§§12.3.1.4. Looking at the definition, this is not surprising, since the sectional curvature yields the value of the curvature operator only on the 2-forms in $\Lambda^2(TM)$ which are of the most degenerate type, namely of the form $x \wedge y$. The complete relations between pinching inequalities for K and for R^* are not understood today. One does not know the optimal pinching of sectional curvature which will imply the positivity of R^*. This pinching probably depends on the dimension. A recent reference is Chen 1990 [360]. However, we know that a 1/4-pinching (pointwise) does not imply $R^* \geq 0$, but the miracle is that it implies (notation below) that

$$K_{\mathbb{C}}^{\text{isotr}} > 0;$$

see §§§12.3.1.4 below for motivation. The proof is just a naive use of the last equality (circular permutation) in 4.28 on page 203, called the first Bianchi identity, see §15.2.

Next we complexify the tangent space (at each point; we do not complexify the whole manifold in general) and then define the complex sectional curvature, denoted by $K_{\mathbb{C}}$. An important notion in §§§12.3.1.4 is that of *complex isotropic planes*, namely those on which the complexification of the metric g vanishes. Then the condition $K_{\mathbb{C}}^{\text{isotr}} > 0$ makes sense. Finally we will see, for Kähler manifolds, the notion of holomorphic and of bisectional curvature. The implications can be summed up in the table 12.1.

Table 12.1. Curvature implications

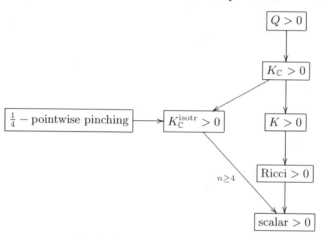

We also introduced into this table the pointwise 1/4-pinching condition; see §§§12.3.1.4. Except the implication of 1/4-pinching, all of the other implications in the table are elementary. Let us just recall the equation 6.9 on page 244 and the fact that the scalar curvature is the trace of the Ricci curvature. For example for any orthonormal basis $\{e_i\}$ of $T_m M$ one has

$$\text{Ricci}(e_1, e_1) = \sum_{i=2}^{d} K(e_1, e_i) \tag{12.1}$$

and

$$\text{scalar} = \sum_{i \neq j} K(e_i, e_j).$$

12.1.2.3 The Problem of Rough Classification

We know that most manifolds cannot carry a metric with curvature of a given sign. So what can a Riemannian geometer ask for? The basic idea is to think of the space of all Riemannian manifolds (up to isometry of course) and try to see if one can see the meaning or the consequences of various bounds for Riemannian invariants. Typically, a rough "classification" would be to prove that some set of conditions imply finiteness for the set of underlying differentiable manifolds which bear a metric satisfying these conditions. This was done in the trail-blazing Cheeger 1970 [330]. If one thinks in a vague sense of the set of all differentiable compact manifolds of a fixed dimension as "discrete," a finiteness theorem would appear like a compactness one. This is naturally linked with possible topologies for the set of all Riemannian metrics of a given dimension. We will try to give a complete account of the most important results obtained in these topics in §12.4.

A special mention should be made of curvature conditions which are not pointwise inequalities but are integral norms (say, roughly of L^2 or L^q type). We will briefly mention results of that kind. One reason for our brevity is that despite their importance, they resort more to analysis than to geometry (you already know the biased temper of the author).

To sum up the structure sketched above we propose table 12.2 on the following page, where one coordinate is the kind of curvature and the other the type of question.

12.1.2.4 References on the Topic, and the Significance of Noncompact Manifolds

We have already explained in the introduction to this book that as we delve into more recent and deeper results, we will give less and less information on the proofs, the tools and the ideas behind the proofs. We also stay in the realm of compact manifolds, for brevity and simplicity, but notable exceptions will be met in §§12.3.1.4 and partly in §§12.3.4. Noncompact manifolds are fundamental, and not only when they appear as the universal covering of a compact one with a "large" fundamental group. Results for noncompact manifolds will be mentioned here and in §14.1. We will of course give various references, both research papers and surveys. The only books on the topic (to the best of our knowledge) are Cheeger & Ebin 1975 [341], Sakai 1996 [1085], Gromov 1999 [633], and Petersen 1997 [1018],[1017]. The book Grove & Petersen 1997 [650] considers quite a few of the topics of this chapter. Petersen 1998 [1019] is very informative and exhibits a deep understanding. The other books on Riemannian geometry already mentioned or those we will quote are more or less specialized. The reader should understand, in case it will not be clear below, that this topic is extremely active today.

12.2 Pinching Problems

12.2.1 Introduction

We saw in §§6.3.2 that the manifolds of constant sectional curvature are locally isometric to spheres, Euclidean spaces or hyperbolic spaces. If compact, they are thus compact quotients of the sphere, Euclidean space or hyperbolic space. Note that the complete classification is still not finished and that such a classification of hyperbolic space quotients is extremely difficult. To solidly

550   12 From Curvature to Topology

Table 12.2. Curvature and topology questions about compact manifolds

Curvature condition	Curvature and topology questions		
	Sectional	Ricci	Scalar
constant	$\mathbb{S}^d(k)$, any k, unique if $\pi_1 = 0$; if not, see table 6.1 on page 287	Einstein manifolds, not classified, no topological restriction known if $d \geq 5$; Ricci flat puzzling	
pinching (almost constant)	$K > 0$: spheres only $K = 0$: needs extra conditions $K < 0$: basically impossible	converges toward Einstein	
> 0	known: only $\mathbb{KP}^n$ and some M^6, M^7, M^{13}. There is a bound on Betti numbers.	no topological restriction known except scalar > 0	complete classification if $\pi_1 = 0$; if $\pi_1 \neq 0$, close to finished
≥ 0	besides products and symmetric spaces, there are examples but no classification	see Ricci flat	see Ricci flat; if exists, no scalar > 0
< 0	universal cover $= \mathbb{R}^d$, so everything is in π_1. No characterization still of those π_1.	Lohkamp: exist on every manifold. More: are C^0 dense among all metrics.	
≤ 0	Exist interesting criteria for separating from < 0 (notion of rank); see ergodic theory in chapter 10		

establish that the space forms are indeed quotients was one of the tasks of Heinz Hopf in the 1930's and one of the motivations for the Hopf–Rinow theorem 52 on page 227. The next natural question is about pinching: assume a compact manifold has a sectional curvature varying not too much (one will say that the manifold is "pinched"). Can one deduce from this that the underlying manifold is, topologically (or perhaps even differentiably) identical to one of the above space forms? After rescaling, we are left with three cases: the pinching question around $+1$, 0, -1 but note that, for the zero case, some normalization is needed, which is done usually by asking the diameter to be not too large, say ≤ 1. Since a trivial but basic remark is that stretching the metric by larger and larger factors will make the curvature go to zero, cf. equation 4.34 on page 207. We will present the positive pinching problem in great detail, because of its historical importance both as a triggering result and as creating new tools. The positive pinching question and its more refined generalizations had an enormous influence, because of its neat statement—a statement which naturally encouraged people to search for the various kinds of generalizations we already mentioned. See the survey Shiohama 2000 [1134].

Returning to the general pinching problem, we will see that the answer, when we only ask that the curvature be pinched, has surprisingly different answers in the three cases. This will begin illustrating the striking differences between space forms of different signs, differences which will met already in §§6.6.2, §§9.5.4, theorem 213 on page 455, and §§11.3.5. There is an answer covering the three cases together, but one needs, besides the pinching of the sectional curvature, both an upper bound on the diameter and a lower bound on the volume. Even if this result is the union of previous results to be met individually below, we mention a general fact, as stated and proved in Fukaya 1990 [531] theorem 15.1:

Theorem 295 *For each d, D, v there is an $\varepsilon = \varepsilon(d, D, v) > 0$ such that if a Riemannian manifold of dimension d has diameter smaller than D, volume larger than v, sectional curvature between bounds $\sigma - \varepsilon < K < \sigma + \varepsilon$ (where $\sigma = -1, 0$ or 1) then this manifold is the underlying manifold of a space form of constant sectional curvature equal to σ.*

The positive case was solved in 1951 by Rauch without Fukaya's extra conditions, see below. The zero case was solved in Gromov 1978 [609] with only an extra upper diameter condition and the negative case was solved also by Gromov (with only an extra upper volume or upper diameter condition); see §§12.2.4. These extra conditions are necessary. For these classical pinching results there is a good survey (up to 1990), namely section 15 of Fukaya 1990 [531]. Informative texts are Gromov 1990 [628] and see Petersen 1996 [1016] as well as the books Grove & Petersen 1997 [650] and Gromov 1990 [628].

There are more general types of pinching, the most general being for the symmetric space forms of §6.6. Such a programme was started in Rauch 1953 [1051]; for references and intermediate results see Min-Oo & Ruh 1979 [927],

1981 [928]. For example in Min-Oo & Ruh 1979 [927], where the proof is mainly analysis, one finds:

Theorem 296 *Let G/K be a compact simply connected irreducible symmetric space and M a Riemannian manifold. Modelled on G/K one can construct on M a principal bundle P and define a suitable norm $\|P\|$ for it. There is a constant ε such that $\|P\| < \varepsilon$ implies that M is diffeomorphic to a finite quotient of G/K.*

12.2.2 Positive Pinching

By some heuristic continuity one might hope to be able to prove that if the sectional curvature is close to a constant, then the underlying manifold (if simply connected) is still the sphere. This was proven in Rauch 1951 [1050] with a pinching constant (i.e. the ratio between the lower and the upper bounds of the sectional curvature) about 3/4. Rauch's paper was seminal in two respects. It was the first controlling the metric on both sides (we presented these two bounds in §§6.4.1). Secondly, he had a nice convexity argument for distance spheres; this argument was systematized by Gromov, see below in the present section. His result was in fact that the manifold is covered by the sphere. Be careful that this does not solve the equivariant pinching problem for manifolds which are not simply connected; see below.

The idea of Rauch's proof was this: we want to compare the manifold with a sphere. In the standard sphere of unit curvature all of the geodesics starting from one given point p go to the antipodal point exactly after time π. Assume now that we have a strong form of pinching like $1 - \varepsilon < K < 1$. Then one can guess that the geodesics will make up a nice ball up to time π. Then some mess will start but the mess will stop soon after time π. Indeed by Myers' theorem 62 on page 243 at the most at length $\pi/\sqrt{1-\varepsilon}$ which is close to π. We know part of this by proposition 67 on page 249 and definition 68 on page 250, since the geodesics diverge no more than those in the sphere of radius $\pi/\sqrt{1-\varepsilon}$ but also as much those in the sphere of radius 1. Rauch made the necessary geometrical analysis to prove first that before time π the exponential ball is covered by a nice topological ball and that what remains in the mess, which he called a "pouch," is contractible to a point. Then a ball with only one point added is a sphere.

12.2.2.1 The Sphere Theorem

Today one has the so-called sphere theorem:

Theorem 297 (Sphere theorem) *If a simply connected manifold has sectional curvature satisfying*
$$\frac{1}{4} < K \leq 1$$
then it is homeomorphic to a sphere.

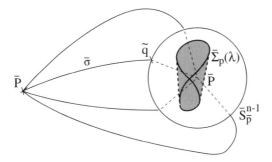

Fig. 12.1. The formation of a pouch

The sphere theorem has an interesting history. First in Klingenberg 1959 [811] the bound of Rauch was improved a little, but principally the notion of injectivity radius was explicitly introduced and the basic theorem discussed in the note 6.5.2.5 on page 276 was proven. Namely the injectivity radius of an even dimensional compact manifold with $0 < K \leq 1$ is larger than or equal to π. Using this result the sphere theorem was proven in even dimensions in Berger 1960 [149]. We will give the proof below. But we just add here that in Klingenberg 1961 [812] the injectivity radius lower estimate was extended to odd dimensions when $1/4 < K \leq 1$ by an argument using Morse theory; see the idea in §§6.5.2. Since Berger's proof was restricted to even dimensions only by the restriction in Klingenberg first result, the sphere theorem follows automatically for any dimension.

For various generalizations there is a recent survey Shiohama 1990 [1133]. We will now present another survey.

The proof of the 1960's consisted in proving that the manifold can be covered with only two topological balls. A classical result from topology, Reeb's theorem, then provides homeomorphism with the sphere, see Milnor 1963 [921]. Note that compactness insures the existence of a $\delta > 1/4$ such that $K \geq \delta$. We pick two points p and q in the manifold M such that $d(p,q) = \text{diam}(M)$. We will show that the open balls $B(p,r)$ and $B(q,r)$ cover M when r is any number strictly between π and $\pi\sqrt{\delta}/2$. Now we need a lemma which we quote explicitly because we will meet it in a more general context in §§§12.2.2.4.

Theorem 298 *If a point q is at maximal distance from a fixed point p then for any tangent vector v at q there is a segment γ from q to p such that the angle between v and $\gamma'(q)$ is less than or equal to $\pi/2$.*

The proof is an easy "*mise en forme*" of the heuristic feeling: if one cannot travel farther away from p than q then in any direction at q one can only travel closer to p, or remain at the same distance. This is essentially (with some refinement) a consequence of the first variation formula 6.3 on page 225.

554   12 From Curvature to Topology

To finish the proof of the sphere theorem, modulo the injectivity radius estimate and Toponogov's theorem 73 on page 258 is now childish: take p and q realizing the diameter, pick any point s which is not in the ball $B(q,r)$ with r defined as above, pick up a segment γ from q to s and a segment η from q to p such that the angle between $\gamma'(q)$ and $\eta'(q)$ is smaller than or equal to $\pi/2$. Look at the standard sphere of radius $r/2$. Then Toponogov's theorem for $K \geq \delta$ implies that $d(p,s) < r$ and one is done.

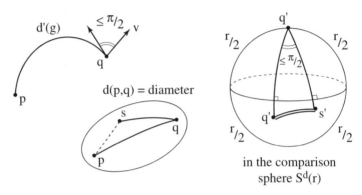

Fig. 12.2. Proving the sphere theorem

There are completely different proofs of this theorem; the most elegant proof is due to Gromov and explained in Eschenburg 1986 [492]. It is in fact very close to Rauch's original proof spirit: just after $\pi/2$ the boundary of a ball of radius R is concave. Then positive curvature shows that the outside of that ball can be contracted to a point just by following the normal field—no trouble arises in the contracting process.

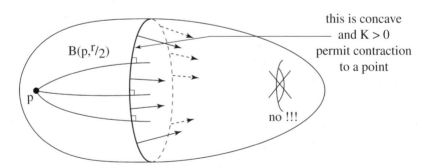

Fig. 12.3. Gromov's proof of the sphere theorem

A careful reader will notice that an abstract pinching constant can be recovered from theorem 295 on page 551 as follows. The diameter condition follows from Myers's theorem (since we are pinching positive curvature) and the volume condition follows easily from Klingenberg's result on the injectivity radius together with one of Rauch's comparison theorems (proposition 75 on page 259).

But there are two dramatic provisos. First, we proved only homeomorphism with the sphere by exhibiting a covering with only two balls. And the proof cannot do better, because there are topological spheres with differentiable structures different from the standard one (the so-called *exotic spheres*). Some exotic spheres are precisely obtained by gluing two half spheres along their equator, but with a suitable identification; see §§4.1.3.6. We are now led to the "differentiable pinching problem" (since of course the simply connected manifolds of constant positive curvature are the standard spheres, each with its standard differentiable structure).

Second there is the problem of what can be done for the nonsimply connected manifolds. We want a comparison result with any space form. The starting manifold being not necessarily simply connected, one desires under a suitable pinching that this manifold is the underlying manifold of a space form of positive constant sectional curvature. This is also called the equivariant pinching problem, the reason being that the proof has to "commute" with the action of the fundamental group on the universal covering (one should realize that this is much more than just asking that the universal cover be homeomorphic or diffeomorphic to the sphere). The standard proof with a two ball covering is clearly not equivariant.

The equivariant question was solved first in Grove, Karcher & Ruh 1974 [647], see also Im Hof & Ruh 1975 [756]. These results also yielded diffeomorphisms. But the pinching constant in this line of investigation depended on the dimension and was not near optimal: sometimes 0.98, even if it goes close to 0.68 when the dimension goes to infinity. For equivariant pinching, homeomorphism or diffeomorphism, there are better constants, but the question of the best constant is open; see the survey Shiohama 1990 [1133]. For the simply connected diffeomorphism problem, after intermediate results appeared, the best constant today is dimensionless and around 0.654, proved in Suyama 1995 [1169], which mixes different techniques.

But nobody knows if $1/4$ is possible or not. Even worse: there is no exotic sphere known which has positive curvature. One knows only of ones with nonnegative sectional curvature: Gromoll & Meyer 1974 [607]. In section 1.6. of Weiss 1993 [1251] it is announced that many exotic spheres cannot be strictly $1/4$-pinched. See also Grove & Wilhelm 1997 [657].

Historically the differentiable pinching problem was solved by Calabi (unpublished) using the center of mass technique (see §§6.1.5 and §§12.4.1.1) and in Shikata 1967 [1131] by methods close to those which will appear in the third part below, namely by an abstract method based on the fact that differentiable manifold structures are in a suitable sense "isolated." This is the place to note that the general theorems of §12.4 can yield almost any pinching type result but without an explicit constant, since the constant is obtained by a contradiction argument of the following type: assume one has a sequence of more and more pinched manifolds. Then one can prove (with additional hypotheses, which are automatically fulfilled in the positive pinching case), that one has a limit Riemannian manifold which is by construction the standard sphere. The isolation implies then that for some (not explicit) pinching constant one is already "on the sphere."

In the other direction, the method of Ruh 1971 [1076] is completely geometric and works equivariantly, which is obviously not the case for the other methods seen in this section. In Ruh's method the sphere will be seen from "outside." The fact that the curvature is very pinched enables us to construct a line bundle τ on the manifold M^d which resembles the normal bundle of the standard sphere in R^{d+1} enough to yield finally an embedding from M into R^{d+1}. What one proves is that the bundle product of the tangent bundle $T_m M$ with τ is trivial. This embedding yields finally a sphere close to the standard one and hence diffeomorphic to it. Such a bundle is called *stabilized*, a name coming from algebraic topology where the technique of making the product of some bundle with some suitable trivial bundle is frequently employed.

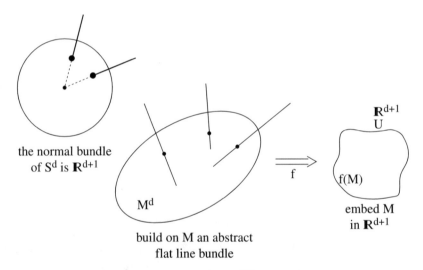

Fig. 12.4. (a) The normal bundle of S^d in $\mathbb{R}^{d+1}$ (b) Build on M an abstract flat line bundle (c) Embed M in $\mathbb{R}^{d+1}$

In Gromoll 1966 [600] there was a very natural and geometric proof, but clearly not equivariant. One starts with the above proof of the sphere theorem by covering with two balls $B(p,r)$ and $B(q,r)$. The result (this is classical differential topology) will be diffeomorphic to the standard sphere if the gluing is close enough to the identity. The gluing takes place ideally in the hypersurface made up by the points equidistant from p and q. This yields a map between the unit tangent spheres at p and q, i.e. a map $\phi : S^{d-1} \to S^{d-1}$.

This map is governed by the behaviour of the geodesics emanating from p and those from q. A refinement of Rauch's comparison theorems on the behaviour of the corresponding Jacobi fields finally permits one to prove that the above map is isotopic to the identity. But Gromoll's pinching constant goes closer and closer to one as the dimension goes to infinity.

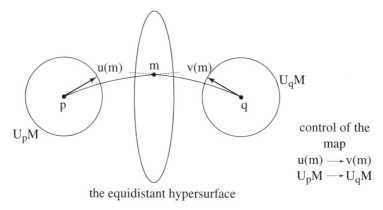

Fig. 12.5. Gromoll's proof of the sphere theorem

12.2.2.2 Sphere Theorems Invoking Bounds on Other Invariants

We will now enter into the spirit explained in the introduction. The game is to replace the one or the two bounds on sectional curvature by bounds on some other curvature or geometric invariants: diameter, volume, etc. A result in this direction is that of Otsu, Shiohama, & Yamaguchi 1989 [988]:

Theorem 299 *If $K > 1$ and the volume of the manifold is close enough to that of the standard sphere, then it is diffeomorphic to a sphere.*

This result was first obtained with an abstract constant by the type of reasoning by contradiction as sketched above. In that text the constant is explicit and the proof uses again a method "from outside:" using geometric arguments one builds up an embedding of the manifold into $\mathbb{R}^{d+1}$ such that the image bounds a ball.

There is also a "differentiable pinching" result yielding manifolds diffeomorphic to the standard sphere. It uses in the hypothesis a pure metric invariant called $\mathrm{pack}_{d+1}(M)$ for a Riemannian manifold M of dimension d. This packing invariant is purely metric—curvature has nothing to do with it. It is equal to one half of the maximum of the minimum nonzero mutual distances between $d+1$ points. The maximum is to be taken over all such sets of points in M. The visual picture is to pack inside the manifold $d+1$ balls of radius as large as possible; the centers of the balls will build up a simplex. A packing number can be defined also for any integer smaller than $d+1$. When one considers only two points instead of $d+1$ one has nothing but the definition of (half) the diameter. Then result is:

Theorem 300 (Grove & Wilhelm 1995 [656]) *If the sectional curvature of a d-dimensional manifold M satisfies $K \geq 1$ and if $\text{pack}_{d+1}(M) > \pi/4$ then this manifold is diffeomorphic to the standard sphere.*

The proof uses Alexandrov geometry (see §§14.5.5) and is a method "from outside" as was Ruh's, except that it just uses the metric and the center of good packing balls to get a suitable embedding. The center of mass technique (see §§6.1.3) is essential here, as in many other places. The use of Alexandrov geometry is natural when working only with the assumption $K \geq 1$. A very informative survey of these sort of results (and of some others below involving the radius, defined in §§12.2.5) is Grove 1992 [645].

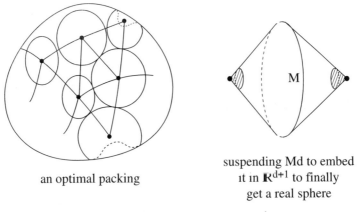

an optimal packing

suspending Md to embed it in $\mathbb{R}^{d+1}$ to finally get a real sphere

Fig. 12.6. (a) An optimal packing (b) Suspending M^d to embed it in $\mathbb{R}^{d+1}$ to finally get a real sphere

12.2.2.3 Homeomorphic Pinching

We come back now to the homeomorphic pinching problem. and recall first that the $\mathbb{KP}^n$ (for $K \neq \mathbb{R}$) have a standard metric with sectional curvature ranging from 1 to 1/4; see §§4.4.3. So the sphere theorem is optimal as far as the pinching constant is concerned. Next it is natural is to ask what happens at 1/4 and also a little below 1/4. Note how mathematicians always proceed— they are never happy, they always want to go further. And if they cannot get the best, they manage to find something interesting on the way. Here the situation is in three respects quite satisfactory. We will give the results and thereafter the ideas for the proofs. First we have the so called *rigidity theorem* which is a perfect answer for the 1/4 question:

Theorem 301 (Rigidity theorem, Berger 1960 [149]) *Assume that a simply connected Riemannian manifold has sectional curvature between 1/4 and 1 and that it is not homeomorphic to a sphere. Then the manifold is isometric to $\mathbb{KP}^n$ (with $K \neq \mathbb{R}$) endowed with its standard metric.*

A pleasant way to look at this rigidity result is to say that any change of metric on a standard $\mathbb{KP}^n$, no matter how small it is, will yield sectional curvature outside a range of 1/4.

Second: one can go a little below 1/4 (how far below perhaps depends on the dimension):

Theorem 302 (Durumeric 1987 [466][1]) *There is an (unknown) constant $\varepsilon(d)$ such that any $1/4 - \varepsilon(d)$ pinched simply connected manifold of even dimension d is homeomorphic to a sphere or some $\mathbb{KP}^n$.*

This is unsatisfactory for the abstractness and the evenness restrictions. The even dimension restriction is due to the fact that at that time it was not known how to go below 1/4 for the injectivity radius in odd dimensions: see the note 6.5.2.5 on page 276. And a lower bound for the injectivity is needed in convergence theorems. Since now we have the Abresch–Meyer theorem 93 on page 276 then theorem 302 extends now to any dimension. But we are still left with the question of finding explicitly the constants $\varepsilon(d)$. A complete answer is still not known but definitive progress is charted below. Note that the results are optimal when the dimension is odd, and that in even dimensions the situation is vaguely similar to that met in §10.10 when studying manifolds all of whose geodesics are periodic:

Theorem 303 (Abresch & Meyer 1996 [7]) *Let M be a simply connected compact manifold with sectional curvature between 1 and $\eta(d)$. There is an explicit $\eta(d)$ such that if d is odd then the manifold is homeomorphic to a sphere, while if d is even one knows only that the cohomology rings $H^*(M,\mathbb{Q})$ and $H^*(M,\mathbb{Z}/2)$ are isomorphic to the corresponding cohomology rings of one of the following manifolds of dimension d: S^d, $\mathbb{CP}^{d/2}$, $\mathbb{HP}^{d/4}$ or $\mathbb{CaP}^2$.*

Also see Zizhou 1999 [1309] and Abresch & Meyer [8]. The best η is unknown; it might well depend on the dimension. We refer to the Abresch–Meyer article for the present best known value which is independent of the dimension and is approximately $\frac{1}{4}(1+\varepsilon_{\text{odd}})^{-2}$ with $\eta = \varepsilon_{\text{odd}} = 10^{-6}$, for d odd, and $\eta = \frac{1}{4}(1+\varepsilon_{\text{even}})^{-2}$ with $\varepsilon = 1/27000$ for d even. We know that one cannot bring these ε constants arbitrarily close to 0, in view of the existing examples of positive curvature manifolds: see §§12.3.1.1. Exceptions are dimension 2 (by the Gauß–Bonnet theorem 28 on page 138) and dimension 3 in view of Hamilton's results on the heat flow presented in equation 11.9 on page 525 and in §§§12.3.1.1. In dimension 4 one finds only S^4 or $\mathbb{CP}^2$ with pinching $0.188 < K < 1$ by Seaman 1989 [1117]. His technique is a refinement of Bochner's, see §§12.3.2 and §15.6. One proves that the second Betti number can be at most one by applying Bochner's formula to two linearly independent harmonic 2-forms. For higher dimensions we have to stop here for the pinching constant. In fact, to go further below $1/4$ will mean merely to look for a classification of all manifolds (since any manifold has some Riemannian structure) or at least for manifolds of positive curvature. See more on this in §§12.3.1. However one should not forget the low dimensional exceptions. Note that for getting a sphere from the above topological condition one needs to use the solution of the Poincaré conjecture in dimension 5 or more.

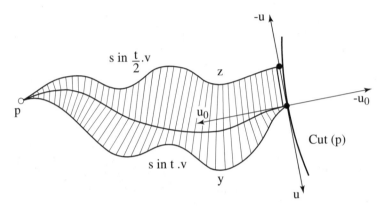

Fig. 12.7. Proof of the rigidity theorem

Let us now give some idea of the proofs. For the rigidity theorem 301 on the preceding page the proof is quite simple and geometrical; it does not need new tools. If one looks at the proof above for the sphere theorem one sees that it still works when the diameter is greater (strictly) than $\pi/2$. Just look at the Toponogov comparison with the sphere of constant curvature equal to 1. So that we can assume that the diameter is equal to $\pi/2$. But this is a very special, say critical, situation which we met in §§6.5.5: our manifold has diameter equal to its injectivity radius. This is because of Klingenberg's theorem 92 on page 276 (which is valid only in even dimensions). We saw many implications of that theorem in §§6.5.5 and §§10.4.2, but could not conclude anything without curvature conditions. But here we have the condition $1/4 \leq K < 1$. Pick any point p and look at its cut-locus Cut-Locus(p) as in §§10.4.2. It is a nice submanifold. If it consists only in one point we are finished—we are on the sphere of constant curvature equal to 1. If we are not in this case, then we might have geodesics still going from p to the same given q in Cut-Locus(p). Then the corresponding Jacobi fields have to satisfy the equality in Rauch's comparison theorem (proposition 75 on page 259) and the curvature condition shows that this field has to be of the form $\sin(t)v$ (for a parallel transported vector field v) and the curvature has to be equal to 1 along it. Then if we look at geodesics from p going to different points in Cut-Locus (p), the tangent vectors to Cut-Locus (p) will correspond this time to the opposite situation for the equality case in the Rauch comparison theorem (proposition 75). Again the curvature condition implies that such a Jacobi field is of the form $\sin(t/2)v$ where v is a fixed (parallel transported) vector and more: the curvature is equal to $1/4$ along it. Finally Jacobi fields can be only of two types : $\sin(t/2)v$ (for curvature $1/4$) or $\sin(t)v$ (for curvature 1). This enables us with some work to completely reconstruct the metric by the geodesics starting from p "à la Cartan" as seen in §§6.3.1. So we get the standard $\mathbb{KP}^n$, including $\mathbb{RP}^n$.

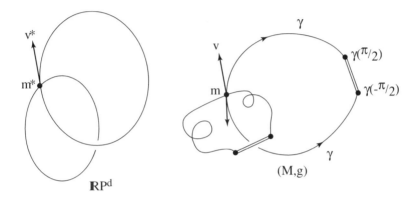

Fig. 12.8. How to build a map from $\mathbb{RP}^n$ to M

For the odd-dimensional result the above will not apply, even if we know by theorem 93 on page 276 that the injectivity radius is not too small. In fact this case is treated, at the same time as that of even dimension, by the horseshoe technique which we explain now. If one appeals to the theorem 304 on the following page, we have a dichotomy for the diameter. If it is larger than $\pi/2$ then we have a sphere. So from now on we suppose that the diameter is closed to $\pi/2$. We hope now that the following heuristic thought will work: when the diameter and the injectivity radius are equal, the geodesics are periodic and of length π. In the present case the diameter being very close to the injectivity radius, one can hope that any geodesic γ starting from p say, will be almost closed, namely that $d(\gamma(\pi/2), \gamma(-\pi/2))$ will be small, hopefully even smaller than the injectivity radius. If this is the case one could then connect them by a unique segment, then making some kind of a horseshoe. The trick is now that of theorem 222 on page 463 and this implies exactly the conclusions of the theorem. These heuristic ideas have been made to work. This is a beautiful geometric achievement carried out in Abresch & Meyer 1994 [8]. But it involves a very technical tool, namely generalized Rauch estimates on

Jacobi fields Y to control not only their norms $\|Y(t)\|$ but also to control their angular velocity $\|Y'(t)\| / \|Y(t)\|$ with curvature estimates.

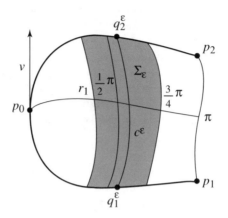

Fig. 12.9. The horseshoe trick

12.2.2.4 The Sphere Theorem with Lower Bound on Diameter, and no Upper Bound on Curvature

But the game is far from being finished, because one would like to get results with weaker hypotheses in the spirit of the introduction. A spectacular event was:

Theorem 304 (Grove & Shiohama 1977 [654]) *Let M be a Riemannian manifold with $K \geq \delta > 0$ and a diameter larger than $\pi\sqrt{\delta}/2$. Then M is homeomorphic to a sphere.*

So the upper bounds for the curvature completely disappear. This is not too surprising today as we saw for Colding's formulas in §§6.4.2 and in the philosophy in §§6.4.3 because of Myers's results. In a manner different from chapter 6 on page 221, we comment here a little bit about the possibility of getting rid of an upper bound for the curvature and still controlling the topology. Very roughly, let us look at the two pictures in figure 12.10 on the next page.

One sees that smaller and smaller fingers give rise to huge positive curvature, and still the topology stays the same. Oppositely if one adds fingers "with holes," then this introduces negative curvature and smaller and smaller fingers give more and more negative curvature. Of course this does not prove anything. We will below introduce techniques to realize this dream.

The first technique was very new; it still uses Toponogov's theorem 73 on page 258, but the main point which is absolutely fundamental is to be able to go beyond the injectivity radius. We comment now in detail because the

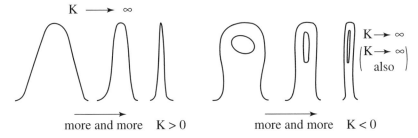

Fig. 12.10. (a) More and more positive curvature (b) More and more negative curvature

authors invented a new tool, the notion of critical point for distance functions, which turned out to be of basic importance thereafter up to the present day.

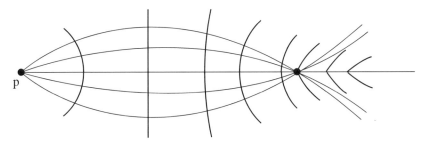

Fig. 12.11. The open balls $B(p,r)$ remain diffeomorphic to $\mathbb{R}^d$ even if their boundaries are not smooth

The trick is to use the distance function $d(p, \cdot)$ from a point p and have the following definition:

Definition 305 *A point q is said to be* critical *for the distance function $d(p, \cdot)$ if for any direction v at q there is a segment (a shortest geodesic) γ from p to q whose speed vector γ' at q makes an angle with v not larger than $\pi/2$.*

The geometer will like to see some other equivalent views of critical distance. The point q will be not critical (for $d(p, \cdot)$, and one can call it *regular*) if there exists a tangent vector v at q such that all the segments from q to p make an angle with v larger than $\pi/2$. Or: all of the directions of the segments from p to q are contained in some open hemisphere of the unit sphere $U_q M$.

As an example, we saw above in theorem 298 on page 553 that is the case when $d(p, \cdot)$ is maxima at q. Another case would be the antipodal point of p on a geodesic loop through p. Is this notion coherent with the notion of a critical

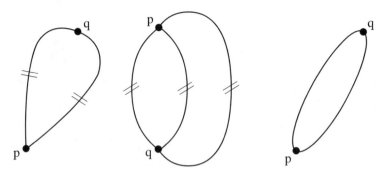

Fig. 12.12. Critical distance

point for a smooth function on a differentiable manifold as given in §§10.3.2? The distance function $d(p,\cdot)$ is differentiable at every point q not in the cut locus of p, and its gradient is the velocity of the unique segment from p to q. So q is certainly not critical. Moreover along this vector, the first derivative of the distance is positive. In the smooth case, at a critical point there is no direction where the first derivative is nonzero. And the definition above for the distance function says exactly that. Conversely, if q is not critical, looking at the subset of the unit sphere at q made of the vectors which are the speed at q of the set of all segments from p to q, one sees immediately that there is some vector w which makes an angle with all of these vectors larger than $\pi/2$. Going along this direction will exactly generalize a positive first derivative.

To prove theorem 304 on page 562 one imitates the proof that a manifold covered by two open balls is a topological sphere (see the discussion following theorem 297 on page 552). One just builds up from these two balls a smooth function with only one maximum. In the Riemannian case one could look at the "equidistant" S^{d-1} and smooth the angles along it.

There is, as suggested just above, a nice smoothing argument based at each noncritical point to find a vector of strictly increasing distance. This is just dittoing the basic fact of Morse theory (see §§10.3.2) that the topology of the sublevel domains do not change when one follows the gradient as long as one does not meet any critical point. Then one has only to remark that the two conditions of the theorem imply that if p and q have distance between them equal to the diameter, then Toponogov's theorem implies immediately that there is no other critical point for $d(p,\cdot)$ besides p and q. The critical point technique is now a basic tool in many situations. Surveys on it are: Cheeger 1991 [334], Grove 1985 [643], Meyer 1989 [916], Grove 1990 [644], Karcher 1989 [780]. See also Gromov's theorem 326 on page 580 on manifolds of nonnegative curvature.

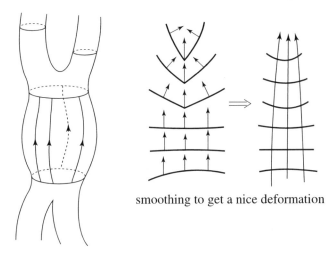

Fig. 12.13. Smoothing to get a nice deformation

12.2.2.5 Topology at the Diameter Pinching Limit

In the same spirit as we did for the rigidity theorem 301 on page 559, we want of course to study what happens in theorem 304 on page 562 at the exact value of diameter $\pi\sqrt{\delta}/2$. The horizons of ignorance were pushed far away with the following diameter rigidity theorem:

Theorem 306 (Gromoll & Grove 1987 [603]) *A simply connected compact manifold M with sectional curvature $K \geq 1$ and $\mathrm{diam}(M) \geq \pi/2$ is either (1) homeomorphic to the sphere or (2) isometric to a compact symmetric space or (3) has the cohomology algebra of $\mathbb{C}a\mathbb{P}^2$.*

Note that the conclusion is close to theorem 301 on page 559, but still not finished in the case of $\mathbb{C}a\mathbb{P}^2$. For the nonsimply connected case the answer is complete: one gets a spherical space form or the very special $\mathbb{CP}^{2k+1}/\mathbb{Z}_2$. No other such space form exists because of Synge's theorem 64 on page 246, as already remarked in §§6.6.3.

But we want more, in the spirit of the introduction and the skeleton of this chapter, i.e. we want to relax the diameter a little bit below the limiting value of $\pi\sqrt{\delta}/2$. This was partially achieved in

Theorem 307 (Durumeric 1987 [466]) *For a suitable ε, any compact Riemannian manifold of diameter larger than $\pi\sqrt{\delta}/2 - \varepsilon$ with sectional curvature $K \geq 1$ is homeomorphic to a spherical space form manifold or a simply connected manifold whose cohomology ring is generated by a single element (which is the case for the $\mathbb{K}\mathbb{P}^n$ but the converse at the topology level is not true).*

So that for theorem 306 on the previous page and theorem 307 on the preceding page the situation is still not finished today and in some sense close to the situation in §§6.5.5. We will now explain the proof of theorem 306 on the previous page and why it fails for manifolds which remind us of the Cayley plane. Very roughly speaking, if our manifold is not a sphere, then we have a critical diameter as in the proof of theorem 301 on page 559. So one has again the situation of a point p and its cut locus Cut-Locus(p). Now one looks at the map $U_pM \to$ Cut-Locus(p), one proves that is is a Riemannian submersion and one studies the corresponding metric foliation of U_pM into spheres. But such foliations are extremely rigid—in Gromoll & Grove 1988 [604] it is proven that such a foliation is metrically congruent to the standard Hopf fibration (see §§4.1.3) except possibly for $S^{15} \to M^8$ which corresponds to the hypothetic Cayley plane.

To conclude as we did in theorem 301 on page 559 one should use the radius:

Definition 308 *The radius of a metric space is the minimum of the radii of balls which contain (hence coincide with) the whole space.*

The radius is a finer invariant than the diameter. For example, think of a very thin ellipsoid. Its radius is close to half its diameter, as shown in figure 12.14

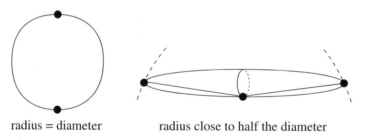

radius = diameter   radius close to half the diameter

Fig. 12.14. (a) Radius = diameter (b) Radius close to half of the diameter

To have a large radius says much more than only a large diameter. And this invariant can have strong effects—just see theorem 321 on page 575 and

Theorem 309 (Wilhelm 1996 [1263]) *Any simply connected compact Riemannian manifold with sectional curvature at least $K \geq 1$ and radius at least $\pi/2$ is either homeomorphic to a sphere or isometric to a $\mathbb{KP}^n$.*

The proof is very geometric again. It is a strong refinement in the case of the 15 dimensional sphere S^{15} of the Gromoll–Grove Riemannian submersion theorem 306 on the preceding page

It is proven in Grove & Petersen 1993 [649] that $K \geq 1$ and radius larger than $\pi/2$ implies homeomorphy to a sphere. For more on this see the very informative Petersen 1998 [1019]. In the spirit of weaker and weaker hypotheses see Petersen & Wei 1996 [1024]. We will meet the radius again in theorem 321 on page 575.

12.2.2.6 Pointwise Pinching

There is also the pointwise pinching problem: instead of pinching the ratio of minimum and maximum curvatures over the whole manifold, we control the ratio of minimum and maximum curvatures at each point:

Definition 310 *At each point $m \in M$ define the* curvature ratio *to be*

$$\frac{\inf_P K(P)}{\sup_P K(P)}$$

where the sup *and* inf *are taken over all 2-planes $P \subset T_m M$ at the point m.*

Theorem 311 *If for every point $m \in M$ on a simply connected compact Riemannian manifold one has curvature ratio less than $1/4$, then the manifold M is homeomorphic to a sphere.*

Of course the Toponogov triangle comparison theorem and methods of §§12.2.2 are hopeless here. The above result is a corollary of theorem 336 on page 588 as we will see in §§§12.3.1.4. Historically the first result concerning pointwise pinching was obtained in Ruh 1982 [1078], but for a pinching depending on dimension and requiring strong pinching, i.e. close to 1. In exchange the proof applies to the equivariant case and yields diffeomorphism. Its proof involves a great deal of analysis and reduces ultimately to using the difficult theorem 296 on page 552 concerning pinching comparison with symmetric spaces, which we mentioned. There are many other results of that vein; see the almost final result of Margerin 1991, 1993, 1994 [895, 897, 898], chapter 5 for detailed statements and the technique of the proof, as well as the nice survey on deformation techniques along Ricci flow introduced already in §§11.4.3 which uses a lot of analysis. Also see section 4 of Besse 1994 [184] and the last chapter of Hebey 1999 [691].

12.2.2.7 Cutting Down the Hypotheses

Again one would like milder hypotheses. Motion in this direction starts with sectional curvature, then one tries to replace it by Ricci curvature, then by scalar curvature and finally by various integral norms for various curvatures. The subject advances so rapidly today that we can only give recent references: Petersen, Shteingold & Wei 1996 [1021], Gallot 1988 [540], Anderson & Cheeger 1991 [49], Gao 1990 [546], and Petersen & Wei 1996 [1024]. Various norms are introduced and though natural they are still very long when written down explicitly. Curvature operator pinching on noncompact manifolds is discussed in Chen & Zhu [359].

12.2.3 Pinching Near Zero

The space forms with zero curvature are flat tori and their finite quotients (see §§6.6.2). Might there be some $\varepsilon > 0$ such that if a manifold has diameter bounded by 1 and curvature bounded on both sides by ε and $-\varepsilon$, then this manifold admits a flat structure? Unlike positive curvature pinching, the answer in the present case is no. There are other manifolds known to be "almost flat." They can be viewed metrically as compact quotients of nilpotent Lie groups. Topologically they are nothing but successive fibrations whose fibers are always a circle and some finite quotient. Heuristically this curvature smallness is possible because a circle (a curve) always has zero curvature, and is a direct consequence of the Riemannian submersion formulas: see equation 15.17 on page 721.

The basic paper Gromov 1978 [609] succeeded in proving the converse: the solution of the pinching problem is true, but the answer consists in not only flat manifolds but the so-called *infranilmanifolds*.

Theorem 312 (Ruh 1982 [1077]) *There is a universal constant $\varepsilon(d) > 0$ such that if a Riemannian manifold M^d satisfies $-1 \leq K \leq 1$ and $\operatorname{diam} M < \varepsilon(d)$ then M^d is an infranilmanifold.*

Note that some normalization is needed, since scale changes $g \mapsto \lambda g$ with λ greater and greater will make the curvature smaller and smaller: $K_{\lambda g} = \lambda^{-1} K_g$. Some readers might prefer normalizing by $|K| < \varepsilon$ and $\operatorname{diam} = 1$, or even better in one shot:
$$|K| \operatorname{diam}^2 < \varepsilon.$$

Gromov proved that under such a condition, a compact Riemannian manifold has a finite covering which is the quotient of a simply connected nilpotent Lie group. Ruh found a little more and his work yielded an infranilmanifold, i.e. for any thusly pinched manifold M there is a nilpotent Lie group N and a discrete subgroup Λ of the semidirect product $N \rtimes \operatorname{Aut}(N)$ of N by its automorphism group $\operatorname{Aut}(N)$ such that M is diffeomorphic to N/Λ and moreover Λ is of finite index in N. This is an analogue of the Bieberbach theorem (see §§6.6.2): compact flat manifolds are finite quotients of tori.

A completely detailed proof of Gromov's result is given in Buser & Karcher 1981 [296]. Their techniques are purely geometric but a complete proof is quite long and involved. Two good sketches of it are: Sakai 1996 [1085] pages 317 and up, and Fukaya 1990 [531] §8 and §9. Using Toponogov's triangle approach, one studies in detail the fundamental group as isometries of the universal cover. The curvature being extremely small, the exponential map is a covering map at very large distances, and then globally not much different from a covering. The elements of the fundamental group almost commute, as can be seen by also controlling the parallel transport (remember the golden triangle §15.2, §15.4, §15.5). The curvature is the parallel transport around infinitesimal parallelograms, so that one has to prove a kind of integrated version of this. The control on the commutators of the fundamental group for its elements generated by small geodesic loops is a generalization of a famous lemma of Margulis which was used for the rigidity of space forms (see §§6.6.4). Then one can boil down to nilpotent groups. Except for the best $\varepsilon(d)$ which is still to be found, this result is then optimal. Gromov's $\varepsilon(d)$ is of the order of

$$\varepsilon(d) \sim e^{-e^{e^{d^2}}}.$$

Also recall theorem 295 on page 551 where one effectively gets only flat space forms as manifolds, but with the extra hypothesis of a lower bound on volume. This was observed in Gromov 1978 [611].

12.2.4 Negative Pinching

Negative pinching provides even less control on topology than zero pinching.

Theorem 313 (Gromov & Thurston 1987 [640], Farrell & Jones 1989, 1993 [505, 509]) For any given larger than 2 and any $\varepsilon > 0$ there are manifolds of dimension d with curvature satisfying $-1 \leq K \leq -1 + \varepsilon$ which bear no metric of constant curvature.

Do not hope even for a general result around 1/4 for the negatively curved $\mathrm{Hyp}_{\mathbb{K}}^n$, since this was ruled out in Farrell & Jones 1994 [510]. The techniques used to construct these manifolds are very geometric. In Gromov & Thurston 1987 [640] one starts with space forms and builds up clever gluings along totally geodesic hypersurfaces inside them. In Farrell & Jones 1989 [505] one makes connected sums with exotic spheres and then controls the curvature of suitable metrics with the formulas for curvature in Riemannian submersions which we record in equation 15.17 on page 721. In the first type of construction the final topology can be quite sophisticated but in the second technique one gets only manifolds homeomorphic to space forms since one makes connected sums with various exotic spheres. Note that from §§§12.4.1.2 ever larger diameters are forced by the Gromov's theorem 315 on the next page. The large variety of these "counterexamples," still not classified, explains the sense in which we wrote above that there is less control on negatively pinched manifolds. In other words, there is no structure theorem as in theorem 312 on the facing page where we had a generalized Bieberbach theorem.

But one has the following related optimal result:

Theorem 314 *Assume that a compact Riemannian manifold M has $-4 \leq K \leq -1$ and that $\pi_1(M)$ is a group isomorphic to the fundamental group of a compact space form of negative curvature which is not of constant curvature (i.e. not real hyperbolic). Then M is isometric to that space form, i.e. is locally symmetric.*

This was proven in Ville 1985 [1217] for four dimensional manifolds. The complex case was proven independently in Hernandez 1991 [707] and Yau & Zheng 1991 [1298]. For the quaternionic and the Cayley case the result follows from Hernandez 1991 [707], Corlette 1992 [404], Gromov 1991 [624] and Aravinda & Farrell 2000 [58, 59].

We also have a good pinching result, optimal for the ingredients if one considers the counterexamples just above:

Theorem 315 (Gromov 1978 [611]) *There is a number ε depending on d, D, V so that a manifold with $-1 < K < -1 + \varepsilon$ of dimension d and with either (1) diameter smaller than D or (2) volume smaller than V carries a metric of constant negative curvature.*

Gromov proved this directly but it is also a consequence of combining theorem 295 on page 551 with the important intermediate result:

Theorem 316 (Gromov 1978 [611], Heintze 1976 [698]) *There is a universal constant $v(d)$ such that any compact manifold of dimension d with $-1 \leq K < 0$ has volume larger than $v(d)$.*

The proof is again very geometric; one plays on the manifold with the smallest possible periodic geodesics generating the fundamental group, as in the proof of theorem 90 on page 273. The diameter condition is missing: this comes from an inequality of Gromov 1978 [611] which bounds the diameter as a function of only the dimension and the volume under the negative curvature condition of theorem 314 on the previous page. The proof is very geometric—volumes are compared in two opposite ways, with curvature estimates and with the special forms that isometries of negative curvature manifolds must take, as in hyperbolic spaces.

Note 12.2.4.1 (The philosophy of negative curvature) Let us consider for the moment manifolds which are complete, but not necessarily compact. The philosophy of Gromov is that it is extremely hard to understand the class of manifolds of negative curvature and, for example, to construct new ones. Recall first that in §§6.6.2 we had examples of Gromov & Piatetski-Shapiro of nonarithmetic space forms, and the examples of theorem 315. Now in Gromov & Thurston 1987 [640] one finds, for any $\varepsilon > 0$ some manifold of negative curvature, but which never admits a metric such that its sectional curvature can be bounded by $-1 \leq K \leq -\varepsilon$. The construction consists in taking a space form M of constant curvature, and finding in it a totally geodesic submanifold of codimension 2. Then one looks at suitably ramified coverings of M branched along N. The final proof is very hard. It uses once again the diffusion of cycles used already on page 334 and theorem 274 on page 518.

It is also hard to construct real analytic examples. One can find some in Abresch & Schroeder 1992 [9]. Do not forget Gromov's finiteness theorem 364 on page 610 for negatively curved analytic manifolds. ♦

Note 12.2.4.2 (Most geometries are negatively curved) Even more embarrassing is that geometric objects (not necessarily manifolds) of negative curvature are very numerous. This contrasts with the rarity of compact manifolds of negative curvature. Gromov even has a "vague" conjecture to the effect that "in high dimensions every hyperbolic manifold is arithmetic."

We explain now why most geometries are of negative curvature. The simplest way to construct geometries is to glue together Euclidean (flat) simplices or more generally polytopes. In dimension two, our claim comes from the fact that when one glues polygons with more than 6 vertices, then at every vertex the sum of the angles will be larger than 2π. For such a piecewise flat geometry, the curvature is concentrated at the vertices (a distribution type of curvature) and its values are

$$K = 2\pi - \text{sum of the angles at that vertex},$$

a value which is negative since one has to glue at least three polygons. For higher dimensions, it is less clear how one can carry out such a construction because the vertex condition is very complicated and still not understood. For more on this, see the category of CAT (k) spaces in §§14.5.6. ♦

12.2.5 Ricci Curvature Pinching

In pinching the Ricci curvature, we face first the question of which Riemannian manifolds have constant Ricci curvature. This is nothing but classifying all Einstein manifolds, the topic of the third part of chapter 11 Even if you want to ignore our ignorance concerning this classification, still the pinching question makes sense and there is the

Theorem 317 (Anderson 1990 [39]) *There is a constant ε depending only on numbers d, i, D, σ such if any compact manifold M with Riemannian metric g satisfies $|\text{Ricci} - \sigma g| < \varepsilon$, $\text{Inj}(M) > i$ and $\text{diam}(M) < D$ then M carries an Einstein metric on the manifold under consideration.*

At least today, such a theorem has not helped in the search for Einstein manifolds. But another type of pinching can be considered if one thinks back to Myers' theorem 63 on page 245 which said that if you assume the Ricci curvature is at least $(d-1)k$, then the diameter is bounded by $\pi/\sqrt{k}$ and equality happens only for the standard sphere. Then one can ask for a Myers-type pinching result. Surprisingly enough it is shown in Anderson 1990 [38] that such a pinching result cannot exist (see also Otsu 1991 [987]). This was achieved by building suitable nonspherical examples, with tools from the theory of Riemannian submersions, where Ricci $> d - 1$ and diameter comes closer and closer to π, even though the examples are not spheres. But if the diameter is replaced by the volume then

Theorem 318 (Perelman 1994 [1008]) *There is a number $\varepsilon(d)$ such that if a Riemannian manifold M of dimension d satisfies both*

$$\text{Ricci} \geq d-1 \text{ and } \text{Vol}(M) > \beta(d) - \varepsilon(d)$$

then M is homeomorphic to the sphere.

This is not inconceivable, because Bishop's theorem 107 on page 310 yields an upper bound for the volume so that the diameter is also automatically pinched. The metric aspect of the proof is very geometrical but quite intricate. It mixes techniques of critical point theory, contractibility of balls in concentric ones of larger radius (see §§§7.2.4.6 and the proof of theorem 376 on page 621) and playing cleverly with algebraic topology (see the notion of controlled topology, on page 622). One proves by induction on k that any embedded sphere S^k can be filled up (Gromov's view of it is "homology generates shadow," and "shadows have volume"). Then all of the homotopy groups below the dimension vanish and one then applies the solution of the Poincaré conjecture, for dimension 3 (just see theorem 279 on page 525). Needless to say, the injectivity radius is ignored. We just give the two pictures of Perelman's text in figure 12.15 on the next page.

But this result is now superseded by:

Theorem 319 (Cheeger & Colding 1997 [337]) *There is a number $\varepsilon(d)$ so that if a Riemannian manifold M^d satisfies both*

$$\text{Ricci} \geq d-1 \qquad \text{and} \qquad \text{Vol} \geq \beta(d) - \varepsilon(d)$$

then M^d is diffeomorphic to the sphere S^d.

The proof is given in appendix 1 of the quoted text. Moreover the volume bound can be explicitly computed. This is only the emerged tip of an iceberg, still not finished to be fathomed. We will see below in theorems 321 on page 575, 350 on page 597 and 351 on page 597 other results of that genre. Together with the Colding L^2 Toponogov theorem 77 on page 264 for positive Ricci curvature, another systematic tool introduced by the authors for the general study of manifolds with a lower bound on Ricci curvature is that of suitable warped products (see the note 4.3.6.1 on page 195). We suggest that the reader read the introduction and appendix 2 of Cheeger & Colding 1997 [337] for a survey of ideas, results and a program concerning manifolds with a lower bound on Ricci curvature. This program is carried out under the banner of *synthetic geometry*, a word—as well as a program—forged in Gromov 1978 [612].

An important intermediate result was

Theorem 320 (Colding [386]) *Under the same hypothesis as in theorem 319 the Gromov–Hausdorff distance (see theorem 380 on page 626) between the manifold and the standard round sphere S^d can be made as small as desired.*

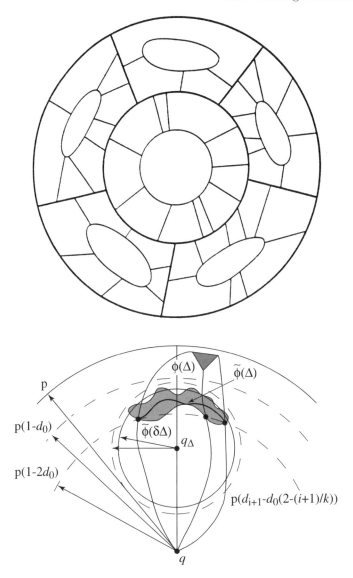

Fig. 12.15. Two pictures from Perelman's article

We will now sketch the proof of theorem 320 on page 572 in detail to help the reader to realize how surprisingly strong metric conclusions can be obtained with only volume estimates. Other theorems with Ricci lower bounds use this technique intensively, so that we are choosing theorem 320 on page 572 as an archetype. What we are going to do now is a deep refinement of what was begun in theorem 108 on page 311. In all this we do not make a precise value for ε, since we are interested only in the ideas of the proof.

The Bishop–Gromov volume estimate theorem 107 on page 310, using the upper bound on ball volumes and the monotonicity, first shows this: if the total volume is close to that of the standard sphere, then all balls in our manifold have volumes close to those of the same radius in the standard metric on a sphere. Then one sees that if two points p, q are such that $d(p,q) > \pi - \varepsilon$, then for any other point x we have $d(p,x) + d(q,x) - d(p,q) < \varepsilon$. If not then one could put into the manifold a ball centered at x which would add too much volume to M, as pictured in figure 12.16.

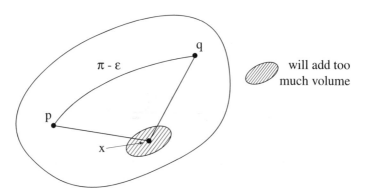

Fig. 12.16. The darkened ball adds too much volume

This shows sphere-like behaviour but is far from enough. Again with theorem 107 on page 310 one can show that the various annuli $B(x, R') \setminus B(x, R)$ also have volume almost equal to that of the same annuli in the standard round sphere. This, at least as regards the measure, shows that the distance functions $d(p, \cdot)$ behave almost like the ones in the round sphere. The aim is now to show the same behaviour but from the perspective of the metric. This is what Colding's inequality theorem 77 on page 264 is good for. Now the almost-isometry with the sphere is constructed as follows. Recall (from §1.9) that in the round sphere, the first spherical harmonics are the cosines of the distance functions from points:

$$\cos d(p, \cdot)$$

and that the standard embedding of the sphere S^d in $\mathbb{E}^{d+1}$ can be realized by taking $d+1$ points $p_1, \ldots, p_{d+1}$ as in figure 12.17 on the facing page and

mapping
$$x \mapsto (\cos d(p_1, x), \ldots, \cos d(p_{d+1}, \cdot)).$$

This embedding is mimicked for our manifold M by choosing two sets of points (almost antipodal) $(p_1, \ldots, p_{d+1})$ and $(q_1, \ldots, q_{d+1})$ such that

$$d(p_i, q_i) > \pi - \varepsilon,$$
$$d(p_i, p_j) = \pi/2,$$
$$|d(p_i, q_j) - \pi/2| < \varepsilon.$$

That choice is possible by various improvements of the metric inequality above. Then one finishes the proof by showing that the map $M \to \mathbb{E}^{d+1}$ given by
$$x \mapsto (\cos d(p_1, x), \ldots, \cos d(p_{d+1}, x))$$
yields the desired metric approximation of S^d. To get a diffeomorphism requires much more work. This was done in Cheeger & Colding 1997 [337] using techniques, due to Reifenberg, originally used to prove existence of minimal submanifolds.

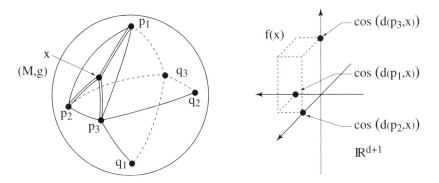

Fig. 12.17. Mapping the sphere into Euclidean space

Another Ricci pinching result, but with "distances" instead of volumes is as follows. It was proven in Grove & Petersen 1993 [649] that $K > 1$ and radius $> \pi/2$ implies homeomorphy to a sphere. In definition 308 on page 566 the radius of a metric space M was defined as the smallest number R such that there is a point m with $B(m, R) = M$. Now we have:

Theorem 321 (Colding 1996 [385]) *If a compact Riemannian manifold satisfies* Ricci $\geq d - 1$ *and its radius is close enough to* π *then it is homeomorphic to the sphere.* [2]

---
[2] How close is close enough can be made explicit.

The idea of the proof is, as before, to show that the Gromov–Hausdorff distance (see §§§12.4.2.3) between the manifold and the standard sphere can be made as small as we wish. As in the work of Grove & Petersen, this is done by squeezing suitable balls in the manifold and defining with them a map to the sphere.

We mention once again the basic philosophy introduced by Gromov, and commented upon in theorem 108 on page 311 and theorem 382 on page 626; namely that the nonincreasing property in Bishop's theorem (as extended by Gromov beyond injectivity radius) for manifolds with lower bounded Ricci curvature, apparently a result concerning the measure, finally can give extremely strong two-sided metric inequalities.

12.3 Curvature of Fixed Sign

We are going to see again here how different the landscape is for positive and for negative curvature. We suggest that the reader take another brief glance at the table 12.1 on page 548.

12.3.1 The Positive Side: Sectional Curvature

Some surveys: Gromoll 1990 [601], Abresch & Meyer 1996 [7], appendix A, and Greene 1997 [591]. We are still in an almost complete mystery. We describe now the only known examples of compact manifolds with positive sectional curvature: because of Myers's theorem, except for a finite classification problem (although difficult), we can afford to look only at the simply connected case.

12.3.1.1 The Known Examples

12.3.1.1.1 Positive Curvature

First come the spheres and the $\mathbb{KP}^n$ met already in §§6.4.3. Now we recall equation 4.38 (or equation 15.15 on page 719) which gives the sectional curvature of normal Riemannian homogeneous spaces G/H:

$$K(v,w) = \frac{1}{4}\|[v,w]_H\|^2 + \|[v,w]_Q\|^2.$$

Hence every such space has nonnegative sectional curvature. Moreover $K(v,w)$ can vanish only if the (total) bracket of v and w vanishes, since its two components in a direct sum have to vanish. So it should not be too hard a job to find the pairs (G, H) for which this vanishing can never happen for two linearly independent vectors. This was done in Berger 1961 [150], with one omission (see just below). Then in Wallach 1972 [1231] and Bérard Bergery 1976 [142] the classification was extended to any (that is to say not necessarily normal) homogeneous space. The final result is:

Theorem 322 *Except for spheres and the $\mathbb{KP}^n$ there are only three such homogeneous "exceptions", two spaces of dimension 7 and 13 and a family $\mathfrak{W}_{p,q}$ of 7 dimensional spaces, nowadays called the* Aloff–Wallach manifolds.

Wallach's manifolds were met in §§§11.3.4.2. The extension needs various tricks, since the sectional curvature is given by the very unexplicit and complicated Nomizu equation 15.15 on page 719. In particular there is geometric trick relating isometries for Riemannian manifolds of even dimension to the positivity of the curvature. This relation is a global version of an infinitesimal one which says:

Theorem 323 (Berger 1966 [152]) *If the curvature is positive and the dimension even, then every Killing vector field (or, the same thing, every one-parameter group of isometries) has a fixed point.*

The proof is a direct application of the Bochner technique and can be seen as the positive analogue of Bochner's nonexistence result for Killing vector field in manifolds of negative Ricci curvature; see §15.6.

A funny and happy coincidence happened. The list of Berger 1961 [150] was incomplete, missing one example. But this missing space in the list turned out to be isometric to the Aloff–Wallach space $\mathfrak{W}_{1,1}$, as discovered in Wilking 1999 [1267]. We now describe in detail the Aloff–Wallach manifolds, because of their importance in other places. As manifolds, the $\mathfrak{W}_{p,q}$ are the 7 dimensional quotients $SU(3)/T(p,q)$ of the Lie group $SU(3)$ by the $T(p,q)$ circle. This means that the direction of the circle (a compact one-dimensional subgroup) is given by the point with coordinates (p,q) in the integral lattice defining a maximal torus of $SU(3)$. Recall that all maximal tori of $SU(3)$ are of dimension 2 and conjugate. Moreover this "universal lattice" is the regular hexagonal one; see figure 11.4 on page 513. In Aloff & Wallach 1975 [26] various homogeneous metrics are constructed on these $\mathfrak{W}_{p,q}$ using Nomizu's formulas. The construction of such a metric depends essentially on two real parameters, because the isotropy action of a $T(p,q)$ on the tangent space at the origin splits into three irreducible parts and then on each irreducible part, all invariant positive definite quadratic forms are proportional; see the end of §§4.3.4. Note that in the semi-desert area of manifolds of positive curvature, Aloff–Wallach manifolds are in fact a fascinating family.

We mention three of their properties. First, by varying the couples (p,q) we find that they have an infinite number of homotopy types (see §§4.1.4). For some couples one can get the same homotopy type among nondiffeomorphic ones. Second, they carry for every (p,q) a metric with a positive lower pinching not much different from $16/(29 \times 37)$; see Huang 1981 [743]. Third they also admit homogeneous Einstein metrics and, if the volume is normalized to 1, the set of their Einstein constants (which are all positive) is infinite: see Wang 1982 [1236], Wang & Ziller 1986 [1239]. Note that these Einstein metrics are not of Aloff–Wallach type. In particular, the Einstein metrics have sectional curvature of both signs. Another of their properties is that the set of positive scalar curvature Einstein metrics, when the volume is normalized to be one, does not converge to a smooth metric. This shows that the Palais–Smale C-condition (see Jost 2002 [768]) is not valid for Riemannian metrics, at least as we understand them today. We met already this type of question in §§§11.4.6.4.

Only four other types of (inhomogeneous) examples are known. They are in dimension 6 and 7: Eschenburg 1992 [493], Taimanov 1996 [1178] and in dimension 13: Bazaikin 1996 [121]. The last ones are in a family closely related to the Aloff–Wallach examples. Pinching constants for these various spaces have funny coincidental properties which are explained in Taimanov 1996 [1178]. Taimanov's study is pursued in Püttmann 1999 [1044] where the pinching constant 1/37 appears and is explained via Taimanov's deformations of Aloff–Wallach metrics. The starting point is to embed the Aloff–Wallach manifolds as totally geodesic submanifolds in the 13-dimensional examples. Moreover there is enough "room" (transversality) to deform the metric quite a lot and still keep the totally geodesic property. One even gets a series of metrics which converge to a smooth one with a pinching constant equal to 1/37. This supports the conjecture that the best pinching constant for Aloff–Wallach manifolds is 1/37 and this was proven in Wilking 1999 [1267]. For best pinching results, also see Püttman 1999 [1044].

No other manifolds are known of positive curvature, except of course small enough deformations of the preceding ones. But there are theoretical restrictions which will be the object of the next section.

12.3.1.1.2 Nonnegative Curvature

If we turn to manifolds of nonnegative curvature, the situation is not really much better and seems as mysterious. As examples of manifolds with nonnegative curvature, we have all symmetric spaces of "positive" type; see §§4.4.3. As well we find the normal homogeneous spaces G/H of compact Lie groups; see above. This does not make a long list, since the maximal Lie subgroups of compact Lie groups were completely classified by Dynkin 1952 [468]. One can also perform Riemannian products of the preceding examples. Besides those, before 1998, the only known examples were some exotic spheres and the connected sum of two symmetric spaces of rank one: see Cheeger 1973 [331] and Gromoll & Meyer 1974 [607]. Those examples are built using a clever mixture of large group actions (low cohomogeneity) of isometry groups and Riemannian submersion techniques to compute the curvature. Now it seems that it might be that manifolds of nonnegative curvature are more numerous than previously thought. In Grove & Ziller 1999 [658], among other examples, it is proven that all $SO(3)$ and $SO(4)$ principal bundles over the sphere S^4, and all S^3 bundles over S^4, admit nonnegative curvature metrics. For almost nonnegative curvature metrics, see Schwachhöfer & Tuschmann 2001 [1113].

This still does not tell us the foggiest about:

Question 324 *Is there any difference—at the level of possible manifolds—between positive and nonnegative sectional curvature?*

12.3 Curvature of Fixed Sign 579

A baffling remark in Yau 1982 [1295], page 670:

Fact 325 *No one knows any compact simply connected manifold with nonnegative curvature for which one can prove that it does not admit a metric of positive curvature.*

For example, Gromov's bound in theorem 326 on the following page on Betti numbers does not make any difference between positive and nonnegative. Yau starts with Hopf's conjecture on $S^2 \times S^2$; see question 294 on page 545. For the nonsimply connected case, Rong's results (see theorem 330 on page 583) provide a partial answer.

It is not surprising that many people tried to address Yau's remark, starting with the Hopf conjecture on $S^2 \times S^2$, by trying to deform such a metric with $K \geq 0$ into one with $K > 0$. This means considering some one parameter family $g(t)$ of metrics and computing the various derivatives at $t = 0$ of the sectional curvature. Technically it is very easy to compute such a derivative for a given tangent plane, but what is difficult is to find a variation for which all the derivatives would be positive. Today this approach still does not work; see Bourguignon 1973 [236] for formulas and reasons why natural approaches do not work. One reason lies in the fact mentioned on page 207: the structure of the sectional curvature as a function on the set of tangent planes (say at a given point) is practically not understood. In particular one does not know where to look for its minimum. Related to this, one should also read Cheeger 1973 [331], and the important Wilking 2002 [1270].

Note 12.3.1.1 (Hopf's questions) Two of the three favourite questions that Hopf was asking since the 1930's are still almost completely open: (1) does a given sign for the curvature imply a given sign for the Poincaré characteristic? (2) Does $S^2 \times S^2$ admit a metric with positive curvature? Hopf's conjecture on the Poincaré characteristic is true for dimensions 2 and 4; for dimension 2 by the Gauß–Bonnet formula, for 4 as follows: (see equation 11.6 on page 518 or equation 15.13 on page 716):

$$8\pi^2 \chi(M) = \int_M \left(|R|^2 - \left| \text{Ricci} - \frac{\text{scalar}}{4} g \right|^2 \right) dV_M$$

in invariant form. But in a nice basis, the integrand becomes

$$K_{12}K_{34} + K_{13}K_{42} + K_{14}K_{23} + R_{1234}^2 + R_{1342}^2 + R_{1423}^2$$

(with the obvious notations for the sectional curvature). This implies Hopf's conjecture as explained in Chern 1955 [365]. But starting in dimension 6 we will see that the formula is so complicated that it cannot say much of use (cf. Bourguignon & Polombo 1981 [246]); for the most part, just some nonvanishing results in the spirit of §§11.3.5. Starting in dimension 6, Hopf's question is still open. There is only a positive result for the Kähler case, which uses L^2-cohomology; see §14.1.

The latest news for $S^2 \times S^2$ is in Kuranishi 1990 [840]. Even more itching is Yau's assertion 325 on the preceding page. This is very irritating, since Synge's theorem trivially excludes $\mathbb{RP}^2 \times \mathbb{RP}^2$ (as well as many other products of manifolds). Rong's theorem 330 on page 583 exclude quite a few more. A recent general list of problems is in Petersen 1996 [1016]. ♦

12.3.1.2 Homology Type and the Fundamental Group

12.3.1.2.1 Homology Type

On the "positive" side one knows at least that not every manifold can carry a metric of nonnegative curvature (a fortiori of positive curvature).

Theorem 326 (Gromov 1981 [613]) *There is universal $n(d)$ such that any compact manifold of dimension d with $K > 0$ has the sum of its Betti numbers (for every field) bounded by $n(d)$.*

For the historian we note here that before Gromov's result in 1981, in the simply connected case at least, believe it or not, the only restriction coming from the condition $K \geq 0$ was that of Lichnerowicz, already valid under only the nonnegativeness of the scalar curvature: see §§12.3.3. For the fundamental group, one has theorem 345 on page 594 of Bochner under the assumption of positive Ricci curvature.

With the above theorem we are left, in a weak sense, with only a finite number of homology types (see §§4.1.4), but an infinite number of homotopy types. Indeed Aloff–Wallach spaces provide examples of infinitely many homotopy types. The tools are somewhat, but not closely, analogous to Morse theory. They require subtle algebraic topology arguments, based on compressibility of balls and topological contents of them. The principal difficulty is that one knows nothing about the injectivity radius (it can be very very small, etc.), since we ask only for $K \geq 0$ without any kind of normalization. One uses in a significant manner the notion of critical point for distance functions; see definition 305 on page 563. The basic lemma is that, by Toponogov's theorem for triangles, a succession of critical points for the distance function to a given point and whose distances are growing at least at a geometric progression, has to be finite (the bounding cardinal being universal). In short: "critical points cannot be too far away."

From this we can easily give the proof. If the function is $d(p, \cdot)$, then let γ and η be segments from p to two critical points q and s. Then the very definition of a critical points enables us to find a segment τ from q to s such that the angle between the speed vectors $\gamma'(q)$ and $\tau'(q)$ is larger than or equal to $\pi/2$. Assume that $d(p, s) \geq \lambda d(p, q)$. Then Toponogov's theorem for $K \geq 0$ as applied to the above triangle shows that the angle between $\gamma'(p)$

and $\tau'(p)$ has to be large (large than some angle function of λ, larger and larger as λ gets larger and larger).

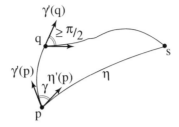

Fig. 12.18. Compressing balls up to meet critical points

Now we have the classical fact:

Proposition 327 *The number of points which one can place on the unit sphere S^d so that their spherical distances are larger than some given number is bounded from above as a function only of the dimension d and that distance bound.*

From this we get the above claim: a set of critical points with respective distances from p larger than a given ratio has cardinality universally bounded. A rough (but not too bad by the way) estimate of the number in proposition 327 can be obtained just by a measure argument, computing the number of disjoint spherical caps. For refined estimates, which is a central problem in pure and applied mathematics, that of *spherical codes*, see the bible Conway & Sloane 1999 [403] or the very recent survey Conway 1994 [402].

Then one has to play with distance functions to different points. The lemma is certainly not enough because we do not know (universally) the injectivity radius. One plays with the topological content of various balls, and always compresses those balls as much as possible without having them meet a critical point. One finds finally a covering by balls with the number of balls universally bounded, and every ball having a topological content (as well as their intersections) which is also universally bounded. The proof is finished by a Mayer–Vietoris argument. The reader can look at the various presentations, which also provide simplifications of the original text: Cheeger 1991 [334], Meyer 1989 [916], Grove 1990 [644] and section 11.5 of Petersen 1998 [1019].

There are two questions concerning theorem 326 on the preceding page which come to my mind. How about an optimal $n(d)$? Gromov's bound was double exponential in d. In Abresch 1985 [4] the author manages to go down to a bound simply exponential in some constant times d^3. The most optimistic conjecture would be 2^d, because this 2^d is the sum of the Betti numbers of the torus T^d. This conjecture is not disproved today; and possibly then equality only for flat tori. Recall that we know so few manifolds of nonnegative curvature.

The second question is in the general spirit of the introduction in §§12.1.1: to what extent can we relax the condition $K \geq 0$?

Theorem 328 (Gromov 1981 [613]) *The sum of the Betti numbers (over any field) of a compact manifold has a bound which is universal in the dimension and exponential with rate of exponential growth given by the positive number* $-\inf K \operatorname{diam}(M)^2$.

So we have partial finiteness for homology types with only a lower bound on the sectional curvature and an upper bound for the diameter; see §§4.1.4. Easy examples, taken among the ones given just after theorem 374 on page 618, show the optimality of these ingredients and the fact that finiteness for homotopy type is hopeless. But of course there is always the problem of optimal constants in the bounds.

12.3.1.2.2 Fundamental group

We turn now to restrictions on the fundamental group. We recall first that by Synge's theorem 64 on page 246, even dimensional compact positively curved manifolds are either simply connected or have simply connected 2-1 covers, so now we concentrate on odd-dimensional manifolds. The fundamental group $\pi_1(M)$ is finite by Myers' theorem 63 on page 245.

Theorem 329 (Gromov 1978 [609]) *There is a universal constant $N(d)$ such that the fundamental group of a manifold with $K > 0$ is generated by at most $N(d)$ generators.*

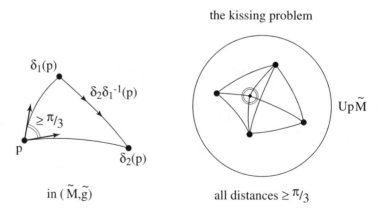

Fig. 12.19. (a) Triangle in the cover $\tilde{M}$ (b) The kissing problem: all distances $\geq \pi/3$

The proof is not a volume of balls argument as in Milnor's theorem 109 on page 311 (to be discussed below) for nonnegative Ricci curvature, but a metric argument. One chooses by induction a "smart" basis δ_i ($i=1,\ldots,n$) for the fundamental group as realized by deck transformations of the universal Riemannian covering of our manifold. This means that the displacement functions are successively as small as possible. Fix a point p in the covering manifold and look at segments γ_i from p to $\delta_i(p)$. Toponogov's theorem 73 on page 258 then shows that any angle between two speed vectors $\delta_i'(p)$ has to be larger than or equal to $\pi/3$. Look at these unit tangent vectors as points of the unit sphere $S^{d-1} \subset T_pM$. Their mutual spherical distances have to be not smaller than $\pi/3$. The number of such points is universally bounded in dimension d. As in proposition 327 on page 581, in the present special case where the spherical distance is equal to $\pi/3$, a poor (but not too poor) bound is easily obtained by a measure argument for spherical caps. For the curious reader we cannot resist mentioning here that this is the famous *kissing number*. Its exact value is known only in dimensions 2, 3, 8 and 24. A good part of the book Conway & Sloane 1999 [403] is devoted to it. One will find there various estimates for it and the current state of the art in techniques for its estimation. The wording "kissing" is explained easily: one looks in $\mathbb{E}^d$ for the maximum number of unit balls which can touch the boundary of one given unit ball without intersecting one another.

But we would like more—namely an upper bound for the possible group structures, and this as a function of the pinching. An old conjecture of Chern was that every Abelian subgroup of the π_1 of a manifold of positive sectional curvature is cyclic. In Rong 1996 [1067], using the very technique used for the proof above, it is proven that this is true, but only up to an index which is bounded by a universal constant $w(d,\delta)$ in the dimension d and the pinching δ. Rong's conjecture today is that $w(d,\delta)$ can be chosen independent of δ. With a much deeper analysis of collapsing, one has the partial result (see also Rong 1997 [1069] where more refined analyses are made).

Theorem 330 (Rong 1996 [1066]) *There are numbers $w(d)$ and $w'(d,\delta)$ so that for any δ-pinched manifold M of dimension d its $\pi_1(M)$ either has a finite cyclic subgroup of index less that $w(d)$ or has order less than $w'(d,\delta)$.*

Finally Chern's conjecture was invalidated in Shankar 1998 [1127]; the counterexamples are suitable quotients of the Aloff–Wallach manifold $\mathfrak{W}_{1,1}$. Note the amusing fact that this was the missing example mentioned in the story of §§§12.3.1.1.

12.3.1.3 The Noncompact Case

In the study of positive and nonnegative sectional curvature, noncompact manifolds also enjoy optimal results, so we will make an exception to our rule of studying only compact manifolds.

Theorem 331 (Gromoll & Meyer 1969 [605], Cheeger & Gromoll 1972 [344]) *A complete noncompact manifold of positive sectional curvature and dimension d is diffeomorphic to $\mathbb{R}^d$. A complete noncompact manifold of nonnegative curvature always admits at least one totally geodesic and totally convex submanifold N called a* soul *such that the manifold is diffeomorphic to the normal bundle of any of its souls N (see figure 12.20).*

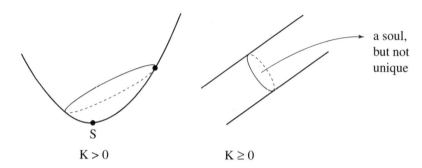

Fig. 12.20. (a) Complete with positive curvature; diffeomorphic to the plane (b) Complete with nonnegative curvature: a soul (which is not unique)

The cylinder shows the optimality of the theorem when $K \geq 0$ and the nonuniqueness of souls in general. The proof of theorem 331 shows that moreover "the isometry group of a complete manifold of positive sectional curvature is itself compact." For the compact case this result is well known, and follows from the classical result which says that the isometry group of a complete manifold is a Lie group. This is an easy fact, since an isometry is clearly determined by its effect on an orthonormal frame at one point. On a compact manifold, the set of orthonormal frames at one point is compact; see Kobayashi 1995 [826] if needed. On the other hand there are examples of manifolds with positive Ricci curvature and whose isometry group is not compact.

How much zero curvature is permitted was recently made precise in

Theorem 332 (Perelman 1994 [1009]) *A single positive sectional curvature not entirely in the soul yields a soul reduced to a point and thereby yields diffeomorphism with $\mathbb{R}^d$.*

Note also that, in a very precise sense, theorem 331 reduces the classification of complete manifolds with positive sectional curvature to that of compact ones, but as we saw above, this classification is in a very primitive state. We sketch now the very geometrical proof of this theorem. It involves nice arguments of convexity mixed with Toponogov's theorem and plays with Busemann functions in disguise. Besides the original texts and the surveys Eschenburg 1994 [494] and Meyer 1989 [916] the theorem is completely proven in the books Sakai 1996 [1085] and Petersen 1997 [1018].

12.3 Curvature of Fixed Sign

Some definitions before we go on. In the noncompact realm, two are basic:

Definition 333 *A ray γ is a geodesic defined on $[0, \infty)$ such that γ is a segment from $\gamma(0)$ to $\gamma(t)$ for any positive t up to infinity. A line is a geodesic defined on $(-\infty, +\infty) = \mathbb{R}$ which is a segment on any interval.*

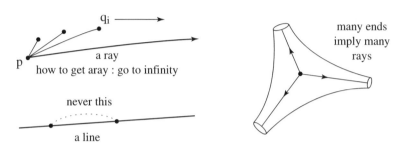

Fig. 12.21. (a) Getting rays to go to infinity (b) Many ends imply many rays

In a primitive sense, on a ray you can go to $+\infty$, and on a line you can go both to $+\infty$ and to $-\infty$. In a (complete) noncompact manifold, there is always a ray starting from any given point p. To build one, just take a limit of segments from p to a sequence of points q_i such that $d(p, q_i)$ tends to infinity with i. There is a limit to this sequence of rays because of the compactness of the unit tangent sphere at p. More: there are at least as many rays as ends of the manifold. The *ends* of a manifold are the connected components which remain distinct during an exhaustion of the manifold by compact domains. For example a paraboloid has one end, a cylinder two ends, etc.

On the other hand, lines do not always exist. Their existence will turn out below to have extremely strong consequences. In term of the cut-locus Cut-Locus (p) of the point p (see §§6.5.4), a ray from p is a geodesic defined on $[0, \infty[$ which never meets Cut-Locus (p). Although they are not directly used here, we mention the *Busemann functions* which will often be used later on in our story:

Definition 334 *The Busemann function b_γ associated to a ray γ is given, for any point q in the manifold, by*

$$b_\gamma(q) = \lim_{t \to \infty} d(q, \gamma(t)) - t$$

The existence of the limit is easy, and we remark that, restricted to γ, the function b_γ reduces to $b_\gamma(t) = -t$. What is important for our intuition is that $b_\gamma(q)$ can be considered as the distance $d(\gamma(\infty), q)$ between q and the point $\gamma(\infty)$, with the translation-normalization placing the origin at $\gamma(0)$. Then one will have distance spheres, with $\gamma(\infty)$ as center, which will be the level sets of b_γ, and metric balls which are the various inverse images $b_\gamma^{-1}\left([r,\infty]\right)$ but with radius r which can also be of any sign. Corresponding here to the rays in standard balls, going from the center to points or the sphere, are here rays, which are also sometimes called *asymptotes*. Beware that they are not unique in general. The corresponding spheres generalize the horospheres of hyperbolic geometry.

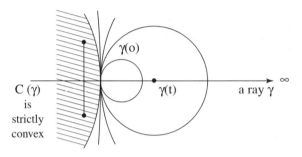

Fig. 12.22. A ray

The amazing idea of Gromoll & Meyer 1969 [605] was this: let γ be any ray and consider the various balls $B\left(\gamma(t),t\right)$, and take their union for all t from 0 to ∞. Now introduce the complement $C(\gamma)$ of this infinite union. The Toponogov theorem 73 on page 258 for positive sectional curvature implies that $C(\gamma)$ is a totally convex set in the manifold, i.e. any segment between any pair of its points is always contained in it; see §§6.1.3. With the definition 334 on the preceding page, the set $C(\gamma)$ is nothing but the complement of the metric ball $b_\gamma^{-1}\left([\gamma(0),-\infty]\right)$. At the same time, we know that on manifolds of nonpositive curvature, these Busemann balls themselves, and not their complements, are totally convex: see §§6.3.3. Such sets can be nicely visualized in the paraboloids

$$z = ax^2 + by^2$$

in $\mathbb{E}^3$.

The next ideas for the proof are as follows. The proof will be finished when one can produce a *pole*, which by definition is a point such that its exponential map is a diffeomorphism. For example, all points are poles in simply connected manifolds of nonpositive curvature (see §§6.3.3). But poles need not exist in noncompact manifolds of positive curvature. One finds examples by suitably modifying paraboloids, as shown in Gromoll & Meyer 1969 [605]. However *simple points* always exist. A point p is said to be *simple* if there is no geodesic loop from p to p. This is equivalent to saying that $\{p\}$ is totally convex. Simple points are uncovered as follows: pick some point p and examine all of the rays γ emanating from p. Look at the intersection Q of all of the associated totally convex sets $C(\gamma)$. Now if Q, which is totally convex, were not compact then one would be able to "escape from it" and then prove the existence of a line.

The positivity of the curvature implies easily the nonexistence of lines. This is (for example) a particular case of theorem 349 on page 596, but can be proven in many ways. It remains to show that Q being totally convex and compact, it must contain at least one simple point. Now differential topology mixed with the existence of simple points shows that the manifold is diffeomorphic to $\mathbb{R}^d$.

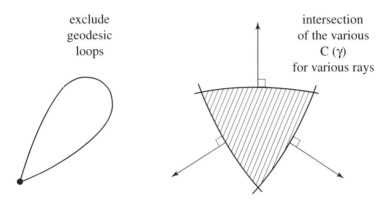

Fig. 12.23. (a) Exclude geodesic loops (b) Intersection of the various $C(\gamma)$ for various rays

We will now cease considering only manifolds of strictly positive curvature, because this old proof is superseded by a proof for manifolds with only nonnegative curvature, which we will now explain. The basic fact, proven with Toponogov's theorem, is that:

Theorem 335 (Cheeger & Gromoll 1972 [344]) *On a compact Riemannian manifold of nonpositive sectional curvature, if γ is any ray, then the associated Busemann function b_γ is a convex function (see proposition 60 on page 234).*

Then one introduces the function which is the minimum of the various Busemann functions associated to the totality of rays issuing from a given fixed point x. This is still a convex function—call it f_x. Its superlevel sets are convex. In fact they are compact since, if not, one could produce a new ray, contradicting the minimal definition of f_x. In particular the maximal superlevel set C_x is compact. If it is a point, we are finished. If C_x is a submanifold, we have found a soul. If not a submanifold, it is still a totally convex set, and then there is some work to do. The work consists in shrinking C_x into a soul, by considering the distance function to the boundary of C_x. All of the details are geometric but quite long and technical; see them in section V.3 of Sakai 1996 [1085] or in section 11.4 of Petersen 1998 [1019], or in Greene's article in Grove & Petersen 1994 [650]. The proof proceeds such that, when the soul is not a point, one meet rectangles with zero curvature, and this finishes the positive curvature case and the whole story.

12.3.1.4 Positivity of the Curvature Operator

12.3.1.4.1 Positive Curvature Operator

Topological spheres can be characterized (up to homeomorphism for dimension at least 5) by a suitable "positive curvature" condition, namely that of the *curvature operator* (see §§12.1.2 for this curvature invariant). Recall that the curvature operator of the standard sphere is the identity. To observe optimality of the results below, one has only to notice that the $\mathbb{KP}^n$ (with $\mathbb{K} \neq \mathbb{R}$) has only a nonnegative curvature operator. For example, on $\mathbb{CP}^n$ the curvature operator vanishes on the Kähler 2-form.

Theorem 336 (Micallef & Moore 1988 [917]) *A simply connected manifold with positive curvature operator is homeomorphic to the sphere.*

The question of a possible diffeomorphism is still open, as well the equivariant question for the nonsimply connected manifolds, except in dimension 4. The case of dimension 2 comes from the Gauß–Bonnet theorem 28 on page 138 and that of dimension 3 is a consequence of the Hamilton's theorem 279 on page 525. For dimension 4 we have an optimal result:

Theorem 337 (Hamilton 1986 [679]) *A manifold of dimension 4 with positive curvature operator is diffeomorphic to S^4 or $\mathbb{RP}^4$.*

The curvature operator seems to have been considered for the first time (at least in relation with topology) in Bochner & Yano 1952 [213] where it was proven that if the curvature operator is 1/2-pinched then all of the real Betti numbers have to vanish, yielding only real homology spheres. The technique used was Bochner's one. It was realized that the apparently complicated term $\text{Curv}_p(R; \omega, \omega)$ (see on page 708) for harmonic p-forms has to look close to that of the curvature operator. Real understanding came in

Theorem 338 (Meyer 1971 [915]) *If the curvature operator is positive, then the terms $\text{Curv}_p(R; \omega, \omega)$ are positive for every p.*

The proof of the homeomorphism (in any dimension) uses strong geometric analysis, namely existence of minimal harmonic maps of the sphere S^2 into our manifold. Then one analyses the second variation, which has to be positive for minimal harmonic maps. This will finally yield the fact that the index cannot be the right one and then one proves that homotopy groups of the manifold vanish in dimension smaller than that of half the dimension of the manifold. It remains only to use Smale's solution of the Poincaré conjecture for dimensions greater than four. In fact the proof above uses only the condition $K_{\mathbb{C}}^{\text{isotr}} > 0$ of positivity for isotropic planes (see the hierarchy in §§12.1.2).

An interesting byproduct of the paper of Micallef and Moore is the solution of the "pointwise" (1/4)-pinching problem already mentioned in §§§12.2.2.6: we assume that at every point $m \in M$ one has $1/4 < K(m) \leq 1$. Then the manifold is homeomorphic to the sphere, because this sectional curvature pinching implies the positivity of the curvature of isotropic planes. This is a nice computation done in Micallef & Moore 1988 [917] and left to the reader as an exercise. Note that such a result is completely out of the reach of triangle comparison theorems and that the result, at least today, is only for simply connected manifolds and yields only homeomorphism.

Hamilton's proof in dimension 4 is completely different. It is via deformation of Riemannian metrics under the Ricci flow; see §§11.4.3. Under such a deformation, the curvature operator defect to be the identity (that of the standard sphere) decreases strictly, and then the positivity and the dimension condition enables Hamilton to show that the deformation equation can be continued up to infinite time so that, at the end, one has constant sectional curvature and, in dimension 4, the positively curved space forms are only the standard S^4 and $\mathbb{RP}^4$ thanks to Synge's theorem 64 on page 246.

The nonsimply connected case, starting dimension 5, is open.

12.3.1.4.2 Nonnegative curvature operator

The classification of manifolds with nonnegative curvature operator is finished:

Theorem 339 *A simply connected manifold with nonnegative curvature operator is a Riemannian product whose components are among*

1. *a manifold homeomorphic to the sphere*
2. *a symmetric space*
3. *a Kähler manifold biholomorphic to $\mathbb{CP}^n$.*

By theorem 336 on the preceding page we can assume that R^* has some zero part. Then one has equality in theorem 338 on the facing page. Bochner's formula, as written by Meyer, shows that there should be a nonzero exterior form with vanishing covariant derivative, i.e. invariant under parallel transport, hence invariant under the holonomy group. Now we apply the holonomy group classification theorem 397 on page 643. It remains only to throw out the holonomy groups $\text{Spin}(7)$ and G_2, to take care of the cases where the holonomy is $Sp(1)\,Sp(n)$, and finally the Kähler holonomy. Various authors contributed to this; see the references in Chen 1990 [360], and a good part of the proof in the book Petersen 1997 [1018].

12.3.1.5 Possible Approaches, Looking to the Future

12.3.1.5.1 Rong's Work on Pinching

Back to a "classification" of the manifolds of positive curvature, we saw that we are still lacking any general statement. We quote here four different approaches to attack the problem. The first and the most promising is that of Rong 1996 [1067]. Thanks to the finiteness theorem 372 on page 616, in a given even dimension there are only finitely many diffeomorphism types of manifolds with $0 < \delta \leq 1$, from Klingenberg's lower bound for the injectivity radius in theorem 92 on page 276. If the dimension is odd, we no longer have Klingenberg's lower bound of π for the injectivity radius. What we quote now uses collapsing theory quite heavily; see §§§12.4.3.1. The first thing to do is to look at the odd-dimensional case. We will assume positive pinching constant δ.

Theorem 340 (Rong 1996 [1067]) *Suppose that M is a δ pinched, positively curved, simply connected, compact Riemannian manifold. Modulo a finite number of cases, one can modify the Riemannian metric into a new one, still of positive curvature with pinching larger than $\delta/2$, but admitting a free circle action of isometries.*

The dimension hypothesis fits well with theorem 323 on page 577. The quotient is sadly enough only an *orbifold* (see §§14.5.2), also of positive curvature by the Riemannian submersion formulas (see equations 4.40 on page 211 and 15.17 on page 721). This almost reduces the classification to that of the even dimension case and the classification of some circle bundles over them (always modulo a finite number). For results along these lines, see also Tuschmann 1997 [1204]. The starting point is to look at the conditions in Cheeger's finiteness theorem 372 on page 616 and to assume that we have a very small volume. But then our manifold should be collapsing; see §§§12.4.3.1. Thereafter one applies structure results for collapsing manifolds and the main point now is to change the metric into a new one which satisfies the conditions above. This is achieved by using Ricci flow (see §§11.4.3). There are also results reducing the possibilities for the dimension of the collapsed manifold; see Rong 1997 [1068].

Of course, in looking for a complete classification, naturally we look at all manifolds M^d of a given dimension and admitting a positive curvature metric and write $\delta(M)$ for the infimum of its pinching constants (see below if in doubt). Define $\delta(d)$ as the infimum of all of these $\delta(M^d)$. Is $\delta(d)$ positive?

12.3.1.5.2 Widths

In another direction, a very geometric approach is to use the purely metric concept of *widths*, introduced by Urysohn in 1924 (see Burago & Zalgaller 1988 [283] for references). These widths can be seen as k-dimensional diameters, but they require the notion of dimension for general metric spaces. For a metric space M its k-width, denoted by $\text{Width}_k(M)$, is the lowest bound of the numbers δ such that there exists a k dimensional metric space P and a continuous map $f : M \to P$ such that all its preimages have diameters at most δ.

For example the ordinary diameter verifies

$$\operatorname{diam} M \in [\text{Width}_0(M), 2\,\text{Width}_0(M)] \ .$$

For convex bodies in $\mathbb{R}^d$ a classical inequality is

Theorem 341 *There is universal constant $c(d)$ such that any convex body C verifies*

$$\frac{\operatorname{Vol} C}{c(d)} \leq \prod_{k=0}^{d-1} \text{Width}_k(C) \leq c(d) \operatorname{Vol} C$$

The reader can look at the trivial case of rectangular parallelepipeds (but they have boundary) and of ellipsoids. Widths in Riemannian geometry were introduced in Gromov 1988 [623], where it was conjectured that

Conjecture 342 *Theorem 341 holds true for manifolds of nonnegative curvature.*

The underlying idea is that such a manifold "looks like" a convex body, even like an ellipsoid. A more illuminating perspective, but more imprecise, being: such a manifold "looks like" the parallelepipedic set

$$\prod_{k=0}^{d-1} [0, \text{Width}_k(C)] \ .$$

Gromov's conjecture was settled affirmatively in Perelman 1995 [1010]. The proof is completely geometric but very involved.

Note 12.3.1.2 (Rational homotopy) A third approach to nonnegative curvature is that of the so-called *elliptic rational homotopy* in Grove & Halperin 1983 [646] which is indirectly linked to the study of manifolds all of whose geodesic are periodic; see §10.10.[3] Rational homotopy is directly linked with Sullivan's fine theory of de Rham cohomology; see §§4.1.4.2. This approach leads to the the so-called "double soul" problem; see more in Petersen 1996 [1016], problem 21.

The main fact of algebraic topology required in this theory is

---
[3] This connection should not be too surprising, considering the $\mathbb{KP}^n$.

Theorem 343 (Felix & Halperin 1982 [512]) *The class of d dimensional compact manifolds is divided into exactly two subclasses: either the homotopy groups $\pi_p(M^d)$ are finite for all $p < 2d - 1$, or the numbers*

$$\sum_{q \leq p} \dim \pi_q(M^d) \otimes \mathbb{Q}$$

grow exponentially in p. Manifolds in the first class are called rationally elliptic, *in the second* rationally hyperbolic.

The exact class, except for the definition, of rationally elliptic manifolds is not known, but for example all compact homogeneous spaces are elliptic. One knows that rationally elliptic spaces satisfy $\dim H_*(M^d, Q) \leq 2^d$. Think of the torus and the comments in §§§12.3.1.2 on Gromov's theorem. This shows that, in a sense defining the assertion itself, most manifolds are rationally hyperbolic. Putting those together motivates (afterwards!) Bott's old conjecture to the effect that a simply connected manifold of nonnegative curvature is rationally elliptic. The second byproduct was used heavily in §§10.3.5 for the existence of infinitely many periodic geodesics. The Betti numbers of the loop spaces of rationally hyperbolic manifolds also grow exponentially. ♦

12.3.1.5.3 The best pinched metric

A last approach would be to look at the best pinched metric on a given manifold admitting a metric of positive curvature. There is an optimal pinching δ (of course $\delta < 1/4$ to be of interest). At least in even dimensions, now the hypothesis $\delta_i \leq K \leq 1$ for a sequence of pinching constants δ_i converging toward the upper bound δ satisfies the hypothesis of convergence in the comments following theorem 384 on page 627: the diameter is bounded above by Myers' theorem 63 on page 245 and the volume from below because of Klingenberg's theorem 92 on page 276 on the injectivity radius. So we have an optimal metric which, as far as we know, is of class $C^{1,\alpha}$. It seems that there is no text studying this approach. The main difficulty was already mentioned in Yau's fact 325 on page 579: it seems very hard, at least today, to use the pinching condition.

Note 12.3.1.3 (Moduli) There is very little known concerning moduli: assume that some manifold admits at least one metric of positive curvature. How does the set of all positively curved metrics look? We know only today that it can fail to be connected; an example being the two Aloff–Wallach manifolds $\mathfrak{W}_{-4638661,582656}$ and $\mathfrak{W}_{-2594149,5052965}$, which are diffeomorphic but do not belong to a connected family of positive scalar curvature metrics; see Kreck & Stolz 1993 [834] and §§12.3.3. ♦

12.3.2 Ricci Curvature: Positive, Negative and Just Below

Historically, results on manifolds with a lower bound for Ricci curvature have appeared continually but slowly, and they are not very complete. In particular, for positive or nonnegative Ricci curvature, they start with Myers' theorem 63 on page 245 which was obtained using the second variation formula. Next came Bochner's theorem 345 on the following page. After that, there was a partial result in Calabi 1958 [300] and Lichnerowicz's bound on λ_1 which we gave in theorem 181 on page 408. Thereafter came the splitting theorem 348 on page 595. During those times there was a great harvest of theorems about sectional curvature; see all of the present chapter. However, the pioneering "little green book" Gromov, Lafontaine & Pansu 1981 [616] triggered an avalanche, to the effect that a lower bound on Ricci curvature can be in some cases be as strong as one on sectional curvature. The essence, the little snowball that started this avalanche, has been explained in §§6.4.3.

But in general the sad fact remains that today, for simply connected manifolds, there is no topological condition known as a consequence of positivity of Ricci curvature besides those to be seen in the next section for scalar curvature. A typical example: can any product $M^d = S^2 \times N^{d-2}$ carry a metric of positive Ricci curvature? This question is motivated by the trivial fact that such a product has a positive scalar curvature metric—just shrink the S^2 factor enough. Its sectional curvature will swallow all of the other terms in the summation. Is there any stable homotopy restriction preventing positive Ricci curvature? Recall also the trichotomy in §§12.3.3.4: Ricci flat manifolds still present a mystery (see fact 281 on page 529). One is only provided with a lot of examples, which we will meet if we have not already met them. However there is a conjecture in Stolz 1996 [1162], which will we just paraphrase since it is quite technical, referring to §§14.2.2 for the precise notions. On spin-manifold M^{4k}, the first Pontryagin class $p_1(M)$ admits a "half," denoted by $\frac{1}{2} p_1(M)$. The conjecture is

Conjecture 344 *Every spin-manifold satisfies*

$$\frac{1}{2} p_1(M) = 0$$

and if it admits a metric with positive Ricci curvature, then its Witten genus $\phi_W(M)$ vanishes too.

Although this condition is only the vanishing of a numerical invariant, this would at least split the class of positive scalar curvature metrics from that of positive Ricci curvature metrics. The game definitely requires loop spaces and uses recent concepts of the geometry of the (infinite dimensional) loop spaces. A very vague and bold idea is that Ricci curvature downstairs on the manifold means positive scalar curvature on its loop space for a suitably defined "metric."

Note 12.3.2.1 (Three Dimensions) It is only in dimension 3 that the situation is completely understood. In theorem 279 on page 525 (originally published in Hamilton 1982 [678]) we saw that, using the deformation technique along the Ricci flow from equation 11.9 on page 525, any positive Ricci metric on a 3-manifold can be deformed into one of constant sectional curvature. Note that in dimension 3, Ricci and sectional curvature have the same number of parameters, namely 6, and determine one another. ♦

Note 12.3.2.2 (Positive Ricci Versus Positive Sectional Curvature) But certainly the topological implications of positive Ricci curvature are not as strong as positive sectional curvature: Gromov's bound on Betti numbers in theorem 326 on page 580 splits the two categories, because there are examples of nonnegative Ricci curvature with arbitrarily large Betti numbers. Proof of the existence of such manifolds was achieved simultaneously in Anderson 1990 [40] and Sha & Yang 1991 [1126]. Sha and Yang built up connected sums, using delicate metric surgery on double warped products. An extra difficulty here is that one needs to spread the modifications all over the manifolds during the gluing operation. In Anderson 1990 [40] one uses models from Gibbons & Hawking 1978 [559]. This construction was put into the more general framework of the Schwarzschild metric in Anderson 1992 [42]. More recently Perelman 1997 [1012] built up examples showing that Betti numbers can still be not bounded when the Ricci curvature is bounded from below, even if the volume is bounded from below, and the diameter from above. Contrast this with the finiteness theorems in §§§12.4.1.1 and §§§12.4.1.2. Also see Boyer & Galicki 2002 [252]. ♦

But at the level of the fundamental groups, we have very strong and optimal recent results for nonnegativity and just below. Before going on, do not forget that Myers' theorem implies that the first real Betti number $b_1(M)$ vanishes.

Theorem 345 (Bochner 1946 [210]) *If* Ricci ≥ 0 *on a compact manifold* M *of dimension* d *then* $b_1(M) \leq d$ *and equality implies that the manifold* M *is isometric to a flat torus.*

This was the first appearance of what is called now *Bochner's technique* which appears in many places throughout this book. There are many other applications, e.g. in the proof of theorem 339 on page 589, §§12.3.3, §15.6, and the very geometric §§6.4.2. The reader might grasp the force of it in the surveys Bourguignon 1990 [239], Bérard 1988 [136], and Wu 1988 [1278]. Because of its importance and of its first appearance in our text, we will give details of the proof. It rests first on the formula in equation 15.7 on page 707, valid for any 1-form ω:

Theorem 346 (Bochner 1946 [210])

$$-\frac{1}{2}\Delta\left(|\omega|^2\right) = |D\omega|^2 - \langle \Delta\omega, \omega \rangle + \text{Ricci}\,(\omega_*, \omega_*)$$

which is straightforward application of the Ricci commutation formulas; see more generally §15.5 and §15.6. In proposition 79 on page 265 we took for ω the differential df of a function, very often a distance function. In theorem 323 on page 577 we used the form ω which is dual to a Killing vector field. Back to our present purpose, we now apply the Hodge–de Rham theorem 406 on page 665 for harmonic 1-forms ω, namely we assume that $\Delta\omega = 0$ and integrate the equation in theorem 346 over the compact manifold. By Stokes formula 34 on page 171 the integral of any Laplacian vanishes, so that if moreover Ricci ≥ 0 we are left on the right-hand side with an integral with $D\omega = 0$ and Ricci$(\omega, \omega) = 0$ at every point. Then if the Ricci curvature is positive, $\omega = 0$ and again $b_1(M, \mathbb{R}) = 0$ by de Rham's theorem 32 on page 168 and we recover part of Myers' theorem. Now we see that the nontrivial elements in $b_1(M, \mathbb{R})$ yield 1-forms ω such that $Dw = 0$, i.e. they have vanishing covariant derivatives. This implies a product decomposition on the universal covering (see theorem 56 on page 229) $M^d = N^{d-k} \times \mathbb{R}^{d-k}$.

The next result was Milnor's theorem 109 on page 311: $\pi_1(M)$ has polynomial growth. But adding the pure group theoretical result of Gromov's theorem 352 on page 598:

Theorem 347 *The fundamental group of a manifold of nonnegative Ricci curvature is almost nilpotent, i.e. contains a nilpotent subgroup of finite index.*

This does not finish the classification of possible groups. Somewhat analogously to theorem 331 on page 584 reducing the study of complete noncompact manifolds of positive sectional curvature to the compact case, we have the following reduction theorem:

Theorem 348 (Cheeger & Gromoll [343]) *If a complete Riemannian manifold (not necessarily compact) has nonnegative Ricci curvature then its universal covering splits isometrically as a Riemannian product $N \times \mathbb{E}^k$ where N has nonnegative Ricci curvature and contains no lines.*

We have to prove that the existence of a line (see definition 333 on page 585) implies a splitting into a Riemannian product with a flat manifold. For the Ricci case, the proof was very new, using Busemann functions (see definition 334 on page 585) in Riemannian geometry and the fact that they are always superharmonic; see note 6.4.2.1 on page 265. this was already noticed in Calabi 1958 [300]. Busemann functions figure in particular in section III.22 of Busemann 1955 [290]. The philosophy, as in §§6.4.3, is that

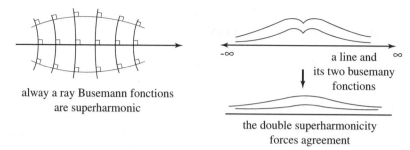

Fig. 12.24. (a) Along a ray, Busemann functions are superharmonic (b) A line and its two Busemann functions (c) The double superharmonicity forces agreement

sectional curvature yields convexity properties for the distance function, while Ricci curvature can only control the Laplacians of distance functions.

Using these two tools and a little bit of analysis (which is needed because distance functions are not necessarily smooth everywhere), the fundamental lemma is

Theorem 349 *If a complete Riemannian manifold contains a line (see defintion 333 on page 585 for the definition of line) and has nonnegative Ricci curvature then it splits isometrically as a Riemannian product $N \times \mathbb{R}$.*

It is interesting to note that such a statement, under the much stronger hypothesis that of nonnegative sectional curvature, was obtained back in Toponogov 1964 [1195] using precisely his comparison theorem. Toponogov's proof went as follows: take a line γ and some point p not on γ. Now consider the two triangles $\{p, q, \gamma(t)\}$ and $\{p, q, \gamma(t')\}$ made up by a foot point q for the distance from p to γ and the points $\gamma(t)$ and $\gamma(t')$ when t goes to $-\infty$ and t' goes to $+\infty$. Apply Toponogov's theorem 73 on page 258 to both. Equality will be forced in Toponogov's theorem by the limit of the triangle inequality, and this in turn forces the sectional curvature in the triangles to be zero. This being valid for any point p in the manifold, one gets exactly the zero curvatures needed to prove the product decomposition (see equation 4.38 on page 209). For manifolds with nonnegative Ricci curvature, one defines two Busemann functions b^+ and b^- with the points $\gamma(-\infty)$ and $\gamma(+\infty)$ of the line under consideration and applies the maximum principle for superharmonic functions to the sum $b^+ + b^-$.

Note 12.3.2.3 (Simply connected compact positive Ricci curvature manifolds) Concerning simply connected Ricci positive manifolds, we should mention a relation with theorem 219 on page 459. The main idea is that the very proof of Myers's theorem, when iterated, proves immediately that Ricci $> (d-1)\delta$ implies that any geodesic of length larger than $k\pi/\sqrt{\delta}$ has index at least $k(d-1)$ (k being any integer). Mixed with theorem 219, this leaves some hope that we might deduce topological implications from positive

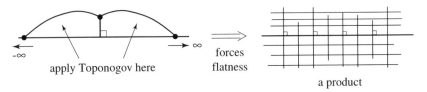

Fig. 12.25. Toponogov's approach to splitting

Ricci curvature via the Betti numbers of the loop spaces of the manifolds. We mentioned already in §§10.3.5 that this line of inquiry has not been fruitful, precisely because no Ricci curvature assumption is needed in theorem 219 nor in theorem 227 on page 465. For general considerations on the positive Ricci curvature question, see pages 75–86 of Gromov 1991 [627], chapter 7 of Gromov 1999 [633] and the references quoted in the introduction of Wilhelm 1997 [1264]. ♦

12.3.2.0.4 Just below zero Ricci curvature

The challenge now is to relax our hypothesis and to go below zero. There is a heuristic hope that we might obtain some results, since Ricci ≥ 0 implies $b_1(M,\mathbb{R}) \leq \dim M$ it is reasonable to expect that some heuristic semicontinuity will yield that $b_1(M,\mathbb{R})$, being an integer, will remain no greater than $\dim M$ when the Ricci curvature is not too large a negative quantity (with some normalization). And moreover we expect equality only for flat tori. Partial results appeared in Gromov 1999 [633] and Gallot 1981 [539]. But a more ambitious semi-continuity would be to control the fundamental group, in view of theorem 347 on page 595. Today both hopes are fulfilled, by

Theorem 350 (Colding 1997 [387]) *There is a constant $\varepsilon = \varepsilon(d) > 0$ such that any compact Riemann manifold of dimension d with $\mathrm{diam}(M)^2$ Ricci $> -\varepsilon(d)$ and $b_1(M) = d$ is homeomorphic to a torus (as long as $d \neq 3$).*

Theorem 351 (Cheeger & Colding 1996 [336]) *There is a constant $\varepsilon = \varepsilon(d) > 0$ such that every compact Riemannian manifold of dimension d with*

$$\mathrm{diam}(M)^2 \, \mathrm{Ricci} > -\varepsilon(d)$$

has a nilpotent subgroup of finite index in its fundamental group.

The second result is incredibly stronger than the "old" pinching around zero result theorem 312 on page 568. In fact this result was conjectured by Gromov and proved in the case of lower bounded sectional curvature in Fukaya & Yamaguchi 1992 [533]. The proof involves a deep study of collapsing under only a lower bound, which is done in Yamaguchi 1991 [1282]. Moreover the authors mentioned there that their technique could be extended to the Ricci curvature lower bound modulo some reasonable "splitting" conjectures. The proof of the Ricci case uses the same outline as the proof of the sectional case, but also new techniques coming from a deep study of the geometry of noncompact Riemannian manifolds with lower bounds on Ricci curvature—in particular an extension to Gromov–Hausdorff limits (see §§§12.4.2.3) of the splitting theorem of Cheeger and Gromoll. In both theorems, an underlying idea is that when ε is small enough, many aspects of the geometry behave as if the Ricci curvature were positive. Another ingredient is to extend the splitting theorems 350 on the previous page to more general metric spaces, namely Alexandrov spaces (see §§14.5.5).

A sketch of the proof of the first result is as follows. One starts with Gromov's proof in Gromov 1999 [633] to the effect that $b_1(M, \mathbb{R}) \leq \dim M$ for small enough ε : on the universal covering of a suitable finite covering of diameter D, one constructs a basis of $\pi_1(M)$ (seen as deck transformations, and in the spirit of the proof of theorem 329 on page 582) so that the generators displace the points not too much, but so that every element displaces them enough (in order to avoid overly squeezed triangles). Independently, on suitably large balls, one constructs d harmonic functions which have L^2 small Hessians and have L^2 almost orthogonal gradients. One mixes this with predictability arguments (which are possible thanks to Colding's formula in theorem 76 on page 262) and a smart basis as above to obtain, when inf Ricci $\times$ diam$(M)^2$ goes to zero, a Gromov–Hausdorff limit (see §§12.4.2) which is a torus.

In the preceding results, one difficult point is to prevent collapsing, since one does not have a bound for the injectivity radius. Kontsevich is developing a description of the closure of the set of nonnegative Ricci curvature manifolds; this is important in theoretical physics.

Note 12.3.2.4 (Structure at infinity) For the proof of theorem 351 on the preceding page, an essential technique is to study the structure at infinity of the manifolds under consideration. This technique is important and was introduced in Gromov 1981 [614] to prove:

Theorem 352 (Gromov, 1981 [614]) *If a finitely generated group has polynomial growth, then it is a subgroup of finite index of a nilpotent Lie group.*

The proof introduced two basic concepts, and was so revolutionary that we will explain its scheme in detail. First one considers a group Γ as a metric homogeneous space by saying that the distance between two elements g, h of Γ is the length of the element $h^{-1}g$, which is defined to be the number of generators (in a fixed generating set) needed to write down $h^{-1}g$. The Cayley graph of Γ is built up by joining two points when their distance is exactly equal to 1. One looks now at the limit, when ε goes to zero, of the sequence of metric spaces $(\Gamma, \varepsilon^{-1}d)$. This limit was solidly defined in Gromov's paper, i.e. we have to find some topology on the space of all metric spaces and hope for convergence. This topology is that of the Gromov–Hausdorff metric which we will define in detail and comment on in §§§12.4.2.3. This metric was the second new concept introduced in the article. The rest of the proof is long and hard. One shows first that the limit is a nilpotent Lie group and thereafter one realizes Γ as a subgroup of it. So one has introduced "the structure at infinity of the group" and considered it as a object in and of itself. This structure at infinity is not be confounded with the sphere at infinity $S(\infty)$ to be met in §§§12.3.4.3. ♦

Back to the proof of theorem 351 on page 597. One studies the sequence of Riemannian metrics on M given by $\varepsilon^{-1}g$ when ε goes to zero, and looks for a nice limit. For a general Riemannian manifold such a limit can be awful. The wonderful point here is that the hypothesis Ricci ≥ 0 implies that the limit is a metric cone, namely a metric product of some compact metric space with $\mathbb{R}^+$. Technically things are much more complicated, because all of the limits we have been considered are taken for the non-compact universal covering of the manifold under consideration. Then limits have to be taken in the pointed category, and the limit can depend on the base point. Moreover to prove theorem 351 one has only a Ricci curvature bound which can be negative, since very small, and all the above has to be applied to a limit in the category of manifolds with Ricci diam $M^2 \geq -\varepsilon$. One needs to introduce generalized Riemannian manifolds, in the spirit of §14.5. All together, the proof is extremely long, hard and subtle. The authors employ some synthetic Ricci curvature Riemannian geometry; see the second appendix of Cheeger & Colding 1997 [337].

When they exist, limit cones are not so simple and their study shows that the wording "the structure at infinity" is a subtle concept. The fact is that those cones need not be unique when the base point varies, even under conditions stronger that the positivity of the Ricci curvature; see Perelman 1997 [1011].

Note 12.3.2.5 (Thin triangles) A new technique for studying complete manifolds with nonnegative Ricci curvature was introduced in Abresch & Gromoll 1990 [5]. One proves that under a diameter growth condition the topology is bounded. The result can also be thought of as a weak quantitative generalization of the splitting theorem. This article was the first appearance of "thin" triangles. Colding's L^1 Toponogov theorem 76 on page 262 is the dramatic generalization of this previous result. The strongest splitting result is the one for Gromov–Hausdorff limits, in Cheeger & Colding 1996 [336], §6.
♦

12.3.3 The Positive Side: Scalar Curvature

We advance to scalar curvature. The story here is one of the most beautiful chapters of recent Riemannian geometry (even if still not finished). However Gromov has some complaint about it, which we will address soon. We assume in this section that dim $M \geq 3$, since the Gauß–Bonnet theorem 28 on page 138 solves all of our questions when dim $M = 2$. The pinching question as well as the negative sign questions are completely ruled out by

Theorem 353 (Kazdan & Warner 1975 [801]) *Every manifold admits a metric with constant negative scalar curvature.*

The proof uses the formula for scalar curvature under a conformal deformation and suitable approximations. We are left with the positive and the nonnegative case. Today we know exactly which are the simply connected manifolds able to carry metrics of positive scalar curvature. We use freely here in advance notions from §§14.2.2:

Theorem 354 (Gromov & Lawson 1980 [635], Stolz 1992 [1161])
Every compact simply connected manifold of dimension at least five, which is not spin, carries a metric of positive scalar curvature. A spin manifold of dimension at least five carries a metric with positive scalar curvature if and only if its α-genus is zero.

This complete classification was achieved through successive efforts of both geometers and topologists. The main geometric tools are those of Gromov & Lawson 1980 [635]. They consist in proving, using Riemannian smoothing functions in a clever way, in conjunction with Riemannian submersion formulas 15.17 on page 721, that any surgery of codimension smaller than 3 will yield a manifold of positive scalar curvature, provided only that the two manifolds used in the surgery (see §§4.3.7) both admit metrics of positive scalar curvature. For more details we refer the reader to the very good recent survey Rosenberg & Stolz 1994 [1071]. The first result goes back to Lichnerowicz 1963 [867], see §§14.2.2 for more on the tools.

12.3.3.1 The Hypersurfaces of Schoen & Yau

A completely different tool has been used in

Theorem 355 (Schoen & Yau 1979 [1109]) *In a manifold M with nonnegative scalar curvature, every homology class in $H_{d-1}(M)$ can be realized by a hypersurface which admits some metric with nonnegative scalar curvature.*

By induction on the dimension, one can see that this result implies topological restrictions. The proof is based on results of geometric measure theory, (see §§14.7.2) to get existence of minimal hypersurfaces. But it works only in dimensions less than seven. This dimension bound is needed in geometric measure theory to be sure to get submanifolds; higher dimensions can give rise to singularities. The authors construct a subtle conformal change of the induced metric on the minimal hypersurface, using essentially the fact that the second variation for the volume along the hypersurface has to be nonnegative. Details can be seen in pages 91–95 of Gromov 1991 [627].

Comments on the advantages of both tools figure on page 246 of Rosenberg & Stolz 1994 [1071] and the section 523 of Gromov 1996 [631]. According to Gromov, the main question is to find the geometric concept unifying both techniques—see more on this soon below, and for the beginning of some unification, see Akutagawa & Botvinnik 2000 [14].

12.3.3.2 Geometrical Descriptions

The above results do not finish the story. First, the non-simply connected case is well advanced but still not finished and connected in part with the famous Novikov conjecture, one of the driving forces of recent topology. See Gromov 1996 [631], the survey Rosenberg & Stolz 1994 [1071], and the references there for more on this. And there are also the low dimensions still to be mastered.

Second, the topologist's description is not really geometrical. Moreover, as we are going to see, the proof of the main result above is everything but geometrical except for the surgery control. A contrario, the product $S^2 \times N$ of any manifold N with the 2-sphere always carries a metric of positive scalar curvature (just shrink the two-sphere enough that its curvature will dominate that of N) by equation 12.1 on page 548. Moreover minimal hypersurfaces of a manifold with positive scalar curvature carry metrics with the same property, at least in low dimensions. This is the nice construction mentioned above of Schoen & Yau 1979 [1109]. So there is a lot of work left to check that these two special examples fit with the hypothesis of theorem 354 on the facing page. A geometric classification in that spirit was started very recently in Gromov 1996 [631].

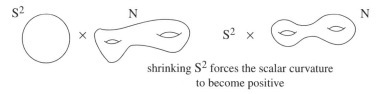

shrinking S^2 forces the scalar curvature to become positive

Fig. 12.26. Shrinking S^2 forces the scalar curvature to become positive

There is a fascinating purely geometric statement in Llarull 1998 [872]:

Theorem 356 (Llarull 1998 [872]) *A Riemannian metric g on the sphere S^d such that $g \geq g_{standard}$ (this means everywhere, i.e. all of the lengths of curves are larger or than equal to their lengths in the standard round metric) has necessarily one point m such that* $\mathrm{scalar}(m) < d(d-1)$ *(the value of the scalar curvature for $g_{standard}$) unless g is the canonical metric.*

But the proof uses all of the spinor theory of the proof sketched below of the above classification. A striking simple example, to illustrate the power of Lichnerowicz' theorem, is the famous $K3$ surface, the algebraic surface in $\mathbb{CP}^3$ cut out by the equation

$$x^4 + y^4 + z^4 + t^4 = 0,$$

which admits a Ricci-flat metric by Yau's theorem 280 on page 527, but for which one cannot prove today in an elementary direct geometric way that it never admits a metric with positive scalar curvature (or even positive sectional curvature).

Note 12.3.3.1 (Volumes of Balls and Divergence of Geodesics)
Bishop's theorem 107 on page 310 on the growth of the volumes of balls does not extend here. One has only an infinitesimal version of it, useless for global purposes. It would be very useful, according to Gromov, to find macroscopic (local) consequences of positive scalar curvature. Another question (which might well be linked with the preceding one) is to decide whether or not a C^0 limit preserves the nonnegativeness of the scalar curvature. The reader can see this question in a more general setting in §§12.4.3. Note that this is false for negative scalar curvature by Lohkamp's result in §§12.3.5. According again to Gromov a starting point for proving this and understanding the theory clearly is the following. The nonnegativeness of sectional curvature is equivalent to the nondivergence of geodesics as compared to the Euclidean case. For scalar curvature one has (in a sense not yet made precise) "nondivergence" but only in (at least) one direction. ♦

12.3.3.3 Gromov's Quantization of K-theory and Topological Implications of Positive Scalar Curvature

In Gromov 1996 [631], Gromov succeeded in extracting some topological implications from the inequality scalar $> \varepsilon^2$. The topology under consideration is K-theory (see §§§14.2.3.4). This is part of Gromov's program to quantize algebraic topology. We saw such an example in his theorem 219 on page 459 on geodesics, and another in the notion of simplicial volume in §§§11.3.5.2. Here K-theory is quantized as follows (we are being very sketchy, skipping the question of the parity of the dimension). A new invariant called the K-area(M) of a Riemannian manifold is defined as

Definition 357 *The K-area(M) is the minimum of the inverse of the largest curvature of all nontrivial vector bundles over M.*

More precisely, one considers all possible complex vector bundles, metrized "à la Riemann" i.e. one puts some metric on the fibre and introduces some connexion which preserve that metric. Such a vector bundle X has an associated "sectional" curvature. As soon as at least one Chern class of X is not zero, these curvatures cannot all vanish. One takes the maximum of the curvature on all tangent planes, denoting it by $\|R(X)\|$. Then K-area(M) is the minimum of the $\|R(X)\|$ when one runs through all such bundles and metrics. The K stands both for the curvature and the K-theory. The fundamental inequality is :

Theorem 358 *If* scalar$(M) > \varepsilon^{-2}$ *then* K-area $\left(M^d\right) < c(d)\varepsilon^2$ *for a constant $c(d)$ depending only on the dimension of M.*

The quoted article also proves the gap theorem 175 on page 404 (as well as many, many other things). The inequality in theorem 358 is a generalization of the theorem 359 on the following page of Gromov and Lawson on the torus.

12.3.3.4 Trichotomy

Third, we have the trichotomy problem if we are interested in the limiting case of nonnegative scalar curvature. On the strictly positive side, things are solved, and on the strictly negative side also, since in fact a long time ago, in Kazdan & Warner 1975 [801, 802], it was proven that any manifold can carry a metric of negative scalar curvature. But what about in between? With a little more than the Kazdan–Warner results (see Besse 1987 [183], section 4.E for details), one knows that for the class of manifolds between these there is only a third possible case: that of manifolds which cannot carry a metric of positive scalar curvature but can carry a metric of zero Ricci curvature. We met this mostly open question in §§12.3.2 and in §§§11.4.6.4: the zero Ricci curvature Riemannian manifolds are still very mysterious today.

Fourth, the total set of positive scalar curvature metrics on a given compact manifold can fail to be connected. This implies a fortiori examples of nonconnectedness for the spaces of positive sectional and Ricci curvature. The first example was Gromov & Lawson 1983 [636] for the sphere S^7. In Kreck & Stolz 1993 [834], the subject is considered further. They make use there of the Aloff–Wallach manifold met before. Particularly nice are the two Aloff-Wallach manifolds $\mathfrak{W}_{-4638661,582656}$ and $\mathfrak{W}_{-2594149,5052965}$, which are diffeomorphic but cannot be connected by a path of positive scalar curvature metrics between the two Aloff–Wallach metrics. For more examples one can consult (among others) the (kind of) survey Lohkamp 1996 [874] (for both signs, by the way) and Lawson & Michelsohn 1989 [850], page 329. Note that nonconnectedness is important but does not say anything about the topological structure of the components; compare with §§12.3.5.

12.3.3.5 The Proof

At this point it is important both historically and for the future to look at the proof of the above results. An expository sketch can be found in pages 95–100 of Gromov 1991 [627]. The tools are of three completely different types. The first is the above surgery, 100% geometrical. The second part is pure algebraic topology, namely cobordism theory. This theory will tell you which manifolds can be built up by surgery with simple building blocks (among them some of the $\mathbb{KP}^n$). One has to show that any manifold satisfying the conditions above can be obtained by surgery with building blocks which do have positive scalar curvature. All of this will insure the sufficiency of the condition.

The necessity is a completely different story and needs a new tool. It started with Lichnerowicz 1963 [867]. This was historically the first condition obtained from the positivity of the scalar curvature (and in those times even of the positivity of the sectional curvature!). Assume the manifold admits a spin structure (see §§14.2.2) and follow the general scheme in the proof of theorem 345 on page 594. Then the generalization of Bochner's technique (see §15.6) is available through the theory of elliptic operators to yield harmonic spinors, namely those which vanish under the Dirac operator $\slashed{D}$. For this Dirac operator the new "Bochner formula" is surprisingly simple and reads in short

$$\slashed{D}^2 = D^*D + \frac{\text{scalar}}{4}.$$

So $scal > 0$ implies that there are no harmonic spinors. Then the Atiyah–Singer index theorem (see §§14.2.3.5) implies that some invariant called the $\hat{A}$-genus has to vanish. To go to the α-genus needed some extra work which was achieved in Hitchin 1974 [719].

12.3.3.6 The Gromov–Lawson Torus Theorem

Last, not least, we cannot resist to quote the very appealing

Theorem 359 (Gromov & Lawson 1983 [636]) *On a torus, a metric of nonnegative scalar curvature has to be flat.*

For surfaces, this result is a direct consequence of the Gauß–Bonnet formula 28 on page 138. The technique of the proof (suitably twisted spinor bundles, using the fact that a torus has larger and larger coverings which are still tori) is put in a general framework in Lawson & Michelsohn 1989 [850], IV.§7. The idea is that the Bochner formula for the generalized Dirac operator of those bundles has two terms; the first is the scalar curvature, the second reflecting more the downstairs metric. Their behaviour, when one considers larger and larger coverings, forces the curvature of the torus to vanish. In one direction at least, Gromov's formula in theorem 358 on the previous page captures the essence of this torus theorem.

12.3.4 The Negative Side: Sectional Curvature

12.3.4.1 Introduction

This topic is an entire world in itself and is quite beautifully developed. In fact, since Hadamard in 1898 and 1901, results on negative sectional curvature have kept flowing in. First by Cartan in the 1920's, by Eberhard Hopf in 1939, Preissmann in the 1940's and then the results described below. Due to this huge amount of results we are forced to choose a very sketchy exposition.

The von Mangoldt–Cartan–Hadamard theorem 72 on page 255 of the 1920's, clarified by the Hopf–Rinow theorem 52 on page 227, implies that complete manifolds of nonpositive sectional curvature have a universal covering which is diffeomorphic to $\mathbb{R}^d$. This follows from the stronger fact that any pair of points is joined by a unique geodesic which is automatically a segment. So the simply connected case "looks like" hyperbolic or Euclidean geometry. Then apparently for the naive observer the entire problem consists in studying the fundamental group: its algebraic nature, classification, differences between the negative and the nonpositive case, etc. But the subject turned out to be not so simple and in particular the characterization of possible groups is still far away. Also one might hope that the compact negatively curved manifolds look close to space forms, see §6.6. But the theory of negative pinching seen in §§12.2.4 might raise suspicions.

In fact the subject has enjoyed many strong results. But at the same time it appears ever more difficult to get a deep understanding of the possible fundamental groups. In particular, the theory of compact negatively curved manifolds is misleading, so we will make another exception to our rule of studying only compact manifolds. We will see below many nice results on the fundamental groups of compact negatively curved manifolds. But to date there is no property known which they satisfy and which is not satisfied by hyperbolic groups. This notion of *hyperbolic group* was introduced in the monumental Gromov 1987 [622]; Ghys & de la Harpe 1988 [558] is a book explaining the subject. The force of this notion sits in its four equivalent definitions, ranging from group theory to Riemannian manifolds. Roughly speaking (and using the word metric for groups) hyperbolic groups enjoy the same asymptotic isoperimetric behaviour and the same large triangle inequalities as in classical hyperbolic spaces. All of this is a strong incentive to leave, at least for the moment, the world of compact (or even finite volume) negatively curved manifolds.

It is important to mention that it is always difficult to construct compact manifolds of negative curvature—we will meet this question again.

12.3.4.2 Literature

There is a recent survey of the field: Eberlein, Hamenstädt, & Schroeder 1990 [472], but note that the authors admit not being complete even in 50 pages.[4]

---

[4] We admit not being complete in quite a few more.

The book Ballmann, Gromov & Schroeder 1985 [106] was quite complete when it appeared. The more recent book Ballmann 1995 [103] is a complete exposition of the "rank rigidity" for manifolds of nonpositive curvature and finite volume, see below. Note that space forms (see §6.6) appear in many instances in this survey. We refer the reader to §6.6 to complete our very partial and biased exposition, which follows Gromov's philosophy. It is impossible not to refer to the basic texts: Gromov 1987 [622], Gromov 1993 [629]. See also the "beginner" text Gromov 1991 [627],

As remarked in the introduction of Eberlein, Hamenstädt & Schroeder 1990 [472], the field involves very different techniques coming from different fields of mathematics. Here we note common ground with dynamical systems in §10.6, and space forms of §6.6. In particular, to simplify the exposition, we have put as much as possible of the dynamical results in §10.6 and the rigidity results for space forms (of negative curvature) in §6.6.

12.3.4.3 Quasi-isometries

To study simply connected negatively curved manifolds, we can begin naively by comparing different metrics of negative curvature on $\mathbb{R}^d$. The most sensible notion of equivalence, ignoring the fine detail, is that of *quasi-isometry*. Two metrics are said to be *quasi-isometric* if there are two constants such that each is squeezed between the other multiplied by these two constants. It turns out that it is very difficult to decide whether two given metrics (say of negative curvature) are quasi-isometric. We will soon appreciate the difficulty. Any diffeomorphism between two compact Riemannian manifolds lifts up trivially to a quasi-isometry between their universal coverings. Now the universal covering of a compact negatively curved manifold has curvature bounded from above and below. So we forget the original compact manifold and start with our first question:

Question 360 *Classify metrics on $\mathbb{R}^d$ with negative sectional curvature K bounded from above and below: $a \leq K \leq b < 0$, up to quasi-isometries.*

The classification turns out to be extremely difficult as soon as $d \geq 3$. For $d = 2$ use the conformal representation theorem and compute the curvature. One gets a Laplacian for the conformal factor, and the bounds on curvature finally yield a bounded function. Hence when $d = 2$ all metrics of uniformly negative curvature are quasi-isometric. The only similar case in higher dimensions is when both metrics under consideration admit an orthogonal one point symmetry. This time the Jacobi field equation gives a conformal factor which is bounded.

But that is the end of the easy part. We will still stay in the negatively pinched range, since without curvature conditions the game is of no interest. It is first difficult to prove that the two space forms which come under consideration, namely Hyp^4 and $\mathrm{Hyp}_{\mathbb{C}}^2$, are not quasi-isometric. This is a corollary of the Mostow rigidity theorem 99 on page 293. In that direction, it seems very difficult to classify spaces satisfying either $-4 < K \leq -1$ or $-4 \leq K \leq -1$. For example, it was already difficult to get an example with curvature which could not be rescaled to lie in the range $-4 \leq K \leq -1$ (see Mostow & Siu 1980 [950]). We meet the difficulty of constructing manifolds of negative curvature (see also §6.6 and §§12.2.4). Even if one performs some geometric manipulations on known examples, the original examples always come from number theory. A contrario, as remarked in the note 12.2.4.1 on page 570, singular objects of negative curvature are in some sense "most" of the natural geometric objects (in dimension 2, glue heptagons or more-gons; in higher dimensions things are more subtle). One now has a quite large family (depending on real parameters) of manifolds such that no two are quasi-isometric. This is in essence in Pansu 1989 [995]. Moreover those metrics are homogeneous.

12.3.4.3.1 The fundamental groups

We continue to consider the possible negatively curved metrics on $\mathbb{R}^d$ but look at it from the other side:

Question 361 *Can any discrete finitely generated group act by isometries with nonbounded orbits on a negatively curved metric on $\mathbb{R}^d$?*

There is no restriction known today besides the trivial one obtained from the fact there must be a quotient which is a $K(\pi, 1)$: our group should have some homology in some dimension. No conjecture seems in view; see Gromov 1993 [629]. The above considerations play a role in Novikov's conjecture, which has become a gargantuan influence on geometry and topology in the last ten years; see the very end of Gromov 2000 [634].

From the other direction, one has a good understanding of how certain special groups can act. In Kleiner & Leeb 1997 [810] it is shown that any quasi-isometry of an irreducible symmetric space of rank at least 2 then it is of bounded distance from an isometry; this is false for the reducible case. This can be seen as a contribution to the understanding of symmetric forms of rank at least two, which are still the source of problems.

12.3.4.3.2 Tools

We look still further at a simply connected manifold of nonpositive or negative curvature. First, any half-geodesic is a ray and every geodesic is a line (see definition 333 on page 585 and theorem 72 on page 255). In particular, we have a lot of Busemann functions (see definition 334 on page 585), which are convex along with all of their sublevel sets. We have all of the consequences listed in §§6.3.3, e.g. the convexity of the distance function on $M \times M$, the distance functions from points, and the norm of Jacobi field. Extremely important is the infinitesimal version of the notion of Busemann function, which played a basic role in Anosov's theorem 243 on page 478: consider a line γ, the point

$p = \gamma(0)$, and any unit tangent vector u orthogonal to $\gamma'(0)$. As drawn in figure 12.27, there is a unique Jacobi field Y (resp. Z) along γ such that $Y(0) = u$ and $\lim_{t \to \infty} Y(t) = 0$ (resp. $\lim_{t \to -\infty} Z(t) = 0$). And of course we have the Toponogov comparison theorem 73 on page 258 for $K < 0$: here there is no restriction on the size of the triangle, so one always has a super-Euclidean geometry, and even more when K is bounded from above by a negative constant.

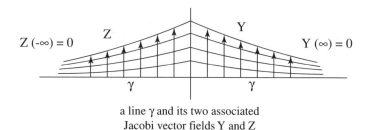

a line γ and its two associated
Jacobi vector fields Y and Z

Fig. 12.27. A line γ and its two associated Jacobi vector fields

Basically the smarter tools all consist in compactifying, since working with the infinite is hard. The main idea, caricaturally, is to actively contemplate the manifold *from infinity* via its compactification to be seen below. This is not to be confused with the structure at infinity met in theorem 352 on page 598. A baffling example of this technique was the proof of Gromov's result on discrete groups with polynomial growth. For example, a lattice, seen from infinity, looks like a vector space, while a hyperbolic structure looks like a tree. We are still on the same manifold $\mathbb{R}^d$ with a nonpositively curved metric, most often with a lower bound for K. We extract a few things from the basic reference Ballmann, Gromov & Schroeder 1985 [106] for detailed proofs and from the survey Eberlein, Hamenstädt & Schroeder 1990 [472].

Definition 362 *One defines the* sphere at infinity $S(\infty)$ *of this situation, by considering on the set of all rays the following equivalence relation: two rays γ and δ are said to be equivalent if the distance $d(\gamma(t), \delta(t))$ is bounded as t goes to infinity. Then $S(\infty)$ is the quotient by this equivalence relation.*

This was already in Hadamard 1898 [674] in studying negatively curved surfaces. One puts a topology on $S(\infty)$, the *cone topology* defined in an obvious way using only rays emanating from a given fixed point. It does not depend on the point chosen and if one varies the metric only by quasi-isometries, one induces homeomorphisms on the different realizations of $S(\infty)$.

We would like to have a measure and a metric on $S(\infty)$. The metric which Gromov calls the *Tits metric* is very special. It is actually a metric, i.e. finite everywhere, when the curvature is negative (more precisely, when the curvature has a uniform negative upper bound). This is linked with the

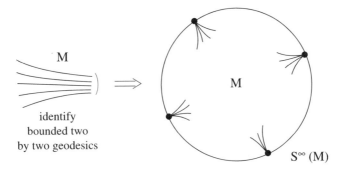

Fig. 12.28. Constructing the sphere at infinity $S(\infty)$

visibility condition introduced in Eberlein & O'Neill 1973 [473]. We say that one point is *visible* from another if there is a geodesic joining them. There are some more subtle families of metrics one can define on $S(\infty)$ which undergo quasi-conformal transformations under quasi-isometric maps.

When the curvature is only nonpositive, then the Tits metric becomes a kind of jigsaw metric, finite for some pairs of points. The typical examples are the symmetric spaces of rank larger than one (see §§4.3.5). The Tits distance between rays is exactly the Euclidean angle. More generally, these finite parts are associated to flat parts which we are going to meet below. The totality is more a jigsaw (or railway) metric than a Carnot–Carathéodory one (see §14.5). We think we can help the reader to get some feeling about the field by mentioning that two equivalent rays, in a simply connected manifold of uniformly negatively bounded curvature, have distance $d(\gamma(t), \delta(t))$ decreasing exponentially with time t. For the many structures on $S(\infty)$ and their relations with various codings of periodic geodesics, see the survey Ledrappier 1995 [859]. Also see Bergeron 1999 [178].

12.3.4.4 Volume and Fundamental Group

We turn now to properties of compact manifolds of negative (resp. nonpositive) curvature and of their fundamental groups.

12.3.4.4.1 Volumes

Let us turn first to their volumes. As was the case for space forms in §6.6, the possible volumes are discrete and in particular isolated from 0 for space forms of any type.

Theorem 363 (Heintze 1976 [698]) *Any manifold of sectional curvature K bounded as $-1 \leq K < 0$ has volume no smaller than a constant $c(d) > 0$ depending only on the dimension.*

This is often called the *Heintze–Margulis lemma,* because it was obtained by Margulis in the special case of space forms to get rigidity results; see §§6.6.2. We used it in studying negative pinching to obtain the negative finiteness theorem 315 on page 570. Both authors provide a purely geometric proof, using a "collar" argument. The strong divergence of geodesics in negative curvature as well as the nature of hyperbolic isometries force a tube around the smallest periodic geodesic to have not too small a diameter and then one applies the comparison theorem as for the tube argument in the proof of theorem 90 on page 273. We naturally hope here that negatively curved compact manifolds have finiteness results with milder conditions than the four ones needed in the general Cheeger finiteness theorem 372 on page 616. This was achieved in Gromov 1978 [611], and is discussed in detail in Ballmann, Gromov & Schroeder 1985 [106] and improved in Fukaya 1984 [527]. We mention here a finiteness theorem of Gromov which is true for real analytic manifolds but false in the differentiable case. It is extremely rare in Riemannian geometry to have a result which distinguishes between these two categories.

Theorem 364 (Gromov) *If a compact Riemannian manifold M has sectional curvature satisfying $-1 \leq K < 0$ and if the Riemannian metric is analytic then the sum of its Betti numbers is bounded by $c(d)\operatorname{Vol}(M)$, where $c(d)$ is a universal constant, depending only on the dimension d.*

This result is optimal, as shown by examples. The examples are obtained by connected sums, a typical smooth but not analytic operation. See Ballmann, Gromov & Schroeder 1985 [106] for details of the proof. Compact space forms of negative curvature were already hard to construct; see §6.6. For variable curvature, we saw examples in §§12.2.4, but real analytic ones are harder to construct, since gluing and surgery cannot be controlled "cheaply" with only Riemannian submersion formulas; see Abresch & Schroeder 1992 [9].

12.3.4.4.2 Fundamental Groups

The first statement describing the structure of fundamental groups of negatively curved compact manifolds, due to Milnor, was already discussed as theorem 104 on page 306. It states that for compact manifolds, negative curvature implies that the fundamental group has exponential growth. In Gromov 1978 [611] (and see 2.5.6 in Buser & Karcher 1981 [296] for details) a bound is obtained for the number of generators of this group. As in proposition 327 on page 581 and theorem 329 on page 582, Gromov uses the triangle comparison theorem and then counts points on the unit sphere whose mutual distances are bounded from below.

The most general statement is in Farrell & Jones 1989 [507], which is a topological version of Mostow's rigidity theorem 99 on page 293:

Theorem 365 (Farrell & Jones 1989 [507]) *Let M be a compact hyperbolic manifold (i.e. a space form) and of dimension larger than 5 and let N*

be any topological manifold which is homotopically equivalent to M. Then M and N are homeomorphic.

This is also presented in the book Farrell 1996 [504].

Theorem 366 (Gromoll & Wolf 1971 [608]) *If the manifold is compact with nonpositive curvature and with a fundamental group which is algebraically a product of groups, with moreover no center, then the manifold itself is a Riemannian product.*

The proof consists in improving the original idea of Preissmann 1943 [1041] which we will now elaborate. Preissmann's result was

Theorem 367 (Preissmann 1943 [1041]) *Abelian subgroups of the fundamental group of a compact negatively curved manifold are cyclic.*

Proof. Consider two commuting deck transformations and look at any parallelogram they generate (see figure 12.29). If this is not a squeezed (degenerate) parallelogram, then you can cut it into two triangles and apply the fact that the sum of the angles is smaller than π by the assertion 2 on page 255. Adding together these angles, one contradicts the angle property of this parallelogram.

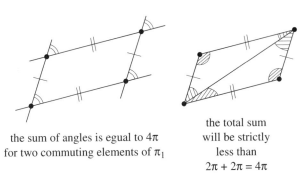

the sum of angles is egual to 4π
for two commuting elements of π_1

the total sum
will be strictly
less than
$2\pi + 2\pi = 4\pi$

Fig. 12.29. (a) The sum of angles in equal to 4π, for two commuting elements of π_1 (b) The total sum will be strictly less than $2\pi + 2\pi = 4\pi$

With more work on this idea, one finds

Theorem 368 (Lawson & Yau 1972 [851]) *In compact manifold with nonnegative sectional curvature, commuting elements of π_1 require flat parts, namely the presence of a free Abelian subgroup in π_1 implies the existence of flat totally geodesic tori in the manifold.*

This can be seen as a converse of the Bieberbach theorem 98 on page 291 for the classification of flat space forms. The idea behind the proof is very natural: extend Preissmann's technique. If two elements commute, seen as deck transformations, this will imply the existence of a parallelogram with equal sides and angles. But then we are in the equality case of Toponogov's theorem which implies that the filled up parallelogram is flat and totally geodesic.

12.3.4.5 Negative Versus Nonpositive Curvature

For negative curvature we saw that, at least today, hyperbolic groups are the "final answer." The big problem now is to appreciate the difference between negative and nonpositive curvature, both in the compact manifold as well as for the fundamental groups. There are extremely strong results which show in essence that the two cases can be dramatically separated. Otherwise put, to go from one to the other you have to make quite a jump. We give the latest result and refer to Eberlein, Hamenstädt & Schroeder 1990 [472] for a survey and to Ballmann 1995 [103] for the proofs.

When thinking about nonpositive curvature (as well analogously as for nonnegative curvature), it is always educational to keep in mind the two completely different types of examples, namely (1) the products of manifolds of negative curvature and (2) the symmetric spaces of rank larger than 1. We met this alternative twice before, for nonnegative curvature in §§§12.3.1.4 and for the simplicial volume of Gromov in §§§11.3.5.2. In both cases there were a lot of planes with zero curvature, but the way they were located in the set of planes was completely different. In a product, (see equation 4.38 on page 209) any plane made by two vectors with one tangent to the first manifold in the product, and the second tangent to the second manifold, has zero sectional curvature. For symmetric spaces (see §§4.3.5), they have a rank which is the common dimension of their totally geodesic flat submanifolds, which are moreover all conjugate by the action of the isometry group. This is interesting only when the rank is at least two—otherwise these totally geodesic submanifolds are just individual geodesics. In particular, every direction is contained in at least one such flat. At the infinitesimal Jacobi field level,

Lemma 369 *In a symmetric space of rank at least two, given any unit vector, along the geodesic it generates there is a nontrivial parallel Jacobi field (orthogonal to the geodesic of course and non zero).*

This never happens for generic manifolds. In a Riemannian manifold, we defined the rank in definition 68 on page 250, and we met it again in §§§12.3.1.4. The strongest result today is the following "tout ou rien" result for compact Riemannian manifolds of nonpositive curvature. We assume any reasonable irreducibility condition.

Theorem 370 *On a compact manifold of nonpositive curvature, either the rank of the space is at least two, and then it is a space form with its symmetric space metric, or the rank is one, and then the geodesic flow is ergodic on the subset of the unit tangent bundle made up by the vectors of rank one.*

Today it is still unknown if this subset of rank one vectors is of full measure. A basic tool used here is theorem 397 on page 643.

Note that the rank, a geometric invariant, can be (at least conceptually) computed from the fundamental group; see Eberlein, Hamenstädt & Schroeder 1990 [472]. The existence of flats and what they really imply is a major issue today; see for example Hummel & Schroeder 1998 [750].

The above dichotomy admits many extensions with weaker hypotheses; for example, compactness could be replaced with completeness and finite volume, or a group "duality condition" and also to metric spaces mimicking Riemannian manifolds of nonpositive curvature; see the CAT (0) spaces in §§ 14.5.6 on page 680.

12.3.4.5.1 Sectional Curvature Just Above Zero

Finally we end with the question systematically treated in this chapter: to relax just a little bit our requirement of nonnegative curvature. It is important to realize that without extra hypotheses, everything is permitted. At least for the simply connected case: a construction of Gromov was published in Buser & Gromoll 1988 [295] and generalized in Bavard 1987 [118]. There examples are built, for any ε, of metrics on spheres and other 3-dimensional manifolds whose curvature satisfies $K < \varepsilon$ and are of bounded diameters. But also see Fukaya & Yamaguchi 1991 [532]. Their results are that one can obtain restrictions on the fundamental group—namely that this group is almost nilpotent, only with an extra hypothesis, like $K > -1$ and an upper bound on the diameter. The above shows again that, in a rough sense, upper bounds for the curvature (if not 0) mean nothing for the topology. This idea will frequently reappear in §§§12.4.1.2

12.3.5 The Negative Side: Ricci Curvature

We have here the strongest possible answer: Lohkamp 1994 [873] belongs to a series of papers where it is shown that, caricaturally, negative curvature "means nothing" or, equivalently, "permits everything." Worse: any metric (typically of positive curvature) can be continuously approximated by a metric of negative Ricci curvature (of course not with too much differentiability). His tools come from analysis, related to the h-principle of Gromov 1986 [620], and also from looking at the formula giving the variation of Ricci curvature under deformation of metrics where one can see that it is quite easy to lower Ricci curvature.

Another very important point fact is that, on any manifold, the total set of metrics of negative Ricci curvature is topologically trivial, i.e. not only connected, but contractible. Note that this contrasts sharply with the positive case: see the note 12.3.1.3 on page 592. We suggest that the reader consult the very informative Lohkamp 1996 [874]. In fact, the freedom left by negative Ricci curvature is even more unbelievable; see theorem 167 on page 400 for the results of Lohkamp 1996 [874].

12.4 Finiteness, Compactness, Collapsing and the Space of Riemannian Metrics

In §§12.4.1 and §§12.4.2 we are going to present ideas in the reverse of the natural order, which would be from weaker to stronger structures. The route we follow is the same as in the surveys Abresch 1990 [4], Fukaya 1990 [531], in appendix 6 of the book Sakai 1996 [1085], in Petersen 1997 [1018], and in the second edition Gromov 1999 [633] of the pioneering book Gromov 1981 [616].

12.4.1 Finiteness

12.4.1.1 Cheeger's Finiteness Theorems

Before introducing Cheeger's work, it might be useful to recall here some facts about the three ratios homotopy/homology, homeomorphy/homotopy, diffeomorphy/homeomorphy which were explained in §§4.1.4. under the heading "the classification of manifolds." Here we will look at such a classification in the sunlight of Riemannian geometry. The two first ratios are in general infinite. We just mention that even the classification of homology spheres is not finished. For the second, the theory of characteristic classes is an ideal tool; a lucid exposition is in Milnor & Stasheff 1974 [925]. The third ratio is always finite in five or more dimensions; basic texts are Hirsch & Mazur 1974 [716], Kirby & Siebenmann 1977 [805]. These finite numbers can be computed explicitly using topological information on the manifold under consideration. Recall that dimension 4 is an exception; examples with infinitely diffeomorphism types with the same homeomorphism type can even be taken among algebraic surfaces.

We note that Gromov's theorem 328 on page 582 is only a weak finiteness result for homology types. The finiteness topic started in fact simultaneously in 1966 with Cheeger 1967 [328] and Weinstein 1967 [1246]. For expository reasons we present first the simplest case, which is the only one treated by Weinstein. In despair of every classifying manifolds of positive sectional curvature (remember how we suffered in §§12.3.1) Weinstein proved

Theorem 371 (Weinstein 1967 [1246]) *If d is even, then there is only a finite number $N(\delta)$ of different homotopy types of d dimensional manifolds whose sectional curvature satisfies $0 < \delta \leq K \leq 1$.*

Unhappily $N(\delta)$ goes to infinity when δ goes to 0, precisely as $c(d)\delta^{-d}$. Note that in the other direction, Gromov's theorem quoted above on Betti numbers of nonnegatively curved manifolds implies a very partial bound but only for (integral) "homology types." The proof of the homotopy finiteness was purely geometrical, covering the manifold with convex balls whose number can be estimated as only a function of δ as follows. Bishop's theorem 107 on page 310 yields an upper bound for the total volume, while Rauch's comparison theorem yields a lower bound for the volume of balls within the injectivity radius: theorem 103 on page 306. Moreover we work within the convexity radius which is larger than or equal to $\pi/4$ thanks to Klingenberg's theorem 92 on page 276 and 95 on page 278; it is here where the restriction to even dimensions enters. Now we use the metric trick of lemma 125 on page 333, packing as many balls of radius $\pi/8$ as we can implies that the corresponding convex balls of radius $\pi/4$ cover. Such packings are called *efficient*. They have also the extra advantage that one can universally control the number of balls which meet a given one. The number of these balls is then controlled by the obvious measure argument. The homotopy type of our manifold is that of a simplicial complex with as many vertices as the above balls, as in figure 12.30

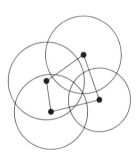

with convex balls covering the whole manifold one gets a simplicial complex with the same topology

Fig. 12.30. With convex balls covering the whole manifold, one gets a simplicial complex with the same topology

We will now look at the two next levels: homeomorphy types and then diffeomorphy types. Note also that in any such finiteness result one can also ask for explicit bounds for the size of those finite sets. We will see that in some cases there is no such explicit bound because the proof is obtained by a contradiction argument based on convergence of some infinite set; but there are cases where one can give explicit bounds.

There is no hope to extend the above result to odd dimensions. For example we saw that there is an infinite set of Aloff–Wallach manifolds $\mathfrak{W}_{p,q}$ (see §§§12.3.1.1) which are all homeomorphic, but not all diffeomorphic. See theorem 3.9 of Kreck & Stolz 1993 [834]. One can still get not only homotopy but even diffeomorphy finiteness theorems, but under stronger hypotheses. Cheeger's ground breaking insight, independent of Weinstein's, was

Theorem 372 (Cheeger 1967 [328], Cheeger 1970 [330]) *Consider the set of Riemannian manifolds of a given dimension $d > 4$ satisfying the four conditions*

1. $K > a$

2. $K < b$

3. $\text{Vol} > v > 0$

4. $\text{diam} < D$

where a, b, v, D are four fixed real numbers such that v has to be positive while a and b can be of any sign). This set of Riemannian manifolds contains only finitely many different manifolds up to diffeomorphism.

Finiteness holds not only for homotopy type but also for diffeomorphism type and without any assumption on the sign of the curvature. In some instances one can effectively get rid of the condition that the volume be greater than some lower bound. For example when one works within the class of manifolds with a non-zero characteristic number, as remarked in Cheeger 1970 [330]. This comes immediately from Chern's formulas used in theorem 272 on page 515 with the spirit of the notion of minimal volume; another instance is that of negatively curved manifolds, because of Heintze's theorem 363 on page 609.

The path of the proof is important for future considerations. First cover the manifold by coordinate balls whose number is universally bounded (universal means of course in a, b, v, D and course the dimension d). This is essentially still thanks to the Bishop volume estimate with Ricci curvature (which comes here from the sectional curvature bound). The next problem is to get a positive lower bound for the injectivity radius. This was the first part of Cheeger's work:

Lemma 373 *The injectivity radius has a universal explicit (positive) lower bound which is a function of a, b, v, D, d*

as we saw in theorems 89 on page 272 and 90 on page 273. Finally, the number of efficient packings as in the proof of theorem 371 on page 614 is bounded, as are their intersection patterns. Remember that the four bounds are needed in Cheeger's control of the injectivity radius.

The second part goes by contradiction: assume that we have an infinite number of manifold M_i satisfying the hypotheses of Cheeger's theorem. Cover them with exponential coordinate ball charts $\phi_{ik} : B(0,r) \to M_i$ as above. One can assume that the number of balls (the number of values that k takes on) is the same for all of them. On each M_i look at the transition functions $\phi_{ik}^{-1} \phi_{ih}$. From the bounds $K \geq a$ and $K \leq b$, the Rauch comparison theorems (propositions 74 on page 259 and 75 on page 259) imply that the transition functions are bi-Lipschitz, and hence converge after passing possibly to a subsequence. So finally for large i the M_i have transition functions which are very close, and they must therefore be diffeomorphic.[5]

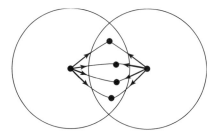

Fig. 12.31. One can control the transitions thanks to the Rauch comparison theorems via exponential maps

In fact Cheeger only proved finiteness for homeomorphisms, and to go from homeomorphism to diffeomorphism types, he used the results from differential topology mentioned above: the number of possible diffeomorphism types on a given simply connected topological manifold is finite when $d > 4$. However later on Cheeger managed to prove diffeo-finiteness for any dimension (under these four bounds) without appeal to differential topology; references are Cheeger & Ebin 1975 [341] theorem 7.37 to be completed by Peters 1984 [1013].

12.4.1.1.1 Improving Cheeger's Proof

There are two ways to work more efficiently at Cheeger's result. The problem is to control with only the sectional curvature (without the metric) some derivatives of the coordinate changes along with their uniform norm. One is to use the compactness theorems; see §§12.4.2. The other is to use the center of mass technique. The center of mass in Riemannian geometry was introduced in §§6.1.5. Recall that it can be obtained in convex balls by two equivalent definitions (hence its usefulness). The first is to minimize the sum of the squares of the distances. The second is to ask for a given linear dependence between the velocity vector from the center of mass to the points under consideration. Then Peters's proof goes as follows: in the sequence of manifolds M_i considered above we will show directly that M_i and M_j are diffeomorphic for i and j large enough, by constructing a diffeomorphism as follows. First we pass to a subsequence such that the domains of the exponential charts ϕ_{ij} and ϕ_{ik} have the same intersection pattern both in M_i and M_j. Now around a point x in M_i we look at all balls containing x and look at the various images $y_1, \ldots, y_N$ of x in M_j given by the corresponding ball charts for M_j of the type $\phi_{ik}\phi_{jk}^{-1}$. The map $f : M_i \to M_j$ is defined as

---

[5] Of course there are many details which we have ignored.

$$f(x) = \text{center of mass of } \{y_1, \ldots, y_N\}.$$

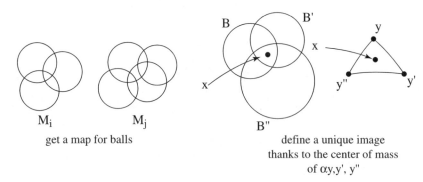

Fig. 12.32. (a) We get a map for balls (b) Define a unique image thanks to the center of mass of $\{y, y', y''\}$

The first being purely metric, and the second involving tangent vectors, there is some C^1 control automatically, as already commented upon in §§6.1.3, via the Rauch comparison theorems (propositions 74 on page 259 and 75 on page 259).

12.4.1.2 More Finiteness Theorems

A priori it seems impossible in Cheeger's results to avoid injectivity radius estimates in order to get contractible (or convex) balls. But remember Gromov's proof for the Betti numbers of nonnegatively curved manifolds: one of the basic facts is that critical points (see definition 305 on page 563) for distance functions cannot "arise too far from the center." Remember also theorem 304 on page 562 where one also works with only a lower bound on sectional curvature, asking that the diameter be large. One ignores the injectivity radius once again. This point of view that "large positive curvature does not modify the topology" was pictured in figure 12.10 on page 563. The critical point technique can be used to get rid of the sectional upper bound in theorem 372 on page 616:

Theorem 374 (Grove & Petersen 1988 [648]) *Given numbers*

$$K \in \mathbb{R}, \quad v > 0, \quad D < \infty$$

the class of manifolds of a given dimension d and with

1. *sectional curvature at least as large as K*
2. *volume greater than v and*
3. *diameter at most D*

contains only finitely many homotopy types (the number of homotopy types is bounded explicitly in d, D, K, v).

Easy examples show that no one of these three bounds can be removed from the statement (see figure 12.33).

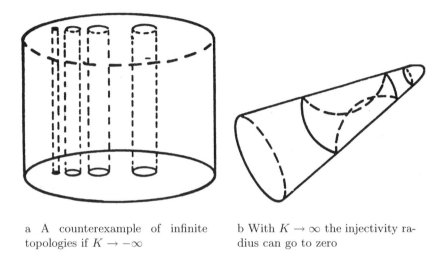

a A counterexample of infinite topologies if $K \to -\infty$

b With $K \to \infty$ the injectivity radius can go to zero

Fig. 12.33. Counterexample manifolds

The idea for the proof is to get rid of the injectivity radius estimate (hence the convexity radius estimate) in Weinstein's proof of theorem 371 on page 614 and to replace it by some "criticality radius" to use the technique of the proof of the Grove–Shiohama theorem 304 on page 562 and finally cover the manifold by balls having the same topological properties as convex balls. But the approach fails because the conditions in theorem 374 on the facing page are not enough to yield a uniform bound for the "one-point criticality radius," as figure 12.33 shows. When the smooth rounding off of the cone (near its vertex) gets smaller and smaller, the curvature still remains positive (but note that it goes to infinity). However the radius of contractible balls goes to 0, since there are smaller and smaller geodesic loops. But there are no periodic geodesics, and this is the clue. Under the conditions in theorem 374 on the facing page there is a uniform bound for the distance of pairs of points which are mutually critical; see figure 12.12 on page 564.

To prove this, if p and q were antipodal points on a periodic geodesics, then follow the proof of theorem 90 on page 273. Remember that when finding Cheeger's bound for the injectivity radius, the upper bound for the curvature was used only to get rid of conjugate points; see the scholium 91 on page 274. But for the proof of theorem 90 only volume and diameter bounds and a lower

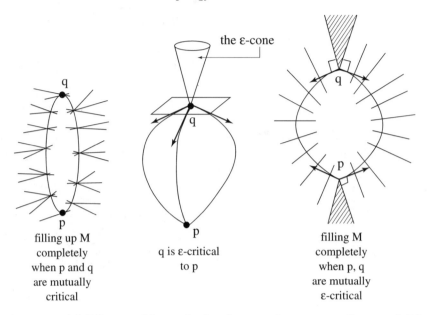

Fig. 12.34. (a) Filling up M completely when p and q are mutually critical (b) q is ε-critical to p (c) Filling M completely when p and q are mutually ε-critical

bound on curvature were needed. This is exactly the situation here. The proof of theorem 90 is generalized as follows.

Consider two mutually critical points p and q and the set Σ of segments between them, and look at the set S of geodesics emanating orthogonally from these segments and of length smaller than or equal to the diameter. As in the proof of theorem 89 on page 272 the set S is in fact the whole manifold, since any point p has in Σ a closest point q and the criticality condition and the first variation formula show that the segment between p and q is orthogonal to Σ. The volume of S is again essentially controlled by the Rauch comparison theorem. To finish the proof one has to imitate the proof of theorem 371 on page 614 but replace contractible balls by contractibility properties in $M \times M$ around the diagonal, plus (among other things) the following improvement, which is essential: there is a universal lower bound for the ε-mutually critical points. The ε-criticality means that, in the definition 305 on page 563 the angle $\pi/2$ is replaced by $\pi/2 + \varepsilon$. The figure below shows how to get a bound: one will get the whole of M by taking the union of the set S above with the cone defined by the normal tangent hyperplanes at p and q having a vertex angle equal to ε. For an ε small enough, this completes the proof. A consequence of this improvement with the ε-criticality is that (see definition 163 on page 367)

Proposition 375 *The manifolds in the class of theorem 372 on page 616 admit a uniform contractibility function.*

Then one works as in §§10.3.4: to each segment between p and q one associates canonically a geodesic in $M \times M$ which is orthogonal in $M \times M$ to the diagonal Δ. Now one uses the uniform contractibility in $M \times M$, and with some tricks one is able to cover the manifold efficiently with suitable metric balls whose radii is precisely controlled by the lower bound on the ε-mutual criticality radius.

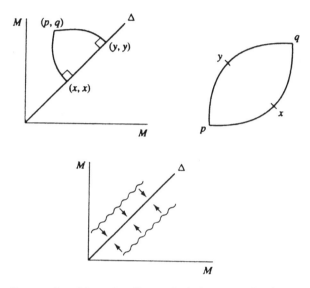

Fig. 12.35. Contracting M on its diagonal Δ for pairs (p,q) which are not ε-mutually critical

12.4.1.2.1 Homeomorphism Finiteness

It seems impossible to do better, namely to obtain homeomorphy finiteness. However the picture of the cone might indicate that large positive curvature does not change the topology, this being certainly not the case for negative curvature. However one has the very strong:

Theorem 376 (Grove, Petersen & Wu 1990–1991) *The class of manifolds of given dimension d, lower bounded sectional curvature, lower bounded volume and upper bounded diameter contains only finitely many homeomorphy types if $d \neq 3$ and finite diffeomorphy types if in addition $d \neq 4$.*

Dimension restrictions when $d = 3$ are due to the status of the Poincaré conjecture and when $d = 4$ where one has to be happy (today at least) only with homeomorphisms. We remark here that it is also open today to decide if the counterexamples (mentioned above) yielding infinitely many differentiable structures on some topological manifolds of dimension 4 can have metrics which satisfy these bounds. Also the numbers of such manifolds cannot be universally bounded today with the kind of proof used.

The proof involves new ideas since those of theorem 372 on page 616. It uses convergence techniques which we will see in §§12.4.2, working in the Gromov–Hausdorff space of theorem 380 on page 626, and a lot of deep algebraic topology. In particular the technique called *controlled topology*, is employed, using in particular the existence of a contractibility function (see definition 163 on page 367). The existence of such a function is ensured by proposition 375 on page 620.

In the same article there is the astonishing curvature-free result:

Theorem 377 (Grove, Petersen & Wu 1990 [651]) *The class of compact manifolds of a given dimension $d \neq 3$, with a fixed upper bound on volume and fixed lower bound on injectivity radius contains only finitely many manifolds up to homeomorphism (hence up to diffeomorphism if in addition $d \neq 4$).*

This means we have finiteness with not a single curvature restriction and with only two numerical invariants. The result follows immediately from the proof of theorem 376 on the previous page if one remarks that the bound on the injectivity radius provides a contractibility function. It remains only to apply Croke's inequality from theorem 149 on page 355. A weaker homotopy type result under the assumptions of theorem 377 was obtained in Yamaguchi 1988 [1281]. The result theorem 377 was mentioned in §§11.2.3, motivating some very natural questions.

Back to the Grove–Petersen–Wu finiteness result; Greene 1994 [591] and Greene & Petersen 1992 [592] throw interesting light on it, in particular making explicit the role of the volumes of tubes around periodic geodesics.

12.4.1.2.2 Integral Curvature Bounds

As mentioned in §§§12.1.2.3 the urge for stronger results, i.e. with weaker hypotheses, did not stop. In particular there are results obtained by replacing some pointwise bounds on the curvatures by various integral bounds. Precise statements are always quite elaborate, and hence not stated here; see Yang 1992 [1286], Petersen & Wei 1996 [1023], Petersen, Wei & Ye 1997 [1025] and the book Petersen 1997 [1018]. One geometric problem encountered in this programme is to control the volume of a tube around a periodic geodesic; this was done in Petersen, Shteingold & Wei 1996 [1021]. The proofs are not completely geometrical but involve a lot of analysis.

12.4.1.3 Ricci Curvature

We next wonder if it is possible to achieve finiteness of topological types with only bounds on Ricci curvature. We saw in §§12.3.2 that Gromov's homology finiteness for lower bounded sectional curvature and upper bounded diameter does not extend to lower bounded Ricci curvature. However this did not stop geometers asking what the infimum of the Ricci curvature means, since controlling Ricci curvature is hopefully a "reasonable" degree of curvature control: see §§6.4.3. Results started with Abresch & Gromoll 1990 [5] and are now very strong. We make a partial choice among them and refer to the very informative survey Anderson 1994 [45] and the references there. Note that some of these results are optimal with respect to the ingredients in view of the examples in Perelman 1997 [1012].

Theorem 378 (Anderson 1990 [39]) *Take any numbers $\lambda, D > 0, I > 0$, where λ can be of any sign. There is a finite number of diffeomorphism types in a given dimension under the assumptions:*

$$\text{Ricci}_g \geq \lambda$$
$$\text{diam}_g(M) \leq D$$
$$\text{Inj}(M, g) \geq I > 0.$$

Since one succeeded for sectional curvature to go beyond the injectivity radius, we naturally wish to do the same here. The story is not finished but:

Theorem 379 (Anderson 1990 [39] and Anderson & Cheeger 1991 [49]) *Pick numbers $\lambda, v > 0, D > 0, \Lambda > 0$. In any odd dimension there are only finitely many diffeomorphism types of manifolds which bear a metric satisfying all of the conditions*

$$|\text{Ricci}| \leq < \lambda$$
$$\text{Vol} \geq v > 0$$
$$\text{diam} \leq D$$

and

$$\int_M |R|^{d/2} \leq \Lambda.$$

When the dimension is even it is compulsory (there are examples) to add orbifolds (see §§14.5.2).

Note that in dimension 4 the $L^{d/2}$-bound on the full curvature tensor is not needed thanks to the Allendoerfer–Weil formula 11.6 on page 518 or equation 15.10 on page 712. It is not clear if it is reasonable to conjecture that this extra condition is not really needed. Note finally that all of these results were obtained by contradiction, and the proofs do not yield explicit bounds for the numbers of diffeomorphism types.

The idea for the above results is to prove $C^{1,\alpha}$-compactness; see §§12.4.2. One needs to know that Ricci curvature in harmonic coordinates, up to first order terms, is equal to the Laplacian of the metric (see equation 6.22 on page 267). But one has to control the harmonic radius (i.e. the largest balls on which harmonic coordinates are well defined and linearly independent, see §§12.4.2 for more on that). When using the bound on $\int_M |R|^{d/2}$ the difficulty is to understand how the curvature concentrates; this is achieved by looking for singular points in sequences converging in the Gromov–Hausdorff space to an orbifold, found by Anderson 1990 [39]. Then one rescales the metrics around the singular points and shows that iteration of this process ends after a finite number of times.

Results on manifolds with Ricci curvature bounds appear continually; we mention Dai, Wei & Ye 1996 [424] which, using the smoothing technique of §§12.4.3, proved many new results.

Note 12.4.1.1 (Scalar Curvature and Finiteness) From theorem 354 on page 600 (for example) it is clear that there is not a single finiteness statement to expect from bounds on scalar curvature. ♦

12.4.2 Compactness and Convergence

12.4.2.1 Motivation

Climbing Jacob's ladder, even in the fog, it is a natural question to ask for some kind of convergence and compactness within the set of Riemannian manifolds satisfying various bounds. This means trying to study the set of all Riemannian manifolds as an object in itself. Moreover we met before many problems where a convergence theorem would have been useful. Among various texts we suggest the surveys Fukaya 1990 [531] and Petersen 1997 [1017].

Look for example at diffeomorphism finiteness theorems or even more simply at the differentiable pinching problem. It is intuitive that in some sense the set of differentiable structures is discrete. Within a compact subset we have finiteness. And for example isolation of the standard sphere. Note that would be a very pleasant proof but not yielding an explicit pinching constant.

Another motivation is that of chapter 11: look at some functional on the set (or some subset) of Riemannian metrics on a given manifold M. Does there exist some Riemannian manifold realizing the infimum of that functional (a best, an extremal Riemannian structure on M)? Sadly enough all of the convergence results we are going to see seem to be unable to give an answer in that domain, see e.g. the minimal volume and the embolic volume in §§11.2.3 and §§11.3.1 as well as Einstein manifolds in §§11.4.1.

12.4.2.2 History

It seems that the first appearance of this direction was in Shikata 1967 [1131], where an isolation result was obtained for differentiable structures. Thereafter considerations of that kind were implicit in Cheeger 1970 [330]. Then the theorem (not numbered) on page 74 of Gromov 1983 [617]. The proof uses implicitly a convergence theorem that the author always took for granted. According to him, since some people doubted it or at least asked for more details, he published a proof in Gromov 1981 [616]. The proof was still a little incomplete, and then Katsuda 1985 [790] offered a complete proof. In 1985 various proofs of the optimal result (see below) started circulating, all using centers of mass and harmonic coordinates. Printed references are Greene & Wu 1988 [593], Peters 1987 [1014] and Kasue 1989 [783]. For the Ricci flow, see Lu 2001 [882]. We now have to make things precise, since the notion of convergence needs a topology.

12.4.2.3 Contemporary Definitions and Results

We now present the notions and the related results in the order which is currently the most sensible, even if it looks as if it is adapted to more general geometric objects than Riemannian manifolds. It is in fact unavoidable to begin in this generality, and moreover simplifies considerably the proofs quoted above. We follow mainly Fukaya 1990 [531] and refer to that paper for precise definitions. Another survey is Petersen 1990 [1015]. And of course the second edition of Gromov 1981 [616] is Gromov 1999 [633].

Contemporary developments start with the Gromov–Hausdorff distance $d_{\mathfrak{G}-\mathfrak{H}}$ between isometry classes of abstract compact metric spaces. We recall a classical "exercise" of general topology. Consider all compact sets of a given Euclidean space $\mathbb{E}^d$ and define a metric on this set as follows: $d_{\mathfrak{H}}(X,Y)$ is the infimum of the numbers ε such that X is contained in the ε-neighborhood of Y and also Y is contained in the ε-neighborhood of X. See figure 12.36.

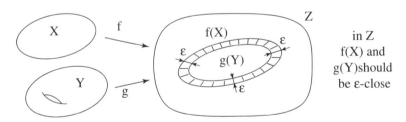

Fig. 12.36. In Z, $f(X)$ and $g(Y)$ should be ε-close

A classical exercise is to show that the set made up of all of these subsets is itself compact. Hausdorff's definition extends trivially to closed subsets of any metric space, but for two isometry classes of abstract metric spaces X and Y, one has to use the following trick: consider all so-called ε-approximations $f : X \to Y$, i.e. Y is equal to the ε-neighborhood of $f(X)$ inside Y and for every $x, x' \in X$ one has

$$|d_X(x, x') - d_Y(f(x), f(x'))| < \varepsilon.$$

Theorem 380 *Define the* Gromov–Hausdorff *distance* $d_{\mathfrak{G}-\mathfrak{H}}$ *between two compact metric spaces X and Y to be the infimum of positive numbers ε such that there is an ε approximation $X \to Y$ and there is also an ε approximation $Y \to X$. The set of isometry classes of compact metric spaces, denoted* $\mathfrak{Met}$, *endowed with the metric* $d_{\mathfrak{G}-\mathfrak{H}}$, *is itself a complete metric space.*

This definition and properties appeared first in Gromov 1981 [614, 616]. This extends the well-known analogous statement for compact subsets of $\mathbb{E}^d$. One might prefer to define $d_{\mathfrak{G}-\mathfrak{H}}$ by the following equivalent definition: look at all of the possible pairs of isometric embeddings $f : X \to Z$ and $g : Y \to Z$ into various metric spaces Z (such triples always exist trivially) and look at the infimum of the Hausdorff distance in Z of the pair $f(X), g(Y)$.

Theorem 381 *For noncompact pointed metric spaces there is an analogous statement, the convergence being now for the corresponding metric balls of any radius.*

We saw Gromov's group-theoretic application of this concept in theorem 352 on page 598. In $\mathfrak{Met}$ we have the Riemannian geometry result from Gromov 1981 [616]:

Theorem 382 *Inside* $(\mathfrak{Met}, d_{\mathfrak{G}-\mathfrak{H}})$ *the subset of Riemannian manifolds of a given dimension d with diameter not larger than D and* Ricci $\geq (d-1)\delta$ *is precompact.*

The metric moral of the story is that one is left with a finite number of "possible geometries." The argument is just efficiently packing as many balls of a given radius as possible, as in lemma 125 on page 333; then doubling the radius one gets a covering. The number of balls can now be universally bounded from above with the radius under consideration. This is quite surprising, since from theorem 107 on page 310 one has a priori only an upper bound for the volumes of balls, while we need a lower bound for the $\operatorname{Vol} B(m,r)$ to control the number of nonintersecting balls in the whole manifold, whose volume is $\operatorname{Vol}(M) \leq \operatorname{Vol} \mathbb{S}^d(\delta)$. The trick is to use the non-increasing property of theorem 107, so that for any r

$$\frac{\operatorname{Vol} B(m,r)}{\operatorname{Vol} B(\mathbb{S}^d(\delta),r)} \geq \frac{\operatorname{Vol} B(m,D)}{\operatorname{Vol} \mathbb{S}^d(\delta) D} \tag{12.2}$$

which yields the desired universal bound for

$$\frac{\operatorname{Vol} B(m,r)}{\operatorname{Vol}(M)}.$$

The desired metric approximations are found by the finite sets made up of the centers of the balls. Maps between these sets are obtained by the pigeon hole principle. But of course the closure of such sets is not in the realm of (smooth) manifolds of dimension d. Just think for example of flat tori. Note also that the assertion "a finite number of geometries" is really purely metric, since we saw in §§12.3.2 that manifolds of Ricci curvature can have arbitrary large Betti numbers.

If one exempts the splitting theorem 348 on page 595, then theorem 381 on the facing page was the first to suggest to Riemannian geometers that a Ricci curvature lower bound could be almost as important as a sectional curvature one.

12.4.2.3.1 The Lipschitz Topology

The next topology to come in our story is the Lipschitz topology. In $\mathfrak{Met}$ we will call it d_L. Its value $d_L(X,Y)$ is the infimum of all numbers $\max\left\{L(f), L\left(f^{-1}\right)\right\}$ where $L(f)$ is the Lipschitz constant of a Lipschitz homeomorphism $f : X \to Y$. (More precisely, one has to take the logarithm to get a metric, but it is useful to work without the logarithm.) Recall that for any metric spaces X and Y, a map $f : X \to Y$ is Lipschitz with constant $k = L(f)$ if
$$d\left(f(x), f(x')\right) \leq k\, d\left(x, x'\right)$$
for every pair of points $x, x' \in X$. In Shikata 1967 [1131] a discreteness result concerning d_L was proven. We work within the set $\mathcal{RM}\left(d, a, b, D, v\right)$ of Riemannian manifolds of given dimension d with $a \leq K \leq b$, diameter at most D and volume at most $v > 0$. If N is the d_L limit of a sequence $\{M_i\}$ then for i large enough M_i is diffeomorphic to N. This is how Shikata solved the differentiable pinching problem of §§12.2.2 (with an abstract constant).

But this is still not a convergence statement for any sequence (or some subsequence) in the set $\mathcal{RM}\left(d, a, b, D, v\right)$. This was achieved in two steps. The first is in Gromov 1981 [616], to be completed with Katsuda 1985 [790]. It tells us of local constancy of diffeomorphism type inside $\mathcal{RM}\left(d, q, b, D, v\right)$.

Theorem 383 (Gromov–Katsuda) *In the space* $\mathcal{RM}\left(d, a, b, D, v\right)$ *for N and a sequence M_i, if*
$$\lim_{i \to \infty} d_{\mathfrak{G}-\mathfrak{H}}\left(N, M_i\right) = 0$$
then
$$\lim_{i \to \infty} d_L\left(N, M_i\right) = 0$$
and M_i is diffeomorphic to N for i large enough.

At this stage we still do not have a convergence statement. This needs more work, which was carried out in the Gromov, Greene–Wu and Peters references mentioned above. The precise statement:

Theorem 384 *From any infinite sequence $\{M_i\}$ in $\mathcal{RM}\left(d, a, b, D, v\right)$ one can extract a subsequence which converges toward a Riemannian manifold (N, g) which is diffeomorphic to M_i for i large enough, but the metric g is known in general to be only of class $C^{1,\alpha}$ (for any $\alpha \in (0, 1)$).*

We did not define the $C^{1,\alpha}$ class nor $C^{1,\alpha}$ convergence in §§4.1.1. A function f defined in some domain of $\mathbb{E}^d$ is said to be of class C^α if it is α-Lipschitz continuous, that it is to say its C^α norm

$$\|f\|_{C^\alpha} = \sup_{x \neq y} \frac{|f(x) - f(y)|}{|x - y|^\alpha}$$

is finite. The class $C^{1,\alpha}$ is the class of continuously differentiable functions whose first derivatives are in C^α. There is then an obvious norm taking into account both f and df. Of course such classes and norms extend to any tensor or to differentiable maps of Riemannian manifolds. In theorem 384 on the previous page the metric g is the $C^{1,\alpha}$ limit of the metrics of the extracted sequence.

It is important to realize that, by nature, one cannot expect C^2 convergence nor a C^2 limit for g. A trivial example is a cylinder with two hemispherical caps, which can be trivially obtained as a limit of of surfaces all inside $\mathcal{RM}(2, 0, b, D, v)$, as in figure 12.37.

a not C^2 - limit

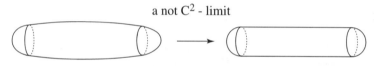

Fig. 12.37. A limit which is not C^2

We saw in §4.5 that the difficulty in proving results like theorem 384 on the previous page lies in the fact that control on the curvature gives control on the metric but not on its derivatives. This is because the Rauch comparison results (propositions 74 on page 259 and 75 on page 259) show only a C^0 behavior. The basic trick is to replace normal (geodesic) coordinates by apparently less geometric ones but in fact better adapted to these questions. Those are the *harmonic coordinates* (see definition 82 on page 267). One chooses d harmonic functions which are linearly independent and satisfy ad hoc boundary conditions. One can do so only within the *harmonic radius* which was first systematically introduced and controlled in Anderson 1990 [39]. Such control is required in many results and we refer for this to the various references which we will meet. With sectional curvature bounds it is easier to control harmonic radius than with Ricci bounds. For the historian we note that harmonic coordinates were used a long time ago by theoretical physicists in Einstein 1916 [484] and in Lanczos 1922 [844]. The founding paper is Jost & Karcher 1982 [769]; also see Hebey & Herzlich 1995 [693]. A recent regularization theorem was proven in Nikolaev [967].

The proof of theorem 383 on the previous page uses coverings with suitable balls whose number is controlled as above in theorem 372 on page 616 and harmonic coordinates on these balls.

If one wants to get smoother limits, one uses smoothing techniques which are important in other instances and were discovered in Bemelmans, Min-Oo & Ruh 1984 [127]. The tool which smooth the metric is the Ricci flow, see §§11.4.3. Thereafter in Abresch 1988 [3] and in Shi 1989 [1130]. See other statements in Fukaya 1990 [531]. The recent Petersen, Wei & Ye 1997 [1025] throws an new light on smoothing, introducing new (optimal) norms to get local geometric control. Smoothness means in particular that the absolute value of the covariant derivatives of the curvature tensor of any order, especially the first one which is technically very useful, can be made as small as required. Smoothing respects most geometric hypotheses and one can for example add smoothness in $\mathcal{RM}(d, a, b, D, v)$, etc.

12.4.2.3.2 Compactness with Ricci curvature bounds

As in instances met above, and in view of theorem 382 on page 626 one is tempted to obtain results of compactness (convergence) with slimmer hypotheses, in particular with only Ricci curvature control. Convergence results where the curvature control is mainly Ricci $\geq -(d-1)$ started in Gao 1990 [546], followed by Anderson 1990 [39]. The main result today is

Theorem 385 (Anderson & Cheeger 1992 [50]) *One has precompactness in the $C^{0,\alpha}$ topology (for any $\alpha \in (0,1)$) under the conditions: Ricci $> r$, injectivity radius $> i > 0$ and volume $< V$.*

The link between finiteness and compactness is explained in Anderson & Cheeger 1992 [50] and is of a quite general nature. If moreover the k-covariant derivatives of Ricci curvature are bounded in absolute value by suitable constants one has precompactness in the $C^{k+1,\alpha}$ topology. See an extremely brief exposition in Hebey & Herzlich 1995 [693]. Note that these extra conditions can be achieved by smoothing as seen above. The proof uses harmonic coordinates (with a lot of analysis, e.g. Sobolev inequalities), and study of the harmonic radius which can be bounded with the metric injectivity radius (this is not too surprising) together with the Ricci bound; see Anderson 1990 [39], and Cheeger, Colding & Tian 1997 [340].

For applications of convergence theorems we refer the reader to the various surveys above. We just note that in most cases, like pinching theorems or "just below" theorems, the convergence theorem yields an unknown bound. One always prefers direct proofs and explicit constants.

Note 12.4.2.1 (Noncompact manifolds) Most of the above results work for (complete) noncompact pointed manifolds. These results are essential for example when one studies the fundamental group and the structure at infinity as seen in various places above. ♦

12.4.3 Collapsing and the Space of Riemannian Metrics

12.4.3.1 Collapsing

If a sequence of Riemannian manifolds does not converge nicely[6] then what is really happening? Do we have some kind of limit space? For example, the limit might be a manifold of smaller dimension, or a reasonable generalization of Riemannian manifolds (see §14.5), or just some metric space. In some sense one is looking for a compactification of the set of Riemannian metrics (of a given dimension) and wondering what goes on when we travel to the boundary. This makes sense only within suitable subsets, namely when we impose some curvature bounds and some metric invariants. When there is no convergence towards a Riemannian manifold (of the same dimension) we say that the manifold is collapsing. This is vague; we will now study things a little more precisely, but we will have to be very brief and incomplete to keep this book of reasonable size. Again the reference we follow here is Fukaya 1990 [531], at least up to 1990. An informative text is Pansu 1985 [994]. The core bible is made up of three parts: Cheeger & Gromov 1985,1986,1990 [346, 347, 348], Fukaya 1987,1989 [528, 530] and by the whole team: Cheeger, Fukaya & Gromov 1992 [342] whose introduction is very helpful. In brief, the general philosophy is that either we have convergence, or collapsing, and that collapsing implies a rich structure. For this philosophy, see Lott 2000 [881].

12.4.3.1.1 Dropping the volume constraint

The first situation we will study is $\mathcal{RM}\,(d,a,b,D,v)$ without the volume condition, working instead in $\mathcal{RM}\,(d,-1,1,D)$ (with obvious notation) after normalizing from (a,b) to $(-1,1)$, which does not hurt when shooting only for general statements. Then we have to study the situation where the volume goes to zero, which by Croke's local embolic theorem 149 on page 355 implies that the injectivity radius goes to zero uniformly. Let us turn to examples. We saw above in the pinching around zero problem that infranilmanifolds are in our class, and not only the obvious tori. Note then that the limit space is the smallest possible one, namely reduced to a point—the big theorem here is the converse of theorem 312 on page 568.

The other basic general example appeared in Gromov 1983 [617]: any manifold admitting a circle action without fixed point, or even just fibered by circles over some other manifold N, collapses to N. This is just an application of Riemannian submersion formulas: take a fixed Riemannian metric invariant under the circle action, and make the fibres smaller and smaller keeping the "horizontal" metric components fixed. Just above we got the infranilmanifolds as a particular case applying this trick to the successive circle fibrations which come from the nilpotent structure. For the Hopf fibrations one sees collapsing of the spheres $S^{2n+1} \to \mathbb{CP}^n$. A fortiori of course this extends to

---

[6] As we have seen this is mainly because the injectivity radius goes to zero.

manifolds admitting any simple torus action. A more sophisticated collapsing of S^3 comes from Clifford tori in S^3 which are parameterized by $[0, 1]$. Each of these tori (except the two circles at the ends) is flat. In each of them consider a geodesic of this flat structure (not a geodesic in the ambient S^3) with an irrational slope, hence everywhere dense. If you shrink the metric more and more along those geodesics, the metric will collapse to the unit interval $[0, 1]$. See figure 4.8 on page 155. Another example of the same vein would be to collapse any symmetric space M onto its quotient by a maximal flat torus.

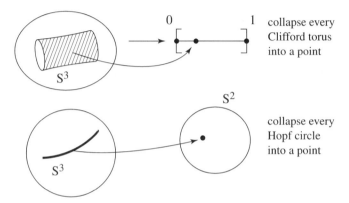

Fig. 12.38. (a) Collapse every Clifford torus to a point (b) Collapse every Hopf circle into a point

Consider a more elaborate example which we met already in figure 11.3 on page 512. Take two surfaces N and N' with boundary (each boundary a union of circles) and look at the products $M = S^1 \times N$ and $M' = S^1 \times N'$. They are manifolds with boundary, and in each the boundary has the topological type of a torus T^2. Then glue M and M' along that T^2 not in the trivial way but interchanging the two circles in T^2 (exchange the parallels and the meridians). The resulting manifold does not in general admit an S^1 action. It does admit one locally on both parts. These two actions agree in the common parts since we have a torus action. It is still possible to define a collapsing structure in $\mathcal{RM}(3, -1, 1, D)$ for this manifold. This example has been put in a very general context in Cheeger & Gromov 1985,1986,1990 [346, 347, 348], where the notions of F and T structures are defined. Other important notions are that of polarized and pure polarized F structures. Detailed definitions are also to be found at the end of the survey Fukaya 1990 [531]. Existence of various structures of those types is linked with the topology of the manifold but how exactly is still an open and important question. See the various references for the state of the art. It is difficult to know when one needs to work within manifolds with bounded diameter.

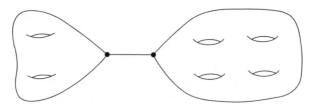

Fig. 12.39. $N \cup [0,1] \cup N'$

For a general F structure one can only build up Riemannian metrics with $-1 \leq K \leq 1$ and injectivity radius going to zero. If the structure is polarized then one can get the volume to go to zero, and if moreover it is pure polarized then one really has collapsing in $\mathcal{RM}(d, -1, 1, D)$. In Cheeger & Gromov 1985,1986,1990 [346, 347, 348] many results are obtained about collapsing and the various Riemannian invariants. The above collapsing theory is used also to study integral formulae for and the existence of characteristic numbers when the manifold is no longer compact but still of finite volume; see Cheeger & Gromov 1985 [346]. In reference to chapter 9, the spectrum during collapsing is considered in Fukaya 1987 [528] and in the recent Kasue & Kumura 1996 [784] where the authors define an interesting notion of spectral distance. The results are good—in a suitable sense one has convergence of eigenvalues and even of eigenfunctions. This is not too surprising if we do not forget that the spectrum is a robust invariant; see why in our discussion of Gromov's *mm* spaces in §14.6.

12.4.3.1.2 A

byproduct of the work of Cheeger & Gromov is a structure statement for any Riemannian manifold:

Theorem 386 (Cheeger & Gromov) *There is a positive number $\varepsilon(d)$ depending only on the dimension d such that given any complete Riemannian manifold M of dimension d with sectional curvature bounded by $-1 \leq K \leq 1$, there is an open set U of M with the following properties: (1) at points in M the injectivity radius is larger than $\varepsilon(d)$ and on U (the thin part) there in an F structure (of positive dimension).*

12.4.3.1.3 Collapsing down to a compact manifold

Consider the structure of the collapsing process when the limit set of $\{M_i\}$ is a compact manifold N. The answer appeared in Fukaya 1987,1989 [528, 530]. We are in $\mathcal{RM}(d, -1, 1, D)$ and assume

$$\lim_{i \to \infty} d_{\mathfrak{G}-\mathfrak{H}}(M_i, N) = 0.$$

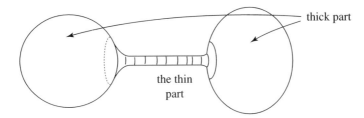

Fig. 12.40. Cheeger & Gromov divide manifolds into "thick" and "thin" parts

For i large enough, there are maps $f_i : M_i \to N$ which are fibrations. The fibers are infranilmanifolds and f_i is almost a Riemannian submersion. In Cheeger, Fukaya & Gromov 1992 [342], the two points of view are united to get very strong statement for collapsing. Finally the bold question of the nature of the $d_{\mathfrak{G}-\mathfrak{H}}$closure of $\mathcal{RM}(d, -1, 1, D)$ was addressed in Fukaya 1988 [529]. When one is interested in various closures on a given manifold things are much easier, see chapter 11 on page 499.

12.4.3.1.4 Alexandrov Spaces

Consider the limiting structure for families of Riemannian metrics involving weaker curvature bounds. The first thing to do is to discard $K \leq 1$ as we did in §§12.4.1 for finiteness results but still keeping $K \geq -1$. It is intuitively clear that limits can now have singularities; think e.g. of small fingers, bubbles, etc. A major discovery is that there is a reasonable generalization of Riemannian geometry which is stable under metric limits subject to the condition $K \geq -1$. This is the notion of an *Alexandrov space*; see §§14.5.5 for more on this. In case of collapsing under only $K \geq -1$ with a smooth limit one has a strong structure result in Yamaguchi 1991 [1282].

12.4.3.1.5 Ricci Curvature Control

As so often before, we would now like control on collapsing with only the Ricci curvature, plus of course various metric invariants. This is an extremely active topic today and hence hard to present concisely. Recall that the initial hope is Gromov's basic view in theorem 382 on page 626. Besides references of the preceding two sections, we mention that Anderson 1992 [42] and Cheeger & Colding 1996 [337] are the beginning of a series of articles taking into account previous results. Recall that Colding's $L^{1,1}$ and L^2 Toponogov theorems (theorems 76 on page 262 and 77 on page 264) play the role of Toponogov's one when one has only a lower Ricci bound. For some insight we mention that those results are part of a programme of Anderson and Cheeger, including the conjecture:

Conjecture 387 *Convergence in $d_{\mathfrak{G}-\mathfrak{H}}$ in the presence of a lower Ricci curvature bound implies volume convergence.*

634   12 From Curvature to Topology

This is now proven in Colding 1997 [387]. For the program of "Ricci synthetic" Riemannian geometry, see appendix 2 of Cheeger & Colding 1997 [337]. 1997a).

12.4.3.1.6 Integral curvature bounds

Next in weakening the assumptions is to use only integral bounds, sometimes mixed with various other ones. We just mention (some of these we have already met): Gallot 1988 [540], Yang 1992 [1286], Gao 1990 [546, 547], Petersen 1997 [1018] and the expository Anderson 1990 [44]. For the behaviour of the spectrum during collapse, see Lott 2000 [881].

12.4.3.2 Closures on a Compact Manifold

On a given compact manifold M, consider the set $\mathcal{RM}(M)$ of all Riemannian metrics. On it we have many topologies—the one already given by the Gromov–Hausdorff metric of theorem 380 on page 626, but also the C^k topologies on $\mathcal{RM}(M)$ coming from the various differentiable structures on M. The C^0 case corresponds for example to looking only at the metrics g, not at any of their derivatives. For example controlling (or defining) the curvature demands working in C^2 at least. Some surveys are Lohkamp 1992 [876] and Lohkamp 1996 [874]. And we now briefly state the results. Some of them are only rephrasing things already seen more or less explicitly. Also see Nikolaev 1991 [967].

One should first remark on the trivial implications from $d_{\mathfrak{G}-\mathfrak{H}}$ closure to the C^k closures. This explains why we will state only the strongest ones. The notations will be obvious. The symbol K will stand for the sectional curvature and the associated inequalities, etc. For each space X with topology T, the notation $\overline{X}^T$ will mean the closure of X in the T topology. Our inequalities start with the sectional curvature:

Theorem 388 *For any compact manifold and any real number* k

$$\overline{K^{\geq k}(M)}^{d_{\mathfrak{G}-\mathfrak{H}}} = K^{\geq k}(M)$$

and

$$\overline{K^{\leq k}(M)}^{d_{\mathfrak{G}-\mathfrak{H}}} = K^{\leq k}(M)$$

These observations come essentially from the Rauch comparison theorem (proposition 74 on page 259). If one passes now to Ricci curvature

Theorem 389

$$\overline{\mathrm{Ricci}^{\geq \alpha}(M)}^{C^0} = \mathrm{Ricci}^{\geq \alpha}(M)$$

and (for the moment the best we know):

$$\overline{Ricci^{\geq\alpha}(M)}^{d_{\mathfrak{G}-\mathfrak{H}}} \neq \mathcal{RM}(M).$$

It is unknown if

$$\overline{Ricci^{\geq\alpha}(M)}^{d_{\mathfrak{G}-\mathfrak{H}}} = Ricci^{\geq\alpha}(M).$$

Lohkamp's result of §§12.3.5 says exactly that

Theorem 390

$$\overline{Ricci^{\leq\alpha}(M)}^{C^0} = \overline{scalar^{\leq\alpha}(M)} = \mathcal{RM}(M)$$

and

$$\overline{scalar^{\leq\alpha}(M)}^{d_{\mathfrak{G}-\mathfrak{H}}} = \overline{Ricci^{\leq\alpha}(M)}^{d_{\mathfrak{G}-\mathfrak{H}}} = \mathcal{RM}(M).$$

As written in Lohkamp's survey, we wrote the four equalities for dramatic effect, although two are consequences of the others.

The complete mystery alluded to in §§12.3.3 is

$$\overline{scalar^{\geq\alpha}(M)}^{C^0}$$

particularly when $\alpha = 0$. Kontsevich is currently developing a picture of the closure of the set of nonnegative Ricci curvature metrics, with applications in mind to theoretical physics.

13 Holonomy Groups and Kähler Manifolds

Contents

13.1	Definitions and Philosophy	637
13.2	Examples	639
13.3	General Structure Theorems	641
13.4	Classification	643
13.5	The Rare Cases	646
	13.5.1 G_2 and Spin(7)	646
	13.5.2 Quaternionic Kähler Manifolds	647
	13.5.3 Ricci Flat Kähler and Hyper-Kähler Manifolds	652
13.6	Kähler Manifolds	654
	13.6.1 Symplectic Structures on Kähler Manifolds	655
	13.6.2 Imitating Complex Algebraic Geometry on Kähler Manifolds	655

In this chapter, up to and including §13.4, **manifolds need not be compact, or even complete, but must have no boundary.** Starting in §13.5, manifolds are once again assumed compact without boundary, unless otherwise stated.

13.1 Definitions and Philosophy

Knowing the importance of groups in mathematics, it is quite natural to try to capture some part of Riemannian geometry in a group. The notion of parallel transport is the key; see proposition 61 on page 240 and §15.4. Given two points p and q in a Riemannian manifold M and a curve c from p to q, the *parallel transport* from p to q along c is an isometric linear isomorphism

$$c_{p \to q} : T_p M \to T_p M$$

between these Euclidean spaces. Let us now consider loops c based at p, and look at

$$c_* = c_{p \to p} : T_p M \to T_p M.$$

This isomorphism is not the identity in general; it is only an element of the orthogonal group $O(T_p M)$ (this group is by definition the set of all Euclidean isomorphisms of T_pM). Such a parallel transport was used in Synge's theorem 64 on page 246. Parallel transport obviously transforms a composition of paths into a product of Euclidean isomorphisms, and therefore:

Definition 391 *As the curve c varies through all possible loops at p, the set of associated c_* in $O(T_pM)$ is a subgroup, denoted by* Hol(p) *and called the* holonomy group *of M at p. If one considers only loops which are homotopic to a point (contractible) one gets a subgroup of* Hol(p) *denoted by* $Hol_0(p)$, *and called the* restricted holonomy group *of M at p.*

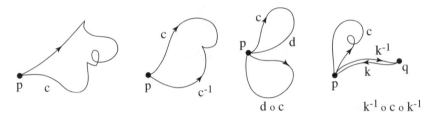

Fig. 13.1. (a) A loop (b) Traveling backwards along a loop c gives the loop c^{-1} (c) Composition of loops (d) Holonomy groups at different points are conjugates

But one is essentially interested in the group structure, and this one does not depend on the point: just connect p and q by any path k, and then the map $c \mapsto k^{-1}ck$ will yield subgroups of the universal orthogonal group $O(d)$ conjugate under an inner automorphism. That is why from now on will speak of *the* holonomy group of a Riemannian manifold M (restricted or not) and write $Hol(M), Hol_0(M)$.

We naturally hope that $Hol(M)$ will reflect the Riemannian structure quite strongly and yield some classification of Riemannian manifolds. Such a classification, at least in a special case, was Élie Cartan's hope in Cartan 1923,1924 [320] where he introduced the notion of holonomy group into the theory of general relativity; then he did calculations up to dimension 3 in Cartan 1926 [313]. Holonomy groups played a crucial role in Cartan's subsequent discovery and classification of symmetric spaces, since he used their holonomy groups to classify them in 1925; see the details of this story in 10.69 and 10.70 of Besse 1987 [183]. One may wonder why such a simple notion was not introduced before. Since Cartan's pioneering work, holonomy groups underwent a scattered history until very recently—in particular they were completely forgotten (or perhaps found too difficult to study) between Cartan's papers and the 1950's. They came back to the fore briefly in Borel & Lichnerowicz 1952 [224]. There are essentially only three surveys of the subject: chapter 10 of Besse 1987 [183], Salamon 1989 [1087], and the remarkable Joyce 2000 [774] (which is devoted to the study of holonomy groups on compact Riemannian manifolds). We mention also Wakakuwa 1971 [1226], but it is hard to find it in libraries outside Japan. The study of holonomy has seen a recent revival in the hands of Bryant, Joyce, Merkulov, Salamon and Schwachhöfer, and exciting applications in string theory.

The philosophy is: a manifold carries a structure invariant under parallel transport if and only if this structure is invariant at a single point under the holonomy group. Another fact is that we will see the curvature coming into holonomy because, following the golden triangle of sections §15.2,§15.4, and §15.5, the curvature is exactly the parallel transport along an infinitesimal parallelogram. After the structure results of §13.3, the main theorem is the classification theorem 397 on page 643. It is very surprising that there are very few possibilities for the holonomy group, in fact only two if one excepts very special manifolds. The corollary is that, with still very exotic structures, only one category of manifolds has a nontrivial structure invariant by parallel transport, namely *Kähler manifolds* which are, by definition, the complex Riemannian manifolds whose complex structure is invariant by parallel transport. We will now briefly describe the main results, after a short acquaintance with some examples.

Recently holonomy groups became important in mathematical physics: see Fröhlich, Granjean & Recknagel 1998 [523, 524] for a systematic presentation of this hierarchy in terms of theoretical physics, and Andersen, Dupont, Pedersen & Swann 1997 [36]. The holonomy group classification is also useful when studying nonpositively curved manifolds; and, on the nonnegative side, for the characterization of symmetric spaces among compact Kähler manifolds by their bisectional curvature see Mok 1988 [933] and theorems 339 on page 589, 402 on page 656 and 403 on page 656.

13.2 Examples

The most natural question is: when is $\mathrm{Hol}(M)$ trivial, i.e. $\mathrm{Hol}(M) = \{1\}$. The answer is easy: $\mathrm{Hol}_0(M) = \{1\}$ if and only if the manifold is flat (see §§4.4.1 and theorem 69 on page 252), proof is left to the reader. This is equivalent to asking that the curvature vanish identically, or that the manifold be locally Euclidean.

Recall that the special orthogonal group $SO(d)$ is the normal subgroup of $O(d)$ consisting of elements of determinant equal to 1. Geometrically, this means the subgroup of elements preserving orientation.

Of course $SO(d)$ is the connected component of the identity of $O(d)$. Therefore one always finds $\mathrm{Hol}_0(M) \subset SO(d)$. It is easy to see that $\mathrm{Hol}_0(M)$ is always a normal subgroup of $\mathrm{Hol}(M)$. Therefore we have a surjective homomorphism

$$\pi_1(M) \to \mathrm{Hol}(M)/\mathrm{Hol}_0(M)$$

from the fundamental group $\pi_1(M)$. Because $\pi_1(M)$ is countable, the image of this homomorphism is also countable. In conclusion, the condition $\mathrm{Hol}(M) \subset SO(d)$ is equivalent to the orientability of the manifold M.

The next example is that of Riemannian products (see §§4.4.3) $M^d \times N^e$. One sees easily that

$$\mathrm{Hol}(M \times N) = \mathrm{Hol}(M) \times \mathrm{Hol}(N),$$

this product group structure being precisely realized by the representation in $O(d+e)$ acting on $\mathbb{R}^{d+e}$ which should be written as

$$\{\mathrm{Hol}(M) \times \mathrm{Id}\,(\mathbb{R}^e)\} \otimes \{\mathrm{Id}\,(\mathbb{R}^d) \times \mathrm{Hol}(N)\},$$

i.e. Hol(M) acts on tangent spaces to N by the identity, and Hol(N) on tangent spaces to M by the identity. Such a reduction of a representation is the strongest possible reducibility, called *complete reducibility*. This comes from the fact that the parallel transport along a path in the product is the direct sum of the parallel transports along the two projections of this path, one in M and the other in N. We will see in §13.3 an amazingly strong converse of this.

By definition, a *Kähler manifold* is one with a complex structure (this means in particular that the coordinates changes are holomorphic for the complex coordinates) together with a Riemannian metric which has with this complex structure the best possible link, namely that multiplication of tangent vectors by unit complex numbers preserves the metric, but moreover the complex structure is invariant under parallel transport. This is equivalent to the condition that the holonomy group be included in the unitary group, hence equivalent also to ask for the existence of a 2-form of maximal rank and of zero covariant derivative. An equivalent definition is this (but the proof is a little tricky): we have a complex structure J with a Riemannian metric g, and the two are compatible:

$$g(Jx, Jy) = g(x, y)$$

for every pair of tangent vectors x, y. Then one gets from the pair g, J an exterior 2-form ω, called the *Kähler form*, defined as

$$\omega(x, y) = g(x, Jy).$$

The manifold is called *Kähler* if and only if ω is closed:

$$d\omega = 0.$$

In Lichnerowicz 1955 [868] pages 258–261 it is proven that the Kähler condition is equivalent to the holonomy group being a subgroup of the unitary group $U(n)$; recall that the notation $U(n) \subset O(2n)$ denotes the Euclidean isomorphisms which preserve the complex structure J. Moreover the condition

$$\mathrm{Hol}(M) \subset SU(n),$$

(where $SU(n)$, the *special unitary group*, denotes the elements in $U(n)$ whose complex determinant is 1), is equivalent to saying that our Kähler manifold is moreover Ricci flat. This is because the Lie algebra of $\mathrm{Hol}_0(p)$ is generated by the curvature endomorphisms, thanks to theorem 396 on page 643. But in the presence of a Kähler metric, taking the derivative of the condition to have complex determinant equal to 1 forces the complex trace to vanish on these endomorphisms, i.e. $R(x, Jy) = 0$. Now use the ideas of §§§13.6.1 to check that the vanishing of those traces is equivalent to the vanishing of the Ricci curvature.

The last example is that of symmetric spaces (see §§4.3.5). For an irreducible, simply connected symmetric space G/H the holonomy group $\mathrm{Hol}(G/H)$ coincides with H; more precisely the adjoint representation of H, i.e. it is the H action associated to the decomposition $\mathfrak{g} = \mathfrak{h} \oplus \mathfrak{m}$ and the relations
$$[\mathfrak{h}, \mathfrak{m}] \subset \mathfrak{m} \quad \text{and} \quad [\mathfrak{m}, \mathfrak{m}] \subset \mathfrak{h}.$$
This is a direct consequence of the formulas in §§15.8.1.

13.3 General Structure Theorems

We will now encounter the surprising fact that holonomy groups are few, and thereby in some sense not too good for classifying Riemannian manifolds. The first astonishing fact is:

Theorem 392 (Borel & Lichnerowicz 1952 [224]) *The restricted holonomy group $\mathrm{Hol}_0(M)$ of any Riemannian manifold (not necessarily complete) is a closed connected subgroup of the orthogonal group, and in particular is compact.*

The proof is very hard; it uses the generation of the holonomy group by *lassos*, see theorem 396 on page 643 and subtle characterizing properties of Lie groups

Next we look at the reducibility of $\mathrm{Hol}(M)$ as it acts on $\mathbb{R}^d$. That is to say, we assume that $\mathrm{Hol}(M)$ leaves invariant some vector subspace $V \subset \mathbb{R}^d$; then it also leaves invariants its Euclidean orthogonal complement W.

Theorem 393 (de Rham) *If $\mathrm{Hol}(M)$ is reducible, then the universal cover of M is a Riemannian product.*

Consider the example of a flat torus; most flat tori are not products, but are locally products.

The proof is not too hard. The invariant subspaces V and W will yield two complementary fields of planes on M, via parallel transport. They turn out to be integrable (in the sense of Frobenius) and finally yield the desired Riemannian product. This is very strong. It yields only completely decomposed representations; for example the obvious double diagonal action of $O(d)$ in $O(2d)$ never appears.

If one now applies a trivial induction, recalling that flatness is equivalent to trivial holonomy, one gets

Theorem 394 *Take any Riemannian manifold M (not necessarily complete) and a point $p \in M$. The tangent space T_pM splits canonically into the orthogonal direct sum*

$$T_pM = T_0 \oplus T_1 \oplus \cdots \oplus T_k$$

and the metric g into a local Riemannian product

$$g = g_0 \times g_1 \times \cdots \times g_k$$

with g_0 is flat and

$$\mathrm{Hol}(p, M) = A_1 \times \cdots \times A_k$$

where every A_i acts irreducibly on T_i and trivially on all the other factors. The universal cover of M splits into a product of k factors.

It might be the place to recall the de Rham theorem 56 on page 229: a locally reducible Riemannian manifold is a global Riemannian product as soon as the manifold is simply connected and complete.

How should one compute the holonomy "theoretically?" We will now introduce the curvature tensor R written here as an antisymmetric 2-form with values in the endomorphism of tangent spaces, the way it appears in the language of the note 15.4.0.1 on page 701. We apply this formula as in the lasso in figure 13.2.

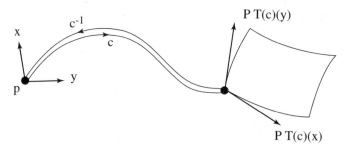

Fig. 13.2. Parallel transport to a point, and there apply the Riemann curvature tensor

We fix a point p and examine the holonomy group $\mathrm{Hol}(p)$ and its Lie algebra $\mathfrak{hol}(p)$, which is the infinitesimal version of $\mathrm{Hol}(p)$, i.e. the tangent space to $\mathrm{Hol}(p)$ at the identity element. We consider any other point q in M and any path c from p to q. Finally, pick a pair of vectors $x, y \in T_pM$. Then we plug in the parallel transport $c_{p \to q}$ from p to q and the formula 15.3 on page 703 implies that

Lemma 395 *The endomorphism*

$$c^{-1}{}_* R(c_* x, c_* y) c_* : T_p M \to T_p M$$

belongs to $\mathfrak{hol}(p)$, for any curve c (not necessarily a loop).

The converse is true, and is only a clever implementation of the Frobenius theorem in the space of orthogonal frames:

Theorem 396 (Ambrose & Singer 1953 [29]) *The holonomy Lie algebra $\mathfrak{hol}(p)$ is exactly the Lie subalgebra of $\mathfrak{so}(d)$ (the Lie algebra of $O(d)$) generated by the elements*

$$c^{-1}{}_* R(c_* x, c_* y) c_* : T_p M \to T_p M$$

where c varies through all curves which start at p, and x and y are any tangent vectors at p.

Despite its esthetic appeal, this is a very funny result. It is clear that one does not need so many elements to compute Hol(p). Moreover in some sense we used all of the holonomy to compute the holonomy. For example look at a generic Riemannian manifold. Take some orthogonal basis $\{v_i\}$ of $T_p M$ and consider the set of endomorphisms $R(v_i, v_j) \subset \mathfrak{so}(d)$. Since there are $d(d-1)/2$ such elements, and since $d(d-1)/2$ is the dimension of $\mathfrak{so}(d)$, the genericity implies equality. This proves two things: first, for a generic metric, $\mathrm{Hol}_0(M) = SO(d)$, the total special orthogonal group. Second, we used only one point, that is to say the "infinitesimal" holonomy group is already the whole group. However theorem 396 is useful in some instances.

For an integral formula using a surface filling a loop, see Nijenhuis 1953 [965] page 54, and the comments on page 704.

13.4 Classification

By the above structure theorem 394 on the preceding page we have only to consider the irreducible case, but we are still on not necessary complete manifolds. Then, with the obvious divisibility dimension conditions, one has only eight possibilities:

Theorem 397 *If Hol_0 is irreducible, then one of the following holds*

- $\mathrm{Hol}_0 = SO(d)$
- *The manifold is locally symmetric.*
- *The manifold is locally Kähler and $\mathrm{Hol}_0 = U(d/2)$*
- *The manifold is locally Kähler, is Ricci flat and $\mathrm{Hol}_0 = SU(d/2)$ and $d \geq 2$*
- $\mathrm{Hol}_0 = Sp(1) Sp(d/4)$ *and* $d \geq 2$
- $\mathrm{Hol}_0 = Sp(d/4)$ *and* $d \geq 2$

- $d = 7$ and $\mathrm{Hol}_0 = G_2$ (an exceptional Lie group of dimension 14 described on page 152)
- $d = 8$ and $\mathrm{Hol}_0 = \mathrm{Spin}(7)$.

It is essential to remark that in this list it is not only the group structure which is given but its precise orthogonal representation. The last six cases being quite special, this explains why we said previously that there are essentially two holonomy groups: there are basically only two cases, the general case where the holonomy is the full (special) orthogonal group and the case of Kähler manifolds. It is important to note that the use of the Bianchi identity for the curvature tensor below is absolutely essential. This is seen in the fact that any compact subgroup of the linear group can be realized as the holonomy group of some Riemannian manifold but with a connection with torsion, in particular not necessarily the Levi–Civita connection of §15.3. This is a local result; for a global one it is just enough to add the topological restriction that the tangent bundle admits the corresponding linear group structure. See the references in one of the books quoted above.

A dramatic consequence: recall first that symmetric spaces are completely know and form quite a short list. Then theorem 397 on the previous page proves that in most Riemannian manifolds there is no nontrivial "object," say for example tensors or spinors of various kinds, invariant under parallel transport. The other way around: the existence of a tensor or spinor, which is not the metric itself or the orientation, and is invariant under parallel transport, implies that the metric is "known," with the exception of the Kähler case.[1]

Where does theorem 397 on the preceding page come from? The result appeared in Berger 1953 [148], but it was really conceptually proven in Simons 1962 [1139]. The key is to prove:

Lemma 398 *If the holonomy group is irreducible but not transitive on the tangent sphere, then the manifold is a a locally symmetric space.*

We look at the orbit $[v]$ of some unit tangent vector $v \in U_p M$, and pick up some vector z orthogonal to the tangent space $T_v[v]$, such a nonzero z exists since the orbit is not full. Note that equation 15.3 on page 703 implies that

$$R(x,y)v \in T_v[v]$$

for all $x, y \in T_p M$, so that $g(R(x,y)v, z) = 0$ for every x, y. The symmetries of R (see equation 4.28 on page 203) enable us to change this into $R(v, z) = 0$. This means, using lemma 395 on page 642, that we have a lot of pairs of vectors (at every point of the manifold) for which $R(v, z) = 0$. This is a situation we know from equation 4.38 on page 209 for products and for symmetric spaces of rank larger than 1 from §§4.3.5. The idea is to organize

---

[1] We say "known" in quotes because in fact the exceptional holonomy groups are excruciatingly difficult to work with, and geometers are nowhere near a classification.

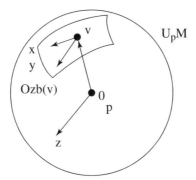

Fig. 13.3. Action of the holonomy on the unit sphere

those v, z planes into totally geodesic submanifolds (see §§6.1.4) by mimicking the geometry of symmetric spaces of rank larger than 1. To go further, we use a saturation, thanks to lemma 395, of all of the curvature tensor in order to get an algebraic object of curvature tensor type which is invariant by parallel transport. But if the curvature tensor is invariant by parallel transport, this one of the characterizing properties of symmetric spaces in theorem 36 on page 190. More details can be found in the two references given above.

To finish with the proof of the list in theorem 397 on page 643 one appeals to the classification of Lie groups acting transitively on spheres. This is by the way a purely topological story—the group action is not asked to be linear. The results are in Montgomery & Samelson 1943 [935], Borel 1949,1950 [221, 222]. The list of possible groups (and respective dimensions of the spheres acted upon) are those of theorem 397 on page 643 plus only two other ones: $S^1 Sp\,(d/4)$ and Spin (9) when in dimension 16. The Bianchi identity eliminates the circle S^1, and the curvature tensor for Spin (9) enjoys so many identities, using equation 4.27 on page 203, that it is algebraically determined at every point and $DR = 0$ using the second Bianchi identity from equation 15.5 on page 705, so that finally see that we are in a symmetric space. This was carried out in Alekseevskii 1968 [17]; also see Brown & Gray 1972 [265] for details.

Direct application of the identity

$$R(x,y)z + R(y,z)x + R(z,x)y = 0$$

(the first Bianchi identity) plugged into the relations defining the respective Lie algebras yields:

Lemma 399 *Manifolds with holonomy $SU\,(d/2)$, $Sp\,(d/4)$, Spin (7) and G_2 are Ricci flat.*

It remains a great mystery that no Ricci flat compact manifolds are known which do not have one of these special holonomy groups.

Symmetric spaces being completely known, and Kähler manifolds (holonomy $U(d/2)$) being treated separately below in §13.6, we have now only to tell the reader what is known about the rare cases.

The picture in figure 13.4 is a good summary; it is taken from Salamon 1989 [1087].

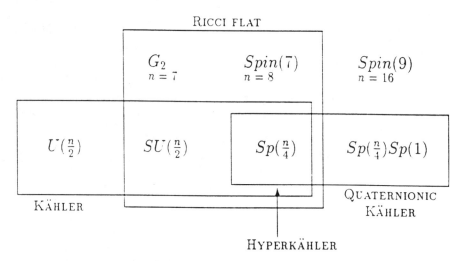

Fig. 13.4. Salamon's illustration of the holonomy groups of Riemannian manifolds

13.5 The Rare Cases

13.5.1 G_2 and Spin(7)

To exhibit a manifold having holonomy group $G_2 \subset SO(7)$ or $\mathrm{Spin}(7) \subset SO(8)$ turned out to be very difficult. Note first that no symmetric space has such a holonomy. It is only in Bryant 1987 [267] that the first examples appeared, but only local. Complete ones appeared in Bryant & Salamon 1989 [271]. Compact ones were proven to exist in Joyce 1996,1997 [772, 771], which is explained in a more leisurely account in Joyce 2000 [774]. Joyce's construction is extremely expensive but a brilliant *tour-de-force* of geometric analysis. It uses deformation of metrics, singular manifolds and the Atiyah–Singer index theorem. For example, for G_2, start with a manifold with a reasonable G_2-structure on its frame bundle. Such a structure has a torsion which measures the defect for the structure under parallel transport. Joyce first constructs manifolds with a G_2-structure whose torsion is small enough. Then Joyce uses a one-parameter deformation technique to get a torsion-free G_2-structure. The same technique applies to the Spin(7) case. Moreover Joyce's technique and results enable him to compute the dimension of the moduli space in every case, and the cohomology of the resulting manifolds. This dimension is always finite by lemma 399 on page 645 and theorem 286 on page 531.

By our philosophy it natural to characterize such holonomies by the objects invariant under parallel transport. This turns out to have an elegant solution: G_2 holonomy is characterized by the existence of a parallel exterior differential form of degree 3 (satisfying a certain nondegeneracy hypothesis).[2] For $\mathrm{Spin}(7)$, the same thing but with a closed parallel 4-form.

There is also a strong relation between these special holonomy groups and parallel spinors: see §§14.2.2. This is due to the fact that both G_2 and $\mathrm{Spin}(7)$ are simply connected (unlike $SO(d)$) and hence induce on a manifold with such a holonomy group a canonical spin structure. Using parallel objects one can define the G_2 (resp. $\mathrm{Spin}(7)$) holonomy manifolds very simply by just asking that the manifold have a parallel spinor; see Moroianu & Semmelmann 2000 [941].

It is not clear if manifolds with these holonomy groups are numerous or scarce. We only recall that on a given compact manifold, the space of moduli for such structures is a finite-dimensional stratified real analytic set: see theorem 286 on page 531.

13.5.2 Quaternionic Kähler Manifolds

Good references for this holonomy group are chapter 14 of Besse's book, chapter 9 of Salamon's and the proceedings Gentili, Marchiafava & Pontecorvo 1998 [554]. Some more recent information will also be added and referenced here. It is tempting to call a Riemannian manifold M^{4n} *quaternionic Kähler* if its holonomy group is $Sp(1)Sp(n) \subset SO(4n)$. But as far as its geometric structure is concerned, the real definition of *quaternionic Kähler* is holonomy contained in $Sp(1)Sp(n)$. This is important because unlike the holonomy groups studied in §§13.5.1, there are irreducible symmetric spaces which offer this inclusion. The complete list is easy to give (and appeared first in Wolf 1965 [1275]). The compact forms are

$$\mathbb{HP}^n, SU(n+2)/U(n), SO(n+4)/SO(n) \times SU(2)$$

and five low dimension exceptions. Note that such manifolds are never Kähler if this inclusion is strict, nor quaternionic in any reasonable sense, but it seems hard to abandon a terminology now standard. There are two symmetric spaces having such a holonomy group, namely $\mathbb{HP}^n$ and $\mathrm{Hyp}_{\mathbb{H}}^n$. They are of rank one. Are there any nonsymmetric examples? We have today the fact:

Fact 400 *There is not a single example known today of a quaternionic Kähler manifold besides the two symmetric spaces $\mathbb{HP}^n$ and $\mathrm{Hyp}_{\mathbb{H}}^n$.*

---

[2] Following our definition of the Cayley numbers on page 154, and the definition of G_2 as the symmetry group of the Cayley numbers, the reader can easily see the invariant 3-form.

Let us see now what we know on the subject. We note first that Bianchi identity, plugged into the Lie algebra relation, implies immediately that these manifolds are Einstein; i.e. the Ricci curvature is proportional to the metric, so that again the moduli space's general structure will follow from Koiso's theorem 286 on page 531. In this chapter sometimes there is the need to take some quotient by $\mathbb{Z}_2 = \mathbb{Z}/2\mathbb{Z}$. We will suppress this complication to simplify visualization. This is seen clearly in the holonomy group $Sp(1)Sp(1)$ which is not $O(4)$, but only

$$SO(4) = (Sp(1)Sp(1))/\mathbb{Z}_2.$$

This is the reason for starting the study of quaternionic Kähler manifolds in dimension 8.

To describe a quaternionic Kähler structure, as opposed to the hyperkähler structure which we will see in §§13.5.3, we start with three endomorphisms I, J, K (think of the quaternions) of TM which preserve the metric g and satisfy the quaternion multiplication table:

$$I^2 = J^2 = K^2 = -1$$
$$IJ = K, JK = I, KI = J$$

Hyperkähler manifolds below will satisfy the same requirements, but have I, J, K invariant under parallel transport. Here we asked only that they are preserved as a whole triple, namely that there are three 1-forms α, β, γ such that

$$\nabla_X I = \gamma(X)J - \beta(X)K$$
$$\nabla_X J = -\gamma(X)I + \alpha(X)K$$
$$\nabla_X K = \beta(X)I - \alpha(X)J.$$

Beware that these I, J, K exist only locally. What exists globally is the two-parameter family $uI + vJ + wK$ with $u^2 + v^2 + w^2 = 1$, and by the above formula this total global set is invariant by parallel transport. For the characterization by invariant objects, there is a 4-form constructed as follows: from I, J, K the metric g "à la Kähler" defines three 2-forms ϕ, ψ, η and the desired invariant 4-form is

$$\omega = \phi \wedge \phi + \psi \wedge \psi + \eta \wedge \eta.$$

It is the existence of this parallel 4-form which explains the wording *quaternionic Kähler*. Moreover the existence of a nontrivial parallel exterior 4-form implies that we are in a quaternionic Kähler structure and with such a structure for this 4-form.

The classification of quaternionic Kähler manifolds is a hard subject. At first glance they look pretty rigid. Quoting Claude LeBrun: "they resemble a symmetric space to an uncomfortable degree." We list now some questions and the results obtained up to now, with an idea of the proofs. The quaternionic Kähler manifolds fall into two different classes: since they are Einstein, the scalar curvature can be positive or negative. If it were zero then the holonomy would be contained in $Sp(n)$. Briefly we will speak of positive (respectively negative) quaternionic Kähler manifolds. By Myers' theorem 62 on page 243 the complete positively curved ones are necessarily compact. We first study the complete negatively curved noncompact case. First there exist nonsymmetric homogeneous negatively curved quaternionic Kähler manifolds, and they were classified by Alekseevskii (see references in Besse's book). It is not known if they can admit compact quotients. Worse: in view of LeBrun 1991 [854] there is an infinite-dimensional moduli space of complete metrics on $\mathbb{R}^{4n}$ with holonomy group equal to $Sp(1)Sp(n)$. They are obtained by deforming the hyperbolic quaternionic space $\mathrm{Hyp}_{\mathbb{H}}^n$. The classification of negatively curved compact case is simple: one knows the arithmetic compact symmetric space forms of $\mathrm{Hyp}_{\mathbb{H}}^n$ found by Borel and described in §§6.6.3, and we saw there that there are no other ones thanks to Gromov and Schoen. There is no guess for the final answer, and the reason for this will be seen in the proof sketched below.

If you turn now to the positive case, one does not know of any nonsymmetric examples, but at least there are partial results, pointing in that direction. First in dimension 8 only $\mathbb{HP}^2$ is quaternionic Kähler by Poon & Salamon 1991 [1038]. And by LeBrun & Salamon 1994 [857] one knows that in a given dimension $4n$ there can exist at most a finite number (up to isometry) of quaternionic Kähler manifolds; also see Herrera & Herrera 2001 [708].

13.5.2.1 The Bérard Bergery/Salamon Twistor Space of Quaternionic Kähler Manifolds

A basic tool was invented independently in 1979-80 by Bérard Bergery and Salamon (see the references in LeBrun & Salamon 1994 [857]) to study quaternionic Kähler manifolds. This is a typical example of the twistor idea (which is only mentioned in §§14.2.3.3), so we will explain it in detail. One constructs a $(4n+2)$-dimensional space $\mathcal{Z}(M)$ fibered over M with fiber S^2. For the standard $\mathbb{HP}^n$ this $\mathcal{Z}(M)$ is nothing but the canonical generalized Hopf fibration

$$\mathbb{CP}^{2n+1} \to \mathbb{HP}^n.$$

The idea is this: remember that the "false Kähler" I, J, K are not well defined, but the complete set of possible choices of them is, i.e. the $uI + vJ + wK$ (with $u^2 + v^2 + w^2 = 1$). We attach to every point $m \in M$ the sphere S^2 of the above (u, v, w). On this fiber space $\mathcal{Z}(M) \to M$ there is a complex structure defined by the picture above. The desired rotation of angle $\pi/2$ on the total tangent space to $\mathcal{Z}(M)$ at the point $\{m; u, v, w\}$ is obtained by combining the canonical rotation of $\pi/2$ on the tangent space to S^2 (orient the spheres, and this takes care of the vertical part) with the action which on

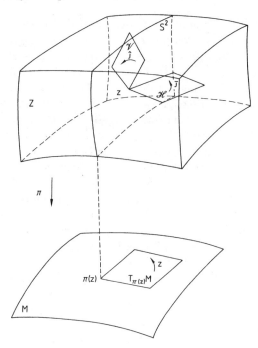

Fig. 13.5. The twistor construction of Bérard Bergery and Salamon

the horizontal part takes the lift of the $\pi/2$ rotation precisely associated to $uI+vJ+wK$. One proves then that this almost complex structure is integrable and so defines a complex structure. The subtle point is to define a nice metric structure on $\mathcal{Z}(M)$ for which the above complex structure will make it a Kähler manifold. This can be done only in the positive case: heuristically because the computations shows that to define a nice metric by lifting one needs positive Ricci curvature downstairs. The conclusion is:

Lemma 401 *When M is a positive scalar curvature quaternionic Kähler manifold, $\mathcal{Z}(M)$ inherits canonically a Kähler–Einstein metric.*

As one might guess, this is a Riemannian submersion with totally geodesic fibers. All the details are in Besse's book.

The rich quaternionic geometry moreover is encoded in $\mathcal{Z}(M)$ and one knows many things more about $\mathcal{Z}(M)$. We have at our disposal the arsenal of Kähler geometry. This is how the above finiteness result is proven. The negative scalar curvature case is untouched by this technique. One can still define a metric of Kähler type for the above complex structure on $\mathcal{Z}(M)$ (which does not need positivity) but the signature of the metric will have to be $(4n, 2)$. Then the geometry of such Kähler objects could be of help, but it seems that there has not been much study of them.

13.5.2.2 The Konishi Twistor Space of a Quaternionic Kähler Manifold

To a quaternionic manifold one can attach canonically another compact twistor type bundle; this was done a long time ago in Konishi 1974/1975 [829]. This time the fiber is three dimensional; it is the group $SO(3)$. The construction is analogous to the above, but instead of the sphere made up of the unit linear combinations of $\{I, J, K\}$, one considers the group $SO(3)$ acting on them. This time the twistor space $\mathcal{S}(M) \to M$ is of dimension $4n + 3$. If M is a positive scalar curvature quaternionic Kähler manifold, one can still define on $\mathcal{S}(M)$ a canonical Riemannian metric making $\mathcal{S}(M) \to M$ a Riemannian submersion, with totally geodesic fibers. Moreover $SO(3)$ acts not only on the fibers but by isometries. This time the special structure on $\mathcal{S}(M)$ inherited from M is called a *3-Sasakian structure*, and $SO(3)$ acts on it by global isometries. We will not define explicitly Sasakian structures; they are some kind of odd-dimensional analogue of Kähler manifolds. The standard reference on them is Yano & Kon 1984 [1291], but also see Boyer, Galicki & Mann 1994 [253]. An example: the unit tangent bundle to the tangent bundle to any Riemannian manifold, or the unit tangent bundle itself when the manifold downstairs is Kähler. The 3-Sasakian manifolds are those on which one has more than one Sasakian structure. Here these "three" isometries replace, in a somewhat weaker sense, the three Kähler structures of the hyperkähler manifolds of the next section. In fact an indirect definition of a 3-Sasakian manifold is to say that the warped product

$$g_{\mathcal{U}} = dr^2 + r^2 g_{\mathcal{S}}$$

on $\mathbb{R}^+ \times \mathcal{S}$ is hyperkähler. These 3-Sasakian manifolds are Einstein and are used these days to construct new examples of Einstein manifolds which are "strongly inhomogeneous," see Boyer, Galicki & Mann 1996 [254]. As for the twistor space $\mathcal{Z}(M)$ above, in the negative scalar curvature case one finds on $\mathcal{S}(M)$ only a Konishi structure of signature $(3, 4n)$.

13.5.2.3 Other Twistor Spaces

Swann 1991 [1171] constructed a third twistor space $\mathcal{U}(M) \to M$ attached to quaternionic Kähler manifolds, with fiber $\mathbb{H}^*$. The fiber is not mysterious—it is basically defined by the quaternionic frames of the basis, quotiented by $Sp(1)Sp(n)$. One has the diagram given in figure 13.6 on the next page.

In the diagram, the upper left arrow is an inclusion as a level set, for any given r as above. The hyperkähler manifold $\mathcal{U}(M)$ is not compact, so most of the machinery of Kähler geometry is not available. As examples show, 3-Sasakian manifolds are much more numerous than quaternionic Kähler ones; in some sense they are weaker structures. For example there are 3-Sasakian manifolds with an arbitrary second Betti number; see Boyer & Galicki 1999 [250].

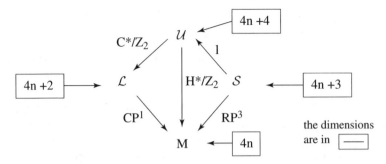

Fig. 13.6. The relations between the twistor spaces of Bérard Bergery/Salamon, Konishi and Swann

Note 13.5.2.1 (Historical mistake) In Berger 1953 [148], the differential forms of respective degrees 3 and 4 determining G_2, Spin (7) and $Sp(1)Sp(n)$ holonomy groups were claimed "not to exist." ♦

13.5.3 Ricci Flat Kähler and Hyper-Kähler Manifolds

Before Yau's 1978 solution of Calabi's conjecture, no compact manifold with holonomy $SU(n)$ was known. Today our knowledge of these manifold is highly advanced. Theorem 280 on page 527 tells us that there are as many $SU(n)$ holonomy manifolds as there are Kähler manifolds with vanishing first Chern class, and thousands of algebraic varieties have vanishing first Chern class. More generally, since Hirzebruch 1966 [717], one knows how to compute quite explicitly the various Betti numbers $b^{p,q}(M,\mathbb{R})$ (see §13.6), and the Chern classes of algebraic (and quite a few other) manifolds. The story of complete noncompact manifolds with $SU(n)$ holonomy is completely different: in LeBrun 1991 [853], for every complex dimension n a complete Ricci–flat Kähler and nonflat metric with holonomy $SU(n)$ is constructed on $\mathbb{C}^n$.

13.5.3.1 Hyperkähler Manifolds

The second group we want to look at now is $Sp(n) \subset SO(4n)$. Manifolds with such a holonomy group are called *hyperkähler*. On such a manifold, we have three different Kähler structures I, J, K defined globally, unlike in the quaternionic Kähler structures of §§13.5.2. These I, J, K are parallel:

$$\nabla I = \nabla J = \nabla K = 0$$

and enjoy the relations

$$I^2 = J^2 = K^2 = -1$$
$$IJ = K$$

and the circular permutations of these. In fact there is a two-parameter family of such I, J, K tensors, namely the $aI + bJ + cK$ with $a^2 + b^2 + c^2 = 1$. Practically they can simply be defined as Kähler manifolds with more than one Kähler structure. The totality of the Kähler structures is necessarily as described above. In fact they really admit a quaternionic structure. An equivalent definition is to ask for a Kähler manifold with a holomorphic parallel exterior 2-form. This 2-form σ is defined by

$$\sigma(x,y) = g(J(x),y) + \sqrt{-1}g(I(J(x)),y)$$

and is holomorphic with respect to the complex structure I. One can also say that one has a complex symplectic structure.

Hyperkähler manifolds are today encountered also in mathematical physics —in particular in the theory of "mirror manifolds": see Voisin 1996 [1223]. Recent references for them are Salamon 1996 [1088], Besse 1987 [183], Biquard & Gauduchon 1996 [194], the books Greene & Yau 1997 [589], Cox & Katz 1999 [408] and Verbitsky 1998 [1211].

Paradoxically it seems that there are more hyperkähler manifolds than quaternionic Kähler manifolds, which is surprising since $Sp(1)\,Sp(n)$ is much smaller that $SU(2n)$.[3] The first complete (noncompact) example appeared in Calabi 1979 [303]. Compact ones can be built using the solution of Calabi's conjecture, and quite deep algebraic geometry: see Beauville 1983 [122], and the holonomy classification result. The basic starting point is this. Assume first that a Kähler manifold has vanishing first Chern class, but also admits a holomorphic 2-form θ of maximal rank. Using the solution of the Calabi conjecture (see theorem 280 on page 527), one first changes the initial Kähler metric into one which is Ricci flat. Then the term $\mathrm{Curv}_2(\sigma,\sigma)$ (see §15.6) in the Bochner formula boils down to $\mathrm{Ricci}(\sigma^*,\sigma^*)$. This was already known to Bochner. It implies that $\nabla \sigma = 0$, so that the form σ is the second Kähler structure we are looking for. The job to be done now is to find algebraic (or more general) manifolds which admit a holomorphic exterior 2-form of maximal rank. There are some examples, but they are quite subtle (it is not enough to know that $b^{2,0}(M,\mathbb{C}) \neq 0$). For their description see 14.B of Besse 1987 [183] or the original Beauville 1983 [122].

Note 13.5.3.1 The naive reader should be warned against thinking that hyperkähler manifolds are the quaternionic analogue of complex manifolds and enjoy coordinates like holomorphic coordinates ("quaternionorphic?"), quaternionic derivatives, etc. This can, sadly enough, happen only in the flat case (then the holonomy is zero and not $Sp(n)$). Fueter had a premonition of that before the 1940's. The main fact is due to Ehresmann: if the derivative $f'(x)$ of a function

$$f : \mathbb{H}^n \to \mathbb{H}^p$$

---

[3] But the story is not finished.

is quaternionic linear at every point x then it is necessarily an affine function. The proof is elementary: the Hessian (second derivative) must be a quadratic form, say $Q(u, v)$, but then plug in noncommuting quaternions $\alpha, \beta \in \mathbb{H}$ and you get

$$Q(\alpha u, \beta v) = \alpha Q(u, \beta v)$$
$$= \alpha \beta Q(u, v)$$

but also, by symmetry

$$= \beta \alpha Q(u, v)$$

This forces $Q(u, v) = 0$, by the assumption that α, β don't commute. However Joyce 1997 [773] managed to develop a theory of "quaternionic functions" on hyperkähler manifolds. A strange but essential aspect of his theory is that the products of quaternionic functions need not be quaternionic; the algebraic side of the story is surprisingly rich and complicated, but appealing. References to quaternionic geometry can be found in modern language in Salamon 1982 [1086] or in Besse 1987 [183] page 410; also see Sommese 1974 [1153], Batson 1992 [117] and Boucksom 2001 [232]. ♦

13.6 Kähler Manifolds

Kähler manifolds were defined in §13.2. Kähler manifolds are the subject of a vast field of research and we will have to be extremely brief. We know of no recent survey. A classic is the book Weil 1958 [1245] and a much respected text is Wells 1980 [1253]. Although motivated by algebraic geometry, Griffiths & Harris 1994 [599] is invaluable for Kähler geometry. Also see chapter 2 of Besse 1987 [183]. Moreover Besse's book contains a great deal of information on Kähler manifolds, in particular a detailed study of the homogeneous ones. The book Amoros, Burger, Corlette, Kotschik & Toledo 1996 [35] is entirely concerned with the fundamental groups of Kähler manifolds (also see Toledo 1997 [1193]) but can be used indirectly, with its references, as a survey or a partial survey of the "Riemann–Kähler" domain. For more general complex geometry one can consult the three books Bedford, d'Angelo, Greene & Krantz 1991 [124].

An analysis of Kähler's historical contribution can be found in Bourguignon 1996 [241]. Of course part of motivation of Riemannian geometers to study Kähler geometry was to be able to add complex analytic techniques in order to attack many problems which are too difficult to be solved in the context of general Riemannian manifolds. However this is not simply a matter of picking the easiest place to work for simplicity's sake, since Kähler manifolds appear entirely naturally in algebraic geometry, so that they are a particularly significant family of Riemannian manifolds. Indeed, Kähler manifolds have recently become extremely important in particle physics and may ultimately play a large role in quantum field theory; see e.g. Voisin 1996 [1223], the references there, and Fröhlich, Granjean, & Recknagel 1998–1999 [523, 524, 525]. One should be careful to remember that Kähler manifolds are almost never to be considered as "complexifications" of "real" manifolds. However they are wonderful to work on, since we have a holomorphic calculus for them.

13.6.1 Symplectic Structures on Kähler Manifolds

The parallelism of the complex structure J implies immediately the curvature relation
$$R(Jx, Jy) = R(x, y)$$
as endomorphisms. Note that a Kähler manifold is then canonically a *symplectic manifold*, so that all of symplectic geometry becomes available. A symplectic structure consists in a even-dimensional manifold M^{2n} endowed with a closed exterior 2-form ω of maximal rank, i.e. the n-th power ω^n is everywhere nonzero. Recent books on the topic are: McDuff & Salamon 1998 [909], Hofer & Zehnder 1994 [724] and also see Audin & Lafontaine 1994 [88]. For contact structures see the book Blair 2002 [202].

13.6.2 Imitating Complex Algebraic Geometry on Kähler Manifolds

Compact Kähler manifolds look like complex algebraic manifolds. There is a good reason for that: a celebrated theorem of Kodaira asserts that this is the case under the sole condition that, via the de Rham theorem, the cohomology class of the Kähler form belongs to the rational cohomology (equivalently has rational periods). For a proof see Griffiths & Harris 1994 [599] or Wells 1980 [1253].

Beware also that the Riemann–Cartan normal coordinates can never be complex, except in the flat case. Also note the heredity: a complex submanifold of a Kähler manifold is also Kähler.

Another strong point is that the Kähler notion is hereditary, not only for (complex) submanifolds but also for many algebraic geometry operations, in particular blowing-up. Note also that ω and its exterior powers ω^k *calibrate*, see the definition on page 667.

Gromov's complaint about Kähler geometry is that the above definition is not very geometric: Gromov 1992 [630]. A geometrical tool, rarely used, is the *diastasis*. Invented in Calabi 1953 [298], it is some kind of adapted metric (in fact it should be viewed more as a potential) but in general defined only locally. Everything concerning it can be found in Hulin 1996 [749] and its bibliography. In this text the diastasis is used to study Einstein manifolds (see more on page 528).

Concerning curvature, besides the sectional curvature, two other notions are natural for Kähler manifolds. First, the *holomorphic curvature*, namely sectional curvature of the real 2-planes which are complex lines. Explicitly they are the numbers $K(x, Jx)$ (where $\|x\| = 1$). Metrics of constant holomorphic curvature are locally isometric to $\mathbb{C}^n, \mathbb{CP}^n$ or $\mathrm{Hyp}_{\mathbb{C}}^n$. This was first stated in Bochner 1947 [211], see Igusa 1954 [755] or Hawley 1953 [685] for a detailed proof. A weaker notion is that of *bisectional curvature*:

$$B(x,y) = R(x, Jx, y, Jy) = K(x,y) + K(x, Jy)$$

for x, y an orthonormal pair of tangent vectors. A very strong result is that of Siu & Yau 1980 [1147] and independently Mori 1979 [940]:

Theorem 402 *Positive bisectional curvature implies that the underlying compact complex manifold is biholomorphic to* $\mathbb{CP}^n$.

This can be seen as an analogue of the Micallef & Moore theorem 336 on page 588 characterizing spheres by the positivity of the curvature operator. Moreover the nonnegative case is settled:

Theorem 403 (Mok 1988 [933]) *Nonnegative bisectional curvature on a compact complex manifold implies biholomorphy with* $\mathbb{CP}^n$ *or isometry with a Hermitian symmetric space, up to products and coverings.*

This is used in theorem 339 on page 589.

For complex manifolds the theory of exterior forms is richer. They enjoy (at least the "pure ones" do) a type (p,q) which refers to how they are written in complex coordinates $z_i, \bar{z}_i$: the "total" degree is $p + q$. For example the Kähler form ω, as well as the Ricci transformed form ρ, are of type $(1,1)$. But one has more: namely that the exterior differentiation can also be split into two pieces ∂ and $\bar{\partial}$. The holomorphic forms are those killed by $\bar{\partial}$, hence of type $(0,q)$. If moreover the manifold is Kähler then the (total) Laplacian $\frac{1}{2}\Delta$ coincides with both the partial Laplacians coming from ∂ and $\bar{\partial}$. Via the Hodge–de Rham theorem and the exterior product with ω (which is of course harmonic), this yields a lot of information. Among that information, let us first mention the growing of Betti numbers up to the middle dimension:

$$b^{p+2}(M, \mathbb{C}) \geq b^p(M, \mathbb{C}).$$

This comes from the fact that because $\nabla \omega = 0$, if any differential form α is harmonic then so is the product $\alpha \wedge \omega$ (in general of course the product of harmonic forms is not harmonic). Another simple example is that positive Ricci curvature forbids the existence of a holomorphic form of any degree. This is a direct consequence of the results in §15.6, because for any $(p, 0)$ form η, the term $\mathrm{Curv}_p(\eta, \eta)$ boils down to only Ricci curvature. The book Hirzeburch 1966 [717] gives enough material to theoretically compute all of

13.6 Kähler Manifolds

the $b^{p,q}(M, \mathbb{C})$ of algebraic manifolds; one can also read Griffiths & Harris 1978 [599].

Second, the Sullivan theory mentioned in §§14.2.1 is much stronger. It is proven in Deligne, Griffiths, Morgan & Sullivan 1975 [438] that the homotopy type of a compact Kähler manifold, over the real numbers, is a formal consequence of the real homology ring. It is also easy to imagine now that for Kähler manifolds, the Bochner vanishing technique will yield a lot of strong results under various assertions on curvature. This is indeed the case and is used heavily for various bundles over Kähler manifolds, especially complex line bundles. There is an immense literature; see the classic Hirzebruch 1966 [717] and Griffiths & Harris 1978 [599]. And thereafter, among others, consider Siu & Yau 1980 [1147] and Sampson 1986 [1092]. Also see the recent Buchdal 2000 [272].

Reconsidering the spectrum of a Riemannian manifold, can one read off whether a manifold is Kähler from its spectrum? We are far from being able to do this. Using the asymptotic expansion there are some special results in Gilkey 1973 [561]. Then Gromov 1992 [630] is more ambitious. There are good results for the first eigenvalue λ_1: see Lichnerowicz 1958 [865] where there is lower bound for λ_1 depending only on a positive Ricci lower bound and in Bourguignon, Li & Yau 1994 [245] an upper bound with the volume for algebraic manifolds.

The power of Kähler manifolds comes in part from their canonical *symplectic geometry*. We hardly touched this very important—though quite recent—topic, as well as that of *contact structures*; see the book Blair 2002 [202]. The unit bundle is a contact manifold in a canonical manner (see section 10.5 on page 471). Many people think that symplectic and contact geometry will be the most important topics in geometry in the next few decades (see Gromov 1999 [633]). McDuff 2000 [908] is a very recent expository text on these topics.

14 Some Other Important Topics

Contents

14.1 **Noncompact Manifolds** **660**
    14.1.1 Noncompact Manifolds
            of Nonnegative Ricci Curvature 660
    14.1.2 Finite Volume 661
    14.1.3 Bounded Geometry 661
    14.1.4 Harmonic Functions 662
    14.1.5 Structure at Infinity 662
    14.1.6 Chopping 662
    14.1.7 Positive Mass 662
    14.1.8 Cohomology and Homology Theories 663
14.2 **Bundles over Riemannian Manifolds** **663**
    14.2.1 Differential Forms and Related Bundles 663
    14.2.2 Spinors 668
    14.2.3 Various Other Bundles 671
14.3 **Harmonic Maps Between Riemannian**
    Manifolds **674**
14.4 **Low Dimensional Riemannian Geometry** **676**
14.5 **Some Generalizations of Riemannian Geometry** **676**
    14.5.1 Boundaries 676
    14.5.2 Orbifolds 677
    14.5.3 Conical Singularities 678
    14.5.4 Spectra of Singular Spaces 678
    14.5.5 Alexandrov Spaces 678
    14.5.6 CAT Spaces 680
    14.5.7 Carnot–Carathéodory Spaces 681
    14.5.8 Finsler Geometry 682
    14.5.9 Riemannian Foliations 683
    14.5.10 Pseudo-Riemannian Manifolds 683
    14.5.11 Infinite Dimensional Riemannian Geometry .. 684
    14.5.12 Noncommutative Geometry 685
14.6 **Gromov's mm Spaces** **685**
14.7 **Submanifolds** **690**
    14.7.1 Higher Dimensions 690
    14.7.2 Geometric Measure Theory
            and Pseudoholomorphic Curves 691

14.1 Noncompact Manifolds

We have already mentioned extensions of various results for compact manifolds to complete manifolds. It is clear that one should restrict oneself to only certain sorts of noncompact Riemannian manifolds to have a hope of obtaining results. Let us mention in particular the possibilities of examining manifolds with finite volume, those with prescribed asymptotic behaviour at infinity, for example quadratic decay,[1] quadratic curvature decay, volume behaviour, Euclidean asymptoticity, etc.

14.1.1 Noncompact Manifolds of Nonnegative Ricci Curvature

A typical example if that of manifolds with nonnegative Ricci curvature. In the case of nonnegative sectional curvature, we saw in §§§12.3.1.3 a perfect structure theorem (splitting and soul) of Cheeger–Gromoll and the bounded topology result of Gromov–Abresch. Note that these results are valid without any extra condition on the geometry (bounded or behaviour at infinity). But if one asks the same question for nonnegative Ricci curvature, the splitting theorem is still valid, but there is no structure theorem and there is no bounded topology result, as seen in examples. This implies that results on nonnegative Ricci curvature on noncompact manifolds should use an extra hypothesis, typically some kind of growth. The results could be of a different nature. Correspondingly, for negative curvature, and its various structures on the sphere at infinity (see 12.3.4.3 on page 606), it seems that an important structure at infinity to be defined (even if not unique in general) is the following. For a given Riemannian manifold (M, g), consider the sequence of the (pointed) manifolds $(M, p, r_i^{-1} g)$ when the "radii" r_i go to infinity. The convergence theorems 385 on page 629 extend easily to the category of pointed manifolds. Petersen 1997 [1018] is a book on this subject. In particular one can extract a subsequence converging toward "some" limit "cone" (a cone at infinity, denoted M_∞ by abuse of notation). This cone need not be unique and also might depend on p. In various instances, one can prove that M_∞ is a volume cone, or better a metric cone, and sometimes even a Euclidean space. Cheeger & Colding 1997 [337] is a nice example of results among many recent ones. Exemplary is the conjecture of Anderson and Cheeger to the effect that if a cone at infinity is isometric to $\mathbb{R}^d$ and the Ricci curvature nonnegative, then the manifold itself is isometric to $\mathbb{R}^d$; this conjecture was proven in Colding 1996 [386]. Also see Cheeger & Colding 1997,1997,1998 [337, 338, 339].

We insist that noncompact manifolds are in many respects more important than compact ones. This is why our partial survey should be completed by the reader in some way or another. The reader could discover more in the references given. Noncompact manifolds appear already among surfaces, especially

---

[1] Quadratic decay means that the curvature decreases as one goes to infinity, at least as r^{-2} where r is the distance to an arbitrary chosen point; there are variations on this definition.

space forms, where the eigenvalue behaviour with respect to the value $1/4$ is even more delicate than for compact surfaces; see for example Luo, Rudnick, & Sarnak 1995 [884].

We saw in §§12.3.2 and §§12.3.4 that noncompact manifolds appear naturally in the analysis of compact manifolds of bounded Ricci curvature, of negative or of nonpositive curvature, when one looks at their universal coverings. But they are also used heavily in different instances, as in collapsing with unbounded diameter or more generally, when one drops the diameter bound in various situations (see §§12.4.3 and references there). See also the very geometric Babenko 1992 [91].

14.1.2 Finite Volume

A natural type of noncompact manifold to study is that of finite volume. Here many results we met above extend, with more or less challenge, depending of the question. To give even a short list of references is impossible. What follows is very biased. Note just to start with the naturality of the topic, that the famous modular domain $SL(2,\mathbb{R})/SL(2,\mathbb{Z})$ is not compact but is of finite volume for its canonical hyperbolic metric (but also has singularities): see Luo, Rudnick & Sarnak 1995 [884]. We also mention the problem of extending the integral Chern formulas from equation 15.14 on page 717 to finite volume manifolds. This turned out surprisingly to be an extremely difficult subject, even for surfaces, where the investigation was initiated in Cohn-Vossen 1935,1936 [382, 383] for the problem of extending the Gauß–Bonnet formula. However a lot of work remains to be done for surfaces. We refer to Shioya 1992 [1136] and the intermediate references there, and just mention that many topics introduced in §§12.3.4 (for negatively curved manifolds) play a role here. The investigation of such formulas in higher dimensions, involving the Euler characteristic χ and also the characteristic (Pontryagin) numbers, was begun in Shioya 1992 [1136]. Noncompact manifolds can have irrational characteristic numbers. More results are to be found recently in Rong 1995 [1065]; also see the survey Lück 1996 [883].

Finite volume is an especially strong condition in the negative curvature realm, where most results valid for compact manifolds extend with often not too much pain to finite volume manifolds. For space forms and Mostow's rigidity see Farrell & Jones 1989 [507]. For the general case see Ballmann, Gromov & Schroeder 1985 [106] and of course Eberlein, Hamenstädt & Schroeder 1990 [472].

14.1.3 Bounded Geometry

Another condition is that of bounded geometry. It occurs naturally for coverings of compact manifolds and homogeneous spaces. The question is studied in Semmes 1996 [1122] in relation with the quite recent notion of complexity. Bounded geometry occurs when, for example, the curvature verifies $|K| < 1$

everywhere. The comparison theorem 73 on page 258 shows that this is really equivalent to the local geometry being bounded.

14.1.4 Harmonic Functions

It is an interesting question when a complete Riemannian manifold admits a nonconstant harmonic function ($\Delta f = 0$) or even more a positive spectrum. This possibility is directly linked with various curvature properties. The founding text is Yau 1975 [1292] where positive Ricci curvature was shown to be enough to forbid nonconstant harmonic functions (this is a "Liouville theorem"). Results continue to appear. We refer only to recent ones and their bibliographys: Colding & Minicozzi II 1998 [388] and Yu 1997 [1300]. Also see Benjamini & Cao 1996 [130] for relations with sectional curvature.

14.1.5 Structure at Infinity

In the noncompact realm various results address the implications of conditions on the structure at infinity, like to be asymptotically Euclidean (with some fixed asymptotic order). See among others Shen 1996 [1128] and references there. There are also gap results, which forbid various compact metrics to be glued with completely flat ones outside, etc. See the introduction of Lohkamp 1996 [874] and Greene & Wu 1990 [595], as well as what was seen in §§12.3.4. Many results can be found in Eberlein, Hamenstädt & Schroeder 1990 [472], like for example asymptotically harmonic manifolds. A deep result is Cheeger & Tian 1994 [357]; this result is used in the results of Cheeger & Colding in §§12.3.2.

14.1.6 Chopping

There are many techniques in noncompact Riemannian geometry. Besides using the sphere at infinity in various ways, we mention the very geometrical *chopping* technique of Cheeger & Gromov 1989 [349]. This is an exhaustion technique where, when one goes to infinity, the boundaries of the successive compact pieces keep a bounded second fundamental form and have controlled volume.

14.1.7 Positive Mass

The technique used for proving the "positive mass conjecture" in Schoen & Yau 1981 [1110] is also very interesting (see the expository Kazdan 1982 [800]). They employed geometric measure theory (see §§14.7.2) and looked at minimal hypersurfaces in the manifolds, studying what they become when they go to infinity. For other viewpoints see the references in Cao 1996 [307].

14.1.8 Cohomology and Homology Theories

On compact manifolds essentially all homology and cohomology theories coincide. On noncompact manifolds, the possible cohomology and homology theories are much more subtle. A good notion for Riemannian geometry is that of L^2 Betti numbers. There are surveys: Pansu 1996/97 [999] and Lück 1996 [883]. We just mention that the Riemannian geometer motivated principally by compact manifolds should contemplate L^2 cohomology with awe. A striking example is in Gromov 1991 [626], where L^2 Betti numbers are used to solve the Hopf conjecture 1 on page 544 for Kähler manifolds:

Theorem 404 (Gromov 1991 [626]) *A compact Kähler manifold M^{2n} of negative curvature has Euler characteristic of sign equal to $(-1)^n$.*

Deep results concerning holonomy groups and volumes of balls appear in Tapp 1999 [1182].

14.2 Bundles over Riemannian Manifolds

We introduce bundles in the order of their Riemannian geometric character. First, the canonical or almost canonical ones (including spinors), then those obtained by twisting canonical ones, then Yang–Mills fields. But before the most natural ones are those of exterior differentials forms met for the first time in §§4.2.2. For spinors, see Gilkey 1995 [564] and Berline, Getzler & Vergne 1992 [179] for the index theorem in a very general context and for more on spinors Lawson & Michelsohn 1989 [850]. Of basic importance are characteristic classes for various bundles over general differentiable manifolds, so one should look at their relations with Riemannian geometry; see §15.7.

14.2.1 Differential Forms and Related Bundles

Exterior differential forms exist on any differentiable manifold, with their degreees extending from 0 to the dimension d of the manifold. The corresponding vector bundles are denoted by $\Lambda^p(T^*M)$, while their spaces of sections are denoted by $\Omega^p(M)$. And the basic link between these spaces of sections is the *exterior derivative*

$$d : \Omega^p(M) \to \Omega^{p+1}(M)$$

with $dd = 0$. Equipped with d, the collection of $\Omega^*(M)$ is called the *differential complex* of M. The closed forms are the ω with $d\omega = 0$. This is leading, as seen in §§4.2.2, to the de Rham theorem 32 on page 168 via Stokes formula 34 on page 171. Although not a Riemannian story, one should know that while de Rham's theorem only yielded the real Betti numbers, more can now be extracted purely from the exterior differential complex, as was discovered in Sullivan 1977 [1166], which yields for example some topological finiteness results (see §§12.4.1). What more can a Riemannian geometer ask for? Of course for interesting relations between the metric and exterior forms.

14.2.1.1 The Hodge Star

A basic fact is the existence (for which one needs the manifold to be oriented) of the Hodge star operator

$$* : \Omega^p(M) \to \Omega^{d-p}(M).$$

It is involutive or antiinvolutive:

$$*^2 = (-1)^{d(p+1)+1}.$$

Given a differential form w, the form $*w$ is defined by

$$*w(x_{p+1}, \ldots, x_d) = w(x_1, \ldots, x_p)$$

for any positively oriented orthonormal basis $\{x_1, \ldots, x_d\}$. For example $*1$ is the volume form, namely the d-form which takes the value $+1$ on any positively oriented orthonormal set of tangent vectors. The existence of the volume form is equivalent to orientability.

14.2.1.2 A Variational Problem for Differential Forms and the Laplace Operator

It is natural to look at the canonical norms $\|w\|$ (see §15.2) and think of the de Rham theorem, to look at the minimum of the L^2 norm

$$\int_M \|w(m)\|^2$$

for a closed form w running through a fixed cohomology class of $H^p_{dR}(M) = H^p(M, \mathbb{R})$ (see §§4.2.2) So we can vary w into $w + d\alpha$ and compute

$$\int_M \|w + d\alpha\|^2 = \int_M \|w\|^2 + 2\int_M \langle w, d\alpha \rangle + \int_M \|d\alpha\|^2.$$

A general theoretical fact is that there is an adjoint operator d^* called the *adjoint* of d, such that

$$\int_M \langle d\sigma, \beta \rangle = \int_M \langle \sigma, d^*\beta \rangle$$

for any pairs of forms σ, β. So that if w not only satisfies $dw = 0$ but also satisfies $d^*w = 0$ then w will be an absolute minimum. The miracle is that this adjoint d^* in our Riemannian manifold is nothing but

$$d^* = (-1)^{d(p+1)} * d *$$

for the Hodge star operation introduced above. There is no need for an orientation since one uses $*$ twice. It is not difficult to transform the paired conditions $d\omega = 0$ (closedness) and $d^*\omega = 0$ (called *co-closedness*)into a single one. This is done by considering the Laplacian

$$\Delta = dd^* + d^*d = (d + d^*)^2.$$

Suppose that the manifold is compact. Then

$$\int_M \langle \Delta\omega, \omega \rangle = \int_M \|d\omega\|^2 + \int_M \|d^*\omega\|^2$$

so that $\Delta\omega = 0$ is equivalent to $d\omega = 0$ and $d^*\omega = 0$. Of course, for functions, which are the forms of degree 0, this Laplacian coincides with that of chapter 9. By definition *harmonic forms* ω are those with $\Delta\omega = 0$. The main result is

Theorem 405 (Hodge–de Rham) *There is precisely one harmonic form in any de Rham cohomology class. Thus the set of harmonic forms of degree p is isomorphic to $H_{dR}^p(M)$, and is a real vector space of dimension equal to the real Betti number $b^p(M, \mathbb{R})$.*

Uniqueness of a harmonic form in each de Rham cohomology class is trivial by the above, but the existence requires hard analysis, and is a typical case of the theory of elliptic operators. The proof does not figure in most Riemannian geometry books, see the classic Warner 1971 [1243] or the recent Jost 2002 [768] for the structure of the proof. It will be important that, contrary to the case of functions, the computation of $\Delta\omega$ when $p > 1$ leads to curvature: see §15.6. But, in exchange, as the revenge of Riemannian geometers, Bochner's approach to the Laplacian will give us tools for relating curvature and topology. See also §9.14 for other types of topological outcomes. Let us also mention the

Theorem 406 (Hodge decomposition theorem) *Any form can be written as a sum*

$$\alpha + d\beta + d^*\gamma$$

where α is harmonic. The forms $\alpha, d\beta$, and $d^*\gamma$ are uniquely determined, although obviously β and γ are not.

Two outcomes of this were mentioned in the spectral analysis in §9.14.

Note 14.2.1.1 (Intrinsically harmonic forms) Concerning harmonic forms one might wonder about the "inverse problem:" if harmonic forms are special or not among closed differential forms. This is a tricky question, attacked first for 1-forms and solved in Calabi 1969 [302]. For higher degrees see Farber, Katz & Levine 1998 [503]. ♦

14.2.1.3 Calibration

One might also wonder about the fact that exterior forms are not good for computing volumes when one restricts them to submanifolds. For example, the lengths of curves in the plane are not measured by integrating a differential form in the plane (but only by "lifting" the curve into the unit tangent bundle). So except for the global point of view of harmonic forms, it seems hopeless to do more geometry with exterior forms. But there are exceptions; the first was discovered in Wirtinger 1936 [1273]. In a Kähler manifold, the Kähler form ω enjoys the following property: for any orthonormal pair of vectors x, y one has

$$\omega(x, y) \leq 1$$

with equality only for complex lines, i.e. $y = \sqrt{-1}x$ and this works also mutatis mutandis for the exterior powers ω^p which match the volume forms precisely on complex p-planes. As we will shortly explain, because ω is closed, applying Stokes formula implies immediately that any complex submanifold of a Kähler manifold has an absolute minimal volume in its homology class (this is much stronger than to be only a minimal submanifold and is called *stability*).

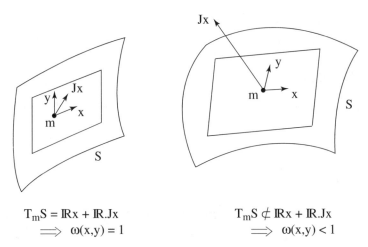

Fig. 14.1. The Wirtinger inequality: (a) $T_m S = \mathbb{R}x \oplus \mathbb{R}\sqrt{-1}x \Rightarrow \omega(x, Jx) = 1$ (b) $T_m S \neq \mathbb{R}x + \mathbb{R}\sqrt{-1}x \Rightarrow \omega(x, y) < 1$

Let us prove the Wirtinger inequality for surfaces (2-dimensional submanifolds) in a complex manifold. We recall first the independence under homology:

$$\int_S \omega = \int_{S'} \omega$$

from Stokes' theorem because $d\omega = 0$ for the Kähler form. Now the basic trick is that $\omega(x, y) \leq 1$ implies

$$\int_S \omega \leq \int_S dA = \mathrm{Area}(S)$$

with equality if and only if the tangent planes of the surface S are stable under $\sqrt{-1}$, i.e. S is a complex curve (which is of course a particular type of real surface) in our complex manifold. Wirtinger used his result for various applications, for example to compute explicitly the volume of any algebraic submanifold of given degree in $\mathbb{CP}^n$. The above inequality can hold for various differential forms of degree p on some Riemannian manifold:

Definition 407 *A p-form ω calibrates (or is a calibration) if it is closed and*

$$\omega(x_1, \ldots, x_p) \leq 1$$

for every orthonormal p-tuple of vectors $x_1, \ldots, x_p$.

The definition comes from Harvey & Lawson 1982 [682]. The straightforward property of a calibrating p-form ω, generalizing Wirtinger's argument, is that if a submanifold N^p has volume equal to $\int_N \omega$ then it is an absolute minimum in its homology class, in particular a stable minimal submanifold. For this it is necessary and sufficient that ω takes the value 1 on each tangent space to N^p. Calibration was used in Berger 1972 [153] to prove the systolic inequality for the standard metric of $\mathbb{KP}^n$ for $\mathbb{K} = \mathbb{H}$ and $\mathbb{K} = \mathbb{C}\mathrm{a}$ (for $\mathbb{K} = \mathbb{C}$ this is Wirtinger's inequality). For the quaternionic case the calibrating form is the form called ω in §§13.5.2. In Harvey & Lawson 1982 [682] a general theory of calibrations is formulated. In particular some generalizations of Wirtinger's inequality are presented, in the sense that calibration is also linked with some PDE the same way complex submanifolds can be defined as holomorphic, namely by a first order operator (the Cauchy–Riemann equation $\bar{\partial} = 0$). In general, even if one has a calibrating differential form ω, the problem of finding submanifolds it calibrates, i.e. submanifolds N on which $\omega = dV_N$, is a highly overdetermined system of partial differential equations, and for most calibrations ω it has no solutions. Recently calibration was used to prove systolic softness (freedom) in Babenko & Katz 1998 [93] (see §6.5) and in Besson, Courtois & Gallot 1995 [189] in a Hilbert manifold framework. See also the pseudo-holomorphic curves in §§14.7.2.

In particle physics, calibrated cycles (not necessarily submanifolds—one allows some sort of singularities, and the precise nature of them is still rather vague) have played a surprising role allowing physicists to break some but not too much supersymmetry from string theory and M-theory.

14.2.1.4 Harmonic Analysis of Other Tensors

There is no reason to stick with exterior forms and not to look at other tensors. Lichnerowicz introduced the theory of the *canonical* and *special Laplacian* in Lichnerowicz 1961 [866]. Particularly important is his Laplacian for symmetric bilinear differential forms, as they can be interpreted as infinitesimal variations of a Riemannian metric. For Hodge–de Rham type theorems concerning these Laplacians, see section 12.C of Besse 1987 [183] where it is used to study deformations of Riemannian structures, i.e. metrics up to isometries (diffeomorphisms). Lichnerowicz's Laplacians are natural, in the sense explained in §15.6.

14.2.2 Spinors

When it exists, the spinor bundle is "almost canonical." For all that follows and more on spin geometry see Lawson & Michelsohn 1989 [850], and for some parts add Gilkey 1995 [564], Berline, Getzler & Vergne 1992 [179] and Friedrich 2000 [522]. To keep this book of reasonable size, we are going to ignore a wealth of detail. The basic idea is: Riemannian geometry has a Euclidean structure on every tangent space, so that, if moreover the space is oriented, the special orthogonal group $SO(d)$ is "the basic group." Besides the fundamental representation on $\mathbb{R}^d$, the special orthogonal group has a basic one on the exterior algebra and, by tensor products, one can build all the other ones. Apparently, the Riemannian geometer should not look further than to those. But the Lie group $SO(d)$ is not simply connected and its (nontrivial, twosheeted when $d > 2$) universal covering group $\mathrm{Spin}(d)$ has an extra representation in a "canonical" complex vector space, the space of spinors of the initial Euclidean space.

14.2.2.1 Algebra of Spinors

First let us point out some purely algebraic facts. The canonical quadratic form on $\mathbb{R}^d$ gives birth to the Clifford algebra $\mathrm{Cliff}(d)$, of dimension $2d$. This algebra is canonically isomorphic as a vector space to the exterior algebra $\Lambda^*\left(\mathbb{R}^d\right)$. There is a simple link between $\mathrm{Spin}(d)$ and $\mathrm{Cliff}(d)$ to the effect that one finally obtains two things in a canonical way. First, a complex vector space $\mathcal{S}$ of dimension $2^{\lfloor d/2 \rfloor}$ (where $\lfloor x \rfloor$ denotes the greatest integer less than a given real number x.) This space shows up when one tries to write $\mathrm{Cliff}(d)$ as an endomorphism algebra $\mathrm{End}(\mathcal{S})$ and is called the (complex) spinor space of $\mathbb{R}^d$. Second, because $\mathrm{Spin}(d)$ can be realized as a subgroup of the group made up by the unit elements of the algebra $\mathrm{Cliff}(d)$, one finds a representation of $\mathrm{Spin}(d)$ on the space $\mathcal{S}$. The theory is complicated by the fact that this representation is irreducible only when d is odd but splits into nonisomorphic representations denoted $\mathcal{S}^+$ and $\mathcal{S}^-$ of half the dimension when d is even.

There are also *real spinors*, but the situation is more complicated with them and is to be considered modulo 8. It is a basic fact that this game cannot be played without having fixed a quadratic form, because the universal cover of the special linear group $SL(d,\mathbb{R})$ does not admit any faithful finite dimensional representation except the ones in the exterior algebra, i.e. those of $SL(d,\mathbb{R})$ itself.

14.2.2.2 Spinors on Riemannian Manifolds

Since all of the above constructions depend only on oriented Euclidean geometry, namely the positive definite quadratic form on $\mathbb{R}^d$, we now carry out this two step construction for any oriented Riemannian manifold M. The (principal) bundle of oriented frames $P_{SO(d)}M$ has $SO(d)$ as structure group, so one first tries to define a universal double covering of this bundle, asking moreover of course for compatibility with the group covering $\text{Spin}(d) \to SO(d)$. This is not always possible and a necessary and sufficient condition for this is the vanishing of the second Stiefel–Whitney class $w_2(M)$ (see Haefliger 1956 [677]). Up to isomorphism, the *spin structures* so obtained are classified by $H^1(M, \mathbb{Z}_2)$. For example, the spin structure is unique on a simply connected manifold. However one usually uses an ambiguous notation, namely $P_{\text{Spin}(d)}M$, to denote any spin bundle covering $P_{SO(d)}M$. Attached to any spin structure on the manifold, (see §§14.2.3) the representation of $\text{Spin}(d)$ on $\mathcal{S}$ automatically produces a (complex) vector bundle denoted by $\mathcal{S}(M)$ and called the (or "a") *spinor bundle* on M. A section of this spinor bundle is what is called a *spinor field* on M. Still using general facts about bundles (see §§14.2.3) the above construction and the Levi-Civita connection on $P_{SO(d)}M$ yield a canonical connection ∇ on $\mathcal{S}(M)$.

The third point, a major one, is that using ∇ one can define on $\mathcal{S}(M)$ a canonical differential operator of degree one, called the *Dirac operator* and denoted by $\slashed{D}$. In even dimensions it exchanges the spinors which are sections of $\mathcal{S}^+$ with those which are sections of $\mathcal{S}^-$. Now for $\slashed{D}^2$ there is a Bochner–Weitzenböck formula which is

$$\slashed{D}^2 = \nabla^*\nabla + \frac{1}{4}\text{scalar}$$

(the *Lichnerowicz formula*) and which was already employed extensively in §§12.3.3. In comparison with formulas for Δ involving the collection of the $\text{Curv}_p(R)$ remainder terms (see theorem 338 on page 588 and §15.6) this looks a priori like a disaster. But it is just the opposite: with much less information on the curvature, one still gets information on the topology with the Hodge theorem for $\slashed{D}$, namely for harmonic spinors σ, i.e. those satisfying $\slashed{D}s = 0$.

14.2.2.3 History of Spinors

The history of spinors is fascinating. It started with Élie Cartan in Cartan 1913 [311] where he completely classified the complex irreducible representations of simple Lie groups. Besides the expected orthogonal and exterior representations, he found an extra one bringing a new space to life. Moreover he prophetically indicates how this representation can generate all others. Then, completely independently, first (in a very primitive form) they appeared in Pauli 1927 [1006], and finally with a clear grasp of their relation to group theory in Dirac 1928 [450], spinors appeared on physical grounds (hence their name) and Dirac defined the Dirac operator; this not for manifolds but just for Minkowski space $\mathbb{R}^{3,1}$. The link between the physicists' Lorentzian spinors and the Euclidean ones was uncovered by Cartan in 1937, but he concluded his book Cartan 1937 [318] by noting the "impossibility" of building a satisfying theory along the lines physicists and geometers were comfortable with: this is so because coordinate changes with general Riemannian metrics on manifolds are only linear (not orthogonal) and we saw above that the universal cover of the special linear group has no finite dimensional representation besides those coming from the genuine linear group. The construction of spinor bundles and of the Dirac operator on them are hard for us to precisely date—it seems that they were more or less folklore. But the year 1963 is of historical importance, being the year when at the same time there appeared (1) the index theorem, which among others yielded the fact that the index for the Dirac operator is the Hirzebruch $\hat{A}$ genus (hence an integer) and (2) the Lichnerowicz scalar curvature formula above. For this history and more on spinors see the introductions of Lawson & Michelsohn 1989 [850] and of Berline, Getzler & Vergne 1992 [179], as well as the postface by Jean-Pierre Bourguignon in Chevalley 1997 [372].

14.2.2.4 Applying Spinors

We saw the fundamental application of spinors in §§12.3.3. The basic idea behind the Lichnerowicz result is that, when one defines much more general spin-type-bundles, one can often prove that $\frac{1}{4}$ scalar is the dominant term. Other applications of spinors are: (1) some proofs of the index theorem and (2) construction of certain special holonomy groups, see §§13.5.1.

14.2.2.5 Warning: Beware of Harmonic Spinors

As opposed to the case of the Laplacian and harmonic differential forms, the kernel of the Dirac operator $\displaystyle{\not}D$ depends on the metric chosen on a given manifold. The dimension of this kernel (the dimension of the space of harmonic spinors) can change, as was first discovered in Hitchin 1974 [719]. Worse: any spin manifold can carry a metric with some nonzero harmonic spinor (at least in dimensions $4n + 3$, and perhaps in all dimensions); see Bär 1996 [116]. However, harmonic spinors are conformal invariants: see Hitchin 1974 [719]. The dependence on the metric is not easy to control because, when one varies the metric, the bundle also changes. Bourguignon & Gauduchon 1992 [243] present a detailed computation for everything in this context, in particular the first variation of the Dirac operator and of its eigenvalues (see references therein for intermediate results).

14.2.2.6 The Half Pontryagin Class

For spin manifolds, the first Pontryagin class $p_1(M)$ (see §15.7) is always the double of some class, denoted then by $\frac{1}{2} p_1(M)$. This class was used in §§12.3.2

as the only known restriction to positive Ricci curvature (besides of course the scalar curvature obstruction).

14.2.2.7 Reconstructing the Metric from the Dirac Operator

Finally, for the very geometrically minded reader, we mention that in Connes 1994 [400] one finds a formula giving the distance between any two points using the Dirac operator, which has the advantage that it carries over to noncommutative geometry.

14.2.2.8 Spin^c Structures

Not much more complicated to define are the Spin^c structures which exist on many more manifolds than have spin structures. In particular they exist on any 4-dimensional manifold, see §14.4 for references to very significant recent applications. Beware of the notation: Spin^c (d) is not the complex group Spin $(n, \mathbb{C})$ but "only"

$$\text{Spin}^c(d) = \left(\text{Spin}(d) \times S^1\right) / \mathbb{Z}_2.$$

14.2.3 Various Other Bundles

The theory of bundles can be found in most books of differential geometry and in detail in Kobayashi & Nomizu 1963-1969 [827, 827]. From the perspective of topology, the classics are Steenrod 1951 [1156] and Husemoller 1975 [751]. Bundles E are differentiable manifolds equipped with a map $E \to B$ onto a base manifold B and a typical fibre F, such that the base B is covered with charts and the preimages in E of those charts are diffeomorphic to products with F. On their intersections, these bundle charts should be smooth on the fibers and most often preserve some given structure. For principal bundles the fiber is acted on simply and transitively by a group, and changes of charts have to be group automorphisms. For vector bundles they should of course be linear maps. When given a principal bundle together with an action of the group on some object (e.g. being a representation in a vector space), one can canonically deduce from it an associated bundle. Such a construction was employed above in order to construct the spinor bundle.

For bundles one has a notion of connection, and then of curvature, of parallel transport and of holonomy. The concept of a connection is the same as the one explained in §15.2, namely one wants to be able to compare (infinitesimally) two fibers and to develop some kind of differential calculus. Of course, the connection one uses should preserve in a reasonable sense the various structures the bundle carries. In particular, Riemannian vector bundles appear when one is given a positive quadratic form on every fiber (with smooth dependence on the base point). Vector bundles have characteristic classes and Chern formulas. Books are Hirzebruch 1966 [717], Berline, Getzler & Vergne 1992 [179], and Gilkey 1995 [564].

14.2.3.1 Secondary Characteristic Classes

We will see in §15.7 that there are no other integral formulas involving curvature besides those of Chern. In Chern & Simons 1974 [370] new subtle invariants were introduced. They take place in various bundles, and are not "downstairs" Riemannian objects. Although typically topologically trivial their connections are not and parallel transport along those connections yields those new invariants. In Cheeger & Simons 1985 [354] they were put downstairs on the manifold where they live as "differential characters." These new invariants became more and more important—in particular recently in mathematical physics and in number theory; see Gillet & Soulé 1992 [567]. Also look at the books Berline, Getzler & Vergne 1992 [179] and Gilkey 1995 [564] section 3.11.

14.2.3.2 Yang–Mills Theory

Yang-Mills fields were "born" in the late seventies by demand of theoretical physicists. A book appeared as soon as Atiyah 1979 [75]. But since then the topic has grown enormously and is now a field of research in itself, with many subtopics. Yang–Mills theory can be roughly outlined as follows. On a given compact Riemannian manifold, most often a "standard" one like the sphere, one picks up some interesting vector bundle over it, smoothly chooses some Hermitian or Euclidean metric on its fibers. Then one looks for the "least twisted" connection preserving this metric, measuring "twist" by the integral of the squared curvature of this connection. This is a problem of the calculus of variations and leads to a condition on the curvature of bundle which looks like harmonicity of the curvature. There is a rich harvest of Yang–Mills theory because what one looks for is like looking for harmonic forms, except that the harmonicity condition is on the curvature. There are results like the Hodge–de Rham theorem 406 on page 665. Its natural differential geometry setting is essential in particle physics (gauge theory). Surveys and recent references are: Donaldson & Kronheimer 1990 [457], Donaldson 1996 [456] and Andersen, Dupont, Pedersen & Swann 1997 [36]. Intermediate references were Freed & Uhlenbeck 1991 [520], and Bourguignon & Lawson 1982 [244].

We just mention here that there is a drastic difference between Yang–Mills fields on various bundles and the condition for the manifold itself to be Einstein (see §11.4). In the Yang–Mills game one keeps the metric downstairs fixed and varies the connection in the bundle, while the Einstein condition couples the Riemannian struture of the basis with the tangent bundle connection. To do both at the same time is a completely different game.

14.2.3.3 Twistor Theory

Twistor theory belongs more to Riemannian geometry than does Yang–Mills theory, even if there are strong links between them. The twistor bundles are various bundles with compact fibers. They are constructed from a Riemannian manifold using its metric and some given additional structure e.g. a complex structure, special holonomy, etc. In dimension 4 every Riemannian manifold has a canonical twistor space above it, with fiber S^2. One example has been explained in some detail in §§13.5.2 for quaternionic-Kähler manifolds. The main point is that twistor spaces have a richer structure that their base spaces, e.g. they may be Kähler. In many cases they have pure geometric applications. One is the classification of minimal surfaces in standard spheres of high dimension carried out in Calabi 1967 [301]. They are also useful for finding holonomy groups of some types (see §§13.5.2), and to construct Einstein manifolds; see for example Hitchin 1995 [720]. Besides a brief appearance in Lawson & Michelsohn 1989 [850], they also appear in the book Besse 1987 [183], and its second edition yet to come. An intermediate reference is Atiyah, Hitchin & Singer 1978 [80].

14.2.3.4 K-theory

General vector bundles are of course only differential geometry at the beginning. If the base manifold is Riemannian, we find also a measure and a canonical connection on the tangent bundle. Vector bundles are interesting because of something called K-theory. This is an algebra made of all complex vector bundles over some given manifold, bringing in an algebraic tool to study differentiable manifolds, the maps between them, etc. For example one of the important facts about the Dirac operator is that it represents the fundamental class in K-theory. For real vector bundles one had to use the more subtle KO-theory.

The quantization of K-theory using Riemannian geometry is just beginning in Gromov 1996 [631], also see definition 357 on page 602 and theorem 358 on page 603.

For general Riemannian real vector bundles, the idea of Atiyah and Singer was to extend to any such bundle the spinor construction of §§14.2.2 (which was sketched for the complex case). The topological construction is the same as for the tangent bundle. One can consider both the complex and the real case. Moreover associated to Riemannian connections there is a canonical Dirac operator and a generalized Lichnerowicz formula which is again surprisingly simple, even for twisted bundles; see II.§8 of Lawson & Michelsohn 1989 [850]. For various vanishing theorems as found in §§12.3.3 an important fact is that the scalar curvature of the base manifold appears separately from the curvature term of the vector bundle under consideration. Finally these generalized Dirac operators became essential in understanding more clearly the index theorem below. This was through various contributions of Atiyah & Bott, McKean & Singer, Patodi & Gilkey; see the introduction of Berline, Getzler & Vergne 1992 [179]. This book is precisely devoted to the index approach with generalized spinors and Dirac operators.

14.2.3.5 The Atiyah–Singer Index Theorem

If one now very broadly generalizes the Laplacian acting on functions and on exterior forms, by considering any elliptic operator on some vector bundle, there is a very deep and universal result, the *Atiyah–Singer index theorem* which was a great event in 1963. Gelfand and Vekua had conjectured the topological invariance of the index. See Atiyah & Singer 1963 [84] for historical references and intermediate results. The *index* of an elliptic operator is the dimension of its kernel minus the dimension of the kernel of its adjoint (its cokernel). The theorem is that this integer, for an elliptic differential operator on a compact manifold M, is equal to a number computed only with two pieces of topological data: (1) the Todd class of M, which is expressible in terms of its Pontryagin classes, (2) an invariant computed with the symbol of the elliptic operator considered as acting on the various Chern classes of the vector bundles on which the operator is defined. In the case of differential forms one just recovers the Euler characteristic via the Hodge–de Rham theorem. Applications of the index theorem are numerous (and far from finished), e.g. yielding integrality of various invariants which a priori were only known to be real numbers. In the world of pure Riemannian geometry, we saw above in §§12.3.3 when we studied scalar curvature. References are Gilkey 1995 [564], Gilkey 2000 [565], Berline, Getzler & Vergne 1992 [179].

14.2.3.6 Supersymmetry and Supergeometry

Starting in 1982 with Witten, and with Quillen and Bismut in 1986, most of the above various notions for bundles were reconsider at a super level, e.g. superspace, superconnection, supersymmetry, etc. In particular in Bismut's work, the Levi-Civita superconnection is fundamental for bringing the "super" concepts into Riemannian geometry. Since then super objects have arisen including connections, Laplacians and Dirac operators, asymptotic expansions of the heat kernel, and index theorems. The book Berline, Getzler & Vergne 1992 [179] is the ideal reference for all of this. For the holomorphic side, see Bismut, Gillet & Soulé 1988 [198, 199, 200].

14.3 Harmonic Maps Between Riemannian Manifolds

It was realized a long time ago that geodesics not only minimize length

$$L = \int \|\gamma'(t)\| \, dt$$

but also (if they are parameterized with constant speed) minimize the energy

$$E = \int \|\gamma'(t)\|^2 \, dt.$$

14.3 Harmonic Maps Between Riemannian Manifolds

The quadratic character, as opposed to the absolute value, makes the computation of the variational derivatives much easier. Moreover this energy can be interpreted as follows: it concerns a map γ from an interval of the real line into the Riemannian manifold and is defined using only the Riemannian structure of the manifold and the interval.

More generally, to a map $\phi : M \to N$ between two Riemannian manifolds (M, g) and (N, h) one can attach an *energy* $E(\phi)$ which is defined as follows: the map ϕ induces on M a pullback quadratic form $\phi^*(h)$. One then take the trace of $\phi^*(h)$ with respect to g.

Definition 408 *The energy of a map* $\phi : M \to N$ *between Riemannian manifolds* (M, g) *and* (N, h) *is*

$$E(\phi) = \frac{1}{2} \int_M \mathrm{tr}_g \, \phi^*(h) \, dV_M$$

for the canonical measure of (M, g). *The map* ϕ *is said to be* harmonic *if its energy* $E(\phi)$ *is critical, i.e. the derivative of the energy vanishes for all variations of* ϕ.

This very general notion was introduced in the founding paper Eells & Sampson 1964 [481]. The theory of harmonic maps from surfaces is quite special because harmonic maps are directly related to minimal surfaces thanks to the conformal representation. But in higher dimensions there is in general no direct connection between minimal submanifolds and harmonic maps, except in some special cases.

Since the Eells & Sampson paper, the subject of harmonic maps has had a rich history, both in answering the natural questions as well as in advancing applications. The reader will find references, in particular to surveys, in the book Eells & Ratto 1993 [480]. At its date of publication, the book Eells & Lemaire 1988 [479] was very systematic, informative and complete. Various regularity results are important. In some cases harmonic maps are defined in a more general context than purely Riemannian manifolds; see the books Jost 2002 [768], Helein 1996 [478] and for applications to Riemann surfaces Jost 2002 [767].

Harmonic maps are a basic tool in contemporary Riemannian geometry as very well illustrated in the Eells & Lemaire report. For example consider (1) their application to space forms in §§6.6.3, where the notion had to be extended to manifolds with singularities, (2) theorem 336 on page 588 for manifolds with positive curvature operator, (3) in Jost & Yau 1990 [770] for nonpositive curvature manifolds (also see §§§12.3.4.5, and the book Jost 1997 [766]). Harmonic maps are essential for studying Kähler manifolds, in particular their fundamental groups, see Amoros, Burger, Corlette, Kotschik & Toledo 1996 [35] and the references in §13.6.

14.4 Low Dimensional Riemannian Geometry

While Riemannian geometry in dimension 2 is wonderful, we have seen that the dimensions 3 and 4 have to be treated with care. For example, many finiteness theorems mentioned in §§12.4.1 hold only in dimension 5 or more. A contrario in some instances one has very strong results in dimensions 3 and 4.

One of the high hopes of Riemannian geometers in low dimension is to use Riemannian tools to solve topology and differential topology problems which are still haunting topologists. Recent reports are Donaldson 1996 [456], the books Donaldson & Kronheimer 1990 [457] and Morgan 1996 [939]. They present in particular the basic tools: Yang–Mills theory, twistors, anti-self-duality in dimension 4 and from Seiberg & Witten 1994 [1118] the Spin^c theory; see §§§14.2.2.8. A classic for the study of 3-dimensional manifolds is Scott 1983 [1116].[2]

For the "best metric" approach based on functionals and their critical points, and the link with the geometrization program of Thurston, see Anderson 1997 [48].

Finally we recall here the two results of Hamilton, using the Ricci flow, for the Ricci curvature in dimension 3 (see theorem 279 on page 525 and §§12.3.2) and for the curvature operator in dimension 4 (see theorem 337 on page 588).

14.5 Some Generalizations of Riemannian Geometry

The book Berestovskij & Nikolaev 1993 [147] is a systematic treatment of some of the following topics.

14.5.1 Boundaries

A first generalization is that of Riemannian manifolds with boundary.[3] Except for the η-invariant and in Besicovitch's results from the note 7.2.3.4 on page 352 (where there are "corners" as well as boundary) we never mentioned boundaries. For the general case one can see Alexander, Berg & Bishop 1990 [20] and the references there. For relations with positive scalar curvature curvature see Lawson & Michelsohn 1989 [850]. For the cut locus see Alexander, Berg & Bishop 1993 [19] and Alexander & Bishop 1998 [21]. For Einstein manifolds with boundary, see a nice rigidity theorem in Schlenker 1998 [1104].

We mention one interesting and practical problem in this area. Assuming some kind of convexity on the boundary ∂M of a Riemannian manifold M, one can define a distance $d(p,q)$ between points of ∂M. To which extent does this distance function determine the interior metric g (up to isometry

---

[2] For further remarks, added 2007, see page 723

[3] Obvious definition: one demands that at the boundary the manifold is smoothly equivalent to one side of an hyperplane in $\mathbb{R}^d$, so that the boundary is smooth.

14.5 Some Generalizations of Riemannian Geometry   677

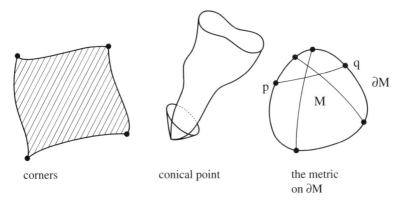

Fig. 14.2. (a) Corners (b) A conical point (c) The metric on the boundary ∂M of a Riemannian manifold M

of the interior). Thinking for a while of earthquakes, tomography, X-rays and scanners one sees the practical importance of such a problem. In an indirect way these ideas were also met in §§12.3.4 and in the "gap" results in §14.1. References are Otal 1990 [986], Michel 1994 [919], Croke 1991 [420], Arconstanzo 1994 [60], and Bourdon 1996 [235]. Also see the Besicovitch results in note 7.2.3.4 on page 352.

Returning to the subject of generalizations of Riemannian geometry, the subject has expanded very rapidly recently and results are flowing now so quickly that we will need to be very concise. Some cases appeared before sporadically; today strong incentives are coming from understanding Riemannian geometry more deeply in various instances, e.g that of §§§12.4.3.1, namely looking at limits of sequences of Riemannian manifolds. Interesting generalizations of Riemannian geometry can be of different types since of course to look at general metric spaces would be too general to obtain reasonably serious results. For example one could define objects playing the roles of tangent vectors, angles, curvature, etc. We will try to guide the reader through this new realm by suitable recent references and of course survey type ones when they exist. Of general type we know only of Berestovskii & Nikolaev 1993 [147].

14.5.2 Orbifolds

Orbifolds were introduced in Satake 1956 [1097] and in Baily 1957 [96] in algebraic geometry, and they were revived in Riemannian geometry by Thurston 1978 [1189]. Essentially they are quotients of Riemannian manifolds by isometries when one permits fixed points with moreover the isotropy group having to be finite. This is typical in the case of space forms. But they came about in more subtle situations, for example as limit spaces of some subsets of Riemannian manifolds with Ricci curvature bounded from below as well as when working on Einstein manifolds, e.g. theorem 379 on page 623; also see Anderson & Cheeger 1991,1992 [49, 50], Anderson 1992 [43], Anderson 1994 [45] and see also Ballmann & Brin 1995 [104]. For orbifolds and quaternionic holonomy see Galicki & Lawson 1988 [538]; for orbifolds in the Kähler realm see Tian 1990 [1190].

14.5.3 Conical Singularities

Spaces with an isolated conical singularity look quite simple, but for applications it is important to see what is happening to the spectrum (see chapter 6). More (and less) generally one can imagining studying Riemannian manifolds by approximating them by PL (piecewise linear) ones, metrically the pieces are all flat and "the curvature is concentrated (distributionally)" at the vertices, the edges, etc. These PL-*manifolds* are the very natural locally Euclidean version of general Riemannian manifolds but with singularities. What is happening to the curvature and to various formulas is the object of Cheeger, Müller, & Schrader 1984 [350] and Cheeger, Müller, & Schrader 1986 [351], see Lafontaine 1985/86 [843] for an expository text. This is in fact a very subtle subject. We saw one case in §§§7.1.2.3 for the isoperimetric profile of nonpositive curved manifolds.

14.5.4 Spectra of Singular Spaces

Cheeger carried out an extensive study of the spectral behaviour of singular spaces in a long series of papers; we refer to Cheeger 1983 [333] and its bibliography. This work came out of the proof in Cheeger 1979 [332] of the Ray–Singer conjecture for the Reidemeister torsion (see the very end of §9.14). One of its byproducts was in analysis, namely results on the diffraction of waves meeting obstacles. Another one was an explicit solution of the heat equation on standard spheres; for both see Cheeger & Taylor 1982 [355, 356]. A good transition to the next topic is found in Kuwae, Machigashira & Shioya 2000–2001 [841, 842].

14.5.5 Alexandrov Spaces

Alexandrov spaces are a wonderful class of metric spaces, because they enjoy two simultaneous properties. First they appear naturally in the boundary of the space of all Riemannian manifolds whose curvature is bounded from below. Second they share many properties with (smooth) Riemannian manifolds. There are many texts of more or less survey type: Perelman 1995 [1010], Berestovskii & Nikolaev 1993 [147], Reshetnyak 1993 [1055], Burago, Gromov & Perelman 1992 [282], Otsu & Shioya 1994 [989] and Yamaguchi 1992 [1283]; also see Kuwae, Machigashira & Shioya 2001 [842] for the Laplacian on Alexandrov spaces. We now give an extremely brief account.

14.5 Some Generalizations of Riemannian Geometry

These spaces appeared first in 1948 in the work of A. D. Alexandrov on convex surfaces, but in 1957 Alexandrov made a more systematic study. Then came the founding paper Burago, Gromov & Perelman 1992 [282]. In §§12.4.3 we wanted to look at the boundary of suitable subsets of the set of all Riemannian manifolds (compactness and convergence results). The description was possible (to some extent) for bounded curvature, say $-1 \leq K \leq 1$. But if all we knew was $K \geq k$ we had only a finiteness theorem and no description of collapsing. The solution lies in the theory of Alexandrov spaces with curvature bounded below.

On the one hand Alexandrov spaces of curvature bounded below are defined very simply. They are locally compact metric spaces which are *length spaces*, namely the metric coincides with the infimum of the length of curves joining the two given points. Note here a question of terminology. In Gromov 1981 [616] the wording was *length space*, while in Gromov 1987 [622] a length space is called a *geodesic space* if moreover any pair of points can be joined by at least one shortest path. Busemann 1955 [290] wrote of these as "intrinsic" metrics. Here such a length space is called *Alexandrov* with curvature at least k if it enjoys (besides completeness of course) everywhere locally a Toponogov comparison theorem (see theorem 73 on page 258) with the standard space form of constant curvature equal to k. One can then prove a global Toponogov theorem but also much more. Even with such a mild condition one has a notion of angles, a notion of a tangent cone which is itself an Alexandrov space of curvature at least zero, and can be defined in many equivalent ways. The points where this cone is not Euclidean are the singular points and the set of such points is of Hausdorff codimension at least 2, so that our spaces are almost everywhere locally Riemannian manifolds. Note that the singular set can be everywhere dense; just think of a suitable limit of convex polyhedra. It is important for generalizations of Riemannian results in the spirit of chapter 12 on page 543 that the notion of critical point for distance functions (see definition 305 on page 563) is still applicable and regions without critical points can be deformed into one another.

On the other hand, the definition of curvature being at least k is purely metric within the metric space $(\mathfrak{Met}, d_{\mathfrak{G}-\mathfrak{H}})$ which was introduced in theorem 380 on page 626. Therefore, limit points of Riemannian manifolds with sectional curvature at least k are automatically Alexandrov spaces with curvature at least k. Hence the results above constitute the answer to our main query and any further results on Alexandrov spaces will provide progress in the study of the boundary of the set of Riemannian manifolds with sectional curvature bounded from below. The following recent references show how active the topic is: Grove 1992 [645] extends some results of the first part of chapter 12. Shiohama & Tanaka 1992 [1135] studies cut loci and distance spheres. Yamaguchi 1992 [1283] generalizes the collapsing structure result of §§§12.4.3.1. Various results generalizing those in chapter 12 can be found in Petersen 1997 [1017], for example the bound for Betti numbers (see theorem 326 on page 580) of nonnegatively curved manifolds (see the references there).

In Cheeger & Colding 1997,1998 [337, 338, 339], many of the above results are generalized to "limit spaces with Ricci curvature bounded below." These spaces occur of course as limit spaces, in the Gromov–Hausdorff sense, of Riemannian manifolds with the same Ricci condition. For various generalized spaces, see Lohkamp 1998 [877]. Recall that Kontsevich is currently developing a picture of the space of Riemannian manifolds with nonnegative Ricci curvature.

14.5.6 CAT Spaces

In the opposite direction, one can look for a large class, but not too large, of spaces generalizing Riemannian manifolds of negative curvature bounded from above. A good notion is that of *hyperbolic space*, which was introduced and studied in Gromov 1987 [622] (not to be confused with hyperbolic space forms which are only a very very special case). For the spirit of the subject see §§12.3.4. After the founding paper Gromov 1987 [622] the topic became an entire world. We will just mention references: Gromov 1993 [629], Ghys & de la Harpe 1990 [558] and the book Bridson & Haefliger 1998 [258].

14.5.6.1 The CAT (k) Condition

Here the analogous condition of Alexandrov spaces of curvature bounded below for an upper bound of the curvature is the CAT (k) condition. Suppose for simplicity that we work in simply connected metric spaces. The CAT (0) condition is that one has, for every triple of points, the triangle inequality discovered by Cartan and rediscovered by Preissmann (see §§6.3.3). For CAT (k), it is the same condition but the comparison is no longer with Euclidean space but with hyperbolic space of constant curvature k. The CAT initials stand for Cartan, Alexandrov, Toponogov. One of the difficulties in these CAT (k) space is that they can have branch points since they can even be graphs. Think also of a one sheet hyperboloid converging toward a cone and look at the origin. A contrario one sheet of the two sheet hyperboloid associated with the same asymptotic cone will converge with a conical point (of positive curvature), a nice Alexandrov space.

Some examples of CAT (k) spaces are found among the PL-manifolds (piecewise linear i.e. Euclidean), although not all PL-manifolds are CAT (k). For PL-surfaces the CAT (0) condition is what you expect: at every vertex the sum of the angles of the triangles meeting there has to be at least 2π. For higher dimensional spaces, is not easy to explicitly describe the CAT (k) condition. The general CAT (k) condition goes the reverse way from the Alexandrov case. What happens to the curvature and to various formulas was mentioned above, see also Bourdon 1996 [235].

14.5.7 Carnot–Carathéodory Spaces

Other recent newcomers in generalizing Riemannian geometry are the Carnot–Carathéodory metric spaces. For them one can say in a caricatural sense that they have very strong singularities at every point. They are the subject of the book Bellaïche & Risler 1996 [126]. A particularly informative passage is the huge portion of this book consisting in the pages 85-323 written by Gromov. A Carnot–Carathéodory space is a manifold M with some vector subspace $V_m \subset T_m M$ of constant dimension smoothly chosen in each tangent space, and moreover on those subspaces one puts a Euclidean structure. Call a curve in M an *admissible curve* if its tangent line at every point m always belongs to V_m. The dimension of V_m is called the *rank*. The metric is then defined by taking the distance between two points to be the infimum of the lengths of admissible curves. Hence the other name for the topic: *sub-Riemannian geometry*. This distance can be infinite (and always will be if the rank is one) but it is always finite when the field of planes V is wild enough, that is to say "completely unintegrable." The game here is the opposite of the Frobenius theorem. Carnot–Carathéodory spaces are used in control theory since in this setting only certain velocites are permitted due to various restrictions imposed by practical situations. In street traffic the geodesics are what you have to do to park and unpark in a narrow slot. See the recent El Alaoui, Gauthier & Kupka 1996 [487] and the introductory text Pelletier & Valère Bouche 1992 [1007]. For the naive reader or someone most at home with partial differential equations, Bryant & Hsu 1993 [270] is extremely informative.

14.5.7.1 Example: the Heisenberg Group

The simplest example is the field of planes (of rank 2) in $\mathbb{E}^3$ which are simply the kernels of the differential 1-form $dz + x\,dy$. The plane at a point (x_0, y_0, z_0) satisfies the equation
$$z + x_0 y = 0.$$
This field of planes is not integrable and yields a metric as defined above. It is interesting to view in group theoretical language, namely as the Lie algebra of the Heisenberg group H, the simplest nilpotent group, made up of the real matrices
$$\begin{pmatrix} 1 & a & b \\ 0 & 1 & c \\ 0 & 0 & 1 \end{pmatrix}.$$
This Lie group has for its Lie algebra the vector space $\mathbb{R}^3$ with the Lie bracket relations
$$[x,z] = [y,z] = 0 \text{ and } [x,y] = z.$$
The field of planes above is the same as the set of left translates of the x,y plane.

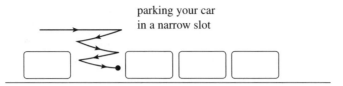

Fig. 14.3. Parking your car in a narrow slot; you can not pick your velocity in any direction other than the one it points in. You control the direction and magnitude of acceleration. The geometry on the unit tangent bundle is slightly more complicated than but similar to Carnot–Carathéodory geometry

14.5.8 Finsler Geometry

As very well explained in Spivak 1970 [1155] where he analyses Riemann's dissertation, Riemann in fact generalized Euclidean geometry by introducing Finsler manifolds. Those are differentiable manifolds M where at every point $m \in M$ is given some Banach structure, i.e. a convex body symmetric around the origin of the tangent space $T_m M$; think of the convex body as the "unit sphere." With this in place, lengths and hence a metric can be defined, as in Riemannian geometry. Riemann writes the prophetic sentence:

> *We will now stick to the case of ellipsoids [quadratic forms] because if not the computations would become very complicated.*
>
> B. Riemann

One can also relax the symmetry condition, e.g. in the Katok type examples mentioned in theorem 221 on page 463.

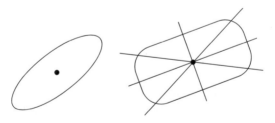

Fig. 14.4. Ellipsoids versus convex bodies. A non-ellipsoidal convex body always has some priviledged directions

There are few geometrical objects as natural as Finsler manifolds, e.g. in classical mechanics they arise whenever the kinetic energy is not quadratic. However Finsler geometry's destiny is not that of Riemannian geometry. It is clear I am biaised concerning the subject but I think that, before very recently, there were very few deep results in Finsler geometry which were not essentially Riemannian. Or, when it makes sense, Riemannian and not true for Finsler metrics. Up to very recently we knew two of them: Leon Green's theorem 257 on page 493 and the Burago–Ivanov theorem 265 on page 497. One counterexample was found in Busemann 1955 [290] where Finsler tori are constructed without conjugate points and not flat. The other was in Skorniakov 1955 [1148] where it is proven that any system of curves on $\mathbb{RP}^2$ satisfying the axioms of projective geometry (with reasonable smoothness) is the system of geodesics of some Finsler (non-Riemannian, usually, thanks to Desargues' theorem) metric; Green's theorem figures in theorem 258 on page 493. The paper Vérovic 1996 [1213] is quite interesting; it is proved there that the result of Besson, Courtois & Gallot 1995 [189] (see theorem 251 on page 484) can be true or false for Finsler manifolds, depending on the choice of different notions of measure in their unit bundle. In the other direction, most of the results in the "filling paper" Gromov 1983 [618] are valid for Finsler metrics.

Here is a personal comment. Finsler spaces look more general than Riemannian metrics. They are in some sense. But in another sense they are more special. In an affine (finite dimensional) real vector space a generic (symmetric) convex body has priviliged directions (points) e.g. those where the affine curvature is critical, etc. But an ellipsoid is isotropic, since all of the directions (the points) are equivalent. This is one reason for the importance of Riemannian geometry.

There are of course many places to find information on Finsler manifolds. Three good recent references are Bao, Chern & Shen 1995 [114], Álvarez Paiva 1991 [27] and Socié-Méthou 2000 [1152]. The note of Foulon 1997 [518] is a perfect blitz introduction to the modern point of view. We mention the fact that Finsler manifolds have a curvature, but the meaning of constancy of this curvature is not clear today. The result of the preceding note is again a rigidity result for compact negative curvature manifold. The positive case remains mysterious, but Bryant 1995 [268] sheds some light on it. Alvarez, Gelfand & Smirnov 1997 [28] characterize Finsler metrics for which the geodesics are the straight lines (Hilbert's fourth problem). Finally we note the use of a Finsler limit metric in the proof of theorem 265 on page 497 to the effect that only flat metrics on a torus can be without conjugate points.

14.5.9 Riemannian Foliations

The study of Riemannian foliations is also a topic in itself as is the study of totally geodesic foliations of Riemannian manifolds. Some books are Tondeur 1988 [1194] and Godbillon 1991 [571]. For the relation with scalar curvature see $1\frac{7}{8}$ of Gromov 1996 [631]. Look to Connes 1994 [400] for a new viewpoint on foliations.

14.5.10 Pseudo-Riemannian Manifolds

Lorentzian manifolds are at first glance similar to Riemannian ones. They are 4-dimensional manifolds equipped with a definite quadratic form in each tangent space, but of signature $(+, +, +, -)$. More generally a *pseudo-Riemannian manifold* may be of any dimensional and with definite quadratic form of any signature in each tangent space. At the beginning they have many things in common with Riemannian manifolds. This is definitely misleading, as was remarked by the physicist C. N. Yang, with the help of a beautiful picture: Yang 1980 [1285]. The picture can also be seen on page 11 of Besse 1987 [183]. The golden triangle is mainly what is common to both subjects. After that, both the questions and the results diverge. Lorentzian "geometry" derives its inspiration mainly from general relativity. Some books on Lorentzian geometry are: Beem, Ehrlich and Easley 1996 [125], Hawking & Ellis 1973 [684], O'Neill 1983 [976], Sachs & Wu 1977 [1082]. A special mention should be given to d'Ambra 1988 [425] (also see d'Ambra & Gromov 1991 [426]) where it is proven that the isometry group of a compact and simply connected Lorentzian manifold is itself compact (for the present, analyticity is needed but this looks like only a small technical point). For surfaces, Weinstein 1996 [1250] studies a Lorentzian conformal concept which is analogous to the concept of Riemann surface as opposed to that of a surface with a Riemannian metric. Of Riemannian flavor is Christodoulou & Klainerman 1993 [379]. See the expository Bourguignon 1991 [240]. Recently pseudo-Riemannian geometry appeared in various contexts, see for example Benoist, Foulon & Labourie 1992 [131] and Kühnel & Rademacher 1995 [837].

14.5.11 Infinite Dimensional Riemannian Geometry

Infinite dimensional Riemannian geometry is still very young. The book Lang 1972 [845] was the first to treat Riemannian geometry systematically in an infinite dimensional setting. Then appeared Klingenberg 1982 [815] and now the third edition Lang 1995 [847], and we suggest strongly that the reader look at it. The most systematic treatment of infinite dimensional Riemannian manifolds available today is Lang 1999 [848]. See also Bourbaki 1971 [234] and Gil-Medrano & Michor 1991 [560]. We mention only a few directions.

The first is to use suitable embeddings of Riemannian manifolds into the standard Hilbert space. This can be done for example by using suitably normalized eigenfunctions of the Laplacian; see Besson, Courtois & Gallot 1991 [188] and Bérard, Besson & Gallot 1994 [140] for the setting and applications to finiteness results.

One can also embed any Riemannian manifold M in the Banach space of continuous functions $C^0(M)$ by the "trivial" map made of the various distance functions, namely
$$m \to d(m, \cdot).$$
This childish map turned out to be essential in the systolic inequality of Gromov 1983 [618]; see theorem 137 on page 345.

The second most natural project is to define a notion of infinite dimensional Riemannian manifolds, starting with some infinite dimensional differentiable manifold M (whatever that means) and then endowing every tangent space

T_mM with some Hilbert space structure. This seems easy. As we see it, this is in fact difficult, at least at the state of the art, for the following reason: from a simple Hilbert space structure one gets in general incomplete metric spaces. With more complicated ones (typically in the Sobolev range) one has completeness but then the geodesics seem to have no good and/or useful geometrical properties; the geodesic equation is too complicated. In Ebin 1970,1972 [474, 475] such a structure is used to solve some fluid mechanics equations. The text Gil-Medrano & Michor 1991 [560] addresses the question of making a manifold with the set of all Riemannian metrics on a given manifold.

In appendix 2 of Arnol'd 1996 [66] the group of diffeomorphisms of a manifold is given an infinite dimensional Riemannian structure which is proven to be always of negative curvature.

For path space Riemannian geometry see Stroock 1996 [1164]. For infinite dimensional Kähler manifolds, see Huckleberry & Wurzbacher 2001 [748].

14.5.12 Noncommutative Geometry

Finally special attention should be given to Connes 1994 [400]: this is the beginning of a complete program to put most (perhaps all) geometries into a very general frame. The frame is that of algebras of operators and the approach is called "noncommutative geometry." In this frame, according to various extra suitable axioms one can recover almost any kind of geometry, including of course Riemannian ones, from families of operators. The next step in the program will be to generalize every concept one could wish, e.g. curvature. For the metric itself this is done and we saw in §§§14.2.2.7 that one can compute the distance with operators on spinors. See also the note 11.4.1.1 on page 523 for the Hilbert functional in this context.

14.6 Gromov's *mm* Spaces

Gromov's *mm* spaces are, in our opinion, the geometry of the future, redefining what we mean by a geometric space, to unify the subjects of probability and metric geometry.

We will sketch here, separately from the preceding section, the new geometric spaces discovered in chapter $3\frac{1}{2}$ of Gromov 1999 [633]. We do this because of their importance now and in the future. We refer of course to Gromov's book for more, but we will try to describe in some detail the contents of this chapter and its spirit. We first quote the author:

> *We humbly hope that the general ambiance of* $\mathcal{X}$ *can provide a friendly environment for treating asymptotics of many interesting spaces of configurations and maps.*
>
> Misha Gromov 1999 [633]

What is the $\mathcal{X}$ in question? The actual metric geometries, even with all their variety, seem not able to provide a basis for research programmes like: (1) studying all the possible configurations of a living organism and its trajectories as a function of time (its "life"); (2) estimating as a function of time the mean diameter of planar nonselfintersecting Brownian motion; improve, both in depth and in rigour, classical results or "facts" of statistical mechanics; (3) construct a geometric theory of probability, i.e. establish a law of large numbers for suitable geometric spaces. A part of this story is known, either formally in the frame of probability theory, or more or less heuristically in statistical mechanics. Still it remains a major task to lay the foundations of an axiomatic theory large enough to treat in one shot all of the above problems, as well as many other unsolved ones.

Here is the answer proposed in the $3\frac{1}{2}$ chapter, which we explain now briefly. One should realize, to understand this program, that the notion of measure is more important than that of metric. In Riemannian geometry, the measure came canonically from the metric, and we used this fact at length in this book, e.g. in the whole of chapter 7, and typically in theorem 318 on page 572. Now, the same way that Riemann dissociated the Euclidean metric from the vector space structure to replace it by a manifold, Gromov has dissociated measure and metric. The new geometry he created is that of *mm spaces* (where *mm* stands for metric and measure). An *mm* space is, by definition, a triple (X, μ, d) where first the measure μ should make (X, μ) a *Lebesgue–Rochlin space*, that is to say that (X, μ) is measure-isomorphic to a real segment $[0, m[$ with a countable set of atoms (points of positive measure). We will only use spaces of finite total measure m. And one only asks that the metric d on (X, μ, d) be a measurable function on $X \times X$. Let us denote by $\mathcal{X}$ the set of all *mm* spaces. For a systematic exposition of analysis on metric spaces, which consists in studying metrics constructed to assist in solving problems of analysis, see the book Heinonen 2001 [697].

Although it is our principal desire to calculate probabilities, it seems that there cannot exist on $\mathcal{X}$ a reasonable measure, but at least there is canonical distance, called by us here d_Γ. This distance is defined only for two spaces of the same total mass. As compared to $d_{\mathfrak{G}-\mathfrak{H}}$(see theorem 380 on page 626) this distance is not very geometric; at any rate it is very hard to visualize it. It is defined as follows: having two *mm* spaces (X, μ, d) and (X', μ', d') one considers all of the measure isomorphisms (which exist by definition)

$$\phi : X \to [0, m[, \phi' : X' \to [0, m[.$$

On the square $[0, m[^2$ one transfers d and d' (and keeps the same notation for them):

$$d, d' : [0, m[^2 \to \mathbb{R}.$$

One then introduces the "almost-distance" $\varepsilon(\phi, \phi')$ between d and d', defined as the smallest ε such that the set of all $t \in [0, m[^2$ with $|\phi(t) - \phi'(t)| < \varepsilon$ is of measure smaller than ε. Then $d_\Gamma(X, X')$ is the infimum, for all of the possible parametrizations ϕ, ϕ' of X, X', of the above $\varepsilon(\phi, \phi')$. It is not easy to verify that $d_\Gamma(X, X') = 0$ implies that X and X' are isomorphic as mm spaces. But it is very pleasant that the proof introduces for mm spaces the following notion which generalizes that of sectional curvature for Riemannian manifolds.

The definition is as follows: for evey integer r one considers the set M_r of all $r \times r$ symmetric matrices. On every M_r there is a natural measure, namely the μ-mass of all r-tuples of points of (X, d, μ) whose mutual distances are precisely given by the elements of the matrix under consideration. The set of these measures, as r runs through the integers, suffices to reconstruct the metric d. This is equivalent to a classical problem of "moments" of functions. A brief reflexion might convince the reader from this, for example after thinking of the extreme case of constant curvature, that there are universal relations for the mutual distances of $(d+2)$-tuples in Hyp^d or S^d, or for Euclidean space $\mathbb{R}^d$ (see section 9.7 of Berger 1994 [167] and Berger 1981 [161]).

The fact that d_Γ is a much better notion, for measure theoretic applications, than $d_{\mathfrak{G}-\mathfrak{H}}$ is seen in figures 14.5: in $d_{\mathfrak{G}-\mathfrak{H}}$ one sees convergence toward a sphere with an interval added ("hair"), but for the d_Γ convergence, only the sphere remains. And of course a physicist, or any person concerned with statistical measurements, will not see the hair.

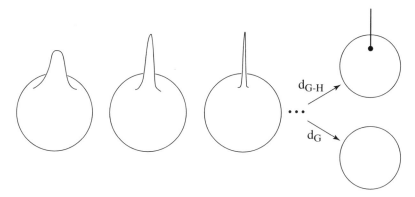

Fig. 14.5. Comparing Gromov and Gromov–Hausdorff limits

The second notion introduced by Gromov is that of *observable diameter* $\mathrm{ObsDiam}\,(X, k)$ (more precisely k-observable with a power of resolution normalized to be equal to 1). By definition $\mathrm{ObsDiam}\,(X, k)$ is the smallest ε such that, for any numerical function f which is 1-Lipschitz on (X, d, μ), there is in $\mathbb{R}$ a subset A with the property that $\mathrm{diam}(A) < \varepsilon$ together with

$$\mu\left(f^{-1}(A)\right) \geq \mu(X) - k.$$

This comes from the desire in physics to define the "visible diameter" of X with regard to various measurements of X. Now, says Gromov, let us just take

$k = 10^{10}$ once and for all for simplicity's sake, and that is what we will do now. The notation is now simply $\mathrm{ObsDiam}(X)$. The geometric law of large numbers which Gromov is after consists in studying the asymptotic behavior of $\mathrm{ObsDiam}(X)$ when the dimension of X tends to infinity. Here is a selection of his results from the $3\frac{1}{2}$ chapter. First:

Theorem 409
$$\mathrm{ObsDiam}\left(S^d\right) = O\left(\frac{1}{\sqrt{d}}\right),$$

although the usual diameter is always equal to π. This is nothing but a reformulation in the present language of a result of Lévy 1951 [861], obtained in fact as early as in 1919. But the new result is that

$$\mathrm{ObsDiam}\left(X^d\right) = O\left(\frac{1}{\sqrt{d}}\right)$$

for the class of Riemannian manifolds X^d with Ricci $> d - 1$ (remember that the Ricci curvature of the standard sphere is equal to $d - 1$). Gromov says that "there is in Paul Lévy everything for this result." One uses for this phenomenon the word *concentration*. Concentration is preserved by d_Γ. To realize the depth of this, one finds in the $3\frac{1}{2}$ chapter examples where there are points of zero Ricci curvature and for which the result fails—there is no longer concentration. These examples consist simply in convex surfaces of revolution with flat points.

Theorem 410 (Gromov's "strong law of large numbers") *For any given compact Riemannian manifold X, write $\prod^n X$ for the n-fold Cartesian product of X with itself. Then*

$$\mathrm{ObsDiam}\left(\prod^n X\right) = O(1).$$

Notice that the diameter of $\prod^n X$ is of order $O(\sqrt{n})$. For a law of large numbers, but where the emphasis is on general functions (Gromov here works with Lipschitz ones) see Talagrand 1995 [1180].

As a first contribution to algebraic geometry in infinite dimension, one has, for any complex algebraic manifold $X \subset \mathbb{CP}^n$ of degree d and of codimension k

$$\mathrm{ObsDiam}(X) \leq \left(\frac{\log n}{n}\right)^{1/2d}$$

when n tends to infinity with k and d fixed. This is the hardest result proved in the $3\frac{1}{2}$ chapter. In short one replaces the positivity of the Ricci curvature by algebraicity. The proof is very long and intricate. It belongs to what Gromov calls "the muddy waters of metric algebraic geometry." If the reader is surprised by this statement, if he doubts that these waters are "muddy," he is advised to consider that one still does not know the behaviour, with respect to their degree, of the intrinsic diameter of the algebraic surfaces of $\mathbb{CP}^3$. For curves the answer is in Bogomolov 1994 [215] and not at all obvious: for algebraic curves in $\mathbb{CP}^2$ the diameter is bounded when the degree is given but there are algebraic curves with arbitrary large diameter. In $\mathbb{CP}^n$ algebraic submanifolds of bounded degree have bounded diameter: we have all of the tools for that, since they are minimal submanifolds by the results from §§§14.2.1.3. Hence every ball within the injectivity radius has a volume which is not too small and the total volume is given by the degree as seen again in §§§14.2.1.3.

Gromov also makes a deep study of concentration in the "fibers." This is the beginning of a study analogous to that of collapsing in §§§12.4.3.1.

Gromov shows that cubes are very mysterious. On one hand, by the above

$$\mathrm{ObsDiam}\left(\prod^n [0,1]\right) = O(1),$$

but if one considers the "void" cube $\prod^n \{0,1\}$, then

$$\mathrm{ObsDiam}\left(\prod^n \{0,1\}\right) = O\left(n^{1/4}\right).$$

"Holes are expensive." For the simplex of dimension n, one has an observable diameter of $O(1/n)$. This is a reformulation of a "classical fact" of statistical mechanics.

Many other accomplishments appeared in the $3\frac{1}{2}$ chapter. An important one consists in defining another metric on $\mathcal{X}$. This new metric looks extremely complicated to the present author, but it has the basic property that if a sequence of metric spaces X_k verify

$$\mathrm{ObsDiam}\,(X_k) \to 0 \text{ as } k \to \infty$$

(e.g. spheres), then the sequence converges for this metric toward a point. This is definitely not the case for d_Γ.

Finally Gromov defines a spectrum for any mm space (remember that the spectrum is a robust invariant). The construction copies the construction of §9.4 and §§1.8.3. It will be enough to show how to define $\lambda_1(X)$, the first eigenvalue of (X, d, μ). The idea is to use Dirichlet's principle, i.e. to minimize the quotient

$$\frac{\int_X \|\nabla f\|^2 \, d\mu}{\int_X f^2 \, d\mu}$$

on the functions of zero integral. This makes sense since we have a measure. It remains only to define $\|\nabla f\|$. But just take the Lipschitz constant of f. This makes sense in any metric space. We end with the beautiful inequality between $\lambda_1(X)$ and the observable diameter (even though the proof is simple):

$$\mathrm{ObsDiam}\,(X, k) \leq \log k^{-1}/2\sqrt{\lambda_1(X)}.$$

14.7 Submanifolds

We are concerned that this topic deserves an entire book, and we are sorry not to discuss it at greater length here. Among other motivations, Riemannian geometry arose is trying to generalize the geometry of surfaces in $\mathbb{E}^3$. There is no reason not to consider the geometry of submanifolds inside a given Riemannian manifold. Because Euclidean geometry is hereditary for subspaces, it follows that Riemannian geometry is also hereditary for submanifolds (but of course the metric is not the induced one in the rough metric sense; that works only for totally geodesic submanifolds), see §§6.1.5. This is one reason more why Riemannian geometry is important. We do not know of any systematic survey for this topic, even in the special and much studied case of submanifolds of euclidean spaces, except for Chen 2000 [358]. Few books give the general equations for a submanifold of any codimension in a Riemannian manifold (the so-called Gauß and Codazzi–Mainardi equations). Among them: chapter VII of Kobayashi & Nomizu 1963-1969 [827, 828], chapter XX (see section 20.14.8) of Dieudonné 1969 [446] and chapter 7 of Spivak 1979 [1155] present the foundations of the theory of submanifolds in a Riemannian manifold. But there is a huge collection of local and of global results. Recent surveys are Hsiang, Palais & Terng 1988 [741], Terng 1990 [1184], Terng & Thorbergsson 1995 [1185], Thorbergsson 2000 [1186], Palais & Terng 1988 [993], see also Willmore 1993 [1272]. In particular, isoparametric hypersurfaces deserve a special mention as relating different topics. The founding text is Cartan 1939 [317]. For isoparametric submanifolds in an infinite dimensional setting, see Heintze & Liu 1999 [700].

14.7.1 Higher Dimensions

An important remark is in order, which was well known around the turn of the century and was already explained in §§4.4.3.5. Surfaces in $\mathbb{E}^3$ are important not only for historical reasons but because of this. When $d \geq 4$ take a generic hypersurface in $\mathbb{E}^d$. Then the sectional curvature (hence the inner metric isometry type) determines the second fundamental form. Classically this implies congruence (this is a purely local statement). The proof is: in a basis which diagonalizes the second fundamental form, the Gauss equations imply that

$$K(e_i, e_j) = h_i h_j$$

where the h_i are the principal curvatures (this is exactly the generalization of the theorema egregium from §§3.1.1). And now, if I know the three products ab, bc and ca, then I know a, b, c as long as none is 0. Thus the metric of a generic hypersurface determines its second fundamental form. The generalized Gauß–Codazzi equations show that together the metric and the second fundamental form determine completely the embedding (see any of the books quoted at the begining of this section). In some sense, there is no "congruence versus isometry" problem left. Of course the above philosophy applies also to hypersurfaces in general Riemannian manifolds M^d for $d \geq 4$.

Things are completely different when $d = 3$. The Gauss curvature (hence the isometry type) yields only the product of the two principal curvatures. Then there is room for a fascinating game, not finished yet. Let us look for example at rigidity: two compact strictly convex hypersurfaces which are abstractly isometric are congruent (matched up by a global isometry of the whole $\mathbb{E}^d$). For $d > 3$ it follows from the above remark, but when $d = 3$ the result remains but the proof is much more sophisticated and is due to Cohn-Vossen and Herglotz. For references and more see Berger & Gostiaux 1988 [175], section 11.4 and Klingenberg 1978 [813] 6.2.8.

This being realized, there remain many topics in the field, especially global questions and we already mention some of them above. Integral geometry is very interesting, but works only (with one exception to be mentioned shortly) in Euclidean spaces and space forms. See the survey Schneider & Wieacker 1993 [1107] and add the classic Santalo 1976 [1093], in particular for Chern's kinematic formula. In Riemannian manifolds, integral geometry works for hypersurfaces and geodesics starting from them. This is basic for Croke's local isoembolic inequality theorem 149 on page 355 and the isoperimetric inequality for nonpositive curvature manifolds in §§§7.1.2.3: see Croke 1980 [411] and Croke 1984 [416].

14.7.2 Geometric Measure Theory and Pseudoholomorphic Curves

Submanifolds enjoying various strong geometric properties are fundamental to many investigations. Two seem to be particularly important today. First the dramatic appearance of *geometric measure theory* (also known as *GMT*) with its harvest of results provided Riemannian geometers with almost all of the existence theorems they could dream of for proving the existence of minimal (or constant mean curvature) objects. This was used above, in the Schoen–Yau approach to positive scalar curvature in theorem 355 on page 600 and for the isoperimetric profile in §§7.1.2 and the classical isoperimetric inequality in §§1.6.8. The theory of "GMT" is an extremely difficult subject. In the founding book Federer 1969 [511] as well as in many other articles on the subject, it very hard to find one's way through in a reasonable time. We suggest reading first Morgan 2000 [937] and then Simon 1983 [1138].

The second application of weird submanifolds appeared in the pioneering paper Gromov 1985 [619]. A recent survey book is Audin & Lafontaine 1994 [88]. Although those pseudoholomorphic curves are mainly used in symplectic geometry, one meets them also in Riemannian geometry; see for example the article of Labourie in the preceeding book and Gromov 1992 [632]. For a two-page sketch, see section 3 of Berger 1998 [169].

15 The Technical Chapter

Contents

15.1	Vector Fields and Tensors	693
15.2	Tensors Dual via the Metric: Index Aerobics	696
15.3	The Connection, Covariant Derivative and Curvature	697
15.4	Parallel Transport	701
	15.4.1 Curvature from Parallel Transport	703
15.5	Absolute (Ricci) Calculus and Commutation Formulas: Index Gymnastics	704
15.6	Hodge and the Laplacian, Bochner's Technique	706
	15.6.1 Bochner's Technique for Higher Degree Differential Forms	708
15.7	Gauß–Bonnet–Chern	709
	15.7.1 Chern's Proof of Gauß–Bonnet for Surfaces	710
	15.7.2 The Proof of Allendoerfer and Weil	711
	15.7.3 Chern's Proof in all Even Dimensions	713
	15.7.4 Chern Classes of Vector Bundles	714
	15.7.5 Pontryagin Classes	715
	15.7.6 The Euler Class	715
	15.7.7 The Absence of Other Characteristic Classes	716
	15.7.8 Applying Characteristic Classes	716
	15.7.9 Characteristic Numbers	716
15.8	Examples of Curvature Calculations	718
	15.8.1 Homogeneous Spaces	718
	15.8.2 Riemannian Submersions	719

We cannot give comprehensive references for this chapter, especially for the generalities. They appear in every book on differential geometry and Riemannian geometry. Only in some special instances will we give references.

15.1 Vector Fields and Tensors

We are on a differentiable manifold M, and it is smooth, i.e. C^∞. In §§4.2.1 we constructed its tangent bundle TM, which is to be seen as a bundle

$$\pi : TM \to M$$

over M. A *vector field* X on M is a (differentiable, as many times as needed, which is always understood) section of TM i.e. a map $X : M \to TM$ such that $\pi X(m) = m$ for any point $m \in M$. This is the modern sophisticated way to say that X is a differentiable choice of a vector $X(m)$ tangent to M at each point m, with m running through M and the map $m \to X(m)$ differentiable. Vector fields can be defined in a short by the way they act on numerical functions $f, g : M \to \mathbb{R}$: they should verify the axiom

$$X(fg) = X(f)g + fX(g)$$

for the product of functions. The operation means, geometrically, differentiating the function f in the direction of X.

We recall from §§4.2.1 the basic fact that without a Riemannian metric there is no second order calculus. A map on functions like $f \to X(Y(f))$ is not a vector field. The reason is that tangent spaces at M at different points cannot be identified. A Riemannian metric will provide an identification for nearby points. However an essential fact is that $f \mapsto X(Y(f)) - Y(X(f))$ verifies trivially the above axiom, so that it is a vector field. It is called the *bracket* (or *Lie bracket*) of X and Y and denoted by $[X, Y]$. For three vector fields X, Y, Z one verifies trivially the Jacobi circular permutation property:

$$[X, [Y, Z]] + [Y, [Z, X]] + [Z, [X, Y]] = 0.$$

The bracket has a basic geometric interpretation: to compute $[X, Y](m_0)$, one follows for a small time ε the integral curve of X, and then, from the attained point, the integral curve of Y, and then in reverse time again the integral curve of X, then of Y. The point you end up at, m_ε, does not coincide with the original point m_0 in general, but the curve described by the point m_ε parameterized by ε has velocity zero at $\varepsilon = 0$, and acceleration $[X, Y](m_0)$. When $[X, Y] = 0$ one says that X and Y commute. The bracket is the defect of commuting; in coordinate systems $\{x_i\}$ the vector fields $\frac{\partial}{\partial x_i}$ commute with one another. Brackets are also a way to know when a distribution of k dimensional subspaces of the tangent space can be integrated into a submanifold; this is the celebrated *Frobenius theorem*, proven in any introductory book on differential geometry.

To every real vector space V of dimension d one attachs canonically many new vector spaces, the *tensor products* of type (r, s). The simplest to define are those of type $(0, s)$. They are nothing but the multilinear forms (valued in the real numbers) over V. And those of type $(1, 0)$ are just V itself. Those of type $(r, 0)$ will be linear combinations of simple tensor products of vectors, like

$$x_1 \otimes \cdots \otimes x_r.$$

A quadratic form is a tensor of type $(0, 2)$, which is moreover symmetric. The tensors of type $(r, 0)$ are just the duals of those of type $(0, r)$, and those of type

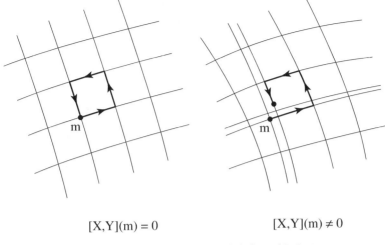

Fig. 15.1. (a) $[X,Y](m) = 0$ (b) $[X,Y](m) \neq 0$

(r,s) are the tensor products of those of type $(r,0)$ with those of type $(0,s)$. The most simple mixed type one is $(1,1)$. Its elements are the endomorphisms of V. A notation can be $\bigotimes^{r,s} V$. It is convenient notation to consider tensors of type (r,s) as multilinear maps valued in the r-th tensor power of V.

The canonicity of the extension $\bigotimes^{r,s}$ enables us to define, for our manifold M, the bundle $\bigotimes^{r,s} TM$ whose fibers are the $\bigotimes^{r,s} T_m M$ when m runs through M. A tensor of type (r,s) on M is just a section of $\bigotimes^{r,s} TM$. In the $(0,s)$, and when moreover one restricts oneself to exterior multilinear forms (i.e. antisymmetric ones), we introduced these as *differential forms* in §§4.2.2 and denoted the spaces of them by $\Omega^s(M)$. The curvature tensor below will be defined as of type $(1,3)$ but often changed into a tensor of type $(0,4)$. For the curvature tensor, $(1,3)$ can be interpreted as an exterior 2-form valued in the endomorphisms of TM since $3 = 1 + 2$.

To check that some object is a tensor one can either employ some conceptual trick like canonicity, or (the hard way) check that it behaves as it should under coordinate changes. For theoretical purposes as below, the following global trick was discovered in the fifties. We state it only for forms. To check that some multilinear form

$$\alpha(X_1, \ldots, X_k)$$

is really a tensor, i.e. that it depends only on the values $X_1(m), \ldots, X_k(m)$ at the point m (and not, for instance, on derivatives), it is enough to check that

$$\alpha(X_1, \ldots, fX_i, \ldots, X_k) = f\alpha(X_1, \ldots, X_i, \ldots, X_k) \qquad (15.1)$$

for every index i and every real valued smooth function f on the manifold.

This simple definition works only in the C^∞ case. If one is not in such a situation, one should proceed directly. For example check that the bracket, although antisymmetric, is not a differential 2-form.

15.2 Tensors Dual via the Metric: Index Aerobics

We are now on a Riemannian manifold M. We ask for neither completeness nor compactness, unless explicitly stated. The first thing to do is to carry over to all tensors the Euclidean structure coming from the tangent bundle. It is classical that a Euclidean structure extends canonically to any tensor power. On any of these tensor spaces will denote the scalar product (the quadratic form) by $\langle , \rangle$ and by $\| \|^2$ the squared norm.

Next we define the canonical musical duality.[1] It starts with the duality between a vector v and 1-form v^*: the form v^* is defined by

$$v^*(x) = g(v, x)$$

for every vector x (where g is the metric) and the vector α_* constructed from a 1-form α is determined by the condition

$$\alpha(v) = g(\alpha_*, v)$$

for every vector v. This duality extends to any Euclidean tensor of type (r, s) and tranforms it at your will into one of type $(r+k, s-k)$. And this extends pointwise trivially to $\bigotimes^{r,s} TM$. We will use this freely. The most important example is the gradient of a function:

$$\nabla f = (df)^*$$

i.e.

$$g(\nabla f, x) = df(x)$$

for any vector x. Recall that for vector fields X

$$df(X) = X(f)$$

by the very definition. The musical operators preserve the norms.

A third operation can be carried out in a Riemannian manifold, which is taking the trace with respect to the metric g of a tensor of type (r, s) to obtain a tensor of type $(r-1, s-1)$. In classical literature, this is called the *contraction* of upper and lower indices. The place where the trace takes place has to be made precise. We do it here for the first two places of a tensor of type $(0, s)$ called α:

---

[1] These dualities are called *musical* because they are often written with symbols like $\flat : V \to V^*$ and $\sharp : V^* \to V$.

$$\mathrm{tr}_g \, \alpha\,(x_3,\ldots,x_s) = \sum_{i=1}^d \alpha\,(x_i, x_i, x_3, \ldots, x_s)$$

for any orthonormal basis $\{x_i\}$. We use the simplest case to define the *Laplacian* of a numerical function f: we take the trace

$$\Delta f = -\,\mathrm{tr}_g \nabla^2 f$$

(with a minus sign) of the $(0,2)$ tensor $\nabla^2 f = \nabla df$, which is sometimes called the *Hessian* of f and sometimes written Hess f.

15.3 The Connection, Covariant Derivative and Curvature

We embark now on describing the *golden triangle* of Riemannian geometry, which is made of three elements: the *curvature* (via the connection), the *parallel transport* and the *absolute calculus* (i.e. an intrinsic differential calculus of all orders). This triangle was first understood by Levi-Civita and Ricci at the start of the XXth century. The basic lemma, which is the key to everything, is the existence and the uniqueness of a canonical connection, called the *Levi-Civita connection*, on any Riemannian manifold. It is important to realize that this lemma is a miracle. Many people have tried to understand it, with more or less sophisticated concepts, but we consider that it remains a miracle.

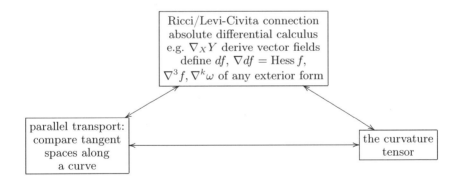

Fig. 15.2. The golden triangle

Let us be more precise. We would like to be able to develop on a Riemannian manifold a differential calculus of all orders. In affine spaces the tangent spaces at various points can be all identified with a single vector space by translations and that is enough. But this is not possible a priori in a smooth manifold since there is no way to compare tangent spaces at different points. Levi-Civita's discovery was that such a comparison is possible infinitesimally in a unique way, if one demands that the Euclidean structures of the tangent spaces be preserved and moreover the second differential of a function be symmetric, see §15.5. The precise notion of such a connection was seen in various ways, mostly in coordinates. Élie Cartan used moving frames to describe it. The notion of connection was introduced in various settings by Élie Cartan, and then Charles Ehresmann put it in the most general frame in Ehresmann 1951 [482]; for more history see note 2 in volume I of Kobayashi & Nomizu 1963-1969 [827, 828]. One can also use the notion of horizontal space in the tangent bundle to the tangent bundle.

Since Koszul's work, the best way of writing the canonical connection (as well as for connections in various bundles with ad hoc modifications) is to define it as an operation $\nabla_X Y$ on pairs of vector fields. Koszul introduced this concept in it 1951. It was used soon in Nomizu 1954 [969] and finally presented in a systematic expository way in Koszul 1960 [831]. Geometrically we want this to enable us to compare infinitesimally closed tangent vectors. This means that we will estimate by $\nabla_X Y$ the defect in the direction X of the vector field Y to be invariant (parallel). See figure 15.3.

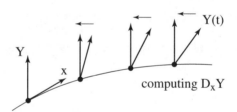

Fig. 15.3. Computing $\nabla_X Y$

Definition 411 *A* linear connection *in a differentiable manifold M is a bilinear map $D_X Y$ from pairs of vector fields X, Y into vector fields which is linear in X and a derivation on Y. Namely, if f is any real valued function, then*
$$D_{fX} Y = f D_X Y$$
but
$$D_X fY = X(f) Y + f D_X Y.$$
The torsion *of D is*
$$T_D(X, Y) = D_X Y - D_Y X - [X, Y].$$

(The reader can check that this is a tensor.) It is said to be torsion-free *if moreover*

15.3 Connection, Derivative and Curvature

$$D_X Y - D_Y X = [X, Y]$$

for every pair X, Y of smooth vector fields.

The axiom $D_{fX}Y = f D_X Y$ ensures for us that, as above, $D_X Y(m)$ depends only on the vector $X(m)$. But for Y one has to know more, for example it is enough to know the restriction of Y to a curve starting from m with velocity $X(m)$. For example, for any curve γ the notation $\gamma'' = \nabla_{\gamma'} \gamma'$ makes sense, and is called the *acceleration* of the curve. It is characteristic of Riemannian manifolds that curves in them have acceleration, something which does not exist in a general differentiable manifold. If for example the speed of γ is constant, i.e. $\|\gamma'\| = 1$, then the acceleration γ'' is always orthogonal to γ'. A crucial fact is that geodesics are exactly the curves whose acceleration vanishes (see equation 4.25 on page 200)

$$\nabla_{\gamma'} \gamma' = 0,$$

and the speed is automatically constant, since

$$\gamma' \|\gamma'\|^2 = 2 \langle \nabla_{\gamma'} \gamma', \gamma' \rangle = 0.$$

Note that $\nabla_X Y$ is not a tensor of type $(1, 2)$. But

$$Z \to \nabla_Z Y$$

is a tensor of type $(1, 1)$.

With a little extra work we can make sense out of applying any connection D to form $D_X Y$ when X and Y are vector fields tangent to a manifold M, but defined only along a curve C (or other immersed submanifold). The resulting $D_X Y$ is also only defined along that curve. Given an oriented curve C on a Riemannian manifold M, and a connection D, we have the obvious choice of vector field defined along C, the unit tangent field; call it ∂_s (using unit speed parameter s). Then we can define the *covariant derivative* of a vector field Y defined along a curve C, written $\dot{Y}$, or $\frac{dY}{ds}$, again forming a vector field defined along C.

On a Riemannian manifold (M, g) a connection will be of interest only if it preserves the metric g:

$$X(g(Y, Z)) = g(D_X Y, Z) + g(Y, D_X Z).$$

This condition is not enough to determine D, but here comes the miracle:

Theorem 412 (The Fundamental Theorem of Riemannian Geometry (Levi-Civita)) There is one and only one torsion-free connection ∇ preserving g and it is called the *Levi-Civita connection* or simply, the connection of (M, g).

700   15 The Technical Chapter

The trick is only to write the three conditions of preservation for three vector fields X, Y, Z and perfom the combination

$$Xg(Y, Z) + Yg(Z, X) - Zg(X, Y).$$

Using the vanishing torsion condition yields immediately the value of $g(\nabla Y, Z)$ is a function of explicit expressions in only g, X, Y, Z. This being true for any Z yields the value of $\nabla_X Y$. In the flat case, i.e. when we are in a Euclidean vector space, the value $\nabla_X Y(m)$ is just the derivation with respect to the vector $X(m)$. This makes sense since Y belongs to a fixed vector space. Let us see what remains, for general Riemannian manifolds, of the rule of commutation of derivatives in the classical calculus in vector spaces. So we look at the possible defect of $\nabla_X \nabla_Y - \nabla_Y \nabla_X$ from its Euclidean value $\nabla_{[X,Y]}$. If the reader does the computation to see the behaviour when X is changed into fX, then the reader will be soon convinced that this expression is a tensor. Then one checks, using the criterion above, that:

Proposition 413 *The expression*

$$\nabla_X \nabla_Y - \nabla_Y \nabla_X - \nabla_{[X,Y]}$$

is in fact a tensor, of type $(1,3)$*, the* curvature tensor *of* (M,g)*.*

By definition, the tensor $R(X,Y)Z$ is given by

$$R(X,Y)Z = \nabla_Y \nabla_X Z - \nabla_X \nabla_Y Z - \nabla_{[Y,X]} Z.$$

Then one can speak of $R(x,y)z$ for tangent vectors x, y, z and note that $R(x,y)z$ is itself a vector. The curvature tensor is identically zero for flat manifolds, and we saw the converse: §§4.4.1 and §§6.3.2. We have put a negative sign (say we exchange X and Y) in order that the curvature of a sphere be positive, or equivalently that the sign be the right one to get back Gauß curvature K (be aware that many authors use a different choice of sign here). The curvature is then a good measure of the defect from being locally Euclidean. Now we use duality to get from R a $(0,4)$ tensor

$$R(x, y, z, t) = g\left(R\left(x, y, z, y\right)\right).$$

The symmetries of this $(0,4)$ tensor claimed in equation 4.28 on page 203 are:

$$R(x, y, z, t) = R(z, t, x, y) = -R(x, y, t, z)$$
$$R(x, y, z, t) = -R(y, x, z, t) \tag{15.2}$$
$$R(x, y, z, t) + R(y, z, x, t) + R(z, x, y, t) = 0 \quad \text{First Bianchi identity}$$

and they stem out of the definition by direct computations. The last one is just a little combinatorics as seen in §§4.4.1. We recall that proposition 413 as it stands is not enough to prove the second variation formula of equation 6.7 on page 242. One needs some extra computation in coordinates or more general stuff about induced connections which we do not really need for applications.

15.4 Parallel Transport

The marvelous thing, even if trivial, is that the connection can be integrated along any curve (but not on a surface in general). Take a curve $\gamma : [0,1] \to M$ and consider some vector $x_0 \in T_{\gamma(0)}M$. For a family of vectors $\{x(t) \in T_{\gamma(t)}M\}$ the equation

$$\nabla_{\gamma'(t)} x(t) = 0$$

is a first order linear ordinary differential equation. It therefore has a unique solution such that $x(0) = x_0$:

Definition 414 *The solution $x(1)$ is called the vector parallel transported from x_0 along the curve γ. Parallel transport is a Euclidean isomorphism*

$$\gamma_{0 \to 1} : T_{\gamma(0)}M \to T_{\gamma(1)}M.$$

For example, if one identifies all of the $T_{\gamma(t)}M$ with $T_{\gamma(0)}M$ by that parallel transport, then for a vector field X along the curve γ the vector $\nabla_{\gamma'(t)}X$ is in fact nothing but

$$\frac{dX}{dt} = X'(t).$$

This is the language we used systematically in the book in various places, starting in proposition 61 on page 240. For the geometric "construction" of the parallel transport, see the figures near proposition 61 on page 240. Another interpretation is given just below.

Note that parallel transport inside flat manifolds, at least locally, is trivial. Since it is the vector space parallelism, along a closed loop, it will be the identity. But in general, if γ is a loop, i.e. if $\gamma(0) = \gamma(1)$, then the Euclidean isomorphism $\gamma_{0 \to 1}$ belongs to the orthogonal group $O\left(T_{\gamma(0)}M\right)$ but is not the identity.

We note here that, although the proof of the first variation formula equation 6.3 on page 225 is straightforward with the above language, the proof of the second variation formula 6.7 on page 242 and that of the Jacobi field equation 6.11 on page 248 are always lengthy and technical. See most books on Riemannian geometry. The difficulty is that one uses proposition 413 on the facing page directly only when the surface of variation is embedded. A typical difficulty arises when some of the curves of the variation have a fixed point, i.e. don't vary.

Note 15.4.0.1 (The horizontal language and the canonical metric on TM) For a geometer it is nice to interpret the connection in the tangent bundle $\pi : TM \to M$. We pick up a vector $v \in TM$ and set $m = \pi(v) \in M$. Now look at the tangent space $T_v(TM)$; here, for any manifold, we have a canonical subspace, namely the space tangent to the fiber T_mM. It is convenient to say that $T_v(T_mM)$ is the *vertical space* of $T_v(TM)$ and to write it $\text{Vert}(v)$.

702   15 The Technical Chapter

The connection provides us with a canonical complement of Vert(v), called the *horizontal space* at v and denoted by Horiz(v). An element $w \in$ Horiz(v) is the initial speed vector w at v of the curve in TM made up of tangent vectors in M which are parallel transported from v along a curve in M with some initial speed vector $x \in T_m M$. Then of course x is equal to the image of w under the differential of π. Conversely, if one is given such a horizontal distribution in TM, the parallel transport along a curve c downstairs in M consists in integrating the horizontal vector field so obtained in the surface $\pi^{-1}(c)$. The direct sum is

$$T_v(TM) = \text{Horiz}(v) \oplus \text{Vert}(v).$$

♦

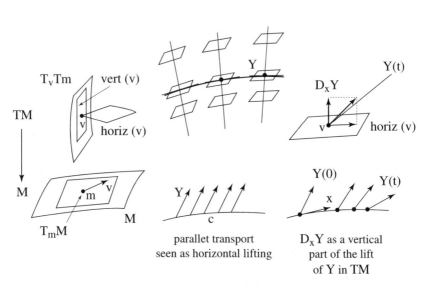

Fig. 15.4. Parallel transport seen as horizontal lifting

Now suppose that we are in a Riemannian manifold. Since the differential $d\pi$ of the projection, as restricted to Horiz(v), is a vector space isomorphism one can define a Riemannian metric on TM as the direct Euclidean sum of g_m, the Euclidean structure on $T_m M$, with the Euclidean structure on Horiz(v) obtained from g_m lifted by $d\pi$ on Horiz(v). This yields the canonical Riemannian metric on TM. For the unit tangent bundle UM one will get an induced metric. This metric played a basic role in chapter 10, one of the main points being that the geodesic flow leaves invariant its canonical measure. This submersion

$$(TM, g_{\text{can}}) \to (M, g)$$

is a particular case of a Riemannian submersion, see §15.8 or §§4.3.6.

15.4.1 Curvature from Parallel Transport

The first arrow in the golden triangle tells us that the curvature is the infinitesimal defect from the identity map of parallel transport along small loops. More precisely, consider two commuting vector fields X and Y. We will look near a point m in a Riemannian manifold M (see §15.1) at the small parallelograms $P(\varepsilon)$ generated by X and Y. Because the vector fields commute, the parallelograms close up at m. If $\varepsilon_* \in O(T_m M)$ denotes parallel transport along the parallelogram $P(\varepsilon)$, the derivative

$$\left.\frac{d}{d\varepsilon}\varepsilon_*\right|_{\varepsilon=0}$$

makes sense—it belongs to the Lie algebra of $O(T_m M)$ (its tangent space at the identity element) and is an endomorphism of $T_m M$. Recall that the Riemann curvature tensor R is in $\Lambda^2(\mathrm{End}(T_m M))$.

$$\left.\frac{d}{d\varepsilon}\varepsilon_*\right|_{\varepsilon=0} = R(X(m), Y(m)) \tag{15.3}$$

or

$$\left.\frac{d}{d\varepsilon}\varepsilon_* z\right|_{\varepsilon=0} = R(X(m), Y(m)) z$$

where $x = X(m), y = Y(m)$ and we pick any $z \in T_m M$.

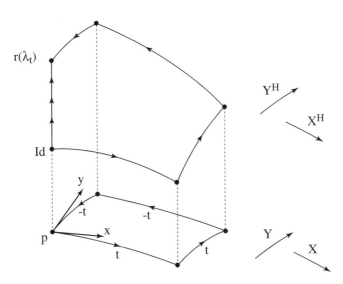

Fig. 15.5. Parallel transport of commuting vector fields

This can be summarized by saying that the curvature measures exactly the defect of the horizontal distribution in TM to be integrable. This was used in chapter 13. We mention now some natural questions, even if not (at least today) fundamental. The first would be to look for a global formula based on note 15.4.0.1, some kind of an integral of the curvature which will yield γ_* for any loop γ, not only infinitesimal ones, but in the spirit of the definition of holonomy groups in §13.1. To our knowledge such a formula exists only in Nijenhuis 1953/1954 [965, 966]. It can be seen as a very geometric explicit realization of the Ambrose–Singer theorem 396 on page 643.

The second question is the *Ambrose problem*. It stems out of Élie Cartan's philosophy seen in §§6.3.1. Cartan proved that a Riemannian metric is determined locally by the knowledge of an exponential map at a given point p and the knowledge of the curvature tensor along the corresponding geodesics. The Ambrose problem asks for a global version of this. For such a result one needs to go up to the cut locus. We saw at the end of §§6.5.4 that this problem has been solved sucessfully only for surfaces.

15.5 Absolute (Ricci) Calculus and Commutation Formulas: Index Gymnastics

We will now introduce the third vertex of the golden triangle, namely the *Ricci (absolute) calculus*. It consists in being able to define, thanks to the canonical connection, a differential calculus of any order which is intrinsic. In particular any tensor T will admit a derivative of any order k, which will be a tensor of type $(r, s+k)$, denoted by $\nabla^k T$, and called the k-th *covariant derivative* of T. The definition is trivial for a linear algebra minded reader. The connection provided us with a first covariant derivative for vector fields X, which is a tensor of type $(1,1)$ written $Z \mapsto D_Z X$. One denotes it by simply ∇X. An important example in computations, as soon as one encounters groups of isometries, is that of a *Killing vector field*. This is the vector field on the manifold made up by the derivative at the origin of a one-parameter group of isometries. Such a field X is Killing if and only if its covariant derivative is antisymmmetric:

$$g\left(\nabla_x \xi, y\right) + g\left(x, \nabla_y \xi\right) = 0$$

for any vectors x and y. We did not meet Killing fields before, because the idea of this book is to skip all computations, but they appear in most, if not all, of the various examples we quoted of manifolds with various remarkable properties.

We have just to extend this notion of covariant derivative to any tensor of type $(r, 0)$ since tensor products of vector fields generate them, and thereafter to multilinear forms by duality, etc. We write the explicit formula only for tensors of type $(0, s)$ and leave to the reader to check that for any tensor α of degree $(0, s)$ the formula

15.5 Absolute Calculus and Commutation Formulas

$$Z \mapsto \nabla_Z (\alpha(X_1, \ldots, X_s))$$
$$= Z(\alpha(X_1, \ldots, X_s)) - \sum_i \alpha(X_1, \ldots, \nabla_Z X_i, \ldots, X_s) \quad (15.4)$$

defines a tensor of degree $(0, s+1)$, which one can denote by $\nabla \alpha$ and write also as $\nabla \alpha (X_1, \ldots, X_s; Z)$.

Then one of course define for any k the k-th derivative $\nabla^k \alpha$, and manage to do the same for any tensor $\mathcal{T}$, e.g. for the curvature tensor ∇R will be a tensor of type $(1, 4)$. A direct computation yields the *second Bianchi identity* for the curvature tensor: For any x, y, z tangent vectors

$$\nabla_x R(y, z) + \nabla_y R(z, x) + \nabla_z R(x, y) = 0 \quad (15.5)$$

which is needed in some problems seen above.

We will now fulfill our promise for the second absolute derivative of a function to be symmetric. Let f be any real valued function and let us look at $\nabla^2 f(X, Y) - \nabla^2 f(Y, X)$. Using only the definitions we get

$$\nabla^2 f(X, Y) - \nabla^2 f(Y, X) = Y(X(f)) - df(\nabla_Y X) - X(Y(f)) + df(\nabla_X Y)$$
$$= [Y, X](f) - df(\nabla_X Y - \nabla_Y X)$$
$$= 0.$$

by the torsion-freedom of the Levi-Civita connection in definition 411 on page 698. Note that the above symmetry is in fact equivalent to this torsion freedom.

Note 15.5.0.1 (Riemannian invariants) Many people think that the curvature and its derivatives are the only Riemannian invariants. This is true and classical when looking at the strong condition of algebraic invariants which stem from the connection, see page 165 of Schouten 1954 [1111] and the references there. In fact the proof easily follows from Epstein 1975 [490]. But things are dramatically different if one asks only for tensors which are invariant under isometries (called "natural"). There is no hope to get any kind of classification of natural tensors as explained in Epstein 1975 [490]. For more see Muñoz Masqué & Valdés 1996 [951]. ♦

We can now complete one missing arrow in the golden triangle: to go from parallel transport to Ricci calculus, i.e. to compute $\nabla \mathcal{T}$ with parallel transport. Pick any curve γ, any basis $\{X_i\}_{i=1}^{i=d}$ of $T_{\gamma(0)} M$ and parallel transport it into $\{X_i(t)\}$ along γ. Then all of the $\nabla_{\gamma'(t)} X_i(t)$ vanish by construction and equation 15.4 tells us that the desired covariant derivative of a $(0, s)$ form α is nothing but the ordinary derivative

$$\frac{d}{dt} \alpha(X_1(t), \ldots, X_s(t)) \Big|_{t=0} = \nabla \alpha(X_1, \ldots, X_s, \gamma'(0)).$$

Only the third arrow remains: the relation between the curvature and the absolute calculus. One can discover it by looking at the third derivative $\nabla^3 f$ of a function. A direct computation yields:

$$\nabla^3 f(x,y,z) - \nabla^3 f(x,z,y) = df(R(y,z)x)$$

so that the curvature can be extracted for the tensors $\nabla^3 f$ for various functions. The commutativity of derivatives stops at the third order. In fact it fails always for any tensor which is not simply a function, i.e. not of order $(0,0)$. But the Ricci commutation formulas enable us to compute the defect of symmetry of $\nabla_X \nabla_Y \mathcal{T} - \nabla_Y \nabla_X \mathcal{T}$ for any tensor. We give the formula for $(0,s)$ tensors:

$$(\nabla_X \nabla_Y \alpha - \nabla_Y \nabla_X \alpha)(X_1,\ldots,X_s) = \sum_i \alpha(X_1,\ldots,R(X,Y)X_i,\ldots,X_s). \tag{15.6}$$

As applied to a closed 1-form and in a global (compact) context, this is the basis of the Bochner technique, as we are going to see post haste. The computations above are straightforward.

15.6 Hodge and the Laplacian, Bochner's Technique

We will work here exclusively on compact manifolds. Because of its importance and the importance of its many generalizations, still not exhausted, we will explain in detail the groundbreaking work of Bochner 1946 [210], even if we have already mentioned it briefly in theorem 346 on page 595, theorem 338 on page 588, and in §§12.3.3.5. Suppose we want to prove that a harmonic function f on a Riemannian manifold is constant, without a clever application of the maximum principle. We could compute $\Delta(f^2)$ as for the second derivative of a square:

$$-\frac{1}{2}\Delta(f^2) = \|df\|^2 - f\Delta f = \|df\|^2$$

since $\Delta f = 0$. We integrate this result using the definition of Δ, to the effect that the integral of the Laplacian of any function vanishes. We are left with

$$\int_M \|df\|^2 = 0$$

so that f is constant.

We now want to do something similar but for exterior forms, ignoring for the moment the Laplacian for them which we introduced in §14.2. Since for a function the Laplacian was

$$-\operatorname{tr}_g \nabla^2 f,$$

15.6 Hodge and the Laplacian, Bochner's Technique

we define the *rough Laplacian* of a form ω (and the same works for a tensor of any type) by
$$\underline{\Delta}\omega = -\operatorname{tr}_g \nabla^2 \omega$$
for the second covariant derivative of ω. Then we compute the Laplacian of the squared norm of ω :
$$-\frac{1}{2}\Delta\left(\|\omega\|^2\right) = \|\nabla\omega\|^2 - \langle\underline{\Delta}\omega, \omega\rangle$$
and integration will yield, for a rough harmonic form $\underline{\Delta}\omega = 0$ the conclusion that
$$\nabla\omega = 0,$$
i.e. ω is a parallel form. But we know from chapter 13 that such parallel transported forms exist almost never, so that our computation leads nowhere in general. We turn to the Hodge–de Rham theorem 405 on page 665 with the Laplacian defined there, with great hopes since these harmonic forms reflect some of the topology of the manifold. But this time $\Delta\omega$ is no longer equal to $-\operatorname{tr}_g \nabla^2 \omega$. At various places in the computation we have to employ equation 15.6 on the preceding page because we are no longer working with mere functions. Remember that the canonical connection has precisely the property that second derivatives commute for functions. The computation for differential forms was done by Bochner, first for 1-forms in Bochner 1946 [210], the result being:
$$-\frac{1}{2}\Delta\left(|\omega|^2\right) = |D\omega|^2 - \langle\Delta\omega, \omega\rangle + \operatorname{Ricci}(\omega_*, \omega_*). \tag{15.7}$$

Now integration over the compact manifold implies immediately that Ricci > 0, forbids any nonzero harmonic form ω (with $\Delta\omega = 0$), and moreover Ricci ≥ 0 permits only parallel transported ones, hence implies a reduction of the holonomy group of the manifold: theorem 345 on page 594 via Hodge–de Rham. It is interesting to note that Bochner was not sure at that time of the solidity of the Hodge theorem, which was only solidly established by de Rham later on, so that his phrasing is quite interesting to look at. This is called the *Bochner vanishing technique*.

Applications of equation 15.7 do not stop here. Let us mention four. The two first consist in taking in equation 15.7 for our form ω the differential of a real-valued function: $\omega = df$, so that we have the formula used in equation 6.20 on page 265 for comparing triangle geometry when Ricci curvature is lower bounded and if one uses the Hessian terminology for $\nabla df = \nabla^2 f$:
$$-\frac{1}{2}\Delta\left(\|df\|^2\right) = \|\operatorname{Hess} f\|^2 - \langle df, \Delta df\rangle + \operatorname{Ricci}(df, df) \tag{15.8}$$

One can use (local) harmonic functions, as was done in §§6.4.2. Let us note that $\nabla(df) = d(\nabla f)$ and take for f an eigenfunction of the Laplacian: $\Delta f = \lambda f$ on an compact manifold. Then one obtains theorem 181 on page 408:

if Ricci $\geq d - 1$ then $\lambda_1 \geq d$. The third application was already in Bochner's paper of 1946: if the Ricci curvature is negative (and the manifold is compact) then there are no Killing vector fields X (infinitesimal isometries). One integrates equation 15.8 on the preceding page over the manifold. The integral of a Laplacian vanishes, and a little computation on Killing fields (see for this and the next theorem Petersen 1997 [1018]) shows that the formula boils down to

$$\int_M \|\nabla X\|^2 - \int_M \mathrm{Ricci}(X, X) \geq 0.$$

This takes care moreover of the case Ricci ≤ 0. The last application is theorem 323 on page 577: on a compact manifold of even dimension with positive sectional curvature, a Killing field has to vanish somewhere. The proof here is not by integration, but looking (with the help of equation 15.8 on the preceding page and the computations just mentioned for Killing fields) at a point where $\|\nabla X\|$ is a minimum.

15.6.1 Bochner's Technique for Higher Degree Differential Forms

For the theory of higher degree exterior forms, it is enough to say that the defect between the rough and the good Laplacians is expressed as

$$\Delta w - \underline{\Delta} w = \mathrm{Curv}_p\left(R; \omega, \omega\right)$$

where $\mathrm{Curv}_p\left(R; \omega, \omega\right)$ is a very complicated expression, namely a quadratic form in ω with coefficients linear in the curvature tensor. A short way to write this is

$$\Delta = dd^* + d^*d = \nabla^*\nabla + \mathrm{Curv}\left(R\right).$$

For historians we note that it was remarked in de Rham 1954 [1056] page 131 that the above formula for Δ with $\mathrm{Curv}_p\left(R\right)$ was already in fact in Weitzenböck 1923 [1252] so that Bochner 1948 [212] rediscovered it; note that Weitzenböck had no global application.

Thereafter people tried to see what results one can get for other degrees of differential forms; see among others Lichnerowicz 1952 [864] and Bochner & Yano 1952 [213]. But it was finally in Meyer 1971 [915] that is was made clear that it is the curvature operator which completely governs the term $\mathrm{Curv}_p\left(R\right)$: see theorem 338 on page 588. A very lucid exposition of this fact is in Lawson & Michelsohn 1989 [850] theorem 8.6. In §§14.2.2 we encountered a Bochner type formula for spinors on a spin manifold. But now, looking at §§14.2.3, it is clear that there is also a Bochner type formula for any Riemannian bundle and its Laplacian Δ. The main point is to find the nontrivial difference term $\Delta - \nabla^*\nabla$. In the above examples only the curvature of the base Riemannian manifold entered. There is no reason that this will be the case in general. What sort of vanishing result could be obtained? The game is to consider suitable bundles and suitable Laplacians. Let us mention some cases. The Bochner vanishing technique is so important that there exist three surveys concerning it: Wu 1988 [1278], Bérard 1988 [136], Bourguignon 1990 [239].

First, for the canonical tensor bundles over the manifold, there a natural Laplacian which was discovered in Lichnerowicz 1961 [866] with a term $\Delta - \nabla^*\nabla$ involving only the curvature tensor. It turned out to be especially useful for symmetric 2-tensors, because they represent the variations of Riemannian metrics; see chapter 12 of Besse 1987 [183]. There is also the complex Laplacian for Kähler manifolds; see §13.6, and it is used a lot in algebraic geometry. In §§§12.3.3.5 we mentioned the use of bundles twisted in some clever way using the spin-bundle in order to study the scalar curvature, the point being that in the $\Delta - \nabla^*\nabla$ term the scalar curvature of the basis is dominant.

There are also special Bochner type formulae when the holonomy group is special (rare, see chapter 13). This is the key for proving the arithmeticity of some space forms, see §§6.6.3 for more and for references.

15.7 Generalizing Gauß–Bonnet, Characteristic Classes and Chern's Formulas

In the late 1920's Hopf wondered about extending the so-called *Gauß–Bonnet theorem* 28 on page 138 which says that for a compact surface M and its curvature K the Euler–Poincaré characteristic is given by

$$\chi(M) = \frac{1}{2\pi} \int_M K.$$

Consequently if K has a given sign then χ has the same sign. Hopf asked about higher dimensions: when the sectional curvature of M has a given sign, does the Euler–Poincaré characteristic (see §§§4.1.4.2) $\chi(M)$ have the expected sign? We said in §12.1 that Hopf managed to prove it in the very special case of space forms (see §§6.6.2) in Hopf 1925 [727] and for hypersurfaces of Euclidean spaces in Hopf 1926 [728]. For hypersurfaces, Hopf's proof consisted in proving equation 3.19 on page 141:

$$\chi(M) = \frac{1}{2} \deg(N) \qquad (15.9)$$

where $N : M \to S^d$ is the generalized Gauß map of the hypersurface M^d in $\mathbb{E}^{d+1}$. Hopf was probably guessing that in the general case it would follow from a generalization of the Gauß–Bonnet formula to higher dimensions. We will see that this is wrong starting in dimension 6. A generalized Gauß–Bonnet formula was indeed obtained in Allendoerfer & Weil 1943 [24]. We comment now a little about this generalization of Gauß–Bonnet, which is not treated in most Riemannian geometry books. Exceptions are: Spivak 1970 [1155] whose chapter 7 is entirely devoted to this, Gray 1990 [583] which has the proof with tubes (which was the intermediate proof of Allendoerfer, using Hermann

Weyls's tube formula and historical remarks), but chapter XII of Kobayashi & Nomizu 1963-1969 [827, 828] covers the whole present section. For the history of characteristic classes, see Dieudonné 1989 [448].

15.7.1 Chern's Proof of Gauß–Bonnet for Surfaces

In Chern 1944 [363] Riemannian geometers were provided with a very conceptual proof of the Allendoerfer–Weil formula. Chern's proof was so nice and also so profound that we will give it in detail. Moreover it gives a conceptual explanation of the local Gauß–Bonnet formula 18 on page 112. We can consider the oriented case, because the characteristic of the oriented two-sheet covering is just twice the characteristic of the nonorientable manifold it covers. A vector field X on a surface M, with isolated zeros m_i for $i = 1, \ldots, k$ yields a 1-1 map $q = X/\|X\| : M\setminus\{m_i\} \to UM$ into the unit tangent bundle. Now on UM there is a canonical 1-form α where $\alpha(x)$ is simply the cosine of the angle of x with the positive unit tangent vector to the fiber (an oriented circle here). The miracle is that the differential $d\alpha$ of α (in UM) is nothing but the inverse image $\pi^* K\, dA$ of the 2-form obtained by multiplying the volume 2-form dA by the Gauß curvature K. Assume first that M is the torus T^2 and pick any unit vector field X; then

$$\begin{aligned}
\int_M K\, dA &= \int_{q(M)} \pi^* K\, dA \\
&= \int_{q(M)} d\alpha \\
&= 0 \\
&= \chi\left(T^2\right)
\end{aligned}$$

by Stokes' theorem since $q(M)$ has no boundary. Now in the general case take very small disks D_i around the zeros x_i of our vector field, and set $D = \cap_i D_i$ and consider $M' = M\setminus D$, a surface with boundary. Apply Stokes' theorem to $q(M') \subset UM$ and note that the boundary of $q(M')$ is the union of the inverse image circles $q(D_i)$.

$$\begin{aligned}
\int_{M'} K\, dA &= \int_{q(M')} \pi^* K\, dA \\
&= \int_{q(M')} d\alpha \\
&= \int_{\partial q(M')} \alpha \\
&= \sum_i \int_{q(\partial D_i)} \alpha.
\end{aligned}$$

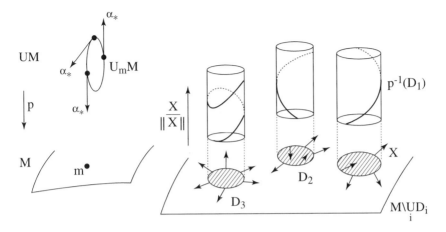

Fig. 15.6. Chern's proof of the Gauß–Bonnet theorem

Now, by the very definition of the form α, the integral

$$\int_{q(\partial D_i)} \alpha$$

tells us how much the vector field X turns around its zero x_i. This is by definition the *index* of X at the point x_i denoted by $\mathrm{Index}\,(X, x_i)$. One can think of it as the Gauß map for plane curves, i.e the number of times that the circle S^1 is covered by the normalized vector field considered as a map from S^1 to S^1. This is in fact true only at the limit, when the radii of the disks go to zero (one need not take a limit in the flat case, but then $K = 0$). Finally, when all the radii go to zero,

$$\int_{M'} K\,dA \to \int_M K\,dA$$

and

$$\sum_i \int_{q(\partial D_i)} \alpha \to \sum_i \mathrm{Index}\,(X, x_i).$$

But by a theorem of Hopf (formula 15.12 on page 713), the latter sum is equal to the Euler-Poincaré characteristic $\chi(M)$. q.e.d.

15.7.2 The Proof of Allendoerfer and Weil

Chern 1944 [363] extends this kind of proof to any (even) dimension. The formula of Allendoerfer and Weil was proven as follows. We are on a Riemannian manifold of dimension $d = 2n$. We again work in the oriented case, as per the remark above. The invariant **K** formed with the curvature which generalizes the determinant of the second fundamental form of a hypersurface in $\mathbb{E}^N$ (see §12.1) can be defined explicitly as follows in any orthonormal oriented basis $\{e_i\}$. We define exterior 2-forms $\Omega_{i,j}$ by

$$\Omega_{i,j}(x,y) = R(e_i, e_j, x, y)$$

(for every pair of tangent vectors x, y). Then $\mathbf{K}$ is defined by a sum of exterior products as

$$\mathbf{K} = \frac{1}{d!} \sum_{i_1...i_d} \varepsilon_{i_1...i_d} \Omega_{i_1 i_2} \wedge \cdots \wedge \Omega_{i_{d-1} i_d}$$

where the sum runs through all permutations $i_1, \ldots, i_d$ of the numbers $\{1, \ldots, d\}$ and $\varepsilon_{i_1,\ldots,i_d}$ denotes the sign of this permutation. And finally we have:

$$\chi(M) = \frac{2}{\operatorname{Vol}(S^d)} \int_M \mathbf{K}. \tag{15.10}$$

(By construction, $\mathbf{K}$ is a d-form.)

Even on unorientable manifolds, one gets an intrinsic scalar $\mathbf{K}$ equal to

$$\mathbf{K} = \mathbf{K}(e_1, \ldots, e_d) \ .$$

It is independent of orientation in the same way as for example for surfaces $\Omega_{1,2}(e_1, e_2)(e_1, e_2) = K$ because exchanging e_1 and e_2 will introduce a negative sign twice. The reader might like to contemplate how to express it with the

$$R_{ijkh} = R(e_i, e_j, e_k, e_h) \ ,$$

namely:

$$\mathbf{K} = \frac{1}{d!} \sum_{i_1...i_d j_1...j_d} \varepsilon_{i_1...i_d} \varepsilon_{j_1...j_d} R_{i_1 i_2 j_1 j_2} \cdots R_{i_{d-1} i_d j_{d-1} j_d} \tag{15.11}$$

where the summation is over all possible choices of two permutations i and j of $\{1, \ldots, d\}$.

In Allendoerfer 1940 [23] this formula is proven for submanifolds of Euclidean space. The Nash embedding theorem 46 on page 218 was not known at that time. Allendoerfer's proof goes as follows: let M^d be a compact submanifold of $\mathbb{E}^N$. The generalized Gauß map this time goes from the normal unit bundle νM to S^{N-1}, where νM is the set of all vectors in $\mathbb{R}^N$ which are orthogonal to TM. For a hypersuface, say oriented, νM is identified with M itself. Now Hopf's formula 15.9 on page 709 is valid again for this generalized Gauß map:

Theorem 415 *For* $N : UM \to S^{N-1}$ *again* $\chi(M) = \frac{1}{2} degree(N)$.

As in §3.4 if we still denote by $d\sigma$ the canonical volume form of the standard sphere S^{N-1}, we still have

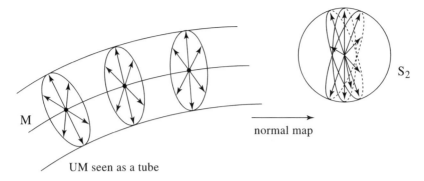

Fig. 15.7. The Gauß map of a submanifold of Euclidean space

$$\deg(N) = \frac{\int_{UM} N^*(d\sigma)}{\int_{S^{N-1}} d\sigma}.$$

It remains only to show that

$$N^*(d\sigma) = K dV_M.$$

This is done by Allendoerfer by using the formula discovered in Weyl 1939 [1256] for computing the volume of a tube generated by a submanifold of $\mathbb{E}^N$. For tubes and relations with this formula, see the book Gray 1990 [583] enterely devoted to tubes or see Berger & Gostiaux 1988 [175], chapter 6.

The definitive proof of Allendoerfer & Weil 1943 [24] was not too very conceptual and used triangulations, local isometric real analytic embedability and approximations, still using embedding and Weyl's tube formula. We note here that Allendoerfer & Weil have a formula for manifolds with boundary, generalizing exactly the Gauß-Bonnet formula 18 on page 112.

15.7.3 Chern's Proof in all Even Dimensions

The above shows that it is extremely desirable to have a proof of the formula 15.10 on the preceding page obtained really by working on the manifold itself. The proof of Chern uses again Hopf's formula (see equation 15.9 on page 709):

$$\chi(M) = \sum_x \text{index}_x(\xi) \tag{15.12}$$

where ξ is a vector field on M with a finite number of singular points x.

We use the same notations as above, take again small balls around singularities, etc. For the even dimensional $d = 2n$ manifold, one considers on the unit tangent bundle $\pi : UM \to M$ the differential form which replaces the α above, that is to say the volume $(2n-1)$-form, noted again by α, of the spheres $U_m M$. But, as soon as $d \geq 4$ then its differential $d\alpha$ is no longer the pullback $\pi^* K\, dV_M$ (where dV_M is again the volume form of M) of some 2n-form on M^{2n}. What Chern does is to prove that there are forms ϕ_i for $i = 1, \ldots, n-1$ such that first

$$\pi^* K\, dV_M = d\left(\alpha + \sum_i \phi_i\right).$$

Second, the integrals of the various ϕ_i vanish when the balls get very small. This concludes the proof. Having obtained such a clever method to compute the Euler–Poincaré characteristic of a Riemannian manifold as a function of the curvature, Chern thought about doing the same for the other invariants of differentiable manifolds M which were known in his time, namely the *Stiefel–Whitney classes* $w_i \in H_i(M, \mathbb{Z}_2)$. But these classes live in the $\mathbb{Z}_2$ cohomology of M, and hence are not accessible through differential forms via de Rham's theorem. One way to define these classes is to use a suitable embedding of the manifold $f : M^d \to \operatorname{Grass}(d, k+d)$ into a Grassmann manifold (see §§4.1.3) where the integer k has to be taken large enough. Then such a map is the exact generalization of the Gauß–Rodrigues map for surfaces in $\mathbb{E}^3$, where the image of a point of M in the Grassmann manifold is the tangent plane $T_m M$. The manifold $\operatorname{Grass}(d, d+k)$ has some cohomology classes over $\mathbb{Z}_2$ and the classes w_i are their pullbacks. Moreover, this is the place to say that Whitney in fact extended Stiefel classes from the case of the tangent bundle to any vector bundle over any compact manifold.

15.7.4 Chern Classes of Vector Bundles

But Chern, reading Ehresmann's work, learned that the complex Grassmann manifolds $\operatorname{Grass}_{\mathbb{C}}(d, d+k)$ have analogous classes, but this time these classes are over the integers, hence also defined over the real numbers. In the pioneering paper Chern 1946 [364] he did two things: he first defined *characteristics classes* for any complex vector bundle V over a manifold. They are now called the *Chern classes*

$$c_i(V) \in H^{2i}(M, \mathbb{Z}) \text{ for } i = 1, \ldots, \operatorname{rk}_{\mathbb{C}} V.$$

Second he proved that those classes can be represented, via the de Rham theorem, by closed exterior differential forms Ch_i of (real) degree $2i$ which depend only (and in a canonical way) on the curvature of any linear connection on V (see §§14.2.3). If the curvature is presented here as an exterior 2-form Ω which is valued in endomorphisms of the bundle V then Ch_i is a linear combination of exterior powers of Ω. It is very simple: one computes the determinant

$$\det\left(\lambda I - \frac{1}{2\pi}\Omega\right) = \sum_i \operatorname{Ch}_i \lambda^i.$$

A lucid exposition is chapter XII of Kobayashi & Nomizu 1963-1969 [827, 828]; for the topological aspects see Milnor & Stasheff 1974 [925] and Husemoller

1994 [751]. Besse has a few pages on them in 2.E. But we should now come back to earth, since we want real objects, not complex ones, at least in general. And also we want to see the connection with the Euler–Poincaré characteristic. We now briefly explain this. But before we get on to that, there is one exception if M is a complex manifold, e.g. Kähler. Then the Chern classes $c_i(M)$ of M are the Chern classes of this tangent bundle, which is here a complex vector bundle. Especially important for us is the first one $c_1(M)$ for Kähler manifolds, since it is essentially the Ricci curvature: see §§11.4.4. In general the form Ch_i is, by the above, a universal polynomial $\text{Ch}_i(R)$ of degree i in the curvature tensor.

15.7.5 Pontryagin Classes

If M has no complex structure, e.g. is of odd dimension, the trick is to complexify its tangent bundle to $TM \otimes_\mathbb{R} \mathbb{C}$ and to look at the classes $c_i(TM \otimes_\mathbb{R} \mathbb{C})$. It turns out that they vanish for all odd values of i. The *Pontryagin classes* $p_i(M)$ are by definition

$$\text{p}_i(M) = (-1)^i c_{2i}(TM \otimes_\mathbb{R} \mathbb{C}) \in H^{4i}(M, \mathbb{Z})$$

and from Chern's formulas they are represented via de Rham theorem by exterior forms $\text{p}_i(R)$ of degree $4i$ as a differential form and a universal polynomial in R of degree $2i$. In all the above *universal* means that the objects depends only on the dimension, as for example the coefficients in the curvature tensor of the Ch_i and the p_i. For a determinant expression in Ω just take

$$\det\left(\lambda I - \frac{1}{2\pi}\Omega\right).$$

15.7.6 The Euler Class

This does not take us back to the Euler–Poincaré characteristic. Even when the dimension $d = 4n$ is a multiple of 4 the characteristic has "nothing to do" with the Pontryagin class $p_n(M)$. The trick is that it is "its square root." More precisely, the determinant $\det(\Omega)$ can be expressed as the square of the expression

$$\sum_{i_1,\ldots,i_d} \varepsilon_{i_1,\ldots,i_d} \Omega_{i_1 i_2} \wedge \cdots \wedge \Omega_{i_{d-1} i_d}.$$

This expression is called the *Pfaffian* of Ω. In multilinear algebra it is nothing but the well known fact that the determinant of an antisymmetric matrix of even order is always a square (in a universal manner in the coefficients).

15.7.7 The Absence of Other Characteristic Classes

It is important to know that there are no other such universal Riemannian integral formulas yielding topological invariants. A heuristic but insufficient reason is that the coefficients of the polynomial

$$\det\left(\lambda I - \frac{1}{2\pi}\Omega\right)$$

are the only invariants of the unitary group for the complex case, and the same for

$$\det\left(\lambda I - \frac{1}{2\pi}\Omega\right)$$

in the real case, except for the Pfaffian. But we might look for a universal polynomial in the curvature tensor and its covariant derivatives, which will have the property that it yields cohomology classes (via de Rham's theorem) independent of the metric on the manifold. This was a query of Gelfand, proved finally in Gilkey 1974 [563] (see also Abrahamov 1951 [1]). See Gilkey 1995 [564] for an up-to-date text. Very recently it was discovered that on Kähler manifolds there are in fact new invariants stemming out of the curvature, the *Rozansky–Witten* invariants, see Kontsevich 1999 [830].

15.7.8 Applying Characteristic Classes

Equipped with Chern and Pontryagin classes and the Allendoefer–Weil formula, can we do a lot in the realm of "curvature and topology?" That is to say, to fulfill Hopf's hope in §12.1. The answer is that one can do no more than the various applications we met above in the book. For the characteristic essentially nothing can be done beyond dimension 4. In dimension 2 Gauß–Bonnet solves everything. In dimension 4 the formula 15.10 on page 712 takes the nice form:

$$\mathbf{K} = 8\pi^2\chi(M) = \|R\|^2 - \left\|\text{Ricci} - \frac{\text{scalar}}{4}g\right\|^2. \tag{15.13}$$

We used it in equation 11.6 on page 518 and in note 12.3.1.1 on page 579 with good results. The curvature formula for the first Pontryagin class is also used heavily in §§11.4.6. But we also saw that, starting in dimension 6, the expression of $\mathbf{K}$ (of degree 3 or more in the curvature tensor) is not linked with the sign of the curvature, as proven in Bourguignon & Polombo 1981 [246].

15.7.9 Characteristic Numbers

To use the above formulas expressing Chern and Pontryagin classes in term of the curvature, one introduces the notion of *characteristic numbers* of a manifold. The first one is the Euler characteristic, valid for any even dimension.

For other ones we have to stick to the case of dimension $d = 4k$ a multiple of 4. Take any sequence of integers $i_1, \ldots, i_s$ such that

$$i_1 + \cdots + i_s = k$$

and consider, for any metric on M, the integral

$$\text{char}(M, i_1, \ldots, i_s) = \int_M p_{i_1} \cdots p_{i_s} \qquad (15.14)$$

This makes sense because we have performed the exterior product of toal degree $4n$ equal to the dimension of the manifold. For the topologist this means nothing but evaluating the cup product of Pontryagin classes over the fundamental class $[M]$ of M:

$$\text{char}(M, i_1, \ldots, i_s) = p_{i_1} \cup \cdots \cup p_{i_s}([M]).$$

Such a real number (in fact an integer) is called a *characteristic number*. We will also add the Euler-Poincaré characteristic to this collection of numbers. Now the universality (in the dimension) of the Chern formulas for Pontryagin classes yields the basic remark of Cheeger seen in theorem 272 on page 515 and used in Cheeger's finiteness theorem 372 on page 616:

Theorem 416 *If the curvature on a compact manifold M verifies $|K| \leq 1$ and if the manifold has some nonzero characteristic number then $\text{Vol}(M) > c(d)$ where $c(d) > 0$ depends only on the dimension d of M.*

Note that for Kähler manifolds one can add more characteristic numbers with the help of Chern classes. But except for these "nonzero" general results, the "integral Chern formulas" are useless, or at least disappointing for the Riemannian geometer who wishes to use only pointwise estimates of the curvature tensor. See more on this in Bourguignon & Polombo 1981 [246], with its revealing title.

For a person less concerned with Riemannian geometry, the above remarks do not prevent Chern classes being considered basic in differential geometry. Recall that they are defined for any bundle, not only the tangent bundle: see Chern 1946 [364]. They are today a building tool in algebraic geometry and in the study of the heat kernel. But what also counts is the fact that there is an integral formula which is polynomial and universal in the curvature. See Berline, Getzler & Vergne 1992 [179], Gilkey 1995 [564].

A basic fact discovered in the first edition (in 1956) of Hirzeburch 1995 [717]:

Theorem 417 (Hirzebruch 1956 [717]) *A suitable combination of Pontryagin classes yields the signature $\sigma(M)$ of a $4n$ dimensional manifold.*

By definition the *signature* is the linear algebra signature of the quadratic form which is defined by the cup product

$$H^{2n}(M,\mathbb{Z}) \otimes H^{2n}(M,\mathbb{Z}) \to \mathbb{Z}$$

of cohomology 2n-classes, i.e. the difference

$$\sigma(M) = b_{2n}^+(M,\mathbb{R}) - b_{2n}^-(M,\mathbb{R})$$

between the number of positive and the number of negative squares. Hirzebruch calls the signature the *index*, because for him the signature is the much more informative pair $b_{2n}^+(M,\mathbb{R}), b_{2n}^-(M,\mathbb{R})$. Hence this important fact: there is an integral formula for the signature given by the integral over the manifold of a universal polynomial in the curvature. See Gilkey 1995 [564] and Berline, Getzler & Vergne 1992 [179]. Also see §§14.2.3 for secondary characteristic classes and the η invariant. The signature is particularly important in dimension 4, since (if the manifold is simply connected) with the signature and the Euler characteristic one can recover $b_{2n}^+(M,\mathbb{R}), b_{2n}^-(M,\mathbb{R})$:

$$\sigma(M) = b_{2n}^+(M,\mathbb{R}) - b_{2n}^-(M,\mathbb{R}), \quad \text{and} \quad 2-\chi(M) = b_{2n}^+(M,\mathbb{R}) + b_{2n}^-(M,\mathbb{R}).$$

Also, in dimension 4, one has simply

$$\sigma(M) = \frac{1}{3}\mathrm{p}_1(M).$$

The integral formulas for both σ and χ were essential ingredients in §§11.4.6.

15.8 Two Examples of Riemannian Manifolds and Calculation of their Curvatures

15.8.1 Homogeneous Spaces

Consider a homogeneous space G/H where G is a Lie group and H is a compact subgroup, and take a Lie algebra decomposition

$$\mathfrak{g} = \mathfrak{h} \oplus \mathfrak{m}$$

with

$$[\mathfrak{h}, \mathfrak{m}] \subset \mathfrak{m}$$

(see the end of §§4.3.4 if needed). Since H is compact, there is a quadratic form on $\mathfrak{m}$ invariant under the action of H, which by translations by G yields a G invariant metric g on the coset space G/H. We give now the formulas for the curvature that we promised in §§4.4.3 and which are extremely useful when one looks for examples of homogeneous spaces with various demands on the curvature. The proof of these formulas, first obtained in Nomizu 1954

[969], are not simple. One can find them in full generality in Sakai 1996 [1085] and Besse 1987 [183]. One has first to compute the canonical connection and then relate it in a workable way with the Lie algebra brackets. The final result is as follows. The notations are: for the Riemannian metric, but preferably read on $\mathfrak{m}$ as a Euclidean structure, one will use the $\langle, \rangle$ notation. And for vectors in $\mathfrak{g}$ the index $v_\mathfrak{m}$ for the corresponding component in $\mathfrak{m}$ in the above direct sum. Then this Euclidean metric on $\mathfrak{m}$ determines uniquely, thanks to the Lie bracket, a map $U : \mathfrak{m} \times \mathfrak{m} \to \mathfrak{m}$ by the condition

$$2 \langle U(x,y), z \rangle = \langle [z,x]_\mathfrak{m}, y \rangle + \langle x, [z,y]_\mathfrak{m} \rangle.$$

The sectional curvature $K(x,y)$ is

$$K(x,y) = -\frac{3}{4} \|[x,y]_\mathfrak{m}\|^2 - \frac{1}{2} \langle [x,[x,y]]_\mathfrak{m}, y \rangle - \frac{1}{2} \langle [y,[y,x]]_\mathfrak{m}, x \rangle \quad (15.15)$$
$$+ \|U(x,y)\|^2 - \langle U(x,x), U(y,y) \rangle$$

for $x, y \in \mathfrak{m}$.

A nice case is when G is compact. On compact groups, bi-invariant metrics (i.e. invariant by both left and right translations) exist, and are the most natural ones. The decomposition $\mathfrak{g} = \mathfrak{h} \oplus \mathfrak{m}$ is orthogonal and $U \equiv 0$. The corresponding homogeneous spaces are called *normal*. Finally,

$$K(x,y) = \frac{1}{4} \|[x,y]_\mathfrak{h}\|^2 + \|[x,y]_\mathfrak{m}\|^2 \quad (15.16)$$

as mentioned in section 11.4.2.2 on page 524, where a geometric proof was provided, not for the exact value, but for the positivity. For symmetric spaces one has $[x,y]_\mathfrak{m} = 0$, but the sign of the curvature varies. We saw in section 11.4.2.2 that the curvature coincides essentially with the Lie algebra bracket up to sign.

15.8.2 Riemannian Submersions

Riemannian submersions were defined in §§4.3.6 and are maps

$$p : M \to N$$

between Riemannian manifolds, with the requirement that around every point n in N there is a neigborhood U such that $p^{-1}(U)$ is diffeomorphic to a product of manifolds: fiber multiplied by base. Now at every point $m \in p^{-1}(n)$ we have the vertical tangent space which is the tangent space $V_m M$ at m to the fiber $p^{-1}(n)$. But the Riemannian metric provides us with a horizontal tangent space, namely the orthogonal complement $H_m M$ of $V m M$ in $T_m M$. Now the differential dp of p restricted to $H_m M$ is a vector space isomorphism between $H_m M$ and $T_n N$ by construction. But both $H_m M$ and $T_n N$ are Euclidean by our data. We say that $p : M \to N$ is a Riemannian submersion when for every n, and every $m \in p^{-1}(n)$, this vector space *isomorphism* is a Euclidean isomorphism. We recall some facts seen in §§4.4.3.

The first interesting point for the geometer is the existence of horizontal geodesics in M which project down to geodesics in N. Let v be any horizontal vector in TM and $dp(v)$ its projection in TN. Then the horizontal lift γ^* of the geodesic γ of N whith initial velocity $dp(v)$ is the geodesic in M with initial velocity v. The proof is as follows: pick any two nearby points m and m' in γ^*. Then there is a unique shortest curve between them, and it has to be part of γ^*. In fact the projection p can only decrease lengths, and strictly, except for completely horizontal curves on which it preserves lengths. Any curve from m to m' will have length larger than or equal to the distance in N between $p(m)$ and $p(m')$.

Using this approach one obtains for the sectional curvatures the equation 4.40 on page 211 which reads $K(P) \geq K(P^*)$ where P is a tangent plane to N and P^* any of its horizontal lifts. This is a direct application of the formula 4.31 on page 205 expressing the curvature as the defect to $2\pi\varepsilon$ of the length of a small circle of radius ε. So that Riemannian submersions can only decrease the curvature.

But this is not an explicit formula like equation 15.16 on the previous page for normal homogeneous spaces. To obtain some explicit formula, we need a general set of formulas for Riemannian submersions. Moreover we have very often mentioned the construction of examples using these formulas of O'Neill, which crucial to construct examples in Riemannian geometry. We will now give only some of those formulas, but will give the starting point to compute anything you want concerning Riemannian submersions. All of those formulas were found in O'Neill 1966 [975]. Book references are: chapter 9 of Besse 1987 [183] and Sakai 1996 [1085].

To get some feeling for submersions, it is good to realize that Riemannian products $M \times N \to M$ are the ideal Riemannian submersions, the fibers N are all totally geodesic (see §§6.1.5) and the horizontal distribution is everywhere integrable since its leaves are the $M \times \{n\}$. The idea is now to introduce two tensors T and A both of type $(2,1)$. The letters H and V will denote respectively the horizontal and the vertical part of vectors. The tensor T expresses the "defect" of the vertical fibers to be totally geodesic. Its general value is:

$$T_E F = H\left(\nabla_{V(E)} V(F)\right) + V\left(\nabla_{V(E)} H(F)\right)$$

and one checks directly that it is a tensor (i.e. does not depends on the way the vectors E and F are extended into vector fields in order to be able to define the various covariant derivatives appearing in the formula). The tensor A expresses the defect of the horizontal distribution to be integrable. Its general value is:

$$A_E F = H\left(\nabla_{H(E)} V(F)\right) + V\left(\nabla_{H(E)} H(F)\right).$$

From now on U and V will be vertical vectors and X and Y horizontal ones. The above interpretation is now justified because:

- $T_U V$ is the second fundamental form of the fibers (and is symmetric)
- for two horizontal vector fields

$$A_X Y = \frac{1}{2} V([X,Y]).$$

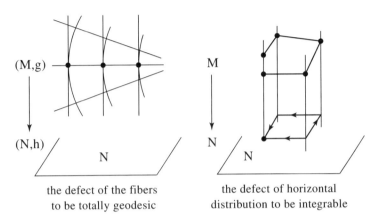

Fig. 15.8. (a) The defect of the fibers to be totally geodesic (b) The defect of the horizontal distribution to be integrable

Here are now the formulas for the three types of sectional curvatures. Both $\{U,V\}$ and $\{X,Y\}$ are orthonormal:

$$K_M(U,V) = K_{\text{fiber}}(U,V) + \|T_U V\|^2 - g(T_U U, T_V V) \qquad (15.17)$$
$$K_M(X,U) = g((D_X T)_U U, X) - \|T_U X\|^2 + \|A_X U\|^2 \qquad (15.18)$$
$$K_M(X,Y) = K_N(X,Y) - 3\|A_X Y\|^2 \qquad (15.19)$$

The last formula expresses quantitatively equation 4.40 on page 211. The above formulas are of course much simpler when one works in the special case of warped products. We met those in §§4.3.6: a warped product of two Riemannian manifolds (M,g) and (N,h) is like a Riemannian product except that one modifies the product metric on $M \times N$ by a real-valued function $f: M \to \mathbb{R}$, that is to say that, for v vertical and w horizontal one sets

$$\|v,w\|^2 = \|v\|_M^2 + f\|w\|_N^2.$$

Then the formula above involves only, besides of course the curvatures of both (M,g) and (N,h), only the function f, its differential df and its Hessian $\nabla^2 f$; see Besse 1987 [183].

16 Poincaré's Conjecture Solved by Riemannian Geometry

One of the hopes mentioned in section 14.4, a fundamental one, has now just become reality. G. Perelman very recently proved Poincaré's conjecture, namely that a simply connected three-dimensional manifold is necessarily homeomorphic to the sphere S3. This proof is contained in the three texts of Perelman deposited in ArXiv : math DG, in 2002 and 2003. The reader interested to see, albeit without details, the scheme of the proof in different degrees of explanation, should look at (Besson 2005), (Milnor 2003), (Bessières 2005), (Morgan 2005), Besson's is the most detailed of these.

Extremely briefly, one starts as in the proof of Theorem 279 following the Ricci flow. But now we have no restriction available on the curvature of the Riemannian manifold with which we start, since any Riemannian metric can be put on our – merely topological – three-dimensonal manifold. The problem is that the Ricci flow eventually cannot be avoided, to develop a singularity, which is a kind of a collapsing (see section 12.4).

Caricaturally, one performs suitable surgeries when a collapsing happens, and then the simply-connectedness of the manifold allows Perelman to show that only a finite number of such surgeries is needed, so that one can, finally, achieve convergence to constant sectional curvature, quod erat demonstrandum. For the moment Perelman's work is still not sufficient to prove Thurston's geometrization program entirely for all three-dimensional manifolds.

Bessières, L. (2005). "Conjecture de Poincaré : la preuve de R. Hamilton et G. Perelman." Gazette des mathématiciens **106**: 7–35.

Besson, G. (2005). "Preuve la conjecture de Poincaré en déformant la métrique par la courbure de Ricci, d'après G. Perel'man." Séminaire Bourbaki **57ème année**: 947–1 à 9457–38.

Milnor, J. (2003). "Towards the Poincaré Conjecture and the Classification of 3-Manifolds." Notices of the AMS **50**: 1226–1233.

Morgan, J. (2005). "Recent progress on the Poincaré conjecture and the classification of 3-manifolds." Bull. of the AMS **42**: 57–78.

References

1. A. A. Abrahamov, *On the topological invariants of Riemannian spaces obtained by integration of tensor fields*, Dokl. Acad. Nauk **81** (1951), 125–128.
2. Uwe Abresch, *Lower curvature bounds, Toponogov's theorem, and bounded topology*, Ann. Sci. École Norm. Sup. (4) **18** (1985), no. 4, 651–670. MR 87j:53058
3. _____, *Über das glatten Riemannscher Metriken*, Bonn, 1988.
4. _____, *Endlichkeitssätze in der Riemannschen Geometrie*, in Geyer [557], pp. 152–176.
5. Uwe Abresch and Detlef Gromoll, *On complete manifolds with nonnegative Ricci curvature*, J. Amer. Math. Soc. **3** (1990), no. 2, 355–374. MR 91a:53071
6. Uwe Abresch and Wolfgang T. Meyer, *Pinching below $\frac{1}{4}$, injectivity radius, and conjugate radius*, J. Differential Geom. **40** (1994), no. 3, 643–691. MR 95j:53053
7. _____, *A sphere theorem with a pinching constant below $\frac{1}{4}$*, J. Differential Geom. **44** (1996), no. 2, 214–261. MR 97i:53036
8. _____, *Injectivity radius estimates and sphere theorems*, in Grove and Petersen [650], Papers from the Special Year in Differential Geometry held in Berkeley, CA, 1993–94, pp. 1–47. MR 98e:53052
9. Uwe Abresch and Viktor Schroeder, *Analytic manifolds of nonpositive curvature*, in Besse [185], En l'honneur de Marcel Berger. [In honor of Marcel Berger], Held in Luminy, France, July 12–18, 1992, pp. 1–67. MR 97j:53039
10. Norbert A'Campo, *Sur la première partie du seizième problème de Hilbert*, Séminaire Bourbaki (1978/79), Springer, Berlin, 1980, pp. Exp. No. 537, pp. 208–227. MR 82g:14049
11. Robert D. M. Accola, *Differentials and extremal length on Riemann surfaces*, Proc. Nat. Acad. Sci. U.S.A. **46** (1960), 540–543. MR 22 #9598
12. Lars V. Ahlfors and Leo Sario, *Riemann surfaces*, Princeton University Press, Princeton, N.J., 1960, Princeton Mathematical Series, No. 26. MR 22 #5729
13. Francesca Aicardi, *Topological invariants of Legendrian curves*, C. R. Acad. Sci. Paris Sér. I Math. **321** (1995), no. 2, 199–204. MR 97d:57002
14. Kazuo Akutagawa and Boris Botvinnik, *Manifolds of positive scalar curvature and conformal cobordism theory*, electronic preprint math.DG/0008139, Los Alamos National Labs, 2000.
15. A. D. Aleksandrov, A. N. Kolmogorov, and M. A. Lavrent'ev (eds.), *Mathematics: its content, methods and meaning*, Izdat. Akad. Nauk SSSR, Moscow, 1956, In Russian. MR 19,520d
16. A. D. Aleksandrov, A. N. Kolmogorov, and M. A. Lavrent'ev (eds.), *Mathematics: Its content, methods and meaning*, Dover Publications Inc., Mineola,

NY, 1999, Translated from the Russian original [15] by Tamas Bartha, Kurt Hirsch and S. H. Gould, Translation edited by Gould, Three volumes bound as one, Reprint of the second 1969 edition, published in three volumes. MR 1 725 757
17. D. V. Alekseevskiĭ, *Riemannian spaces with unusual holonomy groups*, Funkcional. Anal. i Priložen **2** (1968), no. 2, 1–10. MR 37 #6868
18. D. V. Alekseevskiĭ and V. Cortés, *Homogeneous quaternionic Kähler manifolds of unimodular group*, Boll. Un. Mat. Ital. B (7) **11** (1997), no. 2, suppl., 217–229. MR 98i:53062
19. Stephanie B. Alexander, I. David Berg, and Richard L. Bishop, *Cut loci, minimizers, and wavefronts in Riemannian manifolds with boundary*, Michigan Math. J. **40** (1993), no. 2, 229–237. MR 94c:53053
20. _____, *The distance-geometry of Riemannian manifolds with boundary*, in Greene and Yau [597], Proceedings of the AMS Summer Research Institute on Differential Geometry held at the University of California, Los Angeles, California, July 8–28, 1990, pp. 31–36. MR 94f:53083
21. Stephanie B. Alexander and Richard L. Bishop, *Thin Riemannian manifolds with boundary*, Math. Ann. **311** (1998), no. 1, 55–70. MR 99e:53037
22. _____, *Spines and homology of thin Riemannian manifolds with boundary*, to appear, 1999.
23. Carl B. Allendoerfer, *The Euler number of a Riemann manifold*, Amer. J. Math. **62** (1940), 243–248. MR 2,20e
24. Carl B. Allendoerfer and André Weil, *The Gauss-Bonnet theorem for Riemannian polyhedra*, Trans. Amer. Math. Soc. **53** (1943), 101–129. MR 4,169e
25. Norman L. Alling and Newcomb Greenleaf, *Foundations of the theory of Klein surfaces*, Springer-Verlag, Berlin, 1971, Lecture Notes in Mathematics, Vol. 219. MR 48 #11488
26. Simon Aloff and Nolan R. Wallach, *An infinite family of distinct 7-manifolds admitting positively curved Riemannian structures*, Bull. Amer. Math. Soc. **81** (1975), 93–97. MR 51 #6851
27. Juan Carlos Álvarez Paiva, *Some problems in Finsler geometry*, preprint, Université Catholique de Louvain, 1991.
28. Juan Carlos Álvarez Paiva, I. M. Gel'fand, and M. Smirnov, *Crofton densities, symplectic geometry and Hilbert's fourth problem*, in Arnol'd et al. [62], Geometry and singularity theory, pp. 77–92. MR 98a:52005
29. W. Ambrose and I. M. Singer, *A theorem on holonomy*, Trans. Amer. Math. Soc. **75** (1953), 428–443. MR 16,172b
30. _____, *On homogeneous Riemannian manifolds*, Duke Math. J. **25** (1958), 647–669. MR 21 #1628
31. Bernd Ammann, *Spin-Strukturen und das Spektrum des Dirac-Operators*, Ph.D. thesis, Universität Freiburg, Freiburg, Germany, May 1998.
32. _____, *The Willmore conjecture for immersed tori with small curvature integral*, electronic preprint math/9906065, Los Alamos National Labs, 1999.
33. Bernd Ammann and Christian Bär, *The Dirac operator on nilmanifolds and collapsing circle bundles*, Ann. Global Anal. Geom. **16** (1998), no. 3, 221–253. MR 99h:58194
34. _____, *Dirac eigenvalues and total scalar curvature*, to appear, 1999.
35. J. Amorós, M. Burger, Kevin Corlette, D. Kotschick, and Domingo Toledo, *Fundamental groups of compact Kähler manifolds*, American Mathematical Society, Providence, RI, 1996. MR 97d:32037

36. Jørgen Ellegaard Andersen, Johan Dupont, Henrik Pedersen, and Andrew Swann (eds.), *Geometry and physics*, Marcel Dekker Inc., New York, 1997, Papers from the Special Session held at the University of Aarhus, Aarhus, 1995. MR 97i:00013
37. Michael T. Anderson, *Ricci curvature bounds and Einstein metrics on compact manifolds*, J. Amer. Math. Soc. **2** (1989), no. 3, 455–490. MR 90g:53052
38. _____, *Convergence and rigidity of manifolds under Ricci curvature bounds*, Invent. Math. **102** (1990), no. 2, 429–445. MR 92c:53024
39. _____, *Metrics of positive Ricci curvature with large diameter*, Manuscripta Math. **68** (1990), no. 4, 405–415. MR 91g:53045
40. _____, *Short geodesics and gravitational instantons*, J. Differential Geom. **31** (1990), 265–275.
41. _____, *Remarks on the compactness of isospectral sets in low dimensions*, Duke Math. J. **63** (1991), no. 3, 699–711. MR 92m:58140
42. _____, *Hausdorff perturbations of Ricci-flat manifolds and the splitting theorem*, Duke Math. J. **68** (1992), no. 1, 67–82. MR 94k:53054
43. _____, *The L^2 structure of moduli spaces of Einstein metrics on 4-manifolds*, Geom. Funct. Anal. **2** (1992), no. 1, 29–89. MR 92m:58017
44. _____, *Degeneration of metrics with bounded curvature and applications to critical metrics of Riemannian functionals*, in Greene and Yau [597], Proceedings of the AMS Summer Research Institute on Differential Geometry held at the University of California, Los Angeles, California, July 8–28, 1990, pp. 53–79. MR 94c:58032
45. _____, *Einstein metrics and metrics with bounds on Ricci curvature*, in Chatterji [324], pp. 443–452. MR 97g:53046
46. _____, *Scalar curvature and geometrization of 3-manifolds: the nonpositive case*, to appear, 1997.
47. _____, *The static vacuum Einstein equations and degenerations of Yamabe metrics on 3-manifolds*, to appear, 1997.
48. _____, *Scalar curvature, metric degenerations and the static vacuum Einstein equations on 3-manifolds, I*, Geom. and Funct. Anal. **9** (1999), no. 5, 855–967.
49. Michael T. Anderson and Jeff Cheeger, *Diffeomorphism finiteness for manifolds with Ricci curvature and $L^{n/2}$-norm of curvature bounded*, Geom. Funct. Anal. **1** (1991), no. 3, 231–252. MR 92h:53052
50. _____, *C^α-compactness for manifolds with Ricci curvature and injectivity radius bounded below*, J. Differential Geom. **35** (1992), no. 2, 265–281. MR 93c:53028
51. Ben Andrews, *Contraction of convex hypersurfaces by their affine normal*, J. Differential Geom. **43** (1996), no. 2, 207–230. MR 97m:58045
52. _____, *Gauss curvature flow: the fate of the rolling stones*, Invent. Math. **138** (1999), no. 1, 151–161. MR 1 714 339
53. D. V. Anosov, *Geodesic flows on closed Riemann manifolds with negative curvature.*, American Mathematical Society, Providence, R.I., 1969. MR 39 #3527
54. _____, *Generic properties of closed geodesics*, Izv. Akad. Nauk SSSR Ser. Mat. **46** (1982), no. 4, 675–709, 896. MR 84b:58029
55. D. V. Anosov, S. Kh. Aranson, Vladimir I. Arnol'd, I. U. Bronshtein, V. Z. Grines, and Yu. S. Il'yashenko, *Ordinary differential equations and smooth dy-*

namical systems, Springer-Verlag, Berlin, 1997, Translated from the 1985 Russian original by E. R. Dawson and D. O'Shea, Third printing of the 1988 translation [*Dynamical systems. I*, Encyclopaedia Math. Sci., 1, Springer, Berlin, 1988]. MR 99h:58053

56. Vestislav Apostolov, Tedi Drăghici, and Dieter Kotschick, *An integrability theorem for almost Kähler 4-manifolds*, C. R. Acad. Sci. Paris Sér. I Math. **329** (1999), no. 5, 413–418. MR 1 710 103
57. Paul Appell and Émile Lacour, *Principes de la théorie des fonctions elliptiques et applications*, 2. éd. ed., Gauthier–Villars, Paris, 1922, avec le concours de René Garnier.
58. C. S. Aravinda and F. T. Farrell, *Exotic negatively curved structures on cayley hyperbolic manifolds*, unpublished, 2000.
59. _____, *Exotic structures on quaternionic hyperbolic manifolds*, unpublished, 2000.
60. Marc Arconstanzo, *Des métriques finslériennes sur le disque à partir d'une fonction distance entre les points du bord*, Comment. Math. Helv. **69** (1994), no. 2, 229–248.
61. Marc Arcostanzo and René Michel, *Métriques de révolution d'un disque et invariance par rotation de la longueur des géodésiques*, Geom. Dedicata **76** (1999), no. 2, 197–209. MR 1 703 215
62. V. I. Arnol'd, I. M. Gel'fand, V. S. Retakh, and M. Smirnov (eds.), *The Arnol'd–Gelfand mathematical seminars*, Birkhäuser Boston Inc., Boston, MA, 1997, Geometry and singularity theory. MR 97g:00016
63. Vladimir I. Arnol'd, *Catastrophe theory*, third ed., Springer-Verlag, Berlin, 1992, Translated from the Russian by G. S. Wassermann, Based on a translation by R. K. Thomas. MR 93h:58018
64. _____, *Plane curves, their invariants, perestroikas and classifications*, Singularities and bifurcations, Amer. Math. Soc., Providence, RI, 1994, With an appendix by F. Aicardi, pp. 33–91. MR 95m:57009
65. _____, *Topological invariants of plane curves and caustics*, American Mathematical Society, Providence, RI, 1994, Dean Jacqueline B. Lewis Memorial Lectures presented at Rutgers University, New Brunswick, New Jersey. MR 95h:57003
66. _____, *Mathematical methods of classical mechanics*, Springer-Verlag, New York, 1996, Translated from the 1974 Russian original by K. Vogtmann and A. Weinstein, Corrected reprint of the second (1989) edition. MR 96c:70001
67. Vladimir I. Arnol'd, Michael F. Atiyah, P. Lax, and B. Mazur (eds.), *Mathematics: Frontiers and Perspectives*, Inter. Math. Union, Amer. Math. Soc., 2000.
68. Vladimir I. Arnol'd and A. Avez, *Ergodic problems of classical mechanics*, W. A. Benjamin, Inc., New York-Amsterdam, 1968, Translated from the French by A. Avez. MR 38 #1233
69. Vladimir I. Arnol'd and Boris A. Khesin, *Topological methods in hydrodynamics*, Springer-Verlag, New York, 1998. MR 99b:58002
70. Pierre Arnoux, *Ergodicité générique des billards polygonaux (d'après Kerckhoff, Masur, Smillie)*, Astérisque (1988), no. 161-162, Exp. No. 696, 5, 203–221 (1989), Séminaire Bourbaki, Vol. 1987/88. MR 90c:58097
71. Pierre Arnoux, Christian Mauduit, Iekata Shiokawa, and Jun-ichi Tamura, *Complexity of sequences defined by billiard in the cube*, Bull. Soc. Math. France **122** (1994), no. 1, 1–12. MR 94m:11028

72. Mark S. Ashbaugh and Rafael D. Benguria, *Isoperimetric inequalities for eigenvalue ratios*, Partial differential equations of elliptic type (Cortona, 1992), Cambridge Univ. Press, Cambridge, 1994, pp. 1–36. MR 95h:35158
73. _____, *On Rayleigh's conjecture for the clamped plate and its generalization to three dimensions*, Duke Math. J. **78** (1995), no. 1, 1–17. MR 97j:35111
74. Konstantin Athanassopoulos, *Rotation numbers and isometries*, Geom. Dedicata **72** (1998), no. 1, 1–13. MR 99m:58158
75. Michael F. Atiyah, *Geometry of Yang–Mills fields*, Scuola Normale Superiore Pisa, Pisa, 1979. MR 81a:81047
76. Michael F. Atiyah, Raoul Bott, and V. K. Patodi, *On the heat equation and the index theorem*, Invent. Math. **19** (1973), 279–330. MR 58 #31287
77. _____, *Errata to: "On the heat equation and the index theorem" (Invent. Math. **19** (1973), 279–330)*, Invent. Math. **28** (1975), 277–280. MR 58 #31288
78. Michael F. Atiyah, Harold Donnelly, and I. M. Singer, *Eta invariants, signature defects of cusps, and values of L-functions*, Ann. of Math. (2) **118** (1983), no. 1, 131–177. MR 86g:58134a
79. _____, *Signature defects of cusps and values of L-functions: the nonsplit case. Addendum to: "Eta invariants, signature defects of cusps, and values of L-functions"*, Ann. of Math. (2) **119** (1984), no. 3, 635–637. MR 86g:58134b
80. Michael F. Atiyah, Nigel J. Hitchin, and I. M. Singer, *Self-duality in four-dimensional Riemannian geometry*, Proc. Roy. Soc. London Ser. A **362** (1978), no. 1711, 425–461. MR 80d:53023
81. Michael F. Atiyah, V. K. Patodi, and I. M. Singer, *Spectral asymmetry and Riemannian geometry. I*, Math. Proc. Cambridge Philos. Soc. **77** (1975), 43–69. MR 53 #1655a
82. _____, *Spectral asymmetry and Riemannian geometry. II*, Math. Proc. Cambridge Philos. Soc. **78** (1975), no. 3, 405–432. MR 53 #1655b
83. _____, *Spectral asymmetry and Riemannian geometry. III*, Math. Proc. Cambridge Philos. Soc. **79** (1976), no. 1, 71–99. MR 53 #1655c
84. Michael F. Atiyah and I. M. Singer, *The index of elliptic operators on compact manifolds*, Bull. Amer. Math. Soc. **69** (1963), 422–433. MR 28 #626
85. Thierry Aubin, *Métriques riemanniennes et courbure*, J. Differential Geometry **4** (1970), 383–424. MR 43 #5452
86. _____, *Some nonlinear problems in Riemannian geometry*, Springer-Verlag, Berlin, 1998. MR 99i:58001
87. Thierry Aubin and Yan Yan Li, *Sur la meilleure constante dans l'inégalité de Sobolev*, C. R. Acad. Sci. Paris Sér. I Math. **328** (1999), no. 2, 135–138. MR 2000a:46048
88. Michèle Audin and Jacques Lafontaine (eds.), *Holomorphic curves in symplectic geometry*, Birkhäuser Verlag, Basel, 1994. MR 95i:58005
89. Basile Audoly, *Courbes rigidifiant les surfaces*, C. R. Acad. Sci. Paris Sér. I Math. **328** (1999), no. 4, 313–316. MR 99m:53003
90. V. Avakumović, *Über die Eigenfunktionen auf geschlossenen Riemannschen Mannigfaltigkeiten*, Math. Z. **65** (1956), 327–344.
91. Ivan K. Babenko, *Asymptotic invariants of smooth manifolds*, Izv. Ross. Akad. Nauk, Ser. Mat. **56** (1992), no. 4, 707–751. MR 94d:53068
92. _____, *Forte simplex isosystolique de variétés fermées et de polyèdres*, to appear, 2000.

93. Ivan K. Babenko and Mikhail G. Katz, *Systolic freedom of orientable manifolds*, Ann. Sci. École Norm. Sup. (4) **31** (1998), no. 6, 787–809. MR 1 664 222
94. Ivan K. Babenko, Mikhail G. Katz, and Alexander I. Suciu, *Volumes, middle-dimensional systoles, and Whitehead products*, Math. Res. Lett. **5** (1998), no. 4, 461–471. MR 99m:53084
95. Vasilii M. Babič and Vladimir F. Lazutkin, *The eigenfunctions which are concentrated near a closed geodesic*, Problems of Mathematical Physics, No. 2, Spectral Theory, Diffraction Problems (Russian), Izdat. Leningrad. Univ., Leningrad, 1967, pp. 15–25. MR 38 #2708
96. W. L. Baily, *On the imbedding of V-manifolds in projective space*, Amer. J. Math. **79** (1957), 403–430. MR 20 #6538
97. R. Balian and C. Bloch, *Distribution of eigenfrequencies for the wave equation in a finite domain. I. Three-dimensional problem with smooth boundary surface*, Ann. Physics **60** (1970), 401–447. MR 42 #4901
98. _____, *Distribution of eigenfrequencies for the wave equation in a finite domain. II. Electromagnetic field. Riemannian spaces*, Ann. Physics **64** (1971), 271–307. MR 44 #1953
99. _____, *Distribution of eigenfrequencies for the wave equation in a finite domain. III. Eigenfrequency density oscillations*, Ann. Physics **69** (1972), 76–160. MR 44 #7147
100. _____, *Errata: "Distribution of eigenfrequencies for the wave equation in a finite domain. I. Three-dimensional problem with smooth boundary surface" (Ann. Physics **60** (1970), 401–447)*, Ann. Physics **84** (1974), 559. MR 49 #5598
101. _____, *Errata: "Distribution of eigenfrequencies for the wave equation in a finite domain. II. Electromagnetic field. Riemannian spaces" (Ann. Physics **64** (1971), 271–307)*, Ann. Physics **84** (1974), 559–562. MR 49 #5599
102. Werner Ballmann, *Doppelpunktfreie geschlossene Geodätische auf kompakten Flächen*, Math. Z. **161** (1978), no. 1, 41–46. MR 81g:58008
103. _____, *Lectures on spaces of nonpositive curvature*, Birkhäuser Verlag, Basel, 1995, With an appendix by Misha Brin. MR 97a:53053
104. Werner Ballmann and Michael Brin, *Orbihedra of nonpositive curvature*, Inst. Hautes Études Sci. Publ. Math. (1995), no. 82, 169–209 (1996). MR 97i:53049
105. Werner Ballmann and Patrick Eberlein, *Fundamental groups of manifolds of nonpositive curvature*, J. Differential Geom. **25** (1987), no. 1, 1–22. MR 88b:53047
106. Werner Ballmann, Mikhael Gromov, and Viktor Schroeder, *Manifolds of nonpositive curvature*, Birkhäuser Boston Inc., Boston, Mass., 1985. MR 87h:53050
107. Werner Ballmann and Maciej P. Wojtkowski, *An estimate for the measure-theoretic entropy of geodesic flows*, Ergodic Theory Dynamical Systems **9** (1989), no. 2, 271–279. MR 90k:58165
108. Werner Ballmann and Wolfgang Ziller, *On the number of closed geodesics on a compact Riemannian manifold*, Duke Math. J. **49** (1982), no. 3, 629–632. MR 84d:53048
109. Catherine Bandle, *Isoperimetric inequalities and applications*, Pitman (Advanced Publishing Program), Boston, Mass., 1980. MR 81e:35095
110. Victor Bangert, *Geodätische Linien auf Riemannschen Mannigfaltigkeiten*, Jahresber. Deutsch. Math.-Verein. **87** (1985), no. 2, 39–66. MR 87i:58036

111. _____, *Mather sets for twist maps and geodesics on tori*, Dynamics reported, Vol. 1, Teubner, Stuttgart, 1988, pp. 1–56. MR 90a:58145
112. _____, *On the existence of closed geodesics on two-spheres*, Internat. J. Math. **4** (1993), no. 1, 1–10. MR 94d:58036
113. Victor Bangert and Mikhail G. Katz, *Stable systolic inequalities and cohomology products*, electronic preprint math.DG/0204181, Los Alamos National Labs, 2002, To appear in Communications on Pure and Applied Mathematics.
114. David Bao, Shiing-Shen Chern, and Zhongmin Shen (eds.), *Finsler geometry*, Providence, RI, American Mathematical Society, 1996. MR 97b:53001
115. _____, *Preface for "Finsler geometry over the reals"*, Finsler geometry (Seattle, WA, 1995), Amer. Math. Soc., Providence, RI, 1996, pp. 3–13. MR 97c:53030
116. Christian Bär, *Metrics with harmonic spinors*, Geom. Funct. Anal. **6** (1996), no. 6, 899–942. MR 98a:58162
117. Robert J. Baston, *Quaternionic complexes*, J. Geom. Phys. **8** (1992), no. 1-4, 29–52. MR 93h:58145
118. Christophe Bavard, *Courbure presque négative en dimension 3*, Compositio Math. **63** (1987), no. 2, 223–236. MR 88j:53041
119. _____, *Inégalités isosystoliques conformes pour la bouteille de Klein*, Geom. Dedicata **27** (1988), no. 3, 349–355. MR 89k:53012
120. _____, *L'aire systolique conforme des groupes cristallographiques du plan*, Ann. Inst. Fourier (Grenoble) **43** (1993), no. 3, 815–842. MR 95a:53054
121. Ya. V. Bazaĭkin, *On a family of 13-dimensional closed Riemannian manifolds of positive curvature*, Sibirsk. Mat. Zh. **37** (1996), no. 6, 1219–1237, ii. MR 98c:53045
122. Arnaud Beauville, *Variétés Kähleriennes dont la première classe de Chern est nulle*, J. Differential Geom. **18** (1983), no. 4, 755–782 (1984). MR 86c:32030
123. _____, *Le problème de Schottky et la conjecture de Novikov*, Astérisque (1987), no. 152-153, 4, 101–112 (1988), Séminaire Bourbaki, Vol. 1986/87. MR 89e:14029
124. Eric Bedford, John P. D'Angelo, Robert E. Greene, and Steven G. Krantz (eds.), *Several complex variables and complex geometry. Parts 1,2,3*, Providence, RI, American Mathematical Society, 1991. MR 92d:32001
125. John K. Beem, Paul E. Ehrlich, and Kevin L. Easley, *Global Lorentzian geometry*, second ed., Marcel Dekker Inc., New York, 1996. MR 97f:53100
126. André Bellaïche and Jean-Jacques Risler (eds.), *Sub-Riemannian geometry*, Birkhäuser Verlag, Basel, 1996. MR 97f:53002
127. Josef Bemelmans, Maung Min-Oo, and Ernst A. Ruh, *Smoothing Riemannian metrics*, Math. Z. **188** (1984), no. 1, 69–74. MR 85m:58184
128. Radel Ben-Av and Alexander Nabutovsky, *Logical phenomena in quantum gravity*, in preparation, 2002.
129. Riccardo Benedetti and Carlo Petronio, *Lectures on hyperbolic geometry*, Springer-Verlag, Berlin, 1992. MR 94e:57015
130. Itai Benjamini and Jian Guo Cao, *Examples of simply-connected Liouville manifolds with positive spectrum*, Differential Geom. Appl. **6** (1996), no. 1, 31–50. MR 97b:53036
131. Yves Benoist, Patrick Foulon, and François Labourie, *Flots d'Anosov à distributions stable et instable différentiables*, J. Amer. Math. Soc. **5** (1992), no. 1, 33–74. MR 93b:58112

132. Pierre H. Bérard, *On the wave equation on a compact Riemannian manifold without conjugate points*, Math. Z. **155** (1977), no. 3, 249–276. MR 56 #13295
133. _____, *Lattice point problems and eigenvalue asymptotics*, Keio University Department of Mathematics, Yokohama, 1980. MR 83b:10058
134. _____, *Spectres et groupes cristallographiques. I. Domaines euclidiens*, Invent. Math. **58** (1980), no. 2, 179–199. MR 82e:58097a
135. _____, *Spectral geometry: direct and inverse problems*, Springer-Verlag, Berlin, 1986, With appendixes by Gérard Besson, and by Bérard and Marcel Berger. MR 88f:58146
136. _____, *From vanishing theorems to estimating theorems: the Bochner technique revisited*, Bull. Amer. Math. Soc. (N.S.) **19** (1988), no. 2, 371–406. MR 89i:58152
137. _____, *Variétés riemanniennes isospectrales non isométriques*, Astérisque (1989), no. 177-178, Exp. No. 705, 127–154, Séminaire Bourbaki, Vol. 1988/89. MR 91a:58186
138. Pierre H. Bérard and Gérard Besson, *Spectres et groupes cristallographiques. II. Domaines sphériques*, Ann. Inst. Fourier (Grenoble) **30** (1980), no. 3, 237–248. MR 82e:58097b
139. Pierre H. Bérard, Gérard Besson, and Sylvestre F. L. Gallot, *Sur une inégalité isopérimétrique qui généralise celle de Paul Lévy – Gromov*, Invent. Math. **80** (1985), no. 2, 295–308. MR 86j:58017
140. _____, *Embedding Riemannian manifolds by their heat kernel*, Geom. Funct. Anal. **4** (1994), no. 4, 373–398. MR 95g:58228
141. Pierre H. Bérard and David L. Webb, *One can't hear the orientability of surfaces*, preprint 005-95, Math. Sci. Res. Inst., Berkeley, 1995.
142. Lionel Bérard Bergery, *Les variétés riemanniennes homogènes simplement connexes de dimension impaire à courbure strictement positive*, J. Math. Pures Appl. (9) **55** (1976), no. 1, 47–67. MR 54 #6032
143. _____, *Sur de nouvelles variétés riemanniennes d'Einstein*, Institut Élie Cartan, 6, Univ. Nancy, Nancy, 1982, pp. 1–60. MR 85b:53048
144. Lionel Bérard Bergery and C. Boubel, *On the pseudo-Riemannian manifolds whose Ricci tensor is parallel*, to appear, 1999.
145. Lionel Bérard Bergery and A. Ikemakhen, *Sur l'holonomie des variétés pseudo-riemanniennes de signature (n,n)*, Bull. Soc. Math. France **125** (1997), no. 1, 93–114. MR 98m:53087
146. Lionel Bérard Bergery and Mikhail G. Katz, *On intersystolic inequalities in dimension 3*, Geom. Funct. Anal. **4** (1994), no. 6, 621–632. MR 95i:53036
147. V. N. Berestovskij and I. G. Nikolaev, *Multidimensional generalized Riemannian spaces*, Geometry IV, Springer, Berlin, 1993, pp. 165–243, 245–250. MR 1 263 965
148. Marcel Berger, *Sur les groupes d'holonomie des variétés riemanniennes*, Bull. Soc. Math. France **83** (1953), 279–330.
149. _____, *Les variétés riemanniennes $(1/4)$-pincées*, Ann. Scuola Norm. Sup. Pisa Cl. Sci. **14** (1960), 161–170.
150. _____, *Les variétés riemanniennes homogènes normales simplement connexes à courbure positive*, Ann. Scuola Norm. Sup. Pisa Cl. Sci. **15** (1961), 179–246.
151. _____, *Lectures on geodesics in Riemannian geometry*, vol. 33, Tata Institute of Fundamental Research, Bombay, 1965.

152. _____, *Trois remarques sur les variétés riemanniennes à courbure positive*, C. R. Acad. Sci. Paris Sér. A-B **263** (1966), A76–A78. MR 33 #7966
153. _____, *Du côté de chez Pu*, Ann. Sci. École Norm. Sup. (4) **5** (1972), 1–44. MR 46 #8119
154. _____, *Géométrie. Vol. 1*, CEDIC, Paris, 1977, Actions de groupes, espaces affines et projectifs. [Actions of groups, affine and projective spaces]. The publisher is defunct; available in reprint from NATHAN Company, Paris. Also available in translation in English, see Berger 1994 [167]. MR 81k:51001a
155. _____, *Géométrie. Vol. 2*, CEDIC, Paris, 1977, Espaces euclidiens, triangles, cercles et sphères. [Euclidean spaces, triangles, circles and spheres]. The publisher is defunct; available in reprint from NATHAN Company, Paris. Also available in translation in English, see Berger 1994 [167]. MR 81k:51001b
156. _____, *Géométrie. Vol. 3*, CEDIC, Paris, 1977, Convexes et polytopes, polyèdres réguliers, aires et volumes. [Convexes and polytopes, regular polyhedra, areas and volumes]. The publisher is defunct; available in reprint from NATHAN Company, Paris. Also available in translation in English, see Berger 1994 [167]. MR 81k:51001c
157. _____, *Géométrie. Vol. 4*, CEDIC, Paris, 1977, Formes quadratiques, coniques et quadriques. [Quadratic forms, conics and quadrics]. The publisher is defunct; available in reprint from NATHAN Company, Paris. Also available in translation in English, see Berger 1987 [164]. MR 81k:51001d
158. _____, *Géométrie. Vol. 5*, CEDIC, Paris, 1977, La sphère pour elle-même, géométrie hyperbolique, l'espace des sphères. [The sphere for itself, hyperbolic geometry, the space of spheres]. The publisher is defunct; available in reprint from NATHAN Company, Paris. Also available in translation in English, see Berger 1987 [164]. MR 81k:51001e
159. _____, *Volume et rayon d'injectivité dans les variétés riemanniennes.*, C. R. Acad. Sci. Paris Sér. A-B **284** (1977), no. 19, A1221–A1224. MR 55 #11169
160. _____, *Une borne inférieure pour le volume d'une variété riemannienne en fonction du rayon d'injectivité*, Ann. Inst. Fourier (Grenoble) **30** (1980), no. 3, 259–265. MR 82b:53047
161. _____, *Une caractérisation purement métrique des variétés riemanniennes à courbure constante*, E. B. Christoffel (Aachen/Monschau, 1979), Birkhäuser, Basel, 1981, pp. 480–492. MR 83j:53072
162. _____, *Sur les variétés riemanniennes pincées juste au-dessous de 1/4*, Ann. Inst. Fourier (Grenoble) **33** (1983), no. 2, 135–150 (loose errata). MR 85d:53017
163. _____, *Courbure et valeurs propres du laplacien*, Acta convegno studio Roma (Bologna–Pitagora) (Ida Cattaneo Gasparini, ed.), 1985, pp. 1–40.
164. _____, *Geometry. II*, Springer-Verlag, Berlin, 1987, Translated from the French originals [157, 158] by M. Cole and S. Levy. MR 88a:51001b
165. _____, *Systoles et applications selon Gromov*, Astérisque (1993), no. 216, Exp. No. 771, 5, 279–310, Séminaire Bourbaki, Vol. 1992/93. MR 94j:53042
166. _____, *Géométrie et dynamique sur une surface*, Riv. Mat. Univ. Parma (5) **3** (1994), no. 1, 3–65 (1995), Differential geometry, complex analysis (Italian) (Parma, 1994). MR 96g:53051
167. _____, *Geometry. I*, Springer-Verlag, Berlin, 1994, Translated from the French originals [154, 155, 156] by M. Cole and S. Levy. Corrected reprint of the 1987 translation. MR 95g:51001

168. _____, *Encounter with a geometer: Eugenio Calabi*, Manifolds and geometry (Pisa, 1993), Cambridge Univ. Press, Cambridge, 1996, A conference in honour of Eugenio Calabi, pp. 20–60. MR 97k:01023
169. _____, *Rencontres avec un géomètre*, Gaz. Math. (1998), no. 76, 25–45.
170. _____, *Rencontres avec un géomètre, 2ème partie*, Gaz. Math. (1998), no. 77, 29–53.
171. _____, *Riemannian geometry during the second half of the twentieth century*, Jahresber. Deutsch. Math.-Verein. **100** (1998), no. 2, 45–208. MR 99h:53001
172. _____, *Riemannian geometry during the second half of the twentieth century*, University lecture series, vol. 17, American Mathematical Society, Providence, RI, 2000, Enlarged from the 1998 original. MR 1 729 907
173. _____, *Géométrie Vivante*, Cassini, 2003, in French, a work in progress.
174. Marcel Berger, Paul Gauduchon, and Edmond Mazet, *Le spectre d'une variété riemannienne*, Springer-Verlag, Berlin, 1971, Lecture Notes in Mathematics, Vol. 194. MR 43 #8025
175. Marcel Berger and Bernard Gostiaux, *Differential geometry: manifolds, curves, and surfaces*, Springer-Verlag, New York, 1988, Translated from the French by Silvio Levy. MR 88h:53001
176. Marcel Berger, S. Murakami, and T. Ochiai (eds.), *Spectra of Riemannian manifolds*, Kyoto, Kaigai, 1981.
177. Melvyn S. Berger and Enrico Bombieri, *On Poincaré's isoperimetric problem for simple closed geodesics*, J. Funct. Anal. **42** (1981), no. 3, 274–298. MR 82i:58023
178. Nicolas Bergeron, *Sur l'homologie de cycles géodésiques dans des variétés hyperboliques compactes*, C. R. Acad. Sci. Paris Sér. I Math. **328** (1999), no. 9, 783–788. MR 2000g:53038
179. Nicole Berline, Ezra Getzler, and Michèle Vergne, *Heat kernels and Dirac operators*, Springer-Verlag, Berlin, 1992. MR 94e:58130
180. Jürgen Berndt, Franco Tricerri, and Lieven Vanhecke, *Generalized Heisenberg groups and Damek-Ricci harmonic spaces*, Springer-Verlag, Berlin, 1995. MR 97a:53068
181. A. S. Besicovitch, *On two problems of Loewner*, J. London Math. Soc. **27** (1952), 141–144. MR 13,831d
182. Arthur L. Besse, *Manifolds all of whose geodesics are closed*, Springer-Verlag, Berlin, 1978, With appendices by D. B. A. Epstein, J.-P. Bourguignon, L. Bérard Bergery, M. Berger and J. L. Kazdan. MR 80c:53044
183. _____, *Einstein manifolds*, Springer-Verlag, Berlin, 1987. MR 88f:53087
184. _____, *Some trends in Riemannian geometry*, Duration and change, Springer, Berlin, 1994, pp. 71–105. MR 96a:53001
185. Arthur L. Besse (ed.), *Actes de la Table Ronde de Géométrie Différentielle*, Paris, Société Mathématique de France, 1996, En l'honneur de Marcel Berger. [In honor of Marcel Berger], Held in Luminy, France, July 12–18, 1992. MR 97i:53002
186. Laurent Bessières, *Un théorème de rigidité différentielle*, Comment. Math. Helv. **73** (1998), no. 3, 443–479. MR 99m:53085
187. Gérard Besson, *Sur la multiplicité de la première valeur propre des surfaces riemanniennes*, Ann. Inst. Fourier (Grenoble) **30** (1980), no. 1, x, 109–128. MR 81h:58059

188. Gérard Besson, Gilles Courtois, and Sylvestre F. L. Gallot, *Volume et entropie minimale des espaces localement symétriques*, Invent. Math. **103** (1991), no. 2, 417–445. MR 92d:58027
189. ———, *Entropies et rigidités des espaces localement symétriques de courbure strictement négative*, Geom. Funct. Anal. **5** (1995), no. 5, 731–799. MR 96i:58136
190. ———, *A simple proof of the rigidity and the minimal entropy theorem*, preprint 314, Institut Fourier, Grenoble, France, Août 1995.
191. ———, *Minimal entropy and Mostow's rigidity theorems*, Ergodic Theory Dynam. Systems **16** (1996), no. 4, 623–649. MR 97e:58177
192. Gérard Besson, Miroslav Lovric, Maung Min-Oo, and McKenzie Y-K Wang (eds.), *Riemannian geometry*, Fields Institute monographs, no. 4, Providence, R.I., American Mathematical Society, 1996, Papers from a workshop held Aug. 3-13, 1993, in Waterloo, Canada.
193. Edward Bierstone, Boris Khesin, Askold Khovanskii, and Jerrold E. Marsden (eds.), *The Arnoldfest*, Fields Institute Communications, Providence, R.I., Fields Institute, Toronto, Amer. Math. Soc., 1999, Proceedings of a Conference in Honour of the Sixtieth Birthday of V.I. Arnol'd.
194. Olivier Biquard and Paul Gauduchon, *La métrique hyperkählérienne des orbites coadjointes de type symétrique d'un groupe de Lie complexe semi-simple*, C. R. Acad. Sci. Paris Sér. I Math. **323** (1996), no. 12, 1259–1264. MR 97k:53040
195. G. D. Birkhoff, *Dynamical systems with two degrees of freedom*, Trans. Amer. Math. Soc. **18** (1917), 199–300.
196. Richard L. Bishop, *A relation between volume, mean curvature and diameter*, Notices Amer. Math. Soc. **10** (1963), 364.
197. Richard L. Bishop and Richard J. Crittenden, *Geometry of manifolds*, Academic Press, New York, 1964, Pure and Applied Mathematics, Vol. XV. MR 29 #6401
198. Jean-Michel Bismut, Henri Gillet, and Christophe Soulé, *Analytic torsion and holomorphic determinant bundles. I. Bott-Chern forms and analytic torsion*, Comm. Math. Phys. **115** (1988), no. 1, 49–78. MR 89g:58192a
199. ———, *Analytic torsion and holomorphic determinant bundles. II. Direct images and Bott-Chern forms*, Comm. Math. Phys. **115** (1988), no. 1, 79–126. MR 89g:58192b
200. ———, *Analytic torsion and holomorphic determinant bundles. III. Quillen metrics on holomorphic determinants*, Comm. Math. Phys. **115** (1988), no. 2, 301–351. MR 89g:58192c
201. David E. Blair, *Spaces of metrics and curvature functionals*, in Dillen and Verstraelen [449], pp. 153–185. MR 2000h:53003
202. ———, *Riemannian geometry of contact and symplectic manifolds*, Birkhäuser Boston Inc., Boston, MA, 2002. MR 1 874 240
203. Wilhelm Blaschke, *Vorlesungen über Differentialgeometrie und geometrische Grundlagen von Einsteins Relativitätstheorie*, Die Grundlehren der mathematischen Wissenschaften in Einzeldarstellungen, vol. Bd. 1, 7, 29, Julius Springer-Verlag, Berlin, 1921.
204. ———, *Vorlesungen über Differentialgeometrie und geometrische Grundlagen von Einsteins Relativitätstheorie*, second ed., Die Grundlehren der mathematischen Wissenschaften in Einzeldarstellungen, vol. Bd. 1, 7, 29, Julius Springer-Verlag, Berlin, 1924.

205. _____, *Vorlesungen über Differentialgeometrie und geometrische Grundlagen von Einsteins Relativitätstheorie*, third ed., Die Grundlehren der mathematischen Wissenschaften, vol. Bd. 1, 7, 29, Julius Springer-Verlag, Berlin, 1930.
206. Christian Blatter, *Zur Riemannschen Geometrie im Grossen auf dem Möbiusband*, Compositio Math. **15** (1961), 88–107 (1961). MR 25 #3484
207. David Bleecker, *Isometric deformations of compact hypersurfaces*, Geom. Dedicata **64** (1997), no. 2, 193–227. MR 98f:53049
208. Gilbert Ames Bliss, *The geodesic lines on the anchor ring*, Ann. of Math. **4** (1902–1903), 1–21.
209. Jacek Bochnak, Michel Coste, and Marie-Françoise Roy, *Real algebraic geometry*, Springer-Verlag, Berlin, 1998, Translated from the 1987 French original, Revised by the authors. MR 1 659 509
210. Salomon Bochner, *Vector fields and Ricci curvature*, Bull. Amer. Math. Soc. **52** (1946), 776–797. MR 8,230a
211. _____, *Curvature in Hermitian metric*, Bull. Amer. Math. Soc. **53** (1947), 179–195. MR 8,490d
212. _____, *Curvature and Betti numbers*, Ann. of Math. (2) **49** (1948), 379–390. MR 9,618d
213. Salomon Bochner and Kentaro Yano, *Tensor-fields in non-symmetric connections*, Ann. of Math. (2) **56** (1952), 504–519. MR 14,904c
214. Eric Boeckx, Oldřich Kowalski, and Lieven Vanhecke, *Riemannian manifolds of conullity two*, World Scientific Publishing Co. Inc., River Edge, NJ, 1996. MR 98h:53075
215. Fedor A. Bogomolov, *On the diameter of plane algebraic curves*, Math. Res. Lett. **1** (1994), no. 1, 95–98. MR 95a:53109
216. Christoph Böhm, *Inhomogeneous Einstein metrics on low-dimensional spheres and other low-dimensional spaces*, Invent. Math. **134** (1998), no. 1, 145–176. MR 99i:53046
217. Neda Bokan, Peter B. Gilkey, and Rade Živaljević, *An inhomogeneous elliptic complex*, J. Anal. Math. **61** (1993), 367–393. MR 94m:58211
218. Alexey V. Bolsinov and Iskander A. Taĭmanov, *Integrable geodesic flows with positive topological entropy*, Invent. Math. **140** (2000), no. 3, 639–650. MR 1 760 753
219. Bernard Bonnard and Monique Chyba, *Méthodes géométriques et analytiques pour étudier l'application exponentielle, la sphère et le front d'onde en géométrie sous-riemannienne dans le cas Martinet*, ESAIM Control Optim. Calc. Var. **4** (1999), 245–334 (electronic). MR 1 696 290
220. William M. Boothby, *An introduction to differentiable manifolds and Riemannian geometry*, second ed., Academic Press Inc., Orlando, Fla., 1986. MR 87k:58001
221. Armand Borel, *Some remarks about Lie groups transitive on spheres and tori*, Bull. Amer. Math. Soc. **55** (1949), 580–587. MR 10,680c
222. _____, *Le plan projectif des octaves et les sphères comme espaces homogènes*, C. R. Acad. Sci. Paris **230** (1950), 1378–1380. MR 11,640c
223. _____, *Compact Clifford-Klein forms of symmetric spaces*, Topology **2** (1963), 111–122. MR 26 #3823
224. Armand Borel and André Lichnerowicz, *Groupes d'holonomie des variétés riemanniennes*, C. R. Acad. Sci. Paris **234** (1952), 1835–1837. MR 13,986b

225. Max Born and Emil Wolf, *Principles of optics: Electromagnetic theory of propagation, interference and diffraction of light*, revised ed., Pergamon Press, Oxford, 1965. MR 33 #6961
226. Raoul Bott, *On manifolds all of whose geodesics are closed*, Ann. of Math. (2) **60** (1954), 375–382. MR 17,521a
227. _____, *On the iteration of closed geodesics and the Sturm intersection theory*, Comm. Pure Appl. Math. **9** (1956), 171–206. MR 19,859f
228. _____, *Morse theory indomitable*, Inst. Hautes Études Sci. Publ. Math. (1988), no. 68, 99–114 (1989). MR 90f:58027
229. Raoul Bott and Loring W. Tu, *Differential forms in algebraic topology*, Springer-Verlag, New York, 1982. MR 83i:57016
230. Henri Bouasse, *Construction, description et emploi des appareils de mesure et d'observation*, Delagrave, Paris, 1917, Edited by Blanchard.
231. Mohamed Boucetta, *Courbure sectionnelle et intégrales premières symétriques du flot géodésique: généralisation d'un théorème de S. Bochner*, C. R. Acad. Sci. Paris Sér. I Math. **326** (1998), no. 12, 1403–1406. MR 99k:58142
232. Sébastien Boucksom, *Le cône kählérien d'une variété hyperkählérienne*, C. R. Acad. Sci. Paris Sér. I Math. **333** (2001), no. 10, 935–938. MR 1 873 811
233. Nicolas Bourbaki, *Éléments de mathématique. Fasc. XXXIII. Variétés différentielles et analytiques. Fascicule de résultats (Paragraphes 1 à 7)*, Hermann, Paris, 1967, Actualités Scientifiques et Industrielles, No. 1333. MR 36 #2161
234. _____, *Éléments de mathématique. Fasc. XXXVI. Variétés différentielles et analytiques. Fascicule de résultats (Paragraphes 8 à 15)*, Hermann, Paris, 1971, Actualités Scientifiques et Industrielles, No. 1347. MR 43 #6834
235. Marc Bourdon, *Sur le birapport au bord des $CAT(-1)$-espaces*, Inst. Hautes Études Sci. Publ. Math. (1996), no. 83, 95–104. MR 97k:58123
236. Jean-Pierre Bourguignon, *Some constructions related to H. Hopf's conjecture on product manifolds*, Differential geometry (Proc. Sympos. Pure Math., Vol. XXVII, Part 1, Stanford Univ., Stanford, Calif., 1973) (Providence, R.I.), Amer. Math. Soc., 1975, pp. 33–37. MR 52 #1803
237. _____, *Une stratification de l'espace des structures riemanniennes*, Compositio Math. **30** (1975), 1–41. MR 54 #6189
238. _____, *Formes, vibrations et essais non destructifs*, Courier du C.N.R.S. **64** (1986).
239. _____, *The "magic" of Weitzenböck formulas*, Variational methods (Paris, 1988), Birkhäuser Boston, Boston, MA, 1990, pp. 251–271. MR 94a:58181
240. _____, *Stabilité par déformation non-linéaire de la métrique de Minkowski (d'après D. Christodoulou et S. Klainerman)*, Astérisque (1991), no. 201-203, Exp. No. 740, 321–358 (1992), Séminaire Bourbaki, Vol. 1990/91. MR 93d:58164
241. _____, *Eugenio Calabi and Kähler metrics*, Manifolds and geometry (Pisa, 1993), Cambridge Univ. Press, Cambridge, 1996, pp. 61–85. MR 98d:53096
242. _____, *Métriques d'Einstein-Kähler sur les variétés de Fano: obstructions et existence (d'après Y. Matsushima, A. Futaki, S. T. Yau, A. Nadel et G. Tian)*, Astérisque (1997), no. 245, Exp. No. 830, 5, 277–305, Séminaire Bourbaki, Vol. 1996/97. MR 99i:32033
243. Jean-Pierre Bourguignon and Paul Gauduchon, *Spineurs, opérateurs de Dirac et variations de métriques*, Comm. Math. Phys. **144** (1992), no. 3, 581–599. MR 93h:58164

244. Jean-Pierre Bourguignon and H. Blaine Lawson, Jr., *Yang–Mills theory: its physical origins and differential geometric aspects*, Seminar on Differential Geometry, Princeton Univ. Press, Princeton, N.J., 1982, pp. 395–421. MR 83d:53035

245. Jean-Pierre Bourguignon, Peter Li, and Shing-Tung Yau, *Upper bound for the first eigenvalue of algebraic submanifolds*, Comment. Math. Helv. **69** (1994), no. 2, 199–207. MR 95j:58168

246. Jean-Pierre Bourguignon and Albert Polombo, *Intégrands des nombres caractéristiques et courbure: rien ne va plus dès la dimension 6*, J. Differential Geom. **16** (1981), no. 4, 537–550 (1982). MR 83m:53054

247. _____, *Erratum: "Integrands of characteristic numbers and curvature: nothing works from dimension 6 on"*, J. Differential Geom. **21** (1985), no. 2, 309, Correction to Bourguignon and Polombo [246]. MR 87e:53061

248. Brian H. Bowditch, *Some results on the geometry of convex hulls in manifolds of pinched negative curvature*, Comment. Math. Helv. **69** (1994), no. 1, 49–81. MR 94m:53044

249. Werner Boy, *Über die Curvatura integra und die Topologie geschlossener Flächen*, Math. Ann. **57** (1903), 151–184.

250. Charles P. Boyer and Krzysztof Galicki, *3-Sasakian manifolds*, in LeBrun and Wang [858], to appear, p. 59. MR 2001f:53003

251. _____, *New Einstein metrics in dimension five*, electronic preprint math.DG/0003174, Los Alamos National Labs, 2000.

252. _____, *Sasakian geometry, homotopy spheres and positive Ricci curvature*, electronic preprint math.DG/0201147, Los Alamos National Labs, 2002.

253. Charles P. Boyer, Krzysztof Galicki, and Benjamin M. Mann, *The geometry and topology of 3-Sasakian manifolds*, J. Reine Angew. Math. **455** (1994), 183–220. MR 96e:53057

254. _____, *On strongly inhomogeneous Einstein manifolds*, Bull. London Math. Soc. **28** (1996), no. 4, 401–408. MR 97a:53069

255. Charles P. Boyer, Krzysztof Galicki, Benjamin M. Mann, and Elmer G. Rees, *Compact 3-Sasakian 7-manifolds with arbitrary second Betti number*, Invent. Math. **131** (1998), no. 2, 321–344. MR 99b:53066

256. Herm Jan Brascamp and Elliott H. Lieb, *On extensions of the Brunn-Minkowski and Prékopa-Leindler theorems, including inequalities for log concave functions, and with an application to the diffusion equation*, J. Functional Analysis **22** (1976), no. 4, 366–389. MR 56 #8774

257. A. Braunmühl, *Geodätische Linien auf dreiachsigen Flächen zweiten Grades*, Math. Ann. **20** (1882), 557–586.

258. Martin A. Bridson and André Haefliger, *Metric spaces of nonpositive curvature*, Grundlehren der mathematischer Wissenschaften, vol. 319, Springer, 1998.

259. Reinhard Brocks, *Convexité et courbure de Ricci*, C. R. Acad. Sci. Paris Sér. I Math. **319** (1994), no. 1, 73–75. MR 95g:53042

260. _____, *Ein Kompaktheitsatz für Manningfaltigkeiten mit unteren Schranken für Riccikrümmung und Konvexitätsradius*, 1997.

261. Robert Brooks, Carolyn S. Gordon, and Peter Perry (eds.), *Geometry of the spectrum (Seattle WA, 1993)*, Providence, RI, American Mathematical Society, 1994, Proceedings of the 1993 AMS-IMS-SIAM Joint Summer Research Conference on Spectral Geometry held at the University of Washington, Seattle, Washington, July 17–23, 1993. MR 95d:58001

262. Robert Brooks, Peter Perry, and Peter Petersen, V, *Compactness and finiteness theorems for isospectral manifolds*, J. Reine Angew. Math. **426** (1992), 67–89. MR 93f:53034
263. _____, *Finiteness of diffeomorphism types of isospectral manifolds*, in Greene and Yau [597], Proceedings of the AMS Summer Research Institute on Differential Geometry held at the University of California, Los Angeles, California, July 8–28, 1990, pp. 89–94. MR 1 216 613
264. _____, *Spectral geometry in dimension 3*, Acta Math. **173** (1994), no. 2, 283–305. MR 96e:58154
265. Robert B. Brown and Alfred Gray, *Riemannian manifolds with holonomy group* Spin(9), Differential geometry (in honor of Kentaro Yano) (Tokyo), Kinokuniya, 1972, pp. 41–59. MR 48 #7159
266. Jochen Brüning, *Über Knoten von Eigenfunktionen des Laplace-Beltrami-Operators*, Math. Z. **158** (1978), no. 1, 15–21. MR 57 #17732
267. Robert L. Bryant, *Metrics with exceptional holonomy*, Ann. of Math. (2) **126** (1987), no. 3, 525–576. MR 89b:53084
268. _____, *Finsler structures on the 2-sphere satisfying $K = 1$*, in Bao et al. [114], pp. 27–41. MR 97e:53128
269. _____, *Recent advances in the theory of holonomy*, Exposés de le Séminaire Nicolas Bourbaki (1999), no. 861, 1–24, The text of a lecture delivered at the Bourbaki seminar, June 1999. Available from lanl `math/9910059`.
270. Robert L. Bryant and Lucas Hsu, *Rigidity of integral curves of rank 2 distributions*, Invent. Math. **114** (1993), no. 2, 435–461. MR 94j:58003
271. Robert L. Bryant and Simon M. Salamon, *On the construction of some complete metrics with exceptional holonomy*, Duke Math. J. **58** (1989), no. 3, 829–850. MR 90i:53055
272. N. Buchdal, *Compact Kähler surfaces with trivial canonical bundle*, to appear, 2000.
273. Michael A. Buchner, *Simplicial structure of the real analytic cut locus*, Proc. Amer. Math. Soc. **64** (1977), no. 1, 118–121. MR 57 #13783
274. _____, *Stability of the cut locus in dimensions less than or equal to 6*, Invent. Math. **43** (1977), no. 3, 199–231. MR 58 #2866
275. _____, *The structure of the cut locus in dimension less than or equal to six*, Compositio Math. **37** (1978), no. 1, 103–119. MR 58 #18549
276. Emilio Bujalance, José J. Etayo, José M. Gamboa, and Grzegorz Gromadzki, *Automorphism groups of compact bordered Klein surfaces*, Springer-Verlag, Berlin, 1990, A combinatorial approach. MR 92a:14018
277. Leonid A. Bunimovitch, I. P. Cornfeld, R. L. Dobrushin, M. V. Jakobson, N. B. Maslova, Ya. B. Pesin, Yakov G. Sinaĭ, Yu. M. Sukhov, and A. M. Vershik, *Dynamical systems. II*, Springer-Verlag, Berlin, 1989, Ergodic theory with applications to dynamical systems and statistical mechanics, Edited and with a preface by Sinaĭ, Translated from the Russian. MR 91i:58079
278. Ulrich Bunke, *On the gluing problem for the η-invariant*, J. Differential Geom. **41** (1995), no. 2, 397–448. MR 96c:58163
279. Ulrich Bunke and Martin Olbrich, *Selberg zeta and theta functions*, Akademie-Verlag, Berlin, 1995, A differential operator approach. MR 97c:11088
280. D. Burago and Stefan Ivanov, *Riemannian tori without conjugate points are flat*, Geom. Funct. Anal. **4** (1994), no. 3, 259–269. MR 95h:53049
281. Dmitri Burago, Yuri Burago, and Sergei Ivanov, *A course in metric geometry*, American Mathematical Society, Providence, RI, 2001. MR 2002e:53053

282. Yu. D. Burago, Mikhael Gromov, and G. Perel′man, *A. D. Aleksandrov spaces with curvatures bounded below*, Uspekhi Mat. Nauk **47** (1992), no. 2(284), 3–51, 222. MR 93m:53035
283. Yu. D. Burago and V. A. Zalgaller, *Geometric inequalities*, Springer-Verlag, Berlin, 1988, Translated from the Russian by A. B. Sosinskiĭ, Springer Series in Soviet Mathematics. MR 89b:52020
284. Yu. D. Burago and V. A. Zalgaller (eds.), *Geometry. III*, Springer-Verlag, Berlin, 1992, Theory of surfaces, A translation of *Current problems in mathematics. Fundamental directions. Vol. 48* (Russian), Akad. Nauk SSSR, Vsesoyuz. Inst. Nauchn. i Tekhn. Inform., Moscow, 1989, Translation by E. Primrose, Translation edited by Yu. D. Burago and V. A. Zalgaller. MR 95f:53002
285. Keith Burns and Victor J. Donnay, *Embedded surfaces with ergodic geodesic flows*, Internat. J. Bifur. Chaos Appl. Sci. Engrg. **7** (1997), no. 7, 1509–1527. MR 98j:58084
286. Keith Burns and Marlies Gerber, *Real analytic Bernoulli geodesic flows on S^2*, Ergodic Theory Dynamical Systems **9** (1989), no. 1, 27–45. MR 90e:58126
287. Keith Burns and Gabriel P. Paternain, *On the growth of the number of geodesics joining two points*, International Conference on Dynamical Systems (Montevideo, 1995), Longman, Harlow, 1996, pp. 7–20. MR 98g:58137
288. _____, *Counting geodesics on a Riemannian manifold and topological entropy of geodesic flows*, Ergodic Theory Dynam. Systems **17** (1997), no. 5, 1043–1059. MR 98j:58083
289. F. E. Burstall, D. Ferus, K. Leschke, F. Pedit, and Ulrich Pinkall, *Conformal geometry of surfaces in s^4 and quaternions*, Springer-Verlag, Berlin, 2002. MR 1 887 131
290. Herbert Busemann, *The geometry of geodesics*, Academic Press Inc., New York, N. Y., 1955. MR 17,779a
291. Peter Buser, *Über eine Ungleichung von Cheeger*, Math. Z. **158** (1978), no. 3, 245–252. MR 57 #17733
292. _____, *Geometry and spectra of compact Riemann surfaces*, Birkhäuser Boston Inc., Boston, MA, 1992. MR 93g:58149
293. _____, *Inverse spectral geometry on Riemann surfaces*, Recent progress in inverse spectral geometry, Birkhauser, Basel, 1997, pp. 133–173.
294. Peter Buser, John H. Conway, Peter Doyle, and Klaus-Dieter Semmler, *Some planar isospectral domains*, Internat. Math. Res. Notices **9** (1994), 391ff., approx. 9 pp. (electronic). MR 95k:58163
295. Peter Buser and Detlef Gromoll, *On the almost negatively curved 3-sphere*, Geometry and analysis on manifolds (Katata/Kyoto, 1987), Springer, Berlin, 1988, pp. 78–85. MR 90a:53042
296. Peter Buser and Hermann Karcher, *Gromov's almost flat manifolds*, Astérisque (1981), no. 81, 148. MR 83m:53070
297. Peter Buser and Peter Sarnak, *On the period matrix of a Riemann surface of large genus*, Invent. Math. **117** (1994), no. 1, 27–56, With an appendix by J. H. Conway and N. J. A. Sloane. MR 95i:22018
298. Eugenio Calabi, *Isometric imbedding of complex manifolds*, Ann. of Math. (2) **58** (1953), 1–23. MR 15,160c
299. _____, *The space of Kähler metrics*, in Gerretsen [555], pp. 206–207. MR 16,1190j
300. _____, *An extension of E. Hopf's maximum principle with an application to Riemannian geometry*, Duke Math. J. **25** (1957), 45–56. MR 19,1056e

301. _____, *Minimal immersions of surfaces in Euclidean spheres*, J. Differential Geometry **1** (1967), 111–125. MR 38 #1616
302. _____, *An intrinsic characterization of harmonic one-forms*, in Spencer and Iyanaga [1154], pp. 101–117. MR 40 #6585
303. _____, *Métriques kählériennes et fibrés holomorphes*, Ann. Sci. École Norm. Sup. (4) **12** (1979), no. 2, 269–294. MR 83m:32033
304. _____, *Extremal isosystolic metrics for compact surfaces*, in Besse [185], En l'honneur de Marcel Berger. [In honor of Marcel Berger], Held in Luminy, France, July 12–18, 1992, pp. 167–204. MR 97k:53037
305. Eugenio Calabi and Jian Guo Cao, *Simple closed geodesics on convex surfaces*, J. Differential Geom. **36** (1992), no. 3, 517–549. MR 93h:53039
306. Huai-Dong Cao and Bennett Chow, *Recent developments on the Ricci flow*, Bull. Amer. Math. Soc. (N.S.) **36** (1999), no. 1, 59–74. MR 99j:53044
307. Jian Guo Cao, *The Martin boundary and isoperimetric inequalities for Gromov-hyperbolic manifolds*, preprint 251, University of Notre Dame, mathematics dept., Notre Dame, IN, 1996.
308. Jian Guo Cao, Jeff Cheeger, and Xiaochun Rong, *Splittings and CR structures for manifolds with nonpositive sectional curvature*, to appear, 2000.
309. Jian Guo Cao and J. Escobar, *A new 3-dimensional curvature integral formula for pl-manifolds of non-positive curvature*, to appear, 2000.
310. Jian Guo Cao and F. Xavier, *Kähler parabolicity and the Euler number of compact manifolds of non-positive sectional curvature*, to appear, 1998.
311. Élie Cartan, *Les groupes projectifs qui ne laissent invariante aucune multiplicité plane*, Bull. Soc. Math. France **41** (1913), 53–96.
312. _____, *Les groupes réels simples, finis et continus*, Ann. Éc. Norm. **31** (1914), 263–355.
313. _____, *Les groupes d'holonomie des espaces généralisés*, Acta. Math. **48** (1926), 1–42.
314. _____, *Sur certaines formes riemanniennes remarquables des géométries à groupe fondamental simple*, Ann. Éc. Norm. **44** (1927), 345–467.
315. _____, *Groupes simples clos et ouverts et géométrie riemannienne*, J. Math. Pures Appl. **8** (1929), 1–33.
316. _____, *La topologie des espaces représentatifs des groupes de Lie*, Enseignment Math. **35** (1936), 177–200, Exposés de Géometrie VIII.
317. _____, *Sur des familles remarquables d'hypersurfaces isoparamétriques dans les espaces sphériques*, Math. Z. **45** (1939), 335–367. MR 1,28f
318. _____, *The theory of spinors*, Dover Publications Inc., New York, 1981, With a foreword by Raymond Streater, A reprint of the 1966 English translation, Dover Books on Advanced Mathematics. MR 83a:15017
319. _____, *Geometry of Riemannian spaces*, Math Sci Press, Brookline, Mass., 1983, Translated from the French original [321] and with a preface by James Glazebrook. With a preface, notes and appendices by Robert Hermann. MR 85m:53001
320. _____, *On manifolds with an affine connection and the theory of general relativity*, Bibliopolis, Naples, 1986, Translated from the French by Anne Magnon and Abhay Ashtekar, With a foreword by Andrzej Trautman. MR 88b:01071
321. _____, *Leçons sur la géométrie des espaces de Riemann*, Éditions Jacques Gabay, Sceaux, 1988, Reprint of the second (1946) edition. MR 1 191 392

322. _____, *Leçons sur la géométrie projective complexe. La théorie des groupes finis et continus et la géométrie différentielle traitées par la méthode du repère mobile. Leçons sur la théorie des espaces à connexion projective*, Éditions Jacques Gabay, Sceaux, 1992, Reprint of the editions of 1931, 1937 and 1937. MR 93i:01030

323. Fabrizio Catanese and Claude LeBrun, *On the scalar curvature of Einstein manifolds*, Math. Res. Lett. **4** (1997), no. 6, 843–854. MR 98k:53057

324. S. D. Chatterji (ed.), *Proceedings of the International Congress of Mathematicians: August 3-11, 1994, Zürich, Switzerland*, Basel, Birkhauser, 1995.

325. Isaac Chavel, *Eigenvalues in Riemannian geometry*, Academic Press Inc., Orlando, Fla., 1984, Including a chapter by Burton Randol, With an appendix by Jozef Dodziuk. MR 86g:58140

326. _____, *Riemannian geometry—a modern introduction*, Cambridge University Press, Cambridge, 1993. MR 95j:53001

327. J. Chazarain, *Formule de Poisson pour les variétés riemanniennes*, Invent. Math. **24** (1974), 65–82. MR 49 #8062

328. Jeff Cheeger, *Comparison and finiteness theorems for Riemannian manifolds*, Ph.D. thesis, Princeton, 1967.

329. _____, *Finiteness theorems for Riemannian manifolds*, Amer. J. Math. **92** (1970), 61–74. MR 41 #7697

330. _____, *A lower bound for the smallest eigenvalue of the Laplacian*, Problems in analysis (Papers dedicated to Salomon Bochner, 1969) (Princeton, N. J.), Princeton Univ. Press, 1970, pp. 195–199. MR 53 #6645

331. _____, *Some examples of manifolds of nonnegative curvature*, J. Differential Geometry **8** (1973), 623–628. MR 49 #6085

332. _____, *Analytic torsion and the heat equation*, Ann. of Math. (2) **109** (1979), no. 2, 259–322. MR 80j:58065a

333. _____, *Spectral geometry of singular Riemannian spaces*, J. Differential Geom. **18** (1983), no. 4, 575–657 (1984). MR 85d:58083

334. _____, *Critical points of distance functions and applications to geometry*, in de Bartolomeis and Tricerri [436], Lectures given at the First C.I.M.E. Session on Recent Developments in Geometric Topology and Related Topics held in Montecatini Terme, June 4–12, 1990, pp. 1–38. MR 94a:53075

335. _____, *Differentiability of Lipschitz functions on metric measure spaces*, Geom. Funct. Anal. **9** (1999), no. 3, 428–517. MR 1 708 448

336. Jeff Cheeger and Tobias H. Colding, *Lower bounds on Ricci curvature and the almost rigidity of warped products*, Ann. of Math. (2) **144** (1996), no. 1, 189–237. MR 97h:53038

337. _____, *On the structure of spaces with Ricci curvature bounded below. I*, J. Differential Geom. **46** (1997), no. 3, 406–480. MR 98k:53044

338. _____, *On the structure of spaces with Ricci curvature bounded below II*, to appear, 1997.

339. _____, *On the structure of spaces with Ricci curvature bounded below III*, to appear, 1998.

340. Jeff Cheeger, Tobias H. Colding, and Gang Tian, *Constraints on singularities under Ricci curvature bounds*, C. R. Acad. Sci. Paris Sér. I Math. **324** (1997), no. 6, 645–649. MR 98g:53078

341. Jeff Cheeger and David G. Ebin, *Comparison theorems in Riemannian geometry*, North-Holland Publishing Co., Amsterdam, 1975, North-Holland Mathematical Library, Vol. 9. MR 56 #16538

342. Jeff Cheeger, Kenji Fukaya, and Mikhael Gromov, *Nilpotent structures and invariant metrics on collapsed manifolds*, J. Amer. Math. Soc. **5** (1992), no. 2, 327–372. MR 93a:53036
343. Jeff Cheeger and Detlef Gromoll, *The splitting theorem for manifolds of nonnegative Ricci curvature*, J. Differential Geometry **6** (1971/72), 119–128. MR 46 #2597
344. _____, *On the structure of complete manifolds of nonnegative curvature*, Ann. of Math. (2) **96** (1972), 413–443. MR 46 #8121
345. _____, *On the lower bound for the injectivity radius of $1/4$-pinched Riemannian manifolds*, J. Differential Geom. **15** (1980), no. 3, 437–442 (1981). MR 83g:53035
346. Jeff Cheeger and Mikhael Gromov, *On the characteristic numbers of complete manifolds of bounded curvature and finite volume*, Differential geometry and complex analysis, Springer, Berlin, 1985, pp. 115–154. MR 86h:58131
347. _____, *Collapsing Riemannian manifolds while keeping their curvature bounded. I*, J. Differential Geom. **23** (1986), no. 3, 309–346. MR 87k:53087
348. _____, *Collapsing Riemannian manifolds while keeping their curvature bounded. II*, J. Differential Geom. **32** (1990), no. 1, 269–298. MR 92a:53066
349. _____, *Chopping Riemannian manifolds*, Differential geometry, Longman Sci. Tech., Harlow, 1991, pp. 85–94. MR 93k:53034
350. Jeff Cheeger, Werner Müller, and Robert Schrader, *On the curvature of piecewise flat spaces*, Comm. Math. Phys. **92** (1984), no. 3, 405–454. MR 85m:53037
351. _____, *Kinematic and tube formulas for piecewise linear spaces*, Indiana Univ. Math. J. **35** (1986), no. 4, 737–754. MR 87m:53083
352. Jeff Cheeger and Xiaochun Rong, *Collapsed Riemannian manifolds with bounded diameter and bounded covering geometry*, Geom. Funct. Anal. **5** (1995), no. 2, 141–163. MR 96h:53053
353. _____, *Existence of polarized F-structures on collapsed manifolds with bounded curvature and diameter*, Geom. Funct. Anal. **6** (1996), no. 3, 411–429. MR 97f:53074
354. Jeff Cheeger and James Simons, *Differential characters and geometric invariants*, Geometry and topology (College Park, Md., 1983/84), Springer, Berlin, 1985, pp. 50–80. MR 87g:53059
355. Jeff Cheeger and Michael Taylor, *On the diffraction of waves by conical singularities. I*, Comm. Pure Appl. Math. **35** (1982), no. 3, 275–331. MR 84h:35091a
356. _____, *On the diffraction of waves by conical singularities. II*, Comm. Pure Appl. Math. **35** (1982), no. 4, 487–529. MR 84h:35091b
357. Jeff Cheeger and Gang Tian, *On the cone structure at infinity of Ricci flat manifolds with Euclidean volume growth and quadratic curvature decay*, Invent. Math. **118** (1994), no. 3, 493–571. MR 95m:53051
358. Bang-Yen Chen, *Riemannian submanifolds*, in Dillen and Verstraelen [449], pp. 187–418. MR 2000h:53003
359. Bing-Long Chen and Xi-Ping Zhu, *Complete Riemannian manifolds with pointwise pinched curvature*, Invent. Math. **140** (2000), no. 2, 423–452. MR 1 757 002
360. Haiwen Chen, *Manifolds with 2-nonnegative curvature operator*, in Greene and Yau [597], Proceedings of the AMS Summer Research Institute on Differential Geometry held at the University of California, Los Angeles, California, July 8–28, 1990, pp. 129–133. MR 94f:53063

361. Jingyi Chen and Jiayu Li, *Mean curvature flow of surface in 4-manifolds*, Adv. Math. **163** (2001), no. 2, 287–309. MR 2002h:53116
362. Shiu Yuen Cheng, *Eigenvalue comparison theorems and its geometric applications*, Math. Z. **143** (1975), no. 3, 289–297. MR 51 #14170
363. Shiing-Shen Chern, *A simple intrinsic proof of the Gauss-Bonnet formula for closed Riemannian manifolds*, Ann. of Math. (2) **45** (1944), 747–752. MR 6,106a
364. _____, *Characteristic classes of Hermitian manifolds*, Ann. of Math. (2) **47** (1946), 85–121. MR 7,470b
365. _____, *On curvature and characteristic classes of a Riemann manifold*, Abh. Math. Sem. Univ. Hamburg **20** (1955), 117–126. MR 17,783e
366. Shiing-Shen Chern (ed.), *Global differential geometry*, Mathematical Association of America, Washington, DC, 1989. MR 90d:53003
367. _____, *Historical remarks on Gauss-Bonnet*, Analysis, et cetera, Academic Press, Boston, MA, 1990, pp. 209–217. MR 91b:53001
368. Shiing-Shen Chern and Claude Chevalley, *Obituary: Elie Cartan and his mathematical work*, Bull. Amer. Math. Soc. **58** (1952), 217–250. MR 13,810f
369. Shiing-Shen Chern, Philip Hartman, and Aurel Wintner, *On isothermic coordinates*, Comment. Math. Helv. **28** (1954), 301–309. MR 16,622e
370. Shiing-Shen Chern and James Simons, *Characteristic forms and geometric invariants*, Ann. of Math. (2) **99** (1974), 48–69. MR 50 #5811
371. Shiing-Shen Chern and Stephen Smale (eds.), *Global analysis*, American Mathematical Society, Providence, R.I., 1970, Proceedings of Symposia in Pure Mathematics, Vols. XIV-XVI. Edited by Shiing-Shen Chern and Stephen Smale. MR 41 #7686
372. Claude Chevalley, *The algebraic theory of spinors and Clifford algebras*, Springer-Verlag, Berlin, 1997, Collected works. Vol. 2, Edited and with a foreword by Pierre Cartier and Catherine Chevalley, With a postface by J.-P. Bourguignon. MR 99f:01028
373. Ted Chinburg and Alan W. Reid, *Closed hyperbolic 3-manifolds whose closed geodesics all are simple*, J. Differential Geom. **38** (1993), no. 3, 545–558. MR 94k:57020
374. S. Chmutov and S. Duzhin, *Explicit formulas for Arnold's generic curve invariants*, in Arnol'd et al. [62], Geometry and singularity theory, pp. 123–138. MR 97j:57038
375. Yunhi Cho and Hyuk Kim, *On the volume formula for hyperbolic tetrahedra*, Discrete Comput. Geom. **22** (1999), no. 3, 347–366. MR 1 706 606
376. David L. Chopp, *Computing minimal surfaces via level set curvature flow*, J. Comput. Phys. **106** (1993), no. 1, 77–91. MR 94f:53007
377. Gustave Choquet, *Topology*, Academic Press, New York, 1966, Translated from the French by Amiel Feinstein. Pure and Applied Mathematics, Vol. XIX. MR 33 #1823
378. Bennett Chow, *The Ricci flow on the 2-sphere*, J. Differential Geom. **33** (1991), no. 2, 325–334. MR 92d:53036
379. Demetrios Christodoulou and Sergiu Klainerman, *The global nonlinear stability of the Minkowski space*, Princeton University Press, Princeton, NJ, 1993. MR 95k:83006
380. R. V. Churchill, *On the geometry of the Riemann tensor*, Trans. Amer. Math. Soc. **34** (1932), no. 1, 126–152. MR 1 501 632

381. Mónica Clapp and Dieter Puppe, *Critical point theory of symmetric functions and closed geodesics*, Differential Geom. Appl. **6** (1996), no. 4, 367–396. MR 97i:58029
382. Stefan Cohn-Vossen, *Kürzeste Wege und Totalkrümmung auf Flächen*, Compositio Math. **2** (1935), 69–133.
383. _____, *Totalkrümmung und geodätische Linien auf einfach zusammenhängenden vollständingen Flächenstücke*, Recueil Math. Moscou **43** (1936), 139–163.
384. B. Colbois, P. Ghanaat, and E. Ruh, *Curvature and gradient estimates for eigenforms of the Laplacian*, to appear, 1999.
385. Tobias H. Colding, *Large manifolds with positive Ricci curvature*, Invent. Math. **124** (1996), no. 1-3, 193–214. MR 96k:53068
386. _____, *Shape of manifolds with positive Ricci curvature*, Invent. Math. **124** (1996), no. 1-3, 175–191. MR 96k:53067
387. _____, *Ricci curvature and volume convergence*, Ann. of Math. (2) **145** (1997), no. 3, 477–501. MR 98d:53050
388. Tobias H. Colding and William P. Minicozzi, II, *Weyl type bounds for harmonic functions*, Invent. Math. **131** (1998), no. 2, 257–298. MR 99b:53052
389. Yves Colin de Verdière, *Spectre du laplacien et longueurs des géodésiques périodiques. I, II*, Compositio Math. **27** (1973), 83–106; ibid. **27** (1973), 159–184. MR 50 #1293
390. _____, *Quasi-modes sur les variétés Riemanniennes*, Invent. Math. **43** (1977), no. 1, 15–52. MR 58 #18615
391. _____, *Sur le spectre des opérateurs elliptiques à bicaractéristiques toutes périodiques*, Comment. Math. Helv. **54** (1979), no. 3, 508–522. MR 81a:58052
392. _____, *Ergodicité et fonctions propres du laplacien*, Comm. Math. Phys. **102** (1985), no. 3, 497–502. MR 87d:58145
393. _____, *Construction de laplaciens dont une partie finie du spectre est donnée*, Ann. Sci. École Norm. Sup. (4) **20** (1987), no. 4, 599–615. MR 90d:58156
394. _____, *Distribution de points sur une sphère (d'après Lubotzky, Phillips et Sarnak)*, Astérisque (1989), no. 177-178, Exp. No. 703, 83–93, Séminaire Bourbaki, Vol. 1988/89. MR 91k:11089
395. _____, *Comment rendre géodésique une triangulation d'une surface?*, Enseign. Math. (2) **37** (1991), no. 3-4, 201–212. MR 93a:53034
396. _____, *Le spectre du laplacien: survol partiel depuis le Berger–Gauduchon–Mazet et problèmes*, Actes de la Table Ronde de Géométrie Différentielle (Luminy, 1992), Soc. Math. France, Paris, 1996, pp. 233–252. MR 98e:58169
397. Yves Colin de Verdière, Isidoro Gitler, and Dirk Vertigan, *Réseaux électriques planaires. II*, Comment. Math. Helv. **71** (1996), no. 1, 144–167. MR 98a:05054
398. Yves Colin de Verdière and A. Marin, *Triangulations presque équilatérales des surfaces*, J. Differential Geom. **32** (1990), no. 1, 199–207. MR 91f:57001
399. R. Connelly, I. Sabitov, and A. Walz, *The bellows conjecture*, Beiträge Algebra Geom. **38** (1997), no. 1, 1–10. MR 98c:52026
400. Alain Connes, *Noncommutative geometry*, Academic Press Inc., San Diego, CA, 1994. MR 95j:46063
401. _____, *Brisure de symétrie spontanée et géométrie du point de vue spectral*, Astérisque (1997), no. 241, Exp. No. 816, 5, 313–349, Séminaire Bourbaki, Vol. 1995/96. MR 98h:58011a
402. John H. Conway, *Sphere packings, lattices, codes, and greed*, in Chatterji [324], pp. 45–55. MR 97d:11108

403. John H. Conway and N. J. A. Sloane, *Sphere packings, lattices and groups*, third ed., Springer-Verlag, New York, 1999, With additional contributions by E. Bannai, R. E. Borcherds, J. Leech, S. P. Norton, A. M. Odlyzko, R. A. Parker, L. Queen and B. B. Venkov. MR 1 662 447
404. Kevin Corlette, *Archimedean superrigidity and hyperbolic geometry*, Ann. of Math. (2) **135** (1992), no. 1, 165–182. MR 92m:57048
405. Vicente Cortés, *A new construction of homogeneous quaternionic manifolds and related geometric structures*, Mem. Amer. Math. Soc. (to appear). MR 1 708 628
406. Richard Courant and David Hilbert, *Methods of mathematical physics. Vol. I*, Interscience Publishers, Inc., New York, N.Y., 1953. MR 16,426a
407. _____, *Methods of mathematical physics. Vol. II*, John Wiley & Sons Inc., New York, 1989, Partial differential equations, Reprint of the 1962 original, A Wiley-Interscience Publication. MR 90k:35001
408. David A. Cox and Sheldon Katz, *Mirror symmetry and algebraic geometry*, American Mathematical Society, Providence, RI, 1999. MR 2000d:14048
409. H. S. MacDonald Coxeter, *Introduction to geometry*, John Wiley & Sons Inc., New York, 1989, Reprint of the 1969 edition. MR 90a:51001
410. Hallard T. Croft, Kenneth J. Falconer, and Richard K. Guy, *Unsolved problems in geometry*, Springer-Verlag, New York, 1994, Corrected reprint of the 1991 original, Unsolved Problems in Intuitive Mathematics, II. MR 95k:52001
411. Christopher B. Croke, *Some isoperimetric inequalities and eigenvalue estimates*, Ann. Sci. École Norm. Sup. (4) **13** (1980), no. 4, 419–435. MR 83d:58068
412. _____, *The first eigenvalue of the Laplacian for plane domains*, Proc. Amer. Math. Soc. **81** (1981), no. 2, 304–305. MR 82e:35061
413. _____, *An eigenvalue pinching theorem*, Invent. Math. **68** (1982), no. 2, 253–256. MR 84a:58084
414. _____, *Poincaré's problem and the length of the shortest closed geodesic on a convex hypersurface*, J. Differential Geom. **17** (1982), no. 4, 595–634 (1983). MR 84f:58034
415. _____, *Curvature free volume estimates*, Invent. Math. **76** (1984), no. 3, 515–521. MR 85f:53044
416. _____, *A sharp four-dimensional isoperimetric inequality*, Comment. Math. Helv. **59** (1984), no. 2, 187–192. MR 85f:53060
417. _____, *Area and the length of the shortest closed geodesic*, J. Differential Geom. **27** (1988), no. 1, 1–21. MR 89a:53050
418. _____, *An isoembolic pinching theorem*, Invent. Math. **92** (1988), no. 2, 385–387. MR 89e:53061
419. _____, *Rigidity for surfaces of nonpositive curvature*, Comment. Math. Helv. **65** (1990), no. 1, 150–169. MR 91d:53056
420. _____, *Rigidity and the distance between boundary points*, J. Differential Geom. **33** (1991), no. 2, 445–464. MR 92a:53053
421. _____, *Volumes of balls in manifolds without conjugate points*, Internat. J. Math. **3** (1992), no. 4, 455–467. MR 93e:53048
422. Christopher B. Croke and Vladimir A. Sharafutdinov, *Spectral rigidity of a compact negatively curved manifold*, Topology **37** (1998), no. 6, 1265–1273. MR 99e:58191
423. Xianzhe Dai and Guofang Wei, *A comparison-estimate of Toponogov type for Ricci curvature*, Math. Ann. **303** (1995), no. 2, 297–306. MR 96h:53042

424. Xianzhe Dai, Guofang Wei, and Rugang Ye, *Smoothing Riemannian metrics with Ricci curvature bounds*, Manuscripta Math. **90** (1996), no. 1, 49–61. MR 97b:53039
425. Giuseppina D'Ambra, *Isometry groups of Lorentz manifolds*, Invent. Math. **92** (1988), no. 3, 555–565. MR 89h:53124
426. Giuseppina D'Ambra and Mikhael Gromov, *Lectures on transformation groups: geometry and dynamics*, Surveys in differential geometry (Cambridge, MA, 1990), Lehigh Univ., Bethlehem, PA, 1991, pp. 19–111. MR 93d:58117
427. Thibault Damour, Michael Soffel, and Chong Ming Xu, *General-relativistic celestial mechanics. I. Method and definition of reference systems*, Phys. Rev. D (3) **43** (1991), no. 10, 3273–3307. MR 92g:83005
428. Andrew Dancer and McKenzie Y. Wang, *Integrable cases of the Einstein equations*, Comm. Math. Phys. **208** (1999), no. 1, 225–243. MR 2001c:53052
429. Gaston Darboux, *Leçons sur la théorie générale des surfaces et les applications géométriques du calcul infinitésimal. Première partie*, Chelsea Publishing Co., Bronx, N. Y., 1972, Généralités. Coordonnées curvilignes. Surfaces minima, Réimpression de la deuxième édition de 1914. MR 53 #79
430. _____, *Leçons sur la théorie générale des surfaces et les applications géométriques du calcul infinitésimal. Deuxième partie*, Chelsea Publishing Co., Bronx, N.Y., 1972, Les congruences et les équations linéaires aux dérivées partielles. Les lignes tracées sur les surfaces, Réimpression de la deuxième édition de 1915. MR 53 #80
431. _____, *Leçons sur la théorie générale des surfaces et les applications géométriques du calcul infinitésimal. Troisième partie*, Chelsea Publishing Co., Bronx, N. Y., 1972, Lignes géodésiques et courbure géodésique. Paramètres différentiels. Déformation des surfaces, Réimpression de la première édition de 1894. MR 53 #81
432. _____, *Leçons sur la théorie générale des surfaces et les applications géométriques du calcul infinitésimal. Quatrième partie*, Chelsea Publishing Co., Bronx, N. Y., 1972, Déformation infiniment petite et representation sphérique, Réimpression de la première édition de 1896. MR 53 #82
433. _____, *Leçons sur la théorie générale des surfaces. I, II*, Éditions Jacques Gabay, Sceaux, 1993, Généralités. Coordonnées curvilignes. Surfaces minima. [Generalities. Curvilinear coordinates. Minimum surfaces], Les congruences et les équations linéaires aux dérivées partielles. Les lignes tracées sur les surfaces. [Congruences and linear partial differential equations. Lines traced on surfaces], Reprint of the second (1914) edition (I) and the second (1915) edition (II), Cours de Géométrie de la Faculté des Sciences. [Course on Geometry of the Faculty of Science]. MR 97c:01046a
434. _____, *Leçons sur la théorie générale des surfaces. III, IV*, Éditions Jacques Gabay, Sceaux, 1993, Lignes géodésiques et courbure géodésique. Paramètres différentiels. Déformation des surfaces. [Geodesic lines and geodesic curvature. Differential parameters. Deformation of surfaces], Déformation infiniment petite et représentation sphérique. [Infinitely small deformation and spherical representation], Reprint of the 1894 original (III) and the 1896 original (IV), Cours de Géométrie de la Faculté des Sciences. [Course on Geometry of the Faculty of Science]. MR 97c:01046b
435. Guy David and Stephen Semmes, *Fractured fractals and broken dreams*, The Clarendon Press Oxford University Press, New York, 1997, Self-similar geometry through metric and measure. MR 99h:28018

436. Paulo de Bartolomeis and Franco Tricerri (eds.), *Geometric topology: recent developments*, Springer-Verlag, Berlin, 1991, Lectures given at the First C.I.M.E. Session on Recent Developments in Geometric Topology and Related Topics held in Montecatini Terme, June 4–12, 1990. MR 92m:53001
437. W. L. F. Degen, *The cut locus of an ellipsoid*, Geom. Dedicata **67** (1997), no. 2, 197–198. MR 98j:52004
438. Pierre Deligne, Phillip Griffiths, John W. Morgan, and Dennis Sullivan, *Real homotopy theory of Kähler manifolds*, Invent. Math. **29** (1975), no. 3, 245–274. MR 52 #3584
439. Pierre Deligne and G. D. Mostow, *Monodromy of hypergeometric functions and nonlattice integral monodromy*, Inst. Hautes Études Sci. Publ. Math. (1986), no. 63, 5–89. MR 88a:22023a
440. Andrzej Derdzinski, *Einstein metrics in dimension four*, in Dillen and Verstraelen [449], pp. 419–707. MR 2000h:53003
441. Annie Deschamps, *Variétés riemanniennes stigmatiques*, J. Math. Pures Appl. (9) **61** (1982), no. 4, 381–400 (1983). MR 84h:53058
442. Dennis M. DeTurck and Jerry L. Kazdan, *Some regularity theorems in Riemannian geometry*, Ann. Sci. École Norm. Sup. (4) **14** (1981), no. 3, 249–260. MR 83f:53018
443. Antonio J. Di Scala, *On an assertion in Riemann's Habilitationsvortrag*, Enseign. Math. (2) **47** (2001), no. 1-2, 57–63. MR 2002d:53046
444. Ulrich Dierkes, Stefan Hildebrandt, Albrecht Küster, and Ortwin Wohlrab, *Minimal surfaces. I*, Springer-Verlag, Berlin, 1992, Boundary value problems. MR 94c:49001a
445. _____, *Minimal surfaces. II*, Springer-Verlag, Berlin, 1992, Boundary regularity. MR 94c:49001b
446. Jean Dieudonné, *Treatise on analysis*, Pure and Applied Mathematics, Academic Press Inc., Boston, MA, 1969–1993, 8 volumes. Translated from the French by Laura Fainsilber and I. G. Macdonald. The first volume is titled *Foundations of modern analysis*. MR 94b:00001
447. _____, *History of algebraic geometry*, Wadsworth International Group, Belmont, Calif., 1985, An outline of the history and development of algebraic geometry, Translated from the French by Judith D. Sally. MR 86h:01004
448. _____, *A history of algebraic and differential topology. 1900–1960*, Birkhäuser Boston Inc., Boston, MA, 1989. MR 90g:01029
449. Franki J. E. Dillen and Leopold C. A. Verstraelen (eds.), *Handbook of differential geometry. Vol. I*, North-Holland, Amsterdam, 2000. MR 2000h:53003
450. Paul A. M. Dirac, *The quantum theory of the electron, I*, Proc. Roy. Soc. A **117** (1928), 610–624.
451. Manfredo Perdigão do Carmo, *Differential geometry of curves and surfaces*, Prentice-Hall Inc., Englewood Cliffs, N.J., 1976, Translated from the Portuguese. MR 52 #15253
452. _____, *Riemannian geometry*, Birkhäuser Boston Inc., Boston, MA, 1992, Translated from the second Portuguese edition by Francis Flaherty. MR 92i:53001
453. Jozef Dodziuk, *Nonexistence of universal upper bounds for the first positive eigenvalue of the Laplace-Beltrami operator*, in Brooks et al. [261], Proceedings of the 1993 AMS-IMS-SIAM Joint Summer Research Conference on Spectral Geometry held at the University of Washington, Seattle, Washington, July 17–23, 1993, pp. 109–114. MR 95m:53052

454. Peter Dombrowski, *Krümmungsgrössen gleichungsdefinierter Untermannigfaltigkeiten Riemannscher Mannigfaltigkeiten*, Math. Nachr. **38** (1968), 133–180. MR 39 #7536
455. _____, *150 years after Gauss' "Disquisitiones generales circa superficies curvas"*, Société Mathématique de France, Paris, 1979, With the original text of Gauss. MR 80g:01013
456. S. K. Donaldson, *The Seiberg-Witten equations and 4-manifold topology*, Bull. Amer. Math. Soc. (N.S.) **33** (1996), no. 1, 45–70. MR 96k:57033
457. S. K. Donaldson and P. B. Kronheimer, *The geometry of four-manifolds*, The Clarendon Press Oxford University Press, New York, 1990, Oxford Science Publications. MR 92a:57036
458. S. K. Donaldson and C. B. Thomas (eds.), *Geometry of low-dimensional manifolds. 1*, Cambridge, Cambridge University Press, 1990, Gauge theory and algebraic surfaces. MR 93b:57001
459. S. K. Donaldson and C. B. Thomas (eds.), *Geometry of low-dimensional manifolds. 2*, Cambridge, Cambridge University Press, 1990, Symplectic manifolds and Jones-Witten theory. MR 93b:57002
460. Leonbattista Donati and Noëlle Stolfi, *Singularities of illuminated surfaces*, Int. J. of Computer Vision **23** (1997), no. 3, 207–216.
461. Harold Donnelly and Charles Fefferman, *Nodal sets of eigenfunctions on Riemannian manifolds*, Invent. Math. **93** (1988), no. 1, 161–183. MR 89m:58207
462. _____, *Nodal domains and growth of harmonic functions on noncompact manifolds*, J. Geom. Anal. **2** (1992), no. 1, 79–93. MR 93b:58150
463. Olivier Druet and Emmanuel Hebey, *The AB program in goemetric analysis, sharp Sobolev inequalities, and related problems*, to appear, 2000.
464. B. A. Dubrovin, A. T. Fomenko, and S. P. Novikov, *Modern geometry—methods and applications. Part II*, Springer-Verlag, New York, 1985, The geometry and topology of manifolds, Translated from the Russian by Robert G. Burns. MR 86m:53001
465. J. J. Duistermaat and Victor Guillemin, *The spectrum of positive elliptic operators and periodic bicharacteristics*, Invent. Math. **29** (1975), no. 1, 39–79. MR 53 #9307
466. Oguz Durumeric, *A generalization of Berger's theorem on almost $\frac{1}{4}$-pinched manifolds. II*, J. Differential Geom. **26** (1987), no. 1, 101–139. MR 88m:53075
467. Walther Dyck, *Beiträge zur Analysis situs I. Aufsatz. Ein - und zweidimensionale Mannigfaltigkeiten*, Math. Ann. **32** (1888), 457–512.
468. E. B. Dynkin, *Maximal subgroups of the classical groups*, Trudy Moskov. Mat. Obšč. **1** (1952), 39–166, Translation to English in [470]. MR 14,244d
469. _____, *Semisimple subalgebras of semisimple Lie algebras*, Mat. Sbornik N.S. **30(72)** (1952), 349–462 (3 plates). MR 13,904c
470. E. B. Dynkin, M. A. Naimark, P. K. Rasevskii, and N. Ja. Vilenkin, *Five papers on algebra and group theory*, American Mathematical Society Translations–Series 2, vol. 6, Amer. Math. Soc., Providence, RI, 1957.
471. M. G. Eastwood and K. P. Tod, *Local constraints on Einstein-Weyl geometries*, J. Reine Angew. Math. **491** (1997), 183–198. MR 99a:53056
472. Patrick Eberlein, Ursula Hamenstädt, and Viktor Schroeder, *Manifolds of nonpositive curvature*, in Greene and Yau [597], Proceedings of the AMS Summer Research Institute on Differential Geometry held at the University of California, Los Angeles, California, July 8–28, 1990, pp. 179–227. MR 94d:53060

473. Patrick Eberlein and Barrett O'Neill, *Visibility manifolds*, Pacific J. Math. **46** (1973), 45–109. MR 49 #1421
474. David G. Ebin, *The space of Riemannian metrics*, in Chern and Smale [371], Proceedings of Symposia in Pure Mathematics, Vols. XIV-XVI. Edited by Shiing-Shen Chern and Stephen Smale, pp. Vol. XIV: v+367 pp.; Vol. XV: v+307 pp.; Vol. XVI: v+250. MR 41 #7686
475. _____, *Espace des métriques riemanniennes et mouvement des fluides via les variétés d'applications*, Centre de Mathématiques de l'École Polytechnique et Université Paris VII, 1972, Cours de 3ème Cycle, Année Universitaire 1971–1972, Rapport No. M 86.0572. MR 57 #4231
476. David G. Ebin and Gerard Misiołek, *The exponential map on D^s_μ*, in Bierstone et al. [193], Proceedings of a Conference in Honour of the Sixtieth Birthday of V.I. Arnol'd, pp. 153–163.
477. Klaus Ecker, *A local monotonicity formula for mean curvature flow*, Ann. of Math. (2) **154** (2001), no. 2, 503–525. MR 1 865 979
478. James Eells, Jr. and Frédéric Hélein, *Applications harmoniques, lois de conservation et repères mobiles*, Diderot, Paris, 1996.
479. James Eells, Jr. and L. Lemaire, *Another report on harmonic maps*, Bull. London Math. Soc. **20** (1988), no. 5, 385–524. MR 89i:58027
480. James Eells, Jr. and Andrea Ratto, *Harmonic maps and minimal immersions with symmetries*, Princeton University Press, Princeton, NJ, 1993, Methods of ordinary differential equations applied to elliptic variational problems. MR 94k:58033
481. James Eells, Jr. and J. H. Sampson, *Harmonic mappings of Riemannian manifolds*, Amer. J. Math. **86** (1964), 109–160. MR 29 #1603
482. Charles Ehresmann, *Les connexions infinitésimales dans un espace fibré différentiable*, Colloque de topologie (espaces fibrés), Bruxelles, 1950, Georges Thone, Liège, 1951, pp. 29–55. MR 13,159e
483. Paul E. Ehrlich, Yoon-Tae Jung, and Seon-Bu Kim, *Volume comparison theorems for Lorentzian manifolds*, Geom. Dedicata **73** (1998), no. 1, 39–56. MR 99k:53125
484. Albert Einstein, *Nährungsweise Integration de Feldgleichungen der Gravitation [approximative integration of the field equations of gravity]*, Königlich Preußische Akademie der Wissenschaften (1916), 688–696, Also found in [485], pp. 344–357, and translated to English in [486].
485. _____, *The collected papers of Albert Einstein. Vol. 6*, Princeton University Press, Princeton, NJ, 1996, The Berlin years: writings, 1914–1917, Edited by A. J. Kox, Martin J. Klein and Robert Schulmann. MR 97f:01032
486. _____, *The collected papers of Albert Einstein. Vol. 6*, Princeton University Press, Princeton, NJ, 1997, The Berlin years: writings, 1914–1917, English translation of selected texts by Alfred Engel in consultation with Engelbert Schucking, With a preface by Engel and Schucking. MR 98i:01035
487. El-H. Ch. El-Alaoui, J.-P. Gauthier, and I. Kupka, *Small sub-Riemannian balls on $\mathbb{R}^3$*, J. Dynam. Control Systems **2** (1996), no. 3, 359–421. MR 98a:53043b
488. A. El Soufi and S. Ilias, *Riemannian manifolds admitting isometric immersions by their first eigenfunctions*, to appear, 1999.
489. David Elworthy, *Geometric aspects of diffusions on manifolds*, in Hennequin [706], Papers from the 15th–17th Summer Schools held in Saint-Flour, 1985–87, pp. 277–425. MR 89i:60005

490. D. B. A. Epstein, *Natural tensors on Riemannian manifolds*, J. Differential Goemetry **10** (1975), no. 4, 631–645. MR 54 #3617
491. P. Erdös, Peter M. Gruber, and J. Hammer, *Lattice points*, Longman Scientific & Technical, Harlow, 1989. MR 90g:11081
492. Jost-Hinrich Eschenburg, *Local convexity and nonnegative curvature—Gromov's proof of the sphere theorem*, Invent. Math. **84** (1986), no. 3, 507–522. MR 87j:53080
493. _____, *Inhomogeneous spaces of positive curvature*, Differential Geom. Appl. **2** (1992), no. 2, 123–132. MR 94j:53044
494. _____, *Comparison theorems in Riemannian geometry*, Universitá di Trento, 1994.
495. _____, *Lecture notes on symmetric spaces*, to appear, 1997.
496. Jost-Hinrich Eschenburg and John J. O'Sullivan, *Jacobi tensors and Ricci curvature*, Math. Ann. **252** (1980), no. 1, 1–26. MR 81k:53037
497. Jost-Hinrich Eschenburg and McKenzie Y. Wang, *The ODE system arising from cohomogeneity one Einstein metrics*, Geometry and physics (Aarhus, 1995), Dekker, New York, 1997, pp. 157–165. MR 97m:53083
498. Christine M. Escher, *Rigidity of minimal isometric immersions of spheres into spheres*, Geom. Dedicata **73** (1998), no. 3, 275–293. MR 99j:53081
499. Euclid, *The thirteen books of Euclid's Elements translated from the text of Heiberg. Vol. I: Introduction and Books I, II. Vol. II: Books III–IX. Vol. III: Books X–XIII and Appendix*, Dover Publications Inc., New York, 1956, Translated with introduction and commentary by Thomas L. Heath, 2nd ed. MR 17,814b
500. Fr. Fabricius-Bjerre, *On the double tangents of plane closed curves*, Math. Scand **11** (1962), 113–116. MR 28 #4439
501. F. Fang, *Multiplicities of isoparameteric hypersurfaces*, to appear, 1999.
502. F. Fang and Xiaochun Rong, *Positive pinching, volume and second Betti number*, Geom. Funct. Anal. **9** (1999), no. 4, 641–674. MR 1 719 590
503. Michael Farber, Gabriel Katz, and Jerome Levine, *Morse theory of harmonic forms*, Topology **37** (1998), no. 3, 469–483. MR 99i:58026
504. F. T. Farrell, *Lectures on surgical methods in rigidity*, Published for the Tata Institute of Fundamental Research, Bombay, 1996. MR 98j:57050
505. F. T. Farrell and L. E. Jones, *Compact negatively curved manifolds (of dim $\neq 3, 4$) are topologically rigid*, Proc. Nat. Acad. Sci. U.S.A. **86** (1989), no. 10, 3461–3463. MR 90h:57023b
506. _____, *Negatively curved manifolds with exotic smooth structures*, J. Amer. Math. Soc. **2** (1989), no. 4, 899–908. MR 90f:53075
507. _____, *A topological analogue of Mostow's rigidity theorem*, J. Amer. Math. Soc. **2** (1989), no. 2, 257–370. MR 90h:57023a
508. _____, *Rigidity and other topological aspects of compact nonpositively curved manifolds*, Bull. Amer. Math. Soc. (N.S.) **22** (1990), no. 1, 59–64. MR 90h:57023c
509. _____, *Nonuniform hyperbolic lattices and exotic smooth structures*, J. Differential Geom. **38** (1993), no. 2, 235–261. MR 95e:57051
510. _____, *Complex hyperbolic manifolds and exotic smooth structures*, Invent. Math. **117** (1994), no. 1, 57–74. MR 95e:57052
511. Herbert Federer, *Geometric measure theory*, Springer-Verlag New York Inc., New York, 1969, Die Grundlehren der mathematischen Wissenschaften, Band 153. MR 41 #1976

512. Yves Félix and Stephen Halperin, *Rational LS category and its applications*, Trans. Amer. Math. Soc. **273** (1982), no. 1, 1–38. MR 84h:55011
513. Emmanuel Ferrand, *On the Bennequin invariant and the geometry of wave fronts*, Geom. Dedicata **65** (1997), no. 2, 219–245. MR 99c:57061
514. Jacqueline Ferrand, *Les géodésiques des structures conformes*, C. R. Acad. Sci. Paris Sér. I Math. **294** (1982), no. 18, 629–632. MR 83f:53037
515. _____, *Histoire de la réductibilité du groupe conforme des variétés riemanniennes*, Séminaire de théorie spectrale et géométrie (Grenoble), 1999, pp. 9–25.
516. Richard P. Feynman, Robert B. Leighton, and Matthew Sands, *The Feynman lectures on physics. Vol. 1: Mainly mechanics, radiation, and heat*, Addison-Wesley Publishing Co., Inc., Reading, Mass.-London, 1963. MR 35 #3942
517. Livio Flaminio, *Local entropy rigidity for hyperbolic manifolds*, Comm. Anal. Geom. **3** (1995), no. 3-4, 555–596. MR 96k:58170
518. Patrick Foulon, *Locally symmetric Finsler spaces in negative curvature*, C. R. Acad. Sci. Paris Sér. I Math. **324** (1997), no. 10, 1127–1132. MR 98f:53064
519. John Franks, *Geodesics on S^2 and periodic points of annulus homeomorphisms*, Invent. Math. **108** (1992), no. 2, 403–418. MR 93f:58192
520. Daniel S. Freed and Karen K. Uhlenbeck, *Instantons and four-manifolds*, second ed., Springer-Verlag, New York, 1991. MR 91i:57019
521. Michael Freedman, $\mathbb{Z}_2$-*systolic freedom*, electronic preprint math.GT/0002124, Los Alamos National Labs, 2000.
522. Thomas Friedrich, *Dirac operators in Riemannian geometry*, American Mathematical Society, Providence, RI, 2000, Translated from the 1997 German original by Andreas Nestke. MR 1 777 332
523. Jürg Fröhlich, Olivier Grandjean, and Andreas Recknagel, *Supersymmetric quantum theory and differential geometry*, Comm. Math. Phys. **193** (1998), no. 3, 527–594. MR 99k:58015b
524. _____, *Supersymmetric quantum theory, non-commutative geometry, and gravitation*, Symétries quantiques (Les Houches, 1995), North-Holland, Amsterdam, 1998, pp. 221–385. MR 99k:58015a
525. _____, *Supersymmetric quantum theory and non-commutative geometry*, Comm. Math. Phys. **203** (1999), no. 1, 119–184. MR 1 695 097
526. Dmitry Fuchs and Serge Tabachnikov, *More on paperfolding*, Amer. Math. Monthly **106** (1999), no. 1, 27–35. MR 99m:53009
527. Kenji Fukaya, *A finiteness theorem for negatively curved manifolds*, J. Differential Geom. **20** (1984), no. 2, 497–521. MR 86i:53023
528. _____, *Collapsing Riemannian manifolds to ones of lower dimensions*, J. Differential Geom. **25** (1987), no. 1, 139–156. MR 88b:53050
529. _____, *A boundary of the set of the Riemannian manifolds with bounded curvatures and diameters*, J. Differential Geom. **28** (1988), no. 1, 1–21. MR 89h:53090
530. _____, *Collapsing Riemannian manifolds to ones with lower dimension. II*, J. Math. Soc. Japan **41** (1989), no. 2, 333–356. MR 90c:53103
531. _____, *Hausdorff convergence of Riemannian manifolds and its applications*, Recent topics in differential and analytic geometry, Academic Press, Boston, MA, 1990, pp. 143–238. MR 92k:53076
532. Kenji Fukaya and Takao Yamaguchi, *Almost nonpositively curved manifolds*, J. Differential Geom. **33** (1991), no. 1, 67–90. MR 91k:53054
533. _____, *The fundamental groups of almost non-negatively curved manifolds*, Ann. of Math. (2) **136** (1992), no. 2, 253–333. MR 93h:53041

534. P. Funk, *Über Flächen mit lauter geschlossenen geodätischen Linien*, Math. Ann **74** (1913), 278–300.
535. A. Futaki, *An obstruction to the existence of Einstein Kähler metrics*, Invent. Math. **73** (1983), no. 3, 437–443. MR 84j:53072
536. Matthew P. Gaffney, *Asymptotic distributions associated with the Laplacian for forms*, Comm. Pure Appl. Math. **11** (1958), 535–545. MR 20 #5980
537. Michael Gage and Richard S. Hamilton, *The heat equation shrinking convex plane curves*, J. Differential Geom. **23** (1986), no. 1, 69–96. MR 87m:53003
538. Krzysztof Galicki and H. Blaine Lawson, Jr., *Quaternionic reduction and quaternionic orbifolds*, Math. Ann. **282** (1988), no. 1, 1–21. MR 89m:53075
539. Sylvestre F. L. Gallot, *A Sobolev inequality and some geometric applications*, in Berger et al. [176], pp. 45–55.
540. _____, *Inégalités isopérimétriques et analytiques sur les variétés riemanniennes*, Astérisque (1988), no. 163-164, 5–6, 31–91, 281 (1989), On the geometry of differentiable manifolds (Rome, 1986). MR 90f:58173
541. _____, *Volumes, courbure de Ricci et convergence des variétés (d'après T. H. Colding et Cheeger-Colding)*, Astérisque (1998), no. 252, Exp. No. 835, 3, 7–32, Séminaire Bourbaki. Vol. 1997/98. MR 1 685 585
542. Sylvestre F. L. Gallot, Dominique Hulin, and Jacques Lafontaine, *Riemannian geometry*, second ed., Springer-Verlag, Berlin, 1990. MR 91j:53001
543. Sylvestre F. L. Gallot and Daniel Meyer, *Opérateur de courbure et laplacien des formes différentielles d'une variété riemannienne*, J. Math. Pures Appl. (9) **54** (1975), no. 3, 259–284. MR 56 #13128
544. G. A. Gal'perin, *Closed geodesics on tetrahedra*, unpublished, 1995.
545. G. A. Gal'perin and N. I. Chernov, Биллиарды и хаос, "Znanie", Moscow, 1991, With a preface by Yakov G. Sinaĭ. MR 92f:58108
546. L. Zhiyong Gao, *Convergence of Riemannian manifolds; Ricci and $L^{n/2}$-curvature pinching*, J. Differential Geom. **32** (1990), no. 2, 349–381. MR 92f:53049
547. _____, *$L^{n/2}$-curvature pinching*, J. Differential Geom. **32** (1990), no. 3, 713–774. MR 91k:53055
548. Richard J. Gardner, *Geometric tomography*, Cambridge University Press, Cambridge, 1995. MR 96j:52006
549. Jacques Gasqui, *Sur la résolubilité locale des équations d'Einstein*, Compositio Math. **47** (1982), no. 1, 43–69. MR 84f:58115
550. Jacques Gasqui and Hubert Goldschmidt, *The infinitesimal rigidity of the complex quadric of dimension four*, Amer. J. Math. **116** (1994), no. 3, 501–539. MR 96a:53071
551. Paul Gauduchon, *Variétés riemanniennes autoduales (d'après C. H. Taubes et al.)*, Astérisque (1993), no. 216, Exp. No. 767, 4, 151–186, Séminaire Bourbaki, Vol. 1992/93. MR 94k:53060
552. David B. Gauld, *Differential topology*, Marcel Dekker Inc., New York, 1982, An introduction. MR 84k:57013
553. I. M. Gel'fand, *Automorphic functions and the theory of representations*, Proc. Internat. Congr. Mathematicians (Stockholm, 1962), Inst. Mittag-Leffler, Djursholm, 1963, pp. 74–85. MR 31 #273
554. G. Gentili, S. Marchiafava, and M. Pontecorvo (eds.), *Quaternionic structures in mathematics and physics*, Trieste, International School for Advanced Studies (SISSA), Laboratorio Interdisciplinare per le Scienze Naturali ed Umanistiche, 1998, The papers are available electronically at http://www.emis.de/proceedings/QSMP94/. MR 99d:53002

555. Johan C. H. Gerretsen (ed.), *Proceedings of the International Congress of Mathematicians, Amsterdam, 2–9 sept. 1954. Vols. I–III*, Erven P. Noordhoff N. V., Groningen, 1954. MR 16,1190j
556. V. Gershkovich and H. Rubinstein, *Generic cut-loci on surfaces and their generic transformations via singularity theory for Riemannian distance functions*, to appear, 1999.
557. W.-D. Geyer (ed.), *Jubiläumstagung 100 Jahre Deutschen Mathematiker-Vereinigung*, B. G. Teubner, Stuttgart, 1992.
558. É. Ghys and P. de la Harpe (eds.), *Sur les groupes hyperboliques d'après Mikhael Gromov*, Birkhäuser, Boston, MA, 1990, Papers from the Swiss Seminar on Hyperbolic Groups held in Bern, 1988. MR 92f:53050
559. G. W. Gibbons and S. W. Hawking, *Gravitational multi-instantons*, Physics Letters B **78** (1978), no. 4, 430–432.
560. Olga Gil-Medrano and Peter W. Michor, *The Riemannian manifold of all Riemannian metrics*, Quart. J. Math. Oxford Ser. (2) **42** (1991), no. 166, 183–202. MR 92f:58024
561. Peter B. Gilkey, *Curvature and the eigenvalues of the Dolbeault complex for Kaehler manifolds*, Advances in Math. **11** (1973), 311–325. MR 48 #12609
562. _____, *Curvature and the eigenvalues of the Laplacian for elliptic complexes*, Advances in Math. **10** (1973), 344–382. MR 48 #3081
563. _____, *The index theorem and the heat equation*, Publish or Perish Inc., Boston, Mass., 1974, Notes by Jon Sacks, Mathematics Lecture Series, No. 4. MR 56 #16704
564. _____, *Invariance theory, the heat equation, and the Atiyah–Singer index theorem*, second ed., CRC Press, Boca Raton, FL, 1995. MR 98b:58156
565. _____, *The Atiyah–Singer index theorem*, in Dillen and Verstraelen [449], pp. 709–746. MR 2000h:53003
566. Peter B. Gilkey, John V. Leahy, and Jeonghyeong Park, *Spectral geometry, Riemannian submersions, and the Gromov–Lawson conjecture*, Chapman & Hall/CRC, Boca Raton, FL, 1999. MR 1 707 341
567. Henri Gillet and Christophe Soulé, *An arithmetic Riemann–Roch theorem*, Invent. Math. **110** (1992), no. 3, 473–543. MR 94f:14019
568. Ernesto Girondo and Gabino González-Diez, *On extremal discs inside compact hyperbolic surfaces*, C. R. Acad. Sci. Paris Sér. I Math. **329** (1999), no. 1, 57–60. MR 1 703 263
569. Herman Gluck, *The converse to the four vertex theorem*, Enseignement Math. (2) **17** (1971), 295–309. MR 49 #9737
570. Herman Gluck and David A. Singer, *Scattering of geodesic fields. II*, Ann. of Math. (2) **110** (1979), no. 2, 205–225. MR 80k:53073
571. Claude Godbillon, *Feuilletages*, Birkhäuser Verlag, Basel, 1991, Études géométriques. [Geometric studies], With a preface by G. Reeb. MR 93i:57038
572. Alexander Goncharov, *Volumes of hyperbolic manifolds and mixed Tate motives*, J. Amer. Math. Soc. **12** (1999), no. 2, 569–618. MR 99i:19004
573. G. Gong and G. Yu, *Volume growth and positive scalar curvature*, to appear, 1999.
574. V. V. Gorbatsevich, Arkadi L. Onishchik, and È. B. Vinberg, *Foundations of Lie theory and Lie transformation groups*, Springer-Verlag, Berlin, 1997, Translated from the Russian by A. Kozlowski, Reprint of the 1993 translation [*Lie groups and Lie algebras. I*, Encyclopaedia Math. Sci., 20, Springer, Berlin, 1993]. MR 99c:22009

575. Carolyn S. Gordon, *Isospectral closed Riemannian manifolds which are not locally isometric. II*, in Brooks et al. [261], Proceedings of the 1993 AMS-IMS-SIAM Joint Summer Research Conference on Spectral Geometry held at the University of Washington, Seattle, Washington, July 17–23, 1993, pp. 121–131. MR 95k:58166
576. _____, *Survey of isospectral manifolds*, in Dillen and Verstraelen [449], pp. 747–778. MR 2000h:53003
577. Carolyn S. Gordon and Yiping Mao, *Comparisons of Laplace spectra, length spectra and geodesic flows of some Riemannian manifolds*, Math. Res. Lett. **1** (1994), no. 6, 677–688. MR 95i:58181
578. Carolyn S. Gordon and David L. Webb, *Isospectral convex domains in Euclidean space*, Math. Res. Lett. **1** (1994), no. 5, 539–545. MR 95g:58246
579. Carolyn S. Gordon, David L. Webb, and S. Wolpert, *Isospectral plane domains and surfaces via Riemannian orbifolds*, Invent. Math. **110** (1992), no. 1, 1–22. MR 93h:58172
580. Ruth Gornet, *The marked length spectrum vs. the Laplace spectrum on forms on Riemannian nilmanifolds*, Comment. Math. Helv. **71** (1996), no. 2, 297–329. MR 97e:58222
581. _____, *A new construction of isospectral Riemannian nilmanifolds with examples*, Michigan Math. J. **43** (1996), no. 1, 159–188. MR 97b:58143
582. _____, *Continuous families of Riemannian manifolds isospectral on functions but not on 1-forms*, to appear, 1998.
583. Alfred Gray, *Tubes*, Addison-Wesley Publishing Company Advanced Book Program, Redwood City, CA, 1990. MR 92d:53002
584. _____, *Modern differential geometry of curves and surfaces with Mathematica*, second ed., CRC Press, Boca Raton, FL, 1998. MR 1 688 379
585. Matthew A. Grayson, *Shortening embedded curves*, Ann. of Math. (2) **129** (1989), no. 1, 71–111. MR 90a:53050
586. L. W. Green, *Auf Wiedersehensflächen*, Ann. of Math. (2) **78** (1963), 289–299. MR 27 #5206
587. Marvin Jay Greenberg, *Lectures on algebraic topology*, W. A. Benjamin, Inc., New York-Amsterdam, 1967. MR 35 #6137
588. _____, *Euclidean and non-Euclidean geometries*, third ed., W. H. Freeman and Company, New York, 1993, Development and history. MR 94k:51001
589. Brian Greene and Shing-Tung Yau (eds.), *Mirror symmetry. II*, American Mathematical Society, Providence, RI, 1997. MR 97d:00024
590. Robert E. Greene, *Homotopy finiteness theorems and Helly's theorem*, J. Geom. Anal. **4** (1994), no. 3, 317–325. MR 95g:53043
591. _____, *A genealogy of noncompact manifolds of nonnegative curvature: history and logic*, in Grove and Petersen [650], Papers from the Special Year in Differential Geometry held in Berkeley, CA, 1993–94, pp. 99–134. MR 98g:53069
592. Robert E. Greene and Peter Petersen, V, *Little topology, big volume*, Duke Math. J. **67** (1992), no. 2, 273–290. MR 93f:53035
593. Robert E. Greene and Hung Hsi Wu, *Lipschitz convergence of Riemannian manifolds*, Pacific J. Math. **131** (1988), no. 1, 119–141. MR 89g:53063
594. _____, *Addendum to: "Lipschitz convergence of Riemannian manifolds"*, Pacific J. Math. **140** (1989), no. 2, 398, Addendum to Greene and Wu [593]. MR 90j:53061

595. _____, *Non-negatively curved manifolds which are flat outside a compact set*, in Greene and Yau [597], Proceedings of the AMS Summer Research Institute on Differential Geometry held at the University of California, Los Angeles, California, July 8–28, 1990, pp. 327–335. MR 94a:53071

596. Robert E. Greene and Shing-Tung Yau (eds.), *Differential geometry: partial differential equations on manifolds (Los Angeles, 1990)*, Proceedings of Symposia in Pure Mathematics, no. 54, Part 1, Providence, RI, American Mathematical Society, 1993, Proceedings of the AMS Summer Research Institute on Differential Geometry held at the University of California, Los Angeles, California, July 8–28, 1990. MR 94a:00010

597. Robert E. Greene and Shing-Tung Yau (eds.), *Differential geometry: Riemannian geometry (Los Angeles, 1990)*, Proceedings of Symposia in Pure Mathematics, no. 54, Part 3, Providence, RI, American Mathematical Society, 1993, Proceedings of the AMS Summer Research Institute on Differential Geometry held at the University of California, Los Angeles, California, July 8–28, 1990. MR 94a:00012

598. H. P. Greenspan, *Motion of a small viscous drop that wets a surface*, J. Fluid Mech. **84** (1978), 125–143.

599. Phillip Griffiths and Joseph Harris, *Principles of algebraic geometry*, John Wiley & Sons Inc., New York, 1994, Reprint of the 1978 original. MR 95d:14001

600. Detlef Gromoll, *Differenziabare Strukturen und Metriken positiver Krümmung auf Sphären*, Math. Ann. **164** (1966), 353–371.

601. _____, *Spaces of nonnegative curvature*, in Greene and Yau [597], Proceedings of the AMS Summer Research Institute on Differential Geometry held at the University of California, Los Angeles, California, July 8–28, 1990, pp. 337–356. MR 94b:53073

602. Detlef Gromoll and Karsten Grove, *On metrics on S^2 all of whose geodesics are closed*, Invent. Math. **65** (1981/82), no. 1, 175–177. MR 82m:58021

603. _____, *A generalization of Berger's rigidity theorem for positively curved manifolds*, Ann. Sci. École Norm. Sup. (4) **20** (1987), no. 2, 227–239. MR 88k:53062

604. _____, *The low-dimensional metric foliations of Euclidean spheres*, J. Differential Geom. **28** (1988), no. 1, 143–156. MR 89g:53052

605. Detlef Gromoll and Wolfgang T. Meyer, *On complete open manifolds of positive curvature*, Ann. of Math. (2) **90** (1969), 75–90. MR 40 #854

606. _____, *Periodic geodesics on compact Riemannian manifolds*, J. Differential Geometry **3** (1969), 493–510. MR 41 #9143

607. _____, *An exotic sphere with nonnegative sectional curvature*, Ann. of Math. (2) **100** (1974), 401–406. MR 51 #11347

608. Detlef Gromoll and Joseph A. Wolf, *Some relations between the metric structure and the algebraic structure of the fundamental group in manifolds of nonpositive curvature*, Bull. Amer. Math. Soc. **77** (1971), 545–552. MR 43 #6841

609. Mikhael Gromov, *Almost flat manifolds*, J. Differential Geom. **13** (1978), no. 2, 231–241. MR 80h:53041

610. _____, *Homotopical effects of dilatation*, J. Differential Geom. **13** (1978), no. 3, 303–310. MR 82d:58017

611. _____, *Manifolds of negative curvature*, J. Differential Geom. **13** (1978), no. 2, 223–230. MR 80h:53040

612. _____, *Synthetic geometry in Riemannian manifolds*, Proceedings of the International Congress of Mathematicians (Helsinki, 1978) (Helsinki), Acad. Sci. Fennica, 1980, pp. 415–419. MR 81g:53029
613. _____, *Curvature, diameter and Betti numbers*, Comment. Math. Helv. **56** (1981), no. 2, 179–195. MR 82k:53062
614. _____, *Groups of polynomial growth and expanding maps*, Inst. Hautes Études Sci. Publ. Math. (1981), no. 53, 53–73. MR 83b:53041
615. _____, *Hyperbolic manifolds (according to Thurston and Jørgensen)*, Bourbaki Seminar, Vol. 1979/80, Springer, Berlin, 1981, pp. 40–53. MR 84b:53046
616. _____, *Structures métriques pour les variétés riemanniennes*, CEDIC, Paris, 1981, Edited by J. Lafontaine and P. Pansu. MR 85e:53051
617. _____, *Volume and bounded cohomology*, Inst. Hautes Études Sci. Publ. Math. (1982), no. 56, 5–99 (1983). MR 84h:53053
618. _____, *Filling Riemannian manifolds*, J. Differential Geom. **18** (1983), no. 1, 1–147. MR 85h:53029
619. _____, *Pseudoholomorphic curves in symplectic manifolds*, Invent. Math. **82** (1985), no. 2, 307–347. MR 87j:53053
620. _____, *Partial differential relations*, Springer-Verlag, Berlin, 1986. MR 90a:58201
621. _____, *Entropy, homology and semialgebraic geometry*, Astérisque (1987), no. 145-146, 5, 225–240, Séminaire Bourbaki, Vol. 1985/86. MR 89f:58082
622. _____, *Hyperbolic groups*, Essays in group theory, Springer, New York, 1987, pp. 75–263. MR 89e:20070
623. _____, *Width and related invariants of Riemannian manifolds*, Astérisque (1988), no. 163-164, 6, 93–109, 282 (1989), On the geometry of differentiable manifolds (Rome, 1986). MR 90f:53078
624. _____, *Foliated Plateau problem. I. Minimal varieties*, Geom. Funct. Anal. **1** (1991), no. 1, 14–79. MR 92b:53064
625. _____, *Foliated Plateau problem. II. Harmonic maps of foliations*, Geom. Funct. Anal. **1** (1991), no. 3, 253–320. MR 93a:58048
626. _____, *Kähler hyperbolicity and L_2-Hodge theory*, J. Differential Geom. **33** (1991), no. 1, 263–292. MR 92a:58133
627. _____, *Sign and geometric meaning of curvature*, Rend. Sem. Mat. Fis. Milano **61** (1991), 9–123 (1994). MR 95j:53055
628. _____, *Stability and pinching*, Geometry Seminars. Sessions on Topology and Geometry of Manifolds (Italian) (Bologna, 1990), Univ. Stud. Bologna, Bologna, 1992, pp. 55–97. MR 94b:53067
629. _____, *Asymptotic invariants of infinite groups*, Geometric group theory, Vol. 2 (Sussex, 1991), Cambridge Univ. Press, Cambridge, 1993, pp. 1–295. MR 95m:20041
630. _____, *Metric invariants of Kähler manifolds*, Differential geometry and topology (Alghero, 1992), World Sci. Publishing, River Edge, NJ, 1993, pp. 90–116. MR 97f:53111
631. _____, *Positive curvature, macroscopic dimension, spectral gaps and higher signatures*, Functional analysis on the eve of the 21st century, Vol. II (New Brunswick, NJ, 1993), Birkhäuser Boston, Boston, MA, 1996, pp. 1–213. MR 98d:53052
632. _____, *Systoles and intersystolic inequalities*, in Besse [185], En l'honneur de Marcel Berger. [In honor of Marcel Berger], Held in Luminy, France, July 12–18, 1992, pp. 291–362. MR 99a:53051

633. _____, *Metric structures for Riemannian and non-Riemannian spaces*, Birkhäuser Boston Inc., Boston, MA, 1999, Based on the 1981 French original [616]. With appendices by M. Katz, P. Pansu and S. Semmes, Translated from the French by Sean Michael Bates. MR 1 699 320
634. _____, *Spaces and questions*, Geom. Funct. Anal. (2000), no. Special Volume, Part I, 118–161, GAFA 2000 (Tel Aviv, 1999). MR 2002e:53056
635. Mikhael Gromov and H. Blaine Lawson, Jr., *The classification of simply connected manifolds of positive scalar curvature*, Ann. of Math. (2) **111** (1980), no. 3, 423–434. MR 81h:53036
636. _____, *Positive scalar curvature and the Dirac operator on complete Riemannian manifolds*, Inst. Hautes Études Sci. Publ. Math. (1983), no. 58, 83–196 (1984). MR 85g:58082
637. Mikhael Gromov and Pierre Pansu, *Rigidity of lattices: an introduction*, in de Bartolomeis and Tricerri [436], Lectures given at the First C.I.M.E. Session on Recent Developments in Geometric Topology and Related Topics held in Montecatini Terme, June 4–12, 1990, pp. 39–137. MR 93f:53036
638. Mikhael Gromov and I. Piatetski-Shapiro, *Nonarithmetic groups in Lobachevsky spaces*, Inst. Hautes Études Sci. Publ. Math. (1988), no. 66, 93–103. MR 89j:22019
639. Mikhael Gromov and Richard Schoen, *Harmonic maps into singular spaces and p-adic superrigidity for lattices in groups of rank one*, Inst. Hautes Études Sci. Publ. Math. (1992), no. 76, 165–246. MR 94e:58032
640. Mikhael Gromov and William P. Thurston, *Pinching constants for hyperbolic manifolds*, Invent. Math. **89** (1987), no. 1, 1–12. MR 88e:53058
641. Karsten Große-Brauckmann, *Gyroids of constant mean curvature*, Experiment. Math. **6** (1997), no. 1, 33–50.
642. Karsten Große-Brauckmann and Konrad Polthier, *Compact constant mean curvature surfaces with low genus*, Experiment. Math. **6** (1997), no. 1, 13–32.
643. Karsten Grove, *Metric differential geometry*, Differential geometry (Lyngby, 1985), Springer, Berlin, 1987, pp. 171–227. MR 88i:53075
644. _____, *Critical point theory for distance functions*, in Greene and Yau [597], Proceedings of the AMS Summer Research Institute on Differential Geometry held at the University of California, Los Angeles, California, July 8–28, 1990, pp. 357–385. MR 94f:53065
645. _____, *Ramifications of the classical sphere theorem*, in Besse [185], En l'honneur de Marcel Berger. [In honor of Marcel Berger], Held in Luminy, France, July 12–18, 1992, pp. 363–376. MR 97j:53037
646. Karsten Grove and Stephen Halperin, *Contributions of rational homotopy theory to global problems in geometry*, Inst. Hautes Études Sci. Publ. Math. (1982), no. 56, 171–177 (1983). MR 84b:58030
647. Karsten Grove, Hermann Karcher, and Ernst A. Ruh, *Group actions and curvature*, Invent. Math. **23** (1974), 31–48. MR 52 #6609
648. Karsten Grove and Peter Petersen, V, *Bounding homotopy types by geometry*, Ann. of Math. (2) **128** (1988), no. 1, 195–206. MR 90a:53044
649. _____, *A radius sphere theorem*, Invent. Math. **112** (1993), no. 3, 577–583. MR 94e:53034
650. Karsten Grove and Peter Petersen, V (eds.), *Comparison geometry*, Cambridge University Press, Cambridge, 1997, Papers from the Special Year in Differential Geometry held in Berkeley, CA, 1993–94. MR 97m:53001

651. Karsten Grove, Peter Petersen, V, and Jyh-Yang Wu, *Geometric finiteness theorems via controlled topology*, Invent. Math. **99** (1990), no. 1, 205–213. MR 90k:53075
652. _____, *Erratum: "Geometric finiteness theorems via controlled topology"*, Invent. Math. **104** (1991), no. 1, 221–222, Correction to Grove, Petersen and Wu [651]. MR 92b:53065
653. Karsten Grove and K. Shankar, *Rank two fundamental groups of positively curved manifolds*, J. Geom. Anal. (1999).
654. Karsten Grove and Katsuhiro Shiohama, *A generalized sphere theorem*, Ann. Math. (2) **106** (1977), no. 2, 201–211. MR 58 #18268
655. Karsten Grove and G. Walschap, *Transitive holonomy group and rigidity in nonnegative curvature*, to appear, 1999.
656. Karsten Grove and Frederick H. Wilhelm, Jr., *Hard and soft packing radius theorems*, Ann. of Math. (2) **142** (1995), no. 2, 213–237. MR 96h:53054
657. _____, *Metric constraints on exotic spheres via Alexandrov geometry*, J. Reine Angew. Math. **487** (1997), 201–217. MR 98d:53060
658. Karsten Grove and Wolfgang Ziller, *Curvature and symmetry of Milnor spheres*, to appear, 2000.
659. Peter M. Gruber, *A typical convex surface contains no closed geodesic!*, J. Reine Angew. Math. **416** (1991), 195–205. MR 92e:53057
660. Peter M. Gruber and C. G. Lekkerkerker, *Geometry of numbers*, second ed., North-Holland Publishing Co., Amsterdam, 1987. MR 88j:11034
661. Peter M. Gruber and J. M. Wills (eds.), *Handbook of convex geometry. Vol. A, B*, North-Holland Publishing Co., Amsterdam, 1993. MR 94e:52001
662. Pengfei Guan and Yan Yan Li, *The Weyl problem with nonnegative Gauss curvature*, J. Differential Geom. **39** (1994), no. 2, 331–342. MR 95c:53051
663. Luis Guijarro and Peter Petersen, *Rigidity in non-negative curvature*, Ann. Sci. École Norm. Sup. (4) **30** (1997), no. 5, 595–603. MR 98i:53047
664. Victor Guillemin, *The Radon transform on Zoll surfaces*, Advances in Math. **22** (1976), no. 1, 85–119. MR 54 #14009
665. _____, *Wave-trace invariants and a theorem of Zelditch*, Internat. Math. Res. Notices (1993), no. 12, 303–308. MR 95f:58077
666. _____, *Wave-trace invariants*, Duke Math. J. **83** (1996), no. 2, 287–352. MR 97f:58131
667. Victor Guillemin and D. Kazhdan, *Some inverse spectral results for negatively curved 2-manifolds*, Topology **19** (1980), no. 3, 301–312. MR 81j:58082
668. Victor Guillemin and Richard Melrose, *An inverse spectral result for elliptical regions in $\mathbb{R}^2$*, Adv. in Math. **32** (1979), no. 2, 128–148. MR 80f:35104
669. _____, *The Poisson summation formula for manifolds with boundary*, Adv. in Math. **32** (1979), no. 3, 204–232. MR 80j:58066
670. Victor Guillemin and Shlomo Sternberg, *Geometric asymptotics*, American Mathematical Society, Providence, R.I., 1977, Mathematical Surveys, No. 14. MR 58 #24404
671. Carlos Gutierrez and Jorge Sotomayor, *Lines of curvature, umbilic points and Carathéodory conjecture*, Resenhas **3** (1998), no. 3, 291–322. MR 1 633 013
672. Eugene Gutkin, *Billiards in polygons: survey of recent results*, J. Statist. Phys. **83** (1996), no. 1-2, 7–26. MR 97a:58099
673. Martin C. Gutzwiller, *Chaos in classical and quantum mechanics*, Springer-Verlag, New York, 1990. MR 91m:58099

674. Jacques Hadamard, *Les surfaces à courbure opposées et leurs lignes géodésiques*, J. Math. Pures Appl. **4** (1898), 27–73.
675. _____, *Sur le billiard non-euclidien*, Proc. Verb. Soc. Sci. Phys. et Nature Bordeaux **5** (1898).
676. _____, *Sur l'iteration et les solutions asymptotiques des équations différentielles*, Bull. Soc. Math. France **29** (1901), 224–228.
677. André Haefliger, *Sur l'extension du groupe structural d'une espace fibré*, C. R. Acad. Sci. Paris **243** (1956), 558–560. MR 18,920a
678. Richard S. Hamilton, *Three-manifolds with positive Ricci curvature*, J. Differential Geom. **17** (1982), no. 2, 255–306. MR 84a:53050
679. _____, *Four-manifolds with positive curvature operator*, J. Differential Geom. **24** (1986), no. 2, 153–179. MR 87m:53055
680. _____, *The Harnack estimate for the Ricci flow*, J. Differential Geom. **37** (1993), no. 1, 225–243. MR 93k:58052
681. Vagn Lundsgaard Hansen, *Geometry in nature*, A K Peters Ltd., Wellesley, MA, 1993, Translated from the Danish by Tom Artin. MR 94b:00002
682. Reese Harvey and H. Blaine Lawson, Jr., *Calibrated geometries*, Acta Math. **148** (1982), 47–157. MR 85i:53058
683. Joel Hass and Frank Morgan, *Geodesics and soap bubbles in surfaces*, Math. Z. **223** (1996), no. 2, 185–196. MR 97j:53009
684. S. W. Hawking and G. F. R. Ellis, *The large scale structure of space-time*, Cambridge University Press, London, 1973, Cambridge Monographs on Mathematical Physics, No. 1. MR 54 #12154
685. N. S. Hawley, *Constant holomorphic curvature*, Canadian J. Math. **5** (1953), 53–56. MR 14,690d
686. James J. Hebda, *Cut loci of submanifolds in space forms and in the geometries of Möbius and Lie*, Geom. Dedicata **55** (1995), no. 1, 75–93. MR 96e:53048
687. Jens Heber, *Geometric and algebraic structure of noncompact homogeneous Einstein spaces*, preprint, Universität Augsburg, 1997.
688. _____, *Noncompact homogeneous Einstein spaces*, Invent. Math. **133** (1998), no. 2, 279–352. MR 99d:53046
689. Emmanuel Hebey, *Courbure scalaire et géométrie conforme*, J. Geom. Phys. **10** (1993), no. 4, 345–380. MR 94f:53073
690. _____, *Optimal Sobolev inequalities on complete Riemannian manifolds with Ricci curvature bounded below and positive injectivity radius*, Amer. J. Math. **118** (1996), no. 2, 291–300. MR 97c:53058
691. _____, *Nonlinear analysis on manifolds: Sobolev spaces and inequalities*, New York University Courant Institute of Mathematical Sciences, New York, 1999. MR 1 688 256
692. _____, *Scalar curvature type problems in Riemannian geometry*, to appear, 2000.
693. Emmanuel Hebey and M. Herzlich, *Harmonic coordinates, harmonic radius and convergence of Riemannian manifolds*, Rend. Mat. Appl. (7) **17** (1997), no. 4, 569–605 (1998). MR 99f:53039
694. Emmanuel Hebey and Michel Vaugon, *From best constants to critical functions*, Math. Z. **237** (2001), no. 4, 737–767. MR 2002h:58061
695. Gustav A. Hedlund, *Geodesics on a two-dimensional Riemannian manifold with periodic coefficients*, Annals of Math. **33** (1932), 719–739.
696. _____, *On the metrical transitivity of the geodesics on closed surfaces of constant negative curvature*, Annals of Math. **35** (1934), 787–808.

697. Juha Heinonen, *Lectures on analysis on metric spaces*, Springer-Verlag, New York, 2001. MR 2002c:30028
698. Ernst Heintze, *Manningfaltigkeiten negativer Krümmung*, Ph.D. thesis, Universität Bonn, 1976, Habilitationsschrift.
699. Ernst Heintze and Hermann Karcher, *A general comparison theorem with applications to volume estimates for submanifolds*, Ann. Sci. École Norm. Sup. (4) **11** (1978), no. 4, 451–470. MR 80i:53026
700. Ernst Heintze and Xiaobo Liu, *Homogeneity of infinite-dimensional isoparametric submanifolds*, Ann. of Math. (2) **149** (1999), no. 1, 149–181. MR 2000c:58007
701. Sigurdur Helgason, *Differential geometry, Lie groups, and symmetric spaces*, Academic Press Inc. [Harcourt Brace Jovanovich Publishers], New York, 1978. MR 80k:53081
702. _____, *The Radon transform*, Birkhäuser Boston, Mass., 1980. MR 83f:43012
703. _____, *Groups and geometric analysis*, Academic Press Inc., Orlando, Fla., 1984, Integral geometry, invariant differential operators, and spherical functions. MR 86c:22017
704. _____, *Huygens' principle for wave equations on symmetric spaces*, J. Funct. Anal. **107** (1992), no. 2, 279–288. MR 93i:58151
705. D. Hengesh, *Exemples de variétés homogènes d'Einstein à courbure scalaire négative*, to appear, 1998.
706. P. L. Hennequin (ed.), *École d'Été de Probabilités de Saint-Flour XV–XVII, 1985–87*, Springer-Verlag, Berlin, 1988, Papers from the 15th–17th Summer Schools held in Saint-Flour, 1985–87. MR 89i:60005
707. Luis Hernández, *Kähler manifolds and 1/4-pinching*, Duke Math. J. **62** (1991), no. 3, 601–611. MR 92b:53046
708. Haydeé Herrera and Rafael Herrera, *Â-genus on non-spin manifolds with S^1 actions and the classification of positive quaternionic-kähler manifolds*, IHES preprint, 2001.
709. Joseph Hersch, *Quatre propriétés isopérimétriques de membranes sphériques homogènes*, C. R. Acad. Sci. Paris Sér. A-B **270** (1970), A1645–A1648. MR 45 #1444
710. Oussama Hijazi, *Lower bounds for the eigenvalues of the Dirac operator*, J. Geom. Phys. **16** (1995), no. 1, 27–38. MR 96d:58143
711. David Hilbert, *Die Grundlagen der Physik*, Nachr. Akad. Wiss. Göttingen Math.-Phys. Kl (1915), 395–407, Also see the article by the same name in [712], pp. 258–289.
712. _____, *Gesammelte Abhandlungen. Band III: Analysis, Grundlagen der Mathematik, Physik, Verschiedenes, Lebensgeschichte*, Springer-Verlag, Berlin, 1970, Zweite Auflage. MR 41 #8201c
713. David Hilbert and Stefan Cohn-Vossen, *Geometry and the imagination*, Chelsea, New York, 1952, Translated from German to English by P. Nemenyi.
714. Nancy Hingston, *On the growth of the number of closed geodesics on the two-sphere*, Internat. Math. Res. Notices (1993), no. 9, 253–262. MR 94m:58044
715. Morris W. Hirsch, *Differential topology*, Springer-Verlag, New York, 1994, Corrected reprint of the 1976 original. MR 96c:57001
716. Morris W. Hirsch and Barry Mazur, *Smoothings of piecewise linear manifolds*, Princeton University Press, Princeton, N. J., 1974, Annals of Mathematics Studies, No. 80. MR 54 #3711

717. Friedrich Hirzebruch, *Topological methods in algebraic geometry*, Springer-Verlag, Berlin, 1995, Translated from the German and Appendix One by R. L. E. Schwarzenberger, With a preface to the third English edition by the author and Schwarzenberger, Appendix Two by A. Borel, Reprint of the 1978 edition. MR 96c:57002
718. Friedrich Hirzebruch and K. Kodaira, *On the complex projective spaces*, J. Math. Pures Appl. (9) **36** (1957), 201–216. MR 19,1077c
719. Nigel J. Hitchin, *Harmonic spinors*, Advances in Math. **14** (1974), 1–55. MR 50 #11332
720. _____, *Twistor spaces, Einstein metrics and isomonodromic deformations*, J. Differential Geom. **42** (1995), no. 1, 30–112. MR 96g:53057
721. _____, *Einstein metrics and the eta-invariant*, Boll. Un. Mat. Ital. B (7) **11** (1997), no. 2, suppl., 95–105. MR 98g:53085
722. _____, *Integrable systems in Riemannian geometry*, to appear, 1997.
723. Helmut Hofer, Clifford H. Taubes, Alan Weinstein, and Eduard Zehnder (eds.), *The Floer memorial volume*, Birkhäuser Verlag, Basel, 1995. MR 96f:58001
724. Helmut Hofer and Eduard Zehnder, *Symplectic invariants and Hamiltonian dynamics*, Birkhäuser Verlag, Basel, 1994. MR 96g:58001
725. Eberhard Hopf, *Statistik der geodätischen Linien in Mannigfaltigkeiten negativer Krümmung*, Ber. Verh. Sächs. Akad. Wiss. Leipzig **91** (1939), 261–304. MR 1,243a
726. _____, *Closed surfaces without conjugate points*, Proc. Nat. Acad. Sci. U. S. A. **34** (1948), 47–51. MR 9,378d
727. Heinz Hopf, *Die Curvatura integra Clifford–Kleinischer Raumformen*, Nachr. Akad. Wiss. Göttingen Math.-Phys. Kl. II (1925), 131–141.
728. _____, *Über die Curvatura integra geschlossener Hyperflächen*, Math. Ann. **95** (1926), 340–367.
729. _____, *Vektorfelden in n-dimensionalen Mannigfaltigkeiten*, Math. Ann. **96** (1927), 225–250.
730. _____, *Differentialgeometrie und topologische Gestalt*, Jahrbericht der DMV **41** (1932), 209–229.
731. _____, *Über Flächen mit einer Relation zwischen den Hauptkrümmungen*, Math. Nachr. **4** (1951), 232–249. MR 12,634f
732. _____, *Differential geometry in the large*, second ed., Springer-Verlag, Berlin, 1989, Notes taken by Peter Lax and John W. Gray, With a preface by Shiing-Shen Chern, With a preface by K. Voss. MR 90f:53001
733. Heinz Hopf and Willi Rinow, *Über den Begriff der völlständigen differentialgeometrischen Flächen*, Comment. Math. Helv. **3** (1931), 209–225.
734. Lars Hörmander, *The spectral function of an elliptic operator*, Acta Math. **121** (1968), 193–218. MR 58 #29418
735. _____, *The analysis of linear partial differential operators. I*, second ed., Springer-Verlag, Berlin, 1990, Distribution theory and Fourier analysis. MR 91m:35001b
736. _____, *The analysis of linear partial differential operators. II*, Springer-Verlag, Berlin, 1983, Differential operators with constant coefficients. MR 85g:35002b
737. _____, *The analysis of linear partial differential operators. III*, Springer-Verlag, Berlin, 1994, Pseudo-differential operators, Corrected reprint of the 1985 original. MR 95h:35255

738. _____, *The analysis of linear partial differential operators. IV*, Springer-Verlag, Berlin, 1994, Fourier integral operators, Corrected reprint of the 1985 original. MR 98f:35002
739. Hugh Howards, Michael Hutchings, and Frank Morgan, *The isoperimetric problem on surfaces*, Amer. Math. Monthly **106** (1999), no. 5, 430–439. MR 1 699 261
740. Wu-teh Hsiang and Wu-Yi Hsiang, *On the uniqueness of isoperimetric solutions and imbedded soap bubbles in noncompact symmetric spaces. I*, Invent. Math. **98** (1989), no. 1, 39–58. MR 90h:53078
741. Wu-Yi Hsiang, Richard S. Palais, and Chuu-Lian Terng, *The topology of isoparametric submanifolds*, J. Differential Geom. **27** (1988), no. 3, 423–460. MR 89m:53104
742. S. T. Hu, *Differentiable manifolds*, Holt, Rinehart and Winston, Inc., New York, 1969. MR 39 #6343
743. Hua Min Huang, *Some remarks on the pinching problems*, Bull. Inst. Math. Acad. Sinica **9** (1981), no. 2, 321–340. MR 83a:53041
744. Heinz Huber, *Über eine neue Klasse automorpher Funktionen und ein Gitterpunktproblem in der hyperbolischen Ebene. I*, Comment. Math. Helv. **30** (1956), 20–62 (1955). MR 17,603b
745. _____, *Zur analytischen Theorie hyperbolischen Raumformen und Bewegungsgruppen*, Math. Ann. **138** (1959), 1–26. MR 22 #99
746. _____, *Zur analytischen Theorie hyperbolischer Raumformen und Bewegungsgruppen. II*, Math. Ann. **142** (1960/1961), 385–398. MR 23 #A3845
747. _____, *Zur analytischen Theorie hyperbolischer Raumformen und Bewegungsgruppen. II*, Math. Ann. **143** (1961), 463—464. MR 27 #4923
748. Alan Huckleberry and Tilmann Wurzbacher (eds.), *Infinite dimensional Kähler manifolds*, Birkhäuser Verlag, Basel, 2001, Papers from the DMV-Seminar held in Oberwolfach, November 19–25, 1995. MR 2002c:00027
749. Dominique Hulin, *Sous-variétés complexes d'Einstein de l'espace projectif*, Bull. Soc. Math. France **124** (1996), no. 2, 277–298. MR 97h:53047
750. Christoph Hummel and Viktor Schroeder, *Tits geometry associated with 4-dimensional closed real-analytic manifolds of nonpositive curvature*, J. Differential Geom. **48** (1998), no. 3, 531–555. MR 99j:53058
751. Dale Husemoller, *Fibre bundles*, third ed., Springer-Verlag, New York, 1994. MR 94k:55001
752. Martin N. Huxley, *Area, lattice points, and exponential sums*, The Clarendon Press Oxford University Press, New York, 1996, Oxford Science Publications. MR 97g:11088
753. Daniel Huybrechts, *Compact hyper-Kähler manifolds: basic results*, Invent. Math. **135** (1999), no. 1, 63–113. MR 2000a:32039
754. Seungsu Hwang, *A rigidity theorem for Ricci flat metrics*, Geom. Dedicata **71** (1998), no. 1, 5–17. MR 99d:53042
755. Jun-ichi Igusa, *On the structure of a certain class of Kaehler varieties*, Amer. J. Math. **76** (1954), 669–678. MR 16,172c
756. Hans-Christoph Im Hof and Ernst A. Ruh, *An equivariant pinching theorem*, Comment. Math. Helv. **50** (1975), no. 3, 389–401. MR 52 #6610
757. Mikio Ise and Masaru Takeuchi, *Lie groups. I, II*, American Mathematical Society, Providence, RI, 1991, Translated from the Japanese by Katsumi Nomizu. MR 92a:22001

758. Jin-ichi Itoh, *The length of a cut locus on a surface and Ambrose's problem*, J. Differential Geom. **43** (1996), no. 3, 642–651. MR 97i:53038
759. Jin-ichi Itoh and Minoru Tanaka, *On the Hausdorff dimension of the cut locus on a smooth Riemannian manifold*, to appear, 2001.
760. Stefan Ivanov and Irina Petrova, *Riemannian manifold in which the skew-symmetric curvature operator has pointwise constant eigenvalues*, Geom. Dedicata **70** (1998), no. 3, 269–282. MR 99d:53034
761. Victor Ivrii, *Microlocal analysis and precise spectral asymptotics*, Springer-Verlag, Berlin, 1998. MR 99e:58193
762. Felix Jenni, *Über den ersten Eigenwert des Laplace-Operators auf ausgewählten Beispielen kompakter Riemannscher Flächen*, Comment. Math. Helv. **59** (1984), no. 2, 193–203. MR 85i:58118
763. Gary R. Jensen, *Einstein metrics on principal fibre bundles*, J. Differential Geometry **8** (1973), 599–614. MR 50 #5694
764. Alain Joets and Roland Ribotta, *Caustique de la surface ellipsoïdale à trois dimensions*, Experiment. Math. **8** (1999), no. 1, 49–55. MR 2000c:58072
765. Troels Jørgensen, *Compact 3-manifolds of constant negative curvature fibering over the circle*, Ann. Math. (2) **106** (1977), no. 1, 61–72. MR 56 #8840
766. Jürgen Jost, *Nonpositive curvature: geometric and analytic aspects*, Birkhäuser Verlag, Basel, 1997. MR 98g:53070
767. _____, *Compact Riemann surfaces*, second ed., Springer-Verlag, Berlin, 2002, An introduction to contemporary mathematics. MR 1 909 701
768. _____, *Riemannian geometry and geometric analysis*, third ed., Springer-Verlag, Berlin, 2002. MR 2002i:53001
769. Jürgen Jost and Hermann Karcher, *Geometrische Methoden zur Gewinnung von a-priori-Schranken für harmonische Abbildungen*, Manuscripta Math. **40** (1982), no. 1, 27–77. MR 84e:58023
770. Jürgen Jost and Shing-Tung Yau, *Harmonic maps and superrigidity*, in Greene and Yau [596], Proceedings of the AMS Summer Research Institute on Differential Geometry held at the University of California, Los Angeles, California, July 8–28, 1990, pp. 245–280. MR 94m:58060
771. Dominic D. Joyce, *Compact 8-manifolds with holonomy* Spin(7), Invent. Math. **123** (1996), no. 3, 507–552. MR 97d:53052
772. _____, *Compact Riemannian 7-manifolds with holonomy G_2. I, II*, J. Differential Geom. **43** (1996), no. 2, 291–328, 329–375. MR 97m:53084
773. _____, *Hypercomplex algebraic geometry*, Quart. J. Math. Oxford Ser. (2) **49** (1998), no. 194, 129–162. MR 2000a:53082
774. _____, *Compact manifolds with special holonomy*, Oxford University Press, Oxford, 2000. MR 1 787 733
775. Mark Kac, *Can one hear the shape of a drum?*, Amer. Math. Monthly **73** (1966), no. 4, part II, 1–23. MR 34 #1121
776. Georgi I. Kamberov, *Prescribing mean curvature: existence and uniqueness problems*, Electron. Res. Announc. Amer. Math. Soc. **4** (1998), 4–11 (electronic). MR 99c:53066
777. Nikolaos Kapouleas, *Constant mean curvature surfaces constructed by fusing Wente tori*, Invent. Math. **119** (1995), no. 3, 443–518. MR 95m:53008
778. Hermann Karcher, *Remarks on polyhedra with given dihedral angles*, Comm. Pure Appl. Math. **21** (1968), 169–174. MR 36 #5816

779. _____, *A geometric classification of positively curved symmetric spaces and the isoparametric construction of the Cayley plane*, Astérisque (1988), no. 163-164, 6, 111–135, 282 (1989), On the geometry of differentiable manifolds (Rome, 1986). MR 90g:53063
780. _____, *Riemannian comparison constructions*, in Chern [366], pp. 170–222. MR 91b:53046
781. _____, *Submersions via projections*, Geom. Dedicata **74** (1999), no. 3, 249–260. MR 99m:53040
782. Hermann Karcher and Ulrich Pinkall, *Die Boysche Fläche in Oberwolfach*, Mitteilungen der DMV **47** (1997).
783. Atsushi Kasue, *A convergence theorem for Riemannian manifolds and some applications*, Nagoya Math. J. **114** (1989), 21–51. MR 90g:53053
784. Atsushi Kasue and Hironori Kumura, *Spectral convergence of Riemannian manifolds. II*, Tohoku Math. J. (2) **48** (1996), no. 1, 71–120. MR 97f:53075
785. Anatole B. Katok, *The growth rate for the number of singular and periodic orbits for a polygonal billiard*, Comm. Math. Phys. **111** (1987), no. 1, 151–160. MR 88g:58162
786. _____, *Four applications of conformal equivalence to geometry and dynamics*, Ergodic Theory Dynamical Systems **8*** (1988), no. Charles Conley Memorial Issue, 139–152. MR 89m:58165
787. Anatole B. Katok and Boris Hasselblatt, *Introduction to the modern theory of dynamical systems*, Cambridge University Press, Cambridge, 1995, With a supplementary chapter by Katok and Leonardo Mendoza. MR 96c:58055
788. Anatole B. Katok, Jean-Marie Strelcyn, François Ledrappier, and F. Przytycki, *Invariant manifolds, entropy and billiards; smooth maps with singularities*, Springer-Verlag, Berlin, 1986. MR 88k:58075
789. Svetlana Katok, *Coding of closed geodesics after Gauss and Morse*, Geom. Dedicata **63** (1996), no. 2, 123–145. MR 97j:20045
790. Atsushi Katsuda, *Gromov's convergence theorem and its application*, Nagoya Math. J. **100** (1985), 11–48. MR 87e:53067
791. _____, *Correction to: "Gromov's convergence theorem and its application"*, Nagoya Math. J. **114** (1989), 173–174, Erratum for Katsuda [790]. MR 90e:53057
792. Mikhail G. Katz, *The filling radius of two-point homogeneous spaces*, J. Differential Geom. **18** (1983), no. 3, 505–511. MR 85h:53030
793. _____, *The rational filling radius of complex projective space*, Topology Appl. **42** (1991), no. 3, 201–215. MR 93c:53031
794. _____, *Filling volume of the circle*, to appear, 1998.
795. _____, *Systolic freedom in dimension 4*, to appear, 1998.
796. _____, *Local calibration of mass and systolic geometry*, electronic preprint math.DG/0204182, Los Alamos National Labs, 2002, To appear in Geometry and Functional Analysis.
797. Mikhail G. Katz, Matthias Kreck, and Alexander I. Suciu, *Free abelian covers, short loops, stable length, and systolic inequalities*, electronic preprint math.DG/0207143, Los Alamos National Labs, 2002.
798. Mikhail G. Katz and Alexander I. Suciu, *Volume of Riemannian manifolds, geometric inequalities, and homotopy theory*, Tel Aviv Topology Conference: Rothenberg Festschrift (1998), Amer. Math. Soc., Providence, RI, 1999, pp. 113–136. MR 1 705 579

799. _____, *Systolic freedom of loop space*, Geom. Funct. Anal. **11** (2001), no. 1, 60–73. MR 2002c:53067
800. Jerry L. Kazdan, *Positive energy in general relativity*, Bourbaki Seminar, Vol. 1981/1982, Soc. Math. France, Paris, 1982, pp. 315–330. MR 85g:83024
801. Jerry L. Kazdan and Frank W. Warner, *Existence and conformal deformation of metrics with prescribed Gaussian and scalar curvatures*, Ann. of Math. (2) **101** (1975), 317–331. MR 51 #11349
802. _____, *Scalar curvature and conformal deformation of Riemannian structure*, J. Differential Geometry **10** (1975), 113–134. MR 51 #1661
803. C.-W. Kim and J.-W. Yim, *Rigidity of noncompact Finsler manifolds*, Geom. Dedicata (1999), to appear.
804. J. Kim, *8-dimensional Einstein–Thorpe manifolds*, to appear, 1999.
805. Robion C. Kirby and Laurence C. Siebenmann, *Foundational essays on topological manifolds, smoothings, and triangulations*, Princeton University Press, Princeton, N.J., 1977, With notes by John Milnor and Michael F. Atiyah, Annals of Mathematics Studies, No. 88. MR 58 #31082
806. Kazuyoshi Kiyohara, *On infinitesimal $C_{2\pi}$-deformations of standard metrics on spheres*, Hokkaido Math. J. **13** (1984), no. 2, 151–231. MR 86d:58022
807. _____, *Riemannian metrics with periodic geodesic flows on projective spaces*, Japan. J. Math. (N.S.) **13** (1987), no. 2, 209–234. MR 89a:53051
808. Bruce Kleiner, *An isoperimetric comparison theorem*, Invent. Math. **108** (1992), no. 1, 37–47. MR 92m:53056
809. _____, *The local structure of length spaces with curvature bounded above*, Math. Z. **231** (1999), no. 3, 409–456. MR 1 704 987
810. Bruce Kleiner and Bernhard Leeb, *Rigidity of quasi-isometries for symmetric spaces and Euclidean buildings*, C. R. Acad. Sci. Paris Sér. I Math. **324** (1997), no. 6, 639–643. MR 98f:53046
811. Wilhelm Klingenberg, *Contributions to Riemannian geometry in the large*, Ann. of Math. (2) **69** (1959), 654–666. MR 21 #4445
812. _____, *Über Riemannsche Mannigfaltigkeiten mit positiver Krümmung*, Comment. Math. Helv. **35** (1961), 47–54. MR 25 #2559
813. _____, *A course in differential geometry*, Springer-Verlag, New York, 1978, Translated from the German by David Hoffman, Graduate Texts in Mathematics, Vol. 51. MR 57 #13702
814. _____, *Lectures on closed geodesics*, Springer-Verlag, Berlin, 1978, Grundlehren der Mathematischen Wissenschaften, Vol. 230. MR 57 #17563
815. _____, *Lektsii o zamknutykh geodezicheskikh*, "Mir", Moscow, 1982, Translated from the English by A. I. Gryuntal', Translation edited and with a preface by D. V. Anosov. MR 84f:53037
816. _____, *Riemannian geometry*, second ed., Walter de Gruyter & Co., Berlin, 1995. MR 95m:53003
817. Wilhelm Klingenberg and Floris Takens, *Generic properties of geodesic flows*, Math. Ann. **197** (1972), 323–334. MR 46 #6402
818. Tilla Klotz, *On G. Bol's proof of Carathéodory's conjecture*, Comm. Pure Appl. Math. **12** (1959), 277–311. MR 22 #11352
819. Gerhard Knieper, *Spherical means on compact Riemannian manifolds of negative curvature*, Differential Geom. Appl. **4** (1994), no. 4, 361–390. MR 95i:58141

820. _____, *A second derivative formula of the Liouville entropy at spaces of constant negative curvature*, Ergodic Theory Dynam. Systems **17** (1997), no. 5, 1131–1135. MR 98h:58043
821. _____, *The uniqueness of the measure of maximal entropy for geodesic flows on rank 1 manifolds*, Ann. of Math. (2) **148** (1998), no. 1, 291–314. MR 1 652 924
822. _____, *Hyperbolic dynamics and Riemannian geometry*, to appear, 1999.
823. Gerhard Knieper and Howard Weiss, *A surface with positive curvature and positive topological entropy*, J. Differential Geom. **39** (1994), no. 2, 229–249. MR 94m:58170
824. Horst Knörrer, *Geodesics on the ellipsoid*, Invent. Math. **59** (1980), no. 2, 119–143. MR 81h:58050
825. Shoshichi Kobayashi, *On conjugate and cut loci*, in Chern [366], pp. 140–169. MR 90d:53003
826. _____, *Transformation groups in differential geometry*, Springer-Verlag, Berlin, 1995, Reprint of the 1972 edition. MR 96c:53040
827. Shoshichi Kobayashi and Katsumi Nomizu, *Foundations of differential geometry. Vol. I*, John Wiley & Sons Inc., New York, 1996, Reprint of the 1963 original, A Wiley-Interscience Publication. MR 97c:53001a
828. _____, *Foundations of differential geometry. Vol. II*, John Wiley & Sons Inc., New York, 1996, Reprint of the 1969 original, A Wiley-Interscience Publication. MR 97c:53001b
829. Mariko Konishi, *On manifolds with Sasakian 3-structure over quaternion Kaehler manifolds*, Kōdai Math. Sem. Rep. **26** (1974/75), 194–200. MR 51 #13951
830. Maxim Kontsevich, *Rozansky–Witten invariants via formal geometry*, Compositio Math. **115** (1999), no. 1, 115–127. MR 1 671 725
831. Jean-Louis Koszul, *Lectures on fibre bundles and differential geometry*, Published for the Tata Institute of Fundamental Research, Bombay, 1986, With notes by S. Ramanan, Reprint of the 1965 edition. MR 88e:53040
832. Oldřich Kowalski and Zdeněk Vlášek, *Homogeneous Einstein metrics on Aloff-Wallach spaces*, Differential Geom. Appl. **3** (1993), no. 2, 157–167. MR 95a:53069
833. Ekkehard Krätzel, *Lattice points*, Kluwer Academic Publishers Group, Dordrecht, 1988. MR 90e:11144
834. Matthias Kreck and Stephan Stolz, *Nonconnected moduli spaces of positive sectional curvature metrics*, J. Amer. Math. Soc. **6** (1993), no. 4, 825–850. MR 94f:53066
835. Leopold Kronecker, *Über Systeme von Functionen mehrer Variabeln. Zweite Abhandlung.*, Monatsberichte der Königlich Preussischen Akademie der Wissenschaften (1869), 688–698, Also found in Kronecker [836], vol. I, pp. 213–226.
836. _____, *Leopold Kronecker's Werke. Bände I–V*, Chelsea Publishing Co., New York, 1968, Originally published at the behest of the Königlich Preussischen Akademie der Wissenschaften, edited by K. Hensel. MR 38 #5576
837. W. Kühnel and Hans-Bert Rademacher, *Essential conformal fields in pseudo-Riemannian geometry*, J. Math. Pures Appl. (9) **74** (1995), no. 5, 453–481. MR 96h:53079
838. _____, *Essential conformal fields in pseudo-Riemannian geometry. II*, J. Math. Sci. Univ. Tokyo **4** (1997), no. 3, 649–662. MR 98i:53096

839. Ravindra Shripad Kulkarni, *Curvature and metric*, Ann. of Math. (2) **91** (1970), 311–331. MR 41 #2581
840. Masatake Kuranishi, *On some metrics on $S^2 \times S^2$*, in Greene and Yau [597], Proceedings of the AMS Summer Research Institute on Differential Geometry held at the University of California, Los Angeles, California, July 8–28, 1990, pp. 439–450. MR 94b:53068
841. Kazuhiro Kuwae, Yoshiroh Machigashira, and Takashi Shioya, *Beginning of analysis on Alexandrov spaces*, Geometry and topology: Aarhus (1998), Amer. Math. Soc., Providence, RI, 2000, pp. 275–284. MR 2002g:53066
842. _____, *Sobolev spaces, Laplacian, and heat kernel on Alexandrov spaces*, Math. Z. **238** (2001), no. 2, 269–316. MR 1 865 418
843. Jacques Lafontaine, *Mesures de courbure des variétés lisses et des polyèdres*, Astérisque (1987), no. 145-146, 5, 241–256, Séminaire Bourbaki, Vol. 1985/86. MR 88d:53036
844. Cornelius Lanczos, *Ein vereinfachendes Koordinatensystem für die Einsteinschen Gravitationsgleichungen*, Phys. Zeit. **23** (1922), 537–539.
845. Serge Lang, *Differential manifolds*, Addison-Wesley Publishing Co., Inc., Reading, Mass.-London-Don Mills, Ont., 1972. MR 55 #4241
846. _____, *Real and functional analysis*, third ed., Springer-Verlag, New York, 1993. MR 94b:00005
847. _____, *Differential and Riemannian manifolds*, third ed., Springer-Verlag, New York, 1995. MR 96d:53001
848. _____, *Fundamentals of differential geometry*, Springer-Verlag, New York, 1999.
849. Martin Lanzendorf, *Einstein metrics with nonpositive sectional curvature on extensions of Lie algebras of Heisenberg type*, Geom. Dedicata **66** (1997), no. 2, 187–202. MR 98g:53088
850. H. Blaine Lawson, Jr. and Marie-Louise Michelsohn, *Spin geometry*, Princeton University Press, Princeton, NJ, 1989. MR 91g:53001
851. H. Blaine Lawson, Jr. and Shing-Tung Yau, *Compact manifolds of nonpositive curvature*, J. Differential Geometry **7** (1972), 211–228. MR 48 #12402
852. Vladimir F. Lazutkin, *KAM theory and semiclassical approximations to eigenfunctions*, Springer-Verlag, Berlin, 1993, With an addendum by A. I. Shnirel′man. MR 94m:58069
853. Claude LeBrun, *Complete Ricci-flat Kähler metrics on $\mathbf{c}^n$ need not be flat*, Several complex variables and complex geometry, Part 2 (Santa Cruz, CA, 1989), Amer. Math. Soc., Providence, RI, 1991, pp. 297–304. MR 93a:53038
854. _____, *On complete quaternionic-Kähler manifolds*, Duke Math. J. **63** (1991), no. 3, 723–743. MR 92i:53042
855. _____, *On four-dimensional Einstein manifolds*, The geometric universe (Oxford, 1996), Oxford Univ. Press, Oxford, 1998, pp. 109–121. MR 99e:53063
856. _____, *Four-dimensional Einstein manifolds, and beyond*, in LeBrun and Wang [858], Lectures on geometry and topology, sponsored by Lehigh University's Journal of Differential Geometry, pp. 247–285. MR 1 798 613
857. Claude LeBrun and Simon M. Salamon, *Strong rigidity of positive quaternion-Kähler manifolds*, Invent. Math. **118** (1994), no. 1, 109–132. MR 95k:53059
858. Claude LeBrun and McKenzie Wang (eds.), *Surveys in differential geometry: essays on Einstein manifolds*, International Press, Boston, MA, 1999, Lectures on geometry and topology, sponsored by Lehigh University's Journal of Differential Geometry. MR 2001f:53003

859. François Ledrappier, *Structure au bord des variétés à courbure négative*, Séminaire de Théorie Spectrale et Géométrie, No. 13, Année 1994–1995, Univ. Grenoble I, Saint, 1995, pp. 97–122. MR 1 715 960
860. Enrico Leuzinger, *On the trigonometry of symmetric spaces*, Comment. Math. Helv. **67** (1992), no. 2, 252–286. MR 93h:53054
861. Paul Lévy, *Problèmes concrets d'analyse fonctionnelle. Avec un complément sur les fonctionnelles analytiques par F. Pellegrino*, Gauthier-Villars, Paris, 1951, 2d ed. MR 12,834a
862. Leonard Lewin (ed.), *Structural properties of polylogarithms*, American Mathematical Society, Providence, RI, 1991. MR 93b:11158
863. Peter Li and Shing-Tung Yau, *A new conformal invariant and its applications to the Willmore conjecture and the first eigenvalue of compact surfaces*, Invent. Math. **69** (1982), no. 2, 269–291. MR 84f:53049
864. André Lichnerowicz, *Courbure, nombres de Betti, et espaces symétriques*, Proceedings of the International Congress of Mathematicians, Cambridge, Mass., 1950, vol. 2 (Providence, R. I.), Amer. Math. Soc., 1952, pp. 216–223. MR 13,492f
865. _____, *Géométrie des groupes de transformations*, Travaux et Recherches Mathématiques, III. Dunod, Paris, 1958. MR 23 #A1329
866. _____, *Propagateurs et commutateurs en relativité générale*, Inst. Hautes Études Sci. Publ. Math. No. **10** (1961), 56. MR 28 #967
867. _____, *Spineurs harmoniques*, C. R. Acad. Sci. Paris **257** (1963), 7–9. MR 27 #6218
868. _____, *Global theory of connections and holonomy groups*, Noordhoff International Publishing, Leiden, 1976, Translated from the French and edited by Michael Cole. MR 54 #1121
869. Joram Lindenstrauss, *Almost spherical sections: their existence and their applications*, in Geyer [557], pp. 39–61.
870. Joram Lindenstrauss and V. D. Milman, *The local theory of normed spaces and its applications to convexity*, Handbook of convex geometry, Vol. A, B, North-Holland, Amsterdam, 1993, pp. 1149–1220. MR 95b:46012
871. Anders Linnér, *Curve-straightening*, in Greene and Yau [597], Proceedings of the AMS Summer Research Institute on Differential Geometry held at the University of California, Los Angeles, California, July 8–28, 1990, pp. 451–458. MR 94m:58046
872. Marcelo Llarull, *Sharp estimates and the Dirac operator*, Math. Ann. **310** (1998), no. 1, 55–71. MR 98m:53056
873. Joachim Lohkamp, *Metrics of negative Ricci curvature*, Ann. of Math. (2) **140** (1994), no. 3, 655–683. MR 95i:53042
874. _____, *Discontinuity of geometric expansions*, Comment. Math. Helv. **71** (1996), no. 2, 213–228. MR 97f:58134
875. _____, *Global and local curvatures*, in Besson et al. [192], Papers from a workshop held Aug. 3-13, 1993, in Waterloo, Canada, pp. 23–51. MR 97b:53037
876. _____, *Ricci curvature modulo homotopy*, in Besse [185], En l'honneur de Marcel Berger. [In honor of Marcel Berger], Held in Luminy, France, July 12–18, 1992, pp. 437–451. MR 97m:53067
877. _____, *Curvature contents of geometric spaces*, Proceedings of the International Congress of Mathematicians, Vol. II (Berlin, 1998), no. Extra Vol. II, 1998, pp. 381–388 (electronic). MR 2000b:53047

878. Ottmar Loos, *Symmetric spaces. I: General theory*, W. A. Benjamin, Inc., New York-Amsterdam, 1969. MR 39 #365a
879. _____, *Symmetric spaces. II: Compact spaces and classification*, W. A. Benjamin, Inc., New York-Amsterdam, 1969. MR 39 #365b
880. J. Loot, *Collapsing and the differential form Laplacian*, to appear, 1999.
881. John Lott, *Collapsing and the differential form Laplacian*, electronic preprint math.DG/9902111, Los Alamos National Labs, 1999.
882. Peng Lu, *A compactness property for solutions of the Ricci flow on orbifolds*, Amer. J. Math. **123** (2001), no. 6, 1103–1134. MR 2002h:53118
883. W. Lück, L^2 *invariants of regular coverings of compact manifolds and CW complexes (survey)*, preprint, Münster Universität, 1996.
884. Wen Zhi Luo, Zeév Rudnick, and Peter Sarnak, *On Selberg's eigenvalue conjecture*, Geom. Funct. Anal. **5** (1995), no. 2, 387–401. MR 96h:11045
885. Wen Zhi Luo and Peter Sarnak, *Number variance for arithmetic hyperbolic surfaces*, Comm. Math. Phys. **161** (1994), no. 2, 419–432. MR 95k:11076
886. _____, *Quantum ergodicity of eigenfunctions on* $\mathrm{PSL}_2(\mathbb{Z})\backslash \mathbb{H}^2$, Inst. Hautes Études Sci. Publ. Math. (1995), no. 81, 207–237. MR 97f:11037
887. L. A. Lusternik and A. I. Fet, *Variational problems on closed manifolds*, Doklady Akad. Nauk SSSR (N.S.) **81** (1951), 17–18. MR 13,474c
888. L. A. Lusternik and L. G. Schnirelmann, *Sur le problème de trois géodésiques fermées sur les surfaces de genre 0*, C. R. Acad. Sci. Paris **189** (1929), 269–271.
889. Hans Maaß, *Über eine neue Art von nichtanalytischen automorphen Funktionen und die Bestimmung Dirichletscher Reihen durch Funktionalgleichungen*, Math. Ann. **121** (1949), 141–183. MR 11,163c
890. Paul Malliavin and Daniel W. Stroock, *Short time behavior of the heat kernel and its logarithmic derivatives*, J. Differential Geom. **44** (1996), no. 3, 550–570. MR 98c:58164
891. Ricardo Mañé, *Ergodic theory and differentiable dynamics*, Springer-Verlag, Berlin, 1987, Translated from the Portuguese by Silvio Levy. MR 88c:58040
892. _____, *On the topological entropy of geodesic flows*, J. Differential Geom. **45** (1997), no. 1, 74–93. MR 98d:58141
893. Hans von Mangoldt, *Über diejenigen Punkte auf positiv gekrümmten flächen, welche die eigenschaft haben, dass die von ihnen ausgehenden geodätischen Linien nie aufhören, kürzeste Linien zu sein*, J. Reine Angew. Math. **91** (1881), 23–52.
894. Anthony Manning, *Topological entropy for geodesic flows*, Ann. of Math. (2) **110** (1979), no. 3, 567–573. MR 81e:58044
895. Christophe M. Margerin, *Théorie de la déformation en géométrie métrique*, preprint M1005.0890, École polytechnique, Palaiseau, France, 1991.
896. _____, *General conjugate loci are not closed*, Differential geometry: Riemannian geometry (Los Angeles, CA, 1990), Amer. Math. Soc., Providence, RI, 1993, pp. 465–478. MR 94c:53054
897. _____, *Théorie de la déformation en géométrie métrique*, preprint M1100.1293, École polytechnique, Palaiseau, France, 1993.
898. _____, *Théorie de la déformation en géométrie métrique*, preprint M1073.0593, École polytechnique, Palaiseau, France, 1994.
899. G. A. Margulis, *Certain applications of ergodic theory to the investigation of manifolds of negative curvature*, Funkcional. Anal. i Priložen. **3** (1969), no. 4, 89–90. MR 41 #2582

900. _____, *Discrete groups of motions of manifolds of nonpositive curvature*, Proceedings of the International Congress of Mathematicians (Vancouver, B.C., 1974), Vol. 2, Canad. Math. Congress, Montreal, Que., 1975, pp. 21–34. MR 58 #11226
901. F. Marque, *Sur les singularitiés des espaces de cohomogénéité un*, to appear, 1998.
902. William S. Massey, *A basic course in algebraic topology*, Springer-Verlag, New York, 1991. MR 92c:55001
903. Shigenori Matsumoto, *Foundations of flat conformal structure*, Aspects of low-dimensional manifolds, Advanced studies in pure mathematics, vol. 20, Kinokuniya, Tokyo, 1992, pp. 167–261. MR 93m:57014
904. Vladimir S. Matveev and Peter J. Topalov, *Quantum integrability of the Beltrami–Laplace operator for geodesically equivalent metrics*, preprint 120, Max-Planck-Institut für Mathematik, Bonn, Germany, 1999.
905. Rafe Mazzeo, Frank Pacard, and Daniel Pollack, *Connected sums of constant mean curvature surfaces in Euclidean 3 space*, to appear, 2000.
906. John McCleary and Wolfgang Ziller, *On the free loop space of homogeneous spaces*, Amer. J. Math. **109** (1987), no. 4, 765–781. MR 88k:58023
907. _____, *Corrections to: "On the free loop space of homogeneous spaces"*, Amer. J. Math. **113** (1991), no. 2, 375–377, Erratum for McCleary and Ziller [906]. MR 92e:58046
908. Dusa McDuff, *A glimpse into symplectic geometry*, in Arnol′d et al. [67], pp. 175–188.
909. Dusa McDuff and Dietmar Salamon, *Introduction to symplectic topology*, second ed., The Clarendon Press Oxford University Press, New York, 1998. MR 1 698 616
910. H. P. McKean, Jr. and I. M. Singer, *Curvature and the eigenvalues of the Laplacian*, J. Differential Geometry **1** (1967), no. 1, 43–69. MR 36 #828
911. William H. Meeks, III, *A survey of the geometric results in the classical theory of minimal surfaces*, Bol. Soc. Brasil. Mat. **12** (1981), no. 1, 29–86. MR 84d:53007
912. Antonios D. Melas, *On the nodal line of the second eigenfunction of the Laplacian in $\mathbb{R}^2$*, J. Differential Geom. **35** (1992), no. 1, 255–263. MR 93g:35100
913. X. Mengue, *Noncollapsing examples with positive curvature and infinite topological type*, to appear, 2000.
914. Sergei Merkulov and Lorenz Schwachhöfer, *Classification of irreducible holonomies of torsion-free affine connections*, Ann. of Math. (2) **150** (1999), no. 1, 77–149. MR 1 715 321
915. Daniel Meyer, *Sur les variétés riemanniennes à opérateur de courbure positif*, C. R. Acad. Sci. Paris Sér. A-B **272** (1971), A482–A485. MR 43 #5457
916. Wolfgang T. Meyer, *Topogonov's theorem and its applications*, College on Differential Geometry (Trieste), ITP, 1989.
917. Mario J. Micallef and John Douglas Moore, *Minimal two-spheres and the topology of manifolds with positive curvature on totally isotropic two-planes*, Ann. of Math. (2) **127** (1988), no. 1, 199–227. MR 89e:53088
918. J. H. Michael and Leon M. Simon, *Sobolev and mean-value inequalities on generalized submanifolds of R^n*, Comm. Pure Appl. Math. **26** (1973), 361–379. MR 49 #9717
919. René Michel, *Restriction de la distance géodésique à un arc et rigidité*, Bull. Soc. Math. France **122** (1994), no. 3, 435–442. MR 95i:53044

920. V. D. Milman, *Dvoretzky's theorem—thirty years later*, Geom. Funct. Anal. **2** (1992), no. 4, 455–479. MR 93i:46002
921. John W. Milnor, *Morse theory*, Princeton University Press, Princeton, N.J., 1963, Based on lecture notes by M. Spivak and R. Wells. Annals of Mathematics Studies, No. 51. MR 29 #634
922. _____, *Eigenvalues of the Laplace operator on certain manifolds*, Proc. Nat. Acad. Sci. U.S.A. **51** (1964), 542. MR 28 #5403
923. _____, *Whitehead torsion*, Bull. Amer. Math. Soc. **72** (1966), 358–426. MR 33 #4922
924. _____, *A note on curvature and fundamental group*, J. Differential Geometry **2** (1968), 1–7. MR 38 #636
925. John W. Milnor and James D. Stasheff, *Characteristic classes*, Princeton University Press, Princeton, N. J., 1974, Annals of Mathematics Studies, No. 76. MR 55 #13428
926. Tilla Klotz Milnor, *Efimov's theorem about complete immersed surfaces of negative curvature*, Advances in Math. **8** (1972), 474–543. MR 46 #835
927. Maung Min-Oo and Ernst A. Ruh, *Comparison theorems for compact symmetric spaces*, Ann. Sci. École Norm. Sup. (4) **12** (1979), no. 3, 335–353. MR 81e:53027
928. _____, *Vanishing theorems and almost symmetric spaces of noncompact type*, Math. Ann. **257** (1981), no. 4, 419–433. MR 83d:53040
929. S. Minakshisundaram, *Eigenfunctions on Riemannian manifolds*, J. Indian Math. Soc. (N.S.) **17** (1953), 159–165 (1954). MR 15,877d
930. S. Minakshisundaram and Å. Pleijel, *Some properties of the eigenfunctions of the Laplace-operator on Riemannian manifolds*, Canadian J. Math. **1** (1949), 242–256. MR 11,108b
931. G. Misiołek, *The exponential map on the free loop space is Fredholm*, Geom. Funct. Anal. **7** (1997), no. 5, 954–969. MR 98e:58018
932. Edwin E. Moise, *Geometric topology in dimensions 2 and 3*, Springer-Verlag, New York, 1977, Graduate Texts in Mathematics, Vol. 47. MR 58 #7631
933. Ngaiming Mok, *The uniformization theorem for compact Kähler manifolds of nonnegative holomorphic bisectional curvature*, J. Differential Geom. **27** (1988), no. 2, 179–214. MR 89d:53115
934. Bernardo Molina and Carlos Olmos, *Manifolds all of whose flats are closed*, J. Differential Geom. **45** (1997), no. 3, 575–592. MR 98d:53059
935. Deane Montgomery and Hans Samelson, *Transformation groups of spheres*, Ann. of Math. (2) **44** (1943), 454–470. MR 5,60b
936. Frank Morgan, *Riemannian geometry*, second ed., A K Peters Ltd., Wellesley, MA, 1998, A beginner's guide. MR 98i:53001
937. _____, *Geometric measure theory*, third ed., Academic Press Inc., San Diego, CA, 2000, A beginner's guide. MR 1 775 760
938. Frank Morgan, Michael Hutchings, and Hugh Howards, *The isoperimetric problem on surfaces of revolution of decreasing Gauss curvature*, Trans. Amer. Math. Soc. (2000), to appear. MR 1 661 278
939. John W. Morgan, *The Seiberg-Witten equations and applications to the topology of smooth four-manifolds*, Princeton University Press, Princeton, NJ, 1996. MR 97d:57042
940. Shigefumi Mori, *Projective manifolds with ample tangent bundles*, Ann. of Math. (2) **110** (1979), no. 3, 593–606. MR 81j:14010

941. Andrei Moroianu and Uwe Semmelmann, *Parallel spinors and holonomy groups*, J. Math. Phys. **41** (2000), no. 4, 2395–2402. MR 2001f:53099
942. H. Marston Morse, *A one-to-one representation of geodesics on a surface of negative curvature*, Amer. J. Math. **43** (1921), 33–51.
943. _____, *Recurrent geodesics on a surface of negative curvature*, Trans. Amer. Math. Soc. **22** (1921), 84–100.
944. _____, *The calculus of variations in the large*, American Mathematical Society, Providence, RI, 1996, Reprint of the 1932 original. MR 98f:58070
945. Philip M. Morse and Herman Feshbach, *Methods of theoretical physics. 2 volumes*, McGraw-Hill Book Co., Inc., New York, 1953. MR 15,583h
946. Jürgen K. Moser, *Proof of a generalized form of a fixed point theorem due to G. D. Birkhoff*, Geometry and topology (Proc. III Latin Amer. School of Math., Inst. Mat. Pura Aplicada CNPq, Rio de Janeiro, 1976) (Berlin), Springer, 1977, pp. 464–494. Lecture Notes in Math., Vol. 597. MR 58 #13205
947. G. D. Mostow, *Quasi-conformal mappings in n-space and the rigidity of hyperbolic space forms*, Inst. Hautes Études Sci. Publ. Math. No. **34** (1968), 53–104. MR 38 #4679
948. _____, *Strong rigidity of locally symmetric spaces*, Princeton University Press, Princeton, N.J., 1973, Annals of Mathematics Studies, No. 78. MR 52 #5874
949. _____, *On a remarkable class of polyhedra in complex hyperbolic space*, Pacific J. Math. **86** (1980), no. 1, 171–276. MR 82a:22011
950. G. D. Mostow and Yum Tong Siu, *A compact Kähler surface of negative curvature not covered by the ball*, Ann. of Math. (2) **112** (1980), no. 2, 321–360. MR 82f:53075
951. Jaime Muñoz Masqué and Antonio Valdés, *Génération des anneaux d'invariants différentiels des métriques riemanniennes*, C. R. Acad. Sci. Paris Sér. I Math. **323** (1996), no. 6, 643–646. MR 97g:58006
952. Werner Müller, *Analytic torsion and R-torsion of Riemannian manifolds*, Adv. in Math. **28** (1978), no. 3, 233–305. MR 80j:58065b
953. Sumner B. Myers, *Connections between differential geometry and topology*, Duke Math. J. **1** (1935), 376–391.
954. _____, *Riemannian manifolds in the large*, Duke Math. J. **1** (1935), 39–49.
955. _____, *Riemannian manifolds with positive mean curvature*, Duke Math. J. **8** (1941), 401–404. MR 3,18f
956. Alexander Nabutovsky, *Einstein structures: existence versus uniqueness*, Geom. Funct. Anal. **5** (1995), no. 1, 76–91. MR 96e:53061
957. _____, *Disconnectedness of sublevel sets of some Riemannian functionals*, Geom. Funct. Anal. **6** (1996), no. 4, 703–725. MR 97i:58018
958. _____, *Fundamental group and contractible closed geodesics*, Comm. Pure Appl. Math. **49** (1996), no. 12, 1257–1270. MR 98d:53063
959. Alexander Nabutovsky and Regina Rotman, *The length of the shortest closed geodesic on a 2-dimensional sphere*, Int. Math. Res. Not. (2002), no. 23, 1211–1222. MR 1 903 953
960. _____, *Upper bounds on the length of a shortest closed geodesic and a quantitative hurewicz theorem*, unpublished, 2002.
961. Alexander Nabutovsky and Shmuel Weinberger, *Variational problems for Riemannian functionals and arithmetic groups*, electronic preprint math/9711225, Los Alamos National Labs, 1997.

962. _____, *The fractal nature of* Riem/Diff, *I*, preprint, 2000.
963. Nikolai Nadirashvili, *Berger's isoperimetric problem and minimal immersions of surfaces*, Geom. Funct. Anal. **6** (1996), no. 5, 877–897. MR 98f:53061
964. _____, *Hadamard's and Calabi-Yau's conjectures on negatively curved and minimal surfaces*, Invent. Math. **126** (1996), no. 3, 457–465. MR 98d:53014
965. Albert Nijenhuis, *On the holonomy groups of linear connections. IA, IB. General properties of affine connections*, Nederl. Akad. Wetensch. Proc. Ser. A. **56** = Indagationes Math. **15** (1953), 233–240, 241–249. MR 16,171c
966. _____, *On the holonomy groups of linear connections. II. Properties of general linear connections*, Nederl. Akad. Wetensch. Proc. Ser. A. **57** = Indagationes Math. **16** (1954), 17–25. MR 16,172a
967. I. G. Nikolaev, *Bounded curvature closure of the set of compact Riemannian manifolds*, Bull. Amer. Math. Soc. (N.S.) **24** (1991), no. 1, 171–177. MR 91e:53048
968. Johannes C. C. Nitsche, *Lectures on minimal surfaces*, Cambridge University Press, Cambridge, 1989, Translated from the German by Jerry M. Feinberg, With a German foreword. MR 90m:49031
969. Katsumi Nomizu, *Invariant affine connections on homogeneous spaces*, Amer. J. Math. **76** (1954), 33–65. MR 15,468f
970. S. P. Novikov (ed.), *Topology. I*, Springer-Verlag, Berlin, 1996, General survey, Translated from *Current problems in mathematics. Fundamental directions. Vol. 12* (Russian), Akad. Nauk SSSR, Vsesoyuz. Inst. Nauchn. i Tekhn. Inform., Moscow, 1986. Translation by B. Botvinnik and R. G. Burns. Translation edited by S. P. Novikov. MR 96m:57004
971. Morio Obata, *Certain conditions for a Riemannian manifold to be isometric with a sphere*, J. Math. Soc. Japan **14** (1962), 333–340. MR 25 #5479
972. Joseph Oesterlé, *Polylogarithmes*, Astérisque (1993), no. 216, Exp. No. 762, 3, 49–67, Séminaire Bourbaki, Vol. 1992/93. MR 94m:11135
973. Hideki Omori, *Construction problems of Riemannian manifolds*, in Berger et al. [176], pp. 79–90.
974. Barrett O'Neill, *Elementary differential geometry*, Academic Press, New York, 1966. MR 34 #3444
975. _____, *The fundamental equations of a submersion*, Michigan Math. J. **13** (1966), 459–469. MR 34 #751
976. _____, *Semi-Riemannian geometry*, Academic Press Inc. [Harcourt Brace Jovanovich Publishers], New York, 1983, With applications to relativity. MR 85f:53002
977. Arkadi L. Onishchik, *Topology of transitive transformation groups*, Johann Ambrosius Barth Verlag GmbH, Leipzig, 1994. MR 95e:57058
978. B. Osgood, R. Phillips, and Peter Sarnak, *Compact isospectral sets of surfaces*, J. Funct. Anal. **80** (1988), no. 1, 212–234. MR 90d:58160
979. _____, *Extremals of determinants of Laplacians*, J. Funct. Anal. **80** (1988), no. 1, 148–211. MR 90d:58159
980. _____, *Moduli space, heights and isospectral sets of plane domains*, Ann. of Math. (2) **129** (1989), no. 2, 293–362. MR 91a:58196
981. Robert Osserman, *A note on Hayman's theorem on the bass note of a drum*, Comment. Math. Helv. **52** (1977), no. 4, 545–555. MR 56 #17297
982. _____, *The isoperimetric inequality*, Bull. Amer. Math. Soc. **84** (1978), no. 6, 1182–1238. MR 58 #18161

983. _____, *The four-or-more vertex theorem*, Amer. Math. Monthly **92** (1985), no. 5, 332–337. MR 87e:53001

984. Robert Osserman (ed.), *Geometry. V*, Springer-Verlag, Berlin, 1997, Minimal surfaces. MR 98g:53014

985. Jean-Pierre Otal, *Le spectre marqué des longueurs des surfaces à courbure négative*, Ann. of Math. (2) **131** (1990), no. 1, 151–162. MR 91c:58026

986. _____, *Sur les longueurs des géodésiques d'une métrique à courbure négative dans le disque*, Comment. Math. Helv. **65** (1990), no. 2, 334–347. MR 91i:53054

987. Yukio Otsu, *On manifolds of positive Ricci curvature with large diameter*, Math. Z. **206** (1991), no. 2, 255–264. MR 91m:53033

988. Yukio Otsu, Katsuhiro Shiohama, and Takao Yamaguchi, *A new version of differentiable sphere theorem*, Invent. Math. **98** (1989), no. 2, 219–228. MR 90i:53049

989. Yukio Otsu and Takashi Shioya, *The Riemannian structure of Alexandrov spaces*, J. Differential Geom. **39** (1994), no. 3, 629–658. MR 95e:53062

990. D. Page, *Compact rotating gravitational instanton*, Physics Letters B **79** (1978), 235–238.

991. Richard S. Palais, *On the differentiability of isometries*, Proc. Amer. Math. Soc. **8** (1957), 805–807. MR 19,451a

992. Richard S. Palais and Stephen Smale, *A generalized Morse theory*, Bull. Amer. Math. Soc. **70** (1964), 165–172. MR 28 #1634

993. Richard S. Palais and Chuu-Lian Terng, *Critical point theory and submanifold geometry*, Springer-Verlag, Berlin, 1988. MR 90c:53143

994. Pierre Pansu, *Effondrement des variétés riemanniennes, d'après J. Cheeger et M. Gromov*, Astérisque (1985), no. 121-122, 63–82, Seminar Bourbaki, Vol. 1983/84. MR 86j:53063

995. _____, *Métriques de Carnot-Carathéodory et quasiisométries des espaces symétriques de rang un*, Ann. of Math. (2) **129** (1989), no. 1, 1–60. MR 90e:53058

996. _____, *Le flot géodésique des variétés riemanniennes à courbure négative*, Astérisque (1991), no. 201-203, Exp. No. 738, 269–298 (1992), Séminaire Bourbaki, Vol. 1990/91. MR 93a:58136

997. _____, *Sous-groupes discrets des groupes de Lie: rigidité, arithméticité*, Astérisque (1995), no. 227, Exp. No. 778, 3, 69–105, Séminaire Bourbaki, Vol. 1993/94. MR 96a:22021

998. _____, *Differentiability of volume growth*, Proceedings of the Workshop on Differential Geometry and Topology (Palermo, 1996), no. 49, 1997, pp. 159–176. MR 98m:53058

999. _____, *Volume, courbure et entropie (d'après G. Besson, G. Courtois et S. Gallot)*, Astérisque (1997), no. 245, Exp. No. 823, 3, 83–103, Séminaire Bourbaki, Vol. 1996/97. MR 99g:58099

1000. _____, *Sur la régularité du profil isopérimétrique des surfaces riemanniennes compactes*, Ann. Inst. Fourier (Grenoble) **48** (1998), no. 1, 247–264. MR 99i:53035

1001. Gabriel P. Paternain, *On the topology of manifolds with completely integrable geodesic flows*, Ergodic Theory Dynamical Systems **12** (1992), no. 1, 109–121. MR 93g:58117

1002. _____, *Geodesic flows*, Progress in Math., vol. 180, Birkhäuser Boston Inc., Boston, MA, 1999. MR 2000h:53108

1003. Gabriel P. Paternain and Miguel Paternain, *Topological entropy versus geodesic entropy*, Internat. J. Math. **5** (1994), no. 2, 213–218. MR 95h:58101
1004. Gabriel P. Paternain and R. J. Spatzier, *New examples of manifolds with completely integrable geodesic flows*, Adv. Math. **108** (1994), no. 2, 346–366. MR 95g:58166
1005. V. K. Patodi, *Curvature and the eigenforms of the Laplace operator*, J. Differential Geometry **5** (1971), 233–249. MR 45 #1201
1006. Wolfgang Pauli, *Zur Quantenmechanik des magnetischen Elektrons*, Z. Physik **43** (1927), 601.
1007. Fernand Pelletier and Liane Valère Bouche, *The problem of geodesics, intrinsic derivation and the use of control theory in singular sub-Riemannian geometry*, in Besse [185], En l'honneur de Marcel Berger. [In honor of Marcel Berger], Held in Luminy, France, July 12–18, 1992, pp. 453–512. MR 97i:58031
1008. G. Perel′man, *Manifolds of positive Ricci curvature with almost maximal volume*, J. Amer. Math. Soc. **7** (1994), no. 2, 299–305. MR 94f:53077
1009. _____, *Proof of the soul conjecture of Cheeger and Gromoll*, J. Differential Geom. **40** (1994), no. 1, 209–212. MR 95d:53037
1010. _____, *Widths of nonnegatively curved spaces*, Geom. Funct. Anal. **5** (1995), no. 2, 445–463. MR 96c:53063
1011. _____, *A complete Riemannian manifold of positive Ricci curvature with Euclidean volume growth and nonunique asymptotic cone*, in Grove and Petersen [650], Papers from the Special Year in Differential Geometry held in Berkeley, CA, 1993–94, pp. 165–166. MR 98e:53067
1012. _____, *Construction of manifolds of positive Ricci curvature with big volume and large Betti numbers*, in Grove and Petersen [650], Papers from the Special Year in Differential Geometry held in Berkeley, CA, 1993–94, pp. 157–163. MR 98h:53062
1013. Stefan Peters, *Cheeger's finiteness theorem for diffeomorphism classes of Riemannian manifolds*, J. Reine Angew. Math. **349** (1984), 77–82. MR 85j:53046
1014. _____, *Convergence of Riemannian manifolds*, Compositio Math. **62** (1987), no. 1, 3–16. MR 88i:53076
1015. Peter Petersen, V, *Gromov-Hausdorff convergence of metric spaces*, in Greene and Yau [597], Proceedings of the AMS Summer Research Institute on Differential Geometry held at the University of California, Los Angeles, California, July 8–28, 1990, pp. 489–504. MR 94b:53079
1016. _____, *Comparison geometry problem list*, in Besson et al. [192], Papers from a workshop held Aug. 3-13, 1993, in Waterloo, Canada, pp. 87–115. MR 96m:53002
1017. _____, *Convergence theorems in Riemannian geometry*, in Grove and Petersen [650], Papers from the Special Year in Differential Geometry held in Berkeley, CA, 1993–94, pp. 167–202. MR 98k:53049
1018. _____, *Riemannian geometry*, Springer-Verlag, New York, 1998. MR 98m:53001
1019. _____, *Aspects of global Riemannian geometry*, Bull. Amer. Math. Soc. (N.S.) **36** (1999), no. 3, 297–344. MR 1 698 926
1020. _____, *On eigenvalue pinching in positive Ricci curvature*, Invent. Math. **138** (1999), no. 1, 1–21. MR 1 714 334
1021. Peter Petersen, V, S. D. Shteingold, and Guofang Wei, *Comparison geometry with integral curvature bounds*, Geom. Funct. Anal. **7** (1997), no. 6, 1011–1030. MR 99c:53022

1022. Peter Petersen, V and Chadwick Sprouse, *Integral curvature bounds, distance estimates and applications*, J. Differential Geom. **50** (1998), no. 2, 269–298. MR 1 684 981
1023. Peter Petersen, V and Guofang Wei, *Relative volume comparison with integral curvature bounds*, Geom. Funct. Anal. **7** (1997), no. 6, 1031–1045. MR 99c:53023
1024. _____, *Analysis and geometry on manifolds with integral Ricci curvature bounds. II*, Trans. Amer. Math. Soc. (2000), to appear. MR 1 709 777
1025. Peter Petersen, V, Guofang Wei, and Rugang Ye, *Controlled geometry via smoothing*, Comment. Math. Helv. **74** (1999), no. 3, 345–363. MR 1 710 067
1026. A. Petrunin and W. Tuschmann, *Diffeomorphism finiteness, positive pinching, and second homotopy*, Geom. Funct. Anal. **9** (1999), no. 4, 736–774. MR 1 719 602
1027. Ulrich Pinkall and I. Sterling, *On the classification of constant mean curvature tori*, Ann. of Math. (2) **130** (1989), no. 2, 407–451. MR 91b:53009
1028. Mark A. Pinsky, *Feeling the shape of a manifold with Brownian motion— the last word in 1990*, Stochastic analysis (Durham, 1990), Cambridge Univ. Press, Cambridge, 1991, pp. 305–320. MR 93e:58200
1029. _____, *Inverse questions in stochastic differential geometry*, Probability theory (Singapore, 1989), de Gruyter, Berlin, 1992, pp. 3–28. MR 93k:58234
1030. Gilles Pisier, *The volume of convex bodies and Banach space geometry*, Cambridge University Press, Cambridge, 1989. MR 91d:52005
1031. Christophe Pittet, *Systoles on $S^1 \times S^n$*, Differential Geom. Appl. **7** (1997), no. 2, 139–142. MR 98i:53061
1032. _____, *The isoperimetric profile of homogeneous Riemannian manifolds*, J. Differential Geom. **54** (2000), no. 2, 255–302. MR 2002g:53088
1033. A. V. Pogorelov, *Extrinsic geometry of convex surfaces*, American Mathematical Society, Providence, R.I., 1973, Translated from the Russian by Israel Program for Scientific Translations, Translations of Mathematical Monographs, Vol. 35. MR 49 #11439
1034. Henri Poincaré, *Sur les lignes géod'esiques des surfaces convexes*, Trans. Amer. Math. Soc. **6** (1905), 237–274.
1035. Louis Poinsot, *Éléments de statique, suivis de quatre mémoires sur la composition des moments et des aires; sur le plan invariable du système du monde; sur la théorie générale de l'équilibre et du mouvement des systèmes; et sur une théorie nouvelle de la rotation des corps*, Bachelier, Paris, 1842.
1036. Iosif Polterovich, *Heat invariants of Riemannian manifolds*, electronic preprint math/9905073, Los Alamos National Labs, 1999.
1037. G. Pólya and G. Szegö, *Isoperimetric Inequalities in Mathematical Physics*, Princeton University Press, Princeton, N. J., 1951, Annals of Mathematics Studies, no. 27. MR 13,270d
1038. Y. S. Poon and Simon M. Salamon, *Quaternionic Kähler 8-manifolds with positive scalar curvature*, J. Differential Geom. **33** (1991), no. 2, 363–378. MR 92b:53071
1039. I. R. Porteous, *Geometric differentiation for the intelligence of curves and surfaces*, Cambridge University Press, Cambridge, 1994. MR 96b:53006
1040. Gopal Prasad, *Volumes of S-arithmetic quotients of semi-simple groups*, Inst. Hautes Études Sci. Publ. Math. (1989), no. 69, 91–117, With an appendix by Moshe Jarden and the author. MR 91c:22023

1041. Alexandre Preissmann, *Quelques propriétés globales des espaces de Riemann*, Comment. Math. Helv. **15** (1943), 175–216. MR 6,20g
1042. Friedbert Prüfer, Franco Tricerri, and Lieven Vanhecke, *Curvature invariants, differential operators and local homogeneity*, Trans. Amer. Math. Soc. **348** (1996), no. 11, 4643–4652. MR 97a:53074
1043. P. M. Pu, *Some inequalities in certain nonorientable Riemannian manifolds*, Pacific J. Math. **2** (1952), 55–71. MR 14,87e
1044. Thomas Püttmann, *Optimal pinching constants of odd-dimensional homogeneous spaces*, Invent. Math. **138** (1999), no. 3, 631–684. MR 1 719 807
1045. Hans-Bert Rademacher, *The Fadell-Rabinowitz index and closed geodesics*, J. London Math. Soc. (2) **50** (1994), no. 3, 609–624. MR 95k:58040
1046. _____, *On a generic property of geodesic flows*, Math. Ann. **298** (1994), no. 1, 101–116. MR 94j:58040
1047. Ziv Ran, *Universal variations of Hodge structure and Calabi-Yau-Schottky relations*, Invent. Math. **138** (1999), no. 2, 425–449. MR 1 720 188
1048. E. Raphaël, J.-M. di Meglio, Marcel Berger, and Eugenio Calabi, *Convex particles at interfaces*, Journal de physique I **2** (1992), no. 5, 571–579.
1049. John G. Ratcliffe, *Foundations of hyperbolic manifolds*, Springer-Verlag, New York, 1994. MR 95j:57011
1050. H. E. Rauch, *A contribution to differential geometry in the large*, Ann. of Math. (2) **54** (1951), 38–55. MR 13,159b
1051. _____, *Geodesics, symmetric spaces, and differential geometry in the large*, Comment. Math. Helv. **27** (1953), 294–320 (1954). MR 15,744b
1052. D. B. Ray and I. M. Singer, *Analytic torsion*, Partial differential equations (Proc. Sympos. Pure Math., Vol. XXIII, Univ. California, Berkeley, Calif., 1971), Amer. Math. Soc., Providence, R.I., 1973, pp. 167–181. MR 49 #4053
1053. John William Rayleigh and Baron Strutt, *The Theory of Sound*, Dover Publications, New York, N. Y., 1945, 2d ed. MR 7,500e
1054. Christophe Real, *Métriques d'Einstein-Kähler sur les variétés à première classe de Chern positive*, C. R. Acad. Sci. Paris Sér. I Math. **322** (1996), no. 5, 461–464. MR 97c:53108
1055. Y. Reshetnyak (ed.), *Geometry. IV*, Springer-Verlag, Berlin, 1993, Nonregular Riemannian geometry, A translation of *Geometry, 4 (Russian)*, Akad. Nauk SSSR, Vsesoyuz. Inst. Nauchn. i Tekhn. Inform., Moscow, 1989, Translation by E. Primrose. MR 94i:53038
1056. Georges de Rham, *Variétés différentiables. Formes, courants, formes harmoniques*, Hermann, Paris, 1973, Troisième édition revue et augmentée, Publications de l'Institut de Mathématique de l'Université de Nancago, III, Actualités Scientifiques et Industrielles, No. 1222b. MR 49 #11552
1057. _____, *Differentiable manifolds*, Springer-Verlag, Berlin, 1984, Forms, currents, harmonic forms, Translated from the French by F. R. Smith, With an introduction by Shiing-Shen Chern. MR 85m:58005
1058. M. Ritoré, *Constant geodesic curvature curves and isoperimetric domains in rotationally symmetric domains*, to appear, 2000.
1059. Manuel Ritoré and Antonio Ros, *Stable constant mean curvature tori and the isoperimetric problem in three space forms*, Comment. Math. Helv. **67** (1992), no. 2, 293–305. MR 93a:53055
1060. Alain Rivière, *Dimension de Hausdorff de la nervure*, Geom. Dedicata **85** (2001), no. 1-3, 217–235. MR 2002j:28010

1061. Igor Rivin and Jean-Marc Schlenker, *The Schläfli formula in Einstein manifolds with boundary*, Electron. Res. Announc. Amer. Math. Soc. **5** (1999), 18–23 (electronic). MR 2000a:53076

1062. Gilles F. Robert, *Invariants topologiques et géométriques reliés aux longueurs des géodésiques et aux section harmoniques de fibrés*, Ph.D. thesis, Institut Fourier, Grenoble, France, Octobre 1994.

1063. Yann Rollin, *Rigidité d'Einstein du plan hyperbolique complexe*, C. R. Math. Acad. Sci. Paris **334** (2002), no. 8, 671–676. MR 1 903 368

1064. Xiaochun Rong, *The limiting eta invariants of collapsed three-manifolds*, J. Differential Geom. **37** (1993), no. 3, 535–568. MR 94f:57023

1065. _____, *Rationality of geometric signatures of complete 4-manifolds*, Invent. Math. **120** (1995), no. 3, 513–554. MR 97a:53066

1066. _____, *The almost cyclicity of the fundamental groups of positively curved manifolds*, Invent. Math. **126** (1996), no. 1, 47–64. MR 97g:53045

1067. _____, *On the fundamental groups of manifolds of positive sectional curvature*, Ann. of Math. (2) **143** (1996), no. 2, 397–411. MR 97a:53067

1068. _____, *Collapsed manifolds with pinched positive sectional curvature*, to appear, 1997.

1069. _____, *Positive curvature, local and global symmetry, and fundamental groups*, Amer. J. Math. **121** (1999), no. 5, 931–943. MR 1 713 297

1070. A. Ros, *The Willmore conjecture in real projective space*, to appear, 1998.

1071. Jonathan Rosenberg and Stephan Stolz, *Manifolds of positive scalar curvature*, Algebraic topology and its applications, Springer, New York, 1994, pp. 241–267. MR 1 268 192

1072. Regina Rotman, *Upper bounds on the length of the shortest closed geodesic on simply connected manifolds*, Math. Z. **233** (2000), no. 2, 365–398. MR 2001h:53055

1073. L. Rozoy, *Les points ombilics d'une surface*, Séminaire de théorie spectrale et géométrie, Institut Fourier, Grenoble, 1990.

1074. Zeév Rudnick and Peter Sarnak, *Zeros of principal L-functions and random matrix theory*, Duke Math. J. **81** (1996), no. 2, 269–322, A celebration of John F. Nash, Jr. MR 97f:11074

1075. Bernhard Ruh, *Krümmungstreue Diffeomorphismen Riemannscher und pseudo-Riemannscher Mannigfaltigkeiten*, Math. Z. **189** (1985), no. 3, 371–391. MR 86j:53064

1076. Ernst A. Ruh, *Curvature and differentiable structure on spheres*, Comment. Math. Helv. **46** (1971), 127–136. MR 44 #7474

1077. _____, *Almost flat manifolds*, J. Differential Geom. **17** (1982), no. 1, 1–14. MR 84a:53047

1078. _____, *Riemannian manifolds with bounded curvature ratios*, J. Differential Geom. **17** (1982), no. 4, 643–653 (1983). MR 84d:53043

1079. S. Sabourau, *Filling radius and short closed geodesics of the sphere*, unpublished, 2002.

1080. _____, *Global and local volume bounds and the shortest closed geodesic loop*, unpublished, 2002.

1081. _____, *Problèmes isosystolique sur les surfaces*, unpublished, 2002.

1082. Rainer Kurt Sachs and Hung Hsi Wu, *General relativity for mathematicians*, Springer-Verlag, New York, 1977, Graduate Texts in Mathematics, Vol. 48. MR 58 #20239a

1083. E. B. Saff and A. B. J. Kuijlaars, *Distributing many points on a sphere*, Math. Intelligencer **19** (1997), no. 1, 5–11. MR 98h:70011
1084. Takashi Sakai, *Cut loci of Berger's spheres*, Hokkaido Math. J. **10** (1981), no. 1, 143–155. MR 82g:53054
1085. _____, *Riemannian geometry*, American Mathematical Society, Providence, RI, 1996, Translated from the 1992 Japanese original by the author. MR 97f:53001
1086. Simon M. Salamon, *Quaternionic manifolds*, Symposia Mathematica, Vol. XXVI (Rome, 1980), Academic Press, London, 1982, pp. 139–151. MR 84e:53044
1087. _____, *Riemannian geometry and holonomy groups*, Longman Scientific & Technical, Harlow, 1989. MR 90g:53058
1088. _____, *On the cohomology of Kähler and hyper-Kähler manifolds*, Topology **35** (1996), no. 1, 137–155. MR 97f:32042
1089. Andrea Sambusetti, *An obstruction to the existence of Einstein metrics on 4-manifolds*, C. R. Acad. Sci. Paris Sér. I Math. **322** (1996), no. 12, 1213–1218. MR 97c:53073
1090. _____, *On minimal entropy and stability*, to appear, 1999.
1091. Hans Samelson, *On manifolds with many closed geodesics*, Portugal. Math. **22** (1963), 193–196. MR 37 #5819
1092. J. H. Sampson, *Applications of harmonic maps to Kähler geometry*, Complex differential geometry and nonlinear differential equations (Brunswick, Maine, 1984), Amer. Math. Soc., Providence, R.I., 1986, pp. 125–134. MR 87g:58028
1093. Luis A. Santaló, *Integral geometry and geometric probability*, Addison-Wesley Publishing Co., Reading, Mass.-London-Amsterdam, 1976, With a foreword by Mark Kac, Encyclopedia of Mathematics and its Applications, Vol. 1. MR 55 #6340
1094. Peter Sarnak, *Some applications of modular forms*, Cambridge University Press, Cambridge, 1990. MR 92k:11045
1095. _____, *Arithmetic quantum chaos*, The Schur lectures (1992) (Tel Aviv), Bar-Ilan Univ., Ramat Gan, 1995, pp. 183–236. MR 96d:11059
1096. _____, *Extremal geometries*, Extremal Riemann surfaces (San Francisco, CA, 1995), Amer. Math. Soc., Providence, RI, 1997, pp. 1–7. MR 98a:58043
1097. I. Satake, *On a generalization of the notion of manifold*, Proc. Nat. Acad. Sci. U.S.A. **42** (1956), 359–363. MR 18,144a
1098. Alessandro Savo, *Une méthode de symétrization et quelques applications*, C. R. Acad. Sci. Paris Sér. I Math. **322** (1996), no. 9, 861–864. MR 97a:58178
1099. _____, *Uniform estimates and the whole asymptotic series of the heat content on manifolds*, Geom. Dedicata **73** (1998), no. 2, 181–214. MR 1 652 049
1100. _____, *Remarks on the length of the nodes of an eigenfunction and more*, to appear, 1999.
1101. _____, *Lower bounds for the nodal length of eigenfunctions of the Laplacian*, Ann. Global Anal. Geom. **19** (2001), no. 2, 133–151. MR 2002g:58055
1102. H. Scherbel, *A new proof of Hamburger's index theorem on umbilical points*, preprint, ETH Zürich, 1993.
1103. Alexander Schiemann, *Ternary positive definite quadratic forms are determined by their theta series*, Math. Ann. **308** (1997), no. 3, 507–517. MR 98e:11083
1104. Jean-Marc Schlenker, *Métriques sur les polyèdres hyperboliques convexes*, J. Differential Geom. **48** (1998), no. 2, 323–405. MR 2000a:52018

1105. Erhard Schmidt, *Über das isoperimetrische Problem in Raum von n Dimensionen*, Math. Zeit. **44** (1939), 689–788.
1106. Paul Schmutz, *New results concerning the number of small eigenvalues on Riemann surfaces*, J. Reine Angew. Math. **471** (1996), 201–220. MR 97e:58225
1107. Rolf Schneider and John A. Wieacker, *Integral geometry*, in Gruber and Wills [661], pp. 1349–1390. MR 94e:52001
1108. Richard Schoen and Shing-Tung Yau, *On the proof of the positive mass conjecture in general relativity*, Comm. Math. Phys. **65** (1979), no. 1, 45–76. MR 80j:83024
1109. _____, *On the structure of manifolds with positive scalar curvature*, Manuscripta Math. **28** (1979), no. 1-3, 159–183. MR 80k:53064
1110. _____, *Proof of the positive mass theorem. II*, Comm. Math. Phys. **79** (1981), no. 2, 231–260. MR 83i:83045
1111. J. A. Schouten, *Ricci-calculus. An introduction to tensor analysis and its geometrical applications*, Springer-Verlag, Berlin, 1954, 2d. ed, Die Grundlehren der mathematischen Wissenschaften in Einzeldarstellungen mit besonderer Berücksichtigung der Anwendungsgebiete, Bd X. MR 16,521e
1112. Dorothee Schueth, *Continuous families of isospectral metrics on simply connected manifolds*, Ann. of Math. (2) **149** (1999), no. 1, 287–308. MR 2000c:58063
1113. Lorenz J. Schwachhöfer and Wilderich Tuschmann, *Almost nonnegative curvature and cohomogeneity one*, unpublished, 2001.
1114. Richard Schwartz, *Pappus' theorem and the modular group*, Inst. Hautes Études Sci. Publ. Math. (1993), no. 78, 187–206 (1994). MR 95g:57027
1115. _____, *Dynamical versions of Desargues' theorem*, to appear, 1998.
1116. Peter Scott, *The geometries of 3-manifolds*, Bull. London Math. Soc. **15** (1983), no. 5, 401–487. MR 84m:57009
1117. Walter Seaman, *A pinching theorem for four manifolds*, Geom. Dedicata **31** (1989), no. 1, 37–40. MR 90h:53043
1118. N. Seiberg and E. Witten, *Monopoles, duality and chiral symmetry breaking in $N = 2$ supersymmetric QCD*, Nuclear Phys. B **431** (1994), no. 3, 484–550. MR 95m:81203
1119. Herbert Seifert and William Threlfall, *Seifert and Threlfall: a textbook of topology*, Academic Press Inc. [Harcourt Brace Jovanovich Publishers], New York, 1980, Translated from the German edition of 1934 by Michael A. Goldman, With a preface by Joan S. Birman, With "Topology of 3-dimensional fibered spaces" by Seifert, Translated from the German by Wolfgang Heil. MR 82b:55001
1120. Atle Selberg, *Harmonic analysis, part 2*, preprint, Universität Göttingen, 1954.
1121. _____, *Harmonic analysis and discontinuous groups in weakly symmetric Riemannian spaces with applications to Dirichlet series*, J. Indian Math. Soc. (N.S.) **20** (1956), 47–87. MR 19,531g
1122. Stephen Semmes, *Good metric spaces without good parameterizations*, Rev. Mat. Iberoamericana **12** (1996), no. 1, 187–275. MR 97e:57025
1123. Mika Seppälä, *Real algebraic curves in the moduli space of complex curves*, Compositio Math. **74** (1990), no. 3, 259–283. MR 91j:14020
1124. Jean-Pierre Serre, *Homologie singulière des espaces fibrés. Applications*, Ann. of Math. (2) **54** (1951), 425–505. MR 13,574g

782   References

1125. _____, *A course in arithmetic*, Springer-Verlag, New York, 1973, Translated from the French, Graduate Texts in Mathematics, No. 7. MR 49 #8956
1126. Ji-Ping Sha and Da-Gang Yang, *Positive Ricci curvature on the connected sums of $S^n \times S^m$*, J. Differential Geom. **33** (1991), no. 1, 127–137. MR 92f:53048
1127. Krishnan Shankar, *On the fundamental groups of positively curved manifolds*, J. Differential Geom. **49** (1998), no. 1, 179–182. MR 99h:53040
1128. Zhongmin Shen, *Complete manifolds with nonnegative Ricci curvature and large volume growth*, Invent. Math. **125** (1996), no. 3, 393–404. MR 97d:53045
1129. _____, *Curvature, distance and volume in Finsler geometry*, to appear, 1998.
1130. Wan-Xiong Shi, *Deforming the metric on complete Riemannian manifolds*, J. Differential Geom. **30** (1989), no. 1, 223–301. MR 90i:58202
1131. Yoshihiro Shikata, *On the differentiable pinching problem*, Osaka J. Math. **4** (1967), 279–287. MR 37 #2133
1132. Yasushi Shimizu and Akira Shudo, *Polygonal billiards: correspondence between classical trajectories and quantum eigenstates*, Chaos Solitons Fractals **5** (1995), no. 7, 1337–1362. MR 96g:81066
1133. Katsuhiro Shiohama, *Recent developments in sphere theorems*, in Greene and Yau [597], Proceedings of the AMS Summer Research Institute on Differential Geometry held at the University of California, Los Angeles, California, July 8–28, 1990, pp. 551–576. MR 94d:53071
1134. _____, *Sphere theorems*, in Dillen and Verstraelen [449], pp. 865–903. MR 2000h:53003
1135. Katsuhiro Shiohama and Minoru Tanaka, *Cut loci and distance spheres on Alexandrov surfaces*, in Besse [185], En l'honneur de Marcel Berger. [In honor of Marcel Berger], Held in Luminy, France, July 12–18, 1992, pp. 531–559. MR 98a:53062
1136. Takashi Shioya, *Geometry of total curvature*, in Besse [185], En l'honneur de Marcel Berger. [In honor of Marcel Berger], Held in Luminy, France, July 12–18, 1992, pp. 561–600. MR 97m:53063
1137. A. Shnirel′man, *Ergodic properties of eigenfunctions*, Uspekhi Mat. Nauk **29** (1973), no. 6, 181–182.
1138. Leon M. Simon, *Lectures on geometric measure theory*, Australian National University Centre for Mathematical Analysis, Canberra, 1983. MR 87a:49001
1139. James Simons, *On the transitivity of holonomy systems*, Ann. of Math. (2) **76** (1962), 213–234. MR 26 #5520
1140. Yakov G. Sinaĭ, *Introduction to ergodic theory*, Princeton University Press, Princeton, N.J., 1976, Translated by V. Scheffer, Mathematical Notes, 18. MR 58 #28437
1141. _____, *Hyperbolic billiards*, Proceedings of the International Congress of Mathematicians, Vol. I, II (Kyoto, 1990) (Tokyo), Math. Soc. Japan, 1991, pp. 249–260. MR 93c:58163
1142. R. Sinclair and Minoru Tanaka, *Loki: software for computing cut loci*, Experiment. Math. **11** (2002), 1–26.
1143. _____, *The set of poles of a two-sheeted hyperboloid*, Experiment. Math. **11** (2002), 27–36.
1144. I. M. Singer, *Infinitesimally homogeneous spaces*, Comm. Pure Appl. Math. **13** (1960), 685–697. MR 24 #A1100
1145. I. M. Singer and John A. Thorpe, *The curvature of 4-dimensional Einstein spaces*, in Spencer and Iyanaga [1154], pp. 355–365. MR 41 #959

1146. David Singerman, *Symmetries of Riemann surfaces with large automorphism group*, Math. Ann. **210** (1974), 17–32. MR 50 #13505
1147. Yum Tong Siu and Shing-Tung Yau, *Compact Kähler manifolds of positive bisectional curvature*, Invent. Math. **59** (1980), no. 2, 189–204. MR 81h:58029
1148. L. Skorniakov, *Metrization of the projective plane in connection with a given system of curves*, Izv. Akad. Nauk SSSR Ser. Mat. **19** (1955), 471–482.
1149. Stephen Smale, *Mathematical problems for the next century*, in Arnol'd et al. [67], pp. 271–294.
1150. Brian Smyth and Frederico Xavier, *Real solvability of the equation $\partial_{\bar{z}}^2 \omega = \rho g$ and the topology of isolated umbilics*, J. Geom. Anal. **8** (1998), no. 4, 655–671. MR 1 724 211
1151. E. Socié, *Metric properties of Hilbert geometry*, to appear, 1999.
1152. Edith Socié-Méthou, *Caractérisation des ellipsoides de $\mathbb{R}^n$ par leur groupe d'automorphismes*, preprint, Université Louis Pasteur, Strasbourg, 2000.
1153. Andrew John Sommese, *Quaternionic manifolds*, Math. Ann. **212** (1974/75), 191–214. MR 54 #13778
1154. D.C. Spencer and S. Iyanaga (eds.), *Global analysis: Papers in honour of K. Kodaira*, Princeton mathematical series, no. 29, University of Tokyo Press, Tokyo, 1969.
1155. Michael Spivak, *A comprehensive introduction to differential geometry. Vols. I–V*, second ed., Publish or Perish Inc., Wilmington, Del., 1979. MR 82g:53003e
1156. Norman Steenrod, *The topology of fibre bundles*, Princeton University Press, Princeton, NJ, 1999, Reprint of the 1957 edition, Princeton Paperbacks. MR 2000a:55001
1157. Shlomo Sternberg, *Lectures on differential geometry*, second ed., Chelsea Publishing Co., New York, 1983, With an appendix by Sternberg and Victor Guillemin. MR 88f:58001
1158. John Stillwell, *Classical topology and combinatorial group theory*, second ed., Springer-Verlag, New York, 1993. MR 94a:57001
1159. J. J. Stoker, *Geometrical problems concerning polyhedra in the large*, Comm. Pure Appl. Math. **21** (1968), 119–168. MR 36 #5815
1160. _____, *Differential geometry*, John Wiley & Sons Inc., New York, 1989, Reprint of the 1969 original, A Wiley-Interscience Publication. MR 90f:53002
1161. Stephan Stolz, *Simply connected manifolds of positive scalar curvature*, Ann. of Math. (2) **136** (1992), no. 3, 511–540. MR 93i:57033
1162. _____, *A conjecture concerning positive Ricci curvature and the Witten genus*, Math. Ann. **304** (1996), no. 4, 785–800. MR 96k:58209
1163. _____, *Multiplicities of Dupin hypersurfaces*, Invent. Math. **138** (1999), no. 2, 253–279. MR 1 720 184
1164. Daniel W. Stroock, *Some thoughts about Riemannian structures on path space*, Gaz. Math. (1996), no. 68, 31–45. MR 97e:58012
1165. D. J. Struik (ed.), *A source book in mathematics, 1200–1800*, Harvard University Press, Cambridge, Mass., 1969. MR 39 #11
1166. Dennis Sullivan, *Infinitesimal computations in topology*, Inst. Hautes Études Sci. Publ. Math. (1977), no. 47, 269–331 (1978). MR 58 #31119
1167. Dennis Sullivan and Micheline Vigué-Poirrier, *Sur l'existence d'une infinité de géodésiques périodiques sur une variété riemannienne compacte*, C. R. Acad. Sci. Paris Sér. A-B **281** (1975), no. 9, Aii, A289–A291. MR 53 #4133

1168. Toshikazu Sunada, *Riemannian coverings and isospectral manifolds*, Ann. of Math. (2) **121** (1985), no. 1, 169–186. MR 86h:58141
1169. Yoshihiko Suyama, *A differentiable sphere theorem by curvature pinching. II*, Tohoku Math. J. (2) **47** (1995), no. 1, 15–29. MR 95k:53048
1170. A. S. Švarc, *A volume invariant of coverings*, Dokl. Akad. Nauk SSSR (N.S.) **105** (1955), 32–34. MR 17,781d
1171. Andrew Swann, *Hyper-Kähler and quaternionic Kähler geometry*, Math. Ann. **289** (1991), no. 3, 421–450. MR 92c:53030
1172. John L. Synge, *On the connectivity of spaces of positive curvature*, Quart. J. Math. Oxford **7** (1936), 316–320.
1173. Z. I. Szabó, *A short topological proof for the symmetry of 2 point homogeneous spaces*, Invent. Math. **106** (1991), no. 1, 61–64. MR 92f:53055
1174. _____, *Isospectral pairs of metrics on balls, spheres, and other manifolds with different local geometries*, Ann. of Math. (2) **154** (2001), no. 2, 437–475. MR 1 865 977
1175. Serge Tabachnikov, *Billiards*, Panor. Synth. (1995), no. 1, vi+142. MR 96c:58134
1176. Shun-ichi Tachibana, *A theorem of Riemannian manifolds of positive curvature operator*, Proc. Japan Acad. **50** (1974), 301–302. MR 51 #1667
1177. Iskander A. Taĭmanov, *On the existence of three nonintersecting closed geodesics on manifolds that are homeomorphic to the two-dimensional sphere*, Izv. Ross. Akad. Nauk Ser. Mat. **56** (1992), no. 3, 605–635. MR 93k:58051
1178. _____, *On totally geodesic embeddings of 7-dimensional manifolds into 13-dimensional manifolds of positive sectional curvature*, Mat. Sb. **187** (1996), no. 12, 121–136. MR 98c:53070
1179. Masaru Takeuchi, *On conjugate loci and cut loci of compact symmetric spaces. II*, Tsukuba J. Math. **3** (1979), no. 1, 1–29. MR 80k:53082
1180. Michel Talagrand, *Concentration of measure and isoperimetric inequalities in product spaces*, Inst. Hautes Études Sci. Publ. Math. (1995), no. 81, 73–205. MR 97h:60016
1181. Shûkichi Tanno, *A characterization of the canonical spheres by the spectrum*, Math. Z. **175** (1980), no. 3, 267–274. MR 82g:58093
1182. K. Tapp, *Volume growth and holonomy in nonnegative curvature*, Proc. Amer. Math. Soc. **127** (1999), 3035–3041.
1183. Clifford Henry Taubes, *The existence of anti-self-dual conformal structures*, J. Differential Geom. **36** (1992), no. 1, 163–253. MR 93j:53063
1184. Chuu-Lian Terng, *Recent progress in submanifold geometry*, Differential geometry: partial differential equations on manifolds (Los Angeles, CA, 1990), Amer. Math. Soc., Providence, RI, 1993, pp. 439–484. MR 94d:53090
1185. Chuu-Lian Terng and Gudlaugur Thorbergsson, *Submanifold geometry in symmetric spaces*, J. Differential Geom. **42** (1995), no. 3, 665–718. MR 97k:53054
1186. Gudlaugur Thorbergsson, *A survey on isoparametric hypersurfaces and their generalizations*, in Dillen and Verstraelen [449], pp. 963–995. MR 2000h:53003
1187. John A. Thorpe, *Some remarks on the Gauss-Bonnet integral*, J. Math. Mech. **18** (1969), 779–786. MR 41 #963
1188. _____, *Elementary topics in differential geometry*, Springer-Verlag, New York, 1994, Corrected reprint of the 1979 original. MR 95m:53002

1189. William P. Thurston, *Three-dimensional geometry and topology. Vol. 1*, Princeton University Press, Princeton, NJ, 1997, Edited by Silvio Levy. MR 97m:57016
1190. Gang Tian, *On Calabi's conjecture for complex surfaces with positive first Chern class*, Invent. Math. **101** (1990), no. 1, 101–172. MR 91d:32042
1191. _____, *Kähler-Einstein metrics with positive scalar curvature*, Invent. Math. **130** (1997), no. 1, 1–37. MR 99e:53065
1192. J. Tits, *Sur certaines classes d'espaces homogènes de groupes de Lie*, Acad. Roy. Belg. Cl. Sci. Mém. Coll. in 8° **29** (1955), no. 3, 268. MR 17,874f
1193. Domingo Toledo, *Rigidity theorems in Kähler geometry and fundamental groups of varieties*, to appear, 1997.
1194. Philippe Tondeur, *Foliations on Riemannian manifolds*, Springer-Verlag, New York, 1988. MR 89e:53052
1195. V. A. Toponogov, *The metric structure of Riemannian spaces of non-negative curvature containing straight lines*, Sibirsk. Mat. Ž. **5** (1964), 1358–1369. MR 32 #3017
1196. Victor A. Toponogov, *Riemannian spaces containing straight lines*, Dokl. Akad. Nauk SSSR **127** (1959), 977–979. MR 21 #7520
1197. Tatiana Toro, *Doubling and flatness: geometry of measures*, Notices Amer. Math. Soc. **44** (1997), no. 9, 1087–1094. MR 99d:28010
1198. François Trèves, *Introduction to pseudodifferential and Fourier integral operators. Vol. 1*, Plenum Press, New York, 1980, Pseudodifferential operators, The University Series in Mathematics. MR 82i:35173
1199. _____, *Introduction to pseudodifferential and Fourier integral operators. Vol. 2*, Plenum Press, New York, 1980, Fourier integral operators, The University Series in Mathematics. MR 82i:58068
1200. Franco Tricerri and Lieven Vanhecke, *Homogeneous structures on Riemannian manifolds*, Cambridge University Press, Cambridge, 1983. MR 85b:53052
1201. _____, *Variétés riemanniennes dont le tenseur de courbure est celui d'un espace symétrique riemannien irréductible*, C. R. Acad. Sci. Paris Sér. I Math. **302** (1986), no. 6, 233–235. MR 87e:53072
1202. Chiaki Tsukamoto, *Infinitesimal Blaschke conjectures on projective spaces*, Ann. Sci. École Norm. Sup. (4) **14** (1981), no. 3, 339–356. MR 84k:58055
1203. _____, *Infinitesimal Zoll deformations on spheres*, J. Math. Kyoto Univ. **24** (1984), no. 2, 219–224. MR 85j:53048
1204. Wilderich Tuschmann, *On the structure of compact simply connected manifolds of positive sectional curvature*, Geom. Dedicata **67** (1997), no. 1, 107–116. MR 98i:53057
1205. E. R. van Kampen, *The theorems of Gauss-Bonnet and Stokes*, Amer. J. Math. **60** (1938), 129–138.
1206. S. R. S. Varadhan, *Diffusion processes in a small time interval*, Comm. Pure Appl. Math. **20** (1967), 659–685. MR 36 #970
1207. N. Th. Varopoulos, L. Saloff-Coste, and T. Coulhon, *Analysis and geometry on groups*, Cambridge University Press, Cambridge, 1992. MR 95f:43008
1208. William A. Veech, *Teichmüller curves in moduli space, Eisenstein series and an application to triangular billiards*, Invent. Math. **97** (1989), no. 3, 553–583. MR 91h:58083a
1209. _____, *Erratum: "Teichmüller curves in moduli space, Eisenstein series and an application to triangular billiards"*, Invent. Math. **103** (1991), no. 2, 447, Correction to Veech [1208]. MR 91h:58083b

1210. Alain R. Veeravalli, *A rigidity theorem for compact hypersurfaces with an upper bound for the Ricci curvature*, Geom. Dedicata **74** (1999), no. 3, 287–290. MR 1 669 347
1211. Mikhail Verbitsky, *Hyperholomorphic sheaves and new examples of hyper-Kähler manifolds*, to appear, 1998.
1212. Luigi Verdiani, *Invariant metrics on cohomogeneity one manifolds*, Geom. Dedicata **77** (1999), no. 1, 77–111. MR 1 706 504
1213. Patrick Vérovic, *Entropie et métriques de Finsler*, Ph.D. thesis, Institut Fourier, Grenoble, France, Septembre 1996.
1214. _____, *Problème de l'entropie minimale pour les métriques de Finsler*, Ergodic Theory Dynam. Systems **19** (1999), no. 6, 1637–1654. MR 1 738 954
1215. Marie-France Vignéras, *Arithmétique des algèbres de quaternions*, Springer, Berlin, 1980. MR 82i:12016
1216. _____, *Variétés riemanniennes isospectrales et non isométriques*, Ann. of Math. (2) **112** (1980), no. 1, 21–32. MR 82b:58102
1217. Marina Ville, *On $\frac{1}{4}$-pinched 4-dimensional Riemannian manifolds of negative curvature*, Ann. Global Anal. Geom. **3** (1985), no. 3, 329–336. MR 87c:53089
1218. _____, *Sur le volume des variétés riemanniennes pincées*, Bull. Soc. Math. France **115** (1987), no. 2, 127–139. MR 88m:53081
1219. _____, *Vanishing conditions for the simplicial volume of compact complex varieties*, Proc. Amer. Math. Soc. **124** (1996), no. 4, 987–993. MR 96g:32047
1220. È. B. Vinberg, *Absence of crystallographic groups of reflections in Lobachevskiĭ spaces of large dimension*, Trudy Moskov. Mat. Obshch. **47** (1984), 68–102, 246. MR 86i:22020
1221. È. B. Vinberg (ed.), *Geometry. II*, Springer-Verlag, Berlin, 1993, Spaces of constant curvature, A translation of Геометрия. II, Akad. Nauk SSSR, Vsesoyuz. Inst. Nauchn. i Tekhn. Inform., Moscow, 1988, Translation by V. Minachin [V. V. Minakhin], Translation edited by È. B. Vinberg. MR 94f:53002
1222. _____, *Volumes of non-Euclidean polyhedra*, Uspekhi Mat. Nauk **48** (1993), no. 2(290), 17–46, Translation in *Russian Math. Surveys*, 48 (1993), 2, 15–45. MR 94h:52012
1223. Claire Voisin, *Symétrie miroir*, Société Mathématique de France, Paris, 1996. MR 97i:32026
1224. Ya. B. Vorobets, G. A. Gal′perin, and A. M. Stëpin, *Periodic billiard trajectories in polygons: generation mechanisms*, Uspekhi Mat. Nauk **47** (1992), no. 3(285), 9–74, 207. MR 93h:58088
1225. A. W. Wadsley, *Geodesic foliations by circles*, J. Differential Geometry **10** (1975), no. 4, 541–549. MR 53 #4092
1226. Hidekiyo Wakakuwa, *Holonomy groups*, Department of Mathematics, Okayama University, Okayama, 1971, Publications of the Study Group of Geometry, Vol. 6. MR 44 #3240
1227. Arnold Walfisz, *Gitterpunkte in mehrdimensionalen Kugeln*, Państwowe Wydawnictwo Naukowe, Warsaw, 1957, Monografie Matematyczne. Vol. 33. MR 20 #3826
1228. C. T. C. Wall, *Geometric properties of generic differentiable manifolds*, Geometry and topology (Proc. III Latin Amer. School of Math., Inst. Mat. Pura Aplicada CNPq, Rio de Janeiro, 1976), Lecture Notes in Math., no. 597, Springer, Berlin, 1977, pp. 707–774. MR 58 #13144

1229. _____, *Surgery on compact manifolds*, second ed., American Mathematical Society, Providence, RI, 1999, Edited and with a foreword by A. A. Ranicki. MR 2000a:57089

1230. Andrew H. Wallace, *Differential topology: first steps*, W. A. Benjamin, Inc., Reading, Mass.-London-Amsterdam, 1968, Second printing, Mathematics Monograph Series. MR 55 #9098

1231. Nolan R. Wallach, *Compact homogeneous Riemannian manifolds with strictly positive curvature*, Ann. of Math. (2) **96** (1972), 277–295. MR 46 #6243

1232. Peter Walters, *An introduction to ergodic theory*, Springer-Verlag, New York, 1982. MR 84e:28017

1233. Hsien-Chung Wang, *Two-point homogeneous spaces*, Ann. of Math. (2) **55** (1952), 177–191. MR 13,863a

1234. _____, *Topics on totally discontinuous groups*, Symmetric spaces (Short Courses, Washington Univ., St. Louis, Mo., 1969–1970), Dekker, New York, 1972, pp. 459–487. Pure and Appl. Math., Vol. 8. MR 54 #2879

1235. Jun Wang and McKenzie Y. Wang, *Einstein metrics on S^2-bundles*, Math. Ann. **310** (1998), no. 3, 497–526. MR 99b:53072

1236. McKenzie Y. Wang, *Some examples of homogeneous Einstein manifolds in dimension seven*, Duke Math. J. **49** (1982), no. 1, 23–28. MR 83k:53069

1237. _____, *Einstein metrics and quaternionic Kähler manifolds*, Math. Z. **210** (1992), no. 2, 305–325. MR 93h:53046

1238. _____, *Einstein metrics from symmetry and bundle constructions*, to appear, 1999.

1239. McKenzie Y. Wang and Wolfgang Ziller, *Einstein metrics with positive scalar curvature*, Curvature and topology of Riemannian manifolds (Katata, 1985), Springer, Berlin, 1986, pp. 319–336. MR 87k:53114

1240. _____, *Einstein metrics on principal torus bundles*, J. Differential Geom. **31** (1990), no. 1, 215–248. MR 91f:53041

1241. _____, *On isotropy irreducible Riemannian manifolds*, Acta Math. **166** (1991), no. 3-4, 223–261. MR 92b:53078

1242. Mu-Tao Wang, *Long-time existence and convergence of graphic mean curvature flow in arbitrary codimension*, Invent. Math. **148** (2002), no. 3, 525–543. MR 1 908 059

1243. Frank W. Warner, *Foundations of differentiable manifolds and Lie groups*, Springer-Verlag, New York, 1983, Corrected reprint of the 1971 edition. MR 84k:58001

1244. C. Eugene Wayne, *Periodic solutions of nonlinear partial differential equations*, Notices Amer. Math. Soc. **44** (1997), no. 8, 895–902. MR 1 467 653

1245. André Weil, *Introduction à l'étude des variétés kählériennes*, Hermann, Paris, 1958, Publications de l'Institut de Mathématique de l'Université de Nancago, VI. Actualités Sci. Ind. no. 1267. MR 22 #1921

1246. Alan Weinstein, *On the homotopy type of positively-pinched manifolds*, Arch. Math. (Basel) **18** (1967), 523–524. MR 36 #3376

1247. _____, *The generic conjugate locus*, Global Analysis (Proc. Sympos. Pure Math., Vol. XV, Berkeley, Calif., 1968), Amer. Math. Soc., Providence, R.I., 1970, pp. 299–301. MR 42 #6874

1248. _____, *Sur la non-densité des géodésiques fermées*, C. R. Acad. Sci. Paris Sér. A-B **271** (1970), A504. MR 42 #3713

1249. _____, *On the volume of manifolds all of whose geodesics are closed*, J. Differential Geometry **9** (1974), 513–517. MR 52 #11791

1250. Tilla Weinstein, *An introduction to Lorentz surfaces*, Walter de Gruyter & Co., Berlin, 1996. MR 98a:53104
1251. Michael Weiss, *Pinching and concordance theory*, J. Differential Geom. **38** (1993), no. 2, 387–416. MR 95a:53057
1252. Roland Weitzenböck, *Invariantentheorie*, P. Noordhoff, Groningen, 1923.
1253. R. O. Wells, Jr., *Differential analysis on complex manifolds*, second ed., Springer-Verlag, New York, 1980. MR 83f:58001
1254. Henry C. Wente, *Counterexample to a conjecture of H. Hopf*, Pacific J. Math. **121** (1986), no. 1, 193–243. MR 87d:53013
1255. Hermann Weyl, *Über die Bestimmung einer geschlossen konvexen Fläche durch ihr Linienelement*, Vierteljahrsschrift der Naturforschenden Gesselschaft in Zürich **61** (1916), 40–72, Also found in Weyl [1257], band I, pp. 614–644.
1256. _____, *On the volume of tubes*, Amer. J. Math. **61** (1939), 461–472, Also found in Weyl [1257], band III, pp. 658–669.
1257. _____, *Gesammelte Abhandlungen. Bände I, II, III, IV*, Springer-Verlag, Berlin, 1968, Edited by K. Chandrasekharan. MR 37 #6157
1258. Richard L. Wheeden and Antoni Zygmund, *Measure and integral*, Marcel Dekker Inc., New York, 1977, An introduction to real analysis, Pure and Applied Mathematics, Vol. 43. MR 58 #11295
1259. Hassler Whitney, *Differentiable manifolds*, Annals of Math. **37** (1936), no. 2, 645–680.
1260. _____, *On singularities of mappings of Euclidean spaces. I. Mappings of the plane into the plane*, Ann. of Math. (2) **62** (1955), 374–410. MR 17,518d
1261. Frederick H. Wilhelm, Jr., *On the filling radius of positively curved manifolds*, Invent. Math. **107** (1992), no. 3, 653–668. MR 93d:53055
1262. _____, *On radius, systole, and positive Ricci curvature*, Math. Z. **218** (1995), no. 4, 597–602. MR 96d:53040
1263. _____, *The radius rigidity theorem for manifolds of positive curvature*, J. Differential Geom. **44** (1996), no. 3, 634–665. MR 97m:53069
1264. _____, *On intermediate Ricci curvature and fundamental groups*, Illinois J. Math. **41** (1997), no. 3, 488–494. MR 98h:53065
1265. _____, *An exotic sphere with positive curvature almost everywhere*, to appear, 2000.
1266. _____, *Exotic spheres with lots of positive curvature*, to appear, 2000.
1267. Burkhard Wilking, *The normal homogeneous space $(SU(3) \times SO(3))/U^{\bullet}(2)$ has positive sectional curvature*, Proc. Amer. Math. Soc. **127** (1999), no. 4, 1191–1194. MR 99f:53051
1268. _____, *On compact Riemannian manifolds with noncompact holonomy groups*, to appear, 1999.
1269. _____, *On the fundamental groups of complete manifolds of nonnegative curvature*, to appear, 1999.
1270. _____, *Manifolds with positive sectional curvature almost everywhere*, Invent. Math. **148** (2002), no. 1, 117–141. MR 1 892 845
1271. T. J. Willmore, *Mean curvature of Riemannian immersions*, J. London Math. Soc. (2) **3** (1971), 307–310. MR 44 #959
1272. _____, *Riemannian geometry*, The Clarendon Press Oxford University Press, New York, 1993. MR 95e:53002

1273. Wilhelm Wirtinger, *Eine Determinantenidentität und ihre Anwendung auf analytische Gebilde in euklidischer und Hermitescher Maßbestimmung*, Monatshefte für Math. und Physik **44** (1936), 343–365.
1274. Maciej P. Wojtkowski, *Principles for the design of billiards with nonvanishing Lyapunov exponents*, Comm. Math. Phys. **105** (1986), no. 3, 391–414. MR 87k:58165
1275. Joseph A. Wolf, *Complex homogeneous contact manifolds and quaternionic symmetric spaces*, J. Math. Mech. **14** (1965), 1033–1047. MR 32 #3020
1276. _____, *Spaces of constant curvature*, fifth ed., Publish or Perish Inc., Houston, Tex., 1984. MR 88k:53002
1277. David Wraith, *Exotic spheres with positive Ricci curvature*, J. Differential Geom. **45** (1997), no. 3, 638–649. MR 98i:53058
1278. Hung Hsi Wu, *The Bochner technique in differential geometry*, Math. Rep. **3** (1988), no. 2, i–xii and 289–538. MR 91h:58031
1279. Jyh-Yang Wu, *Deformation of asymptotically isospectral metrics*, Geom. Dedicata **76** (1999), no. 1, 31–42. MR 1 699 222
1280. Hidehiko Yamabe, *On a deformation of Riemannian structures on compact manifolds*, Osaka Math. J. **12** (1960), 21–37. MR 23 #A2847
1281. Takao Yamaguchi, *Homotopy type finiteness theorems for certain precompact families of Riemannian manifolds*, Proc. Amer. Math. Soc. **102** (1988), no. 3, 660–666. MR 89d:53088
1282. _____, *Collapsing and pinching under a lower curvature bound*, Ann. of Math. (2) **133** (1991), no. 2, 317–357. MR 92b:53067
1283. _____, *A convergence theorem in the geometry of Alexandrov spaces*, in Besse [185], En l'honneur de Marcel Berger. [In honor of Marcel Berger], Held in Luminy, France, July 12–18, 1992, pp. 601–642. MR 97m:53078
1284. C. T. Yang, *Odd-dimensional wiedersehen manifolds are spheres*, J. Differential Geom. **15** (1980), no. 1, 91–96 (1981). MR 82g:53049
1285. Chen Ning Yang, *Fibre bundles and the physics of the magnetic monopole*, The Chern Symposium 1979 (Proc. Internat. Sympos., Berkeley, Calif., 1979), Springer, New York, 1980, pp. 247–253. MR 82h:81133
1286. Deane Yang, *Convergence of Riemannian manifolds with integral bounds on curvature. I*, Ann. Sci. École Norm. Sup. (4) **25** (1992), no. 1, 77–105. MR 93a:53037
1287. _____, *Convergence of Riemannian manifolds with integral bounds on curvature. II*, Ann. Sci. École Norm. Sup. (4) **25** (1992), no. 2, 179–199. MR 93m:53037
1288. _____, *Rigidity of Einstein 4-manifolds with positive curvature*, Inven. Math. **142** (2000), 435–450.
1289. Paul C. Yang and Shing-Tung Yau, *Eigenvalues of the Laplacian of compact Riemann surfaces and minimal submanifolds*, Ann. Scuola Norm. Sup. Pisa Cl. Sci. (4) **7** (1980), no. 1, 55–63. MR 81m:58084
1290. Kentaro Yano and Salomon Bochner, *Curvature and Betti numbers*, Princeton University Press, Princeton, N. J., 1953, Annals of Mathematics Studies, No. 32. MR 15,989f
1291. Kentaro Yano and Masahiro Kon, *Structures on manifolds*, World Scientific Publishing Co., Singapore, 1984. MR 86g:53001
1292. Shing-Tung Yau, *Harmonic functions on complete Riemannian manifolds*, Comm. Pure Appl. Math. **28** (1975), 201–228. MR 55 #4042

1293. _____, *Calabi's conjecture and some new results in algebraic geometry*, Proc. Nat. Acad. Sci. U.S.A. **74** (1977), no. 5, 1798–1799. MR 56 #9467
1294. _____, *On the Ricci curvature of a compact Kähler manifold and the complex Monge-Ampère equation. I*, Comm. Pure Appl. Math. **31** (1978), no. 3, 339–411. MR 81d:53045
1295. _____, *Problem section*, Seminar on Differential Geometry, Princeton Univ. Press, Princeton, N.J., 1982, pp. 669–706. MR 83e:53029
1296. _____, *Open problems in geometry*, in Greene and Yau [596], Proceedings of the AMS Summer Research Institute on Differential Geometry held at the University of California, Los Angeles, California, July 8–28, 1990, pp. 1–28. MR 94k:53001
1297. _____, *Review of geometry and analysis*, in Arnol'd et al. [67], pp. 353–402.
1298. Shing-Tung Yau and F. Zheng, *Negatively $\frac{1}{4}$-pinched Riemannian metric on a compact Kähler manifold*, Invent. Math. **103** (1991), no. 3, 527–535. MR 92a:53056
1299. Yosef Yomdin, *Volume growth and entropy*, Israel J. Math. **57** (1987), no. 3, 285–300. MR 90g:58008
1300. Guoliang Yu, *Zero-in-the-spectrum conjecture, positive scalar curvature and asymptotic dimension*, Invent. Math. **127** (1997), no. 1, 99–126. MR 97h:58156
1301. Tudor Zamfirescu, *Conjugate points and closed geodesic arcs on convex surfaces*, Geom. Dedicata **62** (1996), no. 1, 99–105. MR 97g:52006
1302. Steven Zelditch, *Uniform distribution of eigenfunctions on compact hyperbolic surfaces*, Duke Math. J. **55** (1987), no. 4, 919–941. MR 89d:58129
1303. _____, *Selberg trace formulae and equidistribution theorems for closed geodesics and Laplace eigenfunctions: finite area surfaces*, Mem. Amer. Math. Soc. **96** (1992), no. 465, vi+102. MR 93a:11047
1304. _____, *The inverse spectral problem for surfaces of revolution*, J. Differential Geom. **49** (1998), no. 2, 207–264. MR 99k:58188
1305. Shunhui Zhu, *The comparison geometry of Ricci curvature*, in Grove and Petersen [650], Papers from the Special Year in Differential Geometry held in Berkeley, CA, 1993–94, pp. 221–262. MR 98c:53054
1306. Günter M. Ziegler, *Lectures on polytopes*, Springer-Verlag, New York, 1995. MR 96a:52011
1307. Wolfgang Ziller, *The free loop space of globally symmetric spaces*, Invent. Math. **41** (1977), no. 1, 1–22. MR 58 #31198
1308. _____, *Geometry of the Katok examples*, Ergodic Theory Dynamical Systems **3** (1983), no. 1, 135–157. MR 86g:58036
1309. T. Zizhou, *Note on cohomology pinching below quarter theorem*, to appear, 1999.
1310. Otto Zoll, *Über Flächen mit Scharen geschlossener geodätischer Linien*, Math. Ann. **57** (1903), 108–133.

Acknowledgements

Springer-Verlag thanks the original publishers of the figures for permission to reprint them in this book.

The figures are identified below by chapter (c), figure (f) and page number (p) in the form c.f (p).

Reprinted from Frank Morgan, *Riemannian Geometry*, A K Peters, Ltd., drawn by James Bredt, © Frank Morgan: 1.17 (15)

Reprinted from Frank Morgan, *Riemannian Geometry*, A K Peters, Ltd., © Frank Morgan: 1.72 (60)

Reprinted from J.-M. di Meglio, "Peut-on faire flotter des troncs d'arbre en apesanteur". Bull. Soc. Franc. Phy., © 1992, Société Française de Physique: 1.26 (23)

Reprinted from Marcel Berger, 200 ans de Science Française: "Maths 89: l'école française 3ème du monde", Science et Vie, © 1989, Montparnasse Multimedia: 1.82 (72), 1.102 (98)

Reprinted from Steven W. McDonald and Allan N. Kaufman, "Wave Chaos in the Stadium: Statistical Properties of Short-Wave Solutions of the Helmholtz Equation". Phys. Rev. A, © 1988, American Physical Society: 1.85 (77)

Reprinted from R. Balian and C. Bloch, "Distribution of Eigenfrequencies for the Wave Equation in a Finite Domain. III: Eigenfrequency Density Oscillations". Annals of Physics, © 1972, Academic Press: 1.86 (78)

Reprinted from Marcel Berger, *Geometry 2*, Springer-Verlag, © Armand Colin Editeur: 1.101 (95)

Reprinted from Manfredo Do Carmo, *Differential Geometry of Curves and Surfaces*, © Prentice Hall: 3.32 (131)

Reprinted from P. Buser, *Geometry and Spectra of Compact Riemann Surfaces*, © Birkhäuser Boston: 4.29 (182)

Reprinted from M. Gromov/S.M. Bates, *Metric Structures for Riemannian and Non-Riemannian Spaces*, © Birkhäuser Boston: 10.23 (459), 10.24 (459)

Reprinted from J.W. Milnor, *Morse Theory*, © Princeton University Press: 4.46 (204), 10.17 (454)

Reprinted from Michael A. Buchner, "The Structure of the Cut Locus in the Dimension Less Than Or Equal To Six". Compositio Math., © 1978, Kluwer Academic Publishers: 6.63 (284)

Reprinted from H.S.M. Coxeter, *Introduction to Geometry*, © Wiley: 7.7 (307)

Reprinted from A.L. Besse (ed.), *Actes de la table ronde de géométrie différentielle, Séminaire et Congrès 1 (1996)*, © Société Mathématique de France: 7.30 (337)

Reprinted from Pierre Bérard, "Variétés riemanniennes isospectrales non isométriques". Sém. Bourbaki, vol. 1988/89, © Société Mathématique de France: 9.10 (411)

Reprinted from Y. Colin de Verdière, "Spectres de variétés riemanniennes et spectre de graphes". Proc. ICM Berkeley, © American Mathematical Society: 9.11 (415)

Reprinted from P. Sarnak, "Arithmetic Quantum Chaos", in: *The Schur Lectures* (1992), edited by Ilya Piatetski-Shapiro and Stephen Gelbart, © Bar-Illan University: 9.13 (425)

Reprinted from Riv. Mat. Univ. Parma, IV. Ser., © Università di Parma: 10.12 (450), 10.13 (450), 10.14 (451), 10.15 (452), 10.16 (452), 10.25 (467), 10.26 (468), 10.27 (469), 10.28 (470), 10.30 (471), 10.32 (479), 10.33 (480), 10.34 (486), 10.35 (487), 10.36 (487), 10.37 (489)

Reprinted from H.E. Rauch, "Geodesics, Symmetric Spaces and Differential Geometry in the Large". Comment. Math. Helv., © 1953, Birkhäuser Basel: 12.1 (553)

Reprinted from Uwe Abresch/Wolfgang T. Meyer, "A Sphere Theorem with a Pinching Constant Below $\frac{1}{4}$". J. Diff. Geometry, © 1996 International Press: 12.9 (562)

Reprinted from Simon Salamon, *Riemann Geometry and Holonomy Groups*, © Pearson Education: 13.4 (646)

Reprinted from Vladimir I. Arnold/André Avez, *Problèmes ergodiques de la mécanique classique*. Monographies internationales de mathématiques modernes, © Dunod Editeur: 10.31 (474)

Reprinted from G. Perelman, "Manifolds of positive Ricci curvature with almost maximal volume", J. Am. Math. Soc, © 1994, American Mathematical Society: cover figure, 12.15 (573)

List of Notation

$*$
: Hodge star, 664

$\langle F \rangle$
: space mean value, 472

$\#$
: connected sum, 198

$\frac{1}{2} p_1(M)$
: half Pontryagin class, 593

Aut
: automorphism group, 568

$B(x,y)$
: bisectional curvature, 656

b_γ
: Busemann function, 585

$b_{2n}^-(M^{4n}, \mathbb{R})$
: anti-self-dual Betti number, 718

$B(m,r)$
: ball of radius r centered at the point m, 222

$\bar{B}(m,r)$
: closed ball, 222

$b_{2n}^+(M^{4n}, \mathbb{R})$
: self-dual Betti number, 718

$B^p(M)$
: coboundaries, 168

$b_p(M, \mathbb{F})$
: Betti numbers, 159

$[X,Y]$
: Lie bracket, 694

$\mathbb{C}a$
: Cayley numbers, 154

$\mathbb{C}a\mathbb{P}^2$
: Cayley projective plane, 154

CAT(k)
: Cartan–Alexandrov–Toponogov hyperbolic spaces, 680

category
: category of a manifold, 505

$CF(L)$
: periodic trajectory counting function, 6

char
: characteristic number, 717

Ch$_i$
: Chern form, 714

$\chi(M)$
: Euler characteristic, 138, 141, 159

Chrom(M)
: chromatic number, 415

$c_i(V)$
: Chern class, 714

C^k
: k times continuously differentiable, 145

Cliff(d)
: Clifford algebra, 668

CvxRad(M)
: convexity radius, 278

$\mathbb{C}\mathbb{P}^n$
: complex projective space, 154

Curv$_p(R; \omega, \omega)$
: curvature terms in the Bochner technique, 708

Cut(γ)
: cut value, 269

Cut-Locus(m)
: cut locus, 278

List of Notation

Δ
    Euclidean Laplacian, 71

$\not{D}$
    Dirac operator, 669

d
    exterior derivative, 167

d^*
    adjoint of the exterior derivative, 664

dA
    area measure of a surface, 112

Δ
    Laplacian, 377, 697

$\underline{\Delta}$
    rough Laplacian, 707

d_Γ
    Gromov distance between mm spaces, 686

$d_{\mathfrak{G}-\mathfrak{H}}$
    Gromov–Hausdorff distance, 625

$d_\mathfrak{H}$
    Hausdorff distance, 625

diam
    diameter of a metric space, 228

Diff(M)
    group of diffeomorphisms, 175

Dirichlet
    Dirichlet quotient, 384

$d(p,q)$
    distance from p to q, 4

ds^2
    Riemannian metric [a.k.a. first fundamental form], 106

dV_M
    Riemannian measure, 299

$D_X Y$
    connection, 698

$E(\phi)$
    energy of a map, 675

$\mathbb{E}^d$
    d dimensional Euclidean space, 2

Emb(M)
    embolic constant, 356

$\eta(s)$
    eta invariant, 428

η
    mean curvature of a hypersurface, 313

exp
    Riemannian exponential map, 222

F^*
    time mean value of F, 472

F_4
    an exceptional Lie group, 155

$f^*\alpha$
    inverse image, 167

FillVal
    filling value, 537

focal(p)
    focal value of a point, 321

$f*g$
    convolution, 395

$\phi_W(M)$
    Witten genus, 593

G_2
    symmetry group of octonions, 152

$\widetilde{GL}(n,\mathbb{R})$
    universal covering group of the general linear group, 152

$GL(n,\mathbb{R})$
    general linear group, 152

GOE
    Gaussian orthogonal ensemble, 425

Grass$_\mathbb{C}(k,n)$
    complex Grassmannian, 153

Grass$_\mathbb{H}(k,n)$
    quaternionic Grassmannian, 154

Grass(k,d)
    real Grassmannian, 152

H
: mean curvature of a surface, 49

$\mathbb{H}$
: quaternions, 146

h
: isoperimetric profile, 315

h_c
: Cheeger's constant, 315

$H^p_{dR}(M)$
: de Rham cohomology, 168

Hess f
: Hessian, 376

$h_{Liouville}$
: Liouville entropy, 476

h_{meas}
: measure entropy, 476

h_{met}
: metric entropy, 476

Hol(p)
: holonomy group at p, 638

$\mathfrak{hol}(p)$
: holonomy Lie algebra at p, 642

Hol$_0(p)$
: restricted holonomy group at p, 638

Sys $H_1(M)$
: homological systole, 338

Sys$_k(M)$
: homotopic systole, 338

$H_p(M, \mathbb{F})$
: homology with $\mathbb{F}$ coefficients, 159

$H^p(M, \mathbb{F})$
: cohomology with $\mathbb{F}$ coefficients, 159

Horiz(v)
: horizontal space, 702

$\mathbb{HP}^n$
: quaternionic projective space, 154

h_{top}
: topological entropy, 475

h_{vol}
: volume entropy, 475

Hyp$^n_\mathbb{C}$
: complex hyperbolic space, 193

Hyp$^2_{Ca}$
: Cayley hyperbolic plane, 193

Hyp$^d(K)$
: hyperbolic space of dimension d and sectional curvature K, 177

Hyp$^n_\mathbb{H}$
: quaternionic hyperbolic space, 193

Hyp$^n_\mathbb{K}$
: hyperbolic space over the field $\mathbb{K}$, 193

Hyp^d
: hyperbolic space of dimension d and sectional curvature -1, 177

II
: second fundamental form, 212

II
: second fundamental form of a surface, 46

inf CF
: infimal asymptotic growth of counting function, 504

inf diam(M)
: minimal diameter of M in any metric with $-1 \leq K \leq 1$, 507

inf $\frac{\text{diam}}{\text{Inj}}(M)$
: infimum over metrics on M of ratio of diameter to injectivity radius, 505

inf $\|R\|_{L^{d/2}(M)}$
: infimum of $L^{d/2}$ norm of the curvature of any metric on M^d, 506

inf Vol(M)
: Gromov's minimal volume, 507

Inj(m)
: injectivity radius, 271

$\int_M \alpha$
    integral, 169
$\operatorname{Isom}(M,g)$
    isometry group, 175
K
    Gauß curvature, 49
K
    curvature of a plane curve, 12
$\mathbf{K}$
    Euler form, 711
k
    algebraic curvature of a plane curve, 15
$K(m,n,t)$
    Euclidean heat kernel, 87
$K(m,n,t)$
    heat kernel, 382
K-area
    K area, 602
$K_{\mathbb{C}}$
    complex sectional curvature (of Kähler manifold), 547
$k_g c$
    geodesic curvature of a curve on a surface, 34
$K_{\mathbb{C}}^{\text{isotr}}$
    complex isotropic curvature, 548
$K_{M,g}(x,y,t)$
    heat kernel, 401
$\mathbb{KP}^n$
    projective space over $\mathbb{K}$, 154, 209
$K(x,y)$
    sectional curvature, 205
λ_1
    first eigenvalue of the Laplacian, 75
$\Lambda(M)$
    space of loops through a point of M, 455
$\Lambda'_{p,q}(M)$
    space of paths in M from p to q, 454

$\operatorname{LGC}(\rho)$
    locally geometrically contractible with contractibility function ρ, 367
$\operatorname{li}(x)$
    logarithmic integral, 447
$\Lambda(M)$
    length of the shortest periodic geodesic, 337
$\Lambda^p(T^*M)$
    exterior tangent bundle, 167
$M(\delta)$
    surface of constant curvature δ, 121
$\mathfrak{Met}$
    the space of metric spaces, 626
$\mathcal{M}_\gamma$
    moduli space of curves of genus γ, 423
$\operatorname{MinSys}(M)$
    systolic quotient, 503
$\operatorname{MinSys}_k(M)$
    higher systolic quotient, 504
mm
    Gromov metric–measure space, 685
∇
    Levi-Civita connection, 699
∇f
    gradient, 234
$\nabla^k \mathcal{T}$
    k-th covariant derivative of $\mathcal{T}$, 704
$N(\lambda)$
    eigenvalue counting function, 73
$\operatorname{ObsDiam}(X,k)$
    k-observable diameter of an mm space X, 687
$\Omega(M)$
    loop space of M, 462

List of Notation   797

$\Omega^p(M)$
    differential forms, 167
$O(n)$
    orthogonal group, 152
$\Omega_*(M)$
    pointed loop space, 462
$\text{pack}_{d+1}(M)$
    packing invariant, 557
c_*
    parallel transport along the entire length of the curve c, 637
$c_{a \to b}$
    parallel transport along curve c from time a to time b, or from point a to point b, 241
π_1
    fundamental group, 150, 160
π_k
    homotopy group, 160
$p_i(M)$
    Pontryagin class, 715
rank
    rank of a vector, 250
Ricci
    Ricci curvature, 244
R_{ijkl}
    Riemann curvature tensor, 202
$\mathcal{RM}(d, a, b, D, v)$
    the set of compact Riemannian manifolds of dimension d, curvature $a \leq K \leq b$, diameter at most D, volume at most v, 627
$\mathcal{RM}^{\text{Vol}=1}_{\text{Inj} \geq \varepsilon}(M)$
    Riemannian metrics with unit volume and bounded injectivity radius, 535
$\mathcal{RM}(M)$
    Riemannian metrics on M, 175
$\mathbb{RP}^2$
    real projective plane, 146
$\mathbb{RP}^d$
    real projective space, 150
$\mathcal{RS}(M)$
    Riemannian structures on M, 175
$\mathcal{RS}^{|K| \leq 1}_{\text{Vol}=1}(M)$
    Riemannian structures with unit volume and sectional curvature $-1 \leq K \leq 1$, 539
$\mathcal{S}(M)$
    Konishi twistor space of a quaternionic Kähler manifold M, 651
$\sigma(M)$
    signature, 717
scalar
    scalar curvature, 398
$S(\infty)$
    sphere at infinity, 608
$S(\infty)$
    sphere at infinity, 182
$SO(n)$
    special orthogonal group, 146, 152
$SO(p, q)$
    special orthogonal group, 152
$\mathbb{S}^d(k)$
    space form of dimension d and sectional curvature k, 208
$\text{Spec}(D)$
    spectrum of Laplacian, 74
$\mathcal{S}(M)$
    spinor bundle, 669
$\text{Spin}^c(d)$
    a "twisted spinor" group, 671
$\text{Spin}(p, q)$
    spin group, 152
$\text{Stab Sys } H_1(M)$
    stable 1-systole, 351
$\text{Stab Sys } H_k(M)$
    stable k-systole, 351
$*$
    Hodge star, 378
$SU(p, q)$
    special unitary group, 152

$\mathrm{Sys}_k(M)$
  k systole of the manifold M, 348
$\mathrm{Sys}(M)$
  systole, 326

T^d
  d dimensional torus, 151, 152
$T_D(X,Y)$
  torsion of connection D, 698
$\tilde{M}$
  universal cover, 150
$T_m M$
  tangent space to M at m, 163
$\otimes^{r,s} V$
  (r,s) tensor product of vector space V, 695
$T_m^* M$
  cotangent space to M at m, 165

$U_m^+ M$
  half tangent sphere, 363
$U\mathbb{K}(n)$
  unitary group over $\mathbb{K}$, 154
UM
  unit tangent bundle, 195
$U(p,q)$
  unitary group, 152

$\mathrm{Vert}(v)$
  vertical space, 702

Vol
  volume, 300
$VRS(d)$
  maximal volume of simplex in Hyp^d, 181

$w_k(M)$
  k-th Stiefel–Whitney characteristic class, 669

$\wedge$
  wedge, 167
$\mathrm{Width}_k(M)$
  k-width of a metric space, 591
$\mathfrak{W}_{p,q}$
  Aloff–Wallach manifold, 577

$\mathcal{X}$
  the space of mm spaces, 686

$\mathcal{Z}(M)$
  Bérard Bergery/Salamon twistor space of a quaternionic Kähler manifold M, 649

ζ
  zeta function of the Laplace–Beltrami operator, 419
$Z_{M,g}(t)$
  spectral function, 401
$Z^p(M)$
  cocycles, 168

List of Authors

Abrahamov, A. A., 716
Abresch, Uwe, 181, 262, 276, 559, 561, 570, 576, 581, 599, 610, 614, 623, 629
Accola, Robert D. M., 330
Ahlfors, Lars V., 139
Aicardi, Francesca, 25
Akutagawa, Kazuo, 601
Alekseevskiĭ, D. V., 645
Alexander, Stephanie B., 676
Alexandrov, Aleksandr Danilovich, 55, 120–125, 217, 258, 678
Allendoerfer, Carl B., 709, 712, 713
Alling, Norman L., 292
Aloff, Simon, 577
Álvarez Paiva, Juan Carlos, 683
Ambrose, W., 216, 284, 643, 704
Ammann, Bernd, 136
Amorós, J., 654, 675
Anderson, Michael T., 161, 216, 267, 419, 500, 501, 509, 520, 522, 523, 532, 533, 567, 571, 594, 623, 624, 628, 629, 633, 634, 676, 678
Andrews, Ben, 17, 24
Anosov, D. V., 465, 472, 478
Appell, Paul, 200
Appolonius of Perga, 233
Aranson, S. Kh., 472
Aravinda, C. S., 570
Arconstanzo, Marc, 677
Arnol'd, Vladimir I., 18, 25, 38, 69, 129, 148, 200, 240, 282, 469, 471, 472, 685
Arnoux, Pierre, 5, 6, 8, 97

Ashbaugh, Mark S., 80
Atiyah, Michael F., 404, 427, 428, 672–674
Aubin, Thierry, 325, 375, 527, 533
Audoly, Basile, 131
Avakumović, V., 374, 403
Avez, A., 471

Babenko, Ivan K., 347–349, 661, 667
Babič, Vasilii M., 408
Baily, W. L., 677
Balian, R., 77, 92, 405, 407
Ballmann, Werner, 255, 257, 466, 471, 477, 478, 482, 483, 606, 608, 610, 612, 661, 678
Bandle, Catherine, 402
Bangert, Victor, 348, 436, 446, 468, 470
Bär, Christian, 670
Baston, Robert J., 654
Bavard, Christophe, 329, 335, 613
Bazaĭkin, Ya. V., 578
Beauville, Arnaud, 341, 653
Beem, John K., 684
Bemelmans, Josef, 629
Ben-Av, Radel, 534
Benedetti, Riccardo, 177, 181, 517
Benguria, Rafael D., 80
Benjamini, Itai, 662
Benoist, Yves, 497, 684
Bérard Bergery, Lionel, 349, 528, 529, 576
Bérard, Pierre H., 71, 75, 78, 79, 83, 87, 89, 90, 96, 319, 320,

375, 383, 385, 386, 401, 402, 408, 417, 594, 684, 709
Berestovskij, V. N., 676–678
Berg, I. David, 676
Berger, Marcel, IX, 2, 10, 17, 18, 20, 22, 25–29, 32, 37, 52, 58, 59, 65, 66, 70, 79, 95, 96, 125, 131–134, 141, 144, 145, 150, 151, 154, 155, 169, 170, 177, 185, 286, 290, 303, 305, 325, 326, 329, 331, 334, 341, 348, 350, 355, 361, 364, 375, 390, 396, 402, 436, 528, 541, 545, 553, 559, 576, 577, 644, 652, 667, 687, 691, 713, 733
Berger, Melvyn S., 467
Bergeron, Nicolas, 609
Berline, Nicole, 375, 396, 398, 663, 668, 670–674, 717, 718
Berndt, Jürgen, 215, 290
Bers, Lipman, 411
Besicovitch, A. S., 335, 336, 352, 353, 416
Besse, Arthur L., 42, 45, 155, 189, 193, 207, 209, 210, 237, 238, 280, 285, 313, 347, 362, 391, 436, 478, 488, 495, 501, 509, 520, 522, 523, 525, 527, 530, 531, 533, 567, 603, 638, 653, 654, 668, 673, 684, 709, 719–721
Bessières, Laurent, 520
Besson, Gérard, 96, 290, 293, 296, 319, 320, 386, 401, 402, 411, 420, 421, 484, 496, 497, 520, 530, 531, 667, 683, 684, 723
Biquard, Olivier, 653
Birkhoff, G. D., 6, 449
Bishop, Richard L., 144, 308, 310, 676
Bismut, Jean-Michel, 674
Blair, David E., 501, 655, 657
Blaschke, Wilhelm, 45, 136, 439, 493
Blatter, Christian, 329, 330

Bleecker, David, 133
Bliss, Gilbert Ames, 37
Bloch, C., 77, 92, 405, 407
Bochnak, Jacek, 161
Bochner, Salomon, 375, 404, 577, 588, 594, 595, 656, 706–708
Boeckx, Eric, 215, 216
Bogomolov, Fedor A., 689
Böhm, Christoph, 529
Bokan, Neda, 529
Bolsinov, Alexey V., 488
Bolyai, Johannes, 101, 176
Bombieri, Enrico, 467
Boothby, William M., 106, 144
Borel, Armand, 293, 295, 296, 638, 641, 645
Born, Max, 493
Bott, Raoul, 160, 427, 453, 463, 491, 673
Botvinnik, Boris, 601
Boucetta, Mohamed, 481
Boucksom, Sébastien, 654
Bourbaki, Nicolas, 684
Bourdon, Marc, 677, 680
Bourguignon, Jean-Pierre, 84, 410–412, 523, 527, 528, 534, 579, 594, 654, 657, 670, 672, 684, 709, 716, 717, 738
Bowditch, Brian H., 231
Bowen, Rufus, 481
Boy, Werner, 25, 136
Boyer, Charles P., 525, 594, 651
Brascamp, Herm Jan, 79
Braunmühl, A., 39
Bridson, Martin A., 680
Brin, Michael, 678
Brocks, Reinhard, 266
Bronshtein, I. U., 472
Brooks, Robert, 419
Brown, Robert B., 645
Brüning, Jochen, 413
Bryant, Robert L., 646, 681, 683
Buchdal, N., 657
Buchner, Michael A., 279, 283, 284
Bujalance, Emilio, 292

Bunimovitch, Leonid A., 7, 436
Bunke, Ulrich, 392, 396, 425, 428
Burago, D., 319, 497
Burago, Yu. D., 4, 26, 28, 180, 305, 590, 678, 679
Burger, M., 654, 675
Burns, Keith, 458, 485, 486
Burstall, F. E., 208
Busemann, Herbert, 595, 679, 683
Buser, Peter, 86, 92, 177, 181, 233, 235, 292, 324, 335, 341, 375, 392, 410, 421–424, 436, 446, 447, 568, 610, 613

Calabi, Eugenio, 22, 233, 262, 329, 335, 336, 347, 375, 451, 527, 593, 595, 653, 655, 665, 673
Cao, Jian Guo, 323, 451, 662
Carathéodory, Constantin, 47, 681
Carnot, Sadi, 681
Cartan, Élie, 144, 155, 190, 193, 208, 214, 215, 233, 249, 253–258, 279, 288, 296, 375, 390, 396, 513, 638, 669, 670, 690, 741
Catanese, Fabrizio, 532
Chavel, Isaac, 75, 79, 87, 95, 96, 120, 136, 144, 246, 272, 278, 303, 305, 308, 325, 326, 353, 362, 363, 375, 396
Chazarain, J., 405
Cheeger, Jeff, 120, 189, 229, 230, 235, 257, 262, 265, 270, 272–278, 315, 324, 396, 409, 429, 445, 512, 514–516, 519, 549, 564, 567, 572, 575, 578, 579, 581, 584, 587, 595, 597, 599, 614, 616, 617, 623, 624, 629–634, 660, 662, 672, 678, 680
Chen, Bang-Yen, 690
Chen, Bing-Long, 567
Chen, Haiwen, 547, 589
Chen, Jingyi, 60
Cheng, Shiu Yuen, 245, 246, 265, 387, 409

Chern, Shiing-Shen, 58, 109, 136, 375, 579, 672, 710, 711, 714, 717
Chernov, N. I., 6
Chevalley, Claude, 375, 670
Chinburg, Ted, 449
Chmutov, S., 25
Cho, Yunhi, 182
Chopp, David L., 59, 60
Choquet, Gustave, 326, 444
Chow, Bennett, 254, 419, 477, 526
Christodoulou, Demetrios, 684
Churchill, R. V., 207
Cohn-Vossen, Stefan, 106, 131, 661
Colding, Tobias H., 262, 264, 572, 575, 597, 599, 629, 633, 634, 660, 662, 680
Colin de Verdière, Yves, 92, 96, 139, 405, 406, 408, 413, 415, 416
Connelly, R., 132
Connes, Alain, 523, 671, 683, 685
Conway, John H., 86, 96, 341, 416, 417, 581, 583
Corlette, Kevin, 296, 570, 654, 675
Cornfeld, I. P., 436
Coste, Michel, 161
Courant, Richard, 38, 75, 77, 78, 95, 96
Courtois, Gilles, 290, 293, 296, 420, 421, 484, 496, 497, 520, 530, 531, 667, 683, 684
Cox, David A., 653
Coxeter, H. S. MacDonald, 2
Crittenden, Richard J., 144
Croft, Hallard T., 11
Croke, Christopher B., 55, 84, 337, 355, 357, 366, 409, 420, 467, 496, 497, 677, 691

Dai, Xianzhe, 266, 624
D'Ambra, Giuseppina, 432, 684
Damek, Ewa, 289
Damour, Thibault, 149
Dancer, Andrew, 525

Darboux, Gaston, 214
Degen, W. L. F., 282
Deligne, Pierre, 295, 657
Derdzinski, Andrzej, 524
Deschamps, Annie, 493
DeTurck, Dennis M., 522
de Rham, Georges, *see* Rham, Georges de
di Meglio, J.-M., 22
Di Scala, Antonio J., 214
Dierkes, Ulrich, 59, 62
Dieudonné, Jean, 157, 171, 212, 690, 710
Dirac, Paul A. M., 669
do Carmo, Manfredo Perdigão, 2, 15, 17, 18, 37, 52, 58, 60, 62, 63, 106, 120, 134, 135, 144, 230, 246, 257, 269, 272, 276
Dobrushin, R. L., 436
Dodziuk, Jozef, 410
Dombrowski, Peter, 106, 109, 111, 123, 212
Donaldson, S. K., 161, 500, 530, 672, 676
Donati, Leonbattista, 70
Donnay, Victor J., 485
Donnelly, Harold, 413, 414, 428
Doyle, Peter, 86
Druet, Olivier, 325
Dubrovin, B. A., 150
Duistermaat, J. J., 405, 406
Durumeric, Oguz, 559, 565
Duzhin, S., 25
Dvoretzky, Aryeh, 63, 305
Dyck, Walther, 136
Dynkin, E. B., 189, 578, 749

Easley, Kevin L., 684
Eberlein, Patrick, 477, 478, 482, 483, 495, 496, 605, 606, 608, 609, 612, 613, 661, 662
Ebin, David G., 120, 189, 229, 230, 257, 270, 272, 276, 278, 534, 549, 617, 685
Ecker, Klaus, 60
Eells, Jr., James, 52, 135, 675

Egorov, Yuri V., 413
Ehresmann, Charles, 653, 698
Ehrlich, Paul E., 684
Einstein, Albert, 267, 628, 750
El-Alaoui, El-H. Ch., 285, 681
Ellis, G. F. R., 684
Elworthy, David, 402
Epstein, D. B. A., 203, 705
Erdös, P., 90, 341, 388
Eschenburg, Jost-Hinrich, 189, 258, 261, 313, 529, 554, 578, 584
Escobar, J., 323
Etayo, José J., 292
Euclid, 101

Fabricius-Bjerre, Fr., 25
Falconer, Kenneth J., 11
Farber, Michael, 665
Farrell, F. T., 293, 496, 569, 570, 610, 611, 661
Federer, Herbert, 691
Fefferman, Charles, 413, 414
Félix, Yves, 592
Ferrand, Emmanuel, 25
Ferrand, Jacqueline, 245
Ferus, D., 208
Feshbach, Herman, 38, 77, 78, 96
Fet, A. I., 463
Feynman, Richard P., 10, 404
Flaminio, Livio, 485
Fomenko, A. T., 150
Foulon, Patrick, 477, 497, 683, 684
Franks, John, 468, 469
Freed, Daniel S., 672
Freedman, Michael, 348
Friedrich, Thomas, 668
Fröhlich, Jürg, 639, 655
Fuchs, Dmitry, 63
Fueter, Rudolf, 653
Fukaya, Kenji, 512, 551, 568, 597, 610, 613, 614, 624, 625, 629–633
Funk, P., 439
Futaki, A., 527

Gaffney, Matthew P., 427
Gage, Michael, 24, 467
Galicki, Krzysztof, 525, 594, 651, 678
Gallot, Sylvestre F. L., 52, 120, 144, 161, 175, 216, 224, 229, 230, 238, 252, 257, 262, 270, 279, 280, 290, 293, 296, 303, 305, 308, 319, 320, 324, 386, 401, 402, 420, 421, 484, 496, 497, 520, 530, 531, 546, 567, 597, 634, 667, 683, 684
Gal′perin, G. A., 6, 451
Gamboa, José M., 292
Gao, L. Zhiyong, 567, 629, 634
Gardner, Richard J., 12
Gasqui, Jacques, 495, 529
Gauduchon, Paul, 95, 96, 375, 390, 396, 530, 653, 670
Gauld, David B., 158
Gauß, Carl Friedrich, 29, 101, 102, 105–120, 176
Gauthier, J.-P., 285, 681
Gel′fand, I. M., 416, 417, 683
Gerber, Marlies, 485, 486
Getzler, Ezra, 375, 396, 398, 663, 668, 670–674, 717, 718
Gibbons, G. W., 594
Gil-Medrano, Olga, 684, 685
Gilkey, Peter B., 375, 396, 398, 427, 428, 529, 657, 663, 668, 671–674, 716–718
Gillet, Henri, 672, 674
Girard, Albert, 32
Gluck, Herman, 20, 128, 283
Godbillon, Claude, 683
Goldschmidt, Hubert, 495
Goncharov, Alexander, 295
Gorbatsevich, V. V., 189
Gordon, Carolyn S., 86, 87, 417, 419
Gornet, Ruth, 417, 418, 429, 497
Gostiaux, Bernard, 2, 10, 17, 18, 20, 25, 29, 37, 52, 58, 59, 65, 66, 131, 134, 141, 144, 145, 150, 151, 169, 170, 691, 713
Grandjean, Olivier, 639, 655
Gray, Alfred, 2, 65, 66, 645, 709, 713
Grayson, Matthew A., 467
Green, L. W., 45, 374, 416, 493
Greenberg, Marvin Jay, 150
Greene, Robert E., 343, 368, 576, 622, 625, 662, 755
Greenleaf, Newcomb, 292
Greenspan, H. P., 71
Griffiths, Phillip, 654, 655, 657
Grines, V. Z., 472
Groemer, Helmut, 27
Gromadzki, Grzegorz, 292
Gromoll, Detlef, 181, 262, 265, 276, 439, 463, 488, 490, 555, 556, 565, 566, 576, 578, 584, 586, 587, 595, 599, 611, 613, 623
Gromov, Mikhael, 4, 26, 56, 63, 160, 181, 218, 228, 255, 257, 261, 286, 293–296, 305, 308, 310–312, 319, 320, 325, 333, 334, 336–338, 341, 342, 345, 347–353, 367, 368, 375, 386, 404, 412, 432, 457, 459, 465, 512, 514, 516–518, 520, 530, 537, 549, 551, 568–570, 572, 580, 582, 591, 593, 597, 598, 600–608, 610, 613, 614, 625–627, 630–633, 655, 657, 661–663, 673, 678–680, 683–685, 691, 758
Große-Brauckmann, Karsten, 135
Grove, Karsten, 439, 457, 488, 490, 505, 555, 558, 562, 564–566, 575, 578, 581, 591, 618, 622, 679, 759
Gruber, Peter M., 90, 341, 388, 452
Guan, Pengfei, 217
Guillemin, Victor, 84, 85, 92, 375, 403, 405, 406, 420, 439
Gutierrez, Carlos, 47
Gutkin, Eugene, 5, 8

List of Authors

Gutzwiller, Martin C., 92, 93
Guy, Richard K., 11

Hadamard, Jacques, 57, 139, 254–257, 375, 434, 444, 478, 480, 608
Haefliger, André, 669, 680
Halperin, Stephen, 591, 592
Hamenstädt, Ursula, 477, 478, 482, 495, 496, 605, 606, 608, 612, 613, 661, 662
Hamilton, Richard S., 24, 216, 467, 525, 588, 594
Hammer, J., 90, 341, 388
Hansen, Vagn Lundsgaard, 161
Harriot, Thomas, 32
Harris, Joseph, 654, 655, 657
Hartman, Philip, 62, 109
Harvey, Reese, 667
Hass, Joel, 467
Hasselblatt, Boris, 436, 471, 475, 478
Hawking, S. W., 594, 684
Hawley, N. S., 656
Heber, Jens, 530
Hebey, Emmanuel, 267, 325, 367, 525, 533, 567, 628, 629
Hedlund, Gustav A., 351, 478
Heinonen, Juha, 686
Heintze, Ernst, 274, 275, 314, 570, 609, 690
Hélein, Frédéric, 675
Helgason, Sigurdur, 189, 237, 396, 439
Hernández, Luis, 570
Herrera, Haydeé, 528, 649
Herrera, Rafael, 528, 649
Hersch, Joseph, 410
Herzlich, M., 267, 628, 629
Hilbert, David, 38, 51, 75, 77, 78, 95, 96, 102, 106, 444, 521, 761
Hildebrandt, Stefan, 59, 62
Hingston, Nancy, 470
Hirsch, Morris W., 158, 161, 614

Hirzebruch, Friedrich, 527, 652, 656, 657, 671, 717
Hitchin, Nigel J., 135, 428, 530, 604, 670, 673
Hofer, Helmut, 655
Hopf, Eberhard, 478, 497
Hopf, Heinz, 47, 134, 226–229, 478, 500, 544, 709
Hörmander, Lars, 90, 374, 375, 403
Howards, Hugh, 319
Hsiang, Wu-teh, 318
Hsiang, Wu-Yi, 318, 690
Hsu, Lucas, 681
Hu, S. T., 144
Huang, Hua Min, 577
Huber, Heinz, 374, 421, 446, 447
Hulin, Dominique, 52, 120, 144, 161, 175, 224, 229, 230, 238, 252, 257, 270, 279, 280, 303, 305, 308, 320, 528, 655
Hummel, Christoph, 613
Hurwitz, Adolf, 27
Husemoller, Dale, 155, 671, 715
Hutchings, Michael, 319
Huxley, Martin N., 90

Igusa, Jun-ichi, 656
Il'yashenko, Yu. S., 472
Im Hof, Hans-Christoph, 555
Itoh, Jin-ichi, 284
Ivanov, Stefan, 216, 319, 497
Ivrii, Victor, 89

Jacobi, Carl Gustav Jacob, 37
Jakobson, M. V., 436
Janet, Maurice, 218
Jenni, Felix, 334
Jensen, Gary R., 524
Jones, L. E., 496, 569, 610, 661
Jørgensen, Troels, 295
Jost, Jürgen, 168, 267, 296, 467, 522, 578, 628, 665, 675
Joyce, Dominic D., 391, 638, 646, 654

Kac, Mark, 88, 89, 374, 404

Kamberov, Georgi I., 135
Kapouleas, Nikolaos, 135
Karcher, Hermann, 25, 132, 136, 196, 233, 235, 258, 261, 267, 274, 275, 289, 306, 314, 555, 564, 568, 610, 628
Kasue, Atsushi, 625, 632
Kato, Tosio, 404
Katok, Anatole B., 5, 7, 8, 97, 436, 447, 471, 475, 478, 483
Katok, Svetlana, 478
Katsuda, Atsushi, 625, 627, 765
Katz, Gabriel, 665
Katz, Mikhail G., 343, 348, 349, 667
Katz, Sheldon, 653
Kazdan, Jerry L., 522, 600, 603, 662
Kazhdan, D., 420
Kelvin, Lord (William Thomson), 171
de Kerékjártó, Béla, 158
Kervaire, Michel A., 156
Kim, Hyuk, 182
Kirby, Robion C., 161, 614
Kiyohara, Kazuyoshi, 488, 493, 495
Klainerman, Sergiu, 684
Klein, Felix, 177, 178
Kleiner, Bruce, 323, 607
Klingenberg, Wilhelm, 2, 37, 106, 131, 189, 269, 272–279, 283, 436, 441, 463, 465, 553, 684, 691
Klotz, Tilla, see Weinstein, Tilla
Klotz, Tilla, 47, 109
Klotz-Milnor, Tilla, see Weinstein, Tilla
Knieper, Gerhard, 436, 481, 483, 485, 488
Knörrer, Horst, 235
Knothe, Herbert, 26
Kobayashi, Shoshichi, 175, 212, 279, 584, 671, 690, 698, 710, 714
Kodaira, K., 527
Koebe, Paul, 254

Kolmogorov, A. N., 473
Kon, Masahiro, 651
Konishi, Mariko, 651
Kontsevich, Maxim, 635, 716
Koszul, Jean-Louis, 698
Kotschick, D., 654, 675
Kowalski, Oldřich, 215, 216, 525
Krätzel, Ekkehard, 388
Kreck, Matthias, 348, 592, 603, 615
Kronecker, Leopold, 136, 767
Kronheimer, P. B., 161, 500, 530, 672, 676
Kühnel, W., 684
Kuijlaars, A. B. J., 541
Kuiper, Nicolaas Hendrik, 218
Kulkarni, Ravindra Shripad, 214
Kumura, Hironori, 632
Kupka, I., 285, 681
Kuranishi, Masatake, 580
Küster, Albrecht, 59, 62
Kuwae, Kazuhiro, 678

Labourie, François, 497, 684
Lacour, Émile, 200
Lafontaine, Jacques, 52, 120, 144, 161, 175, 224, 229, 230, 238, 252, 257, 270, 279, 280, 303, 305, 308, 320, 324, 678
Lanczos, Cornelius, 267, 628
Lang, Serge, 171, 684
Lanzendorf, Martin, 525
Lashof, Richard K., 58
Lawson, Jr., H. Blaine, 136, 600, 603, 604, 611, 663, 667, 668, 670, 672, 673, 676, 678, 708
Lazutkin, Vladimir F., 8, 92, 408
Leahy, John V., 375
LeBrun, Claude, 529–532, 649, 652
Ledrappier, François, 5, 7, 97, 609
Leeb, Bernhard, 607
Leighton, Robert B., 10
Lekkerkerker, C. G., 341, 388
Lemaire, L., 675
Leschke, K., 208
Leuzinger, Enrico, 290

Levine, Jerome, 665
Lévy, Paul, 308, 323, 688
Li, Jiayu, 60
Li, Peter, 136, 411, 412, 657
Li, Yan Yan, 217
Lichnerowicz, André, 216, 374, 404, 408, 600, 604, 638, 640, 641, 657, 667, 708, 709
Lieb, Elliott H., 79
Liebmann, Karl Otto Heinrich, 131
Lindenstrauss, Joram, 305
Linnér, Anders, 24, 29
Liu, Xiaobo, 690
Llarull, Marcelo, 601
Lobachevskii, Nicholas, 101, 176
Löbell, Frank, 292
Loewner, Charles, 327
Lohkamp, Joachim, 400, 416, 488, 603, 613, 614, 634, 662, 680
Loos, Ottmar, 189
Lott, John, 393, 429, 630, 634
Lu, Peng, 526, 625
Lück, W., 661, 663
Luo, Wen Zhi, 407, 425, 661
Lusternik, L. A., 463, 466

Mañé, Ricardo, 436, 456, 471, 475
Maaß, Hans, 374
Machigashira, Yoshiroh, 678
Malliavin, Paul, 400
Mangoldt, Hans von, 40, 254–257
Mann, Benjamin M., 525, 651
Manning, Anthony, 476
Manturov, O., 524
Mao, Yiping, 417
Margerin, Christophe M., 284, 541, 567
Margulis, G. A., 296, 476, 481
Marin, A., 139
Maslova, N. B., 436
Massey, William S., 139, 158
Matsumoto, Shigenori, 208, 534
Matveev, Vladimir S., 497
Mauduit, Christian, 97

Mazet, Edmond, 95, 96, 375, 390, 396
Mazur, Barry, 161, 614
Mazzeo, Rafe, 135
McCleary, John, 461, 771
McDuff, Dusa, 655, 657
McKean, Jr., H. P., 90, 374, 393, 427, 673
Meeks, III, William H., 59
Melas, Antonios D., 86
Melrose, Richard, 84, 85, 92
Meyer, Daniel, 216, 588, 708
Meyer, Wolfgang T., 276, 463, 555, 559, 561, 564, 576, 578, 581, 584, 586
Micallef, Mario J., 588, 589
Michael, J. H., 344
Michel, René, 438, 677
Michelsohn, Marie-Louise, 603, 604, 663, 668, 670, 673, 676, 708
Michor, Peter W., 684, 685
Mills, R. L., 672
Milman, V. D., 305
Milnor, John W., 156, 160, 204, 243, 291, 306, 312, 326, 365, 374, 393, 416, 453, 553, 614, 714
Milnor, Tilla Klotz, 135
Milnor, Tilla Klotz, see Weinstein, Tilla
Min-Oo, Maung, 551, 552, 629
Minakshisundaram, S., 374, 393
Minicozzi, II, William P., 662
Moise, Edwin E., 158
Mok, Ngaiming, 639, 656
Montgomery, Deane, 645
Moore, John Douglas, 588, 589
Morgan, Frank, 29, 171, 319, 320, 467, 691
Morgan, John W., 161, 500, 657, 676
Mori, Shigefumi, 656
Moroianu, Andrei, 647
Morse, H. Marston, 276, 441, 453, 465, 478

Morse, Philip M., 38, 77, 78, 96
Moser, Jürgen K., 169, 464
Mostow, G. D., 293, 295, 296, 607
Muñoz Masqué, Jaime, 203, 705
Müller, Werner, 235, 324, 429, 678
Myers, Sumner B., 129, 226, 245, 262, 283, 312

Nabutovsky, Alexander, 337, 348, 444, 500, 508, 518, 523, 534, 539, 540
Nadirashvili, Nikolai, 135, 410
Naimark, M. A., 749
Nash, John, 218, 445
Nijenhuis, Albert, 643, 704
Nikolaev, I. G., 628, 634, 676–678
Nirenberg, Louis, 62, 217, 375
Nitsche, Johannes C. C., 59
Nomizu, Katsumi, 212, 671, 690, 698, 710, 714, 719
Novikov, S. P., 150

Obata, Morio, 409
Oesterlé, Joseph, 291
Olbrich, Martin, 392, 396, 425
Omori, Hideki, 414
O'Neill, Barrett, 106, 609, 684, 720
Onishchik, Arkadi L., 189
Osgood, B., 86, 418, 419
Osserman, Robert, 20, 26, 84, 485
O'Sullivan, John J., 313
Otal, Jean-Pierre, 496, 677
Otsu, Yukio, 557, 571, 678

Pacard, Frank, 135
Page, D., 528
Palais, Richard S., 108, 174, 522, 529, 690
Pansu, Pierre, 286, 293, 295, 296, 318, 481, 607, 630, 663
Park, Jeonghyeong, 375
Paternain, Gabriel P., 235, 436, 456, 458, 477
Paternain, Miguel, 456
Patodi, V. K., 427, 428, 673

Pauli, Wolfgang, 669
Pedit, F., 208
Pelletier, Fernand, 681
Perel'man, G., 4, 572, 584, 591, 594, 599, 623, 678, 679, 723
Perry, Peter, 419
Pesin, Ya. B., 436
Peters, Stefan, 617, 625
Petersen, V, Peter, 141, 195, 242, 325, 343, 368, 409, 419, 457, 505, 545, 549, 551, 566, 567, 575, 580, 581, 584, 587, 589, 591, 614, 618, 622, 624, 625, 629, 634, 660, 679, 708, 759
Petronio, Carlo, 177, 181, 517
Petrova, Irina, 216
Phillips, R., 86, 418, 419
Piatetski-Shapiro, I., 294
Pinkall, Ulrich, 25, 134, 136, 208
Pinsky, Mark A., 402
Pisier, Gilles, 305
Pittet, Christophe, 325, 349
Planck, Max, 71
Pleijel, Å., 374
Pogorelov, A. V., 134, 217
Poincaré, Henri, 178, 375, 433, 449, 723
Poinsot, Louis, 200, 282
Pollack, Daniel, 135
Polombo, Albert, 579, 716, 717, 738
Polthier, Konrad, 135
Pólya, G., 78, 81
Poon, Y. S., 649
Porteous, I. R., 70
Prasad, Gopal, 295
Preissmann, Alexandre, VII, 611
Prüfer, Friedbert, 215, 216
Przytycki, F., 5, 7, 97
Pu, P. M., 328
Püttmann, Thomas, 578

Radó, Tibor, 139
Rademacher, Hans-Bert, 464–466, 684
Raphaël, E., 22

Rasevskii, P. K., 749
Ratcliffe, John G., 32, 177, 181, 182, 185, 290, 293, 295, 335
Ratto, Andrea, 52, 135, 675
Rauch, H. E., 258, 259, 272, 545, 551, 552
Ray, D. B., 428
Rayleigh, John William, 81
Real, Christophe, 527
Recknagel, Andreas, 639, 655
Reid, Alan W., 449
Rham, Georges de, 159, 160, 168, 170, 226–229, 708
Ricci, Fulvio, 289
Ricci, Giambattista, 245
Riemann, Bernhard, 102, 176, 254
Rinow, Willi, 226–229
Ritoré, Manuel, 318
Robert, Gilles F., 484
Rodrigues, Olinde, 57, 113, 136, 141
Rollin, Yann, 524
Rong, Xiaochun, 428, 512, 514, 515, 519, 583, 590, 661
Ros, Antonio, 318
Rosenberg, Jonathan, 600, 601
Rotman, Regina, 337, 348, 352
Roy, Marie-Françoise, 161
Rozoy, L., 47
Rudnick, Zeév, 407, 425, 426, 661
Ruh, Bernhard, 214
Ruh, Ernst A., 551, 552, 555, 556, 567, 568, 629

Sabitov, I., 132
Sabourau, S., 337, 343
Sachs, Rainer Kurt, 684
Saff, E. B., 541
Sakai, Takashi, 26, 144, 175, 189, 195, 209, 229, 257, 258, 270, 272, 279, 281, 303, 305, 325, 353, 362, 375, 403, 549, 568, 584, 587, 614, 719, 720
Salamon, Dietmar, 655

Salamon, Simon M., 638, 646, 649, 653, 654
Sambusetti, Andrea, 530
Samelson, Hans, 210, 491, 645
Sampson, J. H., 657, 675
Sands, Matthew, 10
Santaló, Luis A., 364, 691
Sario, Leo, 139
Sarnak, Peter, 73, 77, 83, 86, 92, 96, 341, 407, 414, 418, 419, 425, 426, 501, 661
Satake, I., 677
Savo, Alessandro, 324, 387, 400, 414
Scherbel, H., 47
Schiemann, Alexander, 417
Schlenker, Jean-Marc, 134, 676
Schmidt, Erhard, 26, 66
Schmutz, Paul, 424
Schneider, Rolf, 691
Schnirelmann, L. G., 466
Schoen, Richard, 295, 533, 600, 601, 662
Schouten, J. A., 203, 705
Schrader, Robert, 235, 324, 678
Schroeder, Viktor, 255, 257, 477, 478, 482, 495, 496, 570, 605, 606, 608, 610, 612, 613, 661, 662
Schueth, Dorothee, 417
Schwachhöfer, Lorenz J., 578
Schwartz, Richard, 432
Scott, Peter, 676
Seaman, Walter, 560
Seiberg, N., 161, 500, 676
Seifert, Herbert, 158
Selberg, Atle, 374, 446
Semmelmann, Uwe, 647
Semmes, Stephen, 368, 661
Semmler, Klaus-Dieter, 86
Seppälä, Mika, 292
Serre, Jean-Pierre, 417, 455
Sha, Ji-Ping, 594
Shankar, Krishnan, 583
Sharafutdinov, Vladimir A., 420

Shen, Zhongmin, 662
Shi, Wan-Xiong, 629
Shikata, Yoshihiro, 555, 624, 627
Shimizu, Yasushi, 426
Shiohama, Katsuhiro, 285, 551, 553, 555, 557, 562, 679
Shiokawa, Iekata, 97
Shioya, Takashi, 661, 678
Shnirel′man, A., 413
Shteingold, S. D., 325, 567, 622
Shudo, Akira, 426
Siebenmann, Laurence C., 161, 614
Simon, Leon M., 344, 691
Simons, James, 644, 672
Sinaĭ, Yakov G., 5, 436, 472, 475
Sinclair, R., 40
Singer, David A., 128, 283
Singer, I. M., 90, 153, 206, 207, 216, 374, 393, 404, 427, 428, 643, 673, 674
Singerman, David, 292
Siu, Yum Tong, 607, 656, 657
Skorniakov, L., 439, 683
Sloane, N. J. A., 96, 341, 416, 417, 581, 583
Smale, Stephen, 522, 529, 541
Smirnov, M., 683
Smyth, Brian, 47
Sobolev, Sergei L., 325
Socié-Méthou, Edith, 683
Soffel, Michael, 149
Sommese, Andrew John, 654
Sotomayor, Jorge, 47
Soulé, Christophe, 672, 674
Spatzier, R. J., 235
Spivak, Michael, 2, 17, 52, 144, 145, 202, 212, 682, 690, 709
Stasheff, James D., 160, 614, 714
Steenrod, Norman, 671
Stëpin, A. M., 6
Sterling, I., 134
Sternberg, Shlomo, 2, 106, 375, 403
Stillwell, John, 139, 158
Stoker, J. J., 2, 17, 18, 106, 132, 135

Stokes, Sir George Gabriel, 171
Stolfi, Noëlle, 70
Stolz, Stephan, 592, 593, 600, 601, 603, 615
Strelcyn, Jean-Marie, 5, 7, 97
Stroock, Daniel W., 400, 402, 685
Strutt, Baron, 81
Suciu, Alexander I., 348, 349
Sukhov, Yu. M., 436
Sullivan, Dennis, 160, 464, 591, 657, 664
Sunada, Toshikazu, 417
Suyama, Yoshihiko, 555
Švarc, A. S., 306
Swann, Andrew, 651
Synge, John L., 246, 276
Szabó, Z. I., 289, 417
Szegö, G., 78, 81

Tabachnikov, Serge, 5, 7, 63
Tachibana, Shun-ichi, 216
Taĭmanov, Iskander A., 467, 488, 578
Takens, Floris, 465
Takeuchi, Masaru, 280
Talagrand, Michel, 688
Tamura, Jun-ichi, 97
Tanaka, Minoru, 40, 284, 285, 679
Tanno, Shûkichi, 399
Tapp, K., 663
Taubes, Clifford Henry, 530
Taylor, Michael, 396, 678
Teichmüller, Oswald, 293
Terng, Chuu-Lian, 690
Thomson, William (Lord Kelvin), 171
Thorbergsson, Gudlaugur, 690
Thorpe, John A., 106, 144, 153, 206, 207, 530
Threlfall, William, 158
Thurston, William P., 295, 569, 570, 677
Tian, Gang, 527, 629, 662, 678
Tits, J., 289
Toledo, Domingo, 654, 675

Tondeur, Philippe, 683
Topalov, Peter J., 497
Toponogov, V. A., 258, 596
Toro, Tatiana, 311
Trèves, François, 90, 375, 383, 403
Tricerri, Franco, 215, 216, 288, 290
Tsukamoto, Chiaki, 488, 495
Tu, Loring W., 160
Tuschmann, Wilderich, 578, 590

Uhlenbeck, Karen K., 672
Uryson, Pavel Samuilovich, 590

Vafa, Cumrun, 404
Valdés, Antonio, 203, 705
Valère Bouche, Liane, 681
van Kampen, E. R., 109, 137
Vanhecke, Lieven, 215, 216, 288, 290
Varadhan, S. R. S., 400
Vaugon, Michel, 325
Veech, William A., 5, 785
Verbitsky, Mikhail, 653
Vergne, Michèle, 375, 396, 398, 663, 668, 670–674, 717, 718
Vérovic, Patrick, 683
Vershik, A. M., 436
Vignéras, Marie-France, 292, 417, 420, 496
Vigué-Poirrier, Micheline, 464
Vilenkin, N. Ja., 749
Ville, Marina, 518, 520, 570
Vinberg, È. B., 66, 189, 292
Vlášek, Zdeněk, 525
Voisin, Claire, 653, 655
Vorobets, Ya. B., 6

Wadsley, A. W., 490
Wakakuwa, Hidekiyo, 638
Walfisz, Arnold, 388
Wall, C. T. C., 161, 283
Wallace, Andrew H., 158
Wallach, Nolan R., 513, 576, 577
Walters, Peter, 471, 475
Walz, A., 132

Wang, Hsien-Chung, 289, 295
Wang, McKenzie Y., 524, 525, 529, 533, 577
Wang, Mu-Tao, 24, 59
Warner, Frank W., 144, 166, 168, 170, 600, 603, 665
Wayne, C. Eugene, 433, 434
Webb, David L., 86, 87, 417
Wei, Guofang, 266, 325, 567, 622, 624, 629
Weil, André, 323, 654, 709, 713
Weinberger, Shmuel, 500, 508, 518, 534, 539, 540
Weinstein, Alan, 284, 440, 494, 614
Weinstein, Tilla, see Klotz, Tilla
Weinstein, Tilla, 684
Weiss, Howard, 488
Weiss, Michael, 555
Weitzenböck, Roland, 708
Wells, Jr., R. O., 654, 655
Wente, Henry C., 134
Weyl, Hermann, 109, 217, 396, 713, 788
Wheeden, Richard L., 3
Whitney, Hassler, 68, 144, 161
Wieacker, John A., 691
Wilhelm, Jr., Frederick H., 247, 337, 343, 555, 558, 566, 597
Wilking, Burkhard, 577–579
Willmore, T. J., 136, 690
Wintner, Aurel, 109
Wirtinger, Wilhelm, 666
Witten, E., 161, 404, 500, 676
Wohlrab, Ortwin, 59, 62
Wojtkowski, Maciej P., 8, 23, 477
Wolf, Emil, 493
Wolf, Joseph A., 290, 291, 524, 611, 647
Wolpert, S., 86
Wu, Hung Hsi, 594, 625, 662, 684, 709, 755
Wu, Jyh-Yang, 505, 622, 759

Xavier, Frederico, 47
Xu, Chong Ming, 149

Yamabe, Hidehiko, 533
Yamaguchi, Takao, 557, 597, 613, 622, 633, 678, 679
Yang, C. T., 494
Yang, Chen Ning, 672, 684
Yang, Da-Gang, 594
Yang, Deane, 531, 622, 634
Yang, Paul C., 411
Yano, Kentaro, 588, 651, 708
Yau, Shing-Tung, 136, 296, 375, 411–413, 500, 527, 545, 570, 579, 600, 601, 611, 656, 657, 662, 675
Ye, Rugang, 622, 624, 629
Yomdin, Yosef, 457
Yu, Guoliang, 662

Zalgaller, V. A., 26, 28, 180, 305, 590
Zamfirescu, Tudor, 285
Zehnder, Eduard, 655
Zelditch, Steven, 413
Zheng, F., 570
Zhu, Shunhui, 262
Zhu, Xi-Ping, 567
Ziegler, Günter M., 66
Ziller, Wolfgang, 461, 463, 466, 524, 529, 533, 577, 578, 771
Živaljević, Rade, 529
Zizhou, T., 559
Zoll, Otto, 44, 437
Zygmund, Antoni, 3

Subject Index

3-Sasakian manifold, 651

Abresch–Meyer theorem, 276
absolute calculus, 697, 704
acceleration, 12, **699**
action-angle coordinates, 440
adjoint
　of exterior derivative, **664**
Alexandrov space, 285, 598, 633, **678**
Alexandrov's conjecture, 55
algorithmic computability, 538
Allof–Wallach manifold, **513**, 577
Ambrose problem, 216, 250, 284, 704
amenable group, 517
analytic
　real, **44**
angle, 3
antipode, **493**
arithmetic
　hyperbolic manifold, **292**
aspherical, **345**
assay, non-destructive, 84
astronomy, 3
asymptote, 586
asymptotic line, **68**
Atiyah–Singer index theorem, 674
axiom
　Euclidean, 101
　of mobility, **253**, 289
axis
　of a hyperbolic glide, 422

ball bearings, 17

Bangert–Franks–Hingston theorem, 468
Bazaikin manifold, 578
bending, **131**
Bernoulli shift, **472**, 486
Bertrand–Puiseux formula, 110, 205
Besicovitch's 1st theorem, 335, 352
Besicovitch's 2nd theorem, 353
Bessel function, 77
Betti number, 159
Bianchi identity
　first, **203**, 547, **700**
　second, **705**
bicharacteristic, 406
Bieberbach theorem, 291, 568
billiard, 4–8
　table, **4**
biology, 686
biregular
　curve, **11**
Birkhoff canonical form, 406
bisectional curvature, **656**
Bishop theorem, 310
Blaschke manifold, 45, **285–286**, 357
BM measure, 481
Bochner technique, 594, 604, 707
Bochner–Weitzenböck formula, 265
Bonnet theorem, 125
Bonnet–Schoenberg–Myers theorem, 243, 266
bouncing, mirror, **4**
boundary, 676
Bowen–Margulis measure, 481, 483
Brownian motion, 402, 686

814   Subject Index

Brunn–Minkowski theorem, 28
Buchner–Wall theorem, 283
bumpy metric, **465**
bundle
    fiber, 104
    unit tangent, **358**
Bunimovitch stadium, 7
Burago–Ivanov theorem, 497
Busemann function, 265, 479, 584, **585**, 595
Bérard Bergery/Salamon twistor space, 649

calculus of variations, 14
calibration, 484, **667**
camera, 3
canonical Laplacian, 667
canonical transformation, 403
Carathéodory's conjecture, 47
Carnot–Carathéodory metric, 285, 609, **681**
Cartan surface theorem, 214
Cartan–Janet theorem, 218
category
    of a compact manifold, **365**
catenoid, 60
Cauchy's polyhedron theorem, 132
caustic, 8, **92**
Cayley
    numbers, 152, **154**
    plane, **154**, 565
center of mass, **233–235**, 256, 484, 617
characteristic class, 714
    secondary, 428, 672
characteristic number, **716, 717**
chart, 29, **144**
Cheeger finiteness theorem, 445, 616
Cheeger's constant, **315**
Cheeger–Colding theorem, 597
Cheeger–Rong theorem, 514
Chern class, **714**
chopping, 662
chromatic number, 411, 415
circle problem, 76, 388

classical mechanics, 199–200
Clifford algebra, 668
closed
    differential form, **167**
    co-closed, see differential form
Codazzi–Mainardi equations, 212, 690
cohomology, 159
Colding L^1 theorem, 262
Colding L^2 theorem, 264
collapsing, 514
collar, 610
complete
    metric space, **227**
complete reducibility
    of holonomy, **640**
complex isotropic plane, **547**
complex sectional curvature, 547
computable function, **536**
computer programming, 2
concavity, **11**
concentration phenomenon, **305, 688**
condition C
    of Palais–Smale, 522, 523, 529, 578
conductivity, 71
cone
    metric, 660
    topology, 608
    volume, 660
configuration space, 481
conformal, **159**, 180
conformal representation theorem, 254
conformally flat, **208**
conical singularity, 678
conjecture
    Alexandrov's, 55
    Carathéodory, 47
    Willmore, 136
conjugate
    geodesic flows, **495**
    locus, 129
    point, 129, **268**
connected sum, **198**

connection, **698**
  Levi-Civita, 697, **699**
  linear, **698**
contact geometry, 657
contractibility function, **367**, 620, 622
contractible, **445**
  geometrically, **367**
  locally geometrically, **367**
contraction
  of indices, **696**
control theory, 16
controlled topology, 572, 622
convex polygon, 6
convexity, 4, **229–231**
  radius, **278**, 268–286
convolution, **395**
coordinate system, **144**
coordinates
  harmonic, 223, **267**, 623, **628**
  normal, **201**
cotangent bundle, 165
countable at infinity, 145
counting function, 442, 504
covariant derivative, 241, 377, **699**, **704**
covering
  normal, 150
  space, 149, 184–185
    Riemannian, **184**
    universal, **150**
critical
  for the distance function, **563**
  lattice, **341**
  point, **453**
Croke's embolic theorem, 357
cross ratio, 178
crushed ice, 416
cursed exotic $\mathbb{KP}^n$, 492
curvature, VIII, 697
  algebraic, **15**
  bisectional, **656**
  complex sectional, 547
  Gauß, **49**
  geodesic

  of curves on surfaces, **34**
  holomorphic, **656**
  mean, **49**
  of hyperbolic space, 180
  of plane curves, **12**
  of surfaces, 202
  operator, **547**, 588
  principal, of surface, **45–52**
  radius of, **12**
  ratio, **567**
  Ricci, 244
    bounded, 262–266
    lower bounds, 308–314
  Riemann curvature tensor, 103, **200–204**, **700**
  sectional, **204–207**
    bounded, 257–262
    constant, 251–254, 290–295
    nonpositive, 254–257
    upper bounds, 305–308
curve
  biregular, **11**
  heat shrinking, 24, 467
  parallel, **12**
  parameterized, **10**
  plane, 9–29
    simple, **11**
  regular, **10**
curve geometric, **10**
cut
  locus, 39, 125–131, **278**
  point, **127**, **269**
  value, 65, **269**

Dai–Wei theorem, 266
deck transformation, **149**
degree
  of a map, **141**
Delaunay surface, **52**
derivation, **163**
derivative, 164
  covariant, **704**
determinant
  of the Laplacian, 419

diameter, **228**
  observable, 687
diastasis, 528, 655
diffeomorphism, **145**
differential, 164
  character, 672
  complex, **663**
  form, 166–172, 663, **695**
    closed, **167**
    co-closed, **665**
    exact, **168**
    harmonic, **665**
    intrinsically harmonic, 665
diffusion
  of cycles, 334, 518, 570
Dirac
  δ function, 394
  operator, 669
Dirichlet
  principle, 75, **383**, 689
  problem, **72**
  quotient, 75, 371, **384**, 435
double soul problem, 591
doubling property, 311
drum, 78

economics, 200
efficient ball packing, **615**
eigenfunction
  of Δ, **72**, **380**
eigenvalue
  of Δ, **72**, **380**
  small, **422**
Einstein metric, 519, **521**
  sign, **521**
ellipsoid, 37, 128, 235, 269, 281, 282, 440, 486
elliptic
  rational homotopy, 591
  rationally, **592**
elliptic point
  of a surface, **68**
embedded submanifold, 151
embedding, 51
  isometric, 217

embolic
  constant, **356**, 504
  functional, 504, **535**
end, **585**
energy
  of a map, 675
entropy, 473–477
  measure, **475**
  metric, 8, **475**
  topological, 8, **475**
  volume, **475**
equichordal problem, 11
ergodicity, 7, 8, 23, 412, **473**
essential, **345**
estimate
  Weyl eigenvalue, **396**
η invariant, 375, 428, 519
Euclid
  fifth postulate, 101
Euclidean
  geometry, 2–99
  plane, 2
Euler
  angles, 102, 146
  characteristic, 47, 138, 141, **159**, 330, 399, 427, 709, 715, 716
  form, 711
  theorem, 282
exact
  differential form, **168**
excess
  of a triangle, **262**
exotic
  $\mathbb{KP}^n$, 492
    cursed, 492
  quaternionic projective plane, 464
  sphere, **160**, 555
exponential map, **166**, **222**, 248–251
exterior
  algebra, 166
  calculus, 166
  derivative, 167, 663
  product, 167

F structure, 512, 514, 519, 631
  polarized, 512, 519, 631
  pure polarized, 512, 631
Faber–Krahn theorem, 81, 402
fiber bundle, 104, **671**
filling
  radius, **342**
  value, 537
  volume, **343**
finely corrugated wrinkling, 134
first Bianchi identity, **203**, 547, **700**
first fundamental form, 102, 212
first variation, 4, 313
flat, **110**, 291
  conformally, 208
  torus, **185**, 442
Floer homology, 523
flow
  geodesic, *see* geodesic flow
focal value, **321**
focusing, 7
  automatic, 3
foliation
  Riemannian, 683
form
  differential, *see* differential form
Fourier integral operator, 403, 413
Fourier series, 27
Frobenius theorem, 694
fundamental domain, **185**
fundamental solution
  heat equation, **87**, **382**, **394**
fundamental tone, **75**, **79**

gaps
  spectral, **392**
Gauß–Bonnet theorem, 709
Gaussian orthogonal ensemble, **425**
Gauß
  circle problem, 76, 90, 388
  curvature, **49**
  equation, 118, 212
  equations, 690
  lemma, 116
Gauß–Bonnet theorem, 111–115, 519

global, 138
local, 112
generatrix, **48**, 60
generic, **67**
genus, **138**, **329**
geodesic
  closed, 43, 44
  flow, VIII, **359**, 374, 393, 434–436, 438, 440, 446, 447, 449, 456, 471
  loop, 44
  minimal, **222**
  on surfaces, 33–45
  primitive, **445**
  space, **679**
geodesy, 3
geometric curve, **10**
geometric hierarchy, 286–297
geometric measure theory, 171, 320, 600, **691**
geometric number theory, 341
geometrically contractible, **367**
  locally, **367**
geometry
  Euclidean, 2–99
ghost geodesic, 442
glide
  in hyperbolic plane, 422
gluing, 156, 196–198, 514
  of hyperbolic surfaces, 196
golden ratio, 236, 442
golden triangle, 697
gradient, 234
Grassmannian
  complex, **153**
  quaternionic, **154**
  real, **152**
Gromov–Hausdorff distance, 534, 572, 576, 598, 599, 622, 624, 625, **626**, 634
Gromov–Lawson torus theorem, 604
group
  duality condition, 613
  growth of, **306**
  hyperbolic, 605

growth
  of a group, **306**

h-principle, 613
Hadamard's conjecture, 135
half Pontryagin class, 593
Hardy–Littlewood–Karamata theorem, 397, 398
harmonic
  coordinates, 223, **267**, 623, **628**
  differential form, 426, **665**
  manifold, **289**
  map, **675**
  radius, **628**
harmonics, **75**
  spherical, 27
Hausdorff property, 145, 174, 224
heat, 70–99
  equation, **70**, 87–91, **381**
  kernel, **87**, **382**, **394**
  shrinking curve, 24, 467
Heintze theorem, 609
Heintze–Karcher lemma, 309
Heintze–Karcher theorem, 314
Heintze–Margulis lemma, 610
helicoid, 60
helix, 17
Hersch theorem, 410
Hessian, **376**, **697**
hierarchy
  geometric, 286–297
Hilbert hyperbolic surface theorem, 135
Hirzebruch signature theorem, 717
Hodge star, **378**, 664
Hodge–de Rham theorem, 665
holomorphic curvature, **656**
holonomy, **115**, 277, 637–657
  group, 637–657
homeomorphy, **160**
homofocal quadric, **38**
homogeneous space, 152–153, 186–189
  normal, **210**, 719
homological systole, **338**

homology, 159
  Floer, 523
  simplicial, 159
homology sphere, **160**, 538, 540
homophonic, 416
homotopy, **160**
  group, 160
  rational, 160
homowave, 416
Hopf fibration, 155, 350
Hopf–Rinow theorem, 227
horizontal space, **702**
horseshoe, 488, 561
Huber trace formula, 422
hyperbolic
  group, 605
  manifold, **291**
    arithmetic, **292**
  plane models
    hyperboloid, 177
    Klein, 178
    Poincaré, 178
    Poincaré disk, 178
  point
    of a surface, **68**
  rationally, **592**
  space, **177**
    in Gromov's sense, **680**
hyperboloid model
  of hyperbolic plane, 177
hyperelliptic function, 283
hyperkähler, **652**
hyperplane
  hyperbolic, **180**
hypersurface, **151**
  rigidity, 213

ideal triangle, **181**
identity
  first Bianchi, **203**, 547
immersed submanifold, 151
immersion, 51, **164**

index
  of a critical point, **453**
  of an elliptic operator, 674
  of vector field, **711**
induced metric, 30
inequality
  isodiametric, 55
  isoperimetric, *see* isoperimetric inequality
  isosystolic, **327**
infinite dimensional manifold, 684
infranilmanifold, 514, **568**, 630, 633
injectivity, 268–286
  radius, 125–131, **271**, 276
  of surface, 42
inner metric, **4**, 30
  of curves, 9
  of surfaces, 30–33
inscribed square problem, 11
integral, 169
interferometer, 17
intrinsic metric, 30
intrinsically harmonic differential form, 665
inverse image, 167
isodiametric inequality, 55
isometric embedding, 217
isometry, **174**
isoperimetric
  inequality
    in $\mathbb{E}^3$, 63
isoperimetric inequality, 26
  Brunn–Minkowski theorem, 28
  in $\mathbb{E}^3$, 63
  Knothe's proof, 26
  spherical, 305
  Steiner symmetrization, 28
  Steiner's quadrilateral, 27
  Stokes' theorem, 26
isoperimetric profile, VIII, **315**
isosystolic inequality, **327**
isotropy group, **187**

Jacobi
  last theorem, 129

Jacobi vector field, **248**, 402, 478
  trivial, **250**
Janet theorem, 218
jigsaw metric, 609

K-theory, 602, 673
Kac–Feynman–Kato inequality, 404
Kähler
  form, 527, **640**
  geometry, 657
  manifold, 526–528, **639, 640**, 644, 654–657
KAM (Kolmogorov–Arnol'd–Moser) theorem, 408, 464, 480
Katok's theorem, 483
Killing vector field, **704**
kissing number, 582
Klein
  bottle, 317, 329
  model of hyperbolic plane, 178
  surface, **292**
KO-theory, 673
Konishi twistor space, 651

Laplace operator, *see* Laplacian
Laplace–Beltrami operator, *see* Laplacian
Laplacian, **70**, 370, **377**, 697
  canonical, 667
  determinant of, 419
  rough, **707**
  special, 667
lasso, 641
lattice, 185
  critical, **341**
law of large numbers, 686
Lawson's conjecture, 136
le mouvement á la Poinsot, 200, 282
Lebesgue–Rochlin space, **686**
Leibnitz's rule, **163**
lemma
  Heintze–Karcher, 309
  Heintze–Margulis, 610

length, **3**
 counting function, 43
 space, **4**, **679**
 spectrum, **73**, **393**, 405, 407
lens space, **246**, 490
Levi-Civita connection, 697, **699**
Lichnerowicz λ_1 theorem, 408
Lichnerowicz formula, 669
Lie
 algebra, 165
 bracket, 165, **694**
 group, **152**
line, 585, 596
linear connection, **698**
Liouville theorem, 359, 472
Lipschitz, 145, **627**
locally geometrically contractible, **367**
locally symmetric space, **189**
Loewner theorem, 326
long line, 145
loop
 geodesic, **44**
loop space, **462**
Lorentzian manifold, 683

von Mangoldt–Hadamard–Cartan theorem, 255
manifold, **144**
 Allof–Wallach, **513**, 577
 arithmetic hyperbolic, **292**
 hyperbolic, **291**
 infinite dimensional, 684
 Kähler, 526–528, **639**, **640**, 644, 654–657
 quaternionic Kähler, **647**
 real analytic, **145**
 Riemannian, **174**
 topological, **145**
maximum principle
 for parabolic PDE, 402
Maxwell's fisheye, 493
meager, **67**, **465**
mean curvature, **49**, **313**
 flow, 60

measure
 Bowen–Margulis, 481, 483
 entropy, **475**
 isotropy, **289**
 Patterson–Sullivan, 484
mechanics
 classical, 199–200
metric
 ball, **222**
 Carnot–Carathéodory, 609, **681**
 cone, 660
 Einstein, 519, **521**
 entropy, **475**
 induced, 30
 inner, 30
 intrinsic, 30
 jigsaw, 609
 railway, 609
 Ricci flat, **521**
 space, complete, **227**
 Tits, 296
microlocal analysis, 403, 413
Milnor's octahedron, **204**
minimal
 geodesic, **222**
 surface, **59**
  in $\mathbb{E}^3$, 58–62
 volume, 207, **507**
minimax principle, **75**, **384**
mirror bouncing, **4**
mixing, **473**, 486
mm space, 305, 686
mobility
 axiom of, **253**, 289
Möbius group, **182**
Möbius transformation, 294
modular domain, 254, 286, **293**
moduli problem, **502**
Morse
 index, 491
 index theorem, 454
 theorem, 453
 theory, 276, 441, 465, 564
Moser theorem, 169
Mostow rigidity, 293

multiplicity, **380**
musical isomorphisms, 696
Myers–Cheng theorem, 245, 262, 312

Nash embedding theorem, 218, 445
Nash–Kuiper theorem, 218
Nash–Tognoli theorem, 161, 445
natural tensor, 203, **705**
nilmanifold, **260**
nodal
  line, **90**
nodal set, **393**
non-destructive assay, 84
noncommutative geometry, 685
nondegenerate
  critical point, **453**
norm, **173**
normal
  coordinates, **201**
  covering, 150
  homogeneous space, **210**, 719
Novikov conjecture, 601, 607
number theory, 286, 388

observable diameter, 687
octahedron
  Milnor's, **204**
operator
  curvature, 588
orbifold, 590, 623
orientable, **169**
osculating circle, **12**

packing invariant, **557**
pair of pants, 197
Palais theorem, 174
Palais–Smale condition C, 522, 523, 529, 578
panda, 155
pants
  pair of, 197
parabolic point, **68**
parallel transport, **240**, 697, 701
parametrix, 395

parking, 681
Paternain theorem, 477
path, **3**
  shortest, **222**
Patterson–Sullivan measure, 484
pendulum, 102
  double, 199
periodic
  geodesic, VIII
    twist type, 464
  trajectory, 5
Pfaffian, **715**
phase space, 102, 481
PL-manifold, 678
Planck's constant, 71
plane
  Euclidean, 2
  real projective, **146**
plane curve, 9–29
Plateau problem, **59**
Poincaré
  conjecture, 523, 560, 572, 588, 621, 723
  disk model
    of hyperbolic plane, 178
  half-plane model
    of hyperbolic plane, 178
  return map, 405, 464, **469**
Poinsot
  le mouvement á la, 200, 282
point
  simple, **586**
Poisson formula, 389
Poisson type formula, **91**
polarized F structure, 512, 519, 631
pole, **586**
polygon, convex, 6
polytope, **96**
Pontryagin class, **715**
positive mass theorem, 662
pouch, 552
primitive
  geodesic, **445**
principal curvature, **46**, 212
probability theory, 686

problem
  Ambrose, 216, 284, 704
product
  of manifolds, **151**
  of Riemannian manifolds, **183**
  warped, **195**, 572
programming
  computer, 2
projective plane
  real, **146**
projective space, 193, 209, 237, 280, 290–296, 390–391, 565, 566, 576
  complex, **154**
  quaternionic, **154**
  real, **150**, 236, 279
protractor, 18
Prüfer surface, 145
pseudo-Riemannian manifold, 684
Pu theorem, 328
pullback, 167
pure polarized F structure, 512, 631

quadric, **37**
  homofocal, **38**
quantum chaos, 425
quasi-isometry, **606**
quasimode, **92**, 408
quaternionic Kähler
  manifold, **647**
quaternionic projective plane
  exotic, 464
quaternions, **146**
quotient space, 184–185

radius
  convexity, **278**
  filling, **342**
  injectivity, see injectivity radius
  of surface, 42
radius of curvature, **12**
Radon transform, 439
railway metric, 609

rank
  of a map, **164**
  of Riemannian manifold, **250**, 483
  of symmetric space, **193–194**, 232, 289, 461, 476, 483, 578, 607, 609, 612
  of unit vector, **250**
rational homotopy, 160
rationally elliptic, 460, **592**
rationally hyperbolic, 460, **592**
Rauch comparison theorem, 259
ray, 585
Ray–Singer torsion, 428
real analytic, **44**
  manifold, **145**
  Riemannian metric, **145**
real projective plane, **146**, 316, 328
real projective space, 236, 279
real spinor, 668
reducible
  Riemannian manifold, **184**
Reeb's theorem, 553
regular simplex, **181**
Reidemeister torsion, 429
residual, **67**, **465**
resonance, 79, 392
de Rham cohomology, 168
de Rham reducibility theorem, 229
de Rham theorem, 159, 160, 168, 170, 226
Ricci calculus, 704
Ricci curvature, **244**
Ricci flat metric, **521**
Riemann
  curvature tensor, 103
Riemann hypothesis, 423
Riemann surface, **158**, 421
Riemannian
  submersion, **194–196**, 210, 719
Riemannian covering space, **184**
Riemannian manifold, **174**
  reducible, **184**
Riemannian metric, **172**
  real analytic, **145**
Riemannian structure, **175**

rigid body, 102
rigidity theorem, 559
Rodrigues–Gauß map, 57, 113, 136, 141
rolling surface theorem, 56
rough Laplacian, **707**
Rozansky–Witten invariant, 716
ruled
  surface, 60

Santalo's formula, **364**
Sasakian, 403
scalar product, **173**
scar, 92, **426**
Schrödinger equation, 70, 376, **381**, 497
Schwarz symmetrization, **81**
second Bianchi identity, **705**
second fundamental form, **45–52**, 212, **313**
  on a surface, 46
second variation, 119, **242**
secondary characteristic class, 672
segment, **35**, **222**
Seiberg–Witten
  equations, 530
  invariants, 531
Selberg trace formula, 392, 422, 425
semiclassical limit, **376**, 425
series, Fourier, 27
shortest path, **222**
sign
  of Einstein metric, **521**
signature, **717**
simple
  plane curve, **11**
simple point, **586**
simplicial volume, 232, **516**, 540
simply connected, **150**
singularity
  conical, 678
skein, **460**
skinny, **67**
small eigenvalue, **422**
smooth, **144**

soul, 584
space
  covering, 149, 184–185
  homogeneous, 152, 186–189, 296–297
  hyperbolic, **177**
  length, **4**
  lens, **246**
  locally symmetric, **189**
  projective, 193, 209, 237, 290–296, 390–391, 576
    complex, **154**
    quaternionic, **154**
    real, **150**
  quotient, 184–185
  Riemannian covering, **184**
  symmetric, **189**
    rank 1, 295–296
    rank > 1, 296
space form, 148, 251–254, 286–297
space mean value, **472**
space of paths, **454**
special Laplacian, 667
special orthogonal group, 146
spectral gaps, **392**
spectrum, VIII, **73**, **380**
  length, **73**, 405
speed, 10
  of sound, 71
sphere
  exotic, 555
sphere at infinity, 293, 296, 484, **608**
sphere theorem, 552
spherical
  harmonics, 27, **96**
  trigonometry, 31
spherical codes, 581
spin structure, 669
Spin^c structure, 671
spinor, 668
  bundle, 668, **669**
  field, **669**
  real, 668
splitting theorem, 595

stability
  of minimal manifold, **666**
stable, **66**
  systole, **350**
stadium
  Bunimovitch, 7
starfish
  three-legged, 451
Steiner symmetrization, 28
Steiner's quadrilateral, 27
Stiefel–Whitney classes, 669
Stokes' theorem, 171
Stone–Weierstraß theorem, 382, 390
straightedge, 17
strangulation, 451
stroboscope, 72
Sturm–Liouville theory, 121, 314, 402, 441
sub-Riemannian geometry, 681
submanifold, 690
  embedded, 151
  immersed, 151
  totally geodesic, **231–233**
submersion, **164**
  Riemannian, **194–196**, 210, 719
sum
  connected, **198**
supermanifold, 674
superrigidity, 296
supersymmetry, 674
surface, 29–70, 105–142
  constant curvature, 32, 118, 251–254, 500
  Delaunay, **52**
  Klein, **292**
  of revolution, **36**
  Riemann, 158
  Weingarten, 134
  Weinstein, 440
  Zoll, **437**
surgery, 156, 196–198
  trivial, 198
symbol, 378
symmetric space, **189**, 290
  locally, **189**

rank 1, 289
symplectic geometry, 655, 657
  complex, 653
Synge theorem, 246, 276
synthetic geometry, 572
systole, 247, **326**, 503
  k dimensional, 348
  homological, **338**
  stable, **350**
systolic
  (p,q) freedom, **349**
  (p,q) softness, **349**
  inequalities, VIII
  quotient, 328
  ratio (or quotient), **330**

T structure, 512, 631
table, billiard, **4**
tangent
  bundle, 164
  line, **11**
  space, **163**
  vector, **163**
tapestry, **449**
Tarski–Seidenberg theorem, 539
Teichmüller space, 446
telemetry, 3
tensor, 694
  natural, 203, **705**
  product, 694
theorem
  Abresch–Meyer, 276
  Atiyah–Singer index, 674
  Ballmann–Wojtlowski, 477
  Bangert–Franks–Hingston, 468
  Bertrand–Puiseux, 110
  Besicovitch's 1st, 335, 352
  Besicovitch's 2nd, 353
  Bieberbach, 291, 568
  Bishop, 310
  Bonnet, 125
  Bonnet–Schoenberg–Myers, 243, 266
  Brunn–Minkowski, 28
  Buchner–Wall, 283

Burago–Ivanov, 497
Cartan surface, 214
Cartan–Janet, 218
Cheeger finiteness, 445, 616
Cheeger–Colding, 597
Cheeger–Rong, 514
Colding L^1, 262
Colding L^2, 264
conformal representation, 254
Croke's embolic, 357
Dai–Wei, 266
Dvoretzky, 63, 305
Euler, 282
Faber–Krahn, 81, 402
Frobenius, 694
Gauß–Bonnet, 709
Gauß–Bonnet, 111–115, 519
 global, 138
 local, 112
Gromov–Lawson torus, 604
Hardy–Littlewood–Karamata, 397, 398
Heintze, 609
Heintze–Karcher, 314
Hersch, 410
Hilbert hyperbolic surface, 135
Hirzebruch signature, 717
Hodge–de Rham, 665
Hopf–Rinow, 227
hypersurface rigidity, 213
isoperimetric inequality in $\mathbb{E}^3$, 63
Jacobi's last, 129
Janet, 218
KAM (Kolmogorov–Arnol'd–Moser), 408, 464, 480
Katok, 483
Lichnerowicz λ_1, 408
Liouville, 359, 472
Loewner, 326
von Mangoldt–Hadamard–Cartan, 255
Morse, 453
Morse's index, 454
Moser, 169
Mostow rigidity, 293

Myers–Cheng, 245, 262, 312
Nash embedding, 218, 445
Nash–Kuiper, 218
Nash–Tognoli, 161, 445
Palais, 174
Paternain, 477
positive mass, 662
Pu, 328
Rauch comparison, 259
Reeb's, 553
de Rham, 159, 160, 168, 170, 226
de Rham reducibility, 229
rigidity, 559
rolling surface, 56
sphere, 552
spherical isoperimetric inequality, 305
splitting, 595
Stokes', 171
Stone–Weierstrass, 390
Stone–Weierstraß, 382
symmetric space, 190
Synge, 246, 276
Tarski–Seidenberg, 539
Toponogov comparison, 258
triangle comparison, 258
Tsukamoto, 494
Weinstein, 492
Whitney embedding, 161
Wiedersehenmannigfaltigkeit, 493
Wilhelm, 247
Wolpert, 424
theorema egregium, 105
three point transitive, 289
three-legged starfish, 451
time mean value, **472**
Tits metric, 296, **608**
topological
 entropy, 458, **475**, 486, 488
 manifold, **145**
 transitivity, **478**
Toponogov comparison theorem, 258

torsion
  free, 698
  of a connection, **698**, 699
  Ray–Singer, 428
  Reidemeister, 429
torus, **151**
  flat, **185**
totally convex, **230**
totally geodesic submanifold, 198, **231–233**
trajectory
  periodic, 5
transformation
  Möbius, 294
transgression, 455
transitive
  three point, 289
triangle
  on Riemannian manifold, **258**
triangle comparison, 257–267
  Toponogov theorem, 258
trivial
  Jacobi vector field, **250**
Tsukamoto theorem, 494
tube formula, 63–66
Turing machine, 538
turning number, **20**
twist type of periodic geodesic, 464
twistor, 649, 672
twistor space
  Bérard Bergery/Salamon, 649
  Konishi, 651

umbilic, **47**
Umlaufsatz, 19
unit tangent bundle, **358**, 369
unit tangent sphere, **222**
universal covering, **150**

variation
  first, 4, 222–226
  second, **242**
variational principle, **476**

vector field, **694**
  Jacobi, **248**, 478
  trivial Jacobi, **250**
velocity, 10
vertex, **20**
vertical space, **701**
vibrating membrane, **71**
visibility, **609**
volume, 169, 300
  cone, 660
  entropy, **475**
  filling, **343**
  form, **169**, 299
  simplicial, **516**

warped product, **195**, 572
wave equation, **70**, **381**
waves, 70–99
wedge product, 167
Weingarten surface, 134
Weinstein
  surface, 440
  theorem, 492
Weyl eigenvalue estimate, **396**
Weyl problem, 217
whispering gallery, 77
Whitney embedding theorem, 161
widths
  of metric space, **591**
Wiedersehenmannigfaltigkeit theorem, 493
Wilhelm theorem, 247
Willmore conjecture, 136
Wirtinger inequality, **126**, 666
Witten genus, 593
Wolpert theorem, 424
wrinkling
  finely corrugated, 134

Yang–Mills field, 672

Zoll surface, **437**